George Fraser is received in audience by His Holiness Pope Benedict XVI in the presence of His Eminence Cardinal Javier Lozano Barragán at the end of the conference at which George presented the contribution 'Human genetics today: hopes and risks' in this Festschrift.

FIFTY YEARS
OF HUMAN
GENETICS

A Festschrift and *liber amicorum* to celebrate the life and work

GEORGE ROBERT FRASER

FIFTY YEARS OF HUMAN GENETICS

A Festschrift and *liber amicorum* to celebrate the life and work of

GEORGE ROBERT FRASER

Edited by Oliver Mayo and Carolyn Leach

Wakefield
Press

Wakefield Press
1 The Parade West
Kent Town
South Australia 5067
www.wakefieldpress.com.au

First published 2007

Copyright © Oliver Mayo, Carolyn Leach and individual authors, except where otherwise indicated, 2007
All publication has been by permission, these permissions having been obtained by the respective authors

All rights reserved. This book is copyright. Apart from any fair dealing for the purposes of private study, research, criticism or review, as permitted under the Copyright Act, no part may be reproduced without written permission. Enquiries should be addressed to the publisher.

Jacket designed by Liz Nicholson, designBITE
Designed and typeset by Michael Deves, Adelaide
Printed and bound by Hyde Park Press, Adelaide

National Library of Australia
Cataloguing-in-Publication entry

Fifty years of human genetics: a festschrift and liber amicorum to celebrate the life and work of George Robert Fraser.

Bibliography.
Includes index.
ISBN 9781862547537 (hbk).

1. Fraser, George Robert, 1932- . 2. Human genetics. I. Leach, Carolyn, (Carolyn Ruby), 1943- . II. Mayo, Oliver. III. Title.

599.935

Table of contents

Note: the full authorship of each contribution is shown at the beginning of the contribution. This table of contents identifies the colleagues, friends and former students who were invited to contribute to the volume.

STATISTICAL AND POPULATION GENETICS

The origin and purpose of this book

Oliver Mayo

CSIRO Livestock Industries, PO Box 10041, Adelaide BC, SA 5000

This book has come about as a result of a dinner conversation among a small group of people who happened to be friends, colleagues and former students of George Fraser. All those present agreed that he met Dr Johnson's description of Edmund Burke: 'If a man were to go by chance at the same time with Burke under a shed, to shun a shower, he would say – "this is an extraordinary man"'. Furthermore, 'Burke *is* an extraordinary man. His stream of mind is perpetual.' (Boswell 1791)

Thinking further about George, we also agreed that, though his work is well known and influential, he has had perhaps less personal recognition than one would ordinarily have expected. In consequence, it would be a fitting mark of respect and affection if we were able to organise a Festschrift for his 75th birthday, 3 March 2007. Accordingly, I wrote to him with our tentative suggestion, he responded enthusiastically, and we began work.

It is a demonstration of George's nature and achievements that over 60 distinguished friends, former colleagues and former students have provided the contributions which constitute the book. What is perhaps more remarkable is how these contributions cover so many of the advances in human and medical genetics over the fifty years of George's working life.

All of us who have worked with George have come to marvel at his extraordinary facility for languages and for language. At school, he was marked down as a notable classical scholar of the future, and though he followed in his father's footsteps into medicine, he has retained an interest in studying the classical languages, especially Greek, throughout his life. A brief anecdote to illustrate his gift for languages may be permitted from my own experience. When George went to Leiden, he wanted to rent a house together with me, for I was in Holland at his invitation, to work with him for some months. House-renting was not easy for a range of reasons, but after weeks of hunting and living in the Badmotel in Noordwijk aan Zee, we were successful in finding a house and were sitting in a real estate agent's office waiting for the contract to be typed for signature. The 'phone rang, and as the land agent talked in Dutch, I noticed George sit up more and more stiffly. The land agent put the 'phone down and said, in English, 'I am sorry; there has been a mistake; the house is not available.' George replied, politely, 'Yes, there has been a mistake. I suggest you telephone that person and tell him that you made a mistake and the "stupid foreigner" is taking the house. I shall wait to sign the contract.' Astonished, the land agent did as requested. George had not been learning Dutch, but had leafed through an introduction to the language on his 'plane from North America to the Netherlands.

Concern for others has marked George's professional life. He has always admired and tried to follow John Donne (1624) in his well known admonition 'never send to know for whom the bell tolls: it tolls for thee'. Several contributors (Hook, Clarke Fraser, Murday, Yap-Todos …) have noted how exemplary he has been in his clinical practice. As is implicit in this approach, George has always been suspicious of grand schemes in general. Indeed, George has always exhibited the benefits of following Dr Johnson's great maxim, 'clear your *mind* of cant … You tell a man, "I am sorry you had such bad weather the last day of your journey, and were so much wet." You don't care sixpence whether he was wet or dry. You may *talk* in this manner; it is a mode of talking in Society: but don't *think* foolishly.' (Boswell 1791) This clarity of vision and largeness of mind have meant that his outstanding contribution to the taxonomy of human sensory impairment has always had the goal of alleviating human misery and suffering at the level of the individual and the individual family. It has also been conducted as objectively as possible, in the sense of not judging others: 'In the same sense that one kind of order and proportion constitutes beauty, another kind of order and proportion seems to constitute deformity.' (Bayes 1731)

We are fortunate to have contributions (from Opitz and Scambler) dealing with the 21st century's first phase of the molecular elucidation of the pathogenesis of Fraser's syndrome: the discovery and sequencing of the allelic variant associated with the disease and its homologue in the mouse.

Several other contributors (Beighton, Mittwoch, Opitz, Schull: see p. 222) point out that Fraser syndrome was first described in a brief passage in Fraser (1962a), the major paper on genetical load that is the point of departure for Ewens and a reference for Morton and Sved, who address load in three different ways in this volume. George's marvellous paper could in fact have been split into three merely good ones. He has always wanted to convince by the force of his argument, rather than cover his mantelpiece with trophies.

Just as he has mastered at one time or another most major European languages, so through pertinacity and insight George has made significant genetical contributions using mathematics at a level unusual in medical specialists. Testimony to this breadth are the sections on mathematical and population genetics, as indeed are those on the ethics and history of our subject. George has made noteworthy contributions to linkage analysis, the theory of genetical load and the detection of associations between quantitative traits and major genes.

I have already alluded to one of George's favourite texts. Another is the wise advice to medical scientists of William Harvey, discoverer of the circulation of the blood (if one ignores Michael Servetus's claims) and student of Caius, Cambridge, like R. A. Fisher, one of George's mentors. (George himself, of course, was at Trinity.) Over the years, George has sought accurate and sympathetic translations of the passage (which most of us only know in the English of Robert Willis), and we reproduce a number of these after this introduction. Some translations have been made by colleagues and contributors to this volume. In whatever language one reads them, Harvey's words remain good advice for the practising human geneticist.

I wrote above that George is suspicious of grand schemes. This partly reflects the human horrors of the 20th century and their effects on George and those close to him. It was therefore peculiarly fitting that George should have translated into English Benno Müller-Hill's important book on the perversion of science and scientists in Hitler's Germany, *Tödliche Wissenschaft*. We are again fortunate to have in this volume the views of both men on the future, and on how to prevent human outrages from recurring.

In this volume, from their own interests and wide experience and from their association with George Fraser, who has long concerned himself with the intractable ethical problems of genetical and other counselling, many of the authors address the problem here made explicit. It is noteworthy that those authors who favour counselling which can result in preventative intervention generally seek to rest their approach on something like Kant's 'categorical imperative': 'Act in such a way that you always treat humanity, whether in your own person or in the person of any other, never simply as a means, but always at the same time as an end.' However, it is not always clear who is the person who must be treated as an end, or how personal autonomy can be sustained in a medical context, whereas it is certain that workers in public health sometimes address the good of the population as a primary goal. Given that there is no universally accepted right answer, it is appropriate that a diversity of thoughtful ethical views celebrates in this volume one person's life and work, that person someone who practical experience of genetical counselling all his working life (see particularly Fraser 1999 and Fraser 2006b, which, as George has said to me, is a kind of *credo* and *apologia pro vita sua*).

This is a book of friends, and editing it has been a labour of love. I hope that it is a pleasure to read, and for any errors and omissions can only plead, to quote Dr Johnson once more, 'In lapidary inscriptions a man is not upon oath.' (Boswell 1791)

William Harvey's advice in many languages
William Harvey (1578–1657)

Latin   **William Harvey**

Non solet natura usquiam penitiora sua arcana apertius detegere, quam sicubi extra consuetam semitam tenuia sui vestigia monstraverit: nec est ad medicinam recte faciendam tutius iter, quam si quis ex morborum raro contingentium diligenti scrutamine, ad usitatam naturae legem dignoscendam, animum transtulerit. Quippe ita fere in rebus omnibus comparatum est, ut quid illis insit commodi, cuive usui potissimum inserviant, nisi earundem carentia, aut vitiosa constitutione aliqua, vix satis perspicamus.

English   **Robert Willis**

Nature is nowhere accustomed more openly to display her secret mysteries than in cases where she shows traces of her workings apart from the beaten path; nor is there any better way to advance the proper practice of medicine than to give our minds to the discovery of the usual law of Nature by careful investigation of rarer forms of disease. For it has been found in almost all things, that what they contain of useful or applicable nature is hardly perceived unless we are deprived of them, or they become deranged in some way.

Arabic   **Musa Abdelaziz and Nadir Farid**

لا يبدو من المألوف في أي مكان أن الطبيعة تبدو أكثر انفتاحا لعرض أسرار غموضها أكثر مما هو عليه الأمرفي الحالات التي تبدي فيها آثار أعمالها على غير النسق المألوف. كما أنه لا يوجدهناك طريقة مثلى لتطوير المزاولة المناسبة للطب إلا في اطلاق العنان لعقولنا لاكتشاف القانون الاعتيادي للطبيعة عن طريق الاستقصاء الدقيق للأشكال النادرة من المرض. فقد لوحظ في الأشياء جميعها تقريبا بأن ما تحتويه من الطبيعة المفيدة أو المنطبقة لا يتم ادراكه بسهولة الا إذا حرمنا منه أو تم تعطيله بطريقة أو بأخرى.

Chinese Wu Min with the assistance of Shijie Cai

大自然从不轻易地展示她的奥秘，祇在偏离常规时方显露其运转之轨迹。切记靠细心研究罕见疾病所揭示的自然普遍规律，以推动医学的正确实践，除此别无它途。这是因为万物中凡带有实用性的东西，除非失去了它们，或发生了变异，几乎都不被人们所觉察。

Croatian Eduard Klain

Priroda nigdje otvorenije ne otkriva tajne svojih misterija kao u slučajevima kada pokazuje tragove svog djelovanja izvan utabane staze; niti postoji bolji način da se unaprijedi vlastita medicinska praksa nego kad se upustimo u otkrivanje običnih zakona prirode pri istraživanju rijedih oblika bolesti. Jer je otkriveno da se gotovo svi korisni i upotrebljivi sadržaji prirode koji se nalaze u gotovo svim stavrima primjećuju tekar kad su nam uskraćeni ili kad se na neki način pobrkaju.

Dutch Jules Leroy

Nergens is de Natuur meer gewend haar geheime mysteries te openbaren dan in die gevallen waar zij sporen laat zien van haar werkwijzen die sterk van de gebaande weg afwijken; er is trouwens geen betere manier om de goede en juiste geneeskunst te bevorderen dan ons verstand ten volle te geven aan de ontdekking van de gebruikelijke wet der Natuur door zorgvuldig onderzoek van de meer zeldzame vormen van ziekte. Voor bijna alle dingen werd immers vastgesteld dat, wat zij aan nuttigs en toepasselijks inhouden, moeilijk begrijpelijk is, tenzij ons die eigenschappen worden ontzegd of zij op een of andere wijze worden ontregeld.

French Jérôme Lejeune

La Nature n'est jamais plus coutumière de montrer apertement ses mystères secrets que lorsqu'elle laisse voir les traces de son action hors des chemins battus; et il n'est meilleure façon de faire avancer le propre exercice de la médecine qu'en adonnant nos esprits à la découverte des lois ordinaires de la Nature par une soigneuse investigation des formes rares des maladies. Car il se trouve en presque toutes choses, que ce qu'elles contiennent d'utile ou d'applicable ne se perçoit guère que si nous en sommes privés ou si elles sont derangées de quelque manière.

Ga (Ghanaian language) Felix Konotey-Ahulu

Dze nɛng naakpɛ nibii dzeɔ kpo yɛ gbɛdziang ni afɔ-ɔ namɔ kpitio kpitio. Ni nɔni haah tsofah kasemɔh kɛ feemɔ yaa hiɛh krɛdɛɛ lɛ dzi bɔni wɔ kwɛɔ helai ni afɔ-ɔ namɔ mli fitsofitso. Edzaakɛ, wɔ yoo sɛɛ akɛ wɔ nang nokonoko sɛɛnamɔ bɛdza ashɔh nakai nii lɛ yɛ wɔ dɛng.

German

Hartwig Cleve

Die Natur enthüllt ihre Geheimnisse nirgends so sehr als in solchen Zuständen, in denen ihre Arbeitsweise ein wenig vom üblichen Lauf abweicht; es gibt deshalb auch kein besseres Verfahren die Kenntnisse und Praktiken der Medizin voranzutreiben, als das sorgfältige Studium seltene Krankheitszustände. Es ist mit fast allen Dingen so, daß die nützlichen und sinnvollen Einrichtungen oftmals erst erkannt werden, wenn sie uns fehlen oder in irgendeiner Weise gestört sind.

Greek (Modern)

Maria Fraser

Η φύση πουθενά αλλού δεν συνηθίζει πιο ανοικτά να αποκαλύπτει τα απόκρυφα μυστικά της παρά μόνο σε κείνες τις περιπτώσεις που φανερώνει τ'αχνάρια της εργασίας της ξέχωρα απο το πατημένο μονοπάτι. Ούτε υπάρχει καλύτερος τρόπος που συντελεί στη πρόοδο της ιατρικής επιστήμης απο το να αφιερώνουμε τη σκέψη μας στην ανακάλυψη του συνηθισμένου νόμου της φύσης μέσω της προσεκτικής διερεύνησης των σπανιωτέρων μορφών της πάθησης. Γιατί έχει βρεθεί σχεδόν για όλα τα πράγματα ότι οτιδήποτε χρήσιμο ή εφαρμόσιμο έχουν, τότε μόνο γίνεται αντιληπτό όταν το στερηθούμε ή όταν κατά κάποιο τρόπο διαταραχθεί.

Hebrew

Matatiau Glassner with the assistance of Roy Cohen

אין זה ממנהגו של הטבע לחשוף לפנינו באופן בהיר את סודותיו העלומים אלא במקרים ואין שיטה טובה יותר; בהם הוא מציג לעינינו עקבות פעילות הסוטות מן הדרך המוכרת לקדם את מדע הרפואה מאשר מתן הדעת לגילוי חוקי הטבע הפועלים בעולמנו דרך חקר זאת כיוון שהרי גלוי לנו שרוב תופעות הטבע; מעמיק אחר צורות נדירות של מחלה נתפשות על ידינו כמובנות מאליהן עד לאותו רגע בו חסרונן מורגש או שהסדר המוכר לנו משתבש בדרך כולשהי.

Hungarian

Zoltán Papp

A természetnek sehol sem szokása nyíltabban feltárni titkos rejtélyeit, mint azokban az esetekben, ahol a kitaposott úttól eltérő munkálkodásnak nyomait sejteti és nincs jobb mód a helyes orvosi gyakorlatot előbbre vinni, mint elménket a természeti törvények felfedezésének szentelni a betegségek ritkább formáinak gondos kutatása segítségével. Mert majdnem minden dologban azt találjuk, hogy a hasznosat, vagy alkalmazhatót csak akkor értékeljük, ha azoktól megfosztanak bennünket vagy valamely formában zavart szenvednek.

Italian

Angelo Serra

La natura non usa mai svelare nel modo più aperto i suoi segreti se non quando mostra tracce del suo operare al di fuori del sentiero battuto; nè c'é via migliore per progredire in una adeguata pratica medica che applicare la propria mente alla scoperta delle ordinarie leggi della natura mediante una accurata ricerca di forme più rare di malattia. Infatti si sa che, quasi per ogni cosa, ciò che esse contengono di utile o applicabile è difficilmente percepito se non quando ne siamo privati o in qualche modo vengono distrutte.

Italian   Italo Barrai

In nessun caso la natura mostra più apertamente i propri segreti, di quand'ella mostra le tracce dell'opera sua in remoti sentieri; nè v'è modo migliore di avanzar la pratica della medicina, che portare le nostre menti alla scoperta delle usuali leggi della Natura attraverso l'accurata ricerca sulle più rare forme di infermità. Poichè si vede in quasi tutte le cose, che quanto esse contengono di natura utile od applicabile, non è percepito a meno che non ci sia tolto, o venga in alcun modo guastato.

Japanese   Akira Yoshida with the assistance of Katsuko Kasai

自然は常道をはなれたるところにこそ真の秘
めたる妙技の跡を遺すものなれば、希有な病態
の探求を通じて自然法則を発見すべく努めて
こそ医学・医術の進歩はもたらされるのである。
吾人はなべて、その欠失・不調に遇わざれば、事物
の効用を覚らざるものなり。

Norwegian   Pål Møller

Naturen viser fremfor alt sine hemmeligheter der hvor det finnes avvik fra det vanlige. Der er ikke bedre veier til medisinske fremskritt enn ved nøyaktig å vurdere naturens lover ved omhyggelig å undersøke sjeldne sykdommer. For det har vist seg i alle forhold, at det nyttige og hensiktsmessige i naturen er vanskelig å oppdage før det er tatt bort eller ødelagt.

Polish   Marie Ferguson-Smith

Natura nigdzie nie jest bardziej przyzwyczajona do ujawnienia swoich tajemniczych sekretów niż w przypadkach gdzie okazuje ślady swojej czynnosci poza znaną ścieszką; też nie ma lepszego sposobu do postępu w praktyce medycyny niż poddać nasz umysł do odkrycia utartego prawa Natury przez ostrożne badania rzadkich chorób. Jest to uzasadnione że prawie we wszyskich żeczach, to co jest pożyteczne albo do zastosowania jest spostrzegane tylko wtedy, kiedy jesteśmy tego pozbawieni albo kiedy istnieją jakieś zaburzenia.

Portuguese   Francisco Salzano

A Natureza em nenhuma outra situaçao está mais acostumada a mostrar livremente os seus misteriosos segredos do que em casos em que ela apresenta traços de seu trabalho afastados do caminho mais comum; e nem há melhor maneira de avançar a prática apropriada da medicina do que nos concentrarmos na descoberta das leis usuais da Natureza, pela investigação cuidadosa das formas mais raras de doença. Pois tem sido encontrado em quase todas as coisas, que o que elas contém de natureza útil ou aplicável é dificilmente percebido a não ser que sejamos privados delas, ou elas se alterem de alguma maneira.

Romanian   Eva Yap-Todos

Niciunde natura nu e obişnuită să-şi expună mai evident misterele secrete, decât în cazurile în care aceasta ne arată urme ale lucrărilor sale, care se abat de la calea bătătorită; şi nici nu există vreo cale mai bună pentru avansarea potrivită a practicii medicale, decât aceea de a ne dedica minţile descoperirii legilor normale ale Naturii, prin investigaţii meticuloase ale formelor rare de boală. Căci s-a constatat că, aproape în toate lucrurile, ceea ce este folositor sau aplicabil este valorificat cu greu, până în momentul în care suntem deprivaţi de ele sau, dacă acestea suferă anumite modificări.

Russian   Nikolay Bochkov

Нигде больше Природа не готова так открыто демонстрировать свои самые сокровенные таины, как в тех случаях, когда она показывает резултаты трудов своих вдали от протоптанных троп: не существует также более правильново подхода к медицинской практике, чем обратитъ свои разум на открытие обыкновенного закона Природы путем осторожного исследования редких форм болезни. Ибо почти для всех явлений и вещей характерно, что мы почти не воспринемаем то ценное или пригодное что в нихсодержится, пока не лишаемся этого тем или иным образом.

Slovak   Štefan Sršeň

Nikde priroda nezvykne otvorenejšie odhalit' svoje záhady ako v prípadoch, keď sama ukazuje stopy svojho pôsobenja mimo vychodeného chodníka; ani niet lepšieho spôsobu ako pokročit' v medicínskej praxi než venovat' našu pozornost' na objavovanie bežnych zákonov prírody pozorným skúmaním zriedkavejších foriem ochorenia. Pretože sa zistilo temer vo všetkom, že to, čo tam má charakter užitočny a aplikovateľný sa ťažko chápe, ak nie sme o to pripravení, alebo ak sa to nejakým spôsobom nenarušilo.

Spanish   Antonio Velázquez

La Naturaleza no está en ningún sitio más dispuesta a revelar sus misteriosos secretos, que en aquellos casos en los cuales nos muestra las huellas de su obra fuera de los caminos habituales; y no hay mejor manera de avanzar hacia una práctica más adecuada de la medicina, que abriendo nuestras mentes al descubrimiento de las leyes de la Naturaleza mediante la investigación cuidadosa de los casos de enfermedades en sus formas más raras. Porque lo que se ha visto en casi todas las cosas, es que lo que contienen de útil y aplicable apenas se percibe, a menos que estemos faltos de ellas o que se distorsionen en alguna forma.

Welsh   **Evelyn Hughes with the assistance of Gwyn Campbell**

Prin y bydd Natur yn amlygu ei dirgelion cyfrin yn unman yn fwy agored nag y bydd mewn achosion lle mae arlliw o'i phrosesau mwy ymylol i'w weld; nid oes ychwaith well ffordd o symud meddygaeth o'r iawn ryw yn ei blaen na thrwy roi ein bryd ar ddarganfod deddfau arferol Natur wrth archwilio'n fanwl ffurfiau llai cyffredin ar afiechyd. Canys gwelwyd ymron pob peth, nad yw'r hyn ynddo sy'n ddefnyddiol neu'n gymwysadwy prin yn cael ei amgyffred oni bau i ni geal ein hamddifadu ohono neu iddo gael ei afreoleiddio mewn rhyw fodd.

Aspects of development in human genetics since 1956

A. G. Bearn

241 South Sixth Street, Apartment 211, Philadelphia, Pennsylvania 19108, USA

The generous invitation to participate in the Festschrift for our friend and colleague George Fraser requires by long tradition a previously unpublished contribution to human genetics. To my regret, it has been many years since I have been involved, even indirectly, with recent laboratory research. Chagrined by necessity to pen the previous sentences, I must content myself with some general musings on the development of human genetics since I, as a complete tyro, attended the first International Congress of Human Genetics held in Copenhagen in August 1956 under the presidency of Tage Kemp, one of the early contributors to the field. The luminaries of the field that attended the meeting included Haldane, Nachtsheim, Penrose, Fisher, Böök and Muller. It was all rather daunting to a 33-year-old physician who was beginning to be interested in the field. I was there on the very flimsy ground that I had been working on Wilson's disease, a condition which was recessively inherited. As I reflect on the major advances that have been achieved during the past fifty years I am persuaded that one of the axial threads that connect these advances has been the introduction and subsequent exploitation of new methods. And this viewpoint is the thrust of this short piece.

Few would deny that the extraordinary advance in human cytogenetics came in 1955 when Levan learned from Hsu, working with Tjio, that the chromosomes were more easily examined in a squash preparation using hypotonic saline with the addition of colchicine to arrest the chromosomes in metaphase. Using these critical methodological changes, the chromosome number of man was correctly recognised as 46 and human cytogenetics flourished thereafter. The use of fibroblasts in cell culture quickly followed. It is perhaps worth noting that at the 1956 congress, Fuchs reported on successful amniocentesis and led the way to the antenatal detection of hereditary disease, and thereby initiated numerous ethical questions not fully resolved today.

Between 1941 and 1948 Martin and Synge in England had developed a method of partition chromatography to analyse amino acids from protein hydrolysates. It was in 1943 that they concluded that paper was superior to silica gel in their chromatographic experiments and in a groundbreaking paper in 1944 in the *Biochemical Journal* they introduced a methodology soon known as paper chromatography which had as profound effect on human genetics as it did in biochemistry. In human genetics the study of hereditary aminoacidurias was launched. Phenylketonuria is of course one of the earliest conditions studied, when an increased urinary excretion of phenylpyruvic acid was found and, in this way, phenylketonuria was separated from other forms of mental defect. Those interested in ancient history will recall Archibald Garrod's studies on alkaptonuria where he demonstrated that when homogentisic acid is fed to patients with the disease it is excreted, essentially quantitatively, in the urine and coupled with other experiments gave impetus to Garrod's concept of *Inborn Errors of Metabolism*.

Another major advance in human genetics is due to Tiselius, who separated proteins depending on their charge. Whether using electrophoresis in free solution or supported in a medium such as paper or starch an astonishing variety of serum proteins was revealed and a new era in human genetics was founded. The introduction by Smithies showed that starch gel electrophoresis enabled proteins to be separated according to size as well as charge. Perhaps the most illuminating observations were those of Ingram who not only demonstrated that there was a chemical difference between normal and sickle cell haemoglobin but determined that the difference was merely the substitution of the negatively charged glutamic acid being placed by the positively charged valine. It was, in fact, Itano and Pauling who first showed that the normal haemoglobin of sickle cells differed in their electrophoretic mobility, which led to the path breaking studies of Ingram while working in the Perutz laboratory in Cambridge.

These early observations provided powerful proof for the genetic determination of protein structure. The three dimensional structure of haemoglobin was ultimately determined by Perutz using crystallography in 1960. The explosion of knowledge which flowed from methodologies to separate genetic variants of proteins led to a revisiting of the concept of balanced polymorphisms. The finding of dozens of genetic variants without any demonstrable physiological difference led to the view that many were neutral in an evolutionary sense. There is a massive recent endeavour to inquire whether certain allelic variations of a protein may be linked to an increased liability to certain human diseases and response to chemical agents including drugs. The discovery by Nathans and co-workers of restriction enzymes in 1975 revolutionised our understanding in the analysis and the structure of the DNA molecule and disclosed that in higher organisms such as man structural genes are not contiguous but inter-spaced by non-structural regions whose function is still not fully understood. Using this method it was possible to analyse the gene sequence of any segment of DNA which can then be amplified. DNA polymorphisms appear to be considerably greater in number than polymorphisms based on variations in protein polymorphism.

It would of course be inept to fail to mention the epoch making discovery of Watson and Crick in 1952 on the structure of DNA and the cardinal role it plays in molecular biology. Remarkably, it was said at one time that the discovery of the structure of DNA had little impact on human genetics. The importance of Sanger who in 1981 described a method to determine the sequence of DNA can hardly be exaggerated; this ultimately culminated in the triumphant completion of the human genome on 14 April 2003.

Haldane in 1954 made the suggestion that genetic influences might determine the differences observed in man following the use of drugs. Motulsky and Vogel, who introduced the term pharmacogenetics, launched a remarkable avalanche of examples where genes influenced the metabolism of therapeutic agents; polymorphic variation may change the disposition of drugs. Recently it has become apparent that the cytochromes are particularly relevant in this regard. The gene coding for cytochrome CYP2D6 can boast more than 50 single nucleotide polymorphisms. Of more than casual interest is the catalysis of codeine to morphine by the CYP2D6 variants. Specific allelic variations have a different frequency in certain populations. For most of these untoward drug reactions the phenotypic variation is inherited in simple Mendelian fashion. However recent studies suggest that in some instances multifunctional genes may obtain. The well-known difficulty in obtaining satisfactory control in patients on Warfarin may depend on whether patients are either homozygous or heterozygous for the mutant allele of *CYP269*. It is famously dangerous to make predictions but there may come a day when pharmaceutical companies will have to address the genetic individuality of patients to their drugs and will have to consider tailoring them to specific individuals or populations to obtain a satisfactory pharmacological effect with few adverse reactions.

In concluding this slim piece in honour of George, it would be tempting to discuss gene therapy. Thus far, gene therapy has not proved as useful as the theorists had once trumpeted. Even more uncertain is the therapeutic value of embryonic stem cells. Many of the primary studies are behind us but the therapeutic uncertainties that remain are towering. A study to treat spinal cord injuries with embryonic stem cells has already been proposed. At this juncture it seems prudent to remark that all therapeutic uses of embryonic cells lie in the future. Perhaps when George is celebrating a future Festschrift in 2050 these reservations will seem archaic. *Carpe diem*. It has been the burden of this short piece in George's honour to adumbrate, with a few selected examples, the astonishing progress that has been made in human genetics and it is my belief that this has been importantly propelled by the development of new methodologies which in many ways reflect the concepts of paradigmatic shifts of Thomas Kuhn. While my belief of the central role that new methodologies will obtain in the future there is no doubt, as our honouree has consistently demonstrated, there is still room for able and genetically astute clinicians to make significant and enduring discoveries.

Australia antigen and the biology of hepatitis B

Baruch S. Blumberg

Institute of Cancer Research, Fox Chase Cancer Center, 7701 Burholme Avenue, Philadelphia, Pennsylvania 19111, USA

George Fraser has lived a life rich in events and accomplishments. He has dedicated himself to research and scholarship; at the same time he has enriched the friendships that have made the scientific endeavor the fulfilling life it has been for many of us.

The discovery of the infectious agent associated with hepatitis B and the elucidation of new mechanisms for its dissemination are the consequences of a series of studies involving many investigators in our laboratory in Philadelphia. The particular directions the work has followed have been a product of the interests and personalities of the investigators, physicians, technicians and students among others. It has resulted in a complex body of data involving several disciplines. I have been fortunate in having as co-workers dedicated and highly motivated scientists. We have had a congenial atmosphere and I am grateful to my colleagues for this.

Polymorphism and inherited variation

E. B. Ford defined polymorphism as 'the occurrence together in the same habitat of two or more (inherited) discontinuous forms of a species, in such proportions, that the rarest of them cannot be maintained merely by recurrent mutation' (Ford 1956). Examples of polymorphism are the red blood cell groups in which the different phenotypes of a system may occur in high frequencies in many populations. This, in Ford's view, would be unlikely to occur as a consequence of recurrent mutation operating alone to replace a phenotype lost by selection. Another example is the sickle cell haemoglobin system, in which Hb^S genes may be lost from the population each time a homozygote (who has sickle cell disease) fails to contribute to the next generation because of death before the reproductive age. The heterozygotes (Hb^S/Hb^A) appear, however, to be differentially maintained in the population because individuals with this genotype are less likely to succumb to falciparum malaria and consequently survive to contribute to the next generation. The theory implies that there are different selective values to the several forms of polymorphisms. This notion has been questioned recently since it has been difficult to demonstrate selective differences for most (DNA-identified) polymorphisms. Independent of the biological causes for the generation and maintenance of polymorphisms, the concept unifies many interesting biological data. No two people are alike, and polymorphisms probably account for a great deal of variation in humans. There are other important implications of polymorphisms. In some instances, the presence of a small amount of a material may be associated with one effect, and the presence of larger amounts of the same material may be associated with a very different effect. One gene for haemoglobin S protects against malaria, while two genes result in the (often) fatal sickle cell disease. Polymorphisms may produce antigenic differences. Antigenic variants of ABO and other red blood cell groups may cause transfusion reactions. Differences in Rh groups may cause life-threatening antigenic reactions between a mother and her child late in pregnancy and at the time of birth. Polymorphic antigens may have an effect when one human's tissues interact with those of another in blood transfusion, transplantation, pregnancy, intercourse, and possibly, as we shall see, when human antigens are carried by infectious agents.

Oliver Smithies (earlier a graduate student of A.G. Ogston, my mentor at Oxford) developed the starch-gel electrophoresis method that allowed the separation of serum proteins on the basis of complex characteristics of their size and shape. With this, he distinguished several electrophoretically different polymorphic serum proteins (haptoglobins, transferrins, and the like). In 1957 and for several years after, in collaboration with Anthony Allison who was then in Oxford, we studied these variants in Basque, European, Nigerian, and Alaskan (Allison *et al.* 1958, Blumberg *et al.* 1959) populations and found striking variations in gene frequencies. At the same time, I acquired experience and some skill in mounting field studies. Using this and similar techniques in the following years, I studied inherited variants in other populations and regions. These included red blood cell and serum groups in Spanish Basques, in Alaskan and Canadian Indians, and in Eskimos; β-aminoisobutyric acid excretion in Eskimos, Indians, and Micronesians; protein and red blood cell antigens in Greeks, and various variants in North and South American Indians and in U.S. blacks and whites (Alberdi *et al.* 1957, Corcoran *et al.* 1959, Allison *et al.* 1959, Blumberg and Gartler 1959). We identified several 'new' polymorphisms in animals. With Michael Tombs, another of Ogston's pupils, we discovered a polymorphism of alpha lactalbumin in the Zebu cattle of the pastoral Fulani of northern Nigeria (Blumberg and Tombs 1958). Later, Jacob Robbins and I found a polymorphism of the thyroxine binding prealbumin of *Macaca mulatto* (Blumberg and Robbins 1961). From these studies, and those of other investigators, the richness and variety of biochemical and antigenic variation in serum became strikingly apparent.

In the summer of 1960, Allison came to my laboratory at the National Institutes of Health. We decided to test the hypothesis that patients who received large numbers of transfusions might develop antibodies against one or more of the polymorphic serum proteins (either known or unknown) which they themselves had not inherited, but which the blood donors had. We used the technique of double diffusion in agar gel (as developed by Professor Ouchterlony of Göteborg) to see whether precipitating antibodies had formed in the transfused patients which might react with constituents present in the sera of normal persons.

After testing sera from 13 transfused patients (defined as a person who had received 25 units of blood or more), we found a serum that contained a precipitating antibody (Allison and Blumberg 1961). It was an exciting experience to see these precipitin bands and realise that our prediction had been fulfilled. The antibody developed in the blood of a patient (C. de B., male), who had received many transfusions for the treatment of an obscure anaemia. He was extremely cooperative and interested in our research and on several occasions came to Maryland from his home in Wisconsin for medical studies and to donate blood.

We soon found that the antibody in C. de B.'s blood reacted with inherited anti-genic specificities on the low density lipoproteins. We termed this the Ag system; and it has subsequently been the subject of genetic, clinical, and forensic studies (Blumberg *et al.* 1962). We searched for other precipitating systems in the sera of transfused patients on the principle that this approach had resulted in one significant discovery and that a further search would lead to other interesting findings. During my last year at Bethesda, Harvey Alter, a haematologist, came to work with us. We had also been joined by Sam Visnich, a pilot, who, during a slack period in aviation, came to work in our laboratory as a technician.

In 1963, we had been studying the sera of a group of haemophilia patients from Mt. Sinai Hospital in New York City, which had been sent to us by Richard Rosenfield. Antibodies against the Ag proteins were not common in this group of sera, but one day we saw a precipitin band that was unlike any of the others. It had a different configuration, it did not stain readily with Sudan black (suggesting a low lipid content compared to the Ag precipitin), but it did stain red with azocarmine, indicating that protein was a major component. There was a major difference in the distribution of the sera with which the transfused haemophilia patient reacted. Most of the antisera to Ag reacted with a large number (usually about 50 to 90% of the panel sera), but the serum from the haemophilia patient reacted with only one of 24 sera in the panel, and that specimen was from an Australian Aborigine (Blumberg 1964, Blumberg *et al.* 1965). We referred to the reactant as Australia antigen, abbreviated Au. The original Australian sera had been sent to us by Robert Kirk. We subsequently went to Western Australia to collect and test a large number of additional sera.

We then set out to find out why a precipitin band had developed between the serum of a haemophilia patient from New York and that of an Aborigine from Australia. At the outset we had no set views on the explanation, although we were guided by our prior experience with Ag. In preparing this 'history' of the discovery of antigen Au, I constructed an outline, based on a hypothetico-deductive structure, showing the actual events that led to the discovery of the association of Au with hepatitis. It is clear that I could not have planned to find the cause of hepatitis B. This experience does not encourage an approach to basic research that is based exclusively on specific-goal-directed programs for the solution of biological problems.

Our next step was to collect information on the distribution of Au and antibodies to Au in different human populations and disease groups. We had established a collection of serum and plasma samples, later to develop into the blood collection of the Division of Clinical Research of the Institute for Cancer Research, which now numbers more than 200,000 specimens. The antigen was very stable; blood that had been frozen and stored for ten years or more still gave strong reactions for Au. There were some instances in which blood had been collected from the same individual for six or more successive years. If the sera were positive on one occasion, they were in general positive on subsequent testings; if negative initially, they were consistently negative. Presence or absence of Au appeared, at least in the early experiments, to be inherent to an individual.

We were able to use our stored sera for epidemiological surveys and, in a short time, accumulated much information on the worldwide distribution of Au. It was very rare in apparently normal populations of the USA; only one of 1000 sera tested was positive. However, it was quite common in some tropical and Asian populations (for example, 6% in Filipinos from Cebu, 1% in Japanese, and 5–15% in certain Pacific Ocean populations). We will come back to a consideration of the hypothesis that was generated from this set of epidemiological observations after consideration of an interesting disease association discovered at about the same time.

Visnich had been asked to select from our collection the sera of patients who had received transfusions in order to search for more antisera to Au. He decided, however, to use them both as potential sources of antibody and also in the panels against which antisera to Au were tested. Included among the transfused sera were specimens from patients with leukaemia who had received transfusions. A high frequency of Au, rather than antisera to Au, was found in this group. We subsequently tested patients with other diseases and found Au only in transfused patients.

From these observations we made several hypotheses. Although they sound like alternative ones, they in fact are not; and in a sense, all of them have subsequently been supported and are still being tested.

One hypothesis was that, although Au may be rare in normal populations, individuals who have Au are more likely to develop leukaemia than are those without the antigen. That is, there is a common susceptibility factor which makes it more likely for certain people both to have Au and to develop leukaemia. We also suggested that Au might be related to the infectious agent (virus) which is said to be the cause of leukaemia.

A corollary of this hypothesis is that individuals who have a high likelihood of developing leukaemia would be more likely to have Au. Down syndrome patients are more likely to develop leukaemia than are other children; estimation of the increased risk varies from 20 to 2000 times that of children without Down syndrome. I had, in 1964, moved to the Institute for Cancer Research in Philadelphia to start its Division of Clinical Research. There we tested the sera of Down syndrome patients resident in a large institution and found that Au was very common in this group (approximately 30% were Au positive); the prediction generated by our hypothesis was fulfilled by these observations, a very encouraging finding (Blumberg *et al.* 1967). The presence of the antigen in people near Philadelphia also made it possible to study persons with Au more readily. Until this time, everyone with Au who had been identified either lived in Australia, or some other distant place, or was sick with leukaemia.

Down syndrome patients were admitted to the Clinical Research Unit (located in our sister institution, Jeans Hospital) for clinical study. We found again that the presence or absence of Au seemed to be a consistent individual attribute. If Au was present on initial testing, then it was present on subsequent testing; if absent initially, it was not found later. In early 1966 one of our Down syndrome patients, James Bair, who had originally been negative, was found to have Au on a second test. Since this was an aberrant finding we admitted him to the Clinical Research Unit. There was no obvious change in his clinical status. Because he apparently had developed a 'new' protein and since many proteins are produced in the liver we did a series of 'liver chemistry' tests. These showed that between the first testing (negative for Au) and the subsequent testing (positive for Au) this patient had developed a form of chronic anicteric hepatitis.

On 28 June 1966, the day of J.B.'s admission to the Clinical Research Unit, my colleague, Alton Sutnick, wrote the following dramatic note in J.B.'s chart:

> SGOT [serum glutamic oxaloacetic trans-aminase] slightly elevated! Prothrombin time low! We may have an indication of [the reason for] his conversion to Au+.

His prediction proved correct. The diagnosis of hepatitis was clinically confirmed by liver biopsy on 20 July 1966, and we now began to test the hypothesis that Au was associated with hepatitis (London *et al.* 1969). First, we compared the transaminase (SGPT, serum glutamic pyruvic transaminase) levels in males with Down syndrome who had Au and those who did not. The SGPT levels were slightly but significantly higher in the Au(+) individuals. Second, we asked clinicians in Pennsylvania to send us blood samples from patients with acute hepatitis. W. Thomas London and others in our laboratory soon found that many such patients had Au in their blood early in their disease, but the antigen usually disappeared from their blood, after a few days or weeks. Another dramatic incident occurred which added to our urgency in determining the nature of the relation of Au to hepatitis. Barbara Werner (now Dr) was the first technician in our laboratory in Philadelphia. She had been working on the isolation of Au by extensions of the methods developed earlier by Alter and Blumberg. Early in April 1967 she noticed that she was not in her usual good state of health. She was well aware of our observations that Au was related to hepatitis and, one evening, tested her own serum for the presence of Au. The following morning a faint but distinct line appeared, the first case of viral hepatitis diagnosed by the Au test. She subsequently developed icteric hepatitis and, fortunately, recovered completely.

By the end of 1966 we had found that Au was associated with acute viral hepatitis. In our published report (Blumberg *et al.* 1967) we said

> Most of the disease associations could be explained by the association of Au(l) with a virus, as suggested in our previous publications. The discovery of the frequent occurrence of Au(l) in patients with virus hepatitis raises the possibility that he agent present in some cases of this disease may be Australia antigen or be responsible for its presence. The presence of Australia antigen in the thalassaemia and haemophilia patients could be due to virus introduced by transfusions.

That is, we made the hypothesis that Au was (or was closely related to) the aetiological agent of 'viral' hepatitis, and we immediately set about to test it. Our original publication was not widely accepted; there had been many previous reports of the identification of the causative agent of hepatitis so that our claims were naturally greeted with caution. Indeed, an additional paper on Australia antigen and acute viral hepatitis (London *et al.* 1967) which extended our findings published in 1967 was initially rejected for publication on the grounds that we were proposing another 'candidate virus' and there were already many of these.

Confirmation of our findings and the first definitive evidence on the relation of Au to post-transfusion hepatitis came soon. Kazuo Okochi, then at the University of Tokyo, had followed a line of inquiry very similar to ours. He had started with the investigation of antiserum to Ag (lipoprotein), and we had corresponded on this subject. Okochi then found an antiserum in a patient with chronic myelogenous leukaemia which was different from the precipitins in anti-serum to Ag. He also found that it was associated with liver damage. During several field trips to Japan, I had lectured on Australia antigen. Okochi sent the unusual antiserum to us to compare with antiserum to Australia antigen; we found that they were identical. He confirmed our finding of the association of Au with hepatitis and then proceeded to do the first definitive study of transfusion. He found that Au could be transmitted by transfusion and that it led to the development of hepatitis in some of the recipients, and that some transfused patients developed antibody to Au (Okochi and Murakami 1968, Ninomiya and Kaneko 1970). The Au-hepatitis association was also confirmed in 1968 by Alberto Vierucci (Vierucci *et al.* 1968) who had worked in our laboratory and Alfred Prince (Prince 1968).

We had made some preliminary observations in Philadelphia in collaboration with John Senior of the University of Pennsylvania on the transfusion of donor blood which was found to contain Au. We then prepared a controlled, long-term study to determine whether donor bloods which had Au were more likely to transmit hepatitis than those which did not. In 1969 we heard from Okochi that he had already embarked on similar transfusion studies. In June of that year he visited us and showed us his data. These, in his (and our) opinion, demonstrated with a high probability that donor blood containing Australia antigen was much more likely to transmit hepatitis than donor blood which did not contain the antigen. [Similar studies were later done by Dr David Gocke (Gocke and Kavey 1969) and the same conclusions were reached.] We immediately stopped the study and established the practice of excluding donor bloods with Australia antigens in the hospitals where we were testing donor units. This was a dramatic example of how technical information may completely change an ethical problem. Before Okochi's data had become available it was a moral necessity to determine the consequences of transfusing blood containing Australia antigen; and it had to be done in a controlled and convincing manner since major changes in blood transfusion practice were consequent on the findings. As soon as the conclusion of Okochi's well-controlled studies were known to us, it became untenable to administer donor blood containing Australia antigen. *Autres temps, autres moeurs.*

It was, however, possible to do a study to evaluate the efficacy of Au screening on post-transfusion hepatitis using historical controls. Senior and his colleagues had completed an analysis of post-transfusion hepatitis in Philadelphia General Hospital before the advent of screening and found an 18% frequency of post-transfusion hepatitis. In 1969, we started testing all donor blood and excluding Au positive donors. Senior and others undertook a similar follow-up study one year after the screening program began. They found that the frequency of post-transfusion hepatitis had been reduced to six percent, a striking improvement (Senior *et al.*1974).

The practical application of our initially esoteric finding had come about only two years after the publication of our paper on the association between Au and hepatitis (Blumberg *et al.* 1967). In retrospect, one of the major factors contributing to the rapid application of the findings was the simplicity of the immunodiffusion test. Another was our programme of distributing reagents containing antigen and antibody to all investigators who requested them. We did this until this function was assumed by the National Institutes of Health.

After the confirmation of the association of hepatitis with Australia antigen, many studies were published, and, in a relatively short time routine testing in blood banks became essentially universal in the USA and many other countries. It has been estimated that the annual saving resulting from the prevention of post-transfusion hepatitis amounts to about half a billion dollars in the USA.

Virology

Virological methods (e.g., tissue culture and animal inoculation) had been used for many years before our work to search for hepatitis virus, but had not been very productive. Our initial discoveries were based primarily on epidemiological, clinical, and serological observations. Here, I will try to review the early virology work from our laboratory [Robinson and Lutwick have reviewed later work (Robinson and Lutwick 1967)].

Bayer *et al.* (1968), using the isolation techniques initially introduced by Alter and Blumberg (1966), examined isolated Au with the electron microscope. They found particles about 20 nanometres in diameter which were aggregated by antiserum to Au. There were also sausage-like particles of the same diameter, but much elongated (Figures 1 and 2). Subsequently Dane, Cameron, and Briggs identified a larger particle about 42 nm in diameter with an electron-opaque core of about 27 nm (Dane *et al.* 1970). It is probable that this represents the whole virus particle. Both the 20-nm and 42-nm particles contain Australia antigen on their surfaces and this is now termed hepatitis B surface antigen (HBsAg). The surface antigen can be removed from

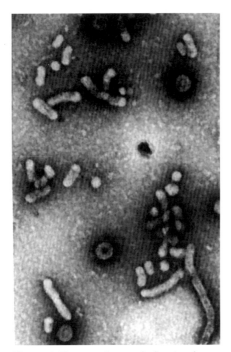

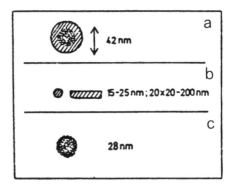

Figure 2. Diagram showing appearance of particles associated with hepatitis B virus, the large or Dane particle (a), the small surface antigen particle, and the sausage-shaped particle (b), and the core of the Dane particle (c).
[Adapted from E. Lycke, *Lakartidningen* **73** 3743 1976]

Figure 1. Electron micrograph showing the several kinds of particles associated with hepatitis B virus (see Figure 2). Magnification, x 90,000. [Electron micrograph prepared by E. Halpern and L. K. Weng]

Dane particles by the action of detergents to reveal the core which has its own antigen, hepatitis B core antigen (HBcAg). Antibodies to both these antigens (anti-HBs, anti-HBc) can be detected in human blood. The surface antigen can be detected in the peripheral blood by the methods we initially introduced and by more recent, more sensitive methods. Anti-HBs is often found in the peripheral blood after infection and may persist for many years. It may also be detected in people who have not had clinical hepatitis. Anti-HBc is usually associated with the carrier state (that is, persistent HBsAg in the blood) but may occur without it. HBcAg itself has not been identified in the peripheral blood. Anti-HBc is also found commonly during the active phase of acute hepatitis, before the development of anti-HBs, but in general does not persist as long as anti-HBs.

DNA has been isolated from the cores of Dane particles and is associated with a specific DNA polymerase. Robinson and Lutwick have shown that the DNA is in the form of double-stranded rings (Robinson and Lutwick 1976). Jesse Summers, Anna O'Connell, and Irving Millman of our Institute have confirmed these findings and provided a model for the molecule, which appears to have double- and single-stranded regions (Figure 3) (Summers *et al.* 1975).

By immunofluorescence and electron microscopy, hepatitis B core particles have been identified in the nuclei of liver cells of infected patients; HBsAg is found in the cytoplasm. It is thought that assembly of the large particles occurs in the cytoplasm and that large and small particles (surface antigen only) emerge from the cells and eventually find their way to the peripheral

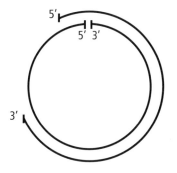

Figure 3. Structure of the DNA extracted from Dane particles proposed by Summers *et al.* (1975). The position of the gaps in the single strands and the location of the 5′ and 3′ ends are shown.

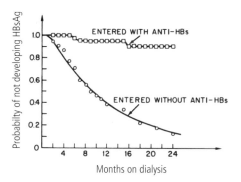

Figure 4. Probability of not developing HBsAg for patients admitted to a renal dialysis unit with and without anti-HBs. The patients with anti-HBs are relatively well protected while those without antibody are very likely to develop infection. [Adapted from Lustbader *et al.* (1976)]

blood. In 1968 we were informed by the US federal government, which provided most of the funds for our work, that they would like to see applications of the basic research they had funded for many years. It occurred to us that the existence of the carrier state provided an unusual method for the production of a vaccine. We presumed that the very large amounts of HBsAg present in the blood could be separated from any infectious particles and used as an antigen for eliciting the production of antibodies. The antibodies in turn would protect against infection with the virus. Irving Millman and I applied separation techniques for isolating and purifying the surface antigen and proposed using this material as a vaccine. To our knowledge, this was a unique approach to the production of a vaccine; that is, obtaining the immunizing antigen directly from the blood of human carriers of the virus. In October 1969, acting on behalf of the Institute for Cancer Research, we filed an application for a patent for the production of a vaccine. This patent was subsequently (January 1972) granted in the USA and elsewhere (Blumberg and Millman 1972).

There are observations which indicate that antibody against the surface antigen is protective. In their early studies, Okochi and Murakami observed that transfused patients with antibody were much less likely to develop hepatitis than those without it (Ninomiya and Kaneko 1970). In a long-term study, London *et al.* (1977b) showed that patients of a renal dialysis unit, and the staff who served them, were much less likely to develop hepatitis if they had antibody than if they did not (Figure 4). Lustbader has used these data to develop a statistical method for rapidly evaluating the vaccine (Lustbader *et al.* 1976).

There have now been several animal and human studies of the vaccine, and the results are promising (Krugman *et al.* 1971, Hilleman *et al.* 1975, Maugh 1975, Purcell and Gerin 1975, Buynak *et al.* 1976, Maupas *et al.* 1976). It should be possible to determine the value of the vaccine within the next few years.

Variation in response to infection with hepatitis B

A physician is primarily interested in how a virus interacts with humans to cause disease. But this is only part of the world of the virus. Our introduction to studies on hepatitis B was not through patients with the disease, but rather through asymptomatic carriers and infected individuals who developed antibody. Therefore, many of our investigations have been of infected but apparently healthy people. There are a variety of responses to infection:

1. Development of acute hepatitis followed by complete recovery. Transient appearance of HBsAg and anti-HBc. Subsequent appearance of anti-HBs which may be persistent.
2. Development of acute hepatitis proceeding to chronic hepatitis. HbsAg and associated anti-HBc are usually persistent.
3. Chronic hepatitis with symptoms and findings of chronic liver disease not preceded by an episode of acute hepatitis. HBsAg and anti-HBc are persistent.
4. Carrier state. Persistent HbsAg and anti-HBc. Carrier is asymptomatic but may have slight biochemical abnormalities of the liver.
5. Development of persistent anti-HBs without detectable HBsAg or symptoms.
6. Persistent HBsAg in patients with an underlying disease often associated with immune abnormalities, that is, Down syndrome, lepromatous leprosy, chronic renal disease, leukaemia, primary hepatic carcinoma. Usually associated with anicteric hepatitis.
7. Formation of complexes of antigen and antibody. These may be associated with certain 'immune' diseases such asperiarteritis nodosa.

Family studies

In our first major paper on Australia antigen (Alter and Visnich 1965) we described family clustering of Au in a Samaritan family from Israel that had been studied by the anthropologist Batsheva Bonne. From it we inferred the hypothesis that the persistent presence of Au was inherited as a simple autosomal recessive trait. The genetic hypothesis has proved to be very useful not in the sense that it is necessarily 'true' [exceptions to the simple hypothesis were noted by us and others very soon (Blumberg 1972)], but because it has generated many studies on the family distributions of responses to infection with hepatitis B. We suggested that hepatitis virus may have several modes of transmission. It can be transmitted horizontally from person to person, i.e. the mode of transmission of 'conventional' infectious agents. This is seen in the transmission of hepatitis B virus (HBV) by transfusion. Other forms of direct and indirect horizontal transmission exist; for example, by sputum, by the faecal-oral route, and, perhaps, by haematophagous insects (see below). It has even been reported that it has been spread by computer cards (Patterson *et al.* 1974), an extraordinary example of adaptation by this ingenious agent! HBV may also be transmitted vertically. If the genetic hypothesis were sustained, then it would imply that the capacity to become persistently infected is controlled (at least in part) as a Mendelian trait. The data are also consistent with the notion that the agent could be transmitted with the genetic material; that the virus could enter the nucleus of its host and in subsequent generations act as a Mendelian trait. The data also suggest a maternal effect. A reanalysis of our family data showed that in many populations more of the offspring were persistent carriers when the mother was a carrier than when the father was a carrier. Many investigators have now shown that women who have acute type B hepatitis just before or during delivery or women who are carriers can transmit HBV to their offspring, who then also become carriers. This may be a major method for the development of carriers in some regions, for example, Japan. Interestingly, this mechanism does not appear to operate in all populations. This suggests that some aspects of delivery and parent-child interaction, differing in different cultures, as well as biological characteristics may affect transmission. The family is an essential human social unit. It is also of major importance in the dissemination of disease. A large part of our current work is directed to an understanding of how the social and genetic relations within a family affect the spread of hepatitis virus.

Host responses to human antigens and HBV: kidney transplantations

London, Jean Drew, Edward Lustbader, and others in our laboratory have undertaken an extensive study of the patients in a large renal dialysis unit in Philadelphia (London *et al.* 1977a, Blumberg *et al.* 1977, London *et al.* 1977b). The renal patients can be characterised on the basis of their responses to infection with hepatitis B. Patients who develop antibody to HBsAg are significantly more likely to reject transplanted kidneys that are not completely matched for HLA antigens than patients who become carriers of HBsAg (Figure 5) (London *et al.* 1977a). Since many of the patients became exposed to hepatitis B while on renal dialysis, their response to

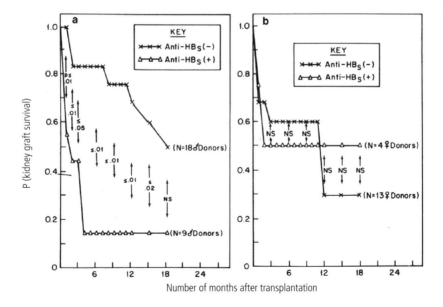

Figure 5. (a) Probability of rejecting a kidney graft by renal dialysis patients who received kidneys from male donors. There is a significant difference in rejection rate between patients who were carriers and those who developed anti-HBs (London *et al.* 1977b).
(b) Probability of rejecting a kidney graft by renal dialysis patients who received kidneys from female donors. There is no difference in the rejection rates between the two groups of patients (London *et al.* 1977b).

infection can be determined prior to transplantation. In this patient population there is a significant correlation between development of anti-HBs and the subsequent development of antibodies to HLA after transplantation. We have also found a correlation between the development of antibody to HLA and anti-HBs in transfused haemophilia patients and in pregnant women. Hence, there appears to be a correlation between the response to infection with HBV and the immune response to polymorphic human antigens in tissue transplants. Further, from preliminary studies, it appears that donor kidneys from males are much more likely to be rejected by patients with anti-HBs than by patients without anti-HBs. These differences were not observed when the kidneys were from female donors. London is now extending his observations to other transplants, in particular, bone marrow, to determine whether a similar relation exists.

Sex of offspring and fertility of infected parents

In many areas of the world, including many tropical regions (for example, the Mediterranean, Africa, southeast Asia, and Oceania) the frequency of HBsAg carriers is very high. In these regions, most of the inhabitants will eventually become infected with HBV and respond in one of the several ways already described. Our family studies and mother-child studies show that there is a maternal effect. Jana Hesser (then a graduate student in anthropology working in our laboratory) and Ioanna Economidou, Stephanos Hadziyannis, and our other Greek colleagues collected information on the sex of newborns in a Greek town in southern Macedonia. In this community the probability of infection with HBV is very high and a majority of the parents had evidence of infection, i.e., detectable HBsAg or anti-HBs (or both) in their blood. It was found that if either parent was a carrier of HBsAg there were significantly more male offspring than in other matings (Hesser *et al.* 1975). Using the Greek data and additional data from Mali in West Africa, London, Drew, and Veronique Barrois (a postdoctoral trainee from Paris) subsequently found that there is a deficiency of male offspring when parents have anti-HBs and that this may be a consequence of differential male mortality during the period in utero (Drew *et al.* 1977). This had led London and his colleagues to test the hypothesis that anti-HBs has specificities in common with Hy or other histocompatibility antigens determined by genes on the Y chromosome. If these observations are supported by additional studies, then HBV may have a significant effect on the composition of populations in places where it is common, which include the most populous regions of the world. The ratio of males to females in a population has a profound effect on population size as well as on the sociology of the population. This connection of anti-HBs with sex selection may also explain why there is a greater likelihood of rejection of male kidneys by renal patients with anti-HBs, and indicate how kidneys can be better selected for transplantation. Pregnancy and transplantation of organs have certain immunological features in common. Rejection of male kidneys and 'rejection' of the male foetus may be mediated by similar biological effects.

Primary hepatic carcinoma

The project with which we are most concerned at present is (i) the relation of hepatitis B to primary hepatic carcinoma (PHC), and (ii) methods for the prevention of the disease. PHC is the most common cancer in men in many parts of Africa and Asia. Investigators in Africa, including Payet *et al.* (1956) and Davies (Steiner and Davies 1957, Davies 1973), had suggested that hepatitis could be the cause of PHC. With reliable tests for Australia antigen we could test

Table 1. Frequency of HBsAg, anti-HBc and an anti-HBs in primary hepatic carcinoma (PHC) and controls in Senegal and Mali, West Africa. Abbreviations: RIA, radioimmunoassay; *P* is the two-tailed probability obtained from Fisher's Exact Test. (Adapted from Larouzé *et al.* (1977))

Test	Patient				Control				*P*
	Number tested	+	−	Percent positive	Number tested	−	+	Percent positive	
Senegal PHC									
HBsAgRIA	39	31	8	79.4	53	6	47	11.3	4×10^{-11}
Anti-HBc	39	35	4	89.7	58	16	42	27.6	1×10^{-9}
Anti-HBs	39	8	31	20.5	58	26	32	44.8	0.02
Total exposed	39	37	2	94.8	58	38	20	65.1	8×10^{-4}
Mali PHC									
HBsAgRIA	21	10	11	47.6	38	2	36	5.2	4×10^{-4}
Anti-HBc	20	15	5	75.0	40	10	30	25.0	5×10^{-4}
Anti-HBs	21	8	13	38.0	40	17	23	42.5	0.95
Total exposed	21	19	2	90.4	40	15	115	62.0	0.02

this hypothesis; there is indeed a striking association of hepatitis B with PHC (Blumberg *et al.* 1975, Larouzé *et al.* 1976) (Table 1). In our studies in Senegal and Mali we found that almost all the patients had been infected with HBV and that most had evidence of current infection (presence of HBsAg or anti-HBc, or both). Ohbayashi and his colleagues (Ohbayashi *et al.* 1972) had reported several families of patients with PHC in which the mothers were carriers. In our study in Senegal (Larouzé *et al.* 1976*)*, Bernard Larouzé and others found that a significantly larger number of mothers of PHC patients were carriers of HBsAg compared with controls, and that none of the fathers of the cases had anti-HBs. In control families, on the other hand, 48% of the fathers developed antibody (Table 2). Our consequent hypothesis is that, in some families, children willbe infected by their mothers, either in utero, at the time of birth, or shortly afterward during the period when there is intimate contact between mother and children. In some cases, the infected child will develop PHC. At each stage, only a fraction of the infected individuals will proceed to the next stage, and this will depend on other factors in the host and in the environment. The stages include retention of the antigen (carrier state), development of chronic hepatitis, development of cirrhosis and finally, development of PHC (Figure 6). We are currently testing this hypothesis in prospective studies in West Africa (Larouzé *et al.* 1977). If it is true, then prevention of PHC could be achieved by preventing infection with HBV, and the vaccine we have introduced, in association with appropriate public health measures, could reduce the amount of infection. This might also involve the use of γ-globulin in the newborn children of carrier mothers, and such studies are now being conducted by Beasley and his colleagues in Taipei. We are now considering the appropriate strategies that might be used to control hepatitis infection and, perhaps, cancer of the liver.

Transmission by insects
HBsAg has been detected by several investigators including Smith *et al.* (1970), Prince *et al.* (1972), Muniz and Micks (1973)*,* and others in mosquitoes collected in the field in areas

Table 2. Frequency of HBsAG, anti-HBc and anti-HBs in patients with primary hepatic carcinoma (PHC) and controls, and the parents of patients and controls. The studies were conducted in Dakar, Senegal, West Africa. Abbreviations: ID, HBsAg by immunodiffusion; RIA, HBsAg by radioimmunoassay. (Adopted from Larouzé *et al.* (1976))

Item	N	+	%+	N	+	%+	P
	Primary hepatic carcinoma (PHC)			*Controls*			
HBsAg(+)ID	28	9	32.1	28	5	17.9	0.35
HBsAg(+)RIA	28	22	78.6	28	16	57.1	0.15
Anti-HBc(+)	28	25	89.2	28	18	64.3	0.05
Anti-HBs(+)	28	7	25.0	28	18	64.3	6×10^{-3}
HBsAg(+), anti-HBc(+) or anti-HBs(+)*	28	27	96.4	28	26	92.9	0.99
	Mothers of PHC			*Mothers of controls*			
HBsAg(+)ID	28	15	53.6	28	3	10.7	1×10^{-3}
HBsAg(+)RIA	28	20	71.4	28	4	14.3	3×10^{-5}
Anti-HBc(+)	28	20	71.4	28	9	32.1	6.9×10^{-3}
Anti-HBs(+)	28	3	10.7	28	15	53.6	1×10^{-3}
HBsAg(+), anti-HBc(+) or anti-HBs(+)*	28	21	75.0	28	19	67.9	0.76
	Fathers of PHC			*Fathers of controls*			
HBsAg(+)ID	27	2	7.4	27	3	11.1	0.99
HBsAg(+)RIA	27	5	18.5	27	5	18.5	1.00
Anti-HBc(+)	27	5	18.5	27	8	29.6	0.52
Anti-HBs(+)	27	0	0	27	13	48.1	3×10^{-5}
HBsAg(+), anti-HBc(+) or anti-HBs(+)*	27	5	18.5	27	18	66.6	7×10^{-6}

* Any evidence of infection with HBV

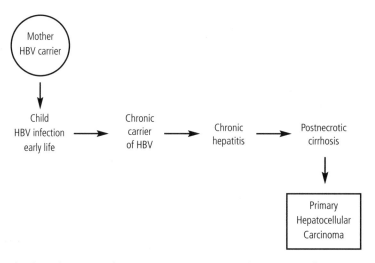

Figure 6. Scheme for the pathogenesis of primary carcinoma, showing the sequence of stages leading to PHC.

where HBsAg is common in the human population. In 1971, we collected mosquitoes in Uganda and Ethiopia and found Au antigen in some (Blumberg *et al.* 1973). In more extensive studies in Senegal, we found a field infection rate of about 1 in 100 for *Anopheles gambiae* and also identified the antigen in several other species of mosquitoes (Wills *et al.* 1976a). It is not known whether HBV replicates in mosquitoes, but it has been reported that it can be detected in mosquitoes many weeks after feeding and it has been found in a mosquito egg. Feeding experiments have been conducted with the North American bedbug *(Cimex lectularius);* these studies show that this insect can also carry the antigen (Newkirk *et al.* 1975).

William Wills, in our laboratory, has found a very high infection rate (~60%) in the tropical bedbug *Cimex hemipterus* collected from the beds of individuals known to be carriers of hepatitis B (Wills *et al.* 1976b). Bedbugs could transfer blood (and the virus) from one occupant of a bed to another. If it is in fact a vector of hepatitis, then it could provide a frequent non-venereal (and unromantic) form of connubial spread. It may also provide a means for transmission from parent to young children who may share the parents' bed in early life; and this would be related to child-rearing practices.

Insect transmission may be important in the programme for the control of hepatitis B infection and for the prevention of chronic liver disease and primary hepatic carcinoma. An understanding of the role of insects in the spread of infection, particularly its transmission from mother to children, would help in designing effective strategies for control.

Hepatitis B as a polymorphism

The original discovery of hepatitis B resulted from the study of serum antigen polymorphisms. Its identification as an infectious agent does not diminish the value of this concept. It is useful to view infection with HBV not only as a 'conventional' infection but also as a transfusion or transplantation reaction; our studies on renal transplantation are a case in point.

HBV appears to have only a small amount of nucleic acid, probably only sufficient to code for a few proteins. Much of the coat (and possibly other portions of the virus) could be produced by the genes of the host. Millman and his colleagues found that the surface antigen contains material with antigenic specificities in common with serum proteins including IgG, transferrin, albumin and β-lipoprotein (Millman *et al.* 1971). If this is true, then the antigenic makeup of the virus would be, at least in part, a consequence of the antigenic characteristics of the host from whence it came; and this, as suggested by our sex studies, may include male antigens. In our discussion of the 'Icron' concept (a name we introduced which is an acronym on the Institute for Cancer Research) (Blumberg *et al.* 1971), we pointed out that the responses of the putative host to HBV may be dictated in part by the nature of the 'match' between the antigens of the host and virus (i.e., the virus acts as if it were a polymorphic human antigen). London *et al.* (1972) and Werner and London (Werner 1975) have described this. If person A is infected with HBV particles that contain proteins antigenically very similar to his own, then he will have little immune response and will tend to develop a persistent infection with the virus. On the other hand, if the proteins of the agent are antigenically different from his, he will develop an immune response to the virus (that is, anti-HBs) and will have a transient infection. During the course of infection in person A, new particles will be synthesised which contain antigenic characteristics of A. In turn, person A can infect person B and the same alternatives present themselves. If the relevant proteins of B are antigenically similar to the antigens of A and the antigens of the HBV produced by A, then B could develop a persistent infection. If they are different, then antibody can form, as described above. (A derivative of this hypothesis is that inflammatory disease of the liver will be associated with the immune response to the infectious agent rather than solely with replication of the agent.) A further possibility is that the virus has complex antigens: some may match the host and some may not and both persistent infection and development of antibodies may occur. The persistent antigens and the antibody in the same individual would have different specificities, and this occurrence has been described (Raunio *et al.* 1970).

This view of the agent as an Icron introduces a complicating element into the epidemiology of infectious agents in which not only the host and virus are factors, but also the previous host or hosts of the agent. If, in fact, the agent does replicate in insects (see above), the antigenic characteristics of previous human hosts may be affected by transmission through another species. This in turn might alter the response of the next host.

Bioethics and the carrier state

During our work a number of bioethical questions have arisen (Blumberg 1976). Experience has shown that these bioethical considerations cannot be separated from 'science,' that answers cannot be provided on a 'purely scientific' ground, and that our technical knowledge is inseparably intertwined with bioethical concerns.

It has been recognised that hepatitis B may be transmitted by means other than transfusion, that is, by contact, faecal-oral spread, insects, and the like. With the introduction of the screening test, many carriers were identified. It is estimated that there are one million such carriers in the USA and more than 100 million in the world. This has led to a situation that may be unique in medicine. Although some carriers may be able to transmit hepatitis by means other than blood transfusion, this is probably not true for many (or most) carriers.

There are studies which show that spread of infection from carriers in health care occupations to patients may not be common. At present there is no satisfactory method of identifying the infectious carriers although it appears that carriers with 'e' antigen, an unusual antigen originally described by Magnius (1975), are much more likely to transmit disease. Despite this, carriers have had professional and social difficulties. Health care personnel who are carriers have been told that they must leave their jobs. In some cases, carriers have changed their pattern of social behaviour because of the fear that they might spread disease to people with whom they come in contact. What appeared to be happening was the development of a class of individuals stigmatised by the knowledge that some member of the 'class' could transmit hepatitis.

The bioethical problems raised from the studies of hepatitis carriers can be viewed as a conflict between public health interests and individual liberty. When the risk to the public is clear, and the restrictions on personal liberties are small, there is little problem in arriving at appropriate regulations. For example, the transfusion of blood containing hepatitis B antigen is a disadvantage to the patient recipient and it has been stopped. The denial of the right to donate blood is not a great infringement of personal activity, and the individuals concerned and society have agreed to accept this moderate restriction. The problems raised by person-to-person transmission are more difficult. The extent of the hazard to the public is not clear, since it is not (now) possible to distinguish carriers who transmit disease from those who do not. On the other hand, if all carriers are treated as infectious, the hazards imposed on the carrier may be enormous, that is, loss of job and ability to continue in the same profession, restriction of social and family contacts, and others. What is clear is that for a very large number of carriers, the risk of transmitting hepatitis by person-to-person contact must be very small. All members of the carrier class should not be stigmatised because some can transmit hepatitis.

On a broader level, an important issue is how far biological knowledge about individuals should impinge on daily lives. Is it appropriate to regulate the risks inherent in people living together and interacting with each other? An issue has been raised with respect to hepatitis because the test is straightforward and because millions of people are tested as part of blood donor programmes. As a consequence of these tests, this particular group of carriers has been identified. There are carriers of other agents, some of them potentially more hazardous (such as staphylococcus otyphoid), and these carriers are not routinely tested and therefore not placed at a disadvantage.

It is hoped that many of these problems can be resolved by continued research into the nature of the hepatitis carrier state, and that carriers who have already been identified will not be jeopardised during this period when necessary information is not available.

A characteristic of many large-scale public health control programmes is the emergence of problems that were not anticipated prior to the institution of the program. For example, the control of malaria has in many areas resulted in a markedly decreased infant mortality and a consequent increase in population. When this has not been accompanied by a concomitant increase in food production, the nourishment and well-being of the population have actually decreased.

With the availability of the serological and environmental tests for hepatitis B, it is now possible to begin the design of control measures for this disease. If the hepatitis B vaccine is found to be effective, then it may also be of value in preventing the development of the carrier state. We are now attempting to investigate the biology of the hepatitis B agent to learn whether some of the consequences of control can be known before the program begins. An example already discussed is the possible effect of HBV infection on sex ratio. The role of the virus in the life of the insects in which it is found is not known, but may be profound; and there may be other effects on the ecology that are not now obvious.

Acknowledgement
Supported by NIH grants CA-06551, RR-05539 and CA-06927 and by an appropriation from the Commonwealth of Pennsylvania.

Copyright © The Nobel Foundation 1977

A tribute to Sir Ronald Fisher

L.L. Cavalli-Sforza

Department of Genetics, Stanford Unversity, USA

I am grateful to George Fraser for stimulating me to contribute to his Festschrift and *liber amicorum*, for kindly supplying me with some of the information below, and for translating my little piece on how I met Sir Ronald Fisher from the original Italian.

George Fraser shares with me the good fortune of having been associated at early formative stages of our respective careers with R. A. Fisher, one of the geniuses of our discipline during the twentieth century. As I shall recount in greater detail below, I first met Fisher in 1948 when I was in my mid-twenties; a few minutes into our initial conversation, I received a totally unexpected offer of a job in his department in order to develop bacterial genetics, and I immediately accepted the proposal. The job I got at the University of Cambridge was not exactly as good as the original offer; Fisher was usually on bad terms with the university administration. But I was in any case very happy to work with him in Cambridge, where he was the Arthur Balfour Professor of Genetics.

George completed Part I of the Natural Sciences Tripos at Cambridge in 1952 at the age of twenty, and needed a subject for Part II in his third undergraduate year. He presented himself to Fisher with a trepidation which turned out to be needless. 'We have been allowed to offer Genetics in Part II since last year but there were no applicants,' he was told. 'This year, you are the only applicant – you are welcome'. So George spent an intellectually very stimulating year listening to four lecture courses from Fisher and from George Owen who was a lecturer in the department, and doing practical work in serology with Robin Coombs. He graduated in June 1953 with the first undergraduate degree in genetics in the history of the University of Cambridge, despite the Arthur Balfour Chair in Genetics having been filled with distinction for 41 years by this date.

In the mean time, I had returned to Italy in 1950 to take up a post at the Istituto Sieroterapico Milanese, where I had been after my degree until moving to Cambridge. Fisher felt that perhaps I was leaving because the original promise he had made had not been fulfilled, but the reality was that the offer looked excellent and I also felt that I would not have any other chance of returning to Italy. In any event, Fisher wanted me to speak to the Master of Gonville and Caius, Fisher's College, who told me to be confident that the original promise by Fisher, which was not honoured by the University, would be kept in due course.

Both George and I have been greatly influenced by these close early contacts with Fisher which we much enjoyed and which changed our lives forever. Thus, in common with all those who knew Fisher well, we could, from that time onwards, keenly appreciate the truth of the words of Hermann Ludwig Ferdinand von Helmholtz:

> Any one who has once come into contact with one or more men of the first rank must have his whole mental standard altered for the rest of his life. Such intercourse is, moreover, the most interesting that life can offer.

Once George had completed his undergraduate degree in Cambridge, Fisher asked me to take him into my laboratory in Milan for a few weeks during the summer of 1953. Over the intervening 54 years, we have kept in touch and seen each other from time to time.

After George left Cambridge in 1953, another contributor to this Festschrift and *liber amicorum* and a good friend and colleague of mine, Walter Bodmer, became a research student of Fisher's before the latter was compelled to retire from his chair at the age of 67 years in 1957.

In 1959, Fisher decided to settle in Adelaide where he was a very welcome guest for some considerable time in the family home of Oliver Mayo, one of the editors of this Festschrift and a contributor, who was seventeen years old at the time of Fisher's arrival. So, there are four persons including myself, who are intimately involved with this book and who have been greatly influenced in the formative stages of their respective careers as geneticists by the genius of Fisher, an intellectual giant. This influence has extended beyond the confines of its lifelong professional implications in the direction of fostering strong long-term links of friendship between us, as manifested in the composition of this Festschrift.

Oliver Mayo maintained close contacts with Fisher until his death in Adelaide in 1962 at the age of 72. Subsequently in 1967, George worked in the Department of Genetics at the University of Adelaide for two years at the invitation of Professor Henry Bennett (long devoted to creating the Fisher archives and responsible for the posthumous publication in five volumes of Fisher's scientific papers, together with two finely edited volumes of Fisher's correspondence on statistics and genetics) whom he had met while Henry had been a postgraduate student in Fisher's department in Cambridge in 1952. Henry Bennett and George were the supervisors, and Walter Bodmer and George Owen were the examiners, of Oliver Mayo's PhD thesis in the Department of Genetics, a thesis thus born of an impeccably Fisherian lineage, as indeed was that of Carolyn Leach, the other editor of this volume, as she was also a PhD student of Henry Bennett's.

In 2005, I wrote an autobiographical description, together with my son, Francesco Cavalli-Sforza, of my career as a scientist. The book, published by Mondadori, Milano, is called *Perché la Scienza: l'avventura di un ricercatore (Why Science: the adventure of a researcher)*. George has translated a few paragraphs from the original Italian, relating the circumstances of my first meeting with Fisher at the Eighth International Congress of Genetics in Stockholm in August 1948.

At the Stockholm Congress, I had one of the most wonderful surprises of my life. I knew Ronald A Fisher, the father of modern statistics, by sight. I had made great efforts to understand this subject from his books, but, genius that he was, he did not exert himself to write in a simple way which would be understandable to all. In spite of this, his books on statistics enjoyed a tremendous success, because he had himself created all the most important and innovative methodology. He was also a very great geneticist.

I had heard him at meetings of the Genetics Society of Great Britain, giving lectures and taking part in discussions. His appearance made him stand out from the crowd; straight white hair, a white moustache and beard, a beautiful open smile. I met him in Stockholm at the door of one of the lecture rooms and I told him that I would like to introduce myself. On hearing my name, he asked me: 'Would you like to come to work in my department as a bacterial geneticist?' It seemed impossible to me that he would make such an unexpected proposal out of the blue in this way. For a moment, I was left speechless with astonishment, especially since I did not understand how he could have known who I was. Probably Kenneth Mather, who had been his student, and in whose laboratory I had spent the first part of the year, had spoken to him of me; perhaps he had seen my name and the topic of my research in the program of the congress.

It was difficult to imagine receiving a more alluring proposal. I do not remember very well how I responded but I believe that I accepted on the spot, without posing any questions.

Thus it was that in the first days of October 1948, my life in Cambridge began.

Some thoughts on early ideas of heredity

Alan E.H. Emery

Green College, Oxford

There will be others contributing to this volume who are better qualified than I am to comment on George Fraser's major contributions to Medical Genetics, particularly in regard to his genetic studies of childhood deafness and blindness. My brief essay is intended to honour a great personal friend of many years who has always had a particular interest in the history of ideas, above all those of the Ancient World, being very much a classical scholar himself.

It seems highly likely that from earliest times there has been an awareness that certain characteristics could be transmitted from one generation to another. Furthermore studies by early settlers in Australia revealed that many Aborigines (whose history on the continent dates back at least 20,000 years) had strong taboos against marriages involving close relatives (Spencer and Gillen 1899). Perhaps they recognised the possible effects such relationships might have on the wellbeing of any offspring. The Ancient Egyptian royal families frequently 'married' close relatives. But it has now been suggested that their children were actually conceived outside such royal relationships.

In some instances the transmission of a trait within families was recognised from earliest times though without any understanding of the mechanism. For instance in the Talmud, the encyclopaedia of Jewish teaching dating from the first century AD, the sons of sisters of the mother of a son with the bleeding disease (haemophilia) were excused from circumcision. The sons of the father's siblings were not so excused. Similarly Gowers recognised in 1879 that like haemophilia, Duchenne muscular dystrophy or Meryon's disease had a predilection for males and when familial was inherited through the mother (Gowers 1879). The cytological basis for X-linkage however, would not be recognised until several decades later.

There are many other instances where a particular trait was recognised to be transmitted in families. Many ideas were advanced to try to explain the mechanisms involved, and some will be explored here, but satisfactory explanations had to await the Mendelian era.

Interestingly the term 'heredity' was only introduced into the English language from the French *hérédité* in the mid-19th century. The term was preferred to 'inheritance' by biologists because the latter had become associated with discredited Lamarckian ideas. Francis Galton claimed to have been the first to use the term but others, including Charles Darwin, had started using the term several years earlier (Blakemore and Jennett 2001).

Aristotle and the early Greeks

It is to Ancient Greece that we can trace back the origins of natural scientific thought (Cruse 2004). Following Hippocrates (c. 460–370 BC) medicine became more a branch of learning rather than philosophy. Nevertheless for a period the faith-healing tradition of Asclepius (the Greek god of healing) and the more rational approach to diagnosis and treatment of Hippocrates continued side by side:

It may be guessed that when physicians of the kind we may with some reservations call secular refused to treat the incurably sick, they were in effect entrusting them to the unexplained powers of Asclepius, as revealed by his priests ... (Phillips 1973)

Among those with a questioning approach to human and animal biology, including heredity, was Aristotle (384–322 BC) whose works include the eight books comprising his *Politics*. It is here that he makes a clear definition (in Books III and V) of heredity as applied to constitutional kingships and the transference of property – a very different matter from the transmission of inherited traits.

In his *Historia Animalium* he discusses in considerable detail the diverse manner in which animals reproduce. He notes that in humans, a man can '... generate up to the age of seventy, a woman up to fifty. Few people at these ages produce children. Generally the limit for men is sixty-five, for women forty-five.' (Aristotle, *Historia Animalium* (Book V: Chapter XIV, 545 b, 30). And he later emphasises '... the excitement connected with sexual appetite and especially with the pleasure derived from sexual intercourse' (Book VI: Chapter XVIII, 571 b, 10). Though he comments on the transmission of coat colour in goats, he clearly believed that the environment (good pasture, water source, wind direction) was important in this case in determining the characteristics of the offspring.

It is in his *De Generatione Animalium* that he considers how offspring come to resemble their parents more than other individuals. Like Anaxagoras (fl. c. 500–428 BC) and Democritus (fl. c. 460 BC), Aristotle believed that male and female (*sic*) 'semen' was formed from every part of the body of each parent (Book I, 721 b, 15), and through sexual intercourse would mix and result in a child resembling its parents. But he noted (Book I, 722 a, 5) that the resemblance of children to their parents is no proof that the 'semen' comes from the whole body because resemblance can occur in voice, nails, hair and way of moving. He clearly was distinguishing congenital and acquired characteristics. Furthermore he realised that if semen is formed from every part of the body of both parents how do we explain parts of a child being only male or only female (Book I, 722 b, 8)? Thus 'semen' could not be secreted from *all* parts of the body (Book I, 723 b, 10). Moreover how could it be explained that sometimes a child resembles an antecedent more than a parent (Book IV, 769 a, 25)? It was also difficult to explain 'monstrosities' (Book IV, 769 b, 10) or as we might refer to them nowadays, congenital abnormalities? Aristotle raised problems concerning heredity which would not find meaningful explanations for another 2000 years.

Maupertuis: The theory of deficiency or excess

Pierre Louis Moreau de Maupertuis (1698–1759) was a French polymath who was a renowned mathematician in his time (Velluz 1969). For example his mathematical analysis of data he collected in northern Finland, as well as elsewhere, proved that the earth was flattened at the poles. He also refuted the theories held by preformationists who believed that a minute human body (a *homunculus*) was present in either the ovum (ovists) after Regnier de Graaf (1641–1673), or sperm (animalculists) after Nicholaas Hartsoeker (1656–1725), which then subsequently developed into a full organism (Lancaster 1995). But it is his ideas about the basis of heredity which are of interest here.

He studied the transmission of polydactyly in a large family as well as in the offspring of a similarly affected bitch from which he bred. These studies led him to postulate that hereditary particles were present in the germ cells, each had an affinity for a like particle from the other parent and each of a corresponding pair was transmitted by a parent to its offspring, and that the particle from either parent may dominate. If there were too few particles then a defect resulting from a deficiency could occur (*monstre par défaut*). Alternatively if there were too many particles then a defect resulting from an excess could occur (*monstre par excès*). Finally if a sudden complete alteration of particles occurred this could account for the origin of a new species (de Maupertuis 1756, Hervé 1912). In some ways these ideas were similar to those of Mendel. But whereas Maupertuis associated dominance or recessiveness with an *excess* or *deficiency* of particles, Mendel associated the inheritance of such traits with the *expression* of one of the two alternative forms of a trait. Other fundamental differences have been pointed out by Sandler (1983). Nevertheless Maupertuis's ideas were certainly original and perhaps because of their very originality were little appreciated at the time.

Hunter: Breeding experiments

John Hunter (1728–1793) was a renowned Scottish anatomist and surgeon in Georgian London (Moore 2005). Among his many interests was the familial occurrence of congenital abnormalities such as spina bifida and harelip. He also carried out a number of breeding experiments on animals and was able to demonstrate which variations and deviations were passed down. His work included crossing a dog with a wolf in order to determine if they belonged to the same species. From his observations on interbreeding, published in the *Philosophical Transactions of the Royal Society* in 1787, he concluded that the wolf, jackal and dog were indeed the same species. Others had carried out similar crossbreeding experiments, such as the French naturalist Buffon, but without success. Hunter was the first to attempt to demonstrate inheritance by such experiments.

Unfortunately Hunter does not appear to have generated any ideas about the mechanisms involved. He was after all a pragmatist. He shunned theorising, preferring to establish facts by careful observation and experiment. In any event the use of animals for breeding experiments would at the time have presented a very major challenge if the aim had been to establish firm ideas of the mechanisms of inheritance.

Dalton: Inheritance of colour blindness

John Dalton (1766–1844) is renowned throughout the scientific world for his enunciation of the Atomic Theory. But he was also interested in his own colour blindness. A story goes that he once chose a scarlet cloth for a suit, not at all befitting someone with his Quaker beliefs!

Though others had commented on colour blindness before him, Dalton gave the first detailed and clear description of the condition (Emery 1988). He presented his observations in a lecture to the Manchester Literary and Philosophical Society on 31 October 1794. This was published in the Society's journal four years later and then in 1831 in the *Edinburgh Journal of Science* (Brockbank 1944, Smyth 1966). Dalton's priority in describing the condition is still recognised in France, where it is known as 'daltonisme'. Dalton noted that his type of colour blindness (now believed to be *protanopia*, an X-linked condition) was relatively common and only affected males (his brother was similarly affected). It did not affect females, nor the parents and children of affected males. He could not offer at the time any explanation for his incredibly perceptive observations. He thought the defect resided in the humours of the eye, not in the retina as we now know.

Adams: Distinguishing patterns of inheritance

Joseph Adams (1756–1818) was the first to write a book in English on hereditary diseases (Adams 1814). This was based on a lifetime of careful clinical observations. This led him to distinguish between congenital disorders (present at birth) which were more often *familial* (by which he meant affected siblings), than *hereditary* (affecting parent and offspring). The former tended to be more serious often resulting in death in early life:

> congenital diseases are more commonly familial, than hereditary: some of them being mortal, cannot indeed be transmitted, of which connate hydrocephalus or watery head is one among other instances. (Adams 1814, p. 14)

Furthermore he observed that in some other congenital disorders which are compatible with survival (congenital cataract, deaf-mutism), the offspring are not affected. He also recognised that inherited diseases often follow a similar course within a family and therefore where there is an early age of onset:

> those of the children who have passed that age without any of the symptoms may be considered as free from the constitutional disposition. (Adams 1814, p. 21)

Thus Adams clearly distinguished between the pedigree patterns of what we would now refer to as dominant and recessive modes of inheritance, though not sex-linked inheritance with transmission through unaffected carrier females. This latter had been clearly delineated in the case of haemophilia by John C. Otto (1774–1844) some years earlier (Otto 1803). Sex-linked inheritance in haemophilia was reviewed in detail and formalised by Nasse a few years later (Nasse 1820).

Haemophilia and its mode of inheritance are not referred to by Adams perhaps because he never saw a case, the disorder being relatively rare, and many cases may also have gone undiagnosed in London at the time.

Darwin: The theory of pangenesis

Charles Darwin (1809–1882), when considering the way in which inherited variations in animals and plants occurred, proposed a provisional hypothesis of *pangenesis*, a term actually derived from the Greek literally meaning 'origin of everything'. In his *The Variation of Animals and Plants under Domestication* in Chapter XXVII (Darwin 1868) he proposed that every cell of the body was capable of producing countless tiny granules, called *gemmules*, which were responsible for the formation of individual organs and features. These particles he believed circulated in the body, passing into the reproductive cells thus ensuring their '… transmission in a dormant state, like seeds in the ground, to successive generations.' (Darwin 1868, p. 350 *et seq.*). He elaborated this theory in considerable detail but was vague as to how variations might actually arise. Interestingly an intimate friend of Darwin, George Romanes (1848–1894), proposed at the time a possible explanation which in some respects was somewhat similar to later genetic concepts. He proposed, at a meeting of the Linnean Society in 1886, that 'physiological' factors, by preventing intercrossing, enabled natural selection to promote diversity and thereby evolution. The paper even provoked an editorial in the *Times* (Forsdyke 2001).

Darwin in his inimitable and methodical way acknowledged the deficiencies of his theory of pangenesis. But the theory was dealt a serious blow by his cousin Francis Galton (1822–1911). The latter carried out blood transfusions between different breeds of rabbits which of course failed to affect the characteristics of their progeny even over several generations.

For this and various other reasons the theory of pangenesis became rejected, to be replaced later by the germ-plasm theory of August Weismann (1834–1914). According to this theory, self-sustaining determinants in the nuclei of the germ cells formed the basis of inheritance (Weismann 1889, 1893). This presaged the later Mendelian idea of discrete hereditary factors (later referred to as genes) as being the basis of inheritance.

Conclusions

These brief notes on some early ideas of heredity have centred on human and, to a lesser extent, animal observations and studies, though of course Darwin had also been interested in plant variation. This emphasis is perhaps understandable because up to Mendel's work, plant breeding had almost always concentrated on hybridisation of different species. At least in humans, by the mid-19th century clear pedigree patterns of different modes of inheritance had become recognised. But none of this work generated any testable hypotheses concerning mechanisms. It is for this reason alone that one realises what a debt we owe to the singular original contributions made by Gregor Mendel.

As much as anyone George Fraser would therefore appreciate the following quotation from Plato's *Republic* (4th century BC), a dialogue where the philosopher considers among other things the nature of ideas and education:

λαμπάδια ἔχοντες διαδώσουσιν ἀλλήλοις.

Those having torches will pass them on to others.

George Fraser, a wonderful friend, and the story of Arthur MacMurrough Kavanagh, an Irish squire born without hands and feet

Ademar Freire-Maia

With the collaboration of George Fraser and with the participation of
Newton Freire-Maia, William J. Schull, Henry Viscardi, Confucius, Karl Popper,
Aristotle, Xenophanes, Isaac Newton, Socrates and Plato

UNESP – Paulista State University, Brazil

I bravely resisted writing something about George Fraser. Not, of course, because I dislike him … On the contrary: I love him! The reason, as I wrote to Oliver Mayo and to George when invited to contribute to this *liber amicorum*, was that I am not involved any more with research projects, and my contacts with George were in any case so brief that I would not normally have been expected to have much to say about him. Incidentally, it was during our collaboration over this little story that I have learned more about him and thus have more to say about him.

To begin with, it is not easy to convince George Fraser when he is convinced that he is right! He strenuously insisted that I could write something, anything. 'When Dértia writes her paper – he wrote me – perhaps you could write a few personal sentences jointly'. Fortunately, Dértia (my wife) wanted to write her sentences alone … But George insisted again: 'If nothing else, you can write *Um abraço* [an embrace, in Portuguese].' Or at least imitate Dértia, who originally wrote to him saying she was too busy, 'But I have time to say that I love you and Maria [George's wife]!' If Dértia, quite active scientifically, could write that, why could you not say at least the same? I almost heard him saying. An irrefutable argument, indeed!

But it seemed insufficient to George to convince him that I was already convinced … So then he quoted some sentences from my previous e-mails: 'We [Dértia and I] certainly would like at least to tell the world that we join your uncountable friends and admirers in this homage'; 'The only thing I could write would be I love George, I love George, I love George, one million times!!!!'; 'What about publishing this love letter?' 'YES, YES, YES, – he enthusiastically replied – as important as a scientific communication; Maria and I would love that'. I was finally convinced, after he promised to help me … And he did!

First of all, I would like to apologise for not being very good at history. Being so young (much less than 80 …), I only think of the present and, at the most, of the near future (no more than a few decades ahead!). So, I don't remember very much about dates and what happened when, except that I know that George and I were born in the same year, 1932, he on 3 March in Užhorod (Uzhhorod), Ruthenia, Czechoslovakia, and I on 14 February, in Boa Esperança, Minas Gerais, Brazil. Actually, 64 years later, George was told by one of his Brazilian friends that in 1932 God had intended to put him on the 14 February earth-shuttle which had been collecting passengers, including myself, destined for Boa Esperança, but that he had been left behind by mistake and the next, and sole, place available for him had been on the shuttle bound for Užhorod, 18 days later (1932 was a leap year). The friend in question had been in a position to learn of (and regret) this sad error because he himself had arrived on a shuttle to Rio de Janeiro some years previously and was engaged at the time in monitoring the passenger lists of all shuttles destined for Brazil. Just to think that George and I might have travelled to Boa Esperança (Good Hope in English – what a beautiful and inspiring name!) together! Užhorod (Town on the Už (River) in English), although situated in Czechoslovakia at the time of George's birth, has moved around so much since then, in the company of the entire province of Ruthenia, first as part of an autonomous republic for a few months, then as part of an independent republic for one day, then to Hungary, then back to Czechoslovakia, then to the USSR, and now to Ukraine. With such a start, George's existence has been even more turbulent and peripatetic than that of Ruthenia, whereas Boa Esperança, together with the entire state of Minas Gerais, has stayed firmly anchored in Brazil, affording me a great measure of stability together with the pleasure of enjoying throughout my life *o jeitinho brasileiro*, an expression which is virtually untranslatable except in that these three words encapsulate the spirit of Brazil and the *joie de vivre* characteristic of my (and George's) beloved country.

Despite the fact that after these dramatic events of 1932, our lives continued to develop at opposite ends of the world, George believed that our paths, as young (and promising!) scientists, crossed both in Montreal in 1958 for the Tenth International Congress of Genetics and then in Rome in 1961 for the Second International Congress of Human Genetics. These congresses occurred long before the beginning of George's Brazilian period, and we did not actually meet each other at either of them. The reason is very simple: I was not there … and the Freire-Maia attending those meetings was my brother Newton who presented our joint work on acheiropodia (a mutilating congenital malformation mainly characterised by complete absence of hands, feet, and forearms, and inherited as an autosomal recessive) in Brazil. Although Newton sadly left us in 2003, he has contributed vicariously, in his capacity as a long-time friend of George, to this article, as will become apparent on reading further.

I have later personal memories of George Fraser on at least three occasions, besides the usual knowledge of his outstanding contribution to science. In this last aspect, of course, I share the feelings of a number of geneticists around the world, admiring George's intelligence and the excellence of his scientific papers.

I remember that on one of these three occasions, we had in fact had a brief scientific 'collaboration', or rather a brief collaborative disagreement, in the pages of *The Lancet* in 1970. And so my contribution to this volume acquired a scientific as well as a personal dimension.

For reasons which were never clear to me or to George himself, one day in the 1960s he happened to be studying the issue of 14 March 1891 of *The Lancet* (Anonymous 1891), where he saw a very interesting book review which he photocopied – a favourite occupation of his. Some years later in 1970, in *The Lancet* of 7 March, he saw a letter from me (Freire-Maia 1970a), and he realised that the photocopy which he had put away in his files so long ago might be relevant. So he wrote a letter to *The Lancet* (30 May: Fraser 1970g), pointing out that acheiropodia was not a malformation restricted to Brazil, as I had written briefly in my letter, and at length in my PhD thesis in 1968 (Freire-Maia 1968). According to George, it had also occurred in a scion of the Irish gentry, Arthur McMurrough Kavanagh, born in 1831, who had suffered from such a malformation according to a biography written by his cousin, Sarah Steele, and who was characterised in the book review of 1891 as one of the most striking personalities of the 19th century. According to a jocular remark in George's letter, although Arthur Kavanagh left many descendants, he was not known to have visited Brazil during his worldwide travels …

Here is an extract from Kavanagh's biography taken from the internet (*http://www.kavanagh-family.com/notable/Art/art.htm*):

Arthur was born at Borris House, Co. Carlow, on 25 March 1831, the third son of Thomas Kavanagh (1767–1837), by his second wife, Lady Harriet Margaret Le Poer Trench, daughter of Richard, second earl of Clancarty. His father was M.P. for Kilkenny in the last Irish parliament and for Co. Carlow in the last two parliaments (of the United Kingdom) under George IV, and the first parliament under William IV.

His family traced its descent to the kings of Leinster. Born with only the rudiments of arms and legs Arthur nevertheless, by indomitable resolution and perseverance, triumphed over his physical defects, and learned to do almost all that the normal man can do, better than most men. Though in general carried on the back of his servant, he had a mechanical chair so contrived that he was able to move about the room without even this assistance.

During 1849–1851 he travelled with his eldest brother, Thomas, and his tutor to India by way of Russia and Persia. Tabriz was reached without notable adventure in November 1849, and the party were introduced to a Persian prince, Malichus Mirza. Arthur fell dangerously ill in December, and was nursed in the prince's harem. On his recovery the travellers crossed Lake Urumiah, and rode through difficult country and blinding sleet and snow to Mosul, passing on the way the scene of the recent murder of Stoddart and Conolly and recovering the latter's prayer-book.

So, this was evidently a very unusual history of a man born without hands and feet. However, I was provoked by George's letter into writing another letter to *The Lancet* (3 October: Freire-Maia 1970b), stating that Arthur Kavanagh's malformation, despite the very sketchy anatomical description available, was clearly different from acheiropodia which, at that time, was thought to be restricted to Brazil. As is usual with scientific knowledge, nothing is black or white, but rather different shades of grey, and, many years later, a true case of acheiropodia was described outside my country, in Puerto Rico (Kruger and Kumar 1994). We Brazilians apparently lost the claim to an exclusive 'privilege', but George admits freely that neither he nor any one else had detailed knowledge of the exact nature of the malformation of Arthur Kavanagh, and that it had been wrong of him to write that it was very likely to have been identical to acheiropodia. He still thinks, however, that the 1891 book review provided a story about the achievements of a man born without hands and feet, which was too extraordinary to be forgotten and too good to keep to himself – hence his provocative (in my eyes at least) letter to *The Lancet*.

No matter! I would be very happy if, like him, I had committed only one scientific mistake in my life … And, better than that, if I had given to science the same magnificent contribution he gave!

But this story of our first contact which had started with an 1891 *Lancet* book review did not end with this exchange of correspondence in 1970 in the same journal. George's long-time friend, Peter Froggatt, was so interested in the 1891 review and the 1970 letter, both of which George sent to him at the time, that, almost three decades later, in 1999, he published together with Norman Nevin, a paper about the 'Incredible Mr Kavanagh' (Froggatt and Nevin 1999). He tells us that the family had taken great care not to allow any detailed descriptions or photographs of Arthur's malformation to survive.

So it could be said that my first encounter with George in *The Lancet* of 1970 was just a very small part of a very long story, from an 1891 book review, which George read by chance, to the present book honouring him. In between these events lie a number of other stories involving handicapped people (the acheiropods) who also showed 'indomitable resolution and perseverance' in overcoming their physical defects. Their stories seemed to me, to my brother Newton Freire-Maia and to our friend William J. Schull, so relevant and inspiring for genetic counselling that we published a joint paper (Freire-Maia *et al.* 1975). It is true that, with respect to risk estimates, acheiropodia offers no special counselling problem, because this anatomically and genetically well-delineated congenital malformation is due to a rare autosomal recessive gene with complete penetrance and relatively little variation in expressivity. As we pointed out, the counselling interest in acheiropodia resides much more in the way in which the affected individuals face their tremendous handicap and try to overcome it, just like Arthur Kavanagh.

It is really remarkable how people born without hands, feet, and forearms can succeed so well and lead almost 'normal' lives. We showed examples of acheiropods capable of feeding and taking care of themselves; of throwing snares and chopping wood; reading a magazine and turning the pages with the stumps; painting pictures; writing; playing billiards; riding a horse; doing housework and sewing clothes; and making toys for children and wooden safes. One of the acheiropods was able 'to use a mattock, chop wood, clean the house, ride a horse, shave, sew shoes, cook and shoot a rifle!'

Our opinion was that the genetic counsellor (like other counsellors) is ethically obliged frankly to admit the possibility that the worst may happen. But – and here comes an extremely important aspect of genetic counselling – even if this worst does happen, some positive aspect can be uncovered. It is here that acheiropodia can contribute to the philosophy and practice of genetic counselling. Of course, this is not a prerogative restricted to acheiropodia. Henry Viscardi, founder of a factory where all the employees were crippled (including himself), was absolutely right when he said, in 1959, that there are no disabled people, merely persons with different degrees of ability for different jobs (Viscardi 1959). Arthur Kavanagh was also one of these persons.

The first acheiropodia family, described in the literature, as early as 1929 by Peacock, was called a 'remarkable family' (Peacock 1929). As Newton Freire-Maia, William Schull and I wrote in 1975, there is no doubt that these people are really remarkable, but not so much for their physical handicap and the extreme rarity of the malformation as for their extraordinary capacity to overcome their deficiency and to live as normal human beings, experiencing victories as well as failures. As Viscardi (1962) wrote: 'We are extraordinary people, Jimmy, we the deformed, the misshapen, we without limbs, or sight, or words, or sound, we are extraordinary to live ordinary lives'. For the benefit of all of us who cannot reconcile ourselves to our deficiencies, we also quote Confucius: 'I used to complain about not having shoes and then I met someone who had no feet'.

The second occasion on which I met George (as far as I remember …) was a very pleasant one, when I had the opportunity to get to know him in his warm person rather than in the cold pages of a journal. This occurred during the Eleventh Latin-American Congress of Human Genetics, held in 1994 in Puerto Vallarta, Mexico. George had given a lecture in Spanish, but read out one slide in Portuguese, since he did not have a Spanish translation.

At the festive party which followed, I embraced him and said: 'Your Portuguese pronunciation is much better than your Spanish!' Actually, this is George's version. Mine is somewhat different. What I told him was: 'Your Portuguese is so good that it is even better than your English … ' At that party, I discovered George Fraser. 'What a nice guy he is' – I certainly thought!

We have a poem in Brazil about friendship, which George likes very much and which he has translated into English. The last line reads

A gente não faz amigos, reconhece-os.

And, in George's translation

We do not make friends, we recognise them.

And this is what George and I felt at that party in Puerto Vallarta.

Incidentally, George's translation of this entire poem about friendship, which embodies for him the spirit of the *liber amicorum*, has been reproduced in the contribution of Eleidi Chautard-Freire-Maia, the widow of my late brother, Newton.

George spent only about one percent of his half-century in our profession in Brazil, between 1970 and 1996, and this represents, in total, only about six months. And I could enjoy being with him no more than one percent of that one percent! He gave a whole course in São Paulo, in 1970, between June and October. He began in English, but soon switched to Spanish, a language which he knew well and which was more easily understandable to his audience because of its similarity to Portuguese, a language of which George knew nothing at that time. This change had the interesting result that the language of his lectures gradually changed to '*Portunhol*' (a mixture of Portuguese and 'Espanhol', or Spanish), often spoken by Brazilians and by Latin Americans in general, when they have some knowledge of the other language in addition to their mother tongue. In this way, his lectures became more appreciated by his audience as each week passed. I was in Brazil at that time, but far from São Paulo. A pity, since we could have discussed personally the controversy then raging between us in the columns of *The Lancet* about acheiropodia in Brazil, and the congenital deformity of an Irish squire, Arthur MacMurrough Kavanagh.

George has also lectured in Curitiba (1974), Rio de Janeiro (1992), and in Curitiba (again), Porto Alegre, and Caxambu in 1996 after his attendance at the Ninth International Congress of Human Genetics in Rio de Janeiro, but I missed him each time! I could not be in any of these places. In 1974, in Curitiba, he visited the Department of Genetics of my late brother, Newton, and of his wife, Eleidi, both long-time friends of his. Several articles about George appeared in the local newspapers during this visit to Curitiba, and he was greatly startled to see himself portrayed as a *Cobrão Internacional* (International Cobra) in one of these articles. So he received an explanation – that in our Brazilian career-structure, if the young initiate eventually reaches the top after having wriggled his way through various intermediate stages, it is as a Cobra. Very logical, thought George, a Big Snake!

During George's later visits to Brazil, I understand that his Portunhol which he had kept up assiduously since 1970, improved to such an extent that for long stretches of his lectures, his language became almost pure unadulterated Brazilian Portuguese.

I had the privilege of meeting him again during his last visit to Brazil in 1996. This third, and last, opportunity I had to meet George was certainly most fascinating and agreeable. He was in São Paulo on his way back to England, and we invited him and his wife, Maria, who was accompanying him, to a supper at our apartment. My wife, Dértia, and I have a wonderful remembrance of that lovely night. This is the image of George and Maria that we vividly keep in our minds and in our hearts.

Both George and I, and both Dértia and Maria, regret very much that such opportunities are unlikely to recur. Although we have long left Užhorod and Boa Esperança, respectively, we still live on different continents and the distance between us, sadly enough, precludes further supper engagements.

I am sure that the best way that I can really honour George Fraser is by mentioning once again another long-time friend of his, and another wonderful person and scientist, my late brother Newton. In his customarily warm manner, Newton used to call George '*Jorginho*', and his wife Maria, '*Mariinha*' (affectionate Portuguese nicknames for Little George and Little Mary). When Newton liked someone in an especially warm manner, he used to say that that person was a '*flor*' (flower). So, *Jorginho* received a surname: *Florzinho* (Little Flower), and Maria, another one: *Rosinha* (Little Rose).

Newton could not, therefore, be missing from this volume. He is also here not only in the few remarks above but also in the translation I made, with George's assistance, of some of his sentiments which concern science and the scientist. Although these are mainly addressed to young students and researchers, every reader can appreciate, in all its generous magnanimity, the inexhaustible and limitless love Newton had for science and for the work of the scientist.

He wrote:

The scientist's imagination is as vivid and as bubblingly effervescent as that of the artist, but his love for the facts drives him to crumble them into their tiniest constituent elements.

Under the surface of apparently disorganised and even chaotic data, the scientist can uncover the resplendent hidden order wherein reigns the peerless beauty which the Universe exhibits.

It is the responsibility of Governments to place the world of Science within the reach of their peoples, not only through free education, but also through museums, scientific centres, and radio and TV programs. Science is a type of entertainment and every one should have the opportunity to enjoy it.

The scientist dedicates to science every single hour of his entire life. Full time means 24 hours a day and seven days a week. Exclusive dedication signifies love.

The limits of science are not the limits of our inquietude, of our ideals, or of our deepest aspirations: these have no limits. Our thirst for knowledge is a thirst for the infinite.' (When I showed George this thought, he remembered Karl Popper, who once said that our knowledge can only be finite, while our ignorance must necessarily be infinite).

The true scientist finds enormous satisfaction in performing his work. He loves science, and, therefore, he likes to study and to investigate. As a matter of fact, he would not know how to live his life in any other way: science *is* his life.

The young person who wants to be a scientist and to dedicate all his time and all his love to science, has at least three certainties:

1. he will die one day (like everybody else),
2. he will not be rich (like almost everybody else),
3. he will enjoy himself a lot (like just a very few people)

Science does not possess many things of its own; it has no country of its own, no language of its own, no ethic of its own. But the scientist has a country of his own, uses a language of his own and has an ethic of his own. All of this is outside the realm of science. Music is also outside the realm of science. In the same way, literature, the cinema, the theatre are outside the realm of science. Painting and sculpture too. Religion is also outside the realm of science. The reality of non-scientific things is as important (or more important) than the reality of scientific things. Love is outside the realm of science.'

As George likes quoting Aristotle and Xenophanes, I shall continue this little story by a quotation from the former, following on directly and logically from my remembrances of my brother Newton, our common friend.

ἡδεῖα δ᾽ ἐστὶ τοῦ μὲν παρόντος ἡ ἐνεργέια,
τοῦ δὲ μέλλοντος ἡ ἐλπίς,
τοῦ δὲ γεγενημένου ἡ μνήμη.

Sweet are the joys of activity in the present,
Of hope for the future,
And of remembrance of things past.

And I shall move towards the end of this little story by referring to George's favourite quotation from the latter

οὗτοι ἀπ᾽ ἀρχῆς πάντα θεοὶ θνητοῖσ᾽ ὑπέδειξαν;
ἀλλὰ χρόνῳ ζητοῦντες ἐφευρίσκουσιν ἄμεινον.

The gods did not reveal all things to mortals in the beginning:
but in long searching man finds that which is better.

This last reflection, one of the few fragments of the writings of the pre-Socratic philosopher, Xenophanes of Colophon, concerning the limitations of human knowledge, which have survived, has, as George told me, guided and informed his entire scientific career which he regards as having been composed solely of a search for that which is better, which he has pursued all his life to the limits of his ability. In the context of this reflection, he regards Xenophanes, in his own mind, as the philosophical and scientific mentor of this *liber amicorum.*

This reflection bears a relationship to Isaac Newton's well-known aphorism: 'I do not know what I may appear to the world, but to myself I seem to have been only like a boy playing on the sea-shore, and diverting myself in finding now and then a smoother pebble or a prettier shell than ordinary, while the great ocean of truth lay all undiscovered before me.' It is especially appropriate to include this aphorism here because, in Portuguese translation, it formed the content of the slide which George showed during his lecture in Spanish when we first met in Puerto Vallarta, Mexico, as long ago as 1974, as I have already related.

Xenophanes' reflection also has much in common with an extract from a book review (Fraser 1970i), written at the time we were exchanging letters in *The Lancet,* about his fellow-protagonist in this account, Arthur Kavanagh, and which is relevant to the conclusion of this little encomium of the great scientist and wonderful friend that George Fraser is:

The concept of the frontiers of knowledge is familiar and readily understandable; that of the boundaries of ignorance is less familiar, although, after a little reflection, no less easy to understand. The matter can perhaps be expressed in terms of an aphorism, possibly original although the thought behind it certainly is not. *Scio ergo nescio* – I know, therefore I know not – or, conversely, *nescio ergo scio* – I know not, therefore I know. This sentiment is somewhat similar to one attributed to Socrates in Plato's Apology: 'As for me, all that I know is that I know nothing.' Paradoxically, to expand the boundaries of ignorance is as much a part of scientific advance as to push them back; the former process is today far more rapid, even though our thinking is geared almost entirely to the latter.

I asked George to write the last paragraph of this story. He did so, and I completely agree with him.

We remain in ignorance of the details of how Arthur Kavanagh, together with our many friends among the acheiropods of Brazil, managed to adapt themselves and their lives so successfully to the limitations imposed by the interruptions in the creation of their bodies. In long searching, we shall learn more about this puzzle, and, during this learning, we shall uncover fascinating new areas of ignorance within which to explore many further aspects of the process of adaptation of fellow-members of our species to the human condition in its myriad variant forms.

Note added in proof

I was very happy to be proved wrong in my prediction in the second paragraph on page 33 very soon after I submitted this paper at the end of 2005. Thus, a few months later, George was unexpectedly invited to give a lecture in Foz do Iguaçu, in Brazil, in September 2006, so that we had the opportunity to arrange continuous further supper engagements with Maria and George during two entire marevllous days which they spent with us in our appartment in São Paolo after the lecture, ten years after their last visit in September 1996. We plan more such reunions at ten-year intervals. The lecture that George gave in Foz do Iguaçu was based on his contribution to this volume, 'Human genetics today: hopes and risks', but, apart from changes in content, the lecture was fundamentally different in that it was delivered in pure unadulterated Brazilian Portuguese, following a very enjoyable collaboration between us to achieve the translation.

'His is een bijzonder mens, dat is hij'

He is a special person, that he is

Hans Galjaard

Emeritus Professor Human Genetics, Erasmus University Rotterdam

In one of the most popular Dutch books for children *Dik Trom* a father repeatedly reacts to the naughty activities of his son by saying 'he is a special child, that he is'. The title of this contribution on the occasion of George Fraser's 75th birthday is a paraphrase which George, impressive polyglot that he is, can easily read; he will also agree with its content.

The period George spent in Holland as a professor and chairman of Anthropogenetics at Leiden University, and hence of our direct professional and personal contacts was short. He was appointed in 1971 and left unexpectedly, even for people close to him, for Newfoundland in 1973. Yet during this brief period he made a variety of impressions some of which I recall after more than thirty years. There was the long search for suitable accommodation in Leiden and Oegstgeest. There were his critical remarks when people or organisations did not perform according to his standards. In a discussion with employees of our National Office of Statistics he asked 'whether they had already adopted the use of the computer' which did not make him friends in that organisation, even though he was in need of their help. George, himself a dedicated worker, also had difficulties accepting the free and easy ways of our Dutch oversocialised society. About one of the population geneticists in his department, he said: 'Yes, he is talented but he is always absent painting his house.'

His successor-to-be, Professor Peter Pearson, worked late hours in the laboratory, doing experiments in cytogenetics. George Fraser, also staying long hours in his office, visited Peter Pearson in his laboratory, and asked him what he was doing. After listening to Peter's explanation, George said: 'Continue to do as I do, Peter, and you will become famous.'

During his Dutch period, George Fraser tried to reorganise the Anthropogenetics Department which had for a long period been supervised by another remarkable foreigner, the Italian geneticist, Marcello Siniscalco. George worked on his book *The causes of profound deafness in childhood*, and wanted to improve Leiden's infrastructure with respect to genetic diagnosis and genetic counselling. This brought us into close contact, because as a professor of Human Genetics at the Erasmus University in Rotterdam, I had started laboratories for postnatal and prenatal diagnosis of inborn errors of metabolism. Together with my colleagues, Dick Bootsma, who was initially responsible for cytogenetics, and Martinus Niermeijer, who focussed on genetic counselling, we built up a separate department of Clinical Genetics. At the same time, together with colleagues from other university hospitals, I tried to convince the Dutch government and health insurers of the future impact of genetics on health care. George was not exactly the type of person whom one would wish to involve in difficult negotiations with civil servants, requiring patience and diplomacy. He was, however, an important source of knowledge and insight for his colleagues.

His contributions varied from advice on individual dilemmas in genetic diagnostics to more general thoughts about ethical issues. Our Crown Princess Beatrix and her husband, Prince Claus, were very much concerned about the first report of the 'Club of Rome', predicting that the earth's natural resources would run out as a result of the ongoing economic and technological growth, and about the predicted environmental damage. They were also very much interested in developments in genetics and biotechnology, and especially in the ethical aspects, including the danger of an increasing gap between wealthy and developing countries. In order to stimulate public information and debate about these issues, they initiated a foundation, *Biosciences and Society*, which organised meetings and edited series of booklets on specific themes and technologies.

One of the first booklets was about Genetics, and had as a title *Geschonden Genen*. George Fraser contributed a chapter (Fraser 1973c) about 'individual choice and social responsibility'. This article (in Dutch!) is still worth reading, 30 years later. He commented in a sensitive way on the difference between selective abortion after prenatal diagnosis and infanticide, on the duty of a society to take optimal care of its handicapped citizens, on selection of gametes and on the minimal effect of medical intervention on the human genetical load.

In 1971, together with Arno Motulsky and Joe Felsenstein (Motulsky *et al.* 1971e), George published an important paper in the Birth Defects Original Article Series of the National Foundation March of Dimes on the long-term implications of intra-uterine diagnosis and selective abortion. In 1972 and 1974, George elaborated on this topic, and broadened it to include long-term effects of recent advances in treatment and prevention of genetic disease (Fraser 1972f, 1974e).

George Fraser was also farsighted in recognising early on the importance of population genetic studies in isolates. He once visited me and asked whether I could organise a study in the Dutch province of Zeeland (where people had hardly migrated for centuries). He also wanted one of my then closest colleagues, Martinus Niermeijer, to direct such a study, not realising that I had my own research projects and responsibilities in Rotterdam.

His departure in 1973 to a really isolated place, Newfoundland, where he hoped, in vain as it turned out, to initiate a study such as he had dreamt of initiating in Zeeland, implied a sudden end of our professional relationship. Had he stayed, George would have been pleased to see that in 1979, the Dutch government and health insurers were the first financially to support a comprehensive network of clinical genetics services with facilities for prenatal and postnatal chromosomal analysis, biochemical diagnosis and genetic counselling. At present, eight departments of Clinical Genetics in a total population of 16 million in the Netherlands, investigate more than 70.000 people annually, and the combination of early diagnosis and genetic counselling offers many couples informed choices about reproduction, contributing significantly to the avoidance of births of children who would otherwise have been seriously handicapped. We now also know that the efforts directed towards early diagnosis and prevention did not have a negative effect on the quality of care of the mentally and/or physically handicapped members of our society. So, two of the major requirements which George Fraser formulated in 1971 have been realised with the help of civil servants of whom at the time George did not have very high expectations.

Since his departure from the Netherlands, George and I have kept in touch during conferences and by exchanging letters. He has always been a faithful colleague and friend, and I would have wished that he had continued his career at Leiden University; in retrospect, he himself would also have wished to do so, as he has told me on several occasions. After his departure from the Netherlands, he has made many other contributions which will be summarised by others in this Festschrift. Recently, George sent me two beautiful pictures of his charming Greek wife and of himself, together with their daughter at her wedding ceremony. George had not changed much during these three decades.

He included with the photographs some reprints about the Fraser syndrome (OMIM 219000). Although George was already famous during his Leiden professorship, for a (clinical) geneticist to have a syndrome named after you must give you an extra feeling of immortality. However, as a geneticist who has primarily worked on the molecular pathogenesis of congenital defects and genetic diseases, I only believed in the lasting fame of George Fraser's name when I read the paper in *Nature Genetics* in 2003 by a multi-institute group with Peter Scambler as the last author, which elucidated various mutations in a gene on chromosome 4q21 which encodes a protein related to the extra-cellular matrix. So now George can relax: there is a gene and also a protein which are named FRAS1 after him (McGregor *et al.* 2003).

Lionel Penrose, mental deficiency and human genetics

Daniel J. Kevles

Yale University, New Haven CT, USA

George Fraser tells us that he came to the Galton Laboratory at University College London in October 1957 and 'thoroughly enjoyed two years of the most stimulating and the most intellectual academic environment which I have ever known.' (Fraser 1998) The Galton was then headed by Lionel Penrose, who quickly became one of Fraser's heroes. Indeed, just a few years before, J. B. S. Haldane, a man not given to overstatement, called Lionel Penrose 'the greatest living authority on human genetics.'[1]

Born in 1898, Penrose was a product of a well-to-do, polymathically capable British family. Lionel possessed the sort of crisp, incisive, pellucid mind that makes for exquisite science. Given a sternly Quaker upbringing, a religious commitment derived from the maternal side of the family, Penrose was sent to Leighton Park, a Quaker School, where he earned a teacher's commendation for declaring that Jesus's message for the Pharisees was 'to do away with their traditions and rites and to look at the things which really mattered.' [2]

To young Penrose, what really mattered were mathematics, science, and chess. The Quakerism counted for a lot, too, especially the pacifism to which his mother's family was unbendingly committed. During World War I, Penrose served in a Friends Ambulance Train Unit. A lecture one evening in France stimulated an interest in Freud's theory of dreams, and by the time he matriculated at Cambridge, in 1919, the knowledge Penrose cared about had gone beyond mathematics and science to include an increasingly intense interest in Freudian psychology. He came to think during this period that religion '*stunts* our mental growth,' that religious belief ought to take a back seat to knowledge. Many years later, his daughter remarked, 'To him, God was simply too vague a concept. It was one of those ideas that you couldn't quantify or test.'[3] Academically, Penrose pursued the Moral Sciences Tripos of psychology, mathematical logic, and philosophy. He did brilliantly at the mathematical logic, disliked the philosophy, and was utterly disappointed by the limited range of studies in psychology.

Graduating in 1922, he went to Vienna vaguely hoping to find intellectual satisfaction in the study of psychoanalysis, but gradually a certain scepticism concerning psychoanalysis set in. Penrose remained fascinated by Freudian insight, but he came to consider psychoanalytic theory too elusive, too slippery for scientific test. His dissatisfaction with it, like that with God, boiled down to the fact that you couldn't quantify its terms.

Increasingly, his interests swung toward the abnormal mind, including the biological role in mental disorder. Feeling the need for a solid grounding in medicine, in 1925 he returned to Cambridge and earned a medical degree while spending some of his time each week as an analyst at the London Clinic of Psycho-Analysis. He took his doctorate at the Cardiff City Mental Hospital with a thesis on a set of schizophrenics.

In 1928, he married Margaret Leathes, a woman of a strongly independent and irreverent cast of mind whom he had met while climbing in the Austrian Alps. He bubbled with enthusiasm to the four children they produced about mathematics, science, Mozart, and chess, especially the mental version with no board but the players' minds – and about wonderful toys and puzzles, physical and mental. He kept a small pedal saw at home with which he constantly fashioned ingenious wooden games and devices.

There was no distinguishing Penrose's playful inventiveness from his character as a human being or as a scientist. In the late nineteen-fifties, he pedalled his pedal saw to produce an ingenious alternative to the Watson-Crick DNA model of genetic reproduction. Resembling interlocking pieces of a puzzle, the wooden units were capable of mechanically reproducing themselves. 'An insult to nucleic acid,' Francis Crick snapped.[4] Penrose was, of course, aware that such a model's function is to suggest ideas, but for him there was no sharp break between devices for play and those for serious science – what started as one, whether the product of mind or saw, might turn into the other.

Penrose was the quintessential anti-religious scientist, but he continued, as Haldane once said of him, to hold Quaker views in everything save theology. His children remember that he always seemed to be writing out cheques for one good cause or another, and the house was often filled with guests, many of them political refugees from various parts of the world.[5] (Harris 1973) A lifetime pacifist, Penrose was generally liberal in his politics – and acidly sceptical toward any sweeping doctrine, particularly eugenics, that pretended to unite theories of biology, medicine, and society. Eugenics offended Penrose's scientifically critical intelligence and his acute moral and social sensibilities.

In 1931, Penrose was appointed to the staff of the Royal Eastern Counties Institution, a home for the mentally disabled in Colchester, England, where his principal task was to investigate the relative roles of heredity and environment in the generation of mental disease and disorder. With his humane Quakerism, precision of mind, and implacably sceptical temperament, Penrose began work at Colchester oriented against the simplistic ideas of many of his predecessors in the field of mental deficiency, finding them marked by class and racial biases, and endowed with considerable sympathy for the unfortunate human beings he was to investigate.

Penrose's attention was early drawn to the type of patients then called 'Mongolian imbeciles.' The first systematic identification of their affliction was made in 1866 by the British physician John Langdon Haydon Down. Down described a syndrome that, along with severe retardation, included an enlarged head and a prolonged, or epicanthic, fold to the eyelid. In Down's time, Western physicians had observed the syndrome only in Caucasians. Down supposed that the disease indicated a biological reversion in its victims to the Mongols of Asia, whom he thought they physically resembled, and who he assumed were a surviving example of an earlier human type. Down interpreted the 'fact' that Caucasians could produce Mongols as evidence for 'the unity of the human species' – a liberal idea running counter to contemporary theories that 'inferior' human races had sprung from separate biological origins. Down believed the disease to be congenital rather than hereditary, and he speculated that the reversion might be caused by parental tuberculosis. (Down 1866)

The identification of the imbeciles with the Mongols of Asia – or, at least, with some general primitive type – persisted. In the nineteen-twenties, in the widely noted book *The Mongol in Our Midst,* the British physician F. G. Crookshank furthered this view by arguing that the syndrome might derive from a recessive 'unit character,' a vestige of man's evolutionary past, and that some Mongol blood no doubt flowed in the veins of many Europeans. (Crookshank 1924)

There were only forty two mongol patients at Colchester; Penrose had to search out others from local and London hospitals and through mental-health organisations, going so far as to track down an afflicted child whom he spotted on the street.[6] To test Crookshank's ideas, Penrose surveyed the blood types of one hundred and sixty six mongols and of a control group of two hundred and twenty five other mental patients. He found that the distribution of blood types in the mongol group was about the same as that in the control group. The results meant, he wrote to a fellow physician, that 'mongolian imbeciles are no more racially Mongolian than other imbeciles.'[7] (Penrose, 1932)

Mongolian imbecility remained a major subject of Penrose's research to the end of his career. Yet from the beginning he judged that Down's syndrome merited special scientific attention, because it seemed so forcefully a product of action on the foetus by the intrauterine environment.

It was noticed early in the century that Down's syndrome births were related to the age of the mother, occurring much more frequently among women over thirty five than among younger women. Nevertheless, there was considerable dispute about the role of maternal age in the origins of the syndrome. Some authorities claimed that what counted was not the mother's age but the father's. Others insisted that the critical factor was the place of the Down's offspring in the family birth order: the mongol was often the last in a long line of children, and it was therefore theorised that the syndrome resulted from the mother's 'reproductive exhaustion.' Then, too, a mother often produced a mongolian imbecile long after the birth of her last previous child, so length of time between births was also advanced as a cause.[8] (Penrose 1934a)

Beginning at Colchester, Penrose worked to extract the truth from the conflicting theories. To choose among the important factors in the birth of a Down's syndrome child, he adopted a simple statistical procedure: calculate the expected number of afflicted offspring on the hypothesis that one factor (for example, maternal age) made a difference while others (for example, birth order) did not; then compare the calculated expectation with the observed incidence. If the two figures matched closely enough, the hypothesis would be demonstrated. ('His statistics are definitely "low brow",' Haldane once remarked, 'but I think effective for the purpose for which they are designed.'[9] See Oliver Penrose's contribution, p. 443) The entire procedure demanded the gathering of complete and accurate family data. In due course, he had extensive information concerning some hundred and fifty families. Analysis of the data revealed that the birth of a Down's syndrome child did not depend upon paternal age. It did not depend upon birth order. It did not depend upon the length of time elapsed since the birth of the last previous child. In most cases, it depended only upon the age of the mother, with the probability of occurrence rising sharply for women over thirty five.[10] (Penrose1934b)

Just why advancing maternal age raised the probability of a Down's syndrome birth, no one at the time could say. Although Penrose, too, was unable to clarify the causes of Down's syndrome, his conclusions about its dependence on maternal age and its likely genetic origins in cases of familial incidence were definitive and rapidly came to be recognised as such.

Though it did not accept the insane, the Royal Eastern Counties' Institution housed, at the time of Penrose's arrival, 'defectives' of all grades and numerous varieties. It was Penrose's task not only to get to know each of the patients but to ascertain everything that might illuminate the nature and causes of their respective deficiencies, especially whether these were primary or secondary in origin.[11] He quickly recognised that, as he put it in 1932, 'there are a great number of different types of retarded mental development, many of which have almost nothing in common with one another except the inability to perform those functional acts which society regards as being an index of intelligence.' But reliable differentiations among the various types required reliable criteria of difference. Penrose rejected out of hand legal grades of mental deficiency, which hinged on social aptitude, as scientifically worthless. ('They are about as much use from the biological standpoint as a classification of aquatic organisms based upon their suitability for consumption as articles of human diet.') He also recognised that legal standards were even less reliable in action than in principle, since he knew that liability to certification as mentally deficient hinged on social class. Penrose insisted upon approaching the study of mental deficiency as 'a branch of human biology.' He preferred a set of criteria expressive more of the patient as such than of the patient's interaction with the social order.[12] (Penrose 1934b)

Penrose was inclined to believe that environment played a major role in the aetiology of defect, but he recognised the possibility that hereditary factors might 'enter significantly into every case of mental deficiency.' If environmental determinants were involved in nearly every case, too, then any attempt to classify mental deficiency as primary or secondary – genetic or acquired -was, in his opinion, 'foredoomed to failure.' He proposed to start from a classificatory *tabula rasa* – to sort out as far as possible all the pure clinical types, then to determine whether any given patient was an example of a pure type or a mixture of more than one.[13] (Penrose 1934b)

To identify the types and untangle their causes, Penrose gathered extensive clinical data on all the Colchester patients. He also oversaw the investigation of each patient's social background and family history. A psychologist administered intelligence tests – designed at Penrose's instigation so that the results would depend as little as possible upon the extent of the test-taker's education – not only to the patients but to members of their families. Penrose knew that 'apart from hereditary likenesses, the child's mentality is, in many ways, copied or modelled on that of the parents' – that 'the parents' social status and ability determine the physical and mental nutrition of the children'[14] (Penrose 1934b)

Penrose was acutely sensitive to the methodological shoddiness that even in the nineteen-thirties continued to plague the field. Although he pursued family medical histories, he understood that queries as to whether the patient seemed to be suffering from a hereditary complaint risked 'a large initial probability of mistake or concealment in the answer.' He laid emphasis on data concerning stillbirths, infant deaths, and the like, realising that neglect of such information could well produce too low an estimate of a given condition's familial incidence, with the consequence that a disease that was really hereditary might seem otherwise. Although he used mental tests, he believed that one could not rely solely on such devices to assess intelligence. For Penrose, test results were just one item in a much larger evidentiary context, and the family histories were to be sifted, resifted, and, if necessary, gone after again to get at the truth. Penrose made it his overall aim 'to understand, as far as possible, the mental outlook of the patients and to relate this to their upbringing, [education], and past emotional experiences.'[15] (Penrose 1934b)

As the survey proceeded, he accumulated evidence confirming the hereditary nature of certain afflictions. Some, including Huntington's chorea, neurofibromatosis, and epiloia, were genetically dominant; others – for example, congenital diplegia, microcephaly, cerebromacular degeneration, and cretinism – were recessive. Penrose found particularly interesting – because of the way it unambiguously announced itself – the recessive condition identified in 1934 by the

Norwegian scientist Ivar Asbjörn Fölling.

Fölling had analysed the urine of four hundred and thirty mentally deficient patients. He detected phenylpyruvic acid in ten of the samples. As soon as Penrose saw Fölling's paper, he analysed the urine of his Colchester patients. If the acid was present, the urine would turn green upon the addition to it of iron trichloride. After four hundred and fifty one samples were treated, the urine of a teenage boy revealed the telltale green colour; it took five or six hundred more before Penrose found a second. The family histories of both these cases strongly suggested that the condition was caused by a rare recessive gene that, when expressed, caused an inborn error of metabolism. It was soon learned that the error occurred in the liver in infancy and that it affected the development of the brain. Juda H. Quastel, a biochemist and a collaborator in the study, coined the word for the disease: 'phenylketonuria,' which was ultimately contracted to PKU.[16]

Penrose recalled in his memoirs that the finding satisfyingly undercut the views held by 'a school of investigators who believed that mental deficiency could almost always be ascribed to inadequate development of the brain, induced by "rotten" heredity.' He was delighted by the idea that 'the mental defect arose because the patient had something wrong with his liver, not his brain' and by 'the possibility in the future of rational treatment for such patients by altering their diet at an early age'.[17]

The expectation was prophetic, but the dietary treatment for PKU lay many years in the future. At the time, only a comparatively small fraction of the mental diseases that Penrose encountered seemed attributable to so definite a genetic, let alone treatable, origin, either dominant or recessive. Although the prevailing wisdom had it that some eighty percent of mental deficiency could be classified as primary amentia, heredity alone seemed to account for only about a quarter of the Colchester cases.[18] A number of the rest seemed to originate from environmental forces, although just what these were was not clear.

In 1938, seven years after he had begun, Penrose published the full results of his Colchester survey in *A Clinical and Genetic Study of 1280 Cases of Mental Defect*. Apart from his conclusions concerning Down's syndrome and phenylketonuria, he reported that, unlike either disorder, most of the Colchester cases were in origin principally neither environmental, pathological, nor genetic but some combination of the three. In his summary view, 'the aetiology of mental defect is multiple, and a facile classification of patients into primary or secondary cases would only have led to a fictitious simplification of the real problems inherent in the data.' (Penrose 1938)

Penrose had hoped from the outset that the Colchester survey would take the field of mental deficiency far beyond the simplicities of then-current eugenics. He in fact succeeded handsomely not least because of his rare arsenal of expertise -the combination of genetics, medicine, psychology, and psychiatry – that he had brought to his task. 'I know of no other investigator who has made such a thorough genetic and clinical analysis of a large group of mentally defective patients,' a sponsoring official wrote to the Medical Research Council. 'Unless I am much mistaken this work by Penrose will be the basis of most researches in mental deficiency during the next few decades. His definite findings on a large number of specific problems are valuable scientific contributions, but it seems to me the chief merit of the work is the new orientation it gives to our genetic approach to this complex problem.'[19]

Penrose, pacifist that he was, spent World War II in Canada, but in 1945 he returned to

Britain to succeed Ronald A. Fisher as Galton Professor of Eugenics at University College London. Haldane had arranged the matter. ('I think that you and I are the British people under 60 who have contributed most to human genetics, and therefore one of us should have the chair. As you have specialised on man and I have not, your claim is somewhat greater.')[20] While Haldane was a brilliant theorist, Penrose, by now a world authority in the genetics of mental deficiency, was also a clinician, not only medically qualified but well versed in psychology as well as psychiatry, a scientist who thrived on direct contact with his human subjects. Neither a master biochemist nor a statistician, he was nevertheless clever enough to invent his own ingenious methods of overcoming ascertainment bias and for performing biochemical assays. The more Penrose branched out into human genetics, the more he came to personify a richly multidisciplinary orientation – statistical, biochemical, medical, and genetic – to the study of human heredity.

While maintaining the Galton's biometric tradition, Penrose shifted the emphasis of the laboratory in a medical and biological direction, and away from the emphasis on eugenics given it by Karl Pearson and Fisher. He told the University College provost in 1961 that since the war, the work of the Galton had been seriously handicapped by 'the stigma of eugenics' and that he found it a 'continual embarrassment' to have to explain that both his laboratory and the professorial chair were 'wrongly named.' In 1954, Penrose had changed the name of the laboratory's principal publication, the *Annals of Eugenics*, to the *Annals of Human Genetics*, and now he succeeded in persuading the authorities of University College to rename his chair the Galton Professorship of Human Genetics.[21] (Penrose 1946)

Penrose established ties with hospitals, medical schools – especially the University College Hospital complex just across Gower Street – and mental institutions, which supplied data on the diverse physiological characteristics and afflictions found among their patients.[22] (Haldane 1970, Box 1978) He also reached out to the overall University College London Department of Biometry, Genetics, and Eugenics, of which the Galton was a part and which was headed by Haldane, who held the Weldon Professorship of Biometry. Haldane was one of the few scientists in the world who enjoyed Penrose's unreserved admiration, and Haldane repaid the compliment; the two were warm friends.

After the Second World War, human geneticists possessed neither the glamour nor the power of physicists, those emperor scientists who had forged radar and the atomic bomb and won the war. Geneticists nevertheless benefited from the general upsurge in the funding of scientific research, especially by governments. Throughout Penrose's tenure, the Galton was well supported by the Rockefeller Foundation and modestly assisted by the Medical Research Council. The permanent staff, including affiliates like Haldane, comprised perhaps eight to ten people. Still, by the standards of post-1945 science, the Galton was neither munificently funded nor heavily staffed. Sylvia Lawler, a blood-group geneticist at the Galton, remembered her experimental equipment: a few deep freezes, some pipettes, and a 'sort of old microscope that Pasteur would have thrown out.' For the most part, people sat at tables and desks working with numbers and papers.[23]

Penrose stretched the available resources to the limit. Positions were funded, usually temporarily, on a catch-as-catch-can basis, with a fellowship here or an assistantship there. A number of the make-do posts were held by women, who of course had been employed in abundance at the Galton since Karl Pearson's day and constituted a relatively cheap supply of

trained – often highly trained – scientific labour. Some of the women at the Galton felt themselves unfairly relegated to positions inferior to those held by men, and a few became lastingly
bitter about it. But at the time only a small number of permanent career opportunities in the laboratory – or in human genetics, for that matter – were available for anybody, male or female.[24]
The Galton was a work-hard place, but it was also lively, congenial, and stimulating. The
geneticist Sylvia Lawler recalled that there were considerable compensations for enduring a
woman's position at the laboratory, not least the sheer excitement of being there.[25]

Penrose was a decidedly laissez-faire director. He did not run the laboratory so much as
preside over it. 'Anyone who managed to get a PhD there had to have a streak of originality,'
Sylvia Lawler later noted. 'There was no spoon-feeding. Penrose would take people in, shut them
in a room, and let them get on with it.' Still, he usually found time for people with results or
problems that interested him.[26] Rarely saying much, he tended to respond to queries with an
intuitive judgment of what was likely to be scientifically right or wrong, and when pressed, he
could be perplexingly elliptical. However, since Penrose did not explain the probable flaw in a
piece of work, people had to figure it out for themselves. In Lawler's judgment, the staff were also
made to use their heads because the technological opportunities were limited by the lack of
sophisticated equipment.[27]

If Penrose inculcated anything explicitly, it was the essential importance of quantification. He
found in measurement, whether of biochemical excretions or of developed physical characteristics, the best possibility of enlarging the scope of certainty in human genetics. No pure Cartesian
rationalism for him. He used to snipe at French scientists: 'The reason they get it wrong is that
they're so logical.' Declining to take anything on pure trust, he always wanted to do his own calculations, in his own way. Still, by example Penrose taught that measurement and mathematics
had to be tempered by scientific experience and judgment, much as he had done at Colchester.[28]
George Fraser has recalled some of Penrose's influence on him: 'The major part of my subsequently published contributions to our profession was to be based on the ideas which I had
acquired from him by diffusion or osmosis. The methodology of two of my contributions is
modelled on the Colchester Survey in that I attempted, in two books, to apply the same techniques to childhood blindness and to childhood deafness or, rather, severe visual handicaps and
severe hearing losses in childhood.' (Fraser 1997)

In many subjects, including biochemical, statistical, and clinical genetics, the Galton was a
groundbreaker. In the postwar years, enlarging upon his long-standing interest in Down's syndrome, Penrose devoted a major part of his own effort to the investigation of fatal malformations,
both congenital and hereditary, and in 1949 he published *The Biology of Mental Defect* (Penrose,
1949), a classic work, widely hailed for giving scientific rigour and credibility to the subject, and
unrivalled in its successive editions on either side of the English-speaking Atlantic.[29]

When the American human geneticist James V. Neel first visited the Galton in the mid-fifties,
he was struck by the fact that the famous laboratory had few experimental facilities and basically
consisted of three offices – one of them Penrose's, ten feet square and lined with books. Neel was
reminded of a proverb his professor liked to quote: 'It's not the size of the cage that determines
how sweetly the canary sings.' The Galton sang the songs of human genetics with exquisite sweet-
ness and power. Its pre-eminence rested on neither size nor money; it hinged, rather, on the high
quality of its diverse staff, above all Penrose and Haldane, and on what both fostered, particularly
an offbeat, sceptical esprit and an incisive style of thought that attracted original men and

women and permitted them to thrive. Between 1945 and 1965, when Penrose left the directorship, the Galton was a mecca for aspiring human geneticists from England, the Empire, the United States, and the Continent, and a list of the postgraduate visitors to Gower Street – among them George Fraser – reads like a later Who's Who in the field.[30]

Notes

1. Haldane to the Principal, Ruskin College, Oxford, 12 March 1953, Haldane Papers, file 2; interview with Shirley Penrose Hodgson.
2. Lionel Penrose, 'Exercise Book, Scriptures, 1913', Penrose Papers, file 3/2.
3. Penrose, 'Why Religion Must Die Out,' n.d. [1920s], Penrose Papers, 35/6.
4. Penrose notes, 'What certain people said when they first encountered the self-reproducing machine in one form or another'; Francis Crick to Penrose, 10 Feb. 1958, Penrose Papers, file 51/16, 51/18.
5. Haldane to the principal, Ruskin College, Oxford, 12 March 1953, Haldane Papers, file 2; Penrose, 'Exercise Book, Scriptures, 1913,' Penrose Papers, file 3/2; interviews with Roland Penrose, Margaret Penrose Newman, and Harry Harris.
6. Penrose, 'Report of the Research Department of the Royal Eastern Counties Institution,' Sept. 1932; Frank Douglas Turner to Penrose, 3 July 1931, 5 Jan. 1932; Penrose to Turner, 4 Jan. 1932; and Essex Voluntary Association for Mental Welfare Materials, Penrose Papers, file 130/7.
7. Penrose to Edmund O. Lewis, Feb. 8, 1932, Penrose to Peter K. McGowan, 13 Feb. 1932, Penrose Papers, file 147/3, file 149/3.
8. Penrose, Draft note to *The Lancet*, 14 March 1938; Penrose Papers, file 61/1.
9. Haldane to Egon Pearson, 27 March 1944, Penrose Papers, file 49/1.
10. Penrose, 'Report of the Research Department of the Royal Eastern Counties' Institution,' Sept. 1932; Penrose to R.A. Fisher, 2 Nov. 1932; Penrose to Frank C. Shrubsall, Aug. 19, 1933; Penrose to R.L. Jenkins, 14 Aug. 1933, Penrose Papers, files 56/2, 61/2, 168/7, 142/5.
11. Penrose to Secretary, Pinsent-Darwin Studentship, 4 Dec. 1931, Lionel S. Penrose Papers, file 149/4; 'Appointment of Research Medical Officer [Colchester], Particulars,' Penrose Papers, file 47.
12. Penrose, 'Report of the Research Department of the Royal Eastern Counties' Institution,' Sept. 1932, Penrose Papers, file 56/2.
13. Penrose, 'For Mental Welfare' [1932], Penrose Papers, file 56/1.
14. Penrose, 'Report of the Research Department, Royal Eastern Counties' Institution,' Sept. 1932; Penrose to Charles Blacker, 2 Oct. 1931; Penrose to Edmund O. Lewis, 30 Jan. 1932; Penrose to Dr. MacCurdy, 23 Dec. 1931,Penrose Papers, files 56/2, 130/9, 147/3, file 149/4.
15. Penrose, 'Report of the Research Department, Royal Eastern Counties' Institution,' Sept. 1932, Penrose Papers, file 56/2..
16. Haldane to Penrose, Nov. 1934; Penrose to Frederick Gowland Hopkins, 23 Nov. 1934; Penrose to Asbjörn Fölling, 15 June 1936; Penrose to Haldane, 3 April 1935; Penrose, 'Memoirs – 1964: Phenylketonuria,' Penrose Papers, files 136, 139/8, 132/4, 72/2
17. Penrose, 'Memoirs – 1964: Phenylketonuria,' Penrose Papers, file 72/2.
18. Penrose, 'Report of the Research Department of the Royal Eastern Counties' Institution,' Sept. 1932, Penrose Papers, file 56/2.
19. Penrose to Ruth Darwin, 31 Aug. 1931, Penrose Papers, file 164/3; E. O. Lewis to David Munro, 18 Sept. 1937, Medical Research Council Records, Mental Disorders, Colchester, file 1588, folder III.
20. Haldane to Penrose, 18 Aug. 1943; Haldane to the Provost, University College London, 9 Aug. 1944, Penrose Papers, files 159, 49/1.
21. Harry Harris interview; Penrose, "Memorandum to Provost," 4 Dec. 1961; Kenneth Ewart to James Henderson, Aug. 15, 1961, and Provost to Penrose, 1 Dec. 1961, Penrose Papers, 175/5.
22. Lionel Penrose, 'The Galton Laboratory Report, 1949–50'; Penrose to Haldane, 26 June 1943; 29 Nov. 1943; Penrose, 'From Eugenics to Human Genetics,' lecture 1965; Penrose, 'Galton Laboratory,' 27 April 1961, Penrose papers, files 159, 77/2, 175/5; interview with C. A. B. Smith.

23. Haldane to Penrose, 17 May 1945; various documents, Lionel Penrose Papers, files 159, 49/6; interviews with Harry Harris, Park Gerald, and Sylvia Lawler.

24. Interviews with Ursula Mittwoch, Harry Harris, Sylvia Lawler.

25. Interviews with Alexander Bearn, Sylvia Lawler.

26. Interviews with Ursula Mittwoch, Park Gerald, CAB Smith, Sylvia Lawler, Barton Childs, Harry Harris, Alexander Bearn.

27. Interviews with Park Gerald, Sylvia Lawler, James V. Neel, Alexander Bearn.

28. Interviews with Harry Harris, CAB Smith, Alexander Bearn, James V. Neel.

29. Interviews with CAB Smith, Ursula Mittwoch; 'Lionel S. Penrose – 1898–1972', *M.A.P.W. Proceedings*, vol. 2, part 5, copy in Penrose Papers, file 20/5.

30. James V. Neel interview; list of postgraduate students and workers in the Galton Laboratory, 1945–65, Penrose Papers, file 49/2.

The Male Procreative Superiority Index (MPSI): its relevance to genetical counselling in Africa

F.I.D. Konotey-Ahulu

Kwegyir Aggrey Distinguished Professor of Human Genetics, Faculty of Science, University of Cape Coast, Ghana

George Fraser, the great encourager as I call him, personifies originality in the discipline of Clinical and Genetic Epidemiology. It is this 'if-it-has-not-been-described-before-it-does-not-mean-it-is-nonsense' attitude, which George always had towards Clinical Research that has led to his significant contributions to Medical Genetics.

I can never forget the encouragement I had when, on asking him whether an Index I had invented to explain certain aspects of African Anthropogenetics was a piece of nonsense, he said after reading the material from his hospital bed in Africa (See below **My first meeting with George Fraser**): 'It sounds reasonable to me. I do not think it is nonsense at all. It needs further study and elaboration.' Without this encouragement, I would never have gone public with the Male Procreative Superiority Index (MPSI) in an international conference on Human Genetics in Hungary (Konotey-Ahulu 1977). I showed that the African husband has a biological fitness on average double that of each of his wives, as reflected by his MPSI, defined as the total number of a man's children divided by the total number of children born to each wife, whether by that husband or by another man (Konotey-Ahulu 1977, 1980). The African male raiding the generation of his children and grandchildren to acquire female partners to pass on his genes, long after the menopause of his first wives, can be persuaded (as I have often done) to limit this propensity and thus lessen the genetic burden of the future (Konotey-Ahulu 1970, Bonney and Konotey-Ahulu 1977).

This, in a way, is not different in my opinion from the 'counselling leading to reproductive restraint' that Fraser and Mayo (1974) mention when they discuss the genetical load. The only difference, I feel, is that the 'reproductive restraint' that they appear to advocate and the approach (Fraser 1972) to 'prevention and treatment of inherited disease', including a discussion of reproductive restraint, differ somewhat from what I have been preaching in Ghana, in that my approach applied exclusively to males, telling them: 'Stop here, please! Do not take any more wives in your old age lest the deleterious genes I have identified in you be passed on to increase the genetic disease burden of the next generation, for one in three of us is a healthy carrier of a recessive gene. One in three of your wives will also be carriers'. The potential of the male for spreading these genes far exceeds that of the female, hence the term 'male procreative superiority.' (Konotey-Ahulu 1977, 1980)

To take just one of many examples from my actual experience, a Ghanaian paramount chief with sickle cell disease caused by a relatively benign 'SC' genotype compatible with an almost normal life span, with 23 wives and 94 children, donated a large number of abnormal beta-globin genes to future generations.

As Africans are living longer, the husband with common diseases compatible with life span of appropriate length, such as essential hypertension, diabetes mellitus, gout, and even prostate cancer, might in the same way account for more genetic pathology in future generations than would be passed on by any of his wives (Konotey-Ahulu 1990, 1996, 2004, Addy 1990, 1992).

I have always been grateful to George Fraser for encouraging me more than 30 years ago to think along these lines.

My first meeting with George Fraser

It was on a Friday in the spring of 1974 that George Fraser arrived from St John's, Newfoundland, where he was Professor of Medical Genetics at the time, in Accra, Ghana, where I was a Physician Specialist at the Korle-Bu and Ridge Hospitals, and the Director of the Ghana Institute of Clinical Genetics. Although I did not know him personally, I was aware of George's impending visit; a colleague of mine, Dr B. B. Edoo, had invited him to investigate the high prevalence of recessively inherited prelingual profound deafness among the inhabitants of the village of Adamarobe about 40 kilometres north of Accra, known to locals as 'the deaf village'. George spent a pleasant Saturday in Accra, and, on the Sunday, he was taken to Adamarobe by some of my colleagues who were familiar with the problems of deafness in the village. While he was there, George was entertained by a large group, executing some native dances. He was very impressed by the fact that even the deaf inhabitants participated fully in the festivities while following the rhythms of the dances as well as their normally hearing relatives.

Monday was to be devoted to sightseeing in Accra, and the scientific investigation in Adamarobe was to start in earnest on Tuesday. I remember that George's request for a street map to guide him in walking around Accra caused some perplexity, since such maps were not readily available at that time. However, a street map was eventually found and George set off in the midday sun on his unaccompanied tour. He took his mission very seriously, and he had traced out on the map in pencil a route to follow.

It was while he was strolling along the pavement at the side of one of the main thoroughfares of Accra that disaster struck. He was looking so intently at his map, taking his bearings and checking his route that he was not looking at the pavement in front of him. Thus, he failed to notice that one of the paving stones, about one metre square, of which the pavement was composed, was missing. As a result, he walked, without the benefit of the slightest trace of a protective reflex, into a hole about one metre deep.

No matter, thought George, as he lifted himself out of the hole. He felt that it was a pity that the watch on his right wrist was shattered, and that his right trouser leg was badly torn. But then, as he looked closely at his right trouser leg, he saw that all was not right under the cloth. In fact, his right foot, instead of pointing forwards, was pointing laterally at a complete right angle to the leg. And then, as the first shock wore off, came the pain, excruciating, agonising … pain and more pain. A passer-by hailed a taxi and accompanied George to the Korle-Bu Hospital whither I was summoned to meet him in person for the first time. A few hours passed and it was evening by the time an orthopaedic surgeon arrived and studied the X-rays, which had been taken. They showed a second-degree Pott's compound fracture of the right ankle due to external rotation, involving both the medial malleolus of the tibia and the lateral malleolus of the fibula. 'Manipulative reduction under general anaesthesia,' pronounced the orthopaedic surgeon. George looked around the emergency room of the Korle-Bu Hospital dubiously. Indeed, there was much to be dubious about; it was that kind of evening.

'Who is going to give the general anaesthetic?' he asked. 'Oh, we will find a nurse.' 'Could it not be done without a general anaesthetic?' Nothing could have pleased the orthopaedic surgeon more; he was anxious to go home after operating throughout a long and busy day. He despatched me to fetch some pethidine and started his preparations. It was late, the hospital pharmacy was closed, and it was difficult to find the pethidine. To cut a long story short, by the time I returned with a loaded syringe, it was all over. George had conducted his part of the transaction stoically, I was told.

From that moment on, he never looked back. A below-knee plaster was applied and he spent the night on the ward in the Korle-Bu Hospital. The next day he was sent back to his hotel, and visiting care was arranged. A couple of days later, he was able to make brief trips outside the hotel by car and even on foot with the aid of crutches. On Saturday, eight days after arriving in Accra, he insisted on flying to London, using his original reservation. So, still with his below-knee plaster, he set off. He did not tell any one that he intended to visit both his mother in the London suburbs and a colleague in the Netherlands *en route* to St John's, Newfoundland, where he arrived safely a full three days after leaving Accra.

The next day, George went to see a local orthopaedic surgeon. 'They let you travel from Accra by plane, wearing that below-knee plaster?!?' exclaimed the surgeon who knew nothing of the three days of stopovers *en route*. 'They must pray to the right gods in Accra,' he said.

During the thirty odd years since the fracture healed, George has had no trouble with his right ankle, which is not deformed in any way. Unless he gives specific thought to the matter, he cannot even remember which ankle was involved.

We do, indeed, have God on our side in Accra!

Later, I moved to the United Kingdom and George returned to the United Kingdom in 1984 after twenty years of life as an itinerant academic around the world. Since 1984, we have seen each other many times, and we have had many meetings and discussions, both social and scientific. How my wife and I enjoyed being invited to the wedding of their daughter in London!

But none of these subsequent encounters was imbued with the drama which characterised our first meeting in the emergency room of the Korle-Bu Teaching Hospital in Accra, except in so far as George's vivacious wife, Maria, has provided us with an aura of Drama, having been born in that celebrated city in Greece.

Unfortunately, George never had an opportunity to return to Adamarobe, 'the deaf village' (David *et al.* 1971), to pursue the investigation he had planned in 1974. Ghanaian colleagues published another paper on the topic a quarter of a century later (Amedofu *et al.* 1999).

Anaximandros: bowing to the fathers of genetics

Michael I. Koukourakis

Department of Radiotherapy – Oncology, Democritus University of Thrace, Alexandroupolis, Greece

Dedicated to Professor George Fraser on the occasion of his 75th birthday. Professor Fraser has for some years been an honorary member of TARG (Tumour and Angiogenesis Research Group) a small research group of ours which has set out on a long journey hoping to understand the beauty and the unspoken pain hidden in the soul of a small cancer cell.

Abstract

The concept of evolution and the existence of specific laws that govern this process through a pre-defined sequence of action and reaction, genesis and decay, were first defined by the pre-Socratic thinkers and are exhaustively analysed in the Cosmo-theory of Anaximandros. This concept is clearly stated in his perception regarding the evolution of life, bringing forward the idea of 'insta-bility of species' (or in modern terms genetic instability and transformation), to serve a predefined necessity, dictated by the law of 'Apeiron', evolution and the creation of Man. Anaximandros, amongst all pre-Socratic philosophers, deserves recognition as the Father of Genetics.

Introduction

The first attempts of the human mind to conceive the principles underlying the function of nature (Cosmos), using methods that transcend the far more ancient mythological approaches, go back to the 6th century BC. In Greece and Minor Asia, schools of 'Thinkers' (stochastes), forerunners of the philosophical schools of Plato and Aristotle, were born. The search for a primary element that gives birth to the physical and psychic world is a fundamental character-istic of the theories developed by these great teachers. Water, fire, earth and air, all four basic ele-ments composing the natural world, were considered as the progenitors of Cosmos, in systematic approaches that never separated matter from the soul, the latter being considered the 'cause' of movement and, therefore, of existence.

Anaximandros (the name 'Anaximander' is more often used in English) of Miletus, successor and apprentice of Thales *(Diogenis Laertios II, 1–2, DK 12a1)*, wrote systematic studies on nature, the periods of the earth, the sphere of the sky and, of course, geometry. Unfortunately, our knowledge on the content of these essays is only fragmentary, still enough to allow a view into the Anaximandrian Cosmology and Science (Diels and Kranz 1964, Kirk *et al.* 1983, 1988).

Anaximandros and the 'Apeiron'

'… the origin and the element of all is "Apeiron"' *(Symplicius, Phys 24, 13 DK 12a9)*
'… from Apeiron Heavens and Cosmoi are secreted' *(Hippolytos, El I, 6 1–2 DK 12a11)*

The 'Apeiron', the 'infinity' (if this is an appropriate term), was considered by Anaximandros the primordial element of existence. Cosmos (plural: Cosmoi) is considered a secretion of 'Apeiron', which also forms the material within Cosmos lies.

The concept of 'Apeiron' is not well understood, even by Aristotle who rejects the idea of an obscure element in addition to the obvious matter that constitutes Cosmos *(Aristotle, Physica, Γ8, 208 α8)*. It is unknown whether this term refers to the infinity of space, as Aristotle believes, or to an internal uncommitted freedom that reaches the concepts of undefined and 'undifferentiated'. This concept brings forward the element of 'Space' (Χωρος), divested from matter, as the 'cause' (the origin of power) and, at the same time, as the primordial element, deprived of qualities, that generates shape and matter. This, certainly, challenges concepts developed by modern physics. 'Apeiron' is the emptiness that possesses the motive/power to secrete matter, shapes and, therefore, Cosmos. Being an element, still a non-tangible material, represents the Cosmic 'soul' itself, the 'Law' that governs the passage from the mathematical probability (Space/Emptiness) to the existence (Cosmos), and back.

The concepts of genesis, decay and evolution

'… Apeiron possesses the whole cause of the genesis and decay of Cosmos … ' *(Pseudoplutarchos Strom. 2 DK 12a10)*
'… the matrix from which things are born is the same with the one to which things decay, according to the "Debt", as they are punished and correct reciprocally the injustice according to the order of "Time" … ' *(Symplicius, Phys 24, 13 DK 12a9)*
'… and refers to "Time", to make evident that genesis, existence and decay are defined (or limited) … ' *(Hippolytos, El I, 6 1–2 DK 12a11)*

Genesis and Decay (*Φθορα – Phthora*) constitute the fundamental motive for the continuous movement of Cosmos, of its continuously changing face, thus 'Evolution' (Loenen 1954). Time is inevitably linked to movement and transformation in the Anaximandrian theory. Shapes are deemed to change through a continuous transformation involving death and genesis of new forms. Unlike Aristotle's view of transformation, where things in Cosmos are innovated by recycling the already existing matter, Anaximandros distinguishes the transformation that occurs within Cosmos, from the process/force of genesis-decay (qualities of Apeiron) that governs evolution through a dialogue between the Space/Law and Cosmos. The course of this crosstalk is defined (or predefined), according to the Law (quality of Apeiron), the 'Debt' (Χρεων, necessity). Cosmos is the product of evolution, a process governed by the very essence of existence and its causes, by the 'Debt' that a result of an action has to pay generating an action, once again. This process certainly spreads in 'Time', as movement and time are identical.

The genesis of animals and men

'… the first animals were born in liquid material, wrapped in peels, and in time, exposed to air and released from the dried broken peel, they lived a different life, for a while … ' *(Aetios V, 19,4)*
'… originally, man was created from other animal species … ' *(Pseudoplutarhos/ Pseudoplutarchos Strom 2)*
'… from hot water and soil creatures were born resembling to fish, and in these, men grew in the

form of embryos, where he stayed till … they were able to feed themselves … ' *(Kinsorinos, de die nat. 4, 7)*

A direct emanation of the concept of evolution is the theory formed by Anaximandros regarding the origins of animals and men. The interplay among the elements of water, earth, air and fire led, through 'Time' and according the 'Debt', to the creation of life. Primitive animals, resembling fish, appeared in the water at first. Obeying to the law of evolution, dryness forced them to transform themselves (through a long series of generations) to creatures living out of water. The same concept is also evident in the thinking of other pre-Socratic philosophers, such as Xenophanes, in whose writing mankind emerges after the drying of the mud, and vanishes when the sea reforms the soil to mud, in a rather periodic manner *(Ippolytos El I 14, 5)*. According to Anaximandros, animals have the power, but more importantly, are obliged to evolve to different forms, as the law of Apeiron imposes the perpetual movement of Cosmos. Anaximandros's concept of evolution is evidently the direct precursor of Darwin's theory.

This concept of a predefined 'instability of species', or in modern terms 'genetic instability', as the cause of the multifarious structure of the living world, is certainly intriguing. The genesis of mammals, and therefore of man, as symbiotic organisms in larger primitive forms of life, has a pre-eminent position in this Pre-Socratic Science and Philosophy. The recognition that man, in contrast to other life forms, needs long support by its parents to survive, certainly puts humans at the top of the life-chain in Cosmos, as the most recent stage of evolution.

The above concepts directly contrast all mythological (outward) approaches regarding the creation of man from Gods. The supernatural is replaced by the concept of the Law that governs the creation and movement of Cosmos, so that the triad Gods/Nature/Life is replaced by the Apeiron/Cosmos/Evolution.

Extending the concepts to modern biology and medicine

The idea that primary forms of life were created in water is not different from the widely accepted biological theory of today regarding the genesis of unicellular organisms in the sea. Genetic material emerged in the water and wrapped with a membrane formed the first eggs for the subsequent development of genetically more stable organisms.

This genetic instability emerges as the blessing that led to human genesis and, at the same time, as the curse of genetic and all kind of diseases. The human genome is prone to mutations, presumably as an epiphenomenon of the underlying 'Law' that pushes existence to lifeforms beyond the human. Gene polymorphisms, one would say life-compatible mutated genes, provide individuality and, at the same time, the individual tendency for a disease. The symbiotic and parasitic presence of life forms (bacteria and viruses) in the human body is a result of the simultaneous genesis (within primitive life forms) of mammals. The viral and bacterial diseases of the present time are the consequence of life forms, partners of our past.

Finally, carcinogenesis emerges as an abortive attempt of the Apeiron to evolve into more 'perfect' (genetically stable) forms, in 'Time' and according to the 'Debt'. Cancer, apart from being a lethal disease, emerges as a remnant of the same law that created life and, presumably, a manifestation of the law that pushes evolution to, as yet, unknown stages. In time we will know whether modern genetics and manipulation of the genome by the human mind itself is the gate to the inevitable next step of Cosmos.

Conclusion

The concept of evolution and the existence of specific laws that govern this process through a pre-defined sequence of action and reaction, genesis and decay, were first defined by the pre-Socratic thinkers and were exhaustively analysed in the Cosmo-theory of Anaximandros. This concept of evolution is evident in his perception regarding the evolution of life, bringing forward, for the first time in the history of the human thinking, the concept of genetic instability and transformation. This is an instability that attends to the most important and indisputable process in Cosmos, thus Evolution. The debt is, therefore, paid off to serve a predefined necessity, the creation of Man.

George Fraser: a unique human and medical geneticist

Arno G. Motulsky

Departments of Medicine (Medical Genetics) and Genome Sciences,
University of Washington Seattle, WA 98195, USA

I had the privilege to have personal interactions with George Fraser since 1957 when I spent eight months in Lionel Penrose's department at the Galton Laboratory at University College in London. George at that time was a PhD candidate in that department. I had never before met someone with his broad fund of knowledge and intellectual brilliance. George had attended one of the UK's top schools (Winchester College) as a scholarship student, and was fluent in many European languages with diplomas in French, Russian and Hungarian from the University of Paris. His training in pre-medical science including mathematics was obtained at Cambridge and was followed by studying genetics and serology with R. A. Fisher and R. R. A. Coombs at that university. Medical school training and house officer service at London Hospital followed. George obtained doctoral degrees (PhD and DSc at the University of London, MB, BChir, and MD at Cambridge) and won a variety of academic prizes for his work. Following medical training he worked at the MRC Population Genetics Research Unit at Oxford and in 1961 came to Seattle (USA) to work as a research fellow in my unit of Medical Genetics at the University of Washington. After working in London as a lecturer on genetics of blindness for three years, he moved to Adelaide (Australia) as reader in genetics. He came back to Seattle as an associate professor and worked in research largely related to human selection of malaria-dependent traits and in clinical genetics (1968–1971). Subsequently, George spent relatively short periods in a variety of positions in medical genetics: Leiden, Holland 1971–1973; St. John's, Newfoundland 1973–1976; Ottawa 1976–1979; Montreal 1979–1980; Bethesda, Maryland 1980–1984; and then found a more permanent working place at the Cancer Genetics Clinic of Churchill Hospital in Oxford from 1984 until his retirement.

George's Fraser's contributions

George has never been a mere biomedical scientist, but a broad scholar in the best traditions of medical and scientific scholarship. His contributions in the area of the genetic causes of profound deafness and blindness in childhood are a solid analysis of these common human conditions which in about 50% of cases have a genetic aetiology. His monographs on blindness (1967) and on deafness (1976) reflect his encyclopaedic knowledge of genetic, biochemical, and medical aspects of these diseases. These books and his many papers that provide more details on blindness and deafness made George a pre-eminent authority in these conditions. But his impact was wider.

In Seattle, George became an indispensable, superb analyst of the large body of data accumulated in population genetic studies by our group on the role of malaria in shaping the distribution of HbS, thalassaemia and G6PD deficiency in Greece, Africa, the Philippines and Taiwan. George's calculations on the long term future effects on gene frequency of recessive diseases following antenatal diagnosis with selective abortion of affected foetuses showed relatively small effects over historical time spans.

Reflecting his broad scholarship and deep knowledge of many other aspects of human and medical genetics, he contributed to many areas mentioned elsewhere in this book. There are few geneticists who combined broad mathematical and statistical ability with the deep biological, clinical and genetic knowledge together with human interests in genetic counselling and a capacity for sympathetic interaction with patients and relatives, particularly in cancer families. I trust that this *liber amicorum* will show all who read the wide-ranging contributions reflecting George's interests how much he is appreciated and respected by many human and medical geneticists all over the world.

George Fraser – the man

George worked on three continents and travelled extensively, getting to know many human and medical geneticists all over the world. He is widely respected for his broad interests that ranged beyond those of most of his peers. His love of the classics, history and poetry was admired but did not always resonate with colleagues who had more mundane interests. George's deep humanity also was demonstrated by another scholarly endeavour not usually carried out by medical academics. He translated Benno Müller-Hill's book *Murderous Science* on the role of German human geneticists, anthropologists and psychiatrists in the elimination of Jews, 'Gypsies' and mentally ill patients in Germany 1933–1945 (Fraser 1988). This book was the first exposition by a knowledgeable geneticist based on interviews with contemporaneous, directly involved physicians and scientists as well as a study of primary sources. To bring this information to the attention of a broader audience, George translated this book from German into English.

Along with his brilliance and humanity, George tended to be a bit absent-minded, leading to mishaps that are amusing only in retrospect. Here are two episodes. First, while passing briefly through Seattle in late 1965 on his way from London to take up an appointment in Adelaide, South Australia, George dined at our house. During the meal, our neighbour rang the bell and enquired whether such-and-such a car belonged to any one of our guests. The car had been parked on the slightly inclined street apparently without properly applying the parking brakes and had smashed into the rear of my neighbour's parked new car with considerable damage. Unfortunately, because of George's status as a visitor to the USA, the insurance for which he had paid at the time when he had bought a second-hand car in order to enjoy a few weeks of travel in North America, was retrospectively declared to have been invalid. Secondly, he and I entered a taxicab at an airport. Shortly after the cab's departure, the driver made a sharp turn and the door next to George flew open. George was about to fall out into the street if I had not held on to him. He apparently had not properly closed the taxi door.

Many of us were deeply affected last year when he sent us a Brazilian poem circulating on the internet, 'Friends', translated from Portuguese by George. This soul-stirring poem reflected George's deep feelings about 'friends who do not know how much they are my friends. They do not perceive the love which I devote to them and the absolute need which I have of them.' The poem appears to convey what George feels about many of his colleagues. 'When I do seek them out, I realise that they do not have any idea of how necessary they are to me.' This poignant poem is a reminder to many of us that we should have spent more time with George than we did over the years. A complete version of George's translation is reproduced in the contribution of Professor E A Chautard-Freire-Maia to this *liber amicorum*.

George R. Fraser: scholar

Jack (W.J.) Schull

Human Genetics Center, University of Texas Health Science Center,
PO Box 20186, Houston, Texas 77025, USA

The year was 1961. It was Rome. The occasion was the Second International Congress of Human Genetics. The decade or so preceding this assembly was one of ferment in genetics. Indeed, it has often been described as the 'golden age' of molecular biology. New technologies, such as moving boundary electrophoresis, made it possible to demonstrate that sickle-cell anaemia was a molecular disease (Pauling *et al.* 1949) and to characterise the specific biochemical differences that distinguish normal from sickle haemoglobin (Ingram 1956, 1957). An even more far-reaching development was the positing by James Watson and Francis Crick (1953a, b) of the helical structure of DNA. When this conjecture withstood all efforts to disprove it, interest moved on to understanding how information was encoded in the double helix. Robert Holley, Govind Khorana and Marshall Nirenberg, through a series of ingenious and independent experiments, set the stage for the ascertainment of the code underlying the structure of the amino acids, the building blocks of all proteins. Illustrative of these experiments was the demonstration of the dependence of cell-free protein synthesis upon naturally occurring or synthetic polyribonucleotides (Nirenberg and Matthaei 1961). But not all of the advances relied upon developments in physical or biological chemistry. Joe-Hin Tjio and Albert Levan (1956), using techniques of a much simpler nature, showed that the number of chromosomes in man was 46, not the 48 commonly held at the time. In 1959, employing these new cytogenetical techniques, Jérôme Lejeune established that Down Syndrome was a chromosomal disorder, a notion that Petrus Waardenburg had advanced a quarter of a century earlier (1932, pp. 47–8), and Charles Ford and his colleagues (1959, see also Patricia Jacobs and J. A. Strong 1959) showed that some variation in human sexuality was chromosomal in origin.

This new knowledge of the molecular and cytogenetical variability existing among human beings had challenging evolutionary implications. How was this variability maintained and what was its evolutionary significance? Two competing hypotheses regarding maintenance existed. One held that variability was maintained through a balance of selection and mutation; whereas the other asserted that variability stemmed from the selective advantage of the heterozygote. The former hypothesis (the 'classical' view) was championed by Herman Muller who, in his presidential address to the American Society of Human Genetics in 1949 introduced the notion of 'loads,' specifically in the context of recurrent mutation. A genetic load, as Muller envisaged it, was the extent to which the Darwinian fitness of the average or optimum genotype at a particular locus was decreased by the presence in a population of deleterious (lethal, semilethal, or subvital) genes. The second hypothesis, that concerning heterotic loci, was arguably the older. The notion of heterosis had been introduced by George Harrison Shull to describe the superiority of the heterozygous genotypes he encountered in corn breeding (1911). This idea was further elaborated by Edmund Brisco Ford (1940, p. 493), the English ecologist, who coined the expression 'balanced polymorphism' to describe those situations wherein two or more genotypes occur simultaneously in the same population at frequencies not accountable by recurrent mutation. In the United States this thesis was championed by Theodosius Dobzhansky and Bruce Wallace as well as Michael Lerner (1954). Unfortunately, the evidence to adjudicate this dispute was fragmentary and inadequate to the needs. However, in 1956, Newton Morton, James Crow, and Muller set forth a possible means of distinguishing between these hypotheses based on data from consanguineous marriages.

Their argument was ingenious and simple mathematically. It went somewhat as follows: if one assumes that different causes of death, genetic and environmental, are independent in action, and that there are many causes all with separate, small probabilities, then the fraction of individuals surviving is

$$S = \Pi(1 - x)\{1 - qFs - q^2(1 - F)s - 2q(1 - q)(1 - F)sh\}$$

where q is the frequency of a given gene, s is the probability of death in the mutant homozygote, h is a measure of dominance, F is the coefficient of inbreeding, and the product, Π, is formed over all causes. Rearranging these terms and collecting those with and without F, and taking the limit of a very large series, one has

$$-\log_e S = A + BF$$

where $\qquad A = \Sigma x + \Sigma q^2 + 2\Sigma q(1 - q)sh$ and $B = \Sigma qs - \Sigma q^2 s - 2\Sigma q(1 - q)sh.$

Summation extends over all environmental causes and all genetic loci having mutant alleles. The former expression, A, represents the damage in a randomly mating population (where F = 0), and the latter, B, is the increased increment of damage that would be fully expressed only in a population where F = 1 (see Schull and Neel 1965, Chapter 15, for a fuller development, and the contributions of Morton, Ewens and Sved for further discussion). This argument suggests that if genetic variability is primarily maintained through a balance of mutation and selection, the ratio, B/A, should be large; whereas if maintenance is through the perpetuation of heterotic loci, the number should be much smaller, equal approximately to the average number of alleles at those loci maintained by a balance of selective forces. Unfortunately, the data then available led to equivocal results. The ratio was neither large enough to establish the one hypothesis nor small enough to support the other.

It was into this environment that George Fraser, a ruddy-cheeked young man with tousled hair and a twinkle in his eyes, stepped. He was well equipped for the task confronting him. He was not only a member of the Population Genetics Research Unit of the Medical Research Council that Alan Stevenson headed, but a former student of R. A. Fisher and Lionel Penrose. As a preamble to his review of the controversy (Fraser 1962), he described the circumstances that prompted his involvement: He wrote 'In March 1961, in the capacity of consultant to the World Health Organization, the author was asked to present, to the IXth session of the United Nations Scientific Committee on the Effects of Atomic Radiation in Geneva, information on the mutational load of human populations, particularly a description of the genetical load and its partition between mutation and other causes (p. 387).' His charge, a formidable one for a 29-year-old, was a departure from his previous scientific involvements, most of which concerned the biology of goitre and of its association with deafness, and indeed from most of his subsequent publications. His review offered no new theoretical insights, that is, he proposed no new components of the load nor did he provide numerical estimates of the importance of those components others had identified. Rather, he emphasised the uncertainty inherent in the assignment of individual human diseases and disabilities to the various components of the genetic load that others had suggested. To put it somewhat differently, he stressed the 'human element' inherent in the genetic load. He obliged us to confront the fact that the ratios so enthusiastically computed and debated masked an enormity of human suffering. Each death or disability to maintain the balance had an immediate human cost – a cost to be reckoned not in morbidity or mortality statistics but in tears.

To George the fact that one is involved in research rather than clinical service does not absolve one of the needs to be compassionate and sensitive in one's dealings with one's fellow human beings. Again, to cite his words he noted: 'While it is not denied that individuals, families and sometimes entire nations have to pay a very heavy socio-economic price, a genetical load is something which no species, least of all man, will ever be without. We can strive only to keep it to the minimum; the ideal homozygote will never be found. This is not a philosophy of despair or a licence to relax all precautions. Thus, for example, a species is adapted to the mutation rate existing at the epoch and an abrupt change can have no consequences for the good. The complexity of the genetical dynamics of a population can be compared with the complexity of the molecular organisation of a cell. Yet just as a cell functions as a beautifully integrated whole so does the total genetical constitution of a population; and just as a cell is sensitive to a variety of insults so is the hereditary material of our species to any uncontrolled changes in its environment' (p. 410).

In 1974, with his colleague Oliver Mayo, Fraser revisited the matter of 'genetical loads,' and in particular, the effects of social organisation in changing or modulating the selective forces that operate on the human community. After a careful and thorough delineation of the strengths and limitations of human intervention, they conclude 'whatever happens to the various components of the genetical load, it should be well within our technological and economic resources to achieve considerable reductions in this burden' (p. 108). This optimistic statement notwithstanding, little concerted effort has been given to achieving this end. As so often happens in science, the pendulum has swung and over time the interest in 'loads' that once obtained in human genetics no longer does. Much of this loss of interest can be ascribed to the hypothesis advanced by Motoo Kimura (1968, 1983), Jack King and Thomas Jukes (1969) and Tomoko Ohta (1973) that most mutations fixed in a population are selectively neutral or nearly so, and thus their frequency in a population is largely a matter of chance rather than selective advantage or disadvantage. Ergo, they impose no burden of significance on a population. Today, research on evolution and the maintenance of genetic variability is rarely pursued at the phenotypic level described here but more commonly at the molecular one where greater precision is possible (see Nei 2005).

It is unlikely that George Fraser will be remembered for his contributions to studies of the maintenance of genetic variability in human populations. But he will be remembered for his significant contributions to the nosology of genetic disease, and in particular those diseases involving the eye and the ear. Here his contributions are seminal. One component was his study of the factors leading to blindness or severe visual disability in childhood which he submitted to Cambridge University in 1966 as his MD doctoral thesis. As a model to emulate in planning his study, he wisely chose Lionel Penrose's famed Colchester Survey (1938). Using records of enrolment in the special schools statutorily established for the visually disabled in England and Wales, Fraser identified 776 children varying in age from 1 to 22 for study.[1] Between 1963 and 1965, more than 90% of these individuals were examined, and extensive social, clinical, and family information were obtained. These basic data were supplemented by family studies and laboratory investigations wherever possible. The laboratory studies included urine analysis, examination of blood smears, detection of rubella antibodies, and assaying the level of the enzyme glucose-6-phosphate dehydrogenase alleged to be associated with cataract formation. Utilising these data, each child's blindness was assigned to one of the following diagnostic categories: (1) mainly due to genetic determination, (2) prenatally acquired, (3) perinatally acquired, and finally, (4) postnatally acquired.

Two of these categories accounted for the bulk of the cases of blindness, specifically, genetic determination and perinatal acquisition. As to the former group, i.e., cases attributed to genetic causation, it included choroido-retinal degenerations, retinoblastoma, pseudoglioma and congenital or infantile retinal detachment, optic atrophy, congenital and infantile cataracts, myopia and childhood retinal detachment, lesions of the cornea, coloboma or microphthalmos or anophthalmos, aniridia, buphthalmos, blindness as a part of a complex syndrome involving other malformations, and blindness secondary to malformations. Among the perinatally acquired instances of blindness, virtually all of the cases were attributable to retrolental fibroplasia stemming from the well-intentioned but misguided oxygen enrichment of the environment of prematurely born infants in vogue in the early post-war years.

Nosologically, the assignment of any given case to one or another of the four categories mentioned above can be challenging, a fact that Fraser and Friedmann clearly acknowledged. Indeed, they observed 'it is evident that this attempt at classification of causes of severe visual disability among the children in this series is only the best possible in the circumstances of the survey, and extrapolation of the results to problems of blindness in children as a whole must be made with caution (p. 136).' We cite their words to underscore the admirable candour and thoroughness of the description of the limitations of their study. These nosological difficulties, however, did not deter George from recognising that among the children he studied was one with an unusual cluster of abnormalities that has since become known as Fraser's Syndrome. The child, a girl (M6 in his study notation), had cryptophthalmos, deafness due to middle and outer ear malformations, high palate, webbing of fingers and toes, atresia of the larynx, wide separation of the symphysis pubis, anal stenosis, and masculinisation of the external genitalia (see also Fraser 1962, page 400). Considerable molecular biological attention has been centred on this syndrome recently because of the recognition that a very similar disorder exists in the mouse, and apparently arises as a consequence of a mutation at the same locus seen in Fraser's syndrome (McGregor *et al.* 2003, Vrontou *et al.* 2003, Takamiya *et al.* 2004, and Jadeja *et al.* 2005)

George's other, and larger and longer (1957–1967), foray into the realm of socially disabling diseases of childhood focused on profound deafness (1976). Although this study followed the paradigm he used in the study of blindness, it also differed in that it involved a series of surveys rather than a single integrated investigation. The bulk of these surveys occurred in the British Isles, including some children of recent migrants of non-European origin, but also involved 520 individuals, both children and adults, whom he studied in South Australia. Again, the cases of deafness ascertained through these studies were divided into those determined by Mendelian inheritance, those being part of a syndrome involving the head and neck primarily, those resulting from acquired causes, and those of 'unknown' cause. This study proved more difficult than the one on blindness in several respects. First, there was not then nor is there now a universally accepted clinical definition of profound deafness. This designation rests primarily on social considerations, such as the inability to profit from conventional schooling. Second, the retrospective nature of the study of several diverse surveys made uniformity of diagnostic criteria impossible. These problems notwithstanding, this study warrants its status as a classic. Indeed, every aspiring epidemiologist or student of deafness should be obliged to read the short final chapter of this monograph. The conclusions set forth here are models of objectivity and offer suggestions to subsequent investigators that are timely and discerning.

In 1979, George's penchant for order and his nosological leanings carried him from Canada, where he was then Chief of the Department of Congenital Anomalies and Inherited Diseases of the Department of National Health and Welfare of the Federal Government of Canada, to the Lister Hill Center for Biomedical Communications at the National Library of Medicine in the United States and to a program headed by Donald Merritt. One aspect of this latter program involved the utilisation of Victor McKusick's Catalogs of Autosomal Dominant, Autosomal Recessive, and X-linked Phenotypes to explore the application of computer techniques to the interactive communication of biomedical knowledge to physicians everywhere. It was this activity, known as the Human Genetics Knowledge Base, that he was to direct. Unfortunately, this program was terminated before it could prove its worth, and George found himself adrift. In 1984, he returned to England to a position as a Senior Research Fellow with the Imperial Cancer Research Fund (ICRF). Specifically he was charged with providing 'a computerised data base to enable the Cancer Epidemiology and Clinical Trials Unit, Oxford, the Mathematics, Statistics and Epidemiology Laboratory, LIF, the Computing Facilities Department and the Cancer Families Study Group and other ICRF workers, to enter and retrieve information on cancer family studies.' Subsequently, in 1990, to him fell the establishment and direction of a Cancer Genetic Clinic. He retained this position until he reached the mandatory retirement age in 1997. As is his wont, George plunged into this new assignment with dedication (Fraser 1999). He undertook a review of the literature on the familial aggregation of cancer as a preliminary to defining the service and research functions he believed the clinic could fulfil. As to service, two functions were of particular importance in his view, namely, (a) to provide realistic assessment of risk to persons who are fearful of cancer because of the presence of this disease among their relatives, and (b) to arrange appropriate screening programs for those individuals and families where the risk of cancer is elevated with a view toward early detection when intervention is apt to be more successful. As to research, he envisaged two functions, i.e., collaboration with institutions throughout the United Kingdom in identifying and characterising cancer susceptibility genes, and furtherance of the study of the clinical pattern of cancer aggregation. Over the first seven years of this clinic's existence, it addressed some 2,165 individual referrals representing 1,771 families. George's assessments of the achievements of the clinic were characteristically muted. Where others might have magnified the importance of their achievements, he held that this was only a beginning in the definition of the hereditary contribution to the pathogenesis of cancer.

Throughout these peregrinations, we continued to correspond and see one another whenever circumstances permitted. Opportunities often arose through service as members of the World Health Organization's Panel of Experts on Human Genetics, or the National Institutes of

Health or some similar agency. Invariably he was not merely a significant addition to any committee on which he served but a source of pithy, often obscure, quotations drawn from his reading. Illustrative of this talent is a remark to be found in a WHO document stemming from a meeting in Mexico at which George served as rapporteur. The committee had grappled with differing levels of uncertainty and doubt about the quality of the data on which its recommendations were to rest. George seeking a politically apt remark offered the following one ascribable to Nicolas Malebranche, a 17th-century French theologian and philosopher: 'One should not think that one has not made a great advance if one has simply learned to doubt. To know when to doubt rationally as a thinking individual is not such a small thing as one might suppose.'

George has been the conscience of human genetics. Throughout his scientific career and his extensive travels, he has never failed to encourage, cajole, and exhort his colleagues to bear in mind the human implications of their actions and research and to eschew notions of superiority or inferiority. He has said, quite forcefully, 'there are no genetically superior human beings and no genetically inferior human beings; there are only human beings who are fellow-members of one species' (Fraser 1997). He has further noted that 'certain inalienable rights inhere in each member of our species, among them being the right to satisfy his or her needs with respect to food, clothing, and shelter, the right to health care and to education, and the right to procreate. The very large, and increasing inequities which obtain today with respect to the fundamental right of adequate access to food, clothing, shelter, health care and education, should be a source of concern and of shame to the more fortunate members of our species.' Like the priest or legal circuit rider of old, George has spread this message of compassion to many places – Adelaide, Brazil, Leiden, Newfoundland, Ottawa, Seattle, Washington, and more recently Oxford.

George's thoughts on the place of genetics in medicine and the role of the geneticist in the purveyance of healthcare are admirably set forth in two presentations, one at a meeting in Dijon in 1997 and the other at a meeting at the Vatican in Rome in 2005. His thoughts and the clarity of their exposition are provocative but not tendentious, and as such stimulate a reader to examine his or her own conscience regarding the economically, socially and numerically disadvantaged peoples of the world. George has taken a courageous stand, one that will be unpopular in some circles, especially among those individuals and groups of individuals who have traditionally achieved and maintained their social stations at the expense of others.

For the reasons adduced above, it is wholly fitting that we, George's admirers, colleagues, and friends, celebrate his seventy fifth birthday, and seize this occasion as an opportunity to disseminate more widely his eloquent defence of the rights of the vulnerable members of the human community. He has enriched our lives in many ways, through his scholarship, his clinical skills, his ethical standards, his good humour, and particularly his capacity to see the folly in the actions of others as well as himself. Indeed, it would be well if, in judging the import of our own actions and accomplishments, we all used the same mirror that George unflinchingly holds before himself.

Note

1. A revised version of his thesis was published in 1967 in collaboration with Allan Friedman by the Johns Hopkins University Press.

'Genetica is als muziek: het uitgangspunt is eenvoudig,
de variaties zijn oneindig en het resultaat is fascinerend'.[1]
Variatie op Leidse stelling

Réflexions sur le Φαινόμενον G. R. Fraser on the occasion van zijn 75ste Geburtstag[2]

Jacques J. P. van de Kamp

Fruinlaan 10, 2313 ER Leiden, The Netherlands

ἄνδρα μοι ἔννεπε, μοῦσα, πολύτροπον, ὃς μάλα πολλὰ πλάγχθη … [3]

Some – a little better educated – individuals will certainly (and rightly) presume that these are the first lines of Homerus' epos *Odyssey*. However these might also be the first lines of a G. R. Fraser biography. Whose birthplace, indeed, except that of George, was situated as well in Russia (Uzhgorod), as in Hungary (Ungvár), as in the dual monarchy Austro-Hungary, as in Czechoslovakia (Užhorod), as in the independent republic of Ruthenia (for one day on 14 March 1939), as in Ukraine (Uzhhorod) (in chronological order)? And who, next, made his actions speak in such famous and far spread places as Adelaide, Baltimore, Bethesda, Brasilia, Bristol, Dartmouth (New Hampshire), Denver, Geneva, Glasgow, Leiden, Liverpool, London (UK), London (Ontario), Montreal, Newcastle, Ottawa, Oxford, Paris, São Paulo, Seattle, Sheffield, St John's (Newfoundland), Toulouse, Uppsala and Washington (in alphabetical order)?

When, in 1970, I had finished my specialisation in paediatrics and desired to work in the field of medical genetics, our 'new to come' professor, George, was even lost in the backlands of Brazil – '*Wie gelukkig wil blijven moet vaak veranderen*'[4] said Confucius (but where and when I have unfortunately forgotten) – and it took weeks before my request reached him and was, luckily, accepted.

Although the University of Leiden was happy and proud to have laid its hands on such a famous professor, his coming was not without difficulties. George appreciated Holland, because a small and stable country was ideal for the population genetic studies he had in mind. But for that goal George needed a COMPUTER and the University Board regretfully let him know that: 1. there was no space for such a machine 2. it was very, very, much too expensive and 3. which was decisive: there was already one such thing in the country. Being a small country.

But population genetics is no longer what it was, due, also, to George's fascination with aviation:

'The jet airplane has already had an incalculably greater effect on human population genetics than any conceivable program of calculated eugenics' (J. Lederberg)

which is – in my view – probably the sole advantage of aircraft and this makes me exclaim to all ministers of integration or immigration or the like:

Immigratie verrijkt het genetisch (èn cultureel!) erfgoed[5]

George always claimed (at least to me, a citizen of that town) that Leiden was the best place to be in the world. However, his Chair of Human Genetics was situated in a laboratory-environment, small kingdoms, sometimes but not always, ruled by excellent kings, which was not an entirely comfortable Umwelt (environment) for one who is more a clinician than a laboratory-man. One day George, detecting a mouse in his room, called for the municipal vermin-destruction service, who could hardly be restrained from destroying all non-human life in the institute to the extreme horror of fruit-fly and mouse geneticists.

During his short stay in Leiden, George strived to bring Clinical Genetics (*avant la lettre*)[6] where it – to my mind also – ought to be: in the hospital. This finally materialised around the turn of the century, thus almost exactly in concordance with one of George's favourite sayings: '*Im Holland geschieht alles vierzig Jahre später*'[7], wrongly attributed to Heinrich Heine.

But life has, according to Eliot, never been easy for a university professor: 'It is a great pity that the life of a professor is so engrossed with executive work and committees, that he has scant time for writing' T. S. Eliot (in 1928 in *Letters of T. S. Eliot*).

And, of course: 'Није сваки дан Божић'[8] (Serbian saying)

Anyway George all his life regretted having left that best of all places, confirming that '*La nature humaine est ainsi fait qu'on s'habitue plus aisément à la détresse qu'au bonheur*'[9] (E. Wiesel. In: *Toutes les fleuves vont à la mer.*)

One thing, a very small one, which George and I had in common, was that we both attended Trinity College; the difference: his was in Cambridge and mine in the less famous place, Haarlem. Far greater was the gap between George's polyglot talents and my very poor pronunciation of the English language, best comparable to Churchill's famous French. So one day I proudly announced the birth of my – first and as it proved later, only – *son*; but George supposed that I was speaking of the weather (*sun*) and muttered 'it is very bad, indeed'. No congratulations!

George's greatest achievements were his extended and still valuable studies on blindness and deafness in childhood. The first achievement betrays the breadth of his interest as: '*De mens is een gevisualiseerd dier, twee ogen op twee poten, (50 milj. hersencellen, de helft van het totaal, zijn betrokken bij de visusfuncties)*'[10] (A.van de Heyden. Commemoration Leiden University, 1996.)

The second achievement – as 'Music is the eye of the ear' Thomas Draxe – points to an interest in music; the only subject on which he never posed (to me at least) as a connoisseur.

I guess that George's deeper feelings are best reflected in the homage he always, in lifelong admiration, paid and pays to his teacher, Lionel Penrose. Penrose's thoughts about the essence of human variance, the value of being 'abnormal', the painful dilemmas in prenatal diagnosis and abortion reappear in many of George's talks and publications.

I remember George saying – and he repeated it in a lecture he held in Dijon in 1997 – that 'In the past, however, we would not have spoken of preventing or eliminating tuberculosis by killing the patient.'

George will surely agree with his teacher's words (and so do I):

'I would rather live in a genetically imperfect society which preserves human standards of life than in one in which technological standards were paramount and heredity perfect'. (Lionel Penrose.)

With perceptible pleasure, George describes Penrose's attitude towards Down-syndrome patients and I suppose that he with me would enjoy seeing Lejeune's vision fulfilled: 'I look forward to the day when a mongolian idiot, treated biochemically, becomes a successful geneticist'.

I was always impressed by the observation that future mothers without experience with Down syndrome and having less than 1% risk of carrying a baby with that disorder almost all opt for prenatal diagnosis and eventually termination of pregnancy, while mothers from families with translocation Down and a considerable higher risk do not ('If my baby turns out to be like my brother or like my nephew, I would be happy'). The lessons are: '*Onbekend maakt onbemind*'[11] and '*Wat de boer niet kent dat eet hij niet*'[12] (Dutch sayings)

Our attitude towards mentally defective people has changed drastically during the last century and there is no reason to be pessimistic; but reason to be careful remains. The feeble-minded have no economical value and economical value seems more than before the only thing that counts. There are still people who, as G. K. Chesterton defined them a century ago, 'have discovered how to combine the hardening of the heart with a sympathetic softening of the head'. And even perhaps Penrose studied the wrong population because, according to the same Chesterton, there are people who are 'clever enough to get all that money and stupid enough to want it'.

I share the fear that Edmond Rostand let Cyrano de Bergerac pronounce:

> *'Je crains tant que parmi notre alchimie exquise*
> *Le vrai du sentiment ne se volatilise*
> …
> *Et que le fin du fin ne soit la fin des fins'.*[13] (Cyrano de Bergerac)

It would not surprise me if George's zeal in the promotion of Benno Müller-Hill's book *Tödliche Wissenschaft: Die Aussonderung von Juden, Zigeunern und Geisteskranken 1933–1945*[14] also found its origin in Penrose's pacifism as a Quaker; no two worlds could be more apart: that of the Nazis and that of the Quakers. On reflection, it is indeed a shame that physicians are not in the forefront in the prevention of war and torture. '*Dulce bellum inexpertis*'[15] (Erasmus.)

But perhaps the medical ethicist was right who taught me that: 'We have physicians to make people healthy not happy'.

I am sure, George, you will at this moment fully agree with Montaigne who complained (long ago): '*Mij lijkt het niet redelijk om mensen voor hun 55^e of 60^e jaar met pensioen te sturen*'[16] (M. de Montaigne, Essays)

But why complain? It must be satisfying to act as an (experienced?) helping hand to Maria, who, undoubtedly has been your help for so many years. And moreover: with knowledge of only Czech, Dutch, English, French, German, Greek, Hungarian, Italian, Portuguese, Russian, Serbo-Croat and Spanish, you can hardly travel around the world, where recently exactly 6,912 known living languages have been registered.

Our common work was restricted to slightly more than two years. A short time, but enough for me to quote your quotation:

> 'He was a man, take him for all in all,
> I shall not look upon his like again'. (*Hamlet,* William Shakespeare)

In conclusion: Lia and I wish you and Maria all the best for many, many, more years. And to supply you with a little homework (as you did to me so often) it is perhaps a bit exaggerated to say '*Mai te mahana e reva tu ai óe ra, aita ia e faito i tou nei mauiui e tau*'[17]. But the citation (from

P. Loti *Le Mariage de Loti*): '*Ia ora na oe i te Atua mau*' [18] is wholeheartedly meant. And finally, meditate on a text that is almost written on your life:

> '*Heureux qui, comme Ulysse, a fait un beau voyage*
> *Et puis est retourné … … …*
> *Vivre entre ses parents le reste de son age*' [19]. (Joachim du Bellay)

Notes

1. Genetics is like music; the starting point is simple, the variations are endless, and the result is fascinating. Variation on a Leiden saying
2. Reflections on the phenomenon 'G R Fraser' on the occasion of his 75th birthday
3. Tell me, O Muse, of the man of many devices, driven far astray
4. They must often change who would be constant in happiness.
5. Immigration enriches the genetic and cultural heritage
6. Before the term existed.
7. In Holland everything happens forty years later.
8. Not every day is Christmas.
9. Human nature is so constituted that Man gets used more easily to distress than to happiness.
10. Man is a visually equipped animal, two eyes on two legs (fifty million brain cells, half the total, are involved in the function of seeing)
11. ' Unknown makes unloved'
12. ' What the farmer does not know, he does not eat'
13. Yet I fear, lest in our exquisite alchemy, true feeling itself might simply cease to be, and the ultimate fineness be merely the end of things.
14. In English translation by George Fraser: *Murderous Science: Elimination by Scientific selection of Jews, Gypsies, and Others Germany 1933–1945.*
15. War is sweet to those who have never taken part in it.
16. It does not seem right to me to pension people off in their 55th or 60th year.
17. Since the day you left, the depths of my sorrow have been without measure
18. Hail and may God protect you!
19. Happy he who like Ulysses has completed a beautiful voyage
    And has then returned …
    To live out the rest of his life surrounded by his family.

Persistent Truncus Arteriosus associated with a homoeodomain mutation of *NKX2.6*

Kirsten Heathcote[1], Claire Braybrook[2], Lulu Abushaban[3], Michelle Guy[1], Maher E. Khetyar[1], Michael A. Patton[1], Nicholas D. Carter[1], Peter J. Scambler[2,3] and Petros Syrris[1]

[1] Department of Clinical Developmental Science, St George's Hospital Medical School, London, SW17 ORE, UK
[2] Molecular Medicine Unit, Institute of Child Health, London WC1N 1EH, UK,
[3] Department of Cardiology, The Chest Hospital, Kuwait City, 13041 Safat, Kuwait

George Fraser – A perspective on heart disease

George had (and has) a unique insight into the genetics of deafness, including the autosomal recessive Jervell and Lange-Nielsen syndrome of profound childhood deafness with a prolonged QT interval of the electrocardiogram, fainting attacks, and sudden death, and this led him, thinking laterally, to speculate in the 1960s on possible causes for unexpected cot deaths. Thus, Fraser and Froggatt (1966) suggested that a proportion, probably small, of sudden unexplained deaths in infancy (cot death, crib death, sudden infant death syndrome (SIDS)) is caused by genetically determined disorders of cardiac conduction, specifically inherited prolongations of the cardiac QT interval. Thirty-five years later, Ackerman *et al.* (2001) showed that two out of 93 hearts from infants with SIDS possessed mutations at the *SCN5A* locus in the genomic DNA extracted from frozen myocardium, consistent with the hypothesis put forward by Fraser and Froggatt (1966). More recently, Hunt (2004) suggested that 4–5% of SIDS deaths are associated with a SCN5A polymorphism. Over the years since 1995, this condition of autosomal dominant 'long QT syndrome' has been associated with a number of cardiac ion-channel gene loci of which SCN5A is just one, not included among those responsible in homozygous or compound heterozygous form for the autosomal recessive Jervell and Lange-Nielsen syndrome, leaving open the question of precisely how big is the small proportion of SIDS deaths which are due to mutations at these loci.

With the advent of gene hunting using information from the Human Genome Programme, many other gene loci for common and rare cardiac conditions have been identified. The following paper, dedicated in this volume to George, describes the identification of such a gene locus by our group at St George's University of London.

Finally, I wish to say how much I have valued friendship and collaboration with George over the past four decades since our first meeting in Seattle in 1968.

Nick Carter

Abstract

Persistent truncus arteriosus (PTA) is a failure of septation of the cardiac outflow tract (OFT) into the pulmonary artery and the aorta. We have used autozygosity mapping of a large consanguineous family segregating for PTA to map the causative locus to chromosome 8p21. An F151L mutation was identified in the homoeodomain of *NKX2.6*, a transcription factor expressed in murine pharyngeal endoderm and embryonic outflow tract myocardium. While expression of *Nkx2.6* during murine embryogenesis is strongly suggestive of a role for this gene in heart development, mice homozygous for a targeted mutation of *Nkx2.6* are normal. However, in these mice it has been shown that *Nkx2.5* expression expands into regions lacking *Nkx2.6* suggesting functional complementation. As transcriptional targets of NKX2.6 are unknown, we investigated functional effects of the mutation in transcriptional and protein interaction assays using NKX2.5 as a surrogate. Introduction of F157L into NKX2.5 substantially reduced its transcription activating function, its synergism with partners at the atrial natriuretic factor (*ANF*) and connexin-40 (*Cx40*) promoters, and its specific DNA binding. We tested NKX2.5 target promoters for NKX2.6 activity. NKX2.6 was inactive at *ANF* but weakly activated transcription of a *Cx40* promoter, whereas the F151L mutant lacked this activity. These findings indicate a previously unsuspected role for NKX2.6 in heart development which should be re-evaluated in more sophisticated model systems.

Introduction

Persistent Truncus Arteriosus (PTA) is a rare congenital heart anomaly accounting for 1% of congenital heart defects (Hoffman and Kaplan 2002). The neonate is well at birth but without surgery heart failure develops and untreated infants rarely survive beyond six months of life. Surgical advances have led to much improved survival and quality of life in recent years (Ferdman and Singh 2003), although even after surgery, PTA is classified as a severe lesion (Hoffman *et al.* 2004). The defect is attributed to incomplete septation of the truncus arteriosus, which is the distal portion of the outflow tract of the embryonic heart tube. Septation occurs when mesenchymal cell proliferation causes endocardial ridges to form within the outflow tract during the fifth week of human development, and cardiac neural crest cells migrate through the primordial pharynx and pharyngeal arches to populate these ridges during the sixth week; this invasion may initiate the fusion of these ridges to form a septum (Webb *et al.* 2003). This spiralling aorticopulmonary septum divides the outflow tract into the two great vessels which arise from the heart. Outflow tract defects occur if this process is deficient. These congenital abnormalities may occur as isolated anatomical defects such as PTA, tetralogy of Fallot, or double outlet right ventricle, or as part of defined syndromes such as deletion of 22q11 (DiGeorge and velocardiofacial syndromes) (Volpe *et al.* 2003).

Several mouse mutations have been associated with PTA. Mutations in transcription factors (e.g. *Pax3, cited2, AP2a, Pitx2*) and signalling proteins (e.g. *sema3C, BmprII, Alk2, Dvl2*) highlight the importance of the neural crest contribution (Creazzo *et al.* 1988) and laterality (Franco and Campione 2003) to outflow tract septation. However, recent work has drawn attention to the role of the pharyngeal endoderm and anterior (secondary) heart field in outflow tract morphogenesis (Kelly and Buckingham 2002). The anterior heart field comprises a population of splanchnic mesoderm cells which contribute to OFT and right ventricular myocardium, as marked by the *islet-1* LIM-homoeodomain transcription factor (Cai *et al.* 2003). *Tbx1* null mice develop PTA, and conditional mutagenesis has defined two distinct roles for *Tbx1* in outflow tract development, neither of which directly involves the neural crest (Xu *et al.* 2004). *Tbx1* regulates proliferation of anterior heart field cells destined for the outflow tract, and is also required in *Nkx2.5* expressing pharyngeal endodermal cells for septation to occur (Xu *et al.* 2004). One role of *Tbx1* is to regulate fibroblast growth factor expression during embryogenesis (Xu *et al.* 2004), including cells in the OFT (Hu *et al.* 2004). Tissue-specific inactivation of *Fgf8* using

Tbx1-Cre results in OFT defects (Brown *et al.* 2004). *TBX1* is hemizygously deleted or heterozygously mutated in human patients with the 22q11 deletion syndrome (Scambler 2003). *NKX2.5* has also been implicated in human congenital heart defects (McElhinney *et al.* 2003), including atrial septal defects, and it interacts at target promoters with two other transcription factors mutated in patients with cardiac septation defects, *TBX5* (mutated in the Holt-Oram syndrome) (Hiroi *et al.* 2001) and *GATA4* (Garg *et al.* 2003). Recently, it has been shown that *Tbx2* is essential for both normal outflow tract development and patterning of the atrioventricular canal. However, *Tbx2* does not appear to be required for an anterior heart field or neural crest contribution, implying mesenchymal cell growth or differentiation defects in the absence of this gene (Harrelson *et al.* 2004). Tbx2 is known to antagonise Tbx5 activity at the *ANF* promoter (Habets *et al.* 2002). Thus it appears that NKX, GATA, and Tbx transcription factors interact to modulate one another's activity and are central to normal cardiac morphogenesis.

Despite the rarity of large human kindreds segregating congenital heart defects, human genetic analyses have been important in elucidating some of these pathways. For instance, linkage and mutation analysis in just two families allowed the identification of *GATA4* mutations in patients with cardiac septal defects (Garg *et al.* 2004). Here, we report linkage analysis of a previously described family comprising six individuals with PTA, where these patients are offspring of double-first cousin marriages. Candidate gene screening identified an amino acid substitution, F151L, at position 20 of the homoeodomain of the transcription factor *NKX2.6*. Functional analysis of this mutation revealed diminished transcriptional activation, and the corresponding mutation of NKX2.5 reduced DNA binding and the synergistic activation of target promoters in conjunction with GATA4 and TBX5. These data indicate a previously unsuspected role for NKX2.6 in cardiac morphogenesis and the importance of the F151 residue for NKX transcription factor activity.

Results

Genetic linkage analysis

The family under study (Figure 1A) has a history of first cousin marriages for generations (Abushaban *et al.* 2003). Six affected children have been born to this family, but three have died during surgery to repair the cardiac lesion so it has not been possible to include them in this genetic study. Karyotype analysis excluded 22q11 microdeletion in the affected subjects, the common genetic cause of truncus arteriosus. Extensive phenotyping identified truncus arteriosus type 1 and no other visible abnormalities. As all affected individuals were the progeny of consanguineous unaffected parents, we initiated autozygosity mapping.

An initial screen of DNA from the three affected subjects using a 10 cM density set of microsatellite markers identified homozygosity common to all three samples at four places in the genome. Further genotyping and haplotype analysis of the whole family excluded three of these loci, but consistent inheritance of markers identical by descent (IBD) was still observed on chromosome 8p21. The region IBD flanked marker D8S1771, and defined an interval of approximately 14 cM as deduced from the Marshfield genetic map (Broman *et al.* 1998), between heterozygous markers D8S1734 and D8S1711. Owing to software constraints we had to analyse the pedigree as three separate nuclear families reducing the power of our analysis. Linkage analysis (HOMOZ) gave max LOD of 2.9 between D8S1771 and D8S1809. We typed addi-

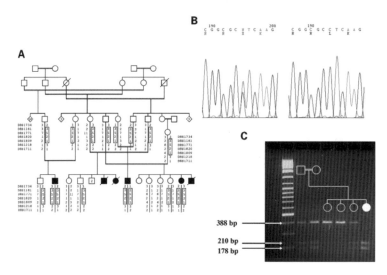

Figure 1. Mutation analysis in the consanguineous family under study. (A) Pedigree of consanguineous family used in this study. Haplotypes are shown for the interval between markers D8S1734 and D8S1711. The proposed linked haplotype is boxed and is homozygous between markers D8S1181 and D8S1809 in the three individuals with PTA. DNA was not available from deceased affected individuals. (B) Electropherograms showing partial sequence of *NKX2.6* exon 2 from an unaffected parent (left) and an affected child (right). The parent is a carrier for the thymine to cytosine transition whereas the affected individual is homozygous for cytosine at position 451. This translates into the F151L missense mutation described in the text. (C) Agarose gel showing *Bsa* HI digest of samples from the Kuwaiti family with partial pedigree above the lanes. The 451T>C mutation creates a *Bsa* HI restriction site in exon 2 of *NKX2.6* which allows the enzyme to cleave the 388 bp PCR product into fragments of 210 bp and 178 bp. The gel shows two heterozygous parents who have children of all three genotypes. The 388 bp PCR product is completely digested in the lane containing the affected child confirming that this individual is homozygous for the mutant sequence.

tional microsatellite markers and single nucleotide polymorphisms within our candidate interval reducing the critical region of homozygosity to approximately 6 Mb between D8S1734 and D8S1218 (Figure 1A).

Mutation screening of positional candidate genes

Genes within the 6 Mb interval not previously implicated in human disease and with reported expression in cardiac tissue or putative roles in heart development were considered positional candidates. Genes including *FZD3, FBX016, BNIP3L, TNFRSF* family members and *NKX3A* were individually sequenced in members of the pedigree with PTA and no mutations were found compared to normal DNA sequence. We also sequenced all novel genes in the region predicted by the Ensembl server (Birney *et al.* 2004) including the predicted human homologue of *Nkx2.6*. *Nkx2.6* encodes a homoeobox transcription factor expressed in murine cardiogenesis (Nikolova *et al.* 1997, Harvey 1996), and as such is a good candidate gene when considering cardiac outflow tract development.

Sequence analysis of *NKX2.6* detected a c.451T>C nucleotide transition in exon 2 of the gene. All three affected individuals were homozygous for C451, parents of affected children were heterozygous (Figure 1B and Figure 1C), and the sequence variant was absent in 100 Kuwaiti,

Nkx2.6 (H. sapiens)   RRKPRVLFSQAQVLALERRFKQQRYLSAPEREHLASALQLTSTQVKIWFQNRRYKCKRQR

Nkx2.6 (M. musculus)   QRKSRVLFSQAQVLALERRFKQQRYLTAPEREHLASALQLTSTQVKIWFQNRRYKSKSQR

Nkx2.5 (H. sapiens)   RRKPRVLFSQAQVYELERRFKQQRYLSAPERDQLASVLKLTSTQVKIWFQNRRYKCKRQR

Nkx2.5 (M. musculus)   RRKPRVLFSQAQVYELERRFKQQRYLSAPERDQLASVLKLTSTQVKIWFQNRRYKCKRQR

Tinman (D. melanogaster) RRKPRVLFSQAQVLELECRFRLKKYLTGAERII-AQKLNLSATQVKIWFQNRRYKSKRGD

Figure 2. Alignment of homoeodomain protein sequences of Nkx2.6, Nkx2.5 and tinman. Identical amino acids between Nkx2.6 and Nkx2.5 are shown in white lettering surrounded by black boxes. The missense mutation F151L at position 20 (F20) in the homoeodomain of Nkx2.6 is marked by an asterisk.

100 Arab and 250 Caucasian control DNA samples. This novel *NKX2.6* mutation is predicted to substitute phenylalanine for leucine at position 151 in the predicted protein sequence (F151L), and position 20 of the homoeodomain (when considering homoeodomains we will refer to the residue as F20). The Nk-2 subgroup of the homoeodomain family comprises six vertebrate homologs (Nkx2.3, 2.5, 2.6, 2.7, 2.8 and 2.9) based on their close structural similarity to *tinman*, a transcription factor essential for development of the heart-like dorsal vessel in *Drosophila* (Newman and Krieg 1998). The F20 residue is conserved in Nk-2 proteins (Harvey 1996) from at least 10 different species (Figure 2).

F151L missense mutation impairs transcriptional activity of NKX2.6 and NKX2.5
In the absence of known downstream genes or a binding site recognition sequence for NKX2.6, we used NKX2.5 as a surrogate in biochemical studies to assess functional consequences of an F151L homoeodomain substitution. NKX2.5 was chosen firstly because of the strong conservation with NKX2.6, especially throughout the homoeodomain (Figure 2), and secondly because biochemical pathways utilising NKX2.5 are relatively well characterised. Thirdly, there are *in vivo* data suggesting Nkx2.5 may compensate for lack of Nkx2.6 during embryogenesis (see discussion), and therefore indicating that some DNA binding and transcriptional regulatory activities will be shared.

Transcriptional activity of wild-type NKX2.6 (NKX2.6[WT]) and NKX2.5 (NKX2.5[WT]) was compared with the mutant proteins (NKX2.6[F157L] and NKX2.5[F157L]); F20 of the homoeodomain was mutated in each case. Transcriptional activation was assessed by transient transfection assays using a 2.6 kb *ANF* promoter or 1.1 kb *Cx40* promoter upstream of a luciferase reporter (*ANF-luc* and *Cx40-luc*).

Over-expression of NKX2.5[WT] led to activation of both *ANF* and *Cx40* promoters consistent with previous studies. However, when equivalent amounts of protein were expressed, NKX2.5[F157L] displayed less transcriptional activation of both *ANF-luc* and *Cx40-luc* compared to NKX2.5[WT] (Figure 3A and Figure 4A). Co-expression of NKX2.5[WT] and either TBX5 or GATA4 resulted in cooperative activation at the *ANF* promoter as previous studies have shown (Hirio *et al.* 2001, Durocher *et al.* 1997), whereas synergistic transcriptional activation was reduced with co-expression of NKX2.5[F157L] and TBX5, and lost with co-transfection of NKX2.5[F157L] and GATA4 (Figure 3A). NKX2.6[WT] did not activate the *ANF* or *Cx40* promoter in U2OS cells, but weakly activated *Cx40-luc* in H9c2 cardiomyocytes. However this transcriptional activity was lost with co-transfection of NKX2.6[F157L.]

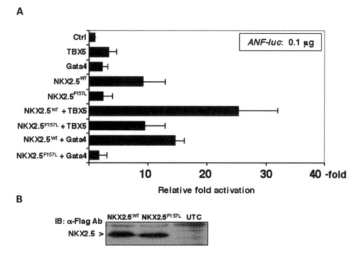

Figure 3. NKX2.5[F157L] shows impaired activation and interactions with cofactors at the *ANF* promoter. **(A)** I n transient transfection assays NKX2.5[WT] led to a 9-fold activation of *ANF-luc* compared to empty pcDNA3 plasmid (CTRL), whereas NKX2.5[F157L] only activated *ANF-luc* 2.4-fold. Co-transfection of TBX5 and NKX2.5[WT] activated *ANF-luc* 25-fold but reporter activation at *ANF* was reduced to 9.5-fold by NKX2.5[F157L]. Co-transfection of GATA4 and NKX2.5[WT] activated *ANF-luc* 15-fold but activation diminished to 1.7-fold with co-expression of NKX2.5[F157L] and GATA4. Differences in transfection efficiency were corrected relative to expression levels of β-galactosidase. Error bars represent the mean ±SD of at least three assays performed in triplicate. (B) NKX2.5[WT] and NKX2.5[F157L] are expressed at equivalent levels as confirmed by immunoblot (IB) of total cell lysate compared to untransfected cells (UTC).

F157L missense mutation reduces DNA binding activity of NKX2.5

Random oligonucleotide selection experiments have determined binding site sequences with highest affinity for several Nk-2 proteins including NKX2.5, and in all cases the binding site conforms to a consensus 5'-T (C/T)AAGTGG-3' (26). Electromobility shift assay (EMSA) of *in vitro* translated NKX2.5[WT] protein efficiently retarded gel migration of radiolabelled probe containing 2 NKX and 1 TBX binding sites (NKX-BS) as previously described (Kasahara *et al.* 2001). However, an equivalent amount of NKX2.5[F157L] protein displayed less binding affinity for NKX-BS probe (Figure 5A). Upon co-incubation of equal amounts of wild type and mutant proteins the presence of NKX2.5[157L] did not affect DNA binding affinity of NKX2.5[WT] for NKX-BS, but DNA binding affinity of NKX2.5[WT] for NKX-BS was reduced in EMSA with 2-fold excess of NKX2.5[F157] (Figure 5B). The NKX-BS probe includes a TBX binding site so we compared DNA binding affinities of NKX2.5[WT] and NKX2.5[F157L] in the presence of TBX5. An increase in amount of shifted NKX-BS probe was reduced when TBX5 was incubated with NKX2.5F157L (Figure 5B). Synthesis of equal quantities of in vitro translated protein from FLAG-NKX2.5WT and FLAG-NKX2.5F157L expression plasmids was confirmed by radiolabelling with ^{35}S-methionine and immunoblotting.

Mutation screening of positional candidate genes

Genes within the 6 Mb interval not previously implicated in human disease and with reported expression in cardiac tissue or putative roles in heart development were considered positional candidates. Genes including *FZD3, FBX016, BNIP3L, TNFRSF* family members and *NKX3A* were individually sequenced in members of the pedigree with PTA and no mutations were found compared to normal DNA sequence. We also sequenced all novel genes in the region predicted by the Ensembl server (Birney *et al.* 2004) including the predicted human homologue of *Nkx2.6*. *Nkx2.6* encodes a homoeobox transcription factor expressed in murine cardiogenesis (Nikolova *et al.* 1997, Harvey 1996), and as such is a good candidate gene when considering cardiac outflow tract development.

Sequence analysis of *NKX2.6* detected a c.451T>C nucleotide transition in exon 2 of the gene. All three affected individuals were homozygous for C451, parents of affected children were heterozygous (Figure 1B and Figure 1C), and the sequence variant was absent in 100 Kuwaiti, 100 Arab and 250 Caucasian control DNA samples. This novel *NKX2.6* mutation is predicted to substitute phenylalanine for leucine at position 151 in the predicted protein sequence (F151L, and position 20 of the homoeodomain (when considering homoeodomains we will refer to the residue as F20). The Nk-2 subgroup of the homoeodomain family comprises six vertebrate homologues (Nkx2.3, 2.5, 2.6, 2.7, 2.8 and 2.9) based on their close structural similarity to *tinman*, a transcription factor essential for development of the heart-like dorsal vessel in *Drosophila* (Newman and Krieg 1998). The F20 residue is conserved in Nk-2 proteins (Harvey 1996) from at least 10 different species (Figure 2).

F151L missense mutation impairs transcriptional activity of NKX2.6 and NKX2.5

In the absence of known downstream genes or a binding site recognition sequence for NKX2.6, we used NKX2.5 as a surrogate in biochemical studies to assess functional consequences of an F151L homoeodomain substitution. NKX2.5 was chosen firstly because of the strong conservation with NKX2.6, especially throughout the homoeodomain (Figure 2), and secondly because biochemical pathways utilising NKX2.5 are relatively well characterised. Thirdly, there are *in vivo* data suggesting *Nkx2.5* may compensate for lack of *Nkx2.6* during embryogenesis (see discussion), and therefore indicating that some DNA binding and transcriptional regulatory activities will be shared.

Transcriptional activity of wild-type NKX2.6 (NKX2.6[WT]) and NKX2.5 (NKX2.5[WT]) was compared with the mutant proteins (NKX2.6[F151L] and NKX2.5[F157L]); F20 of the homoeodomain was mutated in each case. Transcriptional activation was assessed by transient transfection assays using a 2.6 kb *ANF* promoter or 1.1 kb *Cx40* promoter upstream of a luciferase reporter (*ANF-luc* and *Cx40-luc*).

Over-expression of NKX2.5[WT] led to activation of both *ANF* and *Cx40* promoters consistent with previous studies. However, when equivalent amounts of protein were expressed, NKX2.5[F157L] displayed less transcriptional activation of both *ANF-luc* and *Cx40-luc* compared to NKX2.5[WT] (Figure 3A and Figure 4A). Co-expression of NKX2.5[WT] and either TBX5 or GATA4 resulted in cooperative activation at the *ANF* promoter as previous studies have shown (Hiroi *et al.* 2001, Durocher *et al.* 1997), whereas synergistic transcriptional activation was reduced with co-expression of NKX2.5[F157L] and TBX5, and lost with co-transfection of NKX2.5[F157L] and GATA4 (Figure 3A). NKX2.6[WT] did not activate the *ANF* or *Cx40* promoter in U2OS cells, but weakly activated *Cx40-luc* in H9c2 cardiomyocytes. However this transcriptional activity was lost with co-transfection of NKX2.6[F151L].

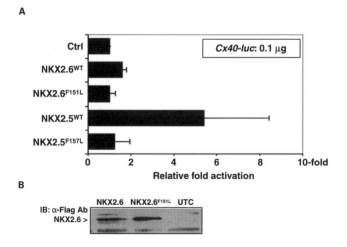

Figure 4. NKX2.6[F151L] and NKX2.5[F157L] show impaired activation at the *Cx40* promoter. (A) NKX2.6[WT] activates *Cx40-luc* by 1.6-fold in H9c2 cells compared to empty pcDNA3 plasmid (CTRL), but NKX2.6[F151L] does not show transcriptional activity at the *Cx40* promoter. NKX2.5[WT] activates *Cx40-luc* 5.4-fold whereas reporter activity is reduced to 1.3-fold by NKX2.5[F157L]. Data represent luciferase activities relative to an internal control and error bars represent the mean ±SD of at least three assays performed in triplicate. (B) Equal protein expression levels for NKX2.6[WT] and NKX2.6[F151L] are indicated by immunoblot (IB) of total cell lysate compared to untransfected cells (UTC).

F157L missense mutation reduces DNA binding activity of NKX2.5

Random oligonucleotide selection experiments have determined binding site sequences with highest affinity for several Nk-2 proteins including Nkx2.5, and in all cases the binding site conforms to a consensus 5'-T (C/T)AAGTGG-3' (Chen *et al.* 1996). Electromobility shift assay (EMSA) of *in vitro* translated NKX2.5[WT] protein efficiently retarded gel migration of radiolabelled probe containing two NKX and one TBX binding sites (NKX-BS) as previously described (Kasahara *et al.* 2001). However, an equivalent amount of NKX2.5[F157L] protein displayed less binding affinity for NKX-BS probe (Figure 5A). Upon co-incubation of equal amounts of wild type and mutant proteins the presence of NKX2.5[F157L] did not affect DNA binding affinity of NKX2.5[WT] for NKX-BS, but DNA binding affinity of NKX2.5[WT] for NKX-BS was reduced in EMSA with 2-fold excess of NKX2.5[F157L] (Figure 5B). The NKX-BS probe includes a TBX binding site so we compared DNA binding affinities of NKX2.5[WT] and NKX2.5[F157L] in the presence of TBX5. An increase in amount of shifted NKX-BS probe was detected when NKX2.5[WT] was incubated with TBX5, but the amount of shifted NKX-BS probe was reduced when TBX5 was incubated with NKX2.5[F157L] Figure 5B). Synthesis of equal quantities of *in-vitro* translated protein from FLAG-NKX2.5[WT] and FLAG-NKX2.5[F157L] expression plasmids was confirmed by radiolabelling with ^{35}S methionine and immunoblotting.

Discussion

We analysed a family with multiple double first-cousin marriages and 6 individuals with PTA as described previously (Abushaban *et al.* 2003). Clinical investigation identified the cardiac defect as PTA type 1, the situation in which the right and left pulmonary arteries arise from a common

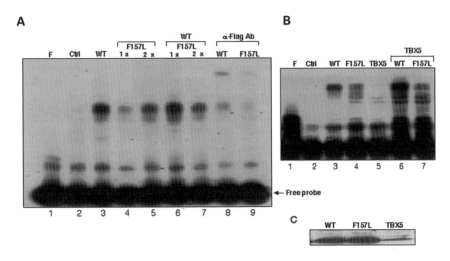

Figure 5. NKX2.5[F157L] has less DNA binding affinity than NKX2.5[WT] for NKX binding sites. (A) Electromobility shift assay of ^{32}P-labelled NKX-BS probe with wild type (NKX2.5[WT]) or mutant protein (NKX2.5[F157L]) reveals decreased DNA binding affinity of NKX2.5[F157L] (lanes 3–5). The presence of an equal amount of NKX2.5[F157L] does not affect DNA binding affinity of wild-type protein (lane 6), but a 2:1 ratio of NKX2.5[WT] to NKX2.5[F157L] leads to a reduction in shifted NKX-BS probe (lane 7). Lane 1 (F) contains free probe and lane 2 (CTRL) contains empty pcDNA3 plasmid. Specificity of DNA binding is shown by supershift EMSAs carried out by pre-incubating NKX2.5[WT] or NKX2.5[F157L] with α-FLAG Ab for 30 minutes prior to addition of ^{32}P-labelled probe (lanes 8 and 9). (B) In the presence of TBX5, the DNA binding affinity of NKX2.5[WT] for NKX-BS is enhanced (lane 6), but this affinity is reduced when Tbx5 is incubated with NKX2.5[F157L] lane 7). Lane 1 (F) contains free probe and lane 2 (CTRL) contains empty pcDNA3 plasmid. Lanes 3–5 contain each of the proteins incubated alone with NKX-BS probe. (C) Protein expression levels were verified by [^{35}S]-labelling and SDS-PAGE.

origin on the truncus arteriosus. Autozygosity mapping identified a region inherited IBD on chromosome 8, within which the *NKX2.6* transcription factor mapped. Mutation screening revealed a T>C nucleotide transition leading to an F20L substitution in the homoeodomain of the protein.

Several considerations indicate F20L is likely to be a pathogenic mutation. The F20 residue in α-helix I of the homoeodomain is conserved in Nk-2 proteins from at least 10 different species (Harvey 1996). F20 is not a predicted direct DNA contact residue, most of which occur in α-helix III (D'Elia *et al.* 2001), but X-ray crystallography on *Drosophila* Engrailed protein (Kissinger *et al.* 1990) and subsequent threading analysis for Pitx2 (Banerjee-Basu and Baxevanis 1999) have identified F20 as one of several hydrophobic residues predicted to stabilise intramolecular interactions between α-helices.

To date, two human genetic disorders are caused by mutation of F20 in homoeodomain proteins. Most notably an identical substitution of F20L has been reported in *SHOX* (Short HOmoeoboX containing gene) in Leri Weill dyschondrosteosis, a short stature syndrome caused by SHOX haploinsufficiency (Falcinelli *et al.* 2002). Additionally F20S in the homoeodomain of *PROP-1* causes Combined Pituitary Hormone Deficiency (Osorio *et al.* 2000). These considerations, combined with the fact that the F20 homoeodomain substitution which segregated with PTA in the pedigree described here was not observed in 900 control chromosomes, are strongly supportive of the notion that F151L is not a rare polymorphism and is indeed causative in our family. We have, however, failed to detect *NKX2.6* mutations in 10 other sporadic cases of PTA type 1, indicating heterogeneity.

Nkx2.6 is expressed in the caudal pharyngeal arches (Nikolova *et al.* 1997) and at opposite poles of the developing heart. Earlier, at E8–8.5, transcripts are found in the progenitors of posterior myocardial cells and later in the sinus venosa and dorsal pericardium. Most notably, at E9.5, *Nkx2.6* is expressed in the OFT itself (Biben *et al.* 1998). Expression is not observed after E10.5. As discussed above, cells in the pharyngeal arches and, self-evidently, cells within the OFT are important for outflow tract septation.

Mutation of *Nkx2.5* in mouse by gene-targeting results in embryonic lethality; although a linear heart tube forms it fails to undergo looping morphogenesis (Lyons *et al.* 1995). In contrast, targeted disruption of *Nkx2.6* results in normal mouse embryos with no obvious heart or pharynx abnormalities (Tanaka *et al.* 2000). Notably, *Nkx2.5* mRNA expression expands to the pharyngeal pouch endoderm, and has been proposed to be indicative of functional compensation for loss of *Nkx2.6* in the pharyngeal pouches of these embryos (Tanaka *et al.* 2000). *Nkx2.6* is also expressed in branchial arch mesectoderm and ectoderm (Nikolova *et al.* 1997), tissues with a known role in cardiac morphogenesis (Macatee *et al.* 2003). Furthermore, overlapping functions for Nkx2.5 and Nkx2.6 have been revealed in double knockout mouse embryos, which lack pharyngeal pouches due to reduced proliferation and survival of pharyngeal endodermal cells (Tanaka *et al.* 2001). Given the high degree of homology between Nkx2.5 and Nkx2.6 over the homoeodomain (90% identity) and overlapping pharyngeal expression patterns, these data make it likely that Nkx2.5 can act at a subset of Nkx2.6 target genes, although there may be less activity in the reciprocal direction.

Using *ANF* and *Cx40* reporters, we demonstrate that a homoeodomain F20L substitution impairs transcriptional activation by NKX2.6 at the *Cx40* promoter, and reduces transcriptional activation function of NKX2.5 at both *Cx40* and *ANF* promoters. Furthermore, synergistic transcriptional activation is reduced between NKX2.5[F157L] and both cofactors GATA4 and TBX5 at the *ANF* promoter. EMSA showed that NKX2.5[F157L] retains some DNA binding activity, but at reduced levels compared to NKX2.5[WT].

Together our results suggest that reduced transcriptional activation function of NKX2.5[F157L] is a consequence of both impaired DNA binding and interaction with cofactor proteins.

A model for synergistic activation of *ANF* by NKX2.5 and GATA4 has been proposed (Sepulveda *et al.* 1998) and describes how hydrophobic interactions between C- and N-terminal domains of NKX2.5 protein could prevent access to target gene binding site recognition sequences. In its inactivated state, NKX2.5 has a low affinity for its target binding sequence but interaction with cofactors such as GATA4 or TBX5 could remove steric hindrance effects, thereby increasing specific DNA binding affinity of NKX2.5. Consistent with this model, we observed an increase in shifted NKX-BS probe when NKX2.5[WT] was incubated in the presence of TBX5 which was reduced with NKX2.5[F157L] and TBX5. These results indicate that DNA binding affinity of NKX2.5[WT] for NKX-BS probe is enhanced in the presence of cofactor TBX5 and that the F20 residue is important for this interaction. To summarise, using NKX2.5 as a surrogate in our functional analyses, we have shown that the F20 residue is important for cooperative protein interactions involved in transcription of downstream genes and is likely to have a role in specific DNA binding.

The genetic and biochemical data presented here implicate *NKX2.6* in the process of outflow tract septation. This role was not evident from the *Nkx2.6* null mouse model, even though *Nkx2.6* is expressed in tissues known to play a vital role in the septation process. This discrepancy might be explained by a species difference in the requirement for *Nkx2.6*, or because this mouse mutant is highly unlikely to produce any Nkx2.6 protein. In patients homozygous for *NKX2.6*[T451C] we predict that stable protein would be produced, and be capable of residual but largely non-functional DNA binding and protein interactions. This presence of NKX2.6[F151L] protein could lead to a diminished ability of NKX2.5 to compensate for loss of function aspects of the *NKX2.6* mutation. Alternatively, the mutant protein could exert a dominant negative effect, but we feel this is unlikely given the recessive nature of the condition. The identification of NKX2.6 targets will be important for further understanding of outflow tract septation, and for testing whether NKX2.6, like NKX2.5, forms complexes with TBX and GATA family proteins. As *Tbx1* null mice have abnormal development of derivatives of the pharyngeal endoderm and PTA (Baldini 2004), and *Tbx2* null mice have outflow tract defects (Harrelson *et al.* 2004), these would be logical candidates to test for such an interaction.

Materials and Methods

PTA family and controls
Informed consent was obtained from participating individuals and the study was approved by the Ministry of Health in Kuwait. Blood samples were collected and DNA was extracted using standard methods. The same procedure was followed for control samples.

Linkage analysis and recombination mapping
Microsatellite analysis excluded the DiGeorge region on chromosome 22q11 (data not shown). The genome-wide screen used the ABI Linkage Mapping Set Version 2.0 (Applied Biosystems) according to protocols supplied by the manufacturer. PCR amplification and separation of amplicons on an ABI PRISM 377 DNA sequencer were performed using standard protocols. GENESCAN and GENOTYPER software (Applied Biosystems) were used to identify alleles. Areas of homozygosity were investigated with unlabelled microsatellite markers analysed by PCR and polyacrylamide gel electrophoresis. These additional markers were identified on the Marshfield genetic map and Ensembl physical map. Linkage analysis was performed using MAPMAKER/HOMOZ software assuming autosomal recessive inheritance, 100% penetrance and trait-allele frequency of 0.0001. One hundred control chromosomes were typed to establish allele frequencies of relevant markers in the Kuwaiti population.

Candidate gene identification and sequence analysis
Positional candidate genes were identified using information from Project Ensembl (Birney et al. 2004) and GenBank (Benson *et al.* 2004). All exons encoding the open reading frame of candidate genes were amplified by PCR using intronic primers. PCR products were sequenced using BigDye terminators v3 (Applied Biosystems) and analysed by ABI PRISM 377 DNA sequencer using standard protocols. All three individuals with a PTA phenotype, parents and an unrelated control DNA sample were screened for sequence variants.

Mutation analysis

We analysed *NKX2.6* for the 451T>C sequence variant in all family members who had provided DNA samples and at least 900 control chromosomes from people of mixed ethnic origins (including 200 Kuwaiti, 200 Arab and 500 Caucasian chromosomes). Mutation analysis was performed using either direct sequencing or restriction fragment length polymorphism (RFLP) analysis of PCR products as the 451T>C mutation creates a *Bsa* HI site (affected = 210 bp + 178 bp fragments; unaffected = 388 bp fragment). Reference sequence for *NKX2.6* was taken from Ensembl, gene number ENSG00000180053. Sequences of all primers and PCR conditions used in this study are available on request. For RFLP-PCR analysis, 10 μl of PCR product was incubated with 0.5 unit *Bsa* H I restriction enzyme according to standard protocols, and digested samples resolved on 2% agarose gels.

Recombinant plasmids and site-directed mutagenesis

For expression constructs, cassettes flanked by *Eco* RI and *Xho* I restriction sites were generated by PCR using *NKX2.6*-specific primers (5'-ATGCTGCTGAGCCCCGT-3' and CCAGGCC-CTGACACCCTGCA-3'), and *NKX2.5*-specific primers (ATGTTCCCCAGCCCT-GCTCTCA-3' and 5'-AGGCTCGGATACCAATGCA-3') incorporating and cloned into pcDNA3-N-FLAG vector (Invitrogen). F157L or F151L missense mutations were introduced in FLAG-Nkx2.5 and FLAG-Nkx2.6 plasmids respectively, by *in vitro* mutagenesis (Altered Sites II System, Promega UK) and verified by sequencing. pGVB-*ANF-luc*, HA-Nkx2.5 and FLAG-Tbx5 were provided by Hiroshi Akasawa and Issei Kumuro (Chiba University Graduate School of Medicine, Japan). N-Myc-Gata4 was provided by Deepak Srivastava (University of Texas Southwestern Medical Center, USA), pGL3-*Cx40-luc* was donated by Birgit Teunissen and Marti Bierhuizen (University Medical Center, Utrecht, The Netherlands).

Reporter gene constructs and transient transfection assays

U2OS (osteosarcoma) or H9c2 (rat cardiomyoblast) cell lines were routinely maintained in Dulbecco's modified Eagle's medium supplemented with 10% foetal bovine serum. Sub-confluent cells in 6-well plates were transfected using Effectene transfection reagent (Qiagen), with 100 ng reporter construct (either *ANF-luc* or *Cx40-luc*) and 100 ng β-gal-CMV as an internal control for transfection variability. The following FLAG epitope-tagged expression constructs were transfected either alone or in combination: FLAG-NKX2.5[WT] FLAG-NKX2.5[F157L] FLAG-TBX5; GATA4-Myc; FLAG-NKX2.6[WT] and FLAG-NKX2.6[F151L]. Total plasmid amount per well was adjusted to 400 ng using empty pcDNA3 plasmid DNA. U2OS cells were harvested 48 hours after transfection using 200 μl lysis buffer, (H9c2 cells 24 hours after transfection), then assayed for luciferase activity using a luminometer (Turner TD-20/20) and β-galactosidase activity. Differences in transfection efficiency were corrected relative to expression levels of β-galactosidase. At least three independent experiments were performed in triplicate. Equal expression levels of protein using FLAG-NKX2.5[WT] and FLAG-NKX2.5[F157L] expression constructs were confirmed by immunoblotting using anti-FLAG monoclonal antibody (α-FLAG Ab) (Sigma).

Antibodies, SDS-PAGE and immunoblotting

FLAG-epitope tagged NKX2.5[WT], NKX2.5[F157L], NKX2.6[WT] and NKX2.6[F151L] expression plasmids were transfected into U2OS cells as described above. Protein expression was evaluated by Western blotting using α-FLAG Ab 48 hours after transfection (Sigma). Protein samples were prepared by lysis of cells in RIPA sample buffer (150 mM NaCl, 1% NP-40, 50 mM Tris, pH7.5, 0.5% DOC, 0.1% SDS), separated on 12% SDS-PAGE gels and electroblotted on to ECL nitrocellulose (Amersham, UK). Following blocking in 5% Marvel/1x TBS /0.1% Tween-20 (blocking buffer) for 2 hours at room temperature or overnight at 4°C. Primary antibody α-FLAG Ab was diluted 1:1500 in blocking buffer and hybridised to the blots for 1 hour at room temperature. Subsequently, anti-mouse polyclonal antibody conjugated to peroxidase (Amersham, UK) was used at 1:5000 for 1 hour at room temperature. Bands were visualised using enhanced chemiluminescence (Amersham, UK).

Electromobility shift assays

EMSAs were used to assess DNA binding properties of NKX2.5^{F157L} compared to NKX2.5^{WT}. The NKX-BS probe used for EMSA contained two NKX binding sites (TNAAGTGG) and an overlapping Tbx5 binding site (TCACACCT) (base pairs -264 to –227 in the *ANF* proximal promoter (Kasahara *et al.* 2001)). Oligo 5'-TCACACCTTTGAAGTGGGGGGCCTCTTGAG-GCAAATC-3' was annealed to 5'-GATTTGCCTCAAGAGGCCCCCACTTCAAAGGT-GTGA-3' and labelled with [^{32}P] dATP using Klenow polymerase. Wild-type and mutant proteins were transcribed and translated using the T^{N}T-coupled reticulolysate system (Promega, UK). Part (2.5 μl) of the reaction was incubated at room temperature for 20 minutes in 20 μl binding buffer (20 mM HEPES, pH 7.9, 60 mM KCl, 1 mM MgCl$_2$, 0.5 mM DTT and 10% glycerol) with 2 μg of polydI.dC and 1 μl of labelled oligo. Protein/DNA complexes were separated on a 6% polyacrylamide gel in Tris borate buffer at 4°C. Gels were dried and exposed to film. Supershift EMSAs were carried out by pre-incubating NKX2.5^{WT} or NKX2.5^{F157L} with 2 μl of α–FLAG Ab for 30 minutes on ice prior to addition of labelled oligo. Equal transcription/translation efficiency of NKX2.5^{WT} and NKX2.5^{F157L} FLAG-tagged fusion proteins was confirmed by transcription/translation of proteins in the presence of ^{35}S-labelled methionine, with subsequent SDS-PAGE analysis and immunoblotting using α-FLAG Ab.

Acknowledgements

We would like to thank the family for participating in this study. Plasmids were kindly provided by Hiroshi Akasawa, Issei Kumuro, Deepak Srivastava, Birgit Teunissen and Marti Bierhuizen. We thank Koen Devriendt, Maria Cristina Diglio and Katrina Prescott who provided DNA samples from sporadic cases of PTA. KH thanks Mark Adams for assistance with preparation of line art for figures. PS is funded by British Heart Foundation, MK, MG are funded by Birth Defects Foundation (UK).

Butyrylcholinesterase: a still mysterious enzyme

Eleidi A. Chautard-Freire-Maia and Ricardo L. R. Souza

Genetics Department, Federal University of Paraná, P. O. Box 19071, 81531–990 Curitiba, PR, Brazil

The discovery of the role of chemical agents in the transmission of nerve impulses opened a path that later revealed the existence of butyrylcholinesterase (BChE; EC. 3.1.1.8; acylcholine acyl-hydrolase). Two Nobel laureates, Sir Henry Dale and Otto Loewi with their respective teams, were responsible for the opening of this path, which began in 1914, when Dale speculated 'on the possible occurrence of acetylcholine in the animal body, and on its physiological significance if it should be found there; and had pointed out the extraordinary evanescence of its action, suggesting that an esterase probably contributed to its rapid removal from the blood' (Dale 1936). This speculation was confirmed in 1926 by Otto Loewi and Ernest Navratil who were able to establish the existence of acetylcholine in animals and also showed that the rapid disappearance of its action was due to its breakdown catalysed by an esterase (apud Loewi 1936).

The debut of butyrylcholinesterase happened when Stedman *et al.* (1932) compared the properties of a horse serum esterase with those from other esterases and proposed the term choline-esterase for this serum enzyme capable of hydrolysing choline esters more rapidly than non-choline esters. Alles and Hawes (1940) found that the cholinesterase in plasma (BChE) was not the same as that present in erythrocytes (acetylcholinesterase; EC. 3.1.1.7; AChE), since they had different relative activities with acetylcholine and its methyl derivatives. The fact that AChE hydrolyses only choline esters and that BChE hydrolyses a variety of non-choline esters as well, led Mendel and Rudney (1943) to propose the denominations of true or specific cholinesterase and pseudo- or non-specific cholinesterase, respectively.

The main reason that impelled scientists to know more about AChE was its function in the termination of cholinergic neurotransmission. In the case of BChE, whose physiological role still remains obscure, the main motivation to know more about this enzyme was its role in metabolism of xenobiotics.

The first report of a xenobiotic involved with BChE concerned a choline derivative, succinylcholine, synthesised by Hunt and Taveau (1906) who studied its action on blood pressure. However, the curare properties of succinylcholine were only recognised in 1949 by the team of Daniel Bovet (again, a Nobel laureate). The relative ease with which succinylcholine is hydrolysed by butyrylcholinesterase and the very low toxicity of the choline and the succinic acid formed after this hydrolysis are responsible for the short curare-like action and the remarkable tolerance of the organism towards this drug (Bovet 1957). Soon, this drug started to be indicated as a simple injection in cases necessitating a particularly short-acting muscle relaxant (endoscopy, electro-convulsive therapy), and as a perfusion for long surgical procedures. The first clinical observations regarding the use of succinylcholine (suxamethonium) in anaesthetics were made in 1949 by Pietro Valdoni and his group in Rome, and in 1951 this drug was introduced in Sweden and Austria (Bovet 1957).

The signal for a muscle to contract starts in the nervous system and reaches the neuromuscular junction by the release of acetylcholine at nerve terminals and its subsequent binding to receptors in the muscle cell membrane. AChE rapidly breaks down this neurotransmitter, enabling the occurrence of a further stimulus to contraction. The agonist succinylcholine binds to the receptor causing muscle twitching at first. However, a few seconds later, the maintained depolarisation causes the muscle cells to become non-excitable, leading to a paralysis of a few minutes. Since succinylcholine is not hydrolysed by AChE, impulse transmission between nerve and muscle is inhibited and lasts until the drug concentration is sufficiently reduced by diffusion and the action of plasma BChE.

Although a short apnoea of two to six minutes is expected after the injection of 1 mg succinylcholine per kg of body weight, cases of prolonged apnoea of several hours were reported soon after this drug started to be administered. Given that Glick (1941) had already shown that succinylcholine was hydrolysed by BChE, the activity of this enzyme had to be examined in patients abnormally susceptible to this muscle relaxant. Bourne *et al.* (1952) and Evans *et al.* (1952) found that susceptible patients had low BChE activity. Forbat *et al.* (1953) noticing low BChE activity in a patient with apnoea and in his brother, inferred that this trait could be genetically determined. This hypothesis was substantiated by Lehmann and Ryan (1956) who observed recurrence of the low BChE activity phenotype in families of suxamethonium-sensitive probands.

Although the enzyme level pointed to a genetic determination, it was not sufficient for discriminating phenotypes because of overlapping values. Phenotype classification became possible after a screening test was designed by Kalow and Genest (1957), using the local anaesthetic dibucaine to inhibit BChE differentially: the usual enzyme is more than 70% inhibited, the atypical enzyme is less than 20% inhibited and the inhibition ranges from 40 to 70% in the case of heterozygous individuals. In the same year, Kalow and Staron (1957) used this method for the screening of nearly 1700 human sera and estimated the frequency of the atypical allele as 1.4% ± 0.4%.

BChE was the second human enzyme – just after glucose-6-phosphate-dehydrogenase – to have its genetic variants revealed by the use of drugs, and contributed to the beginning and development of Pharmacogenetics. Geneticists started to study BChE, previously the province of pharmacologists, and BChE genetic variability was reported for many population samples (review in Whittaker 1986 and for Brazilian populations see Chautard-Freire-Maia *et al.* 1984a, b, Alcântara *et al.* 1995).

BChE genetic variability refers to the *BCHE* gene that encodes its polypeptide chain and also to the result of the interaction of the *BCHE* and *CHE2* genes that determine a heterologous form of BChE (C_5), detected by Harris *et al.* (1963) in electrophoresis. The C_5 variant is formed by linkage of the BChE tetramer to a still unidentified protein (Masson 1991). Besides this genetic variability, BChE is expressed in many molecular forms. The present nomenclature for AChE and BChE molecular forms adopts the letters G and A for globular and asymmetric forms (elongated forms) respectively (Massoulié and Bon 1982). Globular forms G1, G2 and G4 (previously called C_1, C_3 and C_4) contain one, two or four subunits and may be either water-soluble enzymes (secreted into body fluids or occurring inside the cells) or membrane-bound forms attached to cell membranes by a proline-rich membrane anchor called PRIMA (Perrier *et al.* 2002). Asymmetric cholinesterases are formed from one to three tetramers (A4, A8, A12) covalently

Table 1. *BCHE* gene variants[a]

Nucleotide change[b]	Name[c]	Reference
Exon 1		
-116; G→A		Bartels *et al.* 1990
Exon 2		
9 to 11; CATCAT→CAT	*I4del*	Maekawa *et al.* 1997
16; ATT→TT	*I6fs*	Bartels *et al.* 1992b, Primo-Parmo *et al.* 1996
35; AAA→AGA	*K12R*	Mikami *et al.* 2004
45; GGG→GGC	*G15G*	Mikami *et al.* 2004
71; ACG→ATG	*T24M*	Maekawa *et al.* 1997
82; TTT→ATT	*F28I*	Yen *et al.* 2003
98; TAT→TGT	*Y33C*	Primo-Parmo *et al.* 1996
109; CCT→TCT	*P37S*	Primo-Parmo *et al.* 1996
208; GAT→CAT	*D70H*	Boeck *et al.* 2002
209; GAT→GGT	*D70G (A)*	McGuire *at al.* 1989
223; GGC→CGC	*G75R*	Souza *et al.* 2005b
270; GAA→GAC	*E90D*	Souza *et al.* 2005b
286; AAT→TAT	*N96Y*	Yen *et al.* 2003
297; ATT→ATG	*I99M*	Souza *et al.* 2005b
298; CCA→TCA	*P100S*	Maekawa *et al.* 1997, Takagi *et al.* 1997, Lu *et al.* 1997
318; AAT→AAAT	*N106fs*	Yen *et al.* 2003, On-Kei Chan *et al.* 2005
344; GGT→GAT	*G115D*	Primo-Parmo *et al.* 1997
351; GGT→GGAG	*G117fs*	Nogueira *et al.* 1990
355; CAA→TAA	*Q119X*	Sudo *et al.* 1996
375; TTA→TTT	*L125F*	Primo-Parmo *et al.* 1996
383; TAT→TGT	*Y128C*	Hidaka *et al.* 1997a
424; GTG→ATG	*V142M (H)*	Jensen *et al.* 1992
486; GCT→GCC	*A162A*	Souza *et al.* 2005b
510; GAT→GAG	*D170E*	Primo-Parmo *et al.* 1996
514; CAG→TAG	*Q172X*	Gätke *et al.* 2001
551; GCC→GTC	*A184V (SC)*	Greenberg *et al.* 1995
592; AGT→GGT	*S198G*	Primo-Parmo *et al.* 1996
596; GCA→GTA	*A199V*	Sakamoto *et al.* 1998
601; GCA→ACA	*A201T*	Primo-Parmo *et al.* 1996
607; TCA→CCA	*S203P*	Hidaka *et al.* 2001
728; ACG→ATG	*T243M (F-1)*	Nogueira *et al.* 1992
748; ACT→CCT	*T250P*	Maekawa *et al.* 1995
765; GAG→GAC	*E255D*	Primo-Parmo *et al.* 1996
800; AAA→AGA	*K267R*	Maekawa *et al.* 1997
811; GAA→TAA	*E271X*	Primo-Parmo *et al.* 1996
880; GTG→ATG	*V294M*	Mikami *et al.* 2004

Table 1 (continued). *BCHE* gene variants[a]

Nucleotide change[b]	Name[c]	Reference
943; ACC→AACC	*T315fs*	Hidaka *et al.* 1992
943; ACC→TCC	*T315S*	Liu *et al.* 2002
988; TTA→ATA	*L330I*	Sudo *et al.* 1997
997; GGT→TGT	*G333C*	Mikami *et al.* 2004
1062 – 1076	*K355insALU*	Muratani *et al.* 1991, Maekawa *et al.* 2004
1093; GGA→CGA	*G365R*	Hada *et al.* 1992, Hidaka *et al.* 1992
1156; CGT→TGT	*R386C*	Yen *et al.* 2003
1169; GGT→GTT	*G390V (F-2)*	Nogueira *et al.* 1992
1200; TGC→TGA	*C400X*	Hidaka *et al.* 1997b
1253; TTC→TCC	*F418S*	Maekawa *et al.* 1995
1270; CGA→TGA	*R424X*	Yen *et al.* 2003
1273; TCC→CCC	*S425P*	Gnatt *et al.* 1990
1294; GAA→TAA	*E432X*	Levano *et al.* 2005
1303; GGA→AGA	*G435R*	Dey *et al.* 1997
1336; TTT→GTT	*F446V*	Dey *et al.* 1998
1351; GAA→TAA	*E451X*	Dey *et al.* 1998
1378; GAG→AAG	*E460K*	Yen *et al.* 2003
1393; AGA→TGA	*R465X*	Maekawa *et al.* 1995
1408; CGG→TGG	*R470W*	Mikami *et al.* 2004
1411; TGG→CGG	*W471R*	Primo Parmo *et al.* 1996
1420; TTT→CTT	*F474L*	On-Kei Chan *et al.* 2005
Intron 2		
IVS2–8T→G		Primo-Parmo *et al.* 1996
Exon 3		
1490; GAA→GTA	*E497V (J)*	Bartels *et al.* 1992a
1500; TAT→TAA	*Y500X*	Bartels *et al.* 1992b; Primo-Parmo *et al.* 1996
1543; CGT→TGT	*R515C*	Maekawa *et al.* 1995
1553; CAA→CTA	*Q518L*	Primo-Parmo *et al.* 1996
Exon 4		
1615; GCA→ACA	*A539T (K)*	Bartels *et al.* 1992b
1914; A→G		Bartels *et al.* 1990

[a]from Souza *et al.* (2005b) and updated.
[b]Glu 1 is the N-terminal amino acid of the mature BChE protein and nt 1 corresponds to the first nucleotide in the codon for Glu 1.
[c]Trivial name in parenthesis.

associated to a collagen-like tail (Col-Q) which anchors them in the extracellular matrix of the neuromuscular junction (Krejci *et al.* 1997).

Population screening of *BCHE* gene variability had been performed for many years only on the basis of phenotypes, characterised by methodologies that used differential inhibitors (Kalow and Genest 1957, Harris and Whittaker 1961, Morrow and Motulsky 1968, Whittaker *et al.* 1981, Alcântara *et al.* 1991, Picheth *et al.* 1994). The conditions for detection of this gene variability at the DNA level were set by a sequence of studies: description of the 574 amino acid sequence of the catalytic unit of human BChE (Lockridge *et al.* 1987); construction of oligonucleotide probes and isolation of cDNA clones (McTiernan *et al.* 1987, Prody *et al.* 1987); and characterisation of the *BCHE* gene structure (Arpagaus *et al.* 1990). Using PCR and DNA sequencing, McGuire *et al.* (1989) identified the atypical mutation: the first of the 65 non-usual variants described at the DNA level (Table 1).

Although no definite role has yet been assigned to BChE, possible functions have been suggested on the basis of different functional studies. Considering catalytic activity, George and Balasubramanian (1981) showed that BChE behaves not only as an esterase but also as an aryl acylamidase (AAA). This AAA activity – inhibited by 5-hydroxytryptamine (5HT) and stimulated by tyramine – cleaves an acyl-amide bond similar to those found in pain-relieving drugs such as paracetamol and phenacetin. Balasubramanian and Bhanumathy (1993), considering the correlation of an excess of tyramine with migraine, speculated that tyramine may promote the cleavage of an endogenous acyl-amide analgesic through stimulation of the AAA in circulation, leading to the disturbance of a pain relief mechanism. BChE would have a role in the control of the amine-sensitive pain mechanisms, having both 5HT and tyramine modulating its AAA activity.

The hypothesis that BChE is associated with lipid metabolism arose from data showing positive correlation of BChE activity with subcutaneous fat (Berry *et al.* 1954), weight (Kalow and Gunn 1959, Simpson 1966, Cucuianu *et al.* 1968; Stueber-Odebrecht *et al.* 1985), BMI (Alcântara *et al.* 2002, 2003a, b), cholesterol and triglycerides (Cucuianu *et al.* 1968, 1978, Alcântara *et al.* 2002) and also from data on the physical association of BChE with lipoproteins such as LDL (Lawrence and Melnick 1961, Dubbs 1966, Kutty *et al.* 1973, Ryhänen *et al.* 1982), as well as HDL (Ryhänen *et al.* 1982). The findings of positive association between BChE activity and lipid levels were explained by Cucuianu *et al.* (1978) as the result of induction of BChE secretion by an accelerated turnover of serum lipids and lipoproteins.

Association of BChE genetic variants with weight and BMI has also been reported. The CHE2 C5+ phenotype – characterised by the presence of the C_5 complex of BChE – was associated with lower weight (Chautard-Freire-Maia *et al.* 1991) and lower BMI (Alcântara *et al.* 2001). Analysing BMI distribution, Souza *et al.* (2005a) found that heterozygotes for the *K* (*A539T*) variant of the *BCHE* gene presented higher BMI variance than homozygotes for the usual (wild-type) variant.

Some recent findings on ghrelin may possibly explain the connection between BChE variability and BMI. The growth hormone (GH) is released from the anterior pituitary not only by the hypothalamic GH-releasing hormone but also by ghrelin (Kojima *et al.* 2001), an endogenous peptide ligand for the growth-hormone secretagogue receptor (GHS-R). The GHS-R is activated by ghrelin that is *n*-octanoylated on the Ser3 residue while the des-*n*-octanoyl form of ghrelin, designated as des-acyl ghrelin, does not interact with this receptor (Hosoda *et al.* 2000).

Tschöp *et al.* (2000) showed that administration of ghrelin generated weight gain by increasing food intake and reducing fat utilisation in rodents. Plasma ghrelin concentration was significantly lower in obese Caucasians than in lean Caucasians and was higher in Caucasians than in Pima Indians, a population with a very high prevalence of obesity (Tschöp *et al.* 2001), leading these authors to propose that the decrease in plasma ghrelin concentration represents a physiological adaptation to the positive energy balance associated with obesity. In human serum, De Vriese *et al.* (2004) showed that BChE is involved in ghrelin desoctanoylation: ghrelin desoctanoylation is partially inhibited by BChE inhibitors; purified BChE degrades ghrelin; BChE activity and ghrelin hydrolysis are positively correlated. Considering these data on ghrelin, it is possible to explain the positive association of BChE activity with BMI as due to a possible role of BChE in the down-regulation of ghrelin. Furthermore, it would be expected that the CHE2 C5+ phenotype that determines about 30% more BChE activity, in comparison to the CHE2 C5- phenotype, would be associated with weight and BMI decrease as was previously shown (Chautard-Freire-Maia *et al.* 1991, Alcântara *et al.* 2001).

After submitting human plasma to electrophoresis, Juul (1968) obtained 12 BChE bands, some of them representing heterologous forms of BChE. In human serum, besides the physical relation between BChE and lipoproteins, the monomer of BChE links covalently to albumin forming a complex (G1-albumin, previously called C_2; Masson 1989) and its tetramer (G4) links to a still unidentified substance to form the C_5 complex (Masson 1991). A complex formed by BChE and transferrin has also been found in chicken serum (Wietnauer *et al.* 1999). The existence of these complexes, as well as the relation of BChE with proteases (Checler *et al.* 1990, Darvesh *et al.* 2001), poses new queries on BChE function.

A role for BChE in cell proliferation was suggested on the basis of data obtained from the neural tube of chicken embryo (Layer 1983). Layer and Sporns (1987), studying central nervous system and lens development in chicken embryo, verified that BChE is expressed before and during mitosis and that AChE appears about 11 hr after mitosis, indicating a relation of BChE with cell proliferation and of AChE with cell differentiation. Layer *et al.* (1993) also proposed that BChE and AChE may regulate neurite growth by means of a non-enzymatic mechanism, perhaps an adhesion function.

In a review of the neurobiology of BChE (Darvesh *et al.* 2003), other BChE relationships with the nervous systems were reported: 1) BChE is widespread and differentially distributed in human neurons when compared to AChE, particularly within the amygdala, hippocampal formation and thalamus; 2) A regulatory role for BChE in the hydrolysis of acetylcholine seems consistent with the fact that inhibition of BChE leads to a dose-dependent increase in the levels of acetylcholine in the brain; 3) BChE seems to be associated with some aspects of the development of the nervous system.

BChE has also been related to Alzheimer's disease (AD): acetylcholinesterase was significantly reduced and butyrylcholinesterase significantly increased, compared with normal cases, in the hippocampus and temporal cortex of AD patients (Perry *et al.* 1978). The neuritic plaques and neurofibrillary tangles characteristic of AD display acetylcholinesterase and butyrylcholinesterase activities with major shifts in optimum pH and inhibitor sensitivity in comparison to cholinesterases from normal brain (Geula and Mesulam 1989). These cholinesterases are selectively inhibited by indolamines which inhibit the AAA activity of this enzyme, raising the possibility that this activity might be related to AD (Wright *et al.* 1993). In this context, it is worth mentioning the association found between the *K* variant of the *BCHE* gene and late onset AD (Lehmann *et al.* 2000). Like other complex diseases, late-onset AD is conditioned by interacting factors of genetic and environmental origin. These interactions may define a subset of vulnerable people for each 'risk gene'. In the case of the association between the *K* variant of the *BCHE* gene and the onset of late-onset AD, this subset of vulnerable people was provisionally limited to >75-year-old male carriers of the *APOEε 4* allele (Lehmann 2002).

The data related to BChE possible functions indicate that it may have both cholinergic and non-cholinergic roles. Researchers like those of our team, who have been studying BChE for some years, continue to be involved in its mysteries while expecting the day when BChE function(s) will be clearly established.

Acknowledgements
We are very much grateful to Professor Oksana Lockridge for the critical reading of this manuscript and for the valuable suggestions made.

Remembrances of George and Maria

Eleidi A. Chautard-Freire-Maia

I first met George when he was a visiting professor at the Department of Biology of the University of São Paulo (Brazil) in 1970. In the same year, I went to Birmingham where I stayed until December 1972 as a British Council scholar supervised by John Edwards in a linkage analysis of a huge sample of Brazilian families studied by Newton Morton and colleagues. John and George were very good friends and I enjoyed participating in their chats.

In 1974 Newton Freire-Maia and I got married in Brazil, and that same year George visited our home in Curitiba and our University Department of Genetics where he gave two lectures. It is a very nice remembrance associated with a happy and friendly atmosphere. George was greatly amused to feature in the local press as a Cobrão Internacional, the term Cobrão (Cobra) often being used in Brazil to indicate an expert. Newton and George had been good friends and had met many times before our marriage.

In August 1996, we had the pleasure to meet Maria for the first time. She and George came to Rio de Janeiro for the IX International Congress of Human Genetics. After the congress, George and Maria came to visit us in Curitiba and stayed for some days. George gave a lecture on cancer genetics at our University Department. We spent many happy hours together while having meals or going to a fair in the old sector of the city or visiting part of the countryside.

These are all sweet memories for me. They also bring closer to me the presence of my husband Newton who left us in May 2003. Newton's last birthday, when he completed 84 years of age, was commemorated at our apartment together with children, grandchildren and some very good friends. At the end of the party, Newton gave to each one of them a very beautiful Brazilian poem in prose on friendship and also the translation into English made by George who is very fond of the poem. Newton used to say that he liked George's translation even more than the original. So, George and Maria were also with us, commemorating Newton's 84th birthday. George's translation of the poem in prose follows the poem itself:

Amigos

Tenho amigos que não sabem o quanto são meus amigos. Não percebem o amor que lhes devoto e a absoluta necessidade que tenho deles.

A amizade é um sentimento mais nobre do que o amor, eis que permite que o objeto dela se divida em outros afetos, enquanto o amor tem intrínseco o ciúme, que não admite a rivalidade.

E eu poderia suportar, embora não sem dor, que tivessem morrido todos os meus amores, mas enlouqueceria se morressem todos os meus amigos! Até mesmo aqueles que não percebem o quanto são meus amigos e o quanto minha vida depende de suas existências ...

A alguns deles não procuro, basta-me saber que eles existem. Esta mera condição me encoraja a seguir em frente pela vida.

Mas, porque não os procuro com assiduidade, não posso lhes dizer o quanto gosto deles. Eles não iriam acreditar. Muitos deles estão lendo esta crônica e não sabem que estão incluídos na sagrada relação de meus amigos.

Mas é delicioso que eu saiba e sinta que os adoro, embora não declare e não os procure. E às vezes, quando os procuro, noto que eles não tem noção de como me são necessários, de como são indispensáveis ao meu equilíbrio vital, porque eles fazem parte do mundo que eu, tremulamente, construí e se tornaram alicerces do meu encanto pela vida.

Se um deles morrer, eu ficarei torto para um lado. Se todos eles morrerem, eu desabo!

Por isso é que, sem que eles saibam, eu rezo pela vida deles. E me envergonho, porque essa minha prece é, em síntese, dirigida ao meu bem estar. Ela é, talvez, fruto do meu egoísmo.

Por vezes, mergulho em pensamentos sobre alguns deles. Quando viajo e fico diante de lugares maravilhosos, cai-me alguma lágrima por não estarem junto de mim, compartilhando daquele prazer ...

Se alguma coisa me consome e me envelhece é que a roda furiosa da vida não me permite ter sempre ao meu lado, morando comigo, andando comigo, falando comigo, vivendo comigo, todos os meus amigos, e, principalmente os que só desconfiam ou talvez nunca vão saber que são meus amigos!

A gente não faz amigos, reconhece-os.

Friends

I have friends who do not know how much they are my friends. They do not perceive the love which I devote to them and the absolute need which I have of them.

Friendship is a more noble sentiment than love, because it allows the affections of its object to be shared with others, while love contains within itself a jealousy which brooks no rivals.

And I could have borne, even though not without pain, the deaths of all my loves, but I should have been driven mad by the deaths of all my friends! Even of those who do not perceive how much they are my friends and how much my life depends on their existence …

Some of them I do not seek out; it suffices me to know that they exist. And just knowing this gives me the courage to go on with my life.

But because I do not see them with any regularity, I cannot tell them how great a delight I take in them. They would not believe it. Many of them are reading this narrative and they do not know that they are included in the sacred fellowship of my friends.

But the joy of knowing and of feeling that I adore them is sweet, even though I do not declare myself and even though I do not seek them out. And at times, when I do seek them out, I realise that they do not have any idea of how necessary they are to me, of how indispensable they are to the equilibrium of my life, because they form part of the world which I constructed with faltering steps, and because they then became the sources of my enchantment with life.

If one of them were to die, I am distorted. If they were all to die, I am destroyed!

Because of this, without their knowledge, I pray for their lives. And I become ashamed because this prayer of mine is in essence directed towards my own well-being. It is perhaps the fruit of my egoism.

At times, I immerse myself in thoughts about some of them. When I travel and find myself in marvellous places, I let a few tears fall because they are not with me, sharing in this pleasure …

If anything consumes me and makes me feel old, it is that the furious whirligig of life does not allow me to have constantly by my side, staying with me, walking with me, speaking with me, living with me, all my friends, especially those who only have a feeling that they may be my friends or who perhaps will never know that they are my friends.

We do not make friends, we recognise them.

Helicobacter pylori infection and DNA damage

Dertia Villalba Freire-Maia[1] and Marcelo S.P. Ladeira[2]

[1] Genetics Department, Universidade Federal de São Paulo (UNIFESP), São Paulo, Brazil
[2] Internal Medicine Department, Universidade Estadual Paulista (UNESP), Botucatu, São Paulo, Brazil

Honouring George Fraser – by Dertia V. Freire-Maia

I am very happy for being invited to participate in this homage to George Robert Fraser, a good friend and one of the most important human geneticists in the world. George is a brilliant colleague and an extraordinary human being.

From the first time we met, my husband Ademar and I became George's 'old' friends and admirers. And not only of him, but also of his wonderful wife, Maria. We had great pleasure in receiving them at our place, in São Paulo, when George came to Brazil to attend an international congress. The same pleasure we had when we met him at a meeting in Puerto Vallarta, Mexico.

George and Maria are our good inspiration.

Abstract

Helicobacter pylori is believed to predispose carriers to gastric cancer by inducing chronic inflammation. The inflammatory process can result in the generation of reactive oxygen and nitrogen species that damage DNA and could inhibit DNA repair. Given the risk for gastric cancer in *H. pylori*-infected humans, we have used the Comet assay to investigate the relationship between DNA damage and *cagA, vacA* and *iceA* genotypes of *H. pylori*. We also verified whether enzymatic treatment induces DNA damage or inhibits repair. The feasibility of using fresh peripheral blood lymphocytes and whole blood cells for the biomonitoring of DNA damage has also been studied. We observed that DNA damage was significantly higher in *H. pylori* infected patients presenting gastritis than in uninfected patients with normal mucosa. Patients infected by *H. pylori* genotypes *cagA+*, *vacAs1m1* and *iceA1* showed higher levels of DNA damage than patients infected by *H. pylori* strains *cagA-*, *vacAs2m2* and *iceA2* and uninfected patients with normal mucosa. Enzymatic digestion neither induced significant DNA damage nor inhibited the DNA repair system. In fresh peripheral blood lymphocytes, the levels of DNA damage were significantly higher in *H. pylori*-infected patients with moderate and severe gastritis than in uninfected patients. In summary, using the Comet assay with an accurate imaging analysis system, we have demonstrated the relevance of *H. pylori* genotypes for the production of DNA damage in the gastric mucosa. Fresh peripheral blood lymphocytes could therefore be a useful tool for biomonitoring DNA damage in *H. pylori*-infected patients.

Introduction

Gastric cancer (GC) remains the second most frequent cause of cancer-related mortality in Brazil (INCA 2002). The single most common cause of GC is infection with *Helicobacter pylori*, which is able to induce gastritis in practically all the infected individuals (Marshall and Warren 1984). Most of the infected patients however remain asymptomatic, and only a minority of the infected individuals develop gastric cancer. This difference might be linked to genetic diversity among *H. pylori* strains (Ladeira *et al.* 2004a). Several bacterial virulence factors, such as vacuolating cytotoxin (vacA) and cytotoxin-associated antigen (cagA), have been described (Ladeira *et al.* 2004a). The *vacA* gene encodes a vacuolating toxin that damages epithelial cells. This gene is present in all *H. pylori* strains and comprises two variable parts. One is the *s* region located at the 5' end with two alleles, *s1* and *s2*. The *s1* type has three subtypes, *s1a, s1b* and *s1c*. The second part of the *vacA* gene is the m-region (middle) and has two alleles, *m1* and *m2* (Ladeira *et al.* 2004a). The *vacA s1/m1* strains produce a large amount of toxin, *s1/m2* produces moderate amounts, and *s2/m2* produces little or no toxin (Ladeira *et al.* 2004a). Recently, another gene has been described, *iceA*. It has two alleles, *iceA1* and *iceA2* (Ladeira *et al.* 2004a). Although the function of *iceA1* is not clear, it encodes a homologue of the type II restriction endonuclease (*Nla*IIIR) of *Neisseria lactamica*. The expression of *iceA1* is weakly related to the occurrence of peptic ulcers. The iceA2 protein is completely unrelated to iceA1 or other known proteins (Ladeira *et al.* 2004a).

Bacterial products and reactive species of oxygen (ROS) and nitrogen (RNS) released in the inflammation site can damage DNA. Together, genotoxicity and cellular responses could lead to mutations in key genes, chromosomal aberrations and altered DNA repair capacity, which may represent an early step in carcinogenesis (Ladeira *et al.* 2004a). The ability to study DNA damage in gastric epithelial cells of *H. pylori*-infected patients should allow more accurate identification of bacterial virulent risk factors and permit, therefore, preventive strategies (Everett *et al.* 2000).

DNA damage can be assessed by Single Cell Gel Electrophoresis (SCGE) or Comet Assay (CA) at the individual cell level. CA has been used to investigate primary DNA damage, such as double-strand breaks (DSBs), single-strand breaks (SSBs), alkali-labile sites (ALs), incomplete repair sites and crosslinks (Tice *et al.* 2000). Therefore, CA has been applied in a great number of studies to investigate the early biological effects of DNA-damaging agents in environmental, occupational and pathological conditions or exposure to chemicals (Hartmann *et al.* 2003). Smith *et al.* (2003) and Schabath *et al.* (2003) have used CA in lymphocytes to assess the risk for breast and bladder cancer, respectively. Given the risk for gastric cancer in *H. pylori*-infected humans, we have used CA to investigate the relationship between DNA damage and *cagA, vacA* and *iceA* genotypes of *H. pylori*. We have also verified whether enzymatic treatment induces DNA damage or inhibits repair. The feasibility of using fresh peripheral blood lymphocytes and whole blood cells for DNA damage biomonitoring has also been studied.

Materials and Methods

Informed consent to participate was obtained from all the patients or their parents. We have studied, prospectively, 124 volunteer patients (55 males and 69 females), with a mean age of 46.06 ± 19.26 years, and an age range of 5–87 years. All the patients were non-smokers, non-alcoholics, and non-drug users. None have received any medication for at least 30 days before the study. Among the 124 subjects, we randomly selected 40 (20 males: 20 females), with mean age of 50.25 ± 17.51 years, and age range of 19–86 years, for analysis of oxidative DNA damage and efficiency of DNA repair in the antrum, corpus and fresh peripheral blood lymphocytes.

Biopsies were obtained during endoscopy from the antrum and corpus, as previously described (Ladeira *et al.* 2004). *H. pylori* infection was confirmed when positive results were obtained for at least two of the following tests: rapid urease test (Coelho *et al.* 1996), histological analysis (Ladeira *et al.* 2004), and gastric biopsy PCR for ureA (Clayton *et al.* 1992). Peripheral blood samples (5 ml) were obtained from all the patients.

Epithelial cells from gastric mucosa biopsies were isolated as described by Pool-Zobel *et al.* (1994) using proteinase K (Invitrogen, Gaithesburg, MD) and 3 mg collagenase I (Invitrogen).

Cell viability was determined using the fluorescein-diacetate (FDA; Sigma, St Louis, MO)/ethidium bromide (EtBr; Sigma) assay according to Hartmann and Speit (1997).

The slides were stained with haematoxylin and eosin and blindly examined for leukocyte levels.

Peripheral blood lymphocytes were isolated as described by Tomasetti *et al.* (2001), with slight modification (Ladeira *et al.* 2005).

In order to assess the effect of proteinase K and collagenase and incubation temperature on the levels of DNA damage from gastric epithelial cells, we exposed lymphocytes from 8 healthy individuals to the same conditions used for preparation of the gastric single-cell suspension. Moreover, due to possible role of RNS in inhibiting the repair of DNA damage (Yuasa 2003), we also assessed whether proteinase K and collagenase incubation could lead to art factual inhibition of the DNA repair system.

The lymphocyte suspension (500 µl) was resuspended in 500 µl of RPMI medium and simultaneously exposed to the following conditions: Group I – immediate analysis of basal DNA damage levels (fresh lymphocytes); Group II – in 100 µl of the same proteinase K and collagenase type I solution used for the preparation of gastric single cell suspensions at 37°C for 45 min.; Group III – at 37°C for 45 min.; Group IV – in 100 µl of 100 µM H_2O_2 (Merck, Darmstadt, Germany) at 4°C for 45 min. (positive control); Group V – after exposure to 100 µl of 100 µM H_2O_2 at 4°C for 45 min., the aliquots were washed in phosphate buffer solution (PBS) and one aliquot was incubated at 37°C for 45 min. to assess the DNA repair capacity; Group VI – another aliquot was incubated in 100 µl of proteinase K and collagenase type I solution at 37°C for 45 min. to assess the possible inhibition of DNA repair capacity.

The PCR amplification for *ureA* was performed according to Clayton *et al.* (1992). The *cagA* was detected by PCR amplification using the methodology described by Peek *et al.* (1995). For analyses of the *vacA* signal sequences and midregions, PCR amplification was performed as described by Atherton *et al.* (1995). The multiplex-PCR described by Atherton *et al.* (1999) was used to confirm whether reactions for both *m1* and *m2* had been mixed or contained hybrid *m1* and *m2* sequences. For detection of the *iceA1* and *iceA2* alleles, a PCR reaction was performed as described by Van Doorn *et al.* (1998). All PCR products were resolved in agarose gels, stained with EtBr, and detected under a short-wavelength UV-light source. The samples were tested twice. The efficiency of DNA repair was assessed by the H_2O_2 assay.

CA was performed as described by Singh *et al.* (1988), with slight modifications as described by Ladeira *et al.* (2004). For detection of oxidised purines and pirimidines (oxidative DNA damage), CA was performed as described by Ladeira *et al.* (2004b). The slides were stained with 40 µl EtBr (20 µg/ml), and analysed in a fluorescence microscope (Axioplan II – Zeiss, Germany), under green light at 400x, using an image analysis system (Comet Assay II – Perceptive Instruments, Suffolk, UK).

DNA damage, including strand breaks (SBs) apurinic/apyrimidinic (AP) sites, oxidised purines and pyrimidines were detected by the alkaline comet assay, modified with lesion-specific enzymes, FPG and Endo III. The standard alkali comet assay measures SBs, and AP sites. The enzyme-modified comet assay measures oxidative DNA damage as a combination of SBs, AP sites and oxidised bases. Subtraction of mean tail moment on incubation with buffer alone from mean tail moment for incubation with enzymes gives the level of oxidised purines or oxidised pyrimidines.

Statistical Analysis

The Kruskal-Wallis test was used to compare mean tail moment among the groups. The paired-sample t test was used to compare levels of DNA damage in antrum, corpus, and lymphocytes for groups treated with H_2O_2 and after incubation at 37°C for 30 min. Significance level (P) was set as 0.05.

Results

H. pylori infection was detected in 79% (98/124) of the subjects. Histopathology showed that 30% (37/124) of the patients had mild gastritis; 21% (26/124) had moderate gastritis; 20% (25/124) had severe gastritis; 4% (5/124) suffered from gastric ulcers; and 4.0% (5/124) had gastric cancer. All the 98 patients with positive histopathology were infected with *H. pylori*, while the remaining 26 patients with normal mucosa (21% of the patients) were not infected. All the cases of gastric cancer were classified as intestinal type.

Cell viability determined by FDA/EtBr was of 78–96%, with a mean of 89%. The analysis of leukocyte content of cell suspensions prepared from 10 samples of gastric biopsies with gastritis presented less than 12% of leukocyte content with mean of 8.3%.

Figure 1 shows the effect of proteinase K, collagenase and incubation temperature on the levels of DNA damage in blood lymphocytes. Fresh blood lymphocytes had low levels of DNA damage, which did not differ significantly from the levels of DNA damage in lymphocytes exposed to proteinase K and collagenase at 37°C for 45 min. or only incubated at 37°C for 45 min. In addition, when proteinase K and collagenase incubations were checked for eventual inhibition of the DNA repair system, the lymphocytes exposed to H_2O_2 (100 μM), treated or not with enzymes, were able to recover within the repair incubation time of 45 min.

The levels of DNA damage in gastric epithelial cells were significantly higher in *H. pylori*-infected patients in the antrum and corpus than in uninfected patients with normal mucosa.

No statistically significant difference was found in the extent of DNA damage detected in whole blood cells from patients with different *H. pylori* genotypes.

The relationships between DNA damage in gastric epithelial cells from the antrum and corpus evaluated by CA and the different *H. pylori* genotypes are presented in Figure 2. Patients infected by *cagA*⁺ strains had higher levels of DNA damage in both the antrum and corpus than uninfected patients and patients infected by *cagA*⁻ strains. Patients infected by *cagA*⁻ strains had higher levels of DNA damage than uninfected patients. Patients infected with *H. pylori* *vacAs1* strains had significantly higher levels of DNA damage in both the antrum and corpus compared to uninfected patients or to patients infected with *vacAs2* strains (Figure 2). The levels of DNA damage were significantly higher in the antrum and corpus of patients infected by *vacAm1 H. pylori* strains than in uninfected patients or in patients infected by *vacAm2* strains. Patients

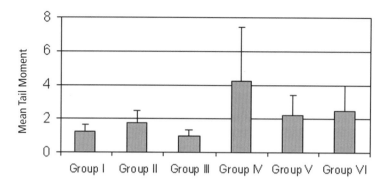

Figure 1. Effect of proteinase K, collagenase and incubation temperature on the levels of DNA damage in blood lymphocytes. *Columns,* mean of tail moment; *bars,* ± SD. Group I = fresh lymphocytes; group II = lymphocytes exposed to proteinase K and collagenase at 37°C for 45 min; group III = lymphocytes incubated at 37°C for 45 min; group IV = lymphocytes exposed to 100 μM H_2O_2 at 4°C for 45 min; group V = lymphocytes of group IV washed in PBS and incubated at 37°C for 45 min; group VI = lymphocytes of group IV washed in PBS and incubated in proteinase K and collagenase type I at 37°C for 45 min. Group IV *vs* group V and group VI (P < 0.05).

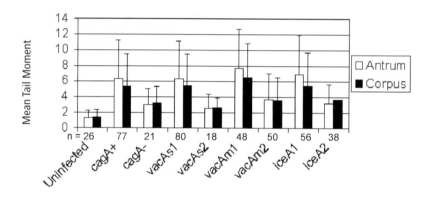

Figure 2. Relationships between DNA damage in gastric epithelial cells from the antrum and corpus measured by the Comet assay and *H. pylori cagA, vacAs1, vacAs2, vacAm1, vacAm2, iceA1* and *iceA2* genotypes. *Columns,* mean of tail moment; *bars,* ± SD. For antrum and corpus: *cagA⁺ vs* non-infected (P < 0.0001), *cagA⁺ vs cagA⁻* (P < 0.05) and *cagA⁻ vs* non-infected (P < 0.05); *vacAs1 vs* non-infected and *vacAs1 vs vacAs2* (P < 0.01); *vacAm1 vs* non-infected (P < 0.0001); *vacAm1 vs vacAm2* (P < 0.001) and *vacAm2 vs* non-infected (P < 0.01); *iceA1 vs* non-infected (P < 0.0001), *iceA1 vs iceA2* (P < 0.05) and *iceA2 vs* non-infected (P < 0.05). Differences between the levels of DNA damage in the antrum and corpus in groups with the same *H. pylori* genotype were not significant.

infected by *vacAm2* strains had higher levels of DNA damage than uninfected patients (Figure 2).

Uninfected patients also had lower levels of DNA damage in both the antrum and corpus than patients infected with *iceA1* or *iceA2* strains. The levels of DNA damage in the antrum and corpus were not significantly different for patients with the same *H. pylori* genotype.

The levels of DSBs, SSBs and ALs and levels of oxidative DNA damage [strand breaks

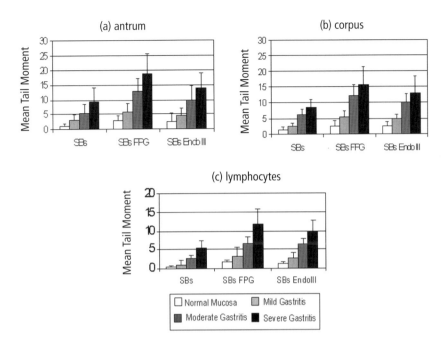

Figure 3. Levels of DNA damage expressed as SBs, SBs + FPG and SBs + endonuclease III in gastric epithelial cells of antrum, corpus and lymphocytes of patients with normal mucosa, and mild, moderate and severe gastritis. SBs strand breaks alone; SBS FPG strand breaks together with FPG sensitive sites (altered purines). SBs Endo III together with oxidised pyrimidines (altered pyrimidines). In the antrum and corpus: levels of SBs, SBs FPG and SBs Endo III from moderate and severe gastritis *vs* normal mucosa and mild gastritis (P < 0.05). In lymphocytes, levels of SBs from moderate and severe gastritis *vs* normal mucosa (P < 0.05) and level of SBs FPG and SBs Endo III from moderate and severe gastritis *vs* mild gastritis and normal mucosa (P < 0.05).

together with FPG sensitive sites, altered purines (SBs FPG) and SBs together with oxidised pyrimidines (SBs Endo III)] were significantly higher in *H. pylori*-infected patients than in uninfected patients with normal mucosa in the antrum, corpus and lymphocytes (data not shown). The levels of SBs were significantly higher in *H. pylori*-infected patients with moderate and severe gastritis than in uninfected patients with normal mucosa and *H. pylori*-infected patients with mild gastritis in the antrum and corpus (Figure 3). In the lymphocytes, the levels of SBs were significantly higher in *H. pylori*-infected patients with moderate and severe gastritis than in uninfected patients. Compared to uninfected patients and *H. pylori*-infected patients with mild gastritis, patients with moderate and severe gastritis had higher levels of SBs FPG and SBs Endo III, in the antrum, corpus and lymphocytes (Figure 3).

We have found a positive correlation between oxidative DNA damage and degree of gastritis in the antrum, corpus and lymphocytes (data not shown).

In order to assess a possible inhibition of DNA damage repair, we exposed gastric epithelial cells and lymphocytes to H_2O_2. The efficiency of DNA repair was lower in patients with moderate and severe gastritis in the antrum and corpus, since the levels of DNA damage did not change with time in the H_2O_2 assay (Figure 4).

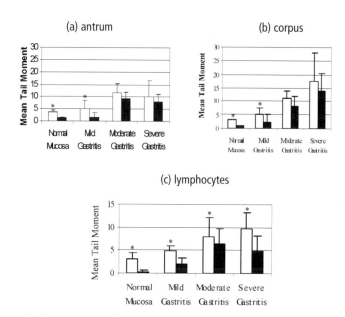

Figure 4. Levels of DNA damage after treatment with 100 μM H_2O_2.* $P < 0.05$.

Discussion

Patients infected by *H. pylori* strains *cagA+*, *vacAs1m1* and *iceA1* showed higher levels of DNA damage than patients infected by *H. pylori* strains *cagA-*, *vacAs2m2* and *iceA2* and uninfected patients with normal mucosa in the antrum and corpus. Part of the DNA damage could be related to bacterial products or derived from the free-radicals generated in the inflammatory process, which could react with the DNA, causing damage in some stem cells and leading to mutations that can favour carcinogenesis (Ladeira *et al.* 2004b).

We have found a positive correlation between oxidative DNA damage and the degree of gastritis, suggesting that oxidative DNA damage increases with increased intensity of gastritis. Hence, the more severe is the inflammatory process greater is the induction of oxidative DNA damage. Moreover, in the antrum and corpus, the efficiency of DNA repair was lower in infected patients with moderate or severe gastritis and the enzymatic digestion did not inhibit DNA repair. That procedure is needed to obtain cellular suspensions from gastric biopsies. So, we suggest that the DNA repair system could be inhibited by RNS in samples from *H. pylori*-infected patients.

In previous studies we have observed a higher risk of extensive DNA damage in patients older than 50 years compared to younger patients with the same degree of gastritis (Ladeira *et al.* 2004b). That risk increases with age (Ladeira *et al.* 2004b). Moreover, the pattern of clinical *H. pylori* disease could be determined by the age wherein the infection is acquired (Veldhuyzen van Zanten *et al.* 1994). Infection in childhood is thought to cause pan gastritis and consequent increased risk for gastric ulcers and gastric cancer, whereas infection as an adult generally leads to antral gastritis and duodenal ulcers (Veldhuyzen van Zanten *et al.* 1994). We believe that early-age establishment of *H. pylori* infection could favour increased accumulation of DNA damage, probably due to inhibition of DNA repair proteins and apoptosis by RNS (Yuasa 2003).

Therefore, patients with the same degree of inflammatory processes and the same *H. pylori* geno-types, but infected at different ages, could show different levels of DNA damage and different risks of developing gastric cancer.

It has been reported that the use of whole blood for the Comet Assay is easier and prevents the possibility of inducing DNA damage during lymphocyte isolation (Speit *et al.* 2003). However, in previous studies (Everett *et al.* 2000, Ladeira *et al.* 2004a), white blood cells were not a good system for detecting differences between *H. pylori*-infected patients and non-infected patients; and fresh peripheral blood lymphocytes were sensitive enough to detect increased levels of DNA damage in *H. pylori*-infected patients with moderate and severe gastritis, when compared to non-infected patients with normal mucosa. As a large migration of free-radicals to plasma can occur during inflammation, they may cause increased levels of oxidative DNA damage in lymphocytes, which are more sensitive than monocytes to oxidative stress (Holz *et al.* 1995). Another possibility is that the higher lifetime of the lymphocytes could lead to DNA damage accumulation.

In summary, using the Comet Assay with an accurate imaging analysis system, we have demonstrated the relevance of *H. pylori* genotypes to induce DNA damage in the gastric mucosa. Fresh peripheral blood lymphocytes could be a useful tool for the biomonitoring of DNA damage from chronic inflammatory exposure in *H. pylori*-infected patients.

Acknowledgements
We are very much grateful to Drs. M.A.M. Rodrigues, D.A. Salvadori, D.A. Queiroz, P. Achilles, P. Padulla, I. Gonçalves, and P.A. Rodrigues, for collaboration in our *H. pylori* research projects; to FAPESP, for financial support; and to Dr. A. Freire-Maia for suggestions to this paper.

Brazilians and the variable mosaic genome paradigm

Sergio D.J. Pena

Departamento de Bioquímica e Imunologia, Universidade Federal de Minas Gerais,
Belo Horizonte, MG 31270–901, Brazil

Xenophanes speaks thus:
And no man knows distinctly anything,
And no man ever will.

Diogenes Laertius 1925

Populations and individuals

There is considerable medical interest in studying human genomic variation and its influence on health and individual drug responses. For that we especially need means to characterise and study the geographical distribution of human genetic diversity. The conventional approach to this question is to divide humanity into populations that can be defined on the basis of race, geography, culture, religion, physical appearance or whatever other criterion that is convenient. It is becoming increasingly clear that such division of humanity into populations may not constitute the most appropriate approach to deal with human variation. Treating people, for instance, of European ancestry and African ancestry, as separate categories for genetic studies tends to contribute to the public perception that the primary difference between these ways of defining populations is biological (Foster and Sharp 2004). This view confounds several issues and obscures the important fact that Europeans are genealogically related to Africans, having evolved as an offshoot of the latter.

Human evolutionary history is remarkably short and the worldwide geographical distribution of genetic traits is basically due to dispersal, with ensuing mutation, selection and genetic drift. In essence, the genetic diversity observable in Europe, Asia, Oceania and the Americas is a merely a subset of the variation found in Africa (Yu *et al.* 2002). As pointed out by Pääbo (2003), from a genomic perspective we are all Africans, either living in Africa or in quite recent exile outside of Africa.

The human genome is composed of hundreds of thousands of genomic blocks of high linkage disequilibrium (Gabriel *et al.* 2002a) each one with its own pattern of variation and genealogy. This forms the basis of a 'Variable Mosaic Genome' (VMG) paradigm, according to which, rather than conceptualising humans as belonging to populations, ethnicities or races, we consider the genome of any particular individual as a mosaic of variable haplotypes (Pääbo 2003). This shifts completely the focus from populations to individuals. As we hope to demonstrate below, the VMG paradigm is especially useful when dealing with genome variation in highly admixed populations, such as Brazilians.

Of course, one cannot envisage the population view and the individual/genealogical view as exclusively antagonistic. Rather, they could be seen as complementary in the description of human genome variation. An analogy with quantum mechanics could be used to illustrate this point: the complementarity principle states that light and electrons have a wave-particle duality. It is impossible to observe both the wave and particle aspects simultaneously. Together, however, they present a fuller description than either of the two taken alone. Most importantly, depending on the experimental arrangement, the light and electrons sometimes behave wavelike and sometimes particle-like, i.e., the model adopted predetermines the kind of properties that will be experimentally observed. Likewise, our scientific interpretations of genetic findings on human genome variation depend on whether we adopt a population or an individual/genealogical model. For instance, under the VMG paradigm certain ideas, such as that of human races, become absolutely meaningless. Also, we cannot overlook the fact that, as pointed out by Avise (2000), strong connections exist between demography and phylogeny. The historical demographies of populations are of profound relevance to phylogeographical patterns over microevolutionary time scales by virtue of their impact on the structure of haplotype genealogies.

The people of Brazil

Brazilians form one of the most heterogeneous populations in the world, the result of five centuries of interethnic crosses among peoples from three continents: Europeans, represented mainly by the Portuguese, Africans and Amerindians. When the Portuguese arrived in 1500, there were roughly 2.5 million indigenous people living in the area of what is now Brazil (Salzano and Freire-Maia 1970). The Portuguese-Amerindian admixture started soon after the arrival of the first colonisers. Mating between European men and indigenous women became commonplace and later (after 1755) was even encouraged as a strategy for population growth and colonial occupation of the country (Mörner 1967). The Amerindian tribes underwent a drastic demographic decline due to conflicts with the European colonisers and diseases to which they were not adapted (Salzano and Freire-Maia 1970). Today there are *circa* 720,000 Amerindians in Brazil, living on land set aside for them by the federal government. From the middle of the 16th century, Africans were brought to Brazil to work on sugarcane farms and, later, in the gold and diamond mines and on coffee plantations. Historical records suggest that between 1551 and 1850 (when the slave trade was abolished), 3.5 million Africans arrived in Brazil (Salzano and Freire-Maia 1970, Curtin 1969). As to the European immigration, it is estimated that *circa* 500,000 Portuguese arrived in the country between 1500 and 1808 (Salzano and Freire-Maia 1970). From then on, after the Brazilian ports were legally opened to all friendly nations, Brazil received increasing numbers of immigrants from several parts of the world. Portugal remained by far the most important source of migrants, followed by Italy, Spain, and Germany. In the 20th century, Asian immigration took place, mainly from Japan, as well as from Lebanon and Syria. According to Callegari-Jacques and Salzano (1999), 58% of the immigrants who arrived in Brazil between 1500 and 1972 were Europeans, 40% were Africans, and 2% were Asians.

In this sense, Brazil might be seen as representing a 'meeting point' for the three major historical geographical components of humanity [Africans, Asians (represented by their Native American descendants) and Europeans]. Of course, one might argue that the same 'meeting point' metaphor could be applied to the USA, as it also has the same three genealogical roots. However, as we will show, the extent of admixture between the three components has been significantly larger in Brazil.

Genetic variation in Brazilians

In the past few years we have been using several different molecular tools to try to characterise the ancestry of Brazilians and the formation of the Brazilian people. We will describe briefly these studies, from which we could unravel evidence of genetic admixture in levels much higher than had previously been suspected. We will start with studies based on uniparental markers because they can be useful to bring forth the concept of haplotype genealogies.

Uniparental genetic markers in Brazilians

There are several types of genetic markers at DNA level, and they can be classified according to their molecular nature and genomic localisation. Autosomal markers are excellent individuality markers because they are diploid and subject to recombination. They can also be useful as ancestry-informative markers (AIMs) as long as the allele frequency difference between two ancestral populations is large (Shriver *et al.* 1997, Parra *et al.* 1998). On the other hand, uniparental maternal (mitochondrial DNA – mtDNA) and paternal (non-recombinant regions of the Y chromosome – NRY) polymorphisms are excellent lineage markers because they are haploid and do not recombine. As such, blocks of genes (haplotypes) transmitted to the next generations remain unaltered in the matrilineages and patrilineages until a mutation supervenes. The mutations that occurred and reached high frequencies after the dispersion of modern man from Africa can be specific to certain regions of the globe and can serve as geographical markers. The mitochondrial DNA and the NRY provide complementary information that can be traced back several generations into the past and allow reconstituting the history of a people through migrations of women and men respectively. It is however relevant to remember that a lineage marker such as mtDNA and NRY provides information about, respectively, a single female or male ancestor of an individual and thus constitutes a small proportion of the genetic constitution of a person.

We examined DNA polymorphisms in the non-recombining portion of the Y chromosome to investigate the contribution of distinct patrilineages to the present-day white Brazilian population. Twelve unique-event polymorphisms were typed in 200 unrelated males from four geographical regions of Brazil and in 93 Portuguese males (Carvalho-Silva *et al.* 2001). In our Brazilian sample, the vast majority of Y chromosomes proved to be of European origin. Only 2% of the Y-chromosome lineages were from sub-Saharan Africa (haplogroup E3a*), and none was Amerindian (haplogroup Q3*). Indeed, there were no significant differences when the haplogroup frequencies in Brazil and Portugal were compared by means of an exact test of population differentiation. Likewise, there was no population differentiation among the four geographical regions of Brazil. Nevertheless, by typing with fast evolving NRY markers we later uncovered a higher within population haplotype diversity in Brazil than in Portugal, explicable by the input of diverse European Y chromosomes (Carvalho-Silva *et al.* 2006).

We also studied a slightly larger sample, with the same basic constitution, for mtDNA. This revealed a very different 'reality' from that yielded by the NRY analysis. Considering Brazil as a whole, 33%, 28% and 39% of matrilineages were of Amerindian, African and European origin, respectively (Alves-Silva *et al.* 2000). As expected, the frequency of different regions reflected their genealogical histories: most matrilineal lineages in the Amazonian region had Amerindian origin, while African ancestrality was preponderant in the Northeast (44%) and the European haplogroups in the South (66%).

In summary, these phylogeographical studies with white Brazilians revealed that the vast majority of patrilineages have European origin, while most matrilineages (>60%) were Amerindian or African. Together, these results configure a picture of strong directional mating between European males and Amerindian and African females, which agrees with the known history of the peopling of Brazil since 1500. Although having little to do with pharmacogenetics *per se*, these studies reveal that the genomes of most white Brazilians are mosaic, having mtDNA and NRY with different phylogeographical origins.

Haplotype blocks and the 'Variable Mosaic Genome' paradigm

We have seen above the genetic properties of non-reticulate uniparental lineage markers. The same kinds of genealogical principles apply in theory to nuclear genes, whose multi-generation transmission routes involve both genders (Avise 2000). In terms of formal theory the major difference from uniparental markers is a four-fold adjustment required to account for the larger effective population size of autosomal alleles. This leads to corresponding four-fold longer coalescent times. A second, less important, difference is that in autosomes, besides mutations, lineages can change because of intragenic (or intrahaplotype, see below) recombination events.

In the past few years it has become evident that much of the human genome is composed of haplotypic blocks ('hapblocks') where polymorphic markers (especially single nucleotide polymorphisms – SNPs) are strongly associated over distances as large as 170 Kb (reviewed in Pääbo 2003, Tishkoff and Verrelli 2003, Wall and Pritchard 2003). The discussion of the origin of these haplotype blocks is beyond the objective of this review. Suffice to say that probably the length of haplotype blocks is influenced by both demographic factors (which is certainly responsible for most of the variation of block sizes among populations) and genomic factors, especially the existence of recombination hot-spots (Zhang *et al.* 2003, Greenwood *et al.* 2004). The existence of such hapblocks has high significance for the feasibility of mapping disease genes by marker association studies, since each block can be defined by typing only 4–5 SNPs. Thus, the number of SNPs needed to achieve fine genomic screening might be reduced from millions to a few hundred thousand (Tishkoff and Verrelli 2003).

We can then envisage the human genome as composed of hundreds of thousands of small genomic blocks of high linkage disequilibrium (like the mtDNA or Y chromosome), each one with its own pattern of variation and genealogical origin. This forms the basis of the VMG paradigm, i.e., that rather than thinking about populations, ethnicities or races, we consider the genome of any particular individual as a mosaic of variable haplotypes (Pääbo 2003). As we already saw this shifts completely the focus from populations to individuals.

Biparental genetic markers, morphological characteristics and ancestry in Brazilians

Using a panel of genetic polymorphisms that display large differences in allelic frequencies (>0.40) between Europeans and Africans, Parra *et al.* (1998) showed that, at a population level, it was possible to estimate with great precision the degree of African and European admixture among Americans. We decided to ascertain whether this same panel of markers would be capable to estimating, on an individual level, the degree of African ancestry in Brazilians. For that, we selected ten of the AIMs used in the American study (Parra *et al.* 2003).

With the purpose of verifying the individual discrimination power of this set of ten AIMs we genotyped a small sample of Portuguese individuals from the Northern part of the country and a sample of individuals from the island of São Tomé, located in the Gulf of Guinea, on the west coast of Africa. These population sources were chosen because they are geographically related to the European and African population groups that participated in the peopling of Brazil. From the genotype information we calculated for each individual an African Ancestry Index (AAI) that is the logarithm of the sum over all alleles of the ratio of the likelihood of a given genotype being from African origin to the likelihood of it being of European origin. There was no overlap between the AAI values obtained for the two groups and 21 logs separated the two medians (9.71 and -11.73, respectively). A complete individual discrimination between the European and African genomes was obtained with an existence of 7.6 logs between the lower African value (AAI5 = 2.86) and the highest European score (AAI5 = 24.86). It was thus clear that the 10-allele set of Parra *et al.* (2003) was highly efficient and provided reliable individual discrimination between European and African genomes.

Our Brazilian sample was composed of 173 individuals from a Southeastern rural community, clinically classified according to their Colour (white, black, or intermediate) with a multivariate evaluation based on skin pigmentation in the medial part of the arm, hair colour and texture, and the shape of the nose and lips. When we compared the AAI values for these individuals, we observed that the groups had much wider ranges than those of Europeans and Africans and that there was very significant overlap between them. However, the comparison of whites versus blacks with the Mann-Whitney U test still showed a modest significant value ($z = 2.62$; $P < 0.01$). On the other hand, comparisons of whites versus Portuguese and blacks versus São Tomé islanders yielded extremely high significance (respectively, $z = 5.08$, $P < 0.0001$ and $z = 25.24$, $P < 0.0001$). Thus, the differences in AAI values of the group of Brazilian blacks compared with Brazilian whites are very distinct and several orders of magnitude smaller than the differences observed between Africans and Europeans. Other studies showed that these data could be reliably extrapolated to the rest of Brazil (Parra *et al.* 2003).

If we consider some peculiarities of Brazilian history and social structure, we can construct a model to explain why Colour should indeed be a poor predictor of African ancestry at an individual level. Most Africans have black skin, genetically determined by a very small number of genes that were evolutionarily selected. Thus, if we have a social race identification system based primarily on phenotype, such as occurs in Brazil, we classify individuals on the basis of the presence of certain alleles at a small number of genes that have impact on the physical appearance, while ignoring all of the rest of the genome. Assortative mating based on Colour, which has been shown by demographic studies to occur in Brazil, will produce strong associations among the individual components of Colour. Indeed, we detected the presence of such positive associations at highly significant levels in a Southeastern Brazilian population (Parra *et al.* 2003). On the other hand, we expect that any initial admixture association between Colour and the AIMs will inevitably decay over time because of genetic admixture. It is easy to see how this combination of social forces could produce a population with distinct Colour groups and yet with similar levels of African ancestry.

In essence, our data indicate that, in Brazil as a whole, Colour is a weak individual predictor of African ancestry. Based on the data generated by the above study, we used the *Structure* software (Pritchard *et al.* 2000) to estimate the number of people in Brazil with African ancestry. In this specific case, *Structure* estimated the proportion of European ancestry in each individual studied. The results, which can be seen in Table 1, are remarkable. More than 75% of self-declared White Brazilians from the North, Northeast or Southeast of the country, present

European ancestry below 90%. Even in Southern Brazil, with its history of strong European immigration, this value was of the order of 49%. For comparison, we calculated, from the data of Shriver *et al.* (2003), the values for European Americans: only 29% present European ancestry below 90%. On the other hand, only 73% of Brazilian Blacks have a predominantly African ancestry, i.e. a proportion of European ancestry below 50%, as compared with 91% in the U.S. (Table 1). Obviously these estimates have been achieved by extrapolating from experimental results with small samples and thus have very wide confidence limits. However, they illustrate cogently the high amount of genetic admixture that has taken place in the formation of the Brazilian people.

Table 1. Proportion of African ancestry in several groups in Brazil and in the United States. (Original data from Shriver *et al.* 2003 and Pena and Bortolini 2004)

European Ancestry	Whites (%)				U.S.	Blacks (%)	U.S.
	Brazil					Brazil	
	South	North	Northeast	Southeast			
>90%	51	24	20	11	71	3	1
>80%	71	43	36	21	92	7	2
>70%	84	53	46	36	99	13	3
>60%	90	73	64	53	100	20	4
>50%	96	82	74	74	100	27	9
>40%	100	92	82	81	100	37	17
>30%	100	98	94	89	100	60	29
>20%	100	98	97	96	100	80	48
>10%	100	100	100	98	100	87	72

Brazilians constitute a trihybrid population with European, African, and Amerindian roots. Thus, we had to ascertain where the Amerindians would be positioned in reference to the AAI scale. For that, we tested DNA samples from 10 individuals from three Amazonian tribes (Karitiana, Surui, and Ticuna) and observed that they fell in the same range as the Europeans. This finding is not unexpected, because several of the population-specific alleles in Parra's set had diverging frequencies in Europe and Africa because of specific selective factors operating in the African environment (Parra *et al.* 1998). Amerindians, who are well known to have a distant Asian ancestry, did not share such environments and would thus, be expected to have frequencies similar to Europeans. To estimate the Amerindian contribution to the Brazilian population we thus needed new polymorphic markers that would be sensitive to the three ancestralities. We then turned to a set of 40 insertion-deletion (indel) polymorphisms that proved to be exquisitely discriminating in that regard (Bastos-Rodrigues *et al.* 2006). Our results showed that White Brazilians have widely varying degrees of European, African and Amerindian ancestry. Even more variable were the results with Black Brazilians sampled in the city of São Paulo (unpublished observation).

It is obvious then that, regardless of their skin colour, the overwhelming majority of Brazilians have a significant degree of African ancestry. Likewise it could be easily demonstrated that, regardless of their skin colour, the overwhelming majority of Brazilians have a significant degree of European ancestry. Finally, although we have not calculated the exact numbers, we can safely predict that regardless of their skin colour, a sizeable proportion of Brazilians have a significant degree of Amerindian ancestry!

It thus makes no sense talking about 'populations' of 'White Brazilians' or 'Black Brazilians' because of the poor correlation between colour and ancestry. Also, it does not make sense talking about African-Brazilians or European-Brazilians because most Brazilians will have significant proportions of African and of European (and of Amerindian) ancestry. Thus, the only possible basis to deal with genetic variation in Brazilians is on a person-by-person basis, according with the VMG paradigm, which allows any individual to have different ancestries in different genomic segments.

Acknowledgements
Research in the authors' laboratory was sponsored by Conselho Nacional de Desenvolvimento Científico e Tecnológico (CNPq) of Brazil.

Dedication
This paper is dedicated to George Fraser on the occasion of his 75th birthday, with admiration and gratitude for his important contributions to human genetics.

George Fraser and the 'malaria hypothesis'

David Weatherall

Weatherall Institute of Molecular Medicine, John Radcliffe Hospital, University of Oxford,
Headington, Oxford, OX3 9DS, UK

Introduction

I first encountered George Fraser in New York at a conference on *Problems of Cooley's Anemia* held by the New York Academy of Sciences in December, 1963. In the late 1950s clinical genetics had started to thrive in the USA, particularly under Victor McKusick at Johns Hopkins Hospital in Baltimore and Arno Motulsky in Seattle. Both George and I were presenting work that we had been carrying out in the USA over the previous years, George in Seattle and I with Victor and later Lock Conley at Hopkins. Since our fields of interest in human genetics went their separate ways after that meeting I did not see much of him for many years, until he returned to Oxford in 1984.

Over the last 20 years I have come to know George well and have come to admire his enormous breadth of scholarship and knowledge right across the field of human biology. Opening one of his letters, often accompanied by a reprint or a copy of some hitherto neglected work in human genetics, always generates excitement and a feeling that one is about to be educated. My other memories of George's more recent Oxford days are the symptoms of impending doom that came over me when I saw him sitting in the audience at one or two of my lectures. As what the late Cyril Clarke described as a 'Sunday morning geneticist', I always found George's appearances at my often superficial talks rather off-putting though, unlike my other Oxford colleagues, he was always too polite to tell me that I was talking rubbish!

It is an enormous pleasure to contribute to this tribute to George. There have been few people who have managed to straddle the fields of clinical and basic genetics with such ease, and with such diverse and often highly original contributions.

Red cell polymorphisms in Greece: George's 1963 lecture

In the period after the Second World War independent studies in Italy and the USA suggested that there is a very high carrier frequency for the genetic blood disease thalassaemia in Mediterranean populations. At the 8th International Congress of Genetics in Stockholm in 1948 the American workers Neel and Valentine suggested that this might reflect a higher mutation rate in certain ethnic groups. J.B.S. Haldane was not impressed with this argument and proposed an alternative mechanism (Haldane 1949). He suggested, in short, that heterozygotes were advantaged because their red cells are smaller than normal and hence they might be more resistant to attack by the sporozoa that cause malaria, a disease that he knew was prevalent in Italy, Sicily and Greece, where the thalassaemia genes occurred at a particularly high frequency. Thus was born what came to be known as the 'malaria hypothesis'.

In subsequent years Haldane's suggestion had mixed fortunes. While there was a growing body of evidence that the high frequency of the sickle cell gene in Africa is related to malaria protection (Allison 1964), the position regarding thalassaemia was much less certain. Population studies in Sardinia, particularly comparing populations at different altitudes, found a reasonably clear correlation between the frequency of malaria and β-thalassaemia (Carcassi *et al.* 1957, Siniscalco *et al.* 1966). On the other hand, when such correlations were sought in other parts of the world they were not always found. During this period, therefore, it was necessary to keep an open mind about whether there might be other mechanisms involved in maintaining the high gene frequencies for thalassaemia.

George Fraser's paper given in New York in 1963 described collaborative studies carried out between Arno Motulsky's group in Seattle and Phaedon Fessas and his colleagues in Greece (Fraser *et al.* 1964j). They had studied almost all the inhabitants of two villages in the Arta region, an area that suffered in the past from a high level of malarial infection. As well as measuring population frequencies for a variety of red cell polymorphisms they were also able to carry out individual family studies. In this region they found a high frequency of β-thalassaemia trait, sickle cell trait, α-thalassaemia, glucose-6-phosphate dehydrogenase deficiency, and a number of other haemoglobin and thalassaemia variants. Since all these traits showed Mendelian segregation ratios it was felt unlikely that a disturbance in segregation had led to the persistence of these variants at a high frequency. Furthermore, there was no evidence of mortality differentials in carriers of these traits. Apart from a slight increase of fertility of female heterozygotes for haemoglobin S, β-thalassaemia and G6PD deficiency, their findings, they felt, were in keeping with Haldane's suggestion that this wide variety of red cell polymorphisms had been maintained at a high frequency by relative protection against *Plasmodium falciparum* malaria. George continued to carry out thorough population studies of this kind, notably in the Philippines, Taiwan, some African populations and Yugoslavia.

The fallow years for the Haldane hypothesis

After these early studies very little progress was made for the next 20 years in trying to obtain population or experimental data in support of the Haldane hypothesis, at least as it applied to the thalassaemias. Perhaps, in retrospect, this should not surprise us. During this time there was increasing evidence that the thalassaemias are a remarkably heterogeneous group of disorders. Furthermore, evidence was amassing that they occur at extremely high frequencies in a line stretching from the Mediterranean, through the Middle East, to the Indian sub-continent and Southeast Asia. Where, it was asked, had they first arisen? Some suggested that they must have had their origin in Greece and been carried east by Alexander the Great's armies. Others had it that they had arisen in the east and been carried west by Marco Polo and his followers. Still others, sitting firmly on the fence, proposed that they had arisen somewhere in the Middle East and moved in both directions! More extensive population studies provided evidence for and against the malaria hypothesis and the field was further disadvantaged by the potential problems of founder effects, drift and the rest.

From the 1980s onwards the thalassaemias became amenable to study at the molecular level and a remarkable picture of their heterogeneity and the complexities of their phenotype/genotype relationships became apparent. At the same time, it became clear that the application of these new techniques to a re-examination of the Haldane hypothesis might help to solve the problems which had bedevilled the field for so many years.

The 'malaria hypothesis' in the molecular era

The studies that have been carried out over recent years that provide solid backing for Haldane's original hypothesis have been reviewed extensively (Weatherall and Clegg 2001, 2002) and will only be outlined briefly here.

The most extensive information about the protection of thalassaemia carriers or those with other mild forms of the disease against *P. falciparum* has been obtained in the case of the α-thalassaemias. Like all the thalassaemias these conditions are very heterogeneous at the molecular level. Normal individuals have two α globin genes per haploid genome; their genotype can be written αα/αα. There are two major classes of α-thalassaemias, the α⁺-thalassaemias in which one of the pair of α-globin genes is lost by deletion or mutation (-α/αα or -α/-α), and the α⁰-thalassaemias in which both the pairs of α genes are lost by deletion (--/αα or --/--). While high frequencies of the α⁰-thalassaemias are restricted to parts of Southeast Asia and the Mediterranean Islands the α⁺-thalassaemias occur at extremely high frequencies throughout the African subcontinent, the Middle East and both South and Southeast Asia. Carrier frequencies in these regions range from 5–70% or more.

It was found that the frequency of α⁺-thalassaemia in the southwest Pacific follows a clinal distribution from North/West to South/East, with the highest frequencies in the north coast of New Guinea and the lowest in New Caledonia. This distribution shows a strong correlation with malaria endemicity. On the other hand, there is no geographical correlation with malaria endemicity and other polymorphic markers in this region (Flint *et al.* 1986). The possibility that α-thalassaemia had been introduced from mainland populations of Southeast Asia, and that its frequency had been diluted as they moved south across the island populations, was excluded when it was found that the molecular forms of α-thalassaemia in Melanesia and Papua New Guinea are quite different to those of the mainland and are set in different α-globin gene haplotypes (Flint *et al.* 1986).

These population studies suggesting a protective effect of α-thalassaemia against *P. falciparum* malaria were augmented by a prospective case-control study of 250 children with severe malaria admitted to hospital in Medang, Papua New Guinea, a region where there is a very high rate of malarial transmission. Compared with normal children, the risk of contracting severe malaria, as defined by WHO guidelines, was 0.4 for α⁺-thalassaemia homozygotes, and 0.66 for α⁺-heterozygotes (Allen *et al.* 1997). These findings have now been confirmed in several other populations.

Since α⁺-thalassaemia homozygotes have no clinical disability, and extremely mild anaemia, at first sight these observations suggest that the high frequency of α-thalassaemia may reflect a transient rather than a balanced polymorphism. However, this view must be regarded with caution. In particular, further detailed analyses of the α⁺-thalassaemia homozygous phenotype are required to be absolutely certain that it is associated with no deleterious features.

Although less progress has been made in the case of the β-thalassaemias, there is some evidence, albeit indirect, that their high frequency also reflects selection. It turns out that every population with a high frequency has a completely different set of β-thalassaemia mutations. In other words, this condition has arisen independently in many different populations and presumably reached its high frequency by selection. There is indirect evidence that this is the case. The β-globin gene haplotype is divided into two regions, 3' and 5' sub-haplotypes, which are separated by a recombination hotspot (see Weatherall and Clegg 2001). However, it turns out that particular β-thalassaemia mutations are closely associated with specific β-globin gene haplotypes, most strongly with the 3' sub-haplotype which contains the β-globin gene, but also with substantial linkage to the 5' sub-haplotype. These findings suggest that a recent cause is responsible for the expansion of the β-thalassaemia mutations; migration has not had sufficient time to disperse them unlike the normal β-globin gene haplotypes, nor has recombination yet disrupted these linkages (Flint *et al.* 1998). In short, it seems clear that β-thalassaemia genes throughout the world have been amplified to high frequencies so recently that none of the other forces, migration, recombination, or drift, has had sufficient time to bring them into genetic equilibrium with their haplotype backgrounds. The timeframes involved with these observations are consistent with the increase in the spread of human malaria following the development of agriculture approximately 5,000 years ago.

The Haldane hypothesis was correct (or almost)

While the evidence that the high gene frequencies of α^+-thalassaemia, and probably β-thalassaemia, have been maintained by heterozygote advantage against *P. falciparum* malaria is now overwhelming, it appears that Haldane was not correct in his idea about the cellular mechanisms of protection; although the notion that the small red cells of thalassaemia carriers might not be a happy environment for the malarial parasite was attractive, it has turned out to be wrong.

Once it became possible to culture malarial parasites a variety of invasion and growth studies suggested that malarial parasites developed through one or more cycles in the red cells of thalassaemia carriers in exactly the same way as they do in normal cells. Thus, while there is increasing evidence that protection by the sickle cell trait against malaria may be mediated through the properties of the red cell, it is becoming clear that in the case of α-thalassaemia this is less likely to be the case. Indeed, studies in both Vanuatu and East Africa are suggesting that an immune mechanism may be more likely (Williams *et al.* 1996, Dr Tom Williams, personal communication). Further evidence of heterogeneity of protective mechanisms comes from studies of the blood group, Duffy, in Africa. The high prevalence of those who do not carry the Duffy antigen appears to reflect a protective effect of this genotype against the milder form of malaria caused by *P. vivax* (Miller *et al.* 1976). It turns out that this variant disrupts the Duffy antigen/chemokine receptor (DARC) promoter and alters a GATA-1 binding site, which inhibits DARC expression on red cells and therefore prevents DARC-mediated entry of *P. vivax* (Tournamille *et al.* 1995).

The picture that is emerging, therefore, is of multiple red cell polymorphisms which protect against different forms of malaria by different mechanisms. And the field has broadened even further. It is now clear that polymorphisms of many other systems, including HLA/DR and a variety of inflammatory mediators, have come under selection in the same way (see Weatherall and Clegg 2002). Recent work, both in human and murine malaria, has underlined the extraordinary diversity of these protective polymorphisms and emphasises how exposure to this single infective agent has modified the human genome quite dramatically over a relatively short time, at least in evolutionary terms.

Postscript

George Fraser's lecture in 1963 was my first exposure to how population geneticists tackle complex problems like the distribution of the thalassaemias and other red cell polymorphisms. For me, it underlined the importance of keeping an open mind and trying to develop the appropriate laboratory technology to tackle the protean possibilities for the distribution of red cell polymorphisms in today's populations. George has done a great deal to bring a critical approach to studies of clinical genetics, particularly in large populations. It has been a great privilege to read his excellent publications over many years and, more recently, to share his friendship.

Acknowledgements
I thank Liz Rose for her help in preparing this manuscript.

Rare variant hypothesis for multifactorial inheritance: susceptibility to colorectal adenomas as a model

Nicola S. Fearnhead[1,2], Bruce Winney[1] and Walter F. Bodmer[1]

[1] Cancer Research U.K. Cancer and Immunogenetics Laboratory, Weatherall Institute of Molecular Medicine, John Radcliffe Hospital, Oxford, OX3 9DS, United Kingdom
[2] Department of Colorectal Surgery, John Radcliffe Hospital, Oxford, OX3 9DU, United Kingdom

Introduction

The earliest success in the identification of the molecular basis of a simple Mendelian disease, and through that the gene responsible, was sickle cell anaemia in 1949 (Pauling *et al.* 1949). However, it was not until late in the last century, following the development of positional cloning technology, that it became possible to identify the genes, and so the molecular basis of clearly inherited diseases, without significant prior knowledge of their molecular basis. As a result, the majority of the genes responsible for classically inherited diseases have been identified, including, for example, cystic fibrosis (Riordan *et al.* 1989), Huntington's chorea (The Huntington's Disease Collaborative Research Group, 1993) and myotonic dystrophy (Carango *et al.* 1993). The focus of much research has now turned to the attempt to identify the genes underlying susceptibility to common complex diseases such as cancer (Pharoah *et al.* 2004), heart disease (Andreotti *et al.* 2002), diabetes (McCarthy *et al.* 2002), obesity (Swarbrick *et al.* 2003) and mental diseases (Kennedy *et al.* 2003).

Two main, but contrasting, hypotheses have emerged regarding the basis for genetic susceptibility to common diseases: the common allele and the rare variant hypotheses. The idea behind the common allele hypothesis comes from early studies on the association between HLA alleles and certain relatively common, mainly autoimmune, diseases such as ankylosing spondylitis, rheumatoid arthritis and type I diabetes (reviewed in Tomlinson and Bodmer 1995). This led to the concept that an association between a relatively common allele and a disease could signal the existence of a variant of a nearby gene on the chromosome due to linkage disequilibrium, which is the tendency for variants of closely linked genes to remain together on the same chromosome or haplotype. It should then be possible to identify the real disease-associated variant by scanning nearby genes for variants that plausibly satisfy the requirement for having an effect on the disease. Though there have been a few notable successes, such as the association between apolipoprotein E (*APOE*) isoforms and hyperlipoproteinaemia (Breslow *et al.* 1982) and that between amyloid precursor protein (*APP*) variants and Alzheimer's disease (Kennedy *et al.* 2003), the majority of such searches have proved disappointing. At best, associations have low relative risks, generally below 1.5, which have been very difficult to follow up (Cardon and Bell 2001, Cardon and Palmer 2003).

The alternative, although not mutually exclusive, hypothesis is that genetic susceptibility to a multifactorial disease is due to summation of the effects of a series of low frequency variants of a variety of different genes, each conferring a moderate, but readily detectable, increase in relative risk. This idea was first clearly suggested by Frayling *et al.* (1998) based on reports of low frequency missense variants in the adenomatous polyposis coli (*APC*) gene that were associated with a significantly increased risk of colorectal cancer or pre-cancerous adenomas (Laken *et al.* 1997, Frayling *et al.* 1998). The *APC* gene, when severely disrupted by nonsense or truncating mutations, causes the dominantly inherited colorectal cancer susceptibility known as familial adenomatous polyposis (FAP), whereas the low frequency missense variants had much lower penetrance effects and were mostly not associated with familial incidence. This led to the hypothesis that 'these types of variants may thus represent a major new facet of the study of multifactorial disease inheritance, representing effects that lie between those of severe clearly inherited susceptibilities and relatively common multifactorial low-penetrance effects, such as are characterised by the many associations between polymorphic HLA variants and autoimmune diseases' (Bodmer 1999).

Recently, as will be discussed below, we have shown that a number of variants in five different genes are sufficient, each on their own, to increase the risk of colorectal adenomas by an average odds ratio of about two (Fearnhead *et al.* 2004). The statistical approach we used to overcome the lack of a very large sample size was to estimate a modified odds ratio using a Mantel-Haenzel approach (Sokal and Rohlf 1995) and combine results for different genes on the same set of cases and controls. Data from the different variants were only combined if there was shown to be no heterogeneity between data sets, using the Mantel-Haenzel test for heterogeneity (Sokal and Rohlf 1995).

Colorectal adenomas as a model multifactorial disease

The adenomatous polyp is the major precursor of colorectal cancer (Morson 1974). The transition of normal epithelium through dysplasia to invasive adenocarcinoma is mediated by a series of genetic and epigenetic events (Fearnhead *et al.* 2002). While any given adenoma may only have a small chance of becoming a carcinoma, the presence in an individual of multiple colorectal adenomas indicates an increased risk of colorectal cancer (Atkin *et al.* 1992), and prophylactic removal of the adenomas substantially reduces the risk of getting colorectal cancer (Winawer *et al.* 1993).

FAP, MYH-associated polyposis (MAP) and hereditary non-polyposis colorectal cancer (HNPCC) are all inherited genetic syndromes that together account for a small fraction, probably less than 4%, of colorectal cancer cases (Fearnhead *et al.* 2002). In addition to these classically inherited syndromes, a significant fraction of colorectal cancers appears to occur in individuals with some form of inherited susceptibility to colorectal cancer (Cannon-Albright *et al.* 1988, Houlston *et al.* 1990). As individuals with multiple colorectal adenomas are at risk of colorectal cancer but mostly do not cluster in families, they provide a valuable group for studying genetic risk factors predisposing to colorectal cancer in this additional category of multifactorial inherited cases.

Following the lead provided by the discovery of rare predisposing variants at the *APC* locus, we decided to search for variants of other candidate genes in a population of individuals with multiple colorectal adenomas. Our criteria for candidate genes were those in which severe disruption of function led to clear-cut inherited syndromes, such as FAP in *APC* mutants, and genes in which mutations or epigenetic changes have been shown to occur somatically in colorectal cancers. These are the genes that are changed as part of the somatic evolutionary pathway leading to colorectal cancer and include, for example, genes involved in Wnt signalling, mismatch repair, cell cycle regulatory mechanisms and cellular adhesion. It was presumed that germ line variants of such genes which gave rise to subtle changes in protein interactions or levels of gene expression, resulting in only a marginal, if any, selective disadvantage to the individual carrying them, might nevertheless give rise to a detectable increased risk of getting colorectal cancer.

Our approach was to screen the germline DNA from 124 multiple adenoma cases for functional variants in five chosen candidate genes, and then to type 483 controls for the identified variants (Table 1). By estimating a modified Mantel-Haenzel odds ratio, we were then able to show that genetic predisposition to colorectal cancer is likely to involve functional variants of multiple genes involved in the progression of the normal epithelium to adenoma to carcinoma sequence. Overall, 30 individuals (25%) carried a rare germline variant as compared to only 12% of random controls (P = 0.0001). Of the 30 multiple adenoma patients with rare variants, four individuals had also developed colorectal cancer(s) and two had had other cancers (Fearnhead *et al.* 2004, 2005).

Table 1. Summary of variants found in germline DNA from 124 patients with multiple adenomas and 483 random controls, derived from Fearnhead *et al.* (2004). *Average number of controls typed for all variants. **Two individuals each had two variants. §Modified Mantel-Haenzel odds ratios. SDS = splice donor site variant.

Gene	Variants	Patients (%)	Controls (%)	Odds ratio	P
APC	*E1317Q*	3/124 (2.4)	6/480 (1.3)	2.0	0.400
CTNNB1	*N287S*	1/124 (0.8)	3/483 (0.6)	1.3	1.000
AXIN1	*P312T, R398H, L445M, D545E, G700S, R891Q*	18/124 (14.5)	37/479.3* (7.7)	2.0§	0.012
hMLH1	*G22A, K618A*	5/124 (4.0)	11/482.5* (2.3)	2.0§	0.175
hMSH2	*H46Q, E808X, ex4SDS*	3/124 (2.4)	0/479.3 (0.0)	11.7§	0.001
Combined		30/124 (24.9)	55/479.8*,** (11.5)	2.2§	0.0001

Discussion

Our work on patients with colorectal adenomas has found rare but functionally important variants of five genes (Fearnhead *et al.* 2004, 2005). Each of the individual variants had an average odds ratio of about 2.0 but, due to the relatively small sample sizes, few of the variants on their own differed significantly in frequency between cases and controls. Analysis using a Mantel-Haenzel approach suggested that the individual data sets were not heterogeneous. When combined, the overall data gave a highly significant modified odds ratio of 2.2 (Table 1, P = 0.0001). Furthermore, 25% (30/124) of patients each had a single variant and none of them had two or more (Fearnhead *et al.* 2004, 2005). The lack of double variants is consistent with the low population frequencies of the variants and the expectation that rare variants act independently.

These data, therefore, support a non-additive rare variant model of susceptibility, whereby it is rare, mainly missense but possibly also nonsense, promoter and splicing, variants that collectively explain a substantial proportion of multifactorial inherited susceptibility to colorectal adenomas. It seems most probable that the same applies to a wide variety of common chronic diseases (Bodmer 1999, Pritchard 2001). The presence of a single variant in an individual appears to be sufficient to increase the risk of susceptibility, although other genetic factors together with the environment will undoubtedly also contribute to development of the disease.

Examples of rare variants, such as we have described for colorectal adenomas, have been described for other diseases. For example, a variant in the *CHEK2* gene, *1100delC* has been found at a frequency of 2% to 14% in *BRCA1/BRCA2*-negative familial breast cancer cases (Mangion *et al.* 2002, Meijers-Heijboer *et al.* 2002, Vahteristo *et al.* 2002). Rare variants in the *BRCA2* gene have been associated with early onset prostate cancer as well as breast cancer. A study of men under the age of 55 years with prostate cancer found truncating germline *BRCA2* mutations in 2.3% (giving a relative risk of 23 for this type of mutation) and *BRCA2* missense and other variants in a further 9.1% (Edwards *et al.* 2003). None of these variants was, however, directly typed in a control population to assess the population frequency.

To reach significance for individual variants at frequencies of 1% or less in disease groups, large sample sizes are needed, especially in the control groups. However an analysis of the functional effect of a variant can add considerably to the confidence in its likely relevance to disease susceptibility (Fearnhead *et al.* 2004, 2005). Evidence is now accumulating that inherited susceptibility to most common complex diseases may be largely attributed to multiple rare variants. Data supporting this hypothesis are emerging in, for example, schizophrenia (Kennedy *et al.* 2003) and venous thrombosis (Franco and Reitsma 2001). Common cancers such as breast cancer are also likely to be caused by multiple rare variants, only a small number of familial cases being associated with mutations in the *BRCA1* and *BRCA2* genes (Meijers-Heijboer *et al.* 2002, Vahteristo *et al.* 2002, Balmain *et al.* 2003).

The sorts of variants we are considering as the basis for susceptibility to common chronic diseases, which are mostly post-reproductive, will either be effectively neutral or very close to neutral. If by chance they drift up in frequency, they will therefore be subject to hardly any constraints on their increase in frequency. What we now see are those variants that happen to have drifted up in frequency. This can occur for many different genes and so can explain why the rare variant hypothesis may well account for susceptibility to a wide range of common diseases and conditions. Such drift effects are very often, if not mostly, population specific, and so the rare variants may often be 'founder' variants that are population specific as in the case of the *APC I1307K* variant in individuals of Ashkenazi Jewish descent. This is consistent with the fact that the *APC E1317Q* variant was not found in the Korean population (Fearnhead *et al.* 2004). The contribution of any given rare variant will be too small to be detected by even quite large sized SNP disease association analysis. The only productive approach is, therefore, to search for DNA variation in carefully chosen candidate genes in selected disease groups. We believe that this approach will ultimately be most rewarding in identifying variants involved in susceptibility to complex multifactorial diseases. The analysis of such variants will contribute both to the understanding of disease aetiology and, at least in some cases, provide the basis for a screening and secondary prevention programme.

It is a pleasure to contribute this paper to a Festschrift in honour of George Fraser, a long-standing friend and colleague who has made many significant contributions to human genetics. We first met through having a common academic heritage from R. A. Fisher. Much later I was fortunate, as Director of the then Imperial Cancer Research Fund, in being able to draw George into an involvement with cancer families and cancer genetics.

Acknowledgement

This paper is extracted from *Cell Cycle* 4 521–525 and is published with permission.

The basal phenotype of *BRCA1*-related breast cancer: past, present and future

William D Foulkes

Program in Cancer Genetics, Departments of Oncology and Human Genetics,
McGill University, Montreal, QC, Canada

The identification of the breast cancer susceptibility gene *BRCA1* was a breakthrough in clinical cancer genetics. For some years, the existence of large pedigrees containing many generations of young women who had been diagnosed with breast cancer, often at a young age, had been known. With the final localisation and identification of *BRCA1*, genetic testing for breast cancer susceptibility came into the clinical realm (Narod and Foulkes 2004). Clinical testing is now widely available, but is expensive and laborious. Several tools have been developed which enable health professionals to predict who is most likely to carry a *BRCA1* or *BRCA2* mutation (Nelson *et al.* 2005). All of these tools use family history as the main variable: some employ a Bayesian approach, whereas other are based on a simpler hand-scoring technique. Each tool has advantages and disadvantages. They all share one drawback: none of them considers the phenotype itself: that is, breast or ovarian cancer.

It has been known for at least 7 years that both *BRCA1*- and *BRCA2*-related breast cancers differ from non-hereditary breast cancers in several aspects, such as a continuous pushing margin. *BRCA1*-related breast cancers are more likely to be ER-negative than are *BRCA2*-related breast cancers (Lakhani *et al.* 1998, Karp *et al.* 1997), and both *BRCA1* and *BRCA2*-related breast cancers are usually HER2-negative. Recently, a more specific, and clinically useful, phenotype has emerged for *BRCA1*-related breast cancers.

In 2000, Perou *et al.* showed that morphologically similar breast cancers could be divided into several groups, based on their gene expression profile (Lakhani *et al.* 2002). One of these molecularly-defined subgroups was found to contain breast cancers that did not stain for either oestrogen receptor (ER) or HER2/neu (also known as erbB-2) proteins in the tumour cells. This is an unusual combination, as ER and HER2 staining is often in opposite directions. As some of the genes that were differentially expressed in these cancers are known to be usually only expressed in the basal cells of the breast (such as cytokeratin 5/6 (CK5/6) and annexin VIII) they called these breast cancers 'basal breast cancers'. 'Luminal' breast cancers were mainly ER-positive, and expressed CK8/18, a protein seen in breast cells adjacent to the lumen of the duct. Most breast cancers express CK8/18. These findings were not meant to imply that basal cancers arose from basal cells (or indeed any type of cell), but that they expressed proteins normally found in these cells in the adult female breast. Immunohistochemical assays using antibodies raised against CK5/6 and CK8/18 showed that these breast cancers tended to have high levels of CK5 and low levels of CK8/18, consistent with the gene expression studies. As *BRCA1*-related breast cancers are often both ER- and HER2-negative, this finding immediately suggested that *BRCA1*-related breast cancers might fall into this so-called basal group. Our group focused on the

ER/HER2-negative sub-group of breast cancers, and showed that within this group of 72 breast cancers diagnosed in Ashkenazi Jewish women under 65 years of age, *BRCA1*-related breast cancers were nine times more likely to express CK5/6 than were tumours that did not arise in *BRCA1/2* mutation carriers (P = 0.002). This observation extended to show that in 247 women (from the same historical cohort) *BRCA1*-related breast cancers were over five times more likely to be 'basal' (i.e. express CK5), irrespective of ER or HER2 status (Foulkes *et al.* 2004). These results published around the same time as a re-analysis by the Stanford group of data from Laura van't Veer and colleagues, which showed that the *BRCA1* cancers in her series were overwhelmingly basal in phenotype (Sorlie *et al.* 2003). The Breast Cancer Linkage Consortium has now confirmed these findings in a much larger series of cases. Importantly, they showed that breast tumours that were ER-negative, CK5/6 and CK14 positive were approximately 30 times more likely to carry a *BRCA1* mutation than were controls: almost one half of all *BRCA1* cancers had this phenotype, whereas it was seen in only 1.6% of the *BRCA1/2*-negative controls (Lakhani *et al.* 2005).

Breast cancers can broadly be divided into those that express luminal keratins, the so-called simple epithelial type keratins such as cytokeratins 7, 8, 18 and 19 and those that feature high levels of expression of genes that are characteristic of the basal epithelial cells of the normal mammary gland, the stratified epithelial cytokeratins, such as cytokeratins 5, 6, 14, 15 and 17 (cytokeratin 6 may not be expressed in breast). Other markers, such as smooth muscle actin, glial fibrillary acidic protein, calponin and P-cadherin may also be present in basal-like breast cancers (Moll *et al.* 1982, Tot 2000, Jones *et al.* 2001, Arnes *et al.* 2005). Basal breast cancers are not frequent, comprising between 3 and 15% of all invasive ductal breast cancers of no special type. Conventional histopathological as well as molecular studies of breast cancers with 'basaloid' cell differentiation have shown that these tumours are often high grade (Jones *et al.* 2001), have areas of necrosis (Tsuda *et al.* 2000), may have a typical or an atypical medullary phenotype (Tot 2000) and have a distinct pattern of genetic alterations (Jones *et al.* 2001), including frequent *TP53* mutations (Sorlie *et al.* 2001). Most studies of outcome have also indicated that basal-like breast cancers often have a poor prognosis, but recent studies show that there may be more than one type of basal breast cancer (Sotiriou *et al.* 2003, Jones *et al.* 2004) and not all of them are associated with a poor prognosis (Jones *et al.* 2004). It may be that the type associated with *BRCA1* mutations are likely to be more aggressive and be associated with a poorer prognosis, as suggested by early studies (Foulkes *et al.* 1997, Ansquer *et al.* 1998). Overall, the clinico-pathological profile of *BRCA1*-related breast cancers is remarkably similar to that of basal breast cancers (Korsching *et al.* 2002, Narod and Foulkes 2004).

Interestingly, some studies have suggested that specifically node-negative (rather than node-positive) *BRCA1*-related breast cancer has a particularly poor outcome (Foulkes *et al.* 2000, Moller *et al.* 2002), and the finding that both *BRCA1* and basal breast cancers in general tend to be large, node-negative tumours associated with a poor outcome suggests that perhaps the mode of spread of these cancers is somewhat different: early haematogenous spread might be favoured (Foulkes *et al.* 2003, Foulkes *et al.* 2004). The successful treatment of ER-positive breast cancer by tamoxifen, and of HER2-positive cancer by HER2, raises the question whether targets such as ER or HER2 might be identified in basal and/or *BRCA1*-related breast cancers, which often lack both these proteins.

In lung cancers, genomic amplification of specific mutations in the Epidermal Growth

Factor Receptor (EGFR), correlate, to a certain extent, with response to gefitinib treatment (Lynch *et al.* 2004, Paez *et al.* 2004, Kobayashi *et al.* 2005, Tsao *et al.* 2005), and in general, the use of small molecules aimed as EGFR has received much attention (Chan *et al.* 2002, Arteaga and Baselga 2004, Ranson and Wardell 2004). Recently, somatic mutations in EGFR were identified in *BRCA1*-related breast cancer (Weber *et al.* 2005), although whether they are truly in excess in such cancers is currently unknown. Notably, *BRCA1* tumours commonly over-express EGFR (Brunet *et al.* 2004, Lakhani *et al.* 2005, van der Groep *et al.* 2004), so it would be interesting to see if EGFR could represent the elusive target in breast cancer.

The concept of the cancer stem cell has be re-visited (Reya *et al.* 2001), and one group claims to have identified 'the' breast cancer stem cell (Al Hajj *et al.* 2003). This very exciting development awaits confirmation by other groups. Böcker and colleagues have developed a model of breast cancer stem cells based on immunohistochemical analysis of breast cancers and normal breast tissue. They argue that the breast stem cell is a CK5/6-positive basally-positioned cell that differentiates into all other breast cell types (Böcker *et al.* 2002), and thus CK5/6-positive cancers could be seen as breast stem cell cancers. Intriguingly, recent immunohistochemical studies have suggested that at least some *BRCA1*-related breast cancers are 'true basal breast cancers', in that some of these cancers express CK5/6 and not the luminal marker CK8/18, whereas most so-called basal cancers also express CK8/18. Indeed, it has been hypothesised that *BRCA1* functions as a stem cell regulator (Foulkes 2004). There is currently no evidence for this hypothesis, and others have pointed out that it is important not to confuse staining patterns in mature cancers with inferences about histogenesis (Gusterson *et al.* 2005). Nevertheless, the aggressive, fast-growing nature (Tilanus-Linthorst *et al.* 2005) of most *BRCA1*-related breast cancers, with their frequent vascular nests (Goffin *et al.* 2003) and invasive tendencies, does fit neatly with the idea of migrating stem cells (Brabletz *et al.* 2005), which would be the necessary target of cancer therapy.

That *BRCA1*-related breast cancers represent a distinct form of breast cancer now seems indubitable. The essence of this is the expression of protein normally seen in basal cells of the breast. This phenotype can be used to better identify *BRCA1* mutation carriers. Turning these findings into practical advantage for the prevention and treatment of breast cancer in these high-risk women is the next challenge.

Dedication and acknowledgements

I first met George Fraser in the early 1990s, when he was running the Cancer Family Clinic in Oxford, while I was doing my PhD at the Imperial Cancer Research Fund at Lincoln's Inn Fields. Therefore, it seems appropriate that my contribution to this Festschrift in honour of George is about hereditary breast cancer. I have learned a lot from George, and regard him as one of the great (medical) geneticists; a direct link to Penrose, Fisher and Haldane and the 'founder group' at the Galton Laboratory. George has travelled widely. Like George, I have travelled to and worked at McGill University. Unlike George, I have stayed here. This short contribution is a gift to George from me.

The work carried out by my group and described in this contribution has been funded by the US Department of Defense, the Susan G. Komen Foundation and the Canadian Breast Cancer Research Alliance.

BRCA1 and BRCA2:
Breaking the cycle and repairing the damage

Julian Barwell[1], Melita Irving [2], Roberto Alonzi[3] and Shirley Hodgson[2]

[1]Clinical Genetics, 7th Floor New Guy's House, Guy's Hospital London SE1 9RT
[2]Clinical Genetics, St George's University of London, Cranmer Terrace, London SW17 ORE
[3]Clinical Oncology, Charing Cross Hospital, Fulham Palace Road, London W6 8RF

Abstract

A family history of breast cancer confers an increased empirical risk of breast cancer. Approximately 5% of all breast cancers are due to an inherited mutation in *BRCA1* or *BRCA2* (Bertwistle and Ashworth 1998). Individuals who have inherited an altered copy of either the *BRCA1* or *BRCA2* gene are at an increased risk of breast and/or ovarian cancer, which may be early onset or bilateral. Pathological features of tumours in carriers of a *BRCA1* or *BRCA2* mutation may help distinguish these individuals from those with sporadic breast cancers. Differences at the histological and protein expression level could provide the opportunity for targeted treatment, and an awareness of the pathogenicity may alter screening practice in those with a mutation, since these mutations cause sensitivity to irradiation due to defects in double stranded DNA repair.

There are differences between *BRCA1* and *BRCA2* with regard to both their cellular action and the clinical phenotype in most carriers. This review examines the evidence for *BRCA1* and *BRCA2* gene functions and the clinical implications of being a *BRCA1* or *BRCA2* mutation carrier.

Introduction

Approximately one women in thirteen worldwide will be affected with breast cancer at some stage during her lifetime, and the rate is one in nine in the United Kingdom (National Statistics 2000). Two highly penetrant breast cancer susceptibility genes, *BRCA1* and *BRCA2*, have been identified but inherited mutations in these genes account for only 15% of familial cases (Peto *et al.* 1999, Anglian Breast Cancer Study Group 2000). However, a recent epidemiological study has suggested that a large proportion of women with breast cancer may have a hereditary susceptibility to this cancer (Peto and Mack 2000). In most breast cancer cases, therefore, other common lower penetrance genes must be involved. Both *BRCA1* and *BRCA2* have multiple roles in the repair of double strand DNA breaks via homologous recombination, cell signalling possibly via a transcription factor action, protein degradation via ubiquination, cell cycle regulation, cell cycle checkpoints and apoptosis (Rosen *et al.* 2003). Abnormalities in any of these pathways caused by other genes might also increase the risk of malignant change (Connor *et al.* 1997).

BRCA1 and *BRCA2* mutation carriers are known to be at a high risk of developing breast cancer and a moderately increased risk of ovarian cancer and are responsible for two-thirds of strongly inherited breast cancer families.

Differences in tumours occurring in women with germ-line mutations of *BRCA1*, and to a lesser extent, *BRCA2* are apparent when compared to sporadic tumours. These pathological features may provide the opportunity to differentiate between mutation carriers and sporadic cases, and further allow targeting of treatment to specific tumour types.

MRI is being assessed as an alternative form of radiological screening for breast cancer in high risk women because of its increased sensitivity and the theoretical risk that *BRCA1* and *BRCA2* mutation carriers may have increased radiosensitivity (Kuhl *et al.* 2000, Stoutjesdijk *et al.* 2001, Warner *et al.* 2001).

This review aims to highlight the pathogenesis and the pathological features of *BRCA1* and *BRCA2* mutation associated tumours, to discuss management in relation to screening modalities, the potential for targeted treatment, and prophylactic surgery, and to explain current opinion on hormonal influences, such as HRT and oral contraceptive use in *BRCA1* or *BRCA2* mutation carriers.

Incidence of *BRCA* mutations

The incidence of *BRCA1* mutation carriers in the general population is estimated to be 0.11%, and for *BRCA2* mutation carriers 0.12% (Peto *et al.* 1999). The proportion of individuals with breast or ovarian cancer with a *BRCA1* or *BRCA2* mutations depends on family history. Families that contained at least four members with either female breast cancer diagnosed before the age of 60 years or male breast cancer diagnosed at any age have been studied by linkage analysis (Ford *et al.* 1998). The disease was linked to *BRCA1* in 52% of families, to *BRCA2* in 32% and to neither in 16%, suggesting the existence of other predisposing genes. In the absence of a family history, a meta-analysis (Whittemore *et al.* 1997) suggested that 4.2% of breast cancer before the age of 70 is due to *BRCA1* germline mutations. Similarly, 5.3% of ovarian cancers were, as estimated by direct mutation analysis, due to germline *BRCA1* mutations.

Clinical penetrance of *BRCA1* and *BRCA2* mutations

The penetrance of *BRCA* mutations does depend on how cancers are ascertained. A meta-analysis of 500 families in which *BRCA1* and *BRCA2* mutations were identified from population based studies have estimated the risks of breast and ovarian cancer in *BRCA1* carriers to be 65% (confidence interval (C.I.) 44–78%) by age 70 and 39% (C.I. 18–54%) respectively. For *BRCA2* the risk of breast cancer was 45% (C.I. 31–56%) for breast cancer and 11% (C.I. 2.4–19%) for ovarian cancer (Antoniou *et al.* 2003). Four hundred and eighty three *BRCA1* mutation carriers ascertained in a breast cancer risk evaluation clinic were followed up to assess breast and ovarian cancer risk (Brose *et al.* 2002). By the age of 70, female breast cancer risk was 72.8% (C.I. 68–78%) and ovarian cancer risk was 40.7% (C.I. 36–46%). The risk of a second primary breast tumour by the age of 70 was 41% (C.I. 34–47%). A study of 173 breast and ovarian cancer families with *BRCA2* mutations (Breast Cancer Linkage Consortium *et al.* 1999) estimated that the cumulative breast cancer risk was 60% (C.I. 44–72%) by age 50, and 77% (C.I. 71–88%) by age 70. The estimated risk of ovarian cancer was 3.3% (C.I. 1–6%) by age 50 and 16% (C.I. 9–23%) by age 70 years.

The lifetime risk of breast cancer in Ashkenazi Jewish *BRCA1* and *BRCA2* mutation carriers identified from index breast cancer cases was shown to be 82% and lifetime ovarian cancer risk was 54% in *BRCA1* and 23% in *BRCA2* mutation carriers (King *et al.* 2003).

Carriers of *BRCA1* and *BRCA2* mutations also have a slightly increased risk of other malignancies (see Table 1). *BRCA1* carriers have a statistically significant increased relative risk of pancreatic cancer of 2.26, endometrial cancer of 2.65, cervical cancer of 3.72, prostate cancer diagnosed under the age of 65 of 1.82. *BRCA2* mutation carriers have a statistically significant increased relative risk of stomach cancer of 2.59, malignant melanoma of 1.43, prostate of 4.65, gallbladder and/or bile duct of 4.97 and pancreas of 3.51 (The Breast Cancer Linkage Consortium 1999, Thompson and Easton 2002a, Brose *et al.* 2002, Thompson and Easton 2002b).

What are the pathological features of breast cancer, which are characteristic of *BRCA1* or *BRCA2* mutation carriers?

The pathological features of *BRCA1* tumours which are significantly different from breast cancers in non-BRCA carriers include an increased total mitotic count, continuous pushing margins and lymphocytic infiltrates, which are characteristic features of medullary cancers. No specific histological type in breast cancer is thought to be associated with *BRCA2* mutation carriers (Lakhani *et al.* 2002).

Approximately 30% of cancers in *BRCA1* mutation carriers have a particular immunohistochemical staining profile, including oestrogen receptor negative, cmyc amplification, HER receptor negative and positive staining for striated epithelial cytokeratin 5 and/or 6. The latter suggests that these tumours may originate from a limited number of epithelial cells with a specific basaloid phenotype (Foulkes *et al.* 2003). Tumours in *BRCA2* mutation carriers tend to be similar to sporadic cancers with respect to hormonal receptor expression.

Hormonal factors

One important question is why individuals with *BRCA1* and *BRCA2* mutations have a predominantly increased risk of breast and ovarian cancer. It has been well established that breast cancer risk is related to the length and timing of oestrogen exposure, such as age at menarche, use of the oral contraceptive pill and hormone replacement therapy, the age at pregnancy and the number of children a woman has had (Madigan *et al.* 1995). Breast and ovarian tissue undergo cycles of cell proliferation, differentiation, involution and apoptosis at different stages of life including birth, puberty, the menstrual cycle, pregnancy, breast feeding and the menopause. There are two possible theories whereby oestrogen responsive tissues may be at an increased risk of cancer. Firstly, that oestrogen increases cellular proliferation by stimulating oestrogen receptor mediated transcription. Secondly, that oestradiol can be metabolised to quinone derivatives causing DNA depurination. In both contexts, any abnormalities in DNA repair (such as *BRCA1* or *BRCA2*) could result in increased somatic mutations (Rzewuska-Lech and Lubinski 2004).

Functions of *BRCA1* and *BRCA2*

BRCA1 was identified in 1994 (Miki *et al.* 1994), and maps to chromosome 17q. It has 22 exons and codes for a protein of 1863 amino acids. The *BRCA2* gene was identified in 1995 (Wooster *et al.* 1995), maps to chromosome 13q and codes for a protein of 3418 amino acids. Many cancer predisposing mutations have been detected in these two breast cancer susceptibility genes (Couch and Weber 1996, Dunning *et al.* 1997), which can result in loss of function of the encoded protein, thereby compromising DNA repair.

The *BRCA1* and *BRCA2* genes encode proteins which are known to have roles in the cell cycle, the DNA repair pathway and cell cycle checkpoint control. DNA repair is a ubiquitous function and both the BRCA genes are expressed in all tissues. Double strand DNA breaks, occurring in the S and G2 phase of the cell cycle, induced by irradiation and cross linking agents are repaired by either non-homologous end joining (NHEJ) and/or homologous recombination (HR). NHEJ does not use the complementary DNA strand as a template, and effectively cuts out the damaged DNA sequence, ligating the two ends of the DNA strands. It is therefore error-prone (Kauff *et al.* 2002, Rebbeck *et al.* 2002, Tutt and Ashworth 2002, Ventikaraman 2002). In HR, the complementary double strand DNA sequence in the other chromatid is used as a template to repair the damaged DNA sequence. This repair procedure is highly conservative and is error free. *BRCA1* and *BRCA2* have particularly important roles in HR, therefore mutation carriers would be expected to exhibit a higher proportion of error prone NHEJ and subsequent mutagenesis.

Can differences be detected between normal cellular function in *BRCA1* and *BRCA2* heterozygotes compared with controls?

The evidence for an abnormality in normal tissue cellular response to DNA damage in *BRCA1* and *BRCA2* mutation carriers is conflicting. The studies have been limited by small numbers and the use of cell lines with altered cellular and apoptotic characteristics.

It is feasible that abnormalities in the cellular response to radiation of *BRCA1* and *BRCA2* heterozygotes may be contributing to the genetic predisposition to breast cancer. It has also been suggested that breast cancer that appears to have a hereditary component, but is not due to *BRCA1* or *BRCA2* mutations, could be due to other inherited types of impaired repair of radiation induced double stranded DNA defects as demonstrated in peripheral blood lymphocytes). This was first suggested by a research group (Scott *et al.* 1996) who determined that a significantly higher percentage of breast cancer patients (40% of 135 cases) had a reduced ability to repair double strand DNA breaks. This was assessed by the *in vitro* lymphocyte G2 radiosensitivity assay, post irradiation, compared with age matched controls (6% of 105 cases). The first-degree relatives of these breast cancer patients also had a slight reduction in repair (Roberts *et al.* 1999), suggesting that the reduced ability to repair these DNA breaks could be inherited. Thus genetic loci controlling chromosome radiosensitivity could be low penetrance familial breast cancer genes. However, these data require confirmation.

The cellular survival of dermal fibroblasts and the number of chromatid breaks in the lymphocytes in *BRCA* mutation carriers has been assessed in vitro post irradiation (Buchholz *et al.* 2002). Results from both assays suggested that cells heterozygous for a mutation in eight *BRCA1* or one *BRCA2* mutation carriers were more radiosensitive than controls. Another study on EBV immortalised lymphoblasts from *BRCA2* mutation carriers also showed a reduction in clonogenic survival, increased micronucleus formation and a DNA repair defect after gamma irradiation (Foray *et al.* 1999). However, there were difficulties with the matching of these cases and the numbers were small.

The repair of X-ray-induced DNA damage in *BRCA1* and *BRCA2* mutation carriers has been reviewed (Nieuwenhuis *et al.* 2002). Fibroblasts and lymphocytes from mutation carriers had no demonstrable gross defects in their ability to rejoin radiation induced DNA breaks.

A recent paper has assessed the G1/S checkpoint in primary fibroblasts from *BRCA1* heterozygotes using UVA, hydrogen peroxide and the DNA cross-linking agent mitomycin C (Shorrocks *et al.* 2004). They showed that release of G1 arrested cells was faster in a *BRCA1* mutation carrier compared with wild type after UVA expression. However, there was no difference in micronuclei, a measure of double strand breaks and chromosome instability suggesting that any checkpoint defect is not associated with genomic instability after oxidative stress or mitomycin C.

A chicken cell line (DT40) heterozygous for a truncated *BRCA2* mutation has been found to have reduced cell proliferation and cell cycle defects (Warren *et al.* 2003). However, there was no difference in cell cycle duration. In particular there was a significant increase in cells blocked prior to entering mitosis at the G2/M check point. The cells had an increased sensitivity to the DNA oxidising agent mitomycin C and cisplatin but not to mitomycin S, UV or X-irradiation. Cells heterozygous for truncated *BRCA2* showed a reduction in the number of rad51 foci. These foci, produced at the points of double strand breaks during S phase, facilitate DNA repair post-irradiation.

Management issues in *BRCA1* and *BRCA2* mutation carriers

Breast cancer screening in BRCA mutation carriers

There are concerns about the use of yearly mammograms in *BRCA1* and *BRCA2* mutation carriers because of the potential risk that irradiation of breast tissue could cause DNA damage, which may be inefficiently repaired and thus potentially carcinogenic. *BRCA1* and *BRCA2* mutation carrier women under the age of 35 should have breast cancer screening but mammograms are less sensitive and specific at this age, due to the fact that younger women have a higher breast density which significantly reduces the efficacy of mammography. There is some evidence that MRI has a superior sensitivity to both mammography and ultrasound (Kuhl *et al.* 2000). The MARIBS study (Leach *et al.* 2005) including 1881 screening years showed that X-ray mammography identified only 40% of the tumours in women at high familial risk, whereas MRI pinpointed 77% and was particularly effective for women known to carry the *BRCA1* gene mutation, detecting 92% of tumours in women carrying this gene, whereas mammography detected only 23%.

Prophylactic measures

There are obviously no randomised controlled trials to assess the benefits of prophylactic bilateral mastectomy. However, there have been two major studies that have assessed individuals at an increased risk of breast cancer who have had prophylactic surgery. The first study (Hartmann *et al.* 1999, Hartmann *et al.* 2001) reviewed 639 patients at the Mayo Clinic, 214 with a high risk of breast cancer and 435 at moderately increased risk. Eighteen of these women at high risk were subsequently shown to be BRCA mutation carriers. Prophylactic surgery, involving total or subcutaneous mastectomy, reduced the risk of breast cancer in both groups by 90%.

The second study (Meijers-Heijboer *et al.* 2001) recruited 139 *BRCA1* and *BRCA2* mutation carriers. Seventy six underwent bilateral prophylactic mastectomy and no breast cancers were identified after a mean follow up of 2.9 years. In the 63 who under went screening without surgery for a median follow up of three years, eight breast cancers were detected.

There is also good evidence that premenopausal oophorectomy halves the risk of breast cancer in women, both in the general population and in *BRCA1* and *BRCA2* mutation carriers (Meijers-Heijboer *et al.* 2002, Rebbeck *et al.* 1999) and is thus a prophylactic option for women who carry *BRCA1* and *BRCA2* mutations.

One study (Kauff *et al.* 2002) assessed the effect of salpingo-oophorectomy compared to ovarian screening on the incidence of breast and ovarian cancer in 170 *BRCA1* and *BRCA2* mutation carriers 35 years of age and older for two years. In the surveillance group, five out of 72 (6.9%) developed ovarian or primary peritoneal cancer and eight of the 72 (12.9%) developed breast cancer. In the surgical group, three out of 98 (3.1%) had ovarian cancer at surgery and one (1.1%) developed primary peritoneal cancer on follow up. Only three developed breast cancer during the follow up period. On average, the time to diagnosis of a breast cancer or *BRCA* related gynaecological tumour was longer in the salpingo-oophorectomy group compared with the surveillance group with a hazard ratio for subsequent breast cancer or *BRCA* related gynaecological tumour of 0.25. Because of the risk of patients developing primary peritoneal cancer, after oophorectomy, *BRCA1* and *BRCA2* mutation carriers are recommended to continue tumour marker (CA 125) surveillance post surgery.

Chemopreventive agents

Tamoxifen has a proven role in the management of oestrogen receptor positive breast cancer. Tamoxifen and newer anti-oestrogens, e.g. raloxifene, are being evaluated for preventing breast cancer in the general female population at increased risk. Initial results with tamoxifen showed that it halved the risk of breast cancer but also increased the risk of adverse cardiovascular events and endometrial cancer (Narod *et al.* 2000, Cuzick *et al.* 2003) and the overall mortality was not reduced. The aromatase inhibitors, e.g. anastrosole, which reduce oestrogen levels, are also being trialled in IBIS II (International Breast Cancer Intervention Study II).

Tumours in *BRCA1* mutation carriers are more likely to be oestrogen receptor negative and tamoxifen may be unhelpful as a treatment in this context (Phillips *et al.* 1999, Lakhani *et al.* 2002). However, it is not well understood whether these tumours start off oestrogen receptor negative or whether they lose their oestrogen receptor during development.

Should *BRCA1* and *BRCA2* mutation carriers use oral contraceptives?

The effect of oral contraceptive use in mutation carriers compared with controls has been assessed. Carriers of a *BRCA1* or *BRCA2* gene mutation who use oral contraceptives for five years have been shown to have a reduced risk of ovarian cancer of approximately 50% (Narod *et al.* 1998). In a retrospective, case/control study of 1311 female *BRCA1* or *BRCA2* mutation carriers, there was no association of oral contraceptive use and breast cancer risk in 330 *BRCA2* mutation carrying individuals (Narod *et al.* 2002). Amongst *BRCA1* carriers, there was an increased risk of breast cancer in those who had used oral contraceptives for at least five years (OR = 1.33; CI 1.11–1.60). Those who began using oral contraceptives after the age of 30 had no increased risk of breast cancer, whereas those who began using oral contraceptives before the age of 30 had an increased risk of breast cancer (OR = 1.29; CI 1.09–1.52). The risk was particularly associated with oral contraceptives used before 1975, which had higher oestrogen concentrations.

HRT has been shown to increase the risk of breast cancer in the general population (Beral *et al.* 2003). However, there is little evidence for the effect of using HRT in BRCA mutation carriers. These questions are being addressed in on-going studies such as EMBRACE (Evaluation of Mutant *BRCA* Carrier Epidemiology). The risk of HRT inducing breast cancer in individuals with a first-degree relative with breast cancer has been assessed (Collaborative Group on Hormonal Factors in Breast Cancer 2001). In those taking HRT, the risk of breast cancer appeared to be greater than the age matched controls but this was not statistically significant.

Ovarian screening

The advent of the CA 125 serum tumour marker measurements and improvements in pelvic ultrasound, along with newer techniques of colour Doppler imaging of ovarian vessels, have led some to advocate the use of these modalities together with bimanual vaginal pelvic examination in the attempt to detect early-stage ovarian cancer. However this remains controversial due to the lack of randomised, controlled evidence and has led to the National Institute of Health consensus statement, which did not advocate the widespread use of ovarian screening (National Institute of Health Consensus Statement 1994). Each of the above modalities, when used alone, would not have the required sensitivity or specificity to be a useful screening tool. A combination of two or three of the techniques is considered appropriate by many clinicians for use in high risk groups but there is no evidence to support this. There are several current studies addressing this issue including the Prostate, Lung Colorectal and Ovarian Cancer Screening Trial (PLCO), the United Kingdom Collaborative Trial for Ovarian Cancer Screening (UKCTOCS) and the United Kingdom Familial Ovarian Cancer Study (UKFOCS).

Prognosis in *BRCA1* and *BRCA2* mutation carriers

There is conflicting evidence regarding the prognosis of cancers in mutation carriers (Kennedy *et al.* 2002). Five-year survival post diagnosis in *BRCA1* mutation carriers with infiltrating breast cancer has been found to be 91% for non-carriers and 63% for *BRCA1* mutation carriers (Moller *et al.* 2002). A poor five-year survival rate in *BRCA* mutation carriers was also suggested by a study from Stoppa-Lyonnet (Stoppa-Lyonnet *et al.* 2000), although other studies have failed to confirm this (Kennedy *et al.* 2002).

The risk of contralateral breast cancer in *BRCA1* and *BRCA2* mutation carriers (Robson *et al*, 1999) has been shown to be increased. A study demonstrated that the contralateral breast cancer risk in mutation carriers was 16.9% five years post diagnosis and 29.5% at ten years (Narod *et al.* 2000). This was significantly decreased with tamoxifen treatment in *BRCA1* carriers (odds ratio 0.38) or with an oophorectomy in *BRCA* mutation carriers under the age of 50 (odds ratio 0.42).

Treatment for cancer in *BRCA1* and *BRCA2* mutation carriers

The oncological management of cancers in *BRCA1* and *BRCA2* mutation carriers is currently the same as for sporadic cases, except offering bilateral mastectomy at primary surgery because of the increased risk of contralateral breast cancer (Robson *et al.* 1999). Since *BRCA1* and *BRCA2* deficient cells repair DNA by error-prone mechanisms such as non-homologous end-joining rather than homologous repair, there is an opportunity to explore therapeutic measures using inhibitors of DNA end-joining (Ventikaraman *et al.* 2003). Breast cancers in gene carriers may be more sensitive to DNA cross-linking drugs, so a combination of cross-linking drugs and inhibitors of DNA end-joining could be synergistic. A trial of the DNA cross-linking agent, cisplatin, for cancer in carriers has been initiated, in view of the exquisite sensitivity of *BRCA2* mutant cells to cross-linking agents (Ashworth, 2003).

New work is being carried out into the use of breast tumour tissue micro-arrays, to try and predict the best treatment regimes for individual patients. An analysis of cDNA arrays (Chang *et al.* 2003) on 24 breast cancer core biopsies showed that they could predict response to taxanes on the basis of the expression of 92 cell cycle, apoptosis and cytoskeletal proteins with a positive predictive value of 92%.

The future

There are still many questions that need to be addressed in the genetics of breast cancer. How specifically are the *BRCA1* and *BRCA2* genes involved in the pathogenesis of breast cancer? What other genes are involved and how important are they? How should individuals at increased risk be identified and managed? Further knowledge of the processes involved in tumorigenesis in mutation carriers may allow better tailoring of chemotherapy and chemoprevention.

Dedication by Shirley Hodgson

George was a good friend of my father, Lionel Penrose, and they worked together fruitfully at a time when human genetics was rapidly developing, exchanging ideas and enthusiasm … George has gone on to develop his work with great success; Lionel would have liked to add something to this, with words of appreciation for this friendship and collaboration.

Genetics in cancer epidemiology at the U.S. National Cancer Institute (NCI): how it began

Robert W. Miller [*]

Scientist Emeritus, Clinical Genetics Branch, DCEG, National Cancer Institute,
EPS-7018, 9000 Rockville Pike, Bethesda, MD, 20892–7231, USA

In 1961 the position, Chief of the Epidemiology Branch, NCI, had been vacant for a year. Epidemiology of acute diseases was well established worldwide, but chronic-disease epidemiology was new, as described by Doll (2001a, b). I was prepared for this position by having trained in paediatrics, epidemiology and radiation effects with experience at the Atomic Bomb Casualty Commission in the study of late radiation effects among children (1956), and as Chief of Pediatrics for the Schull-Neel study of the effects of cousin-marriage on the health of Japanese children (1965).

At my interview, Michael B. Shimkin, Associate Director for Field Studies, greeted me by saying he did not want another nose-counter. I pledged that my studies would be on medical information not being explored by others. Here is a small portion of the subsequent history.

Wilms tumour

My first thought was based on a report by Krivit and Good (1957) of the concurrence of Down syndrome and leukaemia. I wondered what other childhood cancer might be linked to specific birth defects. Wilms tumour (WT) seemed to be a good bet – an embryonal neoplasm with a peak incidence early in life. My wife and I went to what was then called DC Children's Hospital to abstract information from the hospital records. Only a small number of children with WT had been diagnosed. One of them was described by the intern as having widely dilated pupils that did not react to light. The resident physician in ophthalmology diagnosed congenital aniridia (absent irises), a genetic disorder not widely known to paediatricians at that time. Joseph F. Fraumeni, Jr., who had just joined our Branch, Miriam Manning and I then went to five other hospitals where we found five more children with aniridia among 440 with WT (1964). The same study revealed excesses of hemihypertrophy, later linked to Beckwith-Wiedemann (overgrowth) syndrome, and various genitourinary malformations. The opening sentence of our report on WT set the course for our Branch: 'When diseases are found together more often than can be attributed to chance, each can be studied in the light of what is known of the other for clues to aetiology.' We found that hemihypertrophy was also linked to 1) adrenocortical carcinoma as described by Fraumeni and Miller (1967a) and 2) hepatoblastoma, as described by Fraumeni et al. (1968). The association of WT and aniridia was related to a deletion at chromosome 11p13, where the genes were contiguous. The approach, combining new information about the demography of childhood cancers plus multi-hospital review of medical records, was then applied to a series of childhood cancers told in part by Miller (1998).

---

[*] Sadly, Robert Miller died on 23 February 2006.

In 1998 we established the Astute Clinician Lecture given annually at NIH to encourage clinical observations that open new avenues of research. An independent committee selects the lecturer. The first one was J. Bruce Beckwith, cited for his advances in the pathology of WT (Beckwith *et al.* 1990), and for his delineation of the Beckwith-Wiedemann syndrome in 1964 (Beckwith 1963).

In 1972 Louise Strong began her research career as co-author with Alfred Knudson of their report on the two-hit hypothesis applied to WT. In 2003 she wrote on subsequent progress in defining its aetiology, which showed that the tumour is genetically heterogeneous, *WT1* is a tumour suppressor gene that accounts for only a small fraction of WT. The gene plays a key role, however, in urogenital development. Other genes have been identified in the genesis of the tumour with phenotypic differences in pathology, which are related to clinical course, response to treatment and prognosis. Most important in these studies was the establishment in 1969 of the National Wilms Tumor Study Group in the USA, and a comparable group within the International Society of Pediatric Oncology (SIOP) in Europe both of which registered large numbers of children with the tumour for clinical trials and special studies of the disease. Particular attention was given to pathology, as described by Breslow *et al.* (2003). The children continue to be followed into adulthood.

We had noticed that the frequency of WT in Japan was half that in the USA and Europe. Fukuzawa *et al.* (2004) have shown that among US white children, 24% of WT are related to perilobar nephrogenic rests vs only 1% of these tumours in east-Asian children. The absence of this histological type was linked to loss of *IGF2* imprinting. This is an example of the advances made by molecular biology once a genetic syndrome has been clinically identified.

Medical progress

Under this heading, my report on Cancer and Congenital Malformations appeared in the *New England Journal of Medicine* in 1966. In addition to our work, the clastogenic syndromes of Bloom and of Fanconi had just been linked to leukaemia. Case-reports were appearing of cancer in the index case with a genetic syndrome and the same cancer in a close relative who did not have clinical evidence of the syndrome; e.g., leukaemia in non-syndromic relatives of an index case with Fanconi syndrome (FA), as published by Garriga and Crosby (1959). In our review of cancer and congenital malformations, we noted that clusters of leukaemia *and* of congenital heart disease were reported in Niles, Illinois and Orange, Texas, as if the cause was both teratogenic and carcinogenic. These clusters appear now to have occurred by chance.

National childhood cancer death-certificate registry

We went on to study other childhood cancers using data from a registry we had created of death certificates for all cancer deaths from cancer under age 15 years in the U.S., 1960–1964, later updated through 1967. We created it at the suggestion of Brian MacMahon, Professor of Epidemiology at the Harvard School of Public Health. At that time it was easy to get copies of death certificates nationwide. The registry made possible a series of studies that revealed new information about the mortality of type-specific cancers. To make such studies we had to re-code the 22,000 diagnoses according to cell type instead of organ affected. The ICD at that time coded cancers by their anatomical sites, which grouped different cancers affecting the same organ. Thus, teratomas, for example, appeared under about a dozen code numbers.

Dalager and Miller (1974) depicted mortality rates for children by single year of age, which clearly indicated that some cancers originate *in utero*, and others have early childhood peaks that are muted when grouped by the usual 5-year age-intervals. Fraumeni and Miller (1969) studied cancers that were lethal in the first 28 days of age. There were 130, including 44 with leukaemia and 27 with neuroblastoma. In another study, we were able to separate cancers into their histological types, which revealed to Fraumeni and Glass (1970) that Ewing sarcoma is rare in blacks.

The death-certificate registry included home addresses that made possible a study of the time-space distribution of deaths from leukaemia in Los Angeles County, 1960–1964. Glass *et al.* (1968) found no excess of clusters according to a new durable statistical test invented by Nathan Mantel of the Biostatistics Branch, NCI. Sever and Miller (1971) matched the names of 6939 children in Maryland who had been given an experimental measles vaccine and found that none had died of cancer in 1960–1966.

Best of all, we entered the mother's maiden name into the file and could look for siblings who died of cancer by matching on the child's last name, the mother's maiden name and the home address. We added to the literature five more pairs of MZ twins who died of leukaemia before five years of age, a concordance not found in DZ twins. Later it was shown that the explanation was not genetic, but due to transplanted leukaemic cells: MZ twins shared their *in utero* circulation. Five pairs of siblings reported by Glass and Miller (1968) had died before two years of age of histiocytosis X, now called Langerhans-cell histiocytosis, which was not known to be familial or fatal so early in life. It turned up in our data because the code-makers had mistakenly given it a number in the cancer section of the ICD.

The death certificate registry also revealed that the peak for acute lymphocytic leukaemia was at four years of age in whites, but there was no rise at all in the frequency of deaths among blacks. As described by Fraumeni and Miller (1967b), a rise emerged among Japanese children in the 1960s, which followed the elevation in Great Britain in the 1920s and the U.S. in the 1940s. These patterns suggest that the environment was responsible for the peaks in Britain and the U.S., and that genetics accounts for the lack of a peak in blacks.

The most remarkable finding, reported by Miller (1968) was five pairs of siblings in which one child died of a brain tumour another died of soft tissue sarcoma or adrenocortical carcinoma. The following year this array of cancers was proposed as a new syndrome later called the Li-Fraumeni syndrome. Fred Li, who had joined our Branch in 1967, was at a dinner party seated next to a paediatrician who said there was a patient in the hospital with an unusual clustering of several types of cancer in the family. The next morning Fred went to the ward and obtained the family history. Li and Fraumeni (1969) then examined unpublished case-histories we had collected for children with rhabdomyosarcoma. They found three more similar families and described the cancer clusters as a possible new syndrome. At first they named it the breast cancer, soft-tissue sarcoma familial cancer syndrome. I got tired of the long name and called it the Li-Fraumeni syndrome in a paper by Duncan and Miller (1983), an eponym that took hold. Fred, who was later based at the Dana-Farber Cancer Center, spent some time in a molecular biology laboratory there to look for the gene responsible. As reported by Malkin *et al.* (1990), it turned out to be *p53* that was so important in oncogenesis and in maintaining the integrity of the cell cycle that it was selected by *Science* as the molecule of the year in 1993, as described by Culotta and Koshland.

Genetics ignored

Despite these advances, epidemiological studies of genetics in cancer aetiology were held in low regard at NCI and at other cancer research centres in the U.S. and abroad. NCI favoured studies of viruses until about 1972 as detailed in an unpublished book by Baker (2004). In Europe the International Agency for Research on Cancer (IARC) focused almost exclusively on chemical carcinogenesis as exemplified by a volume on the causes of cancer edited by Tomatis *et al.* (1990).

The powers-that-were wanted studies of the most frequent cancers – of adults, not children. Childhood neoplasia was not a public health problem, they said, because it accounted for only 5% of all cancer – and in any event, 'you can't fix genetics'. So, in 1975 the Branch was divided. Joe became the Chief of the Environmental Epidemiology Branch and I headed the Clinical Epidemiology Branch.

Joe and his associates had created a series of atlases on cancer mortality rates for the 3055 counties of the U.S., 1950–1969, which, as summarised by Hoover *et al.* (1975), showed high rates in certain areas which proved to be due to environmental agents, such as bladder cancer at a chemical factory, mesothelioma from asbestos exposure at shipyards during World War II, melanomas in the South, and in the Southeast, oral cancers from dipping snuff among rural women. After 1950–1969 the data were published for comparison at intervals through 1994.

In 1977 I was a member of a National Academy of Sciences Cancer Delegation to the People's Republic of China, where we learned that the Chinese had been making the same sort of cancer maps (Miller 1979). Soon after, Fred Li went to Beijing to help prepare the atlas for publication. It is amazing that the same idea had been implemented on opposite sides of the Earth. In China barefoot doctors had collected the data door-to-door during the Cultural Revolution when research was forbidden.

For 18 years our work at NCI was largely selected by us and approved by an internal committee. Grantees noted that their grant applications were reviewed by expert committees, and project site-visits were made to check on them. They complained that the intramural staff at NIH should do the same. Thereafter, every four years at our project site-visits we were criticised for not focusing on tests of hypotheses.

Books

Branch members convened several first-of-their-kind meetings and published the proceedings: *People at Unusually High Risk of Cancer* (1975) edited by Fraumeni; *Genetics of Human Cancer* edited by Mulvihill, Miller and Fraumeni (1977); and *Neurofibromatosis* edited by Riccardi and Mulvihill (1981). Schottenfeld and Fraumeni edited the monumental *Cancer Etiology and Prevention* now a classic in its third edition. In 1988 Miller and colleagues co-organised a Princess Takamatsu Symposium in Tokyo on *Unusual Occurrences as Clues to Cancer Etiology*, which brought together speakers who had made such observations.

Overview

Malformation syndromes had been described years ago by clinicians whose names live on because of eponyms, such as Down, Fanconi, von Recklinghausen. We added specific cancers to long known, as well as recently defined syndromes. These syndromes were waiting to be studied when molecular biology and other laboratory sciences developed the methods necessary to do so.

As a result, the genes involved were isolated and studied for their actions in oncogenesis and in health; *p53, WT1, RB1* and *NF1* and others were related to our studies. They are involved in the development of cancers of the bone, lung and breast among other cancers of adults. In 1960 very little was known about the somatic-cell genetic origins of cancer. Now it has taken an 800-page book edited by Vogelstein and Kinzler (2002) to tell what has been discovered since then. It is ironic that our work was out of favour because genetics of childhood cancer and malformation syndromes were thought not to be of public health significance. Fraser, who is honoured by this volume celebrating his 75th birthday, should be applauded for bringing attention to William Harvey's advice that is reproduced in different languages beginning on p. xii. It is too poetic to be paraphrased, but in essence he wrote that Nature may best reveal her 'secret mysteries' when she goes off the beaten path – and clinical practice will be advanced by study of rarer diseases. It is as apposite today as when Harvey wrote it shortly before his death in 1657.

George Fraser and the cancer genetic clinic

Victoria A. Murday

Consultant Clinical Geneticist, Department of Medical Genetics, Duncan Guthrie Institute,
Yorkhill Division, Dalnair Street, Glasgow G3 8SJ UK

I first met George when I was a research fellow at the Imperial Cancer Research Fund (ICRF) in London twenty years or so ago. Since then I have always enjoyed meeting up with him, and love to receive his cards, papers and notes which contain all manner of quotes, little bits of Greek, and so on. It amazes me that someone can have such varied knowledge and so many skills. Most of all, however, I enjoy his kindness, concern for others, sense of humour and smile.

In the early 1980s Dr Joan Slack, who was providing a clinical genetic input to St Mark's Hospital in London, took me on as a trainee. Over a period of a couple of years, she started cancer genetic clinics to provide genetic counselling and offer screening to those at risk of common cancers, based on their family histories. Following this pioneering effort, a number of clinics were opened, many of them financed with money from charities such as the ICRF. Finding geneticists interested in this sort of work was not easy at the time; very little was known about the risks and how to manage the families, and there was nothing to offer in terms of genetic testing. George was interested and was put in charge of the clinic in Oxford, and so it was that we got to know each other, through attending the same meetings and struggling over the same sets of clinical problems and issues. I remember all our encounters with pleasure. He was a breath of fresh air, interested in the subject both for its academic and compassionate elements, with little by way of personal ambition.

The cancer genetic clinics were criticised by many; they argued that we were creating anxiety, had no answers for people and had no proof that any of the screening would be of any use. Many doctors at the time had a strong belief that common cancers did not have an inherited component to their aetiology. I remember one of the consultants at St. Mark's saying the work we were doing calculating risks from family studies was pseudo-science! Most clinical geneticists thought cancer genetics uninteresting and irrelevant to the practice of clinical genetics, with the exception of the rare cancer syndromes such as familial polyposis and multiple endocrine neoplasia. The climate was therefore not entirely unchallenging and we trod with care. What we found was extremely exciting at the time. We demonstrated that, far from creating anxiety, people felt better after the clinics, and that the risks of neoplasia that we calculated were real. One of the more revealing and concerning elements was the amount of underlying hidden distress in families and how being able to discuss it helped to alleviate this. Each premature death from cancer recalls the previous ones in the family, and in many cases results in exceptionally tragic periods of bereavement.

Concrete data were thin on the ground, and George has this way of challenging one's over-confidence and certainty about many of the issues with such good humour. He anticipated many of the ongoing problems we have had with managing families. Talking to him would lead to me to ponder on my practice, and I hope I am a better doctor for it. I remember him struggling over a large ovarian cancer family, in which many women had died and women at risk were having prophylactic oophorectomies. '… but Vicky' he said to me, 'what do you *say* to these women, and what is going to happen in the next generation?' There is no answer, and the problem still exists for most of the families we see, despite *BRCA1* and *BRCA2* and mismatch repair and the rest. Twenty years on from our clinics opening we still have no proper data from randomised screening trials, we know very little more than we did then as to who should be screened, at what age and using which method. On the other hand the molecular basis of many of the single gene neoplastic disorders has been identified, and many of these studies would not have been possible without the backup of the cancer genetic clinics. The families tend to be very highly motivated and have provided research material for many groups for the mapping and cloning of the disease susceptibility genes. They continue to provide epidemiological information and, in addition, are a willing group for prevention studies. We know much more about the risks to individuals in families with an identified gene mutation.

I do firmly believe that families benefit from their visits to the clinics and there are many studies to back this up. I know from colleagues in Oxford that George was greatly appreciated by the families for his help and at the end of the day what can be more important?

He loved the many letters he received from grateful patients with whom he had spent an hour or two during the counselling process. He has allowed me to quote from one of them, written in 1990 very soon after he had opened the doors of the Oxford Cancer Genetic Clinic to patients:

> I felt extremely comfortable talking to you … by the end of my visit, I realised that my fears of developing cancer were quite insignificant when compared with other patients. However, you realised just how big a fear I have of cancer and did not think me any less important, and for this I thank you … My sister's death has obviously triggered off my latest worries but you have put my fears and thoughts in perspective and I found great comfort in our talk. I just hope that many of your other patients find similar peace and understanding after talking with you. It is nice to know that you are there to talk to if the need arises.

Genetic deafness and blindness: the contribution of George Fraser and his eponymous colleagues

Peter Beighton

Department of Human Genetics, University of Cape Town Medical School,
7925 Observatory, Cape Town, South Africa

Introduction

George Fraser made major contributions to clinical genetics with his seminal investigations of the causes of blindness and profound deafness in childhood. His books on these subjects, published in 1967 and 1976 respectively, were ahead of their time and together, with his related articles, underpinned future developments in the understanding of these categories of serious disorders.

Several of the genetic conditions mentioned in Fraser's books and articles had previously been documented in the medical literature, although the reports were scanty and often sketchy. In the fashion of the times, some of these conditions had eponymous titles, while in others, Fraser himself established the eponym in his own publications.

Fraser was personally acquainted with certain eponymous authors in the fields of childhood blindness and profound deafness. Some were his collaborators, and he knew others by reputation; all of them had a significant influence on his life and academic contributions. It is fitting, therefore, for the historical record and for the nostalgia inherent in a *liber amicorum*, that Fraser's contributions and brief biographical data pertaining to his eponymous colleagues should form the subject of this chapter.

Deafness and goitre: Pendred syndrome

(Vaughan Pendred 1869–1946)

At the outset of his academic career, Fraser (1959b) gave a presentation on four cases of sporadic goitre and congenital deafness at the Royal Society of Medicine, London. In the following year he published on the same subject (Fraser *et al.* 1960) and subsequently participated in the Fourth International Goitre Conference. His collaborative article in the Congress Transactions identified sporadic goitre with congenital deafness as Pendred's Syndrome (Fraser *et al.* 1961b).

In the years that followed, Fraser published several additional articles on the genetics of thyroid disease and/or deafness (Fraser 1962a, 1963b). He (Fraser 1965a) wrote an account of his investigations in 207 families with profound deafness, including several with goitre, and formally acknowledged Pendred's priority in the nomenclature of the syndrome. Three decades later, Fraser was a co-author in an article which again employed Pendred's eponym (Gausden *et al.* 1997a). Fraser's work on the Pendred syndrome formed the basis of his PhD thesis in Human Genetics successfully submitted in 1960 to the Faculty of Science of the University of London. The title of his thesis was 'Deafness with Goitre (Syndrome of Pendred) and some Related

Aspects of Thyroid Disease' and his supervisor was Professor L.S. Penrose.

Vaughan Pendred (1869–1946) obtained the Fellowship of the Royal College of Surgeons and in 1896 he was appointed as assistant in otorhinolaryngology at the Newcastle-on-Tyne infirmary. He encountered two sisters with large goitres and deaf-mutism; intrigued by this syndromic association, he published an account, remarkable for its pertinence and brevity (Pendred 1896). Pendred obtained a doctorate at the University of Durham in 1901 but renounced academia and entered general practice in Surrey where he had a long and successful career.

Previous reports of goitre and deafness had emanated from Switzerland, Mexico and Ireland but Pendred's report in the English literature remained unique until 1927, when Brain documented five affected sibships in London and postulated autosomal recessive inheritance.

Clinical Notes:
MEDICAL, SURGICAL, OBSTETRICAL, AND THERAPEUTICAL.

———◆———

DEAF-MUTISM AND GOITRE.
BY VAUGHAN PENDRED, M.R.C.S. ENG., L.R.C.P. LOND.,
LATE HOUSE SURGEON TO GUY'S HOSPITAL.

———

THE curious association of deaf-mutism and goitre occuring in two members of a large family has induced me to record these cases. Why this association? Perhaps some readers of THE LANCET may be able to throw some light on the cause of this combination of diseases : Absence of thyroid—cretinism ; overgrowth of thyroid—deaf-mutism. I append the family history as recounted to me by the mother. The family is an Irish one, and the parents have been upwards of forty years resident in Durham. The father, aged sixty-six years, and the mother, aged sixty - seven years, are alive and healthy. They have had ten children, five sons and five daughters. In an epidemic of small-pox twenty-five years ago the whole family was attacked with the exception of the younger of the deaf-mutes, and four males and one female died, although all had been vaccinated, and that recently, as they were children. The remaining son and two of the daughters are healthy and vigorous. The first goitre case is the first-born of the family— a spare woman now aged thirty-eight years. She is deaf and can only mumble indistinctly; little care has been taken to educate her and so she is imbecile. The goitre is a large multilobular hard tumour, the greater part on the right side of the neck ; from time to time she suffers from dyspnœic attacks. The growth was first observed after the small-pox—i.e., at thirteen years of age. The second surviving girl is now aged twenty-eight years, and is the fifth of the family ; she is a small, spare, intelligent woman, her expression being in marked contrast to her sister's. She is not absolutely deaf and can mumble incoherently ; her education has been attended to with so much success that she has been " in service." The tumour is larger than in the other case, but is of the same character ; it has been growing for about fifteen years, and during the last year has caused both dyspnœa and dysphagia, which have become so urgent that I have sent her to-day to Newcastle Infirmary for operation. Durham.

Fig. II–1. An early description of the autosomal recessive syndrome which now bears his name by Dr. Vaughan Pendred; it appeared in *Lancet* 2, 532, 1896. (Reproduced by kind permission of the editor of the *Journal of Medical Genetics*, Fraser 1964c.)

Thirty years after the publication of Brain's article, George Fraser found the records of his patients at the London Hospital, and was able to contact three of the families who were living at their original addresses in the East End of London.

Congenital deafness with cardiac conduction abnormalities: Jervell and Lange-Nielsen syndrome

(Anton Jervell 1901–1987)

In 1964, Fraser, Froggatt and James (Fraser *et al.* 1964c) published their study of several families in which deaf children had experienced fainting attacks and sudden death. Electro-cardiographic investigations had revealed abnormalities of cardiac conduction, which were evidently an important aspect of the pathological process. Later in the same year Fraser and his colleagues published an account of the autosomal recessive transmission of this disorder, employing the eponym 'Jervell and Lange-Nielsen' in the title of their article (Fraser *et al.* 1964i). They also reviewed the pathological changes in the syndrome using autopsy material (Friedmann *et al.*1966d, 1968h). See also Froggatt's contribution to this volume.

Anton Jervell (1901–87) was born and educated in Norway. After post-graduate medical experience in Paris, he obtained his doctorate with a thesis on electrocardiography. He became senior physician at Vestfol County Hospital, near Oslo in 1938, and was active in the Resistance during WWII. In 1956 he was appointed as Professor of Medicine at Ulleval Hospital, Oslo, Norway, where he spent the remainder of his career. In 1957, with his colleague Lange-Nielsen, he published an account of 4 siblings with perceptive deafness, fainting spells and ECG abnormalities.

Jervell remained active in retirement, and in 1983 at the age of 82 years, he addressed a medical society in Oslo, reviewing the syndrome which bears his name.

Treacher Collins syndrome

(Edward Treacher Collins 1862–1932)

The Treacher Collins syndrome is one of the most common heritable conditions in which profound childhood deafness is a component, and Fraser encountered this disorder during his extensive surveys in this field. Treacher Collins died in 1932, the year George Fraser was born, so that his eponym was well established when Fraser commenced his studies. The condition is mentioned in several of Fraser's publications (Fraser 1962e, 1964g).

Edward Treacher Collins was born in 1862 in London, where his father was a medical practitioner. He qualified in medicine at the Middlesex Hospital and followed his elder brother into Ophthalmology, spending the whole of his career at the Royal Eye Hospital, Moorfields, London. In 1900, Treacher Collins documented the association of colobomata of the eyelids and malar hypoplasia, together with malformation of the auricles and variable deafness. These abnormalities constitute the syndrome which bears his name. By the time of his death, Treacher Collins had been associated with Moorfields Hospital for 48 years.

Klein-Waardenburg syndrome

(David Klein 1908–1993; Petrus Waardenburg 1886–1979)
The Klein-Waardenburg syndrome comprises perceptive deafness, a white forelock, heterochromia of the irides and patchy dermal depigmentation. This autosomal dominant disorder is relatively common, and the striking appearance of affected children was noted on several occasions by Fraser during his surveys of profound childhood deafness (Fraser 1964m, 1969e, 1976). The condition had been delineated by Waardenburg in 1951, and although there was controversy concerning eponymic priority, the conjoined eponym was well-established at the time of Fraser's investigations. Klein and Waardenburg collaborated in the genetic eye clinic in Geneva in the middle years of the last century. Fraser had personal contact with Klein, but despite his extensive travels he never met Waardenburg.

David Klein (1908–1993) was born in Falkau, in the Austro-Hungarian Empire. He was educated in Freiburg and qualified in medicine in 1934 at the University of Basel, Switzerland. Klein specialised in ophthalmology and held appointments in Geneva, receiving professorial status in Human Genetics in 1970. He had a distinguished academic career, publishing more than 300 articles and co-authoring a classical treatise 'Genetics and Ophthalmology' with Waardenburg and Franceschetti.

Petrus Waardenburg (1886–1979 was born in Nijeveen, Holland and qualified in medicine at the University of Utrecht. He practised as an ophthalmologist in Arnhem and followed his academic interests in genetic eye disease by collecting data on affected families. In 1932 he published his monograph *Das menschlichen Auge und seine Erbanlangen.* ('s-Gravenhage, Martinus Nijhoff, also published as *Bibliographia Genetica* 1932, VII), and in this book he made the prescient observation that the Down syndrome might be the consequence of a cytogenetic abnormality. This observation is remarkable for the time and its details are virtually unknown almost three-quarters of a century later. A translation from the original German was kindly made by Professor Ursula Mittwoch, a contributor to this volume, and is reproduced in the contribution of Oliver Penrose (p. 443).

Waardenburg was appointed to senior academic status at Leiden, and he continued with his academic interests for many years, publishing his final paper, on albinism, at the age of 84 years.

Goitre and Klinefelter syndrome

(Harry Klinefelter 1912–1990)
Fraser's investigations into the association of deafness and goitre led him into the wider field of thyroid disease. In this way, he encountered a male patient with hypogonadism and hypothyroidism. In 1963 Fraser published a case report on goitre in Klinefelter syndrome, and speculated on the nature of this association (Fraser 1963c).

The Klinefelter syndrome, which affects males, comprises hypogonadism, a Marfanoid habitus and mild intellectual impairment. The condition results from the presence of one or more supernumerary X chromosomes.

Harry F. Klinefelter (1912–1990) was born in Baltimore, Maryland, USA. He qualified in medicine at the Johns Hopkins Hospital in 1937 and after military service during WWII, returned to his alma mater, where he became associate professor of medicine in 1966. The genetics department at the Johns Hopkins Hospital, under Victor McKusick, was at the forefront of the rapid development of clinical genetics in the 1960s. Fraser was a visiting professor with Victor McKusick for six weeks during the summer of 1967 and was associated with him during his appointment at the National Library of Medicine from 1979 to 1984. Despite this proximity, Klinefelter and Fraser were not personally acquainted.

Ocular refraction, twin studies and statistics

(Arnold Sorsby 1902–1980)
George Fraser's interest in genetic ophthalmology extended from rare syndromes to common disorders of visual acuity. While in Seattle (1961–63), Fraser had seen an announcement that Sorsby was looking for someone to conduct a survey of blindness in children. He responded and worked with Sorsby from March 1964 until December 1966 in his Department of Research in Ophthalmology at the Royal College of Surgeons in London. Their collaboration included a statistical and genetical analysis of ocular refraction (Sorsby and Fraser 1964e, Sorsby *et al.* 1966). Sorsby started the Journal of Medical Genetics in 1964, and offered Fraser a place on the editorial board. Fraser wrote eight papers in the first three volumes of this journal (1964–6), including two on the genetics and statistics of ocular refraction, and, in volume 7, several years later, Fraser wrote, at the request of Sorsby, a review article entitled 'Genetical aspects of severe visual impairment in childhood based on his work while he had been in Sorsby's department' (Fraser 1970b).

The main achievement of Fraser's sojourn at the Royal College of Surgeons in London was a dissertation entitled *The Causes of Blindness in Childhood: a Study of 776 Children in Special Schools*. He was awarded the Raymond Horton-Smith Prize for the best MD thesis of the academic year 1965–66; a slightly revised version of this thesis was published in 1967 as a book under the title of *The Causes of Blindness in Childhood: a Study of 776 Children with Severe Visual Handicaps* by the Johns Hopkins Press (Fraser and Friedmann, 1967). During this time, Fraser also worked on various specific research projects including a major study of the Wolfram (DIDMOAD) syndrome, an autosomal recessive condition which involves diabetes insipidus, diabetes mellitus, visual handicap, and even blindness, due to optic atrophy, and moderate deafness (Fraser 1964o,q, Rose *et al.* 1966e). Long after he left the Royal College of Surgeons in 1966, Fraser's links with Sorsby were renewed in 1973 when he contributed a chapter entitled 'Syndromes' to the second edition of Sorsby's monograph *Clinical Genetics* (Fraser 1973).

Arnold Sorsby (1902–1980) was born in Poland and after emigrating to England with his family, graduated in medicine from the University of Leeds in 1921. He trained in ophthalmology, joined the staff of the Royal Eye Hospital, London, and became Dean in 1934. Sorsby's academic contributions received recognition and impetus by his appointment as research professor at the Royal College of Surgeons, London from 1942 until his retirement in 1966. His classical monograph *Genetics in Ophthalmology* first published in 1951 undoubtedly had a strong influence on the direction of Fraser's own research activities. Sorsby's name is used as the title of an autosomal dominant disorder of the retina, termed 'Sorsby fundal dystrophy'.

Visual handicap in skeletal dysplasias

(Pierre Maroteaux 1926-)

During the course of his investigations into syndromic associations of genetic eye disease, Fraser collaborated in the documentation of ocular problems in the dwarfing skeletal dysplasias. In the 1960s Pierre Maroteaux of Paris had made major contributions to the delineation of these skeletal disorders, and an article published with Fraser and other investigators represented the definitive account of visual defects in children with spondyloepiphyseal dysplasias (Fraser *et al.* 1969f).

Pierre Maroteaux was born in Versailles in 1926 and qualified in medicine at the University of Paris in 1952. His career culminated with his appointment as Director of Research at the Hôpital des Enfants Malades, Paris, where his major interest was the delineation and pathogenesis of the heritable skeletal dysplasias. Maroteaux has an international reputation as an originator and contributor in this field, in which his eponym is associated with acromesomelia and pseudoachondroplasia.

Hereditary progressive arthro-ophthalmopathy (Stickler syndrome)
Gunnar B. Stickler (1925-)

Fraser had noted the similarity of the cases of hereditary progressive arthro-ophthalmopathy originally described by Stickler and his collaborators in 1965 (Stickler *et al.* 1965), involving skeletal abnormalities and detached retinas, to a case reported earlier by David (1953), as well as to cases seen by himself (Fraser and Friedmann, 1967). Fraser wrote to Stickler from Adelaide and visited him at the Mayo Clinic in 1966. He pointed out that a hearing defect, previously unrecorded, is a part of this syndrome. Stickler performed the appropriate tests and agreed that a moderate hearing defect, previously unrecognised, formed part of the Stickler syndrome (Stickler and Pugh, 1967).

Gunnar Stickler was born in 1925 and attended the Wilhelmsgymnasium in Munich and studied medicine at the universities of Vienna, Erlangen and Munich. He graduated at Munich in 1949 and two years later emigrated to the USA. Working in the Mayo Clinic, University of Minnesota, he specialised in paediatrics, becoming professor at the Graduate School. In 1973 he was appointed professor and chairman of paediatrics at the Mayo Medical School. In 1960, a twelve year old boy was examined at the Mayo Foundation. The boy had bony enlargements of several joints and was extremely short sighted. His mother was totally blind. Dr Stickler discovered that there were other members of the family with similar symptoms; the first family member having been seen by Charles Mayo (1865–1939) in 1887. This prompted Dr Stickler to study the family. With colleagues he worked to define the condition, the results being published in June 1965. Dr Stickler tentatively named the condition hereditary progressive arthro-ophthalmopathy. Since the 1980s, this condition has been known world-wide as Stickler's syndrome.

Wildervanck syndromes
The Wildervanck (1952) syndrome consists of congenital perceptive deafness, Klippel-Feil anomaly (fused cervical vertebrae with torticollis), and abducens palsy with retractio bulbi (Stilling-Türk-Duane syndrome); cases are described in Fraser (1976). Wildervanck (1963)

also described a syndrome of congenital deafness and split hands and feet, observing the association in two sons of unrelated parents. Birch-Jensen (1949) mentioned a sporadic case of the association. Fraser (1976) saw a brother and sister with this combination. Wildervanck was a general practitioner and human geneticist in Groningen (The Netherlands). Fraser knew him well when working in Leiden in the early 1970s, and often discussed their common interest in deafness with him.

Hallermann-Streiff syndrome (François dyscephalic syndrome)
Jules François (1907–1984)
The features of this syndrome are bird-like facies with hypoplastic mandible and beaked nose, proportionate dwarfism, hypotrichosis, microphthalmia, and congenital cataract. Teeth may be present at birth. Fraser knew Jules François and himself described a case of this syndrome (1967, 1976).

Jules François studied medicine at the University of Louvain, graduating in 1930. He commenced a private practice as an ophthalmologist in Charleroi, while also conducting scientific work. This eventually led to his appointment as professor of ophthalmology at the University of Ghent, where he established a modern ophthalmological department. Jules François published more than 1000 articles. He was an officer of the Order of Leopold II and a Chevalier of the Légion d'Honneur.

Visual handicap and cytogenetics

(John H. Edwards 1928-)
Fraser's interest in the pathogenesis of visual handicap extended to chromosomal abnormalities in affected individuals and he addressed this problem in a multi-authored article (Fraser *et al.* 1970a). One of Fraser's collaborators was John Edwards.

John Edwards was born in 1928 and qualified in medicine at the University of Cambridge. He served as a medical officer in the Falkland Islands dependency survey, and on his return to England he was appointed as lecturer in social medicine at the University of Birmingham. His interests turned to genetics, and in 1958 he joined the MRC Population Genetics Unit, Oxford. Between October 1959 and November 1961 Fraser was also associated with this Unit. Edwards subsequently became Professor of Human Genetics at Birmingham University, before moving to Oxford University in a similar capacity. He was elected FRS in 1979. Fraser was again associated with Edwards at Oxford between 1984 and 1997. The Edwards syndrome or trisomy 18 is a common cytogenetic disorder, which is potentially lethal in the neonate.

Turner syndrome

(Henry Turner 1892–1970; Nikolai Adolphovich Schereschewsky 1885–1961)
Fraser's investigations into deafness and visual handicap involved possible cytogenetic factors, and he had collaborated with Edwards and had been familiar with Klinefelter's work; both these investigators had documented chromosomal abnormalities which subsequently bore their name. In this context, Fraser addressed the issue of the eponymous title of the XO phenotype, arguing

the case for the conjoined format 'Schereschewsky-Turner syndrome' versus the conventional usage 'Turner syndrome' (Fraser 1979).

Henry H. Turner was born in 1892 Harrisburg, Illinois, USA and qualified in medicine at the University of Louisville in 1921. He trained in internal medicine and endocrinology, and was appointed Professor of Medicine at the University of Oklahoma. Turner was a founder of modern endocrinology and was involved in the development of journals, postgraduate education and research in this discipline. In 1938, Turner documented the clinical features of the condition which bears his name, although the cytogenetic basis remained unrecognised until 1954. Turner continued with his clinical contributions during his retirement; he died in 1970 from carcinoma of the lung, which he himself diagnosed.

N.A. Schereschewsky (1895–1961) studied medicine at the Medical Faculty, Moscow University, graduating in 1911. In 1921 he gave a lecture course on endocrinology. In 1925, he first documented the classical phenotype of Schereschewsky-Turner syndrome, in a woman aged 25 years. He worked as chief of the Federal Institute for Experimental Endocrinology (1932–1934) and as director of this Institute (until 1934). Schereschewsky published over 100 articles. His article describing the Schereschewsky-Turner syndrome was written in Russian in *Vestnik Endokrinologii* and entitled 'On the question of malformations in association with an endocrinopathy'; it was translated into French by Dr Gamaleia (*Revue Française d'Endocrinologie* 4 181 1926).

Pituitary dwarfism type III

(Ernst Hanhart 1891–1973)

Fraser met with Hanhart in order to discuss the latter's studies on deaf-mutism in Ayent and Anzère in the Canton Valais, Switzerland. During his surveys in isolated populations, Fraser (1962c, 1964h) studied the Hanhart form of pituitary dwarfism which is present on the island of Krk (Veglia) off the Adriatic cast of the former Yugoslavia.

After qualifying in medicine from the University of Zurich in 1916, Ernst Hanhart (1891–1973) worked as a country practitioner, but in 1921 became an assistant in the Zurich polyclinic. Working under professors Otto Nägeli (1871–1938) and Wilhelm Löffler (1887–1972), Hanhart became interested in human genetics and became a specialist in hereditary disorders.

Hanhart worked extensively on the effects of consanguinity and investigated Swiss families with inherited disorders in isolated communities. He was appointed professor at the University of Zurich in 1942 and was a founding member of the Swiss Society of Genetics. Hanhart achieved an international reputation, but in 1954 he had to resign after having been crippled in a criminal attack.

Comment

George Fraser achieved eponymic immortality with his description in 1962 of two sets of siblings with a disorder comprising cryptophthalmos, and variable additional defects, including skeletal, auricular and genital abnormalities. The Fraser syndrome, which is an autosomal recessive trait, is well known in the fields of ophthalmology and medical genetics. In view of his monumental contributions, it is fitting that Fraser's name should be preserved for posterity in this way. In this context, Fraser stated (personal communication to PB):

> I first met MS, a girl with cryptophthalmos at the age of five years when she was deaf and blind: she was behaving like a wild animal, reacting with a frightened scream to any attempted approach. I noticed that she liked to press loudly ticking clocks to her skull, and suspecting that her deafness was largely conductive, I took her to an otorhinolaryngologist of my acquaintance. He reconstructed her malformed outer and middle ears, and at the age of six years, she heard speech for the first time; more than four decades later, she lives a life which has been a little more tolerable than it would have been without any hearing. Perhaps this small contribution to the improvement in the quality of life of this girl represents a greater achievement than that of my name becoming attached to a syndrome and, in part, to a locus, a gene, and a protein, whether in man or in mouse or in any other living creature.

Acknowledgements

I am grateful to my wife, Greta, my research collaborator and co-author of 40 years standing, for her role in the preparation of the manuscript. Biographical information concerning Fraser's eponymous colleagues is derived from our books *The Man Behind the Syndrome* and *The Person Behind the Syndrome*, published by Springer-Verlag in 1986 and 1997, respectively.

The cardio-auditory syndrome of Jervell and Lange-Nielsen: the anatomy of a population study

Peter Froggatt

3 Strangford Avenue, Belfast BT9 6PG, United Kingdom

Prologue

Many of the contributors to this Festschrift will pay tribute to George Fraser's exceptional talents and seminal researches. Forty-five years ago George and I were brought into a serendipitous collaboration on a pioneering project which was to contribute fundamentally to defining the then rare and arcane cardio-auditory syndrome, now a component of the heterogeneous entity 'the long QT syndromes' (or 'LQTS') which are much in vogue, and which was also to lead to an enduring friendship. This project as Holmes might say 'shows certain points of interest', not all of which could (or should!) have been covered in George's book (Fraser 1976, Ch. III). The following less taut but not entirely unexpurgated version accordingly attempts to illuminate some unlit corners in the published descriptions and so illustrate in microcosm many of George's remarkable and varied credentials. Many of his intellectual ones are well known to a wide audience but some of the others are not, and if I have been able to refer to these even in lighter vein I will be well pleased. I hope George will be well pleased also.

Genesis

In April 1960 Alan Stevenson, my one-time department head at Queen's University Belfast and from 1958 to 1974 Director of the Medical Research Council Population Genetics Research Unit housed at Warneford Hospital in Oxford, telephoned to say that one of his 'young men' was interested in a recently-described (in 1957) syndrome in children who manifested profound perceptive (sensorineural) congenital deafness, fainting attacks of unknown cause, these being frequently fatal, and a prolonged QT interval and other ECG anomalies (Jervell and Lange-Nielsen 1957) – 'screwball' tracings in the later words of the Ann Arbor cardiologist F. P. Wilson (Levine and Woodward 1958). Would I 'nip down' to the Ulster School for the Deaf and Blind and take ECGs on any deaf pupils who had fits, faints or 'turns'? Down I duly nipped and discovered a 12-year-old girl (Irene W.) with well-established fainting fits and the characteristic ECG (Figure 1). This news fairly catapulted George to Belfast and on 11 May we examined the girl and her immediate family in their house in Banbridge, County Down. I arranged for her admission (during August) to the Royal Victoria Hospital Belfast for full assessment. There were no significant abnormal findings other than the presenting signs and symptoms, the bizarre ECG, and a mild hypochromic anaemia. Standard therapies aimed to ameliorate the ECG anomalies were ineffective.

During the next twelve months of close observation her fainting attacks were seen to be clearly related to physical exertion and, especially, to anxiety. They varied in severity and included a near-fatal one triggered by her first visit to her new residential school. She died suddenly during an attack on 5 September 1961 on her first day of term in her new school after the long summer break. Autopsies of the ear and the heart (including the cardiac conduction system) were conducted by noted authorities (respectively, Dr A. I. Friedmann, The Institute of Laryngology and Otology, University of London; and Dr Thomas N. James, then at the Henry Ford Hospital, Detroit). The latter's findings allowed speculation that the 'fainting attacks' were Stokes-Adams-type syncope and that death was probably due to a paroxysmal arrhythmia caused by a conduction disturbance affecting a ventricular myocardium vulnerable on account of its delayed repolarisation time as indicated by the prolonged QT interval. Full descriptions of our cases and details of our surveys are in Fraser *et al.* (1964b), and Fraser *et al.* (1964i).

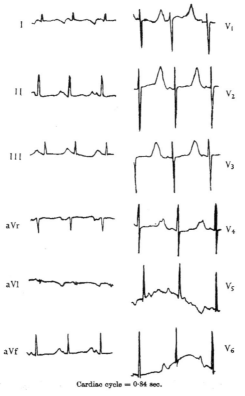

Cardiac cycle = 0·84 sec.

Figure 1. ECG (August 1960) of Irene W aged 12 years. Note the extreme prolongation of the QT interval (indicating delayed ventricular repolarisation) such that the T-wave may be confluent with the following P-wave, and other TU-wave anomalies. (Reproduced from *The Quarterly Journal of Medicine* **33** 363 (1964) by courtesy of Oxford University Press).

Procedure

After we had assessed Irene W. in the autumn of 1960 the question at once arose – what should we do now? George had no doubt: we had to conduct a conventional population genetics survey which would be buoyed by the added incentive of illuminating the mechanism of the often lethal fainting attacks and of saving young life. On this he brooked no argument: this was our medical and moral duty; nothing less would be acceptable. I was less persuaded. I was heavily committed in university and health service duties and to other completely unrelated lines of research (e.g. Cresswell and Froggatt 1963). Furthermore, I felt under no compelling or immediate obligation; I had become involved in the first place simply by helping a researcher from Oxford at the request of my former boss. Three factors finally persuaded me. The first was, quite simply, George's impressive enthusiasm, outstanding intellect, and high probity. I could learn a lot about many things from him. The second was his appeal through precept and example to the more lofty instincts which lurk often well-hidden in anyone brazen enough to call himself or herself a caring scientist. The third, rather less altruistically, was that at the time my research thoughts were turning to include increasingly the looming challenge of sudden unexpected death in infancy ('cot death' or 'crib death') about which little definite was then known but this did not prevent its attracting many speculative though no convincingly coherent theories. Now providence (or serendipity) might allow me to add an aberrant cardiac conduction hypothesis to the rapidly lengthening list! Truly there was a lot of perhaps vital information to be mined from this cardio-auditory syndrome. My reasons may have been less altruistic than George's but it was with genuine interest that I agreed to collaborate, and we planned accordingly.

Based on the known symptomatology, prolongation of the QT interval with (or much less commonly without) a history of fainting attacks among children with congenital perceptive deafness would, strictly, be the diagnostic desideratum. All congenitally deaf children therefore comprised the original survey groups (Ascertainment Method A). As the field-work progressed, however, we relaxed the inclusion criterion to congenitally deaf children (as before) but now confined to those *who already had a history of fits, faints, or 'turns'* (Ascertainment Method B). This was not perhaps impeccably rigorous, but on the then known facts it was reasonable and certainly an attractive expedient; indeed with our limited resources it turned out to be necessary. And our resources certainly were limited, so limited in fact that no funds were available to George to purchase a portable electrocardiograph (ECG) machine. Instead, a machine was constructed in house at (inevitably) greater cost than a commercial equivalent! After the resulting six-month delay, George set off enthusiastically for a school for the deaf where he managed to obtain a few dozen ECG tracings before the machine broke down! Moreover, fully a quarter of these tracings showed gross prolongation of the QT interval – initially fuel for his enthusiasm but which turned out to be a machine-made artefact. Many would there and then have abandoned the study in frustration, but not George, who is nothing if not persistent. He managed to persuade a sympathetic salesman for a medical electronics company to lend him a portable demonstration model ECG machine on 'extended trial', in the event until George's employers in Oxford could afford to buy him a Cambridge Direct Writer ECG.

With his signature drive, enthusiasm, thoroughness and intellectual scruple, George started with Method A taking ECG tracings on 534 boys and 395 girls (some 80% of the enrolments and considered relevantly unbiased) in eight schools for the deaf in England; and using Method B he took tracings on 36 (18 boys and 18 girls) of the 1506 children on the rolls of a further 14 schools for the deaf in England, Wales and Scotland. Together with some other, *ad hoc* ascertainment procedures, probably up to half of the deaf children in Britain as well as many deaf adults were screened. In less populous Ireland we used Method A and obtained ECG tracings from 262 boys and 207 girls (again some 80% of the enrolments and again considered to be relevantly unbiased) at all three schools for the deaf in Northern Ireland and the Republic. We also scrutinised records of deaf adults in Northern Ireland from a previous survey (Stevenson and Cheeseman 1956)

The two schools in the Republic were segregated on gender and run by Religious (Roman Catholic) Orders. They were both in Cabra in North Dublin, and we surveyed their pupils in the spring of 1961. The Dominican school gave a wide, academically-based education; indeed, the Reverend Mother (Mother Mary Nicholas) had a well-justified reputation for having devised a system of single-hand language as an aid to lip-reading. The Christian Brothers (under Brother Esmonde) combined sound, basic education with vocational orientation aimed to prepare their pupils to be self-sufficient in the traditional Irish, mainly rural, economy.

We spent two exemplary days with the nuns followed by two with the Brothers, the latter spiced-up more that somewhat by a generous ration of that lavish hospitality which combines a traditional Irish one (for which my compatriots are justly famous) with the gastronomic *largesse*, always shown by the Brothers to 'travellers', to form, quite literally, a heady brew! This was beyond the experience and perhaps also the comprehension of even George's impressively wide cosmopolitan culture, which had not however previously had a chance to factor-in the *moeurs* of certain Irish Religious Orders! His usual social impeccability was severely tested when Brother Esmonde placed an imposing half-tumbler of John Jameson 10-Year-Old Irish whiskey in front of him as a welcoming lunch-time *aperitif.* By an ingenious subterfuge and some adroit legerde-main which would have done credit to Maskelyne, the ever temperate and resourceful George disposed of his would-be *aperitif* to the doubtful advantage of a nearby potted plant without seemingly being detected, still less inviting comment, least of all causing offence. (I had requested sherry which I knew from previous encounters that the Brothers served in traditional if generous sherry glasses!). After this we tackled heartily (George is an able trencherman) the largest helping of roast beef and trimmings which I have ever seen seriously offered outside some scene of osten-tatious gluttony such as cartooned by Gillray, Cruikshank or Rowlandson! This, or rather these, were followed by a quarter each of a massive rhubarb tart, and the memorable meal was com-pleted by coffee (mercifully unlaced!) before we struggled back, replete *a fortiori*, to the less indul-gent discipline of the Cambridge Direct Writer ECG machine. Something similar was repeated next day (Brother Esmonde said that he had been impressed with our activities leaving it unclear as to whether they were of the researcher or the gourmand!) and we finally but success-fully returned to Belfast contemplating the perils as well as the challenges facing the geneticist in the field. Despite these perils our efforts in Britain and Ireland had in all yielded six sibships with seven propositi and two secondary cases, nine unequivocal cases in all; and as further analysis was to show (Fraser *et al.* 1964i) up to ten further possible cases (in eight sibships), 'possible' in the sense that they could not be confidently categorised on the clinical, family and ECG findings.

Problems (operational)

From the very outset we had been puzzled as to why, in such a bizarre and dramatic condition, the first cases (four in one sibship, three of whom died suddenly) had not been reported in the literature until as recently as 1957 (Jervell and Lange-Nielsen 1957). Surveys of profound childhood deafness over the previous century had uncovered numerous examples of other reces-sively inherited pleiotropic conditions e.g. Pendred syndrome (deafness with goitre) and Usher syndrome (deafness with retinitis pigmentosa). ECG tracings were then of course rarely if ever available and death, even sudden death, in childhood was less remarkable. Furthermore, parental consanguinity was reported in only two of our (six) sibships and in neither of the two in the lit-erature. These facts indicated that the condition could be more common than we had at first supposed and a literature search for further cases could be rewarding. It was, George argued, a scientific and professional imperative. This time I didn't dare to dissent!

We started a literature quest more or less contemporaneous with the fieldwork but to a less disciplined design and time-frame. 'We' flatters my role; it was conducted mostly by George because of his readier access to extensive library stocks (being in Oxford until late 1961, then Seattle until 1963 and then London), but mainly because of his remarkable linguistic facility with complete fluency in several main-stream languages and a high-grade working proficiency in many others, a formidable armoury especially when linked to his phenomenal memory. Hours spent in library stacks and dusty basements (this was in pre-MEDLARS days) had, however, by 1963 uncovered very few additional 'possible' or 'probable' cases (e.g. Morquio 1901, Henning 1928, Danish *et al.* 1963), but they had also uncovered a remarkable German family mentioned almost *en passant* in a book published in 1856. The proposita, from Pappendorf über Mittweida, tragically starred in one of the most dramatic descriptions in the entire literature of this condition and the circumstances, and George's researches into them, are worth recording.

F. L. Meissner, in his classic book *Taubstummheit und Taubstummenbildung* (Meissner 1856) in a passage to illustrate the admirable personality of the well-disciplined 'deaf-and-dumb' in the Leipzig Institute for the Deaf wrote (pp. 119–120) as follows:

> The most outstanding example of this latter feature [remorse] we witnessed in the Leipzig Institute [for the Deaf], in the case of a female deaf-and-dumb patient named Stein who … had apparently pilfered something inconsequential from another pupil … Director Dr Reich … a very fair man … summoned her to appear in front of all the other pupils to find out … what excuse or apology she could provide. When [she] appeared and saw that she was held to be guilty … and saw on the face of her esteemed teacher, naturally, a look of disapproval, she was – without having been in the slightest way punished – seized by such remorse and pain that she suddenly fell to the ground dead and we could do nothing to revive her … it seemed … the tragic result of a Divine judgement, a punishment warning them [the other pupils] never to stray from the path of virtue
>
> … we think that … there must have been some kind of predisposition … followed by an apoplectic fit … at least when Stein's parents were, as delicately as possible, informed of their daughter's unfortunate death they were not at all taken aback but responded directly that this news did not at all surprise them as similar incidents had already occurred in their family – one child had suddenly fallen down dead after a serious fright and another after an extreme fit of rage.

Here was an historical gem: emotional upset a clear trigger for a fatal attack as it has been in most recent unequivocal cases including our first subject (Irene W.). Nothing of course is recorded about the hearing status, age or sex of the unfortunate sibs of the equally unfortunate deaf girl, Stein, and the scent for the hunter after a century would surely be cold. Most researchers would simply have nodded knowingly and shaken their heads in disappointment over the tantalising passage (if they had ever found it!), and moved on. But not George; all his instincts were alerted to the challenge of discovering more about the Steins and he positively revelled in the anticipated excitement of the quest – Holmes's 'The game's afoot' – which is the *sine qua non* of the born researcher and which is such a passport to George's success. Culturally he was also attracted (possibly because of his family's provenance, George has a great interest in central European history and likes to debate the course and issues of World War I), and on 20 August 1965 after the Gregor Mendel Centenary Symposium held in Brno on 4 – 7 August, he bundled his wife and 16-month-old eldest daughter into his red Volkswagen 'Beetle' and the three set off, travelling on his two-day transit visa, to drive to Karl-Marx Stadt (now and originally Chemnitz) in the German Democratic Republic (DDR – 'East Germany') *en route* to the DDR/ Bundesrepublik (BRD – 'West Germany') crossing point at Marienborn-Helmstedt. In the event, however, they cut short their stay in Karl-Marx Stadt and at some risk illegally detoured to Pappendorf über Mittweida and contacted the Pastor of the Evangelical-Lutheran Church of St Wenceslas where the Stein family had worshipped all those years ago. They detoured again towards Berlin now restored as simple tourists, but were stopped and questioned by the Russian military for violating the provisions of the transit visa and could well have been in serious trouble but for George's skills in advocacy and the adequate command of Russian and German to display them. They were nevertheless somewhat fortunate to return safe and well to the BRD.

As with the hospitality of the Christian Brothers in Dublin, the inhospitality of the military in the DDR provided its own peculiar and certainly less congenial perils. But as in Dublin, braving the perils led to success: data from the Pastor in Pappendorf, the site visit, and information from the Director of the Museum of Studies of Deaf-Mutism in Leipzig crucially augmented Meissner's narrative cited above. The proposita in the Stein family was Amalie Friederike Stein, born 1824 and died (as described above) in 1835. Four of her seven sibs died in childhood, two from 'apoplexy' aged respectively two days and three years ('apoplectic fit' was the cause suggested by Meissner 1856 for the sudden death of Amalie), a third died at eight months from 'cough', and the fourth died at three years 'suddenly from tetany', all suggestive trigger mechanisms, be it noted, given the nosology of the day. Unfortunately their hearing status is unknown or unrecorded. The Muses of Learning were satisfied; no Furies would pursue George or haunt his conscience. A 'probably' affected family from the 1830s was now as well documented as it could possibly be.

The nature of the cardio-auditory syndrome requires ECG evidence from a class of children rarely if at all cardiologically surveyed (and then only recently) even if having undiagnosed faints or 'turns' – which among deaf children with their added communication difficulties are commonly ascribed to epilepsy or even psychological causes. Furthermore, the condition is inherited in an autosomal recessive manner. These facts could reasonably explain the delay in recognising examples of the syndrome, and probably to an extent they do. But it is not as simple as that: there has been a similar delay in recognising other 'prolonged QT interval' syndromes without deafness, often with fatal fainting attacks, and usually with a dominant mode of inheritance such as in the cases first fully described independently in Dublin by my compatriot and colleague Conor Ward in 1964 (Ward 1964, 2005) and in Italy by Cesare Romano (Romano *et al.* 1963, Romano 1965). Further discussion on this enigma would be interesting but inappropriate here. Significantly, in the decade (1964–1973) following our papers, 34 further cases of the cardio-auditory syndrome, and over 50 (by 1971) of other 'prolonged QT' syndromes (Kringlebach and Wennevold 1971) were described and many more have been since.

Problems (methodological)

In most cases of the cardio-auditory syndrome the QT prolongation is obvious on simple inspection of the ECG tracing. In some, however, it requires careful measurement and comparator norms: this is essential because a history of fainting attacks while usual is not invariable and moreover the *degree* of QT prolongation and the *frequency* (and severity) of the attacks are not always closely associated. We therefore had to consider whether the syndrome could be diagnosed (among the congenitally sensorineural deaf) *solely by an unequivocal QT interval prolongation.* We were now entering the realms of quantitative, not qualitative, discrimination, a very different (and difficult!) world. Firstly, we had to answer the question – what is a 'normal' QT interval for a given heart rate and age? Existing equations for this calculation were not strictly appropriate for our groups especially in an ambitious pioneer exercise and so we would have had to construct our own. Intellectually I accepted this but nevertheless I shrank from the time and labour involved (I had many departmental commitments) and so tried to persuade George that an existing formula from the literature might be acceptable. George however would have none of this (now based in Seattle he could take a detached operational as well as a detached scientific view!), was adamant that scientific rigour must be preserved and swept aside my excuses. I acquiesced; fortunately I had a very tolerant department head at the time, John Pemberton, still happily thriving at 94, to whom I owe many debts including this one. Eventually I obtained ECG tracings from 195 boys and 175 girls, of the 443 children born on 1 January or 1 July 1946–1955 attending Local Authority (Council) schools in Belfast, which we were able to show was a representative sample of the hearing, age-related population.

Regression equations for the 'predicted' QT interval (as calculated from the appropriate best-fitting quadratic equations) of the form:

$$\text{Predicted QT} = a + b\,(RR) + c\,(RR)^2 + d\,(\text{age in years})$$

where RR is the length of the cardiac cycle in seconds and a, b, c and d are constants, were constructed for each sex for the Irish and the English deaf children surveyed by Method A (as previously described above) and for the two hearing groups, six groups in all. We then calculated for each subject in each group a value for

$$\text{'z'} = \frac{\text{actual QT} - \text{predicted QT}}{\text{standard error of predicted QT}}$$

which in a random sample of the population would be expected to be normally distributed with zero mean, unit variance, and 2.5% falling respectively above +1.96 and below -1.96. This was so for the hearing groups but not for the deaf groups three of which were positively skewed *even after the (four) unequivocal cases of the syndrome ascertained by Method A were excluded.* Forty-eight of these remaining deaf subjects were identified for screening solely on the statistic 'z' > 1.96, but after further rigorous clinical, statistical and family examination, 38 of these could be excluded: as we wrote at the time 'Of the remaining [ten] subjects some may be of the genotype; without a history of fainting attacks it was impossible to make a firm diagnosis' (Fraser *et al.* 1964i). Further ECG measurements and calculations were made but while they added impressively to our statistical and methodological sophistication they did not add to our knowledge of the condition.

Epilogue

Our results were published during 1964 (Fraser *et al.* 1964b, Fraser *et al.* 1964i). George was now back in London in the Godfrey Robinson Unit (Royal National Institute for the Blind) of the Department of Research in Ophthalmology at the Royal College of Surgeons conducting the researches which led to his Doctor of Medicine (MD) thesis at the University of Cambridge in

1966 and the prestigious Raymond Horton-Smith Prize for the best Cambridge MD thesis of the year, a revised version (with A. I. Friedmann) being published as *The Causes of Blindness in Children* (Fraser and Friedmann 1967). His impressive range of genetics publications at this time also showed his intellectual energy and remarkable versatility. I was not myself primarily a geneticist in any meaningful sense but the burgeoning reports of syndromes with prolonged QT intervals and their potential, and often reality, of sudden death including in childhood, strengthened my belief that lethal cardiac conduction mechanisms could contribute to the heterogeneous entity of 'cot death', a belief that I later examined in more detail (Froggatt and James 1973) and perhaps more critically (Froggatt 1977) as further evidence emerged. I had harboured such thoughts on a cardiac role since our Irene W case in 1960, and George and I were I think the first to say so in a letter to the *The Lancet* (Fraser and Froggatt 1964) some few weeks after our first article was published, views which we repeated on several occasions later (Fraser and Froggatt 1966m, 1967). So while George went on to build his remarkable career which justifies this *Festschrift*, I focused at the time some of my (much lower-key) research effort on 'cot death' leading a team which conducted one of the first national surveys (Froggatt *et al.* 1968, 1971a, Froggatt *et al.* 1971b) and continuing my amicable, instructive and rewarding collaboration with the distinguished American cardiologist, Tom James, just as I did, though necessarily more vicariously and intermittently, with the globe-trotting George. It was Tom James's autopsy findings on our first cardio-auditory case (Irene W.) which had originally alerted us to the potentially lethal cardiac dysrhythmic genesis of the fainting attacks and sudden death (Fraser *et al.* 1964b) which with other colleagues I was later to demonstrate unequivocally in another Northern Ireland case (Gerard S.) diagnosed at birth and kept under surveillance for the two years of his life (Froggatt and Adgey 1978 Figs 2 and 3). Tom's autopsy findings in the hearts from our 'cot death' study broke new ground concerning the early post-natal development of the cardiac conduction system, anticipated apoptosis and its consequences in the human heart, and allowed inferences about the potential lethality of significant post-neonatal arrhythmias, and supplied a basis for such a role among the entity 'cot death' (James 1968), ideas which we later synthesised with our survey findings (Froggatt and James 1973) giving, or at least strengthening, in the process leads since energetically followed-up over the past quarter of a century by Peter Schwartz (1983) and David Southall among others (e.g. Towbin and Friedman 1998). It also contributed to an understanding of 'the long QT syndromes' (LQTS). A new dimension has been added by the identification in the LQTS of the several genes involved in encoding ion-channel proteins mainly potassium ones (Ward 2005). George has followed all this very much from the centre of the intellectual field of play but sadly only from the operative side-lines because active involvement from him might well have helped to unravel the tangled skein which is 'cot death'. George is a giant on whose shoulders we would greatly benefit by standing.

There is a common irony in the cardio-auditory syndrome. Intensive study of the condition has taught us very little, if anything, about the aetiology and pathogenesis still less possible treatment of the sensorineural hearing loss which is the syndrome's most unequivocal stigma: even the autopsy findings on the ear remain uninterpreted (Friedmann *et al.* 1966, 1968). It has however taught us much about cardiac conduction, a crucial system in that most crucial of all host organs which initially was only dubiously incriminated in the 'fainting attacks' (later shown to be syncope) and sudden death (later shown to be of cardiac dysrhythmic origin). It has also led to much intensive if so far inconclusive research into the causes of 'cot death' of which up to now the main beneficiary has been the pool of knowledge of cardio-pulmonary physiology in infants which has been, of course, invaluable (the encouraging and impressive reduction in incidence has been obtained mainly by empirical means) and the genome mapping will lead to further scientific findings. What originally interested the population geneticist through surveys of the deaf is now squarely in the very substantial domains of others. The accretion of knowledge often works this way. George would be able to cite an apt quotation!

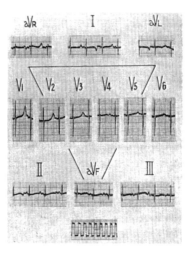

Figure 2. ECG (22 May 1971) of Gerard S aged 4 days. Note the similarity in essential features to Figure 1. The subject was later shown to be profoundly deaf and clearly was a case of the cardio-auditory syndrome, at the time the youngest diagnosed. (Reproduced from the *Ulster Medical Journal* **47** 119 (1978) by courtesy of the editor).

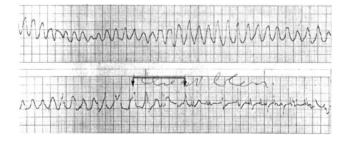

Figure 3. Gerard S (29 September 1973) aged 28 months. An episode in hospital of unequivocal ventricular fibrillation terminated by a precordial 'thump' (arrowed on the second strip). Another episode at home on 9 October proved unresponsive. (Reproduced from the *Ulster Medical Journal* **47** 123 (1978) by courtesy of the editor)

Postscript

One of the most fruitful, fulfilling, and scientifically rewarding periods of my professional life was this collaboration with George. It is seldom that a run-of-the-mill medical academic has had the privilege to work with one so gifted. We had certain temperamental differences which no doubt at times taxed the patience of us both, but we never lost sight of the importance of our mission and vocation and the scientific, professional and humane thrust of our work. I learned a lot from George and much of the modest success I have had in my professional life I attribute to his precept, his example and his philosophy during our collaboration. I am privileged to have been invited to contribute to this worthwhile *liber amicorum*.

Newly emerging concepts in syndromology relevant to deafness

William Reardon

Consultant Clinical Geneticist, Our Lady's Hospital for Sick Children, Crumlin, Dublin 12, Ireland.

Introduction

It is almost fifty years since George Fraser embarked upon his seminal studies on 'Deafness with Goitre (Syndrome of Pendred)' under the guidance of Lionel Penrose and working collaboratively with Margaret Morgans and WR Trotter at University College Hospital, London. For Dr Fraser, whose PhD thesis was based upon his studies of Pendred syndrome cases over the next couple of years, these were the initial, surprisingly assured and purposeful steps for one so young, towards a long and distinguished career in clinical genetics. His subsequent work broadened into epidemiological surveys of deafness, inheritance patterns among deaf x deaf matings and the delineation of syndromes of deafness, and is presented in his remarkable book *The Causes of Profound Deafness in Childhood* (1976). It is difficult now to conceive how groundbreaking this work was. Nowadays even ENT surgeons are waking up to the idea that childhood deafness is likely to be genetic and well over 60 loci for nonsyndromic forms of deafness have been established, in addition to devoted websites, specialist textbooks, treatises and laboratory networks offering testing for several of the more commonly mutated loci. Moreover the syndromology of deafness has developed enormously with allelic syndromes being recognised, genetic heterogeneity for clinically indistinguishable conditions offering new avenues of patient investigation and enhanced evaluation of patients, particularly using neuroradiological approaches first employed by Phelps and Reardon in London in 1988 contributing greatly to patient evaluation and disease recognition. Against this background of ceaseless change in clinical and laboratory practices, it is remarkable that George Fraser's work on deafness continues to have relevance, but such was the rigour and breadth of his work, that his contribution has stood the test of time. A careful reading of Fraser's work shows not only a consistent attention to detail in his assessment of patients and families, but also clinical experience and expertise which never fail to impress. His PhD thesis identified patients with Hashimoto's thyroiditis and deafness as a perfect phenocopy for Pendred syndrome. Forty years later, I learned the same thing! Fraser also recognised the syndrome of cryptophthalmos which bears his name and has led to a series of important gene discoveries in recent times, with the consequence that his name is celebrated eponymously in syndromic, gene and protein nomenclature.

Since this particular term, syndrome, is at the heart of my discussion, a few points of elaboration may be in order. Birth defect syndromes are usually recognised from the report of a single or a few individual cases which bear a resemblance to one another. With the publication of further cases, this emerging new syndrome is expanded by the inclusion of other birth defects not observed in the original reports. Likewise, these follow-on publications tend to throw light on the natural history of the condition, clarify the prognosis and, with luck, establish causation or identify a new investigation which is diagnostic. This is a period of natural tension between aspiring authors, anxious to publish their cases and expand the clinical documentation of the new syndrome and journal editors and referees, who have a duty to keep the literature free of impurities but also an obligation to publish genuine cases which do add to the sum total of knowledge in relation to the newly emerging/emerged condition. However, in the absence of hard objective laboratory investigations, cases which are wrongly attributed can and sometimes do get published resulting in confusion in the emerging literature. One can then understand why it is that for newly emerging, individually rare, conditions based on relatively few cases, the clinical basis of the diagnosis may remain 'soft' for a considerable period. It is worth quoting directly from Aase (1990): 'even after considerable refinement, however, diagnoses based on clinical observations show a great range of latitude and there may be no "gold standard" against which a particular patient can be compared … there is inherent variability in the manifestations of most dysmorphic disorders, both in type and in severity of structural abnormalities … Syndrome diagnosis still relies heavily on the ability of the clinician to detect and to correctly interpret physical and developmental findings and to recognise patterns in them.' Perhaps more than most, George Fraser exemplifies the point which Aase makes.

The impact of gene identification and the altered environment of clinical practice

This contribution written a decade ago might have had a strong emphasis on the need for careful phenotypic examination of patients with a view to gathering together adequate pedigrees to pursue linkage and aspire to gene identification. For many well defined syndromes, these goals have now been attained and current molecular strategies are increasingly turning towards non-Mendelian conditions, often characterised as associations. There is an increasing reliance on molecular cytogenetics to investigate patients whose clinical conditions, occurring sporadically within their families, have previously been unexplained. Much of this work stems from observations of Flint and others in the mid 1990's that up to 7% of unexplained mental retardation could be caused by subtelomeric deletions of chromosomes in patients whose gross chromosomal examination was normal (Knight and Flint 2000, Flint and Knight 2003). As a result of this new focus of research into previously undiagnosable cases, new syndromes are emerging, many of them of relevance to the audiological physican and his/her surgical counterpart.

Meanwhile, rare or poorly defined syndromes continue to be subject to study with a view to identifying the underlying causal mutations and easing diagnostic controversies in cases on the margins of those diagnoses. In parallel with these active research developments, clinicians have worked to apply many of the lessons learned from syndromes and conditions for which diagnostic genetic tests have now become available to enhance clinical management of patients and families with these conditions. It would be impossible in this contribution to allude to all of the advances relevant to syndromology of audiological medicine and otolaryngology practice, so I propose to focus on specific examples which demonstrate the principles above outlined.

1. Identifying a genetic basis for a sporadically occurring condition – CHARGE association becomes a syndrome!

CHARGE association was first described in 1979 by Hall in 17 children with multiple congenital anomalies, who were ascertained because of choanal atresia (Hall 1979). Low-set, small and malformed ears were identified among several of these cases, and associated clinical observations encompassing congenital heart defects, ocular colobomas, deafness, hypogenitalism, facial palsy and developmental delay were noted as inconsistent findings across the patient cohort. Writing in the same year, Hittner and colleagues (Hittner *et al.* 1979) reported 10 children, ascertained for colobomatous microphthalmia, with essentially the same constellation of clinical malformations. The term CHARGE was first proposed by Pagon *et al.* (1981) to reflect the major clinical clues to this diagnosis of Coloboma, Heart Defect, Atresia Choanae, Retarded growth, Ear anomalies/Deafness. As recognised by Graham (2001), the characteristic asymmetry of the clinical findings and the frequent absence of either choanal atresia or coloboma made diagnosis difficult in many cases and several patients looked as if they 'might' have CHARGE association, but without a diagnostic test, the clinical designation of such cases remained dubious. Experienced clinical geneticists often seized upon the ear morphology, the typically cup shaped ear, as a clue to diagnosis in these marginal cases (Figure 1).

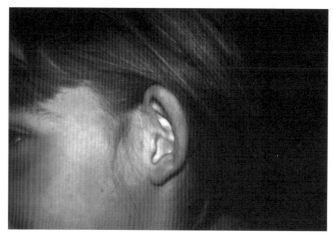

Figure 1. Cupped, prominent ear, in a patient with CHARGE syndrome.

An important clinical landmark was reached in 2001 when Amiel *et al.* (2001) reported absence or hypoplasia of the semicircular canals on temporal bone CT scanning as a core feature of CHARGE association (Figure 2). Likewise, a large scale clinical study by the same group, of clinical characteristics of patients with CHARGE association unsurprisingly showed many other clinical features occurring as uncommon but probably integral features of the syndrome (Tellier *et al.* 1998). In addition to reporting semicircular canal hypoplasia on temporal bone scans in 12 of 12 cases, these authors also drew attention to asymmetric crying facies, oesophageal and laryngeal anomalies, renal malformations and facial clefts among patients with CHARGE association.

Despite these important clinical increments in recognising the totality of the spectrum of associated anomalies, the cause of the condition remained unidentified. A teratogenic aetiology had been proposed but no specific agent had been identified common to women who had had such children (Graham 2001). A few instances of parent to child transmission had been recorded (Graham 2001), suggesting, in this subpopulation of CHARGE cases at least, a genetic, autosomal dominant basis. Other evidence for a genetic basis was drawn from observation of concordance of the condition in monozygotic twins and discordance in dizygotic twins (Graham 2001). Although it was routine clinical practice for clinical geneticists to undertake chromosomal analysis in CHARGE association cases, this was generally seen as an exercise in hope rather than a realistic investigation likely to give an abnormality. Most such cases resolutely showed normal chromosomal analysis. Hurst and colleagues (Hurst *et al.* 1991) had drawn attention to a de novo chromosomal rearrangement, a seemingly balanced whole arm chromosomal rearrangement between chromosomes 6 and 8 in a child with typical clinical features of CHARGE, but there being no further evidence to substantiate this as an important finding, it was equally likely that it was a red herring and not of aetiological significance.

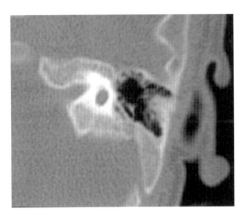

Figure 2. CHARGE syndrome – Axial CT of the petrous bone at the level of the internal auditory meatus at expected level of the horizontal semicircular canal which is absent. The crus of the posterior semicircular canal should also be seen at this level indicating complete absence of the semicircular canals (with thanks to Dr. E. Phelan).

All of this changed, however, when Vissers *et al.* (2004) used the comparative genome hybridisation approach (CGH) to screen CHARGE patients for submicroscopic copy number changes with a view to identifying microdeletions or duplications in patients with CHARGE association. They identified a CHARGE case with a deletion of approximately 2Mb. Recognising the possible value of Hurst's report and obtaining DNA from her case, these researchers then hybridised genomic DNA from Hurst's patient onto the chromosome 8 BAC array and identified two submicroscopic deletions overlapping with the earlier 5Mb 8q12 critical region. Proceeding from these important initial data, no deletions were identified in 17 other cases of CHARGE. Nine genes were identified within this critical region and sequencing of these genes identified mutations in ten of the 17 patients with CHARGE association not related to 8q12 submicroscopic deletions within a specific locus, *CHD7*. The CHD genes are a family of genes encoding Chromodomain Helicase DNA binding proteins, a family of proteins thought to have pivotal roles in early embryonic development by affecting chromatin structure and gene expression. The findings of Vissers *et al.* (2004) clearly establish that haploinsufficiency of the *CHD7* gene results in CHARGE features. Interestingly, and as might have been predicted, the individual with the microdeletion has relatively severe mental retardation in association with the core clinical features of CHARGE – presumably this represents the haploinsufficiency of genes adjacent to *CHD7*, whose specific absence accounts for the typical CHARGE features.

What of the seven individuals for whom neither deletions nor mutations were identified within this locus? It is already known that CHARGE can be associated with chromosome 22q

11.2 deletion syndrome – like phenotype, a cytogenetic deletion syndrome more readily associated with clinical stigmata of Di George sequence, Velo-Cardio-Facial syndrome and Cayler syndrome (de Lonlay-Debeney *et al.* 1997). Consequently the emerging data confirm that CHARGE is a genetically heterogeneous condition, most cases being caused by haploinsufficiency of *CHD7*, but some other cases may possibly represent an extended chromosome 22q11.2 microdeletion syndrome and other cases an as yet unidentified genetic causation. However, it is now fair to recognise that most cases of CHARGE do share an underlying genetic basis, irrespective of variability in clinical signs, and that the condition might correctly be termed a syndrome under the distinction outlined above.

2. Improved cytogenetics identifies new syndromes with specific audiological and ENT relevance. Clinical and cytogenetic interaction can result in recognition of abnormal chromosomes.

One of the questions most posed to geneticists relates to the origins of 'new' syndromes. Of course the conditions referred to as 'new' are not new, but have always existed but not been previously recognised as distinct clinical entities. 'New' syndromes emerge through the medical literature all the time. In the past, these have frequently comprised clinical reports of instructive families or individuals but a particular trend of the last few years has been the identification of syndromes with specific chromosomal abnormalities and which are deemed to be clinically recognisable. Consequently, seeing a patient in whom one is reminded of one of these 'new' cytogenetic syndromes, the clinician has a definite idea of what investigations he/she might request of the laboratory in seeking to establish the underlying diagnosis in that particular patient.

Deletions of chromosome 1p36 represent a good instance of special relevance to clinicians dealing with deafness in the context of developmental delay. Shapira and colleagues presented clinical details of 14 patients with deletion of chromosome 1p36 and identified that the condition was much more common than previously recognised by the then prevailing cytogenetic techniques. Exploiting FISH and other advances in cytogenetic technology facilitated the identification of the syndrome in cases where this had not previously been possible. Moreover, the clinical phenotype described was strongly suggestive of a pattern of malformations which should be clinically recognisable and that the clinician, by redirecting the attention to cytogeneticists towards this area of the karyotype might assist in identification of the underlying chromosomal abnormality and thus solve the diagnostic issue for the patient (Shapira *et al.* 1997). In fact it is clear from reading this paper that the clinicians were able to make the diagnosis clinically once they had become accustomed to the phenotype from the first few cytogenetically positive cases. Following that breakthrough paper, there were several other reports of this syndrome being recognised by clinicians elsewhere and these are well summarised by Slavotinek *et al.* in a major review article (1999).

The clinical profile of affected individuals which has been crystallised from these reported experiences is one of motor delay and hypotonia (90% +), moderate to severe mental retardation (90% +), pointed chin (80%), seizures (70% +), clinodactyly and/or short fifth finger (60% +), ear asymmetry (55% +), low set ears (55% +), hearing deficits (55%+) and other variable features, including congenital heart disease and cleft lip and/or palate. Some have commented on a horizontality of the eyebrows which they find clinically valuable in alerting them to this syndromic diagnosis but that is inconstant, as any examination of published photographs shows. If present, it is a valuable clue. However, for this author at least, the prime clue is often the shape of the

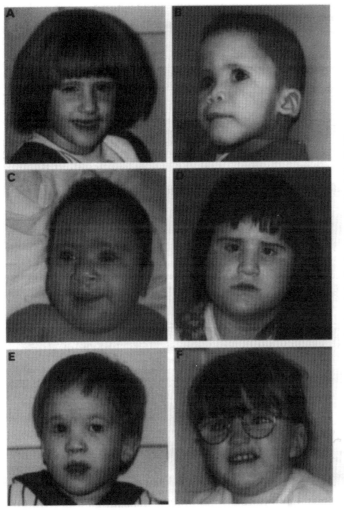

Figure 3. Facial characteristics seen in six children with chromosome 1p36 deletion. Note especially the horizontality of the eyebrows, which is a good clinical sign but not universal. The pointed chin, cases B, D and E especially, is another good clinical clue. (Reproduced with permission from Slavotinek *et al.* 1999 *Journal of Medical Genetics* **36** 657–63.)

chin, which is pointed and often rather prominent (Figure 3).

While the low set ears and ear asymmetry may be noted in audiology or ENT clinics, the main concern will often relate to hearing abnormalities. These have been characterised as high frequency bilateral sensorineural hearing loss in eight of 18 cases in one report, a further two cases having conductive loss characterised as severe degree (Heilstedt *et al.* 1998). It is valuable to know that experienced dysmorphologists will often recognise children with this syndrome clinically, despite a normal karyotype report, and discussion with cytogeneticist colleagues will often lead to re-evaluation of the original chromosome report and the identification of the underlying deletion.

A further example of this clinical-cytogenetic interaction proving valuable in identification of an underlying causative chromosomal abnormality occurs in relation to chromosome 4qter deletion. The existence of a syndrome comprising developmental delay, hypertelorism, often cleft palate and palatal dysfunction, low-set ears, poor growth and abnormal fifth finger nails has been

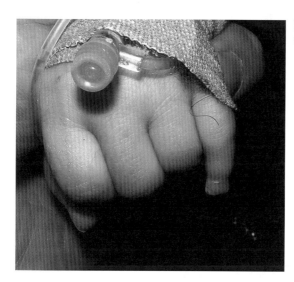

Figure 4. Tale of a nail sign in chromosome 4q– syndrome.

known for many years (Lin *et al.* 1998). Indeed, this latter sign has been recognised by Flannery as to main clue to the diagnosis and led him to coin the term 'tale of a nail' syndrome for the condition (Flannery 1993). However, the deletion can be subtle cytogenetically and, the patient's clinical condition being mild, be missed. Such a case arose in this author's own practice recently.

The patient was the youngest of three sisters born to unrelated parents. She presented aged one year with an acute respiratory arrest. Laryngo-tracheo-bronchoscopy showed multiple haemorrhagic regions in the trachea and main bronchi, consistent with acute respiratory arrest. No obstructive or other cause for this was identified. Routine investigations including basic chromosomes were normal. A genetics referral led to some new points being established – specifically there was no facial dysmorphism but the developmental history was suggestive of slight parental concern in that milestones were not being achieved at the same rate as had occurred in the older siblings. Specifically, as she got older, her speech was clearly delayed. The only clinical sign was an abnormal fifth fingernail unilaterally (Figure 4), which prompted the clinical geneticists to ask for cytogenetic re-evaluation with specific reference to chromosome 4q. A tiny deletion was shown on extended banding review of the chromosomes (Figure 5).

Subsequently this child developed severe palatal insufficiency, with little evidence of gag reflex on video fluoroscopy (Figure 6), which led to gastrostomy and direct feeding. Following fundoplication, airway function improved greatly and eventually it was possible to re-instigate oral feeding. Oropharyngeal hypotonia and palatal dysfunction are a well established feature of the 4q minus syndrome, frequently leading to need for tracheostomy and gastrostomy. Several such cases are described. Considering the numbers of children who have these surgical procedures, it ought to be worth clinically examining the nails for 'tale of the nail' sign and reviewing the chromosomes for evidence of 4q minus abnormality, which can be familial and asymptomatic in some individuals.

3. New clinical signs and associations are crystallised which are relevant to deaf patients and their physicians.

It is difficult to conceive that the practice of medicine can, after all the generations of our antecedents, still throw up new clinical signs. Perhaps it is not so much the clinical sign itself

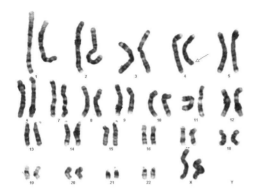

Figure 5. The karyotype of the patient with 'tale of a nail' sign is shown. Note the arrowed deletion of chromosome 4q ter. (with thanks to Mr A Dunlop).

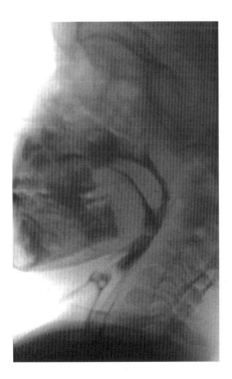

Figure 6. Swallowing study of 4q- showing the aspiration from the pharynx into the trachea, which is often seen in children with this chromosomal abnormality and can lead to life-threatening consequences. This child also shows nasal regurgitation of the swallowed contrast.

which is new, as the recognition of that sign as a marker for a specific genetic disease or syndrome. A case in point with particular relevance to the clinical examination of ears has come to light over the last few years and now bears the eponym Mowat-Wilson syndrome, after the pair of principal observers, Drs David Mowat and Meredith Wilson.

Mowat *et al.* (1998) describes a series of six children bearing a distinctive facial phenotype, in association with mental retardation, microcephaly, short stature and, in four of the six, neonatal Hirschsprung disease. Severe constipation was present in all six. Having established a deletion of chromosome 2q22–23 in one of these patients, the authors then proceeded to review the literature of clinical data from published cases with visible deletions in this region of chromosome 2q and felt there were strong facial resemblances between the features of the six

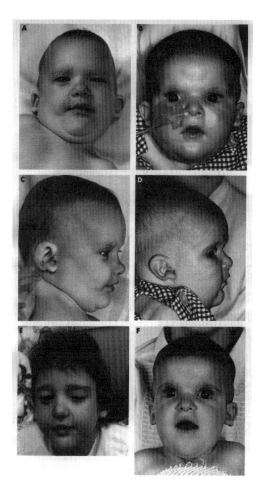

Figure 7. The facial appearance of two children with Mowat-Wilson syndrome is shown in infancy and childhood. (Reproduced, with permission, from Mowat *et al.* 2003 *Journal of Medical Genetics* **40** 305–310.)

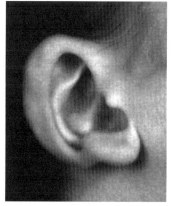

Figure 8. The characteristic ear appearance in Mowat-Wilson syndrome is shown. (Reproduced with permission from Mowat *et al.* 2003 *Journal of Medical Genetics* **40** 305-310.)

cases under report and the case previously identified by Lurie *et al.* (1994). Mowat *et al.* observed that following the recognition of the phenotype in the first two cases in their series, the next three cases were recognised within a six-month period. This phenomenon exemplifies the important learning process which dysmorphologists often comment upon and call 'getting your eye in' – essentially a learning period during which one recognises the phenotype and, having so done, recognises the pattern in future consultations with other patients. This learning stage is an important process in the emergence of any new dysmorphic syndrome. It also follows that if the original authors identified five patients within a few months, then the syndrome must be relatively common so that these cases were unlikely to form a unique set.

Subsequent events have shown that such is indeed the correct conclusion – a review by the original authors in 2003 recorded 45 cases from several continents (Mowat *et al.* 2003). In the interim period between the publication of the original observations and the review, the genetic basis of the syndrome had been established as involving the *ZFHX1B* gene on chromosome 2q22-q23. Some patients, as in the case reported by Lurie *et al.* (1994) and the original observation by Mowat *et al.* (2003) had large deletions encompassing this locus and surrounding regions, occasionally resulting in cytogenetically visible deletions down the microscope. Most patients however had intragenic mutations of *ZFHX1B* and the clue to undertaking this confirmatory test in these individuals lay in the phenotype.

Reflecting on the fundamental facial features, Mowat *et al.* (2003) drew attention to a prominent chin, hypertelorism, deep-set eyes, a broad nasal bridge, saddle nose, prominent rounded nasal tip, posteriorly rotated ears and large uplifted ear lobes. They noted that the configuration of the ear lobes, which they described as 'like orechiette pasta or red blood corpuscles in shape, is a consistent and easily recognised feature.' (Figures 7 and 8).

The point which needs to be established is that a new syndrome has emerged, which is identifiable on the basis of clinical features and that the recognition of those clinical features is the key to directing investigation towards the *ZFHX1B* gene mutation analysis. Perhaps the clinical sign itself is not 'new' – indeed it is likely that this condition has always existed, but the relevance of the clinical signs and their specific causal association with *ZFHX1B* have only recently emerged.

A similar phenomenon is beginning to emerge in relation to some cases of choanal atresia. Most cases of this malformation arise as isolated clinical findings and many patients are never investigated beyond a brief consideration of whether the choanal atresia may represent a presentation of CHARGE syndrome. Most cases arise as new events in the family and, if taken, the family history is unremarkable. One aspect of history which is frequently not sought is the history of the pregnancy and, in particular a history of maternal medication. In fact, a trickle of cases since Greenberg first observed choanal atresia in the offspring of a woman exposed to carbimazole in pregnancy (Greenburg *et al.* 1986, Greenburg 1987) have supported a likely causal effect for choanal atresia in several cases of carbimazole exposure (Myers and Reardon 2006, Lann *et al* 2004), which may be associated with oesophageal atresia, nipple aplasia or hypoplasia and dysmorphic facial features in some instances (Figure 9).

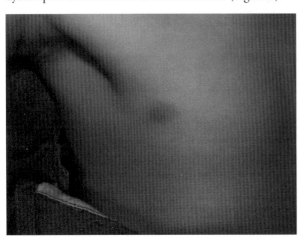

Figure 9. Nipple hypoplasia in Carbimazole exposure is demonstrated. Some children have had complete absence of nipple formation in association with choanal atresia in this teratogenic syndrome.

As with the ear abnormalities in 2q deletion syndrome, choanal atresia related to the ingestion of Carbimazole during pregnancy has long existed but the association been overlooked by failure to take the history of the pregnancy. The lesson is that for cases of isolated choanal atresia it is worth taking a detailed history of the pregnancy and looking carefully at the nipples of the baby.

Often the diagnostic significance of a specific clinical sign can be obscured by lack of reports or failure to observe the sign in cases with the condition. It certainly seems that this observation

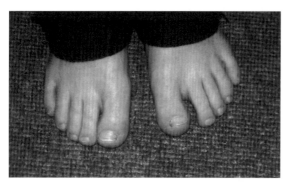

Figure 10. Broad halluces seen in association with deafness in a case of Keipert syndrome. (Reproduced with permission from Reardon and Hall 2003 *American Journal of Medical Genetics* **118A** 86–89.)

is true in respect of Keipert syndrome, in which condition deafness is associated with broad thumbs and halluces (Figure 10).

Only a handful of reports have recognised this rare diagnosis, but the author is aware of at least three further cases which have been brought to his attention following a report of a classical case (Reardon and Hall 2004). Apart from the broad thumbs, there was general reduction in the terminal phalanges on radiology with a large thumb epiphysis (Figure 11). It is likely that there are many other cases of this syndrome currently unrecognised for want of clinical examination.

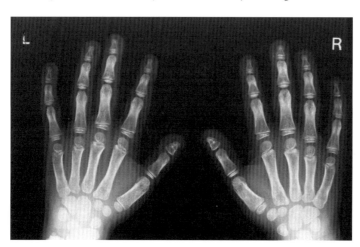

Figure 11. Short terminal phalanges in Keipert syndrome. Note also the abnormally large epiphysis in the thumbs. (Reproduced with permission from Reardon and Hall 2003 *American Journal of Medical Genetics* **118A** 86–89.)

4. Molecular genetics of known syndromes informs clinical classification and explains some previous contradictions

(i) Antley-Bixler syndrome

Antley-Bixler syndrome is a condition deriving from the eponymous 1975 report of a patient with craniosynostosis, radio-humeral synostosis and femoral bowing (Antley and Bixler 1975). Over 30 subsequent cases have been described, sometimes as single events, often as sibships. In common with other children with severe craniosynostosis, many of these children have significant audiological problems, complicating a clinical profile which already encompasses craniofacial, dental, orthopaedic and endocrine elements. Genital abnormality is an inconstant element of the condition. However, the syndrome is very difficult to distinguish from two other clinical disorders – Pfeiffer syndrome with large joint synostosis, in which the genital malformations are absent, and fluconazole embryopathy. In the latter condition, mothers taking fluconazole have given birth to children with a clinical picture which closely resembles Antley-Bixler syndrome and may be indistinguishable (Aleck and Bartley 1997). The observation of clitoromegaly in a single case of Antley-Bixler syndrome led one group of authors to pursue this line of research enquiry further. They observed abnormalities of steroid biogenesis in seven out of 16 patients with a clinical presentation consistent with Antley-Bixler syndrome, finding mutations of the Fibroblast Growth Factor Receptor 2 gene (*FGFR2*) in a further seven cases. This led to the authors postulating that there were different forms of Antley-Bixler syndrome – those associated with steroid biogenesis abnormalities and those whose clinical phenotype might reflect *FGFR* mutation only (Reardon *et al.* 2000). This suggestion has been developed further and mutations in cytochrome P450 oxido-reductase identified in children with disordered steroidogenesis, ambiguous genitalia and Antley-Bixler syndrome, this condition segregating as an autosomal recessive disorder in contrast with the new dominant mutation of *FGFR2*, which gives a similar phenotype, but for the absence of genital ambiguity (Flück *et al.* 2004). Not only has the molecular genetics resolved the differences between the overlapping clinical phenotypes but has also given an understandable reason for the genital ambiguity in some families which was not apparent in the *FGFR2* related forms. Finally the fluconazole embryopathy phenotype can be readily understood in the context of considering the mode of action of that antifungal agent. Fluconazole acts through the cytochrome P450 enzyme C-14 α-demethylase, principally inhibiting the demethylation of lanosterol, the predominant sterol of the fungal cell wall. Although one of the therapeutic advantages of fluconazole is the improved specificity it shows for the fungal cytochrome P450 enzyme complex, the embryopathy is likely to reflect relative adrenal insufficiency in infants who develop features of the embryopathy in mothers exposed to fluconazole during pregnancy. The identification of mutations in the *POR* gene consolidates this likely mechanism of action as the basis of the fluconazole embryopathy and the phenotypic overlap with *FGFR2* mutation and *POR* mutation. Thus clinical observations, in this instance the identification of ambiguous genitalia in a single case, which can initially seem rather disparate, can be crucial to the ultimate understanding of the pathological spectrum in all its variations and the apparent contradictions can be elided.

(ii) Pendred syndrome

There are several other good examples of this in conditions which are considered more 'mainstream' with respect to deafness syndromology. If we look at Pendred syndrome, the classical diagnostic triad of deafness, goitre and a positive perchlorate discharge test have been shown to be relatively poor identifiers of affected individuals and that the substitution of radiological malformation in the form of Mondini malformation or of dilatation of the vestibular aqueduct greatly enhance diagnosis and identification of affected individuals (Reardon *et al.* 2000, Phelps *et al.* 1998). Indeed, in clinical practice, the use of the perchlorate discharge test has largely been supplanted. Likewise the identification of mutations in the *PDS* gene on chromosome 7q has greatly added to the investigative tools available in recognising this syndrome in all its manifestations (Everett *et al.* 1997). Over 100 mutations of the gene are now known, though a small number are much more prevalent than others, some of which have only been observed on a single occasion. The deployment of these new forms of investigation has facilitated the resolution of diagnostic conundrums posed by particular interesting cases and families. For instance,

Gill *et al.* presented a case in which the proband had been dead for 35 years (Gill *et al.* 1999). The patient had been congenitally deaf and hypothyroid, the deafness having been assumed to be secondary to the hypothyroidism. Temporal bone sections had been stored and on review 35 years later, a grossly dilated vestibular aqueduct was identified. An affected younger sibling was identified and investigated with typical clinical and radiological findings of Pendred syndrome. The developmental delay in the index case was clearly attributable to the congenital hypothyroidism, a very rare complication in the profile of Pendred syndrome.

Likewise there have been perplexing families reported, whose clinical conditions have been resolvable by molecular approaches. The best example of such is the Brazilian family recorded by Billerbeck *et al.* (1994). Goitre associated with deafness and a positive perchlorate discharge test was observed in at least two affected individuals in the highly consanguineous pedigree under consideration. To complicate matters, the family emanated from a region of endemic goitre. The likely diagnosis of Pendred syndrome was offset by the observation of positive perchlorate test in the absence of hearing loss in other individuals in the pedigree, while others were recorded with deafness alone or goitre as a sole finding. The identification of mutations in the *PDS* gene facilitated the wider exploration of the underlying pathology in this confusing pedigree. It transpired that the index case, satisfying all typical diagnostic parameters for Pendred syndrome, did harbour a homozygous deletion in exon 3 of the gene, resulting in a frameshift and premature stop. An additional two individuals in the pedigree also shared this genotype, and thus had Pendred syndrome. However, several deaf individuals in the pedigree were not homozygous for the *PDS* mutation, suggesting that they were homozygous for a autosomal recessive for deafness. Moreover six individuals in the family with goitre did not have *PDS* gene mutations and the likely cause for the goitre in these was the endemic iodine deficiency (Kopp *et al.* 1999). Accordingly the clinical classification of this family has been established as comprising three distinct conditions – Pendred syndrome, endemic iodine-deficiency-related goitre and nonsyndromic deafness. Similar phenomena have been observed and formally established in another confusing family (Vaidya *et al.* 1999).

(iii) Waardenburg syndrome

Waardenburg syndrome and the various subtypes of this condition provide one of the most elegant examples of how good clinical observation, careful family studies and integration of molecular data can powerfully combine to enhance understanding of clinical observations which, initially at least, seemed to be at variance with received wisdom, ultimately resulting in the recognition of new disease processes. It is worth briefly reviewing the progress which has been made relating to this group of disorders.

The original observation of deafness with heterochromia iridium, white forelock and white skin patches dates from Waardenburg (1951). Some twenty years later, it was the observations of Arias that the dystopia canthorum segregated with deafness in some families but not in others which led to the delineation of type I (with dystopia) from type II (Arias 1971). Subsequent reports were less amenable to classification – the observation of Shah *et al.* (1981) of infants with Hirschsprung disease and white forelock, seemingly inherited in autosomal recessive manner, and the report from Klein in 1983 of a patient with features of Waardenburg syndrome type I (WSI) associated with severe arm hypoplasia, and arthrogryposis of the wrists and hands.

Aided by careful attention to phenotype and, in particular, to dystrophia canthorum, linkage studies on WSI led to identification of mutations in the *PAX3* gene on chromosome 2 (Tassabehji *et al.* 1992, Baldwin 1992). Subsequent studies have established that almost all cases conforming to the WSI phenotype have mutations at this locus and there is no substantial evidence for genetic heterogeneity. However deafness is a variable feature of the syndrome among WSI individuals – Read and Newton citing a prevalence of 52% in their experience (Read and Newton 1996). This seems not to be related to the nature of the mutation and the exact cause of this variation in penetrance remains unclear. However, identification of mutations in *PAX3* has considerably aided our understanding of clinically confusing situations outlined above. Klein-Waardenburg syndrome, also known as WSIII, has proven to be another phenotype of *PAX3* mutation – in some instances this is due to a contiguous gene deletion syndrome involving the *PAX3* locus and adjacent regions of chromosome 2q35, but in others, intragenic mutations of *PAX3* have been found, either in the homozygous or in the heterozygous state. A good example is the family reported by Wollnik *et al.* (2003), in which parents with WSI shared a mutation for *PAX3*, the offspring being homozygous for the mutation, Y90H, and having the WSIII phenotype. In mice, homozygosity of *Pax-3* mutations results in severe neural tube defects and lethality. Likewise a phenotype of exencephaly and severe contracture and webbing of the limbs in humans has been speculated to be consistent with a severe *PAX3* mutation in homozygous form (Ayme and Philip 1995). Indeed screening patients with neural tube defects for *PAX3* mutation led to the identification of one patient with a myelomeningocoele who, on close examination, was shown to have mild features of WSI, which were also seen to segregate with the mutation in several family members (Hol *et al.* 1995). Less predictable was the finding that Craniofacial-Deafness-Hand syndrome is allelic to WSI, being caused by an exon 2 missense mutation in affected members of this unique family. The phenotype comprises autosomal dominant deafness, with hypoplasia of the nasal bones, telecanthus, nasolacrimal duct absence/obstruction, ulnar deviation of the hands and flexion contractures of the ulnar digits (Sommer and Bartholomew 2003). Notably there are no features of pigmentary disturbance in this pedigree. The mutation in this family is a missense mutation, resulting in substitution of asparagine by lysine (N47K). However, alternative mutation of this N residue, to histidine, in another family results in a more typical clinical outcome of WSIII in affected individuals. What is more, unlike the *PAX3* mutations seen in homozygous form in some WSIII patients, this N47H mutation causes WSIII in heterozygous form (Hoth *et al.* 1993). More commonly, however, the WSIII phenotype is seen as a consequence of compound heterozygosity for *PAX3* mutation. The likely explanation for these seemingly contradictory observations lies in the effects of the mutation on the function of the *PAX3* mutant protein. Indeed, there is evidence for this from the work of DeStefano *et al.* (1998), who studied the relationship between mutation type and clinical sequelae in 271 WS individuals, representing 42 unique *PAX3* mutations. Deletions of the homoeodomain were most significantly correlated with significant clinical findings and were seen to correlate especially strongly with white forelock.

In parallel with these emerging insights into the clinical phenotypes attributable to genetic mutation at the *PAX3* locus, there has been considerable advance in the understanding of the genetic basis of WSII and related disorders. Mutations in the *MITF* gene on chromosome 3p have been established in several families conforming to the clinical definition of WSII. Other clinically interesting phenomena associated with mutation at this locus have also been observed. Deafness is more common as a clinical finding of WSII than is the case in WSI, being observed in approximately 80% of cases (Read and Newton 1996). However, the absence of pigmentary abnormalities in many patients, or the presence of such features in only very subtle form, does lead to difficulty in discrimination between WSII and patient with nonsyndromic deafness. Indeed, Read has confirmed this clinical observation at molecular level with his observation that 10% of cases with a clinical diagnosis of WSII in fact have mutations at the *Connexin 26* locus and do not have WSII at all (Read 2005). As often happens once the molecular genetics of a syndrome is established, conditions which had been considered to represent clinically distinct syndromes have been recognised as allelic forms, due to the identification of a mutation at the same locus. Tietz syndrome is a case in point. The syndrome dates from the 1963 report of Tietz of a family in which deafness segregated as a dominant trait over six generations but always in association with albinism. The irides were blue, with albinoid fundi, the hair being blonde and the skin very fair. *MITF* mutation was shown as the basis of this syndrome of albinism and deafness (Amiel *et al.* 1998). Foremost among the clinical observations which underlie this syndrome was the co-segregation of albinism and deafness in affected individuals, as autosomal dominant traits. Albinism is more often and classically observed as an autosomal recessive trait in clinical genetics. However, families were also known in whom WSII and ocular albinism (OA) existed but in which the pattern of cosegregation was not as clear-cut. This was known as the WSII-OA phenotype. Such pedigrees are rare, but Morrell *et al.* (1997) studied one such pedigree, establishing an intragenic deletion within the *MITF* locus. Individuals with the OA phenotype were shown to have homozygosity or heterozygosity for the R402Q mutation in the tyrosinase gene, which functionally reduces the catalytic activity of the tyrosinase enzyme, in addition to the *MITF* mutation. These observations led the authors to propose that the WSII-OA phenotype is consequent on digenic interaction between *MITF* and tyrosinase, a gene regulated by *MITF*.

Not all cases of WSII phenotype have mutation of the *MITF* locus. Indeed, OMIM currently lists four loci for WSII, respectively termed WSIIA-D. However, only the *MITF* locus is confirmed and to date, this represents the sole locus for WSII at which mutations have been established which result in the WSII phenotype. *MITF* is a key activator for tyrosinase, a major enzyme in melanogenesis and critical for melanocyte differentiation. *PAX3* transactivates the *MITF* promoter and is assisted in doing so by another gene *SOX10*. Not surprisingly, this latter is also an important gene in WS phenotypes and specifically in the WS4 clinical spectrum.

The term WS4 relates, as outlined above, to the observations of Hirschsprung disease in association with other phenotypic characteristics of WS. In a study of a large Mennonite family, many of whose members had Hirschsprung disease, sometimes associated with low grade features of WS (white forelock in 7.6% of cases), Puffenberger *et al.* (1994) identified a causative mutation in the Endothelin Receptor B Gene (*EDNRB)* on chromosome 13. This was an interesting mutation, which showed dose sensitivity, homozygotes having a 74% chance of showing Hirschsprung disease against 21% in heterozygotes. This was a seminal finding, leading not only to the identification of a genetic basis for many cases of nonsyndromic Hirschsprung disease, but also to mutation of the *EDNRB* locus in families conforming to the Waardenburg-Shah phenotype (WS4) (Syrris *et al.* 1999) as well as the recognition of an allelic condition ABCD syndrome (albinism, black lock, cell migration disorder of the neurocytes of the gut and deafness) (Verheij *et al.* 2002). The latter refers to a child with deafness, albinism, a black lock in the right temporo-occipital region and spots of retinal depigmentation, in whom severe intestinal innervation defects were established. These clinical findings were causally attributed to homozygosity for a C to T transition in the *EDNRB* gene resulting in a stop codon with no production of normal protein.

Prompted by these observations and encouraged by the knowledge that mutation of the Endothelin 3 gene in mouse results in a phenotype similar to WS4, Edery *et al.* (1996) searched for and reported mutations of the *EDN3* gene in patients with Waardenburg-Shah syndrome. Subsequently mutation at this same locus have been found in other cases of WS4 but also in isolated cases of Hirschsprung disease and even in a patient with Hirschsprung disease associated with central hypoventilation syndrome.

It is now known that *EDNRB* has a strictly defined role in governing migration of the precursor cells of the enteric nervous system into the colon. Binding sites for SOX 10 enhance the migration of these enteric nervous system cells. Not surprisingly, then, the *SOX10* gene, on chromosome 22q13, is also associated with WS phenotypes. Specifically several patients with WS4 phenotype have been described due to mutation at this locus. Moreover, a patient thought to have a separate condition, Yemenite Deaf-Blind hypopigmentation syndrome, was also reported with *SOX10* mutation (Bondurand *et al.* 1998). Likewise, *SOX10* mutations have been recorded in patients with pigmentary disturbance and deafness suggestive of WS but in whom rectal biopsy is normal. Nonetheless, the patients have persistent bowel symptoms suggestive of bowel obstruction. This establishes that aganglionosis is not the only mechanism associated with intestinal dysfunction in *SOX10* mutation (Pingault *et al.* 2000). Other clinically important phenomena have been observed in the spectrum of *SOX10* associated disease. Donnai (2004) presented details of an adult deaf female with raindrop pigmentation of the skin, in whom a *SOX10* mutation was established. A further set of patients was identified with WS4 features associated with a peripheral neuropathy and/or central hypomyelinating neuropathy associated with *SOX10* mutation (Pingault *et al.* 2000, Inoue *et al.* 1999). This neurological variant is now known as PCWH (Peripheral demyelinating neuropathy, Central demyelinating leukodystrophy, Waardenburg syndrome and Hirschsprung disease) and recent work has established that this more severe phenotype occurs because truncated, mutant SOX10 proteins with potent dominant-negative activity escape the nonsense mediated decay pathway (Inoue *et al.* 2004).

To summarise, mutations at 5 distinct gene loci, *PAX3, MITF, EDNRB, EDN3* and *SOX10*, have been described in association with Waardenburg syndrome phenotypes. However, careful attention to clinical examination and investigation in these patient groups have contributed enormously to an enhanced understanding of the molecular mechanisms, the mutational spectrum and the embryological events which underlie the differing presentations of Waardenburg syndrome.

5. Phenotypic studies of syndromes with an already established genetic basis enhances clinical data, patient management and drives further research.

The cloning of a gene together with the establishment of causative mutations at that locus for various phenotypes is sometimes seen as an end in itself. To researchers engaged in such work, this does represent a momentous milestone. However, to clinicians, families with the condition and those charged with delivery of medical services to such patients and families, the identification of mutations does not usually change patient care other than by facilitating identification of others in the kindred who themselves have inherited the mutation and might benefit from specific screening measures for covert disease. What the identification of mutations underlying a spe-

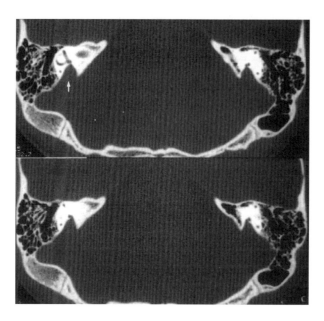

Figure 12. Dilatation of the Vestibular Aqueduct is shown in a typical case of Pendred syndrome. (Reproduced with permission from Reardon *et al.* 2000 *Quarterly Journal of Medicine* **93** 99–104.)

cific syndrome does allow is more detailed phenotypic studies of that condition and encourages the clinical 'teasing out' of clinically overlapping conditions, so that it can become clearly established as to what particular pathology applies in an individual patient or family.

A good example of this is provided by the Dilated Vestibular Aqueduct syndrome (Figure 12). Dilatation of the vestibular aqueduct has been known since 1978. Several series of deaf patients with this radiological phenomenon had been published, resulting in over 200 cases being identified and reported in radiological and ENT literatures. It appears that none of these patients had been recognised as having an underlying genetic syndrome and indeed it is never addressed in any of these publications as to whether any of the patients included in the various series were related. Phelps recognised that almost all cases of Pendred syndrome manifest dilatation of the vestibular aqueduct (Phelps 1996, Phelps *et al.* 1998) and there has since been a mushrooming of interest in this radiological marker of deafness.

This interest in investigating deaf patients more systematically and in seeking to identify the precise basis of the deafness has established that dilatation of the vestibular aqueduct is not confined to Pendred syndrome. Indeed, it is not at all surprising, considering the shared pathology of ion transporter defects seen in both conditions, that renal tubular acidosis and deafness, a distinct autosomal recessive condition, should share this characteristic with Pendred syndrome (Berrettini *et al.* 2001). There are now suggestions that there may be a genetically distinct autosomal recessive syndrome of dilatation of the vestibular aqueduct and deafness separate from Pendred syndrome and for which the locus remains to be established (Pryor *et al.* 2000). Such claims, whether validated in time or not, are only possible because of detailed phenotypic work which has continued following the identification of the genetic basis of Pendred syndrome and the incorporation of such mutational studies into clinical practice. The best current estimate is that Pendred syndrome mutation accounts for about 86% of cases of vestibular aqueduct dilatation (Reardon *et al.* 2000).

Likewise with respect to Branchio-oto-renal (BOR) syndrome, the cloning and identification of mutation at the *EYA1* gene has shown that there are other clinically overlapping phenotypes which are not due to mutation at this locus. Among families, comprising the majority, which do owe their clinical phenotype to *EYA1* mutation, there has been enhanced incorporation of genetic data into patient care and management. Chang *et al.* (2004) have furnished their data incorporating 40 families with 33 distinct mutations segregating and have identified the major features as deafness in 98.5%, preauricular pits in 83.6%, branchial anomalies in 68.5% and renal anomalies in 38.2%. However, other phenotypes have also been associated with mutation at this locus, including cataract and anterior ocular defects (Azumu *et al.* 2000), oto-facio-cervical syndrome (Rickard *et al.* 2001) and a contiguous gene deletion syndrome which is clinically characterised by BOR syndrome but with additional clinical features of Duane eye-retraction syndrome, hydrocephalus and aplasia of the trapezius muscle (Vincent *et al.* 1994). In addition to these allelic diseases emerging form BOR-related studies, there has also been clarification of those families whose clinical phenotype appears to suggest a likely diagnosis of BOR but for whom mutation at *EYA1* was not established and linkage data suggested that the disease phenotype was a function of mutation at another locus. A good example of this is provided by the large kindred forming the basis of the report of another *BOR* locus on chromosome 14q (Ruf *et al.* 2003). This pedigree, comprising over forty affected individuals, differs from the classical BOR syndrome profile in that only approximately 25% had branchial arch related defects, the age of onset of deafness was much later and more variable than is generally seen in *EYA1* related deafness and no renal malformations or anomalies are reported in the clinical data furnished on the family. Purists might argue with the nosology of the syndrome as BOR3, but it is difficult to argue against this in light of the clinical finding of branchial defects in 25% of affected individuals. The mutational basis of this, to date unique, family remains unresolved at this time, but it is worth noting that other 'nonsyndromic deafness' loci map to the same region (*DFNA23* and *DFNB35*). The designation *BOR2* has been given to another hitherto unique dominant pedigree mapping to chromosome 1q (Kumar *et al.* 2000). Such a designation is certainly more contentious as the family had always been considered clinically distinct from BOR by the absence of cervical fistulae, renal anomalies and the presence of lip pits. However, there is no doubt that this pedigree represents another form of autosomal dominant deafness associated with preauricular sinuses.

BOR syndrome also represents an example of how learning that a member of a specific gene family can cause a particular phenotype extends the opportunities for establishing molecular pathology in clinically related situations. *EYA1* is one of four related human loci and mutations at another of the genes in this family, *EYA4*, has been reported in deafness of autosomal dominant nonsyndromic type (Wayne *et al.* 2001).

The genetics of Fraser Syndrome and the blebs mouse mutants

Ian Smyth[1,2] and Peter Scambler[2]

[1]Cancer Research UK, London Research Institute, 44 Lincoln's Inn Fields, London WC2A 3PX
[2]Molecular Medicine Unit, Institute of Child Health, 30 Guilford Street, London WC1N 1EH UK

Abstract

Fraser Syndrome is a recessive multisystem disorder characterised by embryonic epidermal blistering, cryptophthalmos, syndactyly, renal defects and a range of other developmental abnormalities. More than 17 years ago four mapped mouse blebs mutants were proposed as models of this disorder, given their striking phenotypic overlaps. In the last few years these loci have been cloned, uncovering a family of three large extracellular matrix (ECM) proteins and an intracellular adapter protein which are required for normal epidermal adhesion early in development. The proteins have also been shown to play a crucial role in the development and homeostasis of the kidney. We review the cloning and characterisation of these genes and explore the consequences of their loss.

Fraser syndrome

According to Warkany, during the first century AD Pliny the Elder described a family with three children, each of which was born with skin covering the eyes (Warkany and Schraffenberger 1944). This may well be an early description of autosomal recessive cryptophthalmos, a major feature of a syndrome comprehensively described for the first time by George Fraser (Fraser 1962a). He detailed then most of the features now known to be associated with the syndrome bearing his name, and in more recent times Slavotinek and Tifft compiled an excellent summary of the condition. They list four major and eight minor characteristics, two major and one minor, or one major and four minor features being required to make a diagnosis of Fraser syndrome (Slavotinek and Tifft 2002) (FS; MIM219000).

FS is rare, the incidence having been estimated at one in 10,000 stillbirths (Martinez-Frias *et al.* 1998). Of the major features, cryptophthalmos occurs in approximately 90% of cases, half of those being bilateral. Posterior structures of the eye are usually spared, and surgical intervention can result in the ability to distinguish light and dark, and some movements. Microphthalmia or anophthalmia are found in a few cases. The facial appearance is well described by the author Lucinda Franks in a *New Yorker* piece concerning the care of her nephew: '… like a child's unfinished drawing. He had only one unnaturally small eye, on the right side of the face. On the other side, there was concave blankness beneath the brow. His nostrils were separated by a deep cleft, and the nasal ridge was squashed. [His mother] took his curled fist and felt for fingers, but none were there.' (from *http://www.midbio.org/mbc-newethics5.html*)

Of the remaining major features cutaneous syndactyly occurs in nearly 60% of cases, the severity varying enormously. Abnormal genitalia are detected in approximately 20% of cases and ambiguous genitalia can make gender assignment problematic at birth. An affected sib is the final major feature; all cases described have been compatible with recessive inheritance, the incidence being increased in consanguineous parents.

As a group, the minor features occur frequently, and we feel that laryngeal stenosis and renal a/dysgenesis could be considered as major features. Patients may also have skeletal defects, pulmonary hyperplasia, ear malformations with conductive deafness, orofacial clefting, gastrointestinal malformations and heart defects. It has been estimated that 45% of cases are stillborn or die within the first year, primarily because of pulmonary or renal complications (Boyd *et al.* 1988). However, there is no reason to suspect that those patients without these life-threatening complications cannot live a normal lifespan.

In the absence of DNA testing prenatal diagnosis is difficult. Second trimester ultrasonography may detect oligohydramnios and contrastingly voluminous hyperechogenic lungs (Fryns *et al.* 1997). Detection of cryptophthalmos by examination of palpebral fissure slant may be possible, but is complicated by any concomitant oligohydramnios; the variable expressivity means that at least some diagnoses will be missed (Berg *et al.* 2001). A parent support group has been formed and information is available at *http://frasersyndrome.info/*.

The mouse 'blebbing' mutants

In 1988 Robin Winter first speculated that the mouse 'bleb' mutants might be a murine equivalent of FS (Winter 1988; Winter 1990). At that time there were four mapped 'bleb' mutants which localised to different chromosomes. Each line gave affected homozygotes with combinations of eye abnormalities, renal anomalies and syndactyly, with varying severity. The common embryological phenotype of affected embryos was the presence of fluid filled blebs arising at about twelve days of gestation and occurring over the extremities, eyes or hindbrain (Figure 1). These blisters subsequently became haemorrhagic and disappeared during late gestation leaving the characteristic covering of the eye with skin.

The first 'bleb' mutant to be described was *myelencephalic blebs* (*my*) which arose in an X-irradiation screen conducted by Little and Bagg (Grüneberg 1952). The *blebbed* mutation (*bl*) arose from offspring of a male irradiated with neutrons and, at least on the original undefined genetic background, appears to have the most severe phenotype with poor post-natal survival. It was suggested that a different mechanism might be operating during renal and skin development (Darling and Gossler 1994). The *eye blebs* (*eb*) mutation arose spontaneously although histology gave little clue as to the basis for the abnormalities observed (Center and Emery 1997). In severe cases mice failed to form the retina and lens (Swiergiel *et al.* 2000). The final mutant, *head blebs* (*heb*), arose spontaneously and was somewhat milder in severity than *eb* and *my* (Varnum and Fox 1981).

Identification of FS/blebbing genes

Early linkage analyses had established the approximate chromosomal location of all the extant blebbing loci, although these mapping experiments were not at a sufficiently high resolution to give a confident indication of the respective locus in man. Subsequent autozygosity mapping and gene sequencing in FS kindreds revealed loss of function mutations in a novel gene *FRAS1* (chromosome 4), the murine homologue being mutated in *bl* (McGregor *et al.* 2003, Vrontou *et al.* 2003).

Targeted mutation of the glutamate receptor interacting protein (*Grip1*) gene, resulted in a

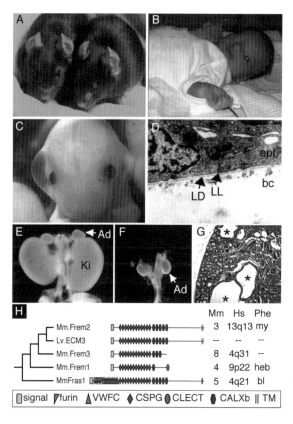

Figure 1. Fraser Syndrome and the blebs mutants

The mouse blebs mutants were identified in a number of genetics screens. The most obvious feature of these animals is cryptophthalmia in which skin covers the globe of the eyes (A; wild type left, *heb* right). Many of the defects such as cryptophthalmia and syndactyly observed in these mice are mirrored in human Fraser Syndrome (B). Many of these defects are caused by loss of epidermal adhesion during embryonic development. These blisters, or blebs, tend to form around the head, limbs and trunk and can often become haemorrhagic late in gestation. The majority resolve prior to birth. The loss of adhesion occurs below the level of the basement membrane lamina densa (D; LD, lamina densa; LL, lamina lucida; epi, epidermis; bc, blister cavity). Three of the blebs mutants are caused by mutations in Fras1, Frem1 and Frem2, complex multidomain extracellular matrix proteins (E; vWFC, von Willebrand factors C; CSPG, chondroitin sulphate proteoglycan domain; CLECT, C-type lectin; CALXb, calcium exchange domain beta; TM, transmembrane domain).

phenotype closely resembling *bl* (Bladt *et al.* 2002, Takamiya *et al.* 2004). *Grip1* mapped to mouse chromosome 10, which suggested that it might be the *eb* gene, and an intragenic deletion of exons 10 and 11 confirmed this suspicion (Takamiya *et al.* 2004). Concurrently, another blebbing phenotype had been discovered in an ENU mutagenesis screen and shown to be allelic with the *heb* locus. Mutations in both alleles were found in a gene encoding a protein with similarity to Fras1 and the *heb* gene was christened *Frem1* (for *Fras-related, extracellular matrix*). While analysis of families not linked to *Fras1* has identified a few pedigrees compatible with linkage to the human homologues *GRIP1* (chromosome 12) and *FREM1* (chromosome 9), we have not discovered any mutations within these genes.

The *Frem2* and *Frem3* genes were identified by genome database interrogation, *Frem2* mapping precisely to a 1cM interval defined by mapping the *my* locus on mouse chromosome 3. Subsequent complementation analysis showed that a gene trap mutation of *Frem2* is allelic to the *my* mutation (Jadeja *et al.* 2005). Three FS families segregated a mutation in human *FREM2* (E1972K) which substitutes a strongly conserved residue within the CALXβ domain, a residue predicted to be important for calcium binding (see below) (Jadeja *et al.* 2005). The *Frem3* gene is not linked to any known blebbing locus, no mouse mutant is available, and no mutations of *FREM3* (chromosome 4) have been found in FS.

Structure and evolution of the 'bleb' proteins

The extracellular blebs genes – *Fras1*, *Frem1* and *Frem2* – encode large multi-domain proteins between 2172 and 4010 amino acids in length (Figure 1). Fras1 and Frem2 are transmembrane proteins while Frem1 instead contains a lectin type C domain at its C-terminus. The hallmark feature of this family is the repeated ~130 amino acid chondroitin sulphate proteoglycan (CSPG) domain, whose structure is thought to be similar to a cadherin fold (Staub *et al.* 2002). The proteins also incorporate one or more copies of the calcium exchange β domain (CALXβ), a calcium chelating motif present in a variety of calcium transporter proteins as well as in other cell adhesion molecules (Schwarz and Benzer 1997). These calcium binding motifs are also structurally similar to cadherin domains (our unpublished data) and their likely functional importance is highlighted by our identification of a disease causing *FREM2* CALXβ missense mutation in Fraser Syndrome patients.

In addition to these protein elements, Fras1 also bears six von Willebrand Factor Type C (vWFC) and 14 cysteine rich partial furin motifs at its N-terminus. The established involvement of these domains in interaction with members of the TGF-β growth factor family (Thomas 2002), has led to the suggestion that Fras1 might modulate their activity within the ECM (McGregor *et al.* 2003). The *Fras* and *Frem* genes are a relatively recently evolved family and form a monophyletic clade found only in deuterostomes. The most ancient member of the family is the sea urchin *ECM3* gene which, on the basis of domain organisation and sequence conservation, is an orthologue of *Frem2*. ECM3 is a major component of fibres which lie on the basal surface of the embryonic ectoderm and it is thought to orchestrate primary mesenchyme cell migration during gastrulation (Hodor *et al.* 2000). Finally, *Grip1* encodes a 7 PDZ domain cytoplasmic protein which has been shown to interact with the most C-terminal residues of both Fras1 and Frem2. In the case of Fras1 at least, loss of Grip1 results in failure of the protein to localise correctly to the basal membrane of keratinocytes in culture (Takamiya *et al.* 2004).

Clues to the function of the *Frem* and *Fras* genes are largely provided by studies of *NG2*, a gene widely expressed in partially differentiated or differentiating cells (Levine and Card 1987, Pluschke *et al.* 1996, Stallcup 2002, Legg *et al.* 2003, Fukushi *et al.* 2004). The CSPG domains in NG2 are capable of diverse interactions with other extracellular components, many of which may have direct implications for Frem and Fras protein interactions. Various studies have demonstrated their interaction with collagen V and VI (Burg *et al.* 1996, Tillet *et al.* 2002), basic fibroblast growth factor and platelet-derived growth factor (Goretzki *et al.* 1999). There is also mounting evidence to suggest that a soluble form of the protein can be generated by matrix metalloproteases (Asher *et al.* 2005, Larsen *et al.* 2003). All of these studies raise the intriguing prospect that the Fras and Frem proteins may have the ability to regulate not just the assembly of structural components of the ECM but also subtly modulate the activity of growth factors in the extracellular milieu.

Gene expression

The blebs genes are expressed in a diverse pattern during embryonic development. As a general rule, they are all expressed at high levels in tissues in which a differentiating and remodelling epidermis is interacting with an underlying mesenchyme (Figure 2), lending credence to the concept that the proteins play both structural and developmental roles during tissue differentiation. With only a few exceptions, the expression of *Frem1* is restricted to the mesenchymal components of these structures, while *Frem2/Fras1* and *Grip1* are expressed in the epithelial special-isations. In the developing skin, *Fras1* is expressed very strongly throughout the developing epidermis (McGregor *et al.* 2003, Vrontou *et al.* 2003), while *Frem1* and *Frem2* expression in the interfollicular regions tends to be markedly reduced in comparison (Smyth *et al.* 2004, Jadeja *et al.* 2005). *Frem2* is also expressed in a highly dynamic pattern in the endoderm and ectoderm of the branchial arches and in a number of important neural signalling centres including the midline of the dorsal forebrain and midbrain/hindbrain boundary (Jadeja *et al.* 2005). *Grip1* is expressed in a variety of other locations, probably reflecting its diverse role in modulating protein trafficking (Jourdi *et al.* 2003, Stegmuller *et al.* 2003, Takamiya *et al.* 2004).

The role of the *Frem* and *Frem* genes in epidermal adhesion
Cryptophthalmos and syndactyly are the most striking physical features of FS and the blebs mice. We reason that they both arise as a consequence of loss of epidermal adhesion leading to inter-rupted epidermal/mesenchymal interactions between the eyelid epithelia or limb AER and the underlying mesenchyme. In all of the mutants, the separation between the epidermis and the dermis occurs below the level of the lamina densa, the lowermost, collagen IV rich, component of the basement membrane (BM). Given the established interaction between the NG2 protein and collagens V and VI (which also contribute to the BM) we and others have studied the dep-osition of these proteins in the blebs mutants. *Grip1* and *Fras1* mutants display defects in dep-osition of these proteins and this may contribute to epidermal delamination (McGregor *et al.* 2003, Vrontou *et al.* 2003, Takamiya *et al.* 2004), but the same is not true in *Frem1* and *Frem2* null mice (Smyth *et al.* 2004, Jadeja *et al.* 2005). Thus the steps leading to blistering are only partly understood.

What is perhaps most striking about all of the mutants is the ability of the epidermis to repair itself, often after suffering delamination which is frequently sufficient to kill the embryo. This observation, and the absence of postnatal skin blistering in the blebs mice, suggests that there is a temporal developmental window when these proteins are required for epidermal adhesion, but that later in gestation other components of the ECM can functionally compensate for their loss. The blistering defects in the blebs mice mirror the effects of loss of collagen VII, which causes dystrophic epidermolysis bullosa (DEB) in humans (Christiano *et al.* 1993). ColVII contributes to the anchoring fibrils which consolidate the interaction between the dermis and epidermis; however, its loss results primarily in postnatal rather than *in utero* blistering (Christiano *et al.* 1993). Additionally, ColVII deposition is unaffected in *Fras1* mutants and is only partly defec-tive in mice lacking *Grip1*, suggesting that epidermal separation in DEB is distinct from that of FS or the blebs mice.

Organogenesis
Renal defects are a hallmark of FS and all of the 'bleb' alleles identified thus far. The agenesis apparent in these models is triggered very early during the development of the kidney, with reduced and apoptotic mesenchymal condensations surrounding the ureteric bud as early as E11.5 (McGregor *et al.* 2003, Takamiya *et al.* 2004). *Frem2* and *Fras1* are both strongly expressed in the nephric epithelium, especially in the tips of the buds, while *Frem1* is expressed at lower levels in the stroma, a pattern which continues as the ureteric tree branches and differ-entiates. Cystic disease in the blebs mice has been best studied in *Frem2* and *Fras1* mutants.

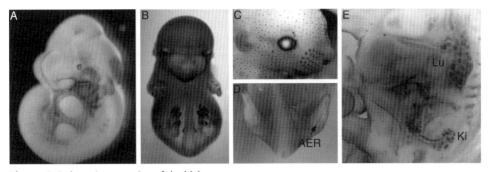

Figure 2. Embryonic expression of the blebs genes
Frem1, -2 and *Fras1* are widely expressed during embryonic development. While *Fras1* is broadly and strongly expressed in the embryonic epidermis, expression of *Frem1* and *Frem2* are comparatively reduced in the inter-follicular epidermis. Expression of both of these genes is restricted to site of epithelial/mesenchymal interactions including the hair follicle, whisker vibrissae, mammary glands and apical ectodermal ridge (AER) (A, C, D, *Frem2*; B, *Frem1*). The genes are also strongly expressed in developing organs affected in both the blebs mutants and in Fraser Syndrome patients including the kidney and lung (E, *Frem2* expression in Lu, lung; Ki, kidney). In most cases, *Frem1* expression in the mesenchyme mirrors *Frem2* and *Fras1* expression in the overlying epithelium.

Animals homozygous for mutations in either or both of these genes develop cortical renal cysts by 12 weeks of age (Jadeja *et al.* 2005). The hyper-proliferative and hyper-apoptotic cysts express markers of both the collecting ducts and thick ascending loops of Henle. In wild type mice, *Frem2* is expressed strongly in adult kidneys in the collecting ducts, proximal convoluted tubules and arterioles which, in combination with the development of renal cysts in the mutant animals, suggest that the protein is required for the maintenance of the differentiated state of the mature renal epithelia. *Fras1* also has a role in regulating the normal lobular development of the lung and the development of vascular tissue in the terminal air sacs (Petrou *et al.* 2005).

Functional analysis

Given their multidomain structure it is highly likely that the Frem and Fras proteins interact with many different components of the ECM, and that some of these interactions are required for normal epidermal adhesion (Figure 3). Their similarity to NG2, and the observation of some defects in collagen deposition in the basement membrane, indicate that one of these interactions is with collagens V and VI, although mice null for these proteins do not display a 'bleb'-like phenotype (Andrikopoulos *et al.* 1995, Bonaldo *et al.* 1998). Given the complementary expression pattern of *Frem1* and *Frem2/Fras1*, and the similarity of the CSPG domains to cadherin, we have proposed that the proteins may form homo- or heterodimeric associations. Engagement of ligands to the ectodomain of NG2 induce cell spreading and alterations in the actin cytoskeleton mediated by Rac1, cdc42, Ack1 and p130cas (Eisenmann *et al.* 1999, Majumdar *et al.* 2003), raising the possibility of signalling *via* engagement of Fras1/Frem2. A study by Kiyozumi *et al.* has also demonstrated that Frem1 is capable of mediating cellular adhesion *in vitro* through interactions with αv- and α8-containing integrins (Kiyozumi *et al.* 2005). Mice null for these integrin subunits display no obvious skin defects which means it is unlikely to be the sole mechanism by which Frem1 mediates epidermal adhesion. However, mice lacking the α8 subunit are char-

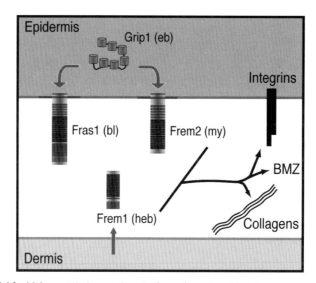

Figure 3. A model for blebs protein interactions in the embryonic epidermis
Mouse studies have shown that the PDZ domain protein Grip1 can bind to the intracellular C-terminus of the transmembrane domain proteins Fras1 and Frem2 and is required for normal localisation of the proteins to the basal side of keratinocytes. Loss of Grip1, Fras1 and Frem2 gives rise to the *eye blebs* (*eb*), *blebbed* (*bl*) and *myelencephalic blebs* (*my*) mutants respectively. Loss of the dermally expressed secreted protein Frem1, results in the *head blebs* (*heb*) phenotype. Frem1 has been shown to interact with integrins αv and α8 while all of the proteins are predicted to interact with collagens and other components of the basement membrane zone (BMZ). Mutations in FRAS1 and FREM2 have been shown to cause Fraser Syndrome in humans.

acterised by severe renal dysgenesis (Muller *et al.* 1997), suggesting that its interaction with Frem1 may be of functional importance.

Future directions and outstanding questions

The cloning of the blebs mutants and the identification of FS mutations has finally proven a genetic association first proposed some 25 years ago. This has given us a new insight into the early events in epidermal differentiation and basement membrane assembly and nephrogenesis. The proteins involved are largely uncharacterised and their interactions with components of the ECM are only just beginning to be understood. Various aspects of human disease are also puzzling. Why are the majority of mutations found in Fras1 when the phenotypes of the different blebs mutants are so similar? Does loss of Frem1 and Grip1 in humans lead to more severe phenotypes, and if this is the case, are we aware of the full influence of these proteins on early development? While the large size and complex domain organisation of these proteins will make dissecting their function challenging, knowledge of the human and mouse phenotypes which result from their loss are already providing avenues by which to pursue these studies.

Dedication

When beginning work on Fraser syndrome in 1999 I had little idea about the history of the condition and who had described it. The original articles were not kept by our library and, as usual, my first stop was OMIM. First contact with George was via my then clinical fellow Lesley

McGregor, whom George introduced to the very first 'Fraser syndrome' patient he had seen and who contributed to his initial description of the condition. We met in 2003 at the European Society of Human Genetics meeting in Birmingham and it was such a pleasure to hear how work on the clinical delineation of the syndrome had developed. George provided all his original articles and has always shown an active interest in what we were doing. This volume will contain many dedications to George testifying to his friendship and clinical acumen dating back decades. I hope our contribution demonstrates the value of the carefully crafted case description and speaks to George's continued contribution to medical research well into retirement. Thank you George, and all best wishes for the future.

Acknowledgements

We would like to thank all the families and clinicians who have helped us in our work, and the Wellcome Trust and British Heart Foundation for support. Also, thanks to Shalini Jadeja and Mieke van Haelst for their efforts in the lab, and to Sue Darling for her work on the blebbed mice. Oxford University Press allowed us to reproduce much of the text and figures from our previous review for this special occasion. The final word must be one of dedication to the late Professor Robin Winter, who was also a great inspiration for this work.

Visual impairment and learning disabilities in children
Plus ça change, plus c'est la même chose

M. Warburg[1] and R. Riise[2]

[1]University of Copenhagen Hospital, Glostrup, DK-2600 Glostrup, Denmark
[2]University of Oslo, Rigshositalet, Norway

It is a great pleasure to honour George Fraser on his 75th birthday. With his important books on *The Causes of Profound Deafness in Childhood* (1976) and *The Causes of Blindness in Childhood* (1967) and his later studies on malformations George has inspired many paediatric ophthalmologists and shown that clinical genetics is an integral part of clinical ophthalmology. It was a new approach to ophthalmology at a time when paediatric ophthalmology was in its very early beginnings. Since the monographs appeared, a number of other studies on the prevalence of childhood blindness have been published. It is interesting to follow how the registration of causes of blindness in children has changed during the last 50 years. These changes can roughly be divided into changes in notification or registration requirements or procedures and changes due to the surprising extent of genetic and clinical heterogeneity. In the following, we shall discuss Nordic registration of visual impairment in childhood (VI), disorders due to non-genetic causes, some heritable disabilities with VI and examples of very rare syndromes seen in children with learning disabilities (LD). The aim of the study is to stress that ophthalmological examinations of children with LD will disclose new heritable syndromes and simultaneously offer treatment for ordinary eye disorders which might not otherwise have been detected. We do not pretend to cover the whole spectrum of VI in childhood.

Notification and registration

Blindness and VI in childhood are rare in industrialised countries where infections, traumas and poor nutrition have been reduced to a minimum, but blindness is involved in a substantial part of genetic disorders and syndromic malformations in these populations, and the affected children are faced with many years of blindness. Blind and VI persons are notified in many countries. In Denmark notification of VI in children is compulsory.

Until the 1980s, children with VI were educated in competitive schools for the blind, and it was an unwanted surprise when in 1976 (Warburg *et al.* 1979) we examined children in all special education schools and discovered that there were as many children with VI among the children with learning disabilities (LD) as there were already in the blind schools. In other words, one half of all children with VI in Denmark had LD, but it was only their LD that had been registered in the National Medical Registers.

De-institutionalisation and decentralisation of care for all children and adults with LD, hearing loss, epilepsy and other congenital or juvenile disabilities were introduced in the 1970–1980s in the Nordic countries. Institutions for people with LD were closed down, and itinerant teachers went to mainstream schools to teach the teachers teaching the visually impaired pupils. Children with visual impairment had the opportunity of acquiring telescopes, magnaprint textbooks and other educational material. Registration of VI and blindness was therefore possible in spite of the decentralisation of care. Children with LD are given special education and parents receive an allowance to meet expenses caused by the impairment. A paediatric examination is usually necessary to obtain this. The former practice of classifying some children as ineducable has been abandoned and all children receive education whether they are able to learn reading and arithmetic or only to learn moving, eating and babbling.

In 1990–92 a Nordic study on VI in childhood based on National Registers was performed in Denmark, Norway, Finland and Iceland. There were 2527 visually impaired children aged 0–17 years or 70 per 100,000 age-specific population, and it was shown that 30–50% (Riise *et al.* 1992 Table 9) had additional impairments, the commonest being learning disabilities. A Swedish study of 2373 visually impaired children aged 0–19 years (Blohmé and Tornqvist 1997, Table 1) showed a prevalence of 109 per 100,000. Among them, 60.4% had additional impairments; learning disabilities comprised 49.6% of the VI children. The Nordic study showed that prenatal conditions were prevalent in 66%, perinatal in 22% and juvenile in 7%. The list of rare genetic conditions comprised the majority of known genetic disorders in paediatric ophthalmology (Rosenberg *et al.* 1992, Tables 2, 3). The Nordic and Swedish studies confirmed that approximately one half of children with low vision had LD. A similar study of VI children in Finland (Rudanko and Laatikainen 2004) found that the incidence of VI in children born at full term had not declined during the years 1972–89.

A doubling of the number of children with VI compared to surveys 50 years ago is mainly caused by inclusion of children with VI associated with moderate and severe LD.

Clinical classification

The clinical classification of causes of disorders with VI derives from a classical meticulous description and illustration of features, and the naming of disorders is based on these features or eponyms. In the industrialised countries, the main causes of VI in childhood are retinopathy of prematurity (ROP), cerebral visual impairment (CVI) and various genetic disorders (Fraser and Friedmann 1967, Rosenberg *et al.* 1992, Hansen *et al.* 1992, Blohmé and Tornqvist 1997). Patients with very rare syndromes are much more common in children with LD than in the background population. Individuals with LD often show VI caused by uncorrected refractive errors. In fact, one half of the adult patients with VI simply needed glasses. Similar investigations have not been described in children.

Retinopathy of Prematurity and Cerebral Visual Impairment

Children with LD now grow up with their families, the majority are examined by paediatricians and many are referred to ophthalmologists because in roughly 50% of children with severe LD squint, uncorrected errors of refraction are common and give rise to mild and moderate VI.

The majority of acquired causes of blindness and VI in children from the industrialised countries are retinopathy of prematurity (ROP) and cerebral visual impairment (CVI). Fifty years ago the incidence of ROP was much higher (Fraser and Friedmann 1967), but when it was understood that oxygen treatment of the premature babies during a short window of time was deleterious, the incidence and prevalence dropped; a rise in incidence is now seen in less affluent but developing middle income countries and in extremely premature neonates in the affluent countries. It is possible that rare mutations in the Norrie gene (Hiraoka *et al.* 2001, Talks *et al.* 2001) play a role in the development of ROP in a few severely premature infants.

CVI is the other common perinatally acquired cause of VI. It is preferentially seen in infants with developmental delay and thus was not registered previously. The diagnosis requires ultrasound exam in infancy and magnetic resonance imagining (MRI) at later ages. In Sweden CVI was diagnosed in 22.9% of children with VI and additional impairments were seen in 96% of them. The important features are ischaemic or haemorrhagic periventricular leucomalacia identified on MRI (Pike *et al.* 1994).

Optic atrophy

In the Nordic series, non-genetic optic atrophy (OA) was the cause of VI in 19% of all cases (n = 483), and 85.5% of them (n = 365) were present in children with additional impairments. There were 39 children with hereditary optic atrophy but the majority of those affected had OA secondary to other causes. In Sweden, OA was present in 17% of the VI and 88% had additional impairments. VI due to OA is a very heterogeneous group, and the diagnoses requires interdisciplinary examinations.

Congenital cataract

Congenital and infantile cataracts are the commonest heritable malformations leading to VI (Hansen *et al.* 1992, Blohmé and Tornqvist 1997). In PubMed, 168 items refer to congenital cataract, and Geneeye has 171 entries of syndromes with congenital, infantile and juvenile cataract. In a survey of 2.6 million children based on the Danish National Register of Patients, Haargaard *et al.* (2004a) found an incidence of congenital or infantile cataract of 72 per 100,000 of which 64% were bilateral. Surgery within the first two years of life was performed in 55% of children with congenital or infantile cataract. Almost all young Danish women are vaccinated against rubella, traumatic cataract has declined over the last 20 years (Haargaard *et al.* 2004a) and hence, the majority of the cases are presumably genetic, although this has been proven in only 29% of the Danish cases (Haargaard *et al.* 2004b). Additional impairments were found in 29% of patients with VI caused by congenital cataract in the Nordic survey (Hansen *et al.* 1992), and in 34% of the Swedish cases (Blohmé and Tornqvist 1997). LD is the commonest complication as already shown by François (1959). Many cataract genes have been mapped and a number have been cloned. Until recently, almost all cases with identified mutations coding for congenital cataract were those involved in systemic disorders. The genetic basis of uncomplicated cases is now being identified in increasing numbers (Reddy *et al.* 2004, Addison *et al.* 2005, Forshew *et al.* 2005); rare genetic cases are found in children with LD, such as for instance in the MICRO syndrome (MIM 600118) (Microcornea, Microcephaly, Cataract, Retardation, Optic atrophy, and various cerebral malformations (Warburg *et al.* 1993)). A gene causing this syndrome has been mapped and cloned (Aligianis *et al.* 2005). Few ophthalmological studies have been performed in children with LD, and rare syndromes not yet described, are bound to be present among them.

Retinal dystrophy

The Nordic study (Hansen *et al.* 1992) showed that childhood retinal disorders – ROP excluded – comprised retinal dystrophy and syndromes with retinal dystrophy or retinitis pigmentosa (RP) are one of the leading causes of VI. RP was diagnosed in 339 patients in the Nordic countries, or 13% of VI children. It has been known for more than 15 years that rare cases of disorders associated with RP were more frequent among children with LD than in the background population and in a survey of the first 3000 patients from the eye clinic for children with LD we found 26 with Bardet-Biedl syndrome, 20 with Batten disease, six with RP and hearing loss and a variety of other identified syndromes with retinal dystrophy (Warburg *et al.* 1991a). Five patients had chromosomal aberrations, two had Down syndrome, there was a boy with a cone dystrophy and a t(1;6)(q44q27) (Tranebjaerg *et al.* 1986), a boy with a cone rod dystrophy and a del(18q21.1) (Warburg *et al.* 1991b), a young man with mosaic trisomy 9 and rod-cone dystrophy and a case of Kearn Sayre syndrome. In the Nordic series, additional impairments were found in 124 (36.6%), the majority being LD. This shows the heterogeneity of RP, but is not useful for calculation of prevalence.

RP is more prevalent among adult patients with LD than in the general population. In a clinical study of all adults with LD in a Danish county, a total of 837 people with moderate, severe and profound LD, RP was found in six persons (Warburg 2001), as compared to a prevalence of 22 per 100,000 in Denmark (Haim 2002).

Molecular genetic studies of RP have led to an understanding of the immense genetic heterogeneity of RP disorders, thus the Geneeye tabulates 55 clinical entries of disorders with RP. The molecular genetics and the biochemical machinery of the rods and cones of the retina have been decoded and classified into proteins encoding visual transduction, visual cycle, transcription factors, structural proteins, the spliceosome complex and cellular traffic (Maubaret and Hamel (2005), and a protein database of genes involved in retinal degenerations is provided by Retina International's Scientific Newsletter protein database (*http://www.retina-international.org/sci-news/protdat.hum*).

The diagnoses are often delayed in children with additional impairments because the patients are difficult to examine. Few care persons will observe reduced visual fields and difficult ambulation in the dark and VA is usually retained for a long time. Children with LD and RP should have karyotype testing or a comparative genomic hybridisation (CGH) exam because a translocation or deletion may pinpoint the locus and accordingly the mutated gene.

Electroretinography (ERG) under anaesthesia and extensive family studies of patients suspected of having with RP are necessary in spite of the difficulties, because a correct diagnosis will improve the psychological and educational development. Carers will understand the situation and can be informed of tube vision, night blindness and progressive visual loss.

Clearly, any counselling and future treatment must take the heterogeneity of molecular genetics of RP into account, however, the differential molecular diagnosis relies on rather expensive laboratory materials and a cheap genetic retina kit has not yet been produced.

Walker-Warburg Syndrome

The Walker-Warburg [1] syndrome is an example of the clinical and genetic heterogeneity that has recently been disclosed in many (most?) congenital and dystrophic disorders.

It presents in severely impaired infants with hydrocephaly, lissencephaly type II, hypoplasia of the vermis, agenesis of the corpus callosum, microphthalmia, cataract, anterior chamber malformation with glaucoma and retinal dysplasia and retinal folds (Warburg *et al.* 1978). The expectancy of life is usually only a few years. Inheritance is autosomal recessive, and the syndrome is caused by a mutation in *POMT1*[2] and *POMT2*[3] (Van Reeuwijk *et al.* 2005), leading to defective O-mannosylation of α-dystroglycan which destroys the pathway traffic between intra- and extra-cellular proteins necessary for muscular function. A different mutation in the *POMT1* gene gives rise to limb-girdle muscular dystrophy with LD[4] (Balci *et al.* 2005, Reeuwijk *et al.* 2005, Topaloglu 2005). There are other clinical syndromes similar to the WWS, viz. Muscle-Eye-Brain[5] due to mutations in the *POMGNT1*[6] and Fukuyama muscular dystrophy[7] in which the *FKRP*[8] and *FCMD*[9] genes are mutated. In most cases, the MEB and the FMD are milder than the WWS, but variation is great and the features of these four disorders can be so like each other that a molecular genetic analysis is necessary to distinguish between them. Although treatment is so far hypothetical, it is possible that boosting or bypassing the genes expressed downstream from the dystroglycan complex may be beneficial (Engvall and Wewer 2003, Rando 2004). The identification of these disorders has resulted in improved understanding of the role of dystrophin and its place in the dystrophin-dystroglycan pathway. It may eventually explain (to the ophthalmologists) why Duchenne muscular dystrophy[10], the LARGE mouse muscular dystrophy, and the WWS share electroretinographic anomalies (Jensen *et al.* 1995).

Heterogeneity

The clinical identification of a disorder is the initial approach to the patient's problems, the pathogenesis and prognosis, while the unfolding of molecular genetic aetiology and heterogeneity leads to an understanding of the fundamental physiology and embryology of tissues and cells.

The hereditary disorders leading to VI in children have not changed much during the last 50 years, but new diagnoses have been included in the registration.

Genetic studies of children with a combination of visual and intellectual impairment have been instrumental in delineating new clinical and genetic disorders and thus have a chance of leading to new a biological understanding and concomitantly offering treatment to the impaired children.

Knowledge of the considerable clinical and genetic heterogeneity has changed the classification of genetic syndromes. Disorders that were previously viewed as singular diseases are now understood to be caused by mutations in different genes. There are, for instance, at least four different loci associated with familial exudative vitreoretinopathy (Toomes *et al.* 2005) and eight loci associated with the Bardet-Biedl[11] syndrome (Mykytun *et al.* 2003, Beales 2005). By contrast, mutations in the same gene can give rise to great variation in clinical expression as mentioned with the WWS. This is a general taxonomic problem which proteomics will presumably solve. Much is known about the embryology of the eye and adnexa; many of the heritable causes of VI will probably be found to exert their action in the prenatal period. The *PAX6* gene which is instrumental in aniridia and holoprosencephaly is an example of this. There is a need for new instruments which can analyse mutations of the DNA of eye disorders rapidly and cheaply.

Summary

Persons with learning disabilities (LD) in Nordic countries have recently been resettled in the main population; the large inmstitutions for people with hearing loss, learning disabilities, epilepsy, visual impairment, etc. have been closed. The children with LD are now being subjected to interdisciplinary examinations whereby it has been observed that roughly 40–50% of all visually impaired children have LD. There are 'new' and rare, but identified diseases and syndromes among these children. We discuss how molecular genetic aetiologies may improve the biological understanding of biochemical pathways in cells and disease.

Notes

1. WWS MIM 209900
2. MIM 607423
3. MIM 607439
4. MIM 609308
5. MEB MIM 253280
6. MIM 606822
7. FMD MIM 253800
8. MIM 606596
9. MIM 607440
10. DMD MIM 310200
11. BBS MIM 209900

Michele, Florina, and George Fraser:
the cryptophthalmos story at the human level

Eva Yap-Todos and Damian Yap

BC Children's Hospital, Vancouver, British Columbia V6H 3V4, Canada

We make a living by what we get; we make a life by what we give.
Sir Winston Churchill (1875–1964)

The goal of life is living in agreement with nature
Zeno (335 BC – 264 BC), from Diogenes Laertius, Lives of Eminent Philosophers

I will not spend my life. I will invest my life.
Helen Keller

Was mich nicht umbringt, macht mich stärker.
That which does not kill me, makes me stronger.
Ce nu mă distruge mă face puternic. (Romanian)
Friedrich Nietzsche

George Fazekas (Hungarian for 'potter') was born in Czechoslovakia of Hungarian parents. After living in England for many years, he changed his name to Fraser. And so that is how we know the eminent scholar, physician, scientist and friend, George Fraser.

It seems to be no coincidence that one of my first patients that I (EY-T) accompanied for eye surgeries in Hungary had Fraser syndrome. The particular reason I was chosen for this task was that, although I was born and lived in Romania, Hungarian was also spoken in my family and I speak the language fluently myself.

My patient, and later foster-daughter, Florina, was born in Romania in March 1999. Due to lack of knowledge with regard to her multiple congenital malformations, none of the doctors knew what to expect regarding her survival. As a result, her parents left Florina in the hospital and, having no hope, they later decided to abandon her legally, and to leave her under constant medical supervision in the Children's Hospital, where I was to do my house officer training. Even after her geneticist was able to put Florina's congenital malformations under the name of Fraser syndrome (aided by the McKusick Catalogues), when Florina was nearly one year old, there was still nothing known about her life expectancy.

However, things started changing for Florina during her eye operations in Hungary in 2000, when she proved that despite her multiple severe malformations and sensory impairments, she was an ambitious and determined tiny human being. It was for the first time that I realised that there was hope for her. There was hope for children born like her! Although the three operations in Hungary did not have a good outcome, they did work out eventually for Florina's good, because she managed, innocently, simply, but very surely, to gain my heart for her and for children like her. Entering our family's care soon afterwards, she started developing, and constantly inspiring those around her.

Later, I learned about Fraser syndrome by meeting on the internet a few families from different parts of the world who have children born with Fraser syndrome. They confirmed that there is definitely hope for these children and became real models for Florina. It was through this personal benefit from individual experiences with families with children born with Fraser syndrome, from around the world, that we started the Fraser syndrome website as an international medical resource and support network for medical professionals and families.

In the spring of 2004, Florina had two eye operations in Great Ormond Street Hospital for Children, in London, UK. It was then that we had the great honour and privilege of having Dr George Fraser himself visit Florina at our residence in Cambridge, UK. I was astonished to find him addressing me in Hungarian, having recognised my accent in English. What could be a better way to learn about the history of Fraser syndrome than from George himself? We learned of George's role in it and his contribution to the life of Michele, the longest-known surviving patient with Fraser syndrome (who is now 51 years old). Although it was an apparently small observation that George made with regard to the conductive rather than sensorineural nature of Michele's deafness when he first met her in 1960 at the age of five years, George's professionalism contributed to a real direct change in the life of this patient. Thus, he immediately referred her to an otorhinolaryngologist who performed surgery which enabled her to hear for the first time soon after meeting George; this also proved later to become an indirect influence in the lives of many others too.

Furthermore, George Fraser's intervention in Michele's life also remains an inspiration for those medical professionals involved in the lives of these patients. Due to their multiple congenital malformations, many patients, particularly from countries where medical resources are scarce, are automatically being left to a life without any sense or hope. However, paying attention to small things regarding these children can bring real changes in their lives as well as in our own humanity.

Our support group, the *Fraser Syndrome Support Group* (*http://frasersyndrome.info*), formed by a few families scattered around the world, is constantly learning valuable lessons that we would not learn if we did not have these children ourselves. George Fraser is a co-patron, together with Dr W.P. Reardon, who has made a contribution to the Festschrift and *liber amicorum*. Some of the main lessons are the following:

There is always hope. Even despite all adverse appearances, there is hope. However, it is our choice to acknowledge that, and hence our responsibility whether we are going to do something about it.
To acknowledge that there is hope and that it is worthwhile to invest in these children, one needs to be informed, therefore it is important to share and exchange one's experience in order to extend knowledge.
Doing just a little can bring about a great and real change.
The more defects and needs a child has, the more help it requires in order to reach its potential.

The physician attending the family needs to consider the fact that all the many components of the Fraser syndrome are very variable in their expression and that this leads to important consequences for the survival of affected individuals and for their integration into society. A large number, probably the majority, do not survive to infancy, being miscarried, or stillborn, or dying in the first days of life. Often this failure to survive is due to extreme degrees of renal aplasia and/or laryngeal atresia, incompatible with life. These could be termed the *severe* cases.

Among those who survive the neonatal period, some are affected more seriously than others. The main variation influencing integration into society is in the sense organs, leading to varying degrees of loss of sight and/or hearing. In the *moderate* cases, one or both of these senses are seriously compromised bilaterally, and educational measures applicable to the deaf or the blind or the deaf-blind should be available. In the case of deaf-blind children especially, there is a tendency to assume co-existent mental subnormality because of the many additional physical handicaps which may be prominent.

It should be emphasised, however, that there is no evidence that mental subnormality is a component of the syndrome. These children, therefore, respond to educational measures just like other deaf, blind or deaf-blind children, albeit with additional difficulties due to the physical handicaps associated with the syndrome. It should be noted that their sensory handicaps, especially deafness, may be amenable to surgical correction.

There is also a substantial proportion of *mild* cases, in whom the eyes and ears are spared on one or both sides, leading to normal hearing and sight.

While these marked variations do not correspond to any distinct biological sub-entities within Fraser syndrome in that severely and mildly affected children may co-exist in the same sibship, the serious social and educational effects of pronounced involvement, especially of the eye and ear, should be borne in mind by the attending physician who should direct the family to the appropriate educational authorities on the basis of an *a priori* assumption that mental development is normal.

This preliminary attempt at classification could perhaps help physicians when they are faced with a case of Fraser syndrome, and when they need to advise the parents with regard to what they should expect. Being the first to advise parents, they can determine their attitude and perhaps the way that they will deal with their child, avoiding inappropriate treatments.

By helping to correct Michele's hearing, George Fraser enabled Michele's parents to send her to a special school and the rest can be read in an excellent book written by her mother, which we hope to publish soon. Both of us were inspired to be able to accompany George on a very heartwarming trip to the East of England to visit Michele in October 2005 when the photograph of the two of them, reproduced here as Figure 1, was taken. They had not seen each other for 40 years.

Although it might seem logical not to invest resources in children with low mental ability or those whom we suspect of being of low mental ability, it is, nevertheless, uplifting and enriching to invest in somebody from whom we expect to get nothing back in return. A total lack of any return is rarely the case, since people invariably find that investment in such instances is returned many fold.

How much more investment then, is required for these children with Fraser syndrome who have normal mental ability but who, without appropriate help, will be classified as retarded. The truth is that due to their multiple sensory impairments, they will never learn just by listening or watching – they have to be actively and constantly taught and stimulated.

That represents our great challenge and herein lies our very humanity: in *helping those who cannot help themselves* (our motto), not because they do not have the capability to do so, but because they are prevented from doing so (i.e. the children with Fraser syndrome who present multiple sensory impairments). The reward is great for those who see that and do something about it. The truth is that most frequently very little is needed. No one asks for great sacrifices.

Figure 1. Michele and George, aged 51 and 74 respectively. (*reproduced with permission*)

George Fraser proved himself willing to pay attention and do something about and for these patients. In the foreword on our Fraser Syndrome Support Group website, George writes, '*Perhaps this small contribution to the improvement in the quality of the life of this girl represents a greater achievement in my life than that of my name becoming attached to a syndrome.*' Indeed, we believe that George not only bequeaths the scientific and medical communities with a wealth of literature and scholarly articles which will no doubt be material for further work for generations, but also an imprint of a man who had his hand on the pulse of creation. We are grateful not only to have known of George but also to have been able, and to be able, to work together with him and build a friendship. We wish him many happy returns on the occasion of his 75th birthday!

Our gratitude, too, remains forever, and will be refreshed by each encounter with a child with Fraser syndrome and their loving families.

Mendelian disorders deserve more attention

Stylianos E Antonarakis[1] and Jacques S Beckmann[2]

[1]Department of Genetic Medicine and Development, University of Geneva Medical School,
and University Hospitals of Geneva, 1 rue Michel-Servet, 1211 Geneva, Switzerland
[2]Department of Medical Genetics, University of Lausanne and Centre Hospitalier Universitaire Vaudois,
2 Ave Pierre Decker, 1011 Lausanne, Switzerland

Fotis Kontoglou, a contemporary iconographer, painter, and writer, once wrote about Leonid Ouspensky (another famous iconographer and writer of the last century): 'Leonid was my best friend, but I never met him'. Last year I received a stunning email from George Fraser entitled 'We do not make friends, we recognise them!' George is a dear friend, and yet we have only met a few times. He is the type of scholar, thinker, polymath, and clinical researcher that I have always admired in my thoughts without ever having had the opportunity to find all these qualities together in one person. Until I met George … I was therefore delighted to learn that besides the friendship, I am also scientifically related to George; a kind of a first cousin if you wish. In the 1960s he worked in Greece with people who later were either my teachers or colleagues with whom I had the opportunity to co-author papers: G. Stamatoyannopoulos, C. Kattamis, D. Loukopoulos, P. Fessas, C. Choremis, D. Ikkos. This chapter is thus dedicated to G.R. Fraser, the friend and relative, by Stylianos Antonarkis.

Abstract

The study of inherited 'monogenic' diseases contributed greatly to our understanding of the mechanisms of pathogenic mutations and gene regulation, and to the development of powerful investigation tools offering new prospects in disease management and treatment. But interest gradually shifted away from monogenic diseases that collectively affect a small fraction of the world's population, to focus on the more 'complex', 'multifactorial', common diseases with genetic predisposition. This quest for the genetic variability associated with common traits should not be done at the expense of Mendelian disorders, since these disorders could still reveal new genetic principles, and clues into gene function.

Introduction

At the latest count there are barely 25,000 protein-coding genes in our genome (Anonymous 2004a), encoding perhaps an order of magnitude more distinct gene products. Although this number will probably be many times revised in the near future, a major and challenging question is currently in the minds of biomedical researchers, funding agencies, patient groups, and bio-technology and drug companies: what is(are) the function(s) of each one of these genes, and how genetic variation of each one contributes to health and disease? Indeed, understanding the gene product interaction and the complex and hierarchical network of biochemical, signal transduc-tion, and development pathways, is necessary for the introduction of intelligent and rationalistic treatment modalities to the myriad of rare and common inherited disease phenotypes.

Herein we focus on current trends in genetic research (from simple to complex traits), highlighting lessons we have learned from the study of Mendelian disorders; we emphasise that the latter represent not only a necessary and indispensable treasure trove for the elucidation of gene function and for the reconstruction of normal and pathological pathways, but could also contribute significantly to our understanding and solving of the common, complex diseases.

Drifting away from Mendelian disorders?

What is the best way to begin to understand gene function? Fortunately, there is not just a single answer to this. Genetics, biochemistry, bioinformatics, and utilisation of model organisms, each provide complementary ways and opportunities for functional annotation of human genes. Major efforts are currently under way to identify all functional elements in the human genome (Anonymous 2004b).

We argue that a timely and low-risk option for elucidating the gene function is to link naturally occurring pathogenic mutations with the so-called monogenic (or monolocus) disorders. The process is well known to medical geneticists; in the past 20 years pathogenic allelic variants of a total of 1811 genes – less than 10% of our gene repertoire – have been found to cause highly penetrant 'monogenic', Mendelian phenotypes (*http://www.ncbi.nlm.nih.gov/entrez/query.fcgi?db =OMIM* (19 October 2005) comprehensive knowledge-based database of human genes and Mendelian phenotypes (Antonarakis *et al.* 2000, Hamosh *et al.* 2005). Most of these discoveries have been achieved before the completion of the euchromatic sequence of the human genome, before the introduction of methods for genome-wide genotyping and gene expression analyses, and before the discovery of the extraordinary variability among human genomes. The paradox is that in the first few years after the acquisition of the complete sequence of the human and other genomes the interest for 'monogenic' disorders is in the decline and most of the current attention is focused on the complex common, multifactorial, polygenic phenotypes. Figure 1A shows the number of novel gene-disease matches in the last 15 years; the data are from OMIM (*http://www.ncbi.nlm.nih.gov/entrez/query.fcgi?db=OMIM*). Figure 1B shows the pathogenic mutations recorded in the HGMD (Stenson *et al.* 2003) (human gene mutation database; *http://archive.uwcm.ac.uk/uwcm/mg/hgmd0.html*). Paradoxically, and contrary to expectations, after the completion of the sequence of the euchromatic part of the human genome the discovery of new gene-disease matches, and additional pathogenic mutations are in the decline compared to the previous decade in which primitive tools for gene discovery were available. A comparison of the number of gene identification reports published in *Nature Genetics* during the first six months of 2000 (n=37) and 2005 (n=14, and mostly as brief communications) is but another illustration of this tendency. One of the reasons for this new trend is that most molecular genetics laboratories and high impact journals have turned their attention to the polygenic, complex, multifactorial phenotypes and traits. In addition, availability of funding resources, privileging the study of common (also commercially attractive) targets over that of rare orphan diseases, further reinforced this trend.

Besides genuine interest, editorial and funding policies, additional factors contributed to this effect. Most low hanging fruits (e.g. highly prevalent, highly penetrant, easily diagnosed monogenic entities) were, as usual, elucidated first. Consequently, we are now predominantly faced with more rare Mendelian disorders, the number of which is hard to estimate (we predict that there may be well over several thousands; perhaps at least one per gene), and for which the usual

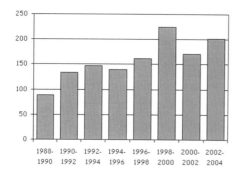

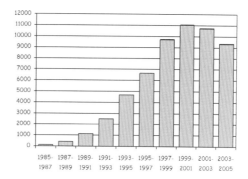

Figure 1A. Number of OMIM entries with allelic variants (pathogenic mutations that cause 'monogenic disorders') created during the indicated time intervals.

Figure 1B. Number of mutation entries in HGMD (http://archive.uwcm.ac.uk/uwcm/mg/hgmd0.html) published in the indicated time intervals.

conventional strategies to map and identify a pathogenic locus or variant may not apply. Elucidation of these entities calls for the development of alternative powerful strategies and tools.

From gene to phenotype

We have learned a great deal from studying these monogenic traits, in terms of mutation processes and pathophysiology; in addition we became aware of particular phenomena such as parental imprinting and epistatic interactions. However, despite the impressive progress made, for most of these diseases there is a fundamental lack of understanding of their biological cause. Identifying cause-effect associations is not a trivial issue: even when we know the exact molecular defect, the link between it and the resulting phenotype can often be elusive: for instance, we cannot explain how alterations in ubiquitously expressed essential splicing factors manifest exquisite tissue selectivity as seen in retinitis pigmentosa (Baehr and Chen 2001). Mendelian disorders are far from having delivered all their secrets yet, and their study remains therefore of great pertinence to medical genetics.

Although research on the genetic predisposition to complex phenotypes should be a major focus of activities in the years to come, we would wish to emphasise the tremendous power of 'monogenic' phenotypes to understand human nosology. Hence we argue that we have now (with the development of the genomic infrastructure and the advancement of the technological platforms) the best opportunity to link, directly or indirectly, the majority of genes with aberrant phenotypes. Some of the reasons for this are discussed below.

From phenotype to gene

Genome-wide linkage analysis to map the critical genomic interval containing a morbid allele is currently much easier because of the development of novel genotyping methods (Affymetrix; *www.affymetrix.com/index.affx*, Sequenom; *www.sequenom.com*, and Illumina; *www.illumina.com* platforms for example) and the results of the HapMap projects that provided information on common SNP variants and their allele frequencies in human populations (Anonymous 2003b, Altshuler *et al.* 2005, Hinds *et al.* 2005). Using these enhanced high-resolution genotyping capacities, it should be possible, even for rare diseases, to narrow onto short shared genomic intervals; common 'disease' haplotypes could be identified in unrelated patients carrying a common ancestral founder mutation, thereby pinpointing the location of the disease allele.

The results of the genome sequence and comparative genome analysis uncovered additional functional genomic elements that could be targets of pathogenic mutations; these include the functionally conserved sequences that are not protein-coding (Dermitzakis *et al.* 2005), the miRNAs and other non-coding RNAs (Eddy 2001, He and Hannon 2004), and the numerous insertion-deletion/copy number variations of the different human genomes (Eddy 2001, Sebat *et al.* 2004, Sharp *et al.* 2005). Hence, positional cloning is now much easier because of the precise knowledge of the human sequence (Anonymous 2004a) and the availability of rapid methods for mutation analysis, both in terms of point mutations or larger genomic lesions. Determination of gene expression differences can also pinpoint candidate genes. What used to be a laborious effort could now be greatly facilitated by merging a variety of enabling methodologies and bioinformatics tools.

The use of sequence conservation and experimentation in model organisms provides ways for validation of dysfunction of mutant proteins. It is worth noting that the major bottleneck is not the mapping of a disease locus or the identification of candidate pathogenic alleles, but the demonstration of the pathogenicity of the true disease-related allele and its subsequent pathophysiological characterisation.

The increased ease of generation of conditional, or tissue-specific (knocked-in or knocked-out) animal or cellular models is a great adjunct to understand and validate the functions of the suspected culprit genes. Indeed, linking a mutation within a defined gene or functional genomic element to a particular phenotype remains essential to assign a function to this genomic feature.

Databases of protein-protein interactions enrich the space of functional candidate molecules for phenotypically related disorders (von Mering *et al.* 2002, Gavin and Superti-Furga 2003, Brunner and van Driel 2004). Using these as Ariadne's thread allows one to track additional genes involved in related pathologies, thereby progressively reconstructing the normal and pathophysiological pathways in which these genes intervene.

Are we finished with Mendelian disorders?

Some investigators may argue that there are not enough 'monogenic' disorders left to study. This is probably not correct; there are currently more than 1500 known monogenic phenotypes for which the defective gene remains elusive (*http://www.ncbi.nlm.nih.gov/entrez/query.fcgi?db =OMIM*). Many of these are rare and present in a few families only. Adequate clinical and technological resources need to be set up to allow and foster sharing of these precious samples as part of collaborative efforts.

In addition, there are large, poorly studied population groups with inbreeding and consanguinity, and it is likely that many unknown recessive 'monogenic' phenotypes exist among them. As much as 20–50% of marriages are consanguineous in approximately one billion people (Bittles 2001).

Furthermore, there are numerous as yet uncharted isolated populations that also harbour high frequency recessive alleles due to founder effect and inbreeding. Finally most of the 'monogenic' causes of foetal wastage and subfertility are currently unexplored (Hackstein *et al.* 2000). Investigation of the former calls for the collection and preservation, whenever possible, of the corresponding tissues, which would then allow complementation of genetic analyses by determination of (RNA or protein) expression profiles (Strachan *et al.* 1997). As many of these foetal events are also characterised by identifiable sub-microscopic chromosomal abnormalities (detectable by the new high-resolution tiling path CGH arrays), wouldn't it be great if a fraction of the existing large sequencing capacity were used to systematically sequence around these rearrangements? This might uncover new pathological mutations as well as to provide a basis for a rationale explanation of dysmorphological features.

Besides the recessive traits, an unknown number of dominant and X-linked 'monogenic' phenotypes may exist in African and other populations that are not adequately studied. A concerted and well-organised effort needs to be initiated (preferably led by international organisations or local interested groups), towards the collection of samples from well-diagnosed, well-informed individuals and their families from selected communities. This is actually the most difficult, yet essential part of the discovery process, since the laboratory and genetic analysis parts are now almost off-the-shelf endeavours.

Human or mouse genetic models

Other investigators argue that the gene knockout project of all open- reading-frames in mice and the precise assessment of the resulting phenotypes (Auwerx *et al.* 2004) will provide similar information in a much more controlled, systematic, and comprehensive way. There are several shortcomings of this approach; not all human genes have a mouse orthologous gene and vice versa, and second, several clinically relevant phenotypes have human-specific manifestations (e.g. mental retardation, cognitive behaviour), which may be difficult to evidence or reproduce in mice. Even when phenotypes are easily scorable and the corresponding genes are present in the respective genomes of both species, it is important to stress that the understanding of a gene dysfunction is better accomplished by an allelic series of (naturally occurring) mutations and not just with sizable gene deletions or other inactivating mutations. Furthermore, in numerous instances, similar phenotypes can be caused by mutations in any of a number of distinct genes (*genetic heterogeneity*), and in other cases, different mutations within the same gene can give rise to distinct phenotypes; the Lamin A/C gene, in which allelic series of mutations result in up to six different clinical phenotypes (Ben Yaou *et al.* 2005), constitutes a particularly illustrative example. In other instances, the same pathological mutation, depending on the genetic background in which it occurs, may display different expressivity, penetrance or even distinct diseases (e.g. in LGMD2B and Miyoshi myopathy Liu *et al.* (1998), Bashir *et al.* (1998)). Yet in other diseases, such as in hereditary prion disease, the disease phenotype is determined by the combined effect of pathogenic mutations and polymorphic variants (Gambetti *et al.* 2003). Mouse models present an additional disadvantage because phenotypic assessment in this species is not as complete and well-developed and sophisticated as it is in humans. One essential advantage in human biology is the billions of patient-physician contacts and trillions of laboratory and imaging tests done in humans that uncover genetically-related phenotypes. Finally, there are considerable differences in the phenotypic spectra between mice and humans: mice have a shorter lifespan and some of the late-onset human phenotypes are not apparent in them; foetal development also differs between man and mice; mutations with severe consequences in humans, could be mild in mice e.g. APP and Alzheimer disease, HPRT and Lesch-Nyhan syndrome (Kuehn *et al.* 1987, Phinney *et al.* 2003) environmental challenges are different in mice and fewer too.

The benefits of studying monogenic disorders

Much of what we know today on the different types of pathogenic mutations, their consequences and pathogenetic mechanisms has been discovered as an on going process in this quest for Mendelian entities (Antonarakis and McKusick 2000, Hamosh *et al.* 2005). For example, mutations involving changes in triplet repeat copy numbers are now part of the common genetic knowledge (Pearson *et al.* 2005), but they were totally unexpected when first reported.

Another major contribution of 'monogenic' disorders in the elucidation of gene function is their genetic and allelic heterogeneity, as well as phenotypic variability. In cystic fibrosis, subtle variations in intronic simple nucleotide repeat number can result in phenotypic and clinical variability (Buratti *et al.* 2004). There is a plethora of examples where the penetrance, expressivity and disease severity differ among mutation carriers, even within single families, pointing towards the involvement of additional genetic factors or modifier genes (Weatherall 2001, Cutting 2005), as if monogenic entities too could reflect a continuous phenotypic distribution. The nature and influence of most modifier loci modulating the clinical phenotypes of Mendelian, monogenic entities remain unknown. Such loci could be among the first beneficiaries of the recently completed HapMap project (Anonymous 2003b, Hinds *et al.* 2005, Altshuler *et al.* 2005). Cystic fibrosis for example represents a primary target for such an exhaustive search for modifiers. Knowledge and statistical tools developed in the course of this endeavour will in return benefit the exploration of complex traits.

The examples above demonstrate that the notion of a 'monogenic' disorder can no longer be understood *stricto sensu*. These disorders provide instructive examples of oligogenic inheritance, for example digenic or triallelic inheritance (Kajiwara *et al.* 1994, Beckmann1996, Katsanis *et al.* 2001), that in turn provide clues to the understanding of polygenic inheritance. Allelic variation in genes or other functional DNA sequences that modify the phenotypic severity of monogenic disorder, or control gene expression variation provide links to additional genomic causes related to phenotypic variability. Mutations affecting regulatory sequences such as promoters, or mRNA stability and turnover rate (Wang *et al.* 2005) can modulate variation in expression levels and thus result in phenotypic alterations. Furthermore, regulated co-expression or biochemical interactions of gene products also result in more gene-phenotype links. Eventually, starting from an allelic variant that causes a 'monogenic' disorder, one could end up with several other genes linked to the same phenotype, a situation similar to the expected molecular genetic causes of polygenic complex phenotypes. Prime examples are thalassaemias, sickle cell disease and methaemoglobulinaemias. Allelic variation involved in the phenotype variability in these disorders is not only within the β-globin gene on chromosome 11 (HBB) but also the HBG (cis to HBB), HBA (chromosome 16), X-linked ATRX, and other genes (Weatherall 2001). Furthermore, more attention to the phenotypic variability of a Mendelian trait could unravel novel functions or interactions of the mutant gene. A striking recent example is that malaria protection by the HBS heterozygotes also involves the enhancement of acquired immunity to the parasite (Williams *et al.* 2005).

More research on Mendelian disorders is needed

The classic argument for more research on 'monogenic' disorders is also truer than ever: the 'monogenic' mutant gene usually provides a window of opportunity to understand the allelic variability for susceptibility to a similar common polygenic phenotype. For example the discovery of

APC mutations in hereditary colon cancer led to the discovery of additional genes, mutant alleles of which also cause or predispose to colon cancer. Additional gene interactions may reveal the full spectrum of mutant alleles related to this common phenotype. Furthermore, multiple rare allelic variants in three candidate genes (ABCA1, APOA1, and LCAT) that cause Mendelian forms of low plasma HDL-cholesterol levels, were found in people with the common complex trait of low HDL-cholesterol (Cohen *et al.* 2004). Along the same vein, the identification of one of the best-documented loci for susceptibility to schizophrenia, mapped by studying 'monolocus' entities (Lindsay *et al.* 1995), might contribute significantly to the unravelling of this severe condition.

It is widely accepted that most of the lessons learnt from the Mendelian disorders will help us in the search of the susceptibility alleles for complex phenotypes. Observations on mutation mechanisms and phenomena such as anticipation (Harper *et al.* 1992), gene dosage effects (Ibanez *et al.* 2004), uniparental disomy (Ledbetter and Engel 1995), imprinting (da Rocha and Ferguson-Smith 2004) variation in copy number repeats (such as contraction of the D4Z4 repeats in FSHD (Tupler and Gabellini 2004)), effects of allelic series not only in dichotomous but also in quantitative traits (e.g. different haemoglobin levels from different beta+ mutations in the HBB gene (Weatherall 2001)) are but some additional examples of the complex issues that emerged from the seemingly simple Mendelian traits.

There is no reason a priori to exclude the possibility that any of these diverse and rich mutation, gene regulation, and epigenetic processes seen in monogenic diseases do not also contribute to common multifactorial traits. It is likely that additional clues for the understanding of complex phenotypes will first come from a fresh and more sophisticated look of these instructive Mendelian traits. Thus all lessons from the latter are more than relevant for the new genetic challenge. Moreover, the exploration of rare orphan diseases may necessitate the development of new identification strategies, which once validated will also benefit the dissection of complex traits.

In summary we argue that the time is now opportune for more research on the genetic dissection/elucidation (of the complex) monogenic disorders (that are teachers of molecular mechanisms, leaders in the understanding complex traits); that effort will provide clues for gene function and enrich the annotation of the human genome. This is not to say that efforts for the molecular analysis of polygenic, complex phenotypes is not appropriate or timely; simply the 'monogenic' phenotypes, themselves being polygenic and complex, are likely to continue to provide unexpected and ground-breaking knowledge towards the common goal: the early diagnosis and treatment of all genetic disorders.

Acknowledgements

We thank all the past and present members of our laboratories for ideas, debates, and discussions. SEA's laboratory is currently supported by the SNF, EU, NIH and the Lejeune and ChildCare Foundations, and JSB's by grants from the SNF and the University of Lausanne.

This article was originally published in 2005, *Nature Reviews Genetics* 6 151–157.

GENATLAS

Jean Frézal, in collaboration with Marline Le Merrer and Claude Mugnier

Service de Génétique Médicale, Hôpital des Enfants Malades, 149 rue de Sèvres, 75743 Paris Cédex 15, FRANCE

Following his PhD thesis presented in 1960 to London University, George Fraser devoted a number of important works, including original articles, reviews, books and his MD thesis (University of Cambridge, 1966) on sensory defects, either blindness or deafness in children. Half a century later, these works keep their currency and their interest, simply because they are founded on an acute clinical sense associated with a great familiarity with formal, epidemiological and population genetics. For example, George was clearly aware of the heterogeneity of the disorders he studied and his estimates of the relative prevalence of dominant, recessive and sex-linked forms of retinopathies are not so different from current estimates, taking into account the differences of classification (Fraser 1966b). Among the syndromic forms of the diseases on which he worked, everybody knows his contribution to the Pendred syndrome and his description of the eponymous Fraser syndrome (Fraser 1962a). However, his role in the delineation of the Wolfram syndrome (Rose *et al.* 1966e) is probably not so well known.

Considering the interest of George in sensory disorders and the involvement of one of my former collaborators, Josseline Kaplan, in that field (Dufier and Kaplan 2005) I first thought to devote my contribution to such disorders. Following a suggestion of George's, I changed my mind. Accordingly, I am delighted and very honoured to present the GENATLAS database (Frézal et al. 1989) in this Festschrift, in honour of George Fraser. Before closing this dedication, I should like to express my great esteem, not only for the distinguished geneticist but also for a man endowed with a vast culture. For us, George definitely is *un homme des Lumières* (a man of the Enlightenment). He is also a marvellous friend and as has been said in a Brazilian poem on friendship of which George is very fond and which he has translated into English: 'We do not make friends, we recognise them'. – one of George's favourite quotations. A complete version of George's translation is reproduced in the contribution of Professor E A Chautard-Freire-Maia to this Festschrift and *liber amicorum*.

A brief history of GENATLAS

GENATLAS was created in 1986 after I was mandated, at the Helsinki meeting in 1985, to organise the Ninth Human Gene Mapping Workshop which was planned to be held in 1987, in Paris. While accrediting me, the Organizing Committee, in consideration of the growing tide of data, insisted that the management of the meeting be computerised. I remember that I was invited by Professor George Cahill, from the Howard Hughes Medical Institute, to a meeting held in Bethesda, in spring or summer 1986, where the question was debated by a brilliant assembly of 'mappers'. It is at this very meeting that I announced the birth of the database, already christened as GENATLAS, trying to convince the audience that GENATLAS actually did exist and presenting, as a proof, a rolling document, something like a phylactery on which a handful of genes had been reliably listed.

Despite this modest beginning, we were just about ready to carry out our mission. Anyway, HMGW9 was the first of the series where the number of assigned genes exceeded one thousand. Although our personal experience was not entirely conclusive, it was a very exciting challenge. So we decided to go ahead.

In 1986, GENATLAS was meant to record the genes and markers localised on the human chromosomes, and this cartographic approach continued to hold a cardinal importance in the course of time, remaining an obligatory requisite for including a gene in the database. However, it became obvious, with time, that we had to add complementary information to the basic mapping information. We were supposed to add pertinent data on the products of the genes, their structure and their function. In parallel we had to put more emphasis on the phenotypes associated with genes. In the mid-nineties, to cope with the overwhelming increase in the volume of data, we engaged ourselves in a thorough revision of the database. A new system of informatics was devised, and the structure of information profoundly modified. The information about the genes and the phenotypes, which was stored until then in a mixed directory, was divided into two separate and specific directories, the GENE and the PHENOTYPE directories, linked together. It is also at that time that GENATLAS was introduced on the net: *http://genatlas.org*.

Bioinformatics

In order to achieve this plan, a new application was developed which is maintained at the computer Centre of the René Descartes University in Paris. So GENATLAS could benefit from the network, system and exploitation competences of the team at the Centre which is a node of the Academic Parisian network connected on Renater. The data are managed by SGBD Oracle.

The relational GENATLAS database

Conceptually, the restructuring aimed at providing answers to questions raised by scientists and clinicians with respect to heritable disorders, that is to say:

Which genes for which pathology?

The identification of a gene corresponding to an illness often relies on the search for mutations in a gene whose presence at an identified locus makes it a candidate for a causal role. This identification may be facilitated by the heterogeneity of numerous pathologies. In fact, the involvement of a gene may be suggested by the co-expression and/or the association with genes already implicated in a pathology (for instance: the dystrophin complex in muscular dystrophy), or a structural homology suggesting a similar function (e.g., ion channels). GENATLAS provides easy access to this information and specifies for each gene: its function, motifs and domains characterising its product, the complex to which it belongs, and the process in which it participates.

Which pathology for which gene?

As the number of identified genes continues to increase, GENATLAS seeks to direct searches so that a link can be made with a pathology. This is done by comparing the description of illnesses localised in the genome, with the expression and function of genes. In order to answer these questions, ontologies have been constructed which allow a rational classification of gene expression, structure and function. In addition, a commentary by an expert helps to channel the information into 300 topics. This attribute of GENATLAS therefore opens up a great number of research possibilities.

The source of information

Either in the old version or in the new one, the information is extracted from the literature, in printed journals, in reviews such as *Current Contents* or online. Furthermore, data recorded in GENATLAS are regularly updated and compared to GENEW, the official database for nomenclature maintained by HUGO which serves as reference point. This leads to new entries in GENATLAS or to a change of symbols and nomenclature. Comparisons with OMIM, Locuslink (Gene) and other databases enable a cross-reference and correction of the previously saved data, when necessary. Relevant references are recorded from MEDLINE. The data obtained electronically are inserted into a window and once validated, they can be consulted on line. All new modifications are dated and validated by an expert. Other than references, a number of features can be automatically documented, such as the OMIM number, nucleotide sequences, number of exons, etc. Chromosome localisation is verified by reference to the original article and compatibility with the mapping information.

External links

GENATLAS has mostly reciprocal links with a number of core or specialised databases such as GeneCards (Weizmann Institute), GENELYNX (Pfizer Corp), Source (Stanford University), SwissProt (Geneva), GENEW for gene nomenclature (London). GENATLAS is on a par with GeneCards and the Gene/Locuslink database (NCBI, USA) which provide similar information but can only be queried through symbol or free text. Therefore, their query potential is lower and subject to background noise. Finally, these databases are concerned with genes. As a rule, they are not associated with illnesses. The latter are simply signalled through a link to OMIM. On the contrary, GENATLAS puts emphasis on the pathologies themselves with the aim of collating all information on the gene and providing a useful tool for clinicians and researchers interested in genetic diseases.

The new directories

The new GENATLAS is made up of three directories. The first one, the GENE file, compiles the genes. This directory includes almost 22,000 entries. It is organised in five sections.

In the first section, every gene is identified by its official symbol (HUGO) and its aliases, which refer to the other names used in the literature. In addition, a short definition is associated with every entity. Otherwise, GENATLAS has been linked to the human sequence data, which will allow the correction of existing information or enable the precise localisation of genes. It is possible to obtain information relevant to the choice of primers for use in positional cloning for monogenic diseases or the study of regulatory regions upstream of genes of interest.

In the same way, certain information deserves to be listed, notably the structure of the gene and its anatomy, by indicating its size and its composition in coding and non-coding sequences and by describing the flanking regions which may contain promoter or regulatory regions. Finally, the repetitive sequences which are situated within the limits of a gene or in its neighbourhood, are also mentioned. The cluster organisation of genes is included, for cases where there is a string of genes and/or gene-like sequences and/or pseudogenes, physically linked together, structurally and functionally related and co-ordinately regulated. Examples are given by the haemoglobin clusters on chromosomes 11 and 16 or by the inflammatory cytokine clusters on chromosomes 17 and 4. Interestingly, some features such as the genes within the genes of which the conservation or the loss can contribute to phenotypic variability, or breakpoints revealed in numerous malignant disorders, are reported.

The second section deals with the DIFFERENT TYPES of RNAs. The largest part refers to unstable RNAs, i.e. to gene transcripts and their isoforms. In a relatively high proportion which may reach one third of the genes according to certain estimates, genes have several transcripts, notably by alternative splicing of the exons. These various alternative forms and/or variants are of interest as they are supposed to account for the phenotypic diversity produced by a limited number of genes. Other types of RNAs are also categorised and stored, such as the stable, s and tRNAs. This chapter will, no doubt, take on more importance with the discovery of the microRNAs and the unravelling of their biological meaning.

The third section, EXPRESSION and SUBCELLULAR LOCALISATION, accounts for one of the major parameters of gene activity and consists in specific transcription according to cell type, tissue, organ, viscera etc. as well as time expression (cell cycle phase) or developmental stage (embryo, foetal, perinatal, adult, etc.). It also refers to the subcellular compartments in which their protein products are present.

The fourth section, PROTEIN, helps to produce an integrated view of protein features. Accordingly, genes are classified by type (functioning, pseudo …) and category of gene product (enzyme, structural protein, etc.). The products act either individually or as components of complex(es). Their action may be an organised set of operations, a process to do or produce something or to accomplish a specific result. Alternatively, the product may direct or control a specific step in a chain of reactions (pathway), aiming at the transformation of an organic substance (metabolism). Besides the metabolic chains, the genes control the way by which a signal of any kind is transferred in order to elicit an appropriate response (signal transduction).

The fifth section is dedicated to VARIANT(S) of the gene, either mutations or polymorphisms. These variants are commonly associated with pathologies; these may rarely be Mendelian characters. In the current organisation of the database, these Mendelian phenotypes are stored in an entry separated from the gene's entry. More often, it is a chromosomal rearrangement linked to a pathology, either hereditary or, generally, acquired. As a rule, they are associated to the susceptibility (or resistance) to a common, non-Mendelian character, such as diabetes or atherosclerosis, or allergy, etc.

Phenotype directory

We would like to stress again the strong emphasis progressively put in the course of time, on the medical aspects of genetics in GENATLAS, as exemplified by the 3500 phenotypes individually categorised. As already mentioned, it is the reason why the information on diseases which was gathered in the GENE directory is now brought together in a specific directory endowed with its own structure. For this purpose, a detailed thesaurus has been compiled for clinical pathologies. This thesaurus is compatible with existing diagnostic tools such as LDDB or POSSUM. The aim of the thesaurus is to enable a more detailed description of the pathologies. It aims at providing a reference tool for the clinician, whether the diagnosis is clear or uncertain. In addition, the database may provide some help for scientists involved in the molecular identification of new phenotypes. A new PHENOTYPE directory, which is still under development, will be organised under a combinatorial model and founded upon the use of a controlled vocabulary and a rational classification of the information, i.e. an ontology (Ashburner *et al.* 2000) including a symbol which is usually different from the gene symbol, synonyms, OMIM number (gateway to numerous databases), the responsible gene, if known, its product, and its mutations

Citation directory

Every item in either the GENE directory or the linked PHENOTYPE directory is associated with at least one reference pertaining to map information, linkage, associated diseases, and mutations. Reference is also made to articles, either original or review, relevant to the expression, structure, and function of the genes and their products, in keeping with the ontologies developed in the database. Currently, more than 62,000 citations are recorded from MEDLINE. Furthermore, abstracts from relevant articles are directly accessible from the references in MEDLINE.

Querying GENATLAS

One of the main interests in developing a relational database model is to deal with complex information in multi-criterion querying strategies. In this respect, the new relational GENATLAS database implements multi-criterion approaches according to the five core sections (GENE, DNA, RNA, Expression, Protein, Variant/Pathology), and several hundred criteria can be used to query for gene and protein structure, expression, function, and interactions, as well as for related diseases. A wide variety of criteria is then available for the research of an individual gene or of a group of genes, either individually or in combination. For example, it is easy to retrieve the genes which code for the enzymes expressed in the liver with subcellular localisation in mitochondria and involved in organic acid metabolism. Starting with 3484 enzymes, one retrieves six genes in a few seconds, by the multi-criterion query. Otherwise, a system enabling the search for genes associated or not with a pathology, is now available.

The future

In conclusion, we aimed at giving a new dimension to GENATLAS in order to promote it as a reference database for the annotation of genetic diseases and genes in man. Since its introduction on the net, the circulation and diffusion of the database have steadily increased. It can be objectively assessed by the Apache server at the national and international levels, using the software awstats (http://awstats.scourceforge.net/). During 2004, 2,100,000 pages were consulted during 25,000 visits yo the web site, per month, by people in France and abroad. In 2005, the number of visits has been close to 30,000 a month. This undertaking had been founded on a phenotypic approach, i.e. the function of a gene, the symptoms of a disease, in order to reciprocally enlighten each other and to try to integrate both clinical and biological information. We do not ignore the question which has been raised about the usefulness of a database such as GENATLAS or its equivalents with the advent of genomics. All things considered, we think that the two approaches are not at all exclusive from each other. Indeed, we think they are complementary, first of all because they allow a cross-control of the information, the usefulness of which being demonstrable by the single example of the relative position of the opsin genes at the distal end of the long arm of the X chromosome. It is well known from biological and molecular studies of colour blindness that the genes for the red and for the green opsins lie, in a close proximity, the recombination fraction being estimated as about 0.08 by George Fraser (Fraser 1970h), a feature which explains the rearrangement observed in most cases of daltonism. However in the Draft of the Genome the two genes are wrongly situated at a great distance from each other, i.e. at several hundreds of thousands of bases at a position incompatible with clinical and biological information (Patrinos and Brookes 2005). In fact, there are still many inaccuracies in the assignment of genes which fully justifies the cross-control (Ashburner *et al.* 2000). Moreover, as stated by Jean Weissenbach (Weissenbach 2000, 2002), 'the very analysis of sequence data remains a major problem … notably in superior eukaryotes. Furthermore, the prediction of a phenotype which could be conferred by the alteration of a function, even when characterised at the biochemical and cellular levels, remains hardly feasible.' Therefore, experimental and phenotypic approaches currently remain the best ways to break through these difficulties.

Currently, some three quarters of the human genes are stored and classified in GENATLAS. It is our conviction that GENATLAS will retain its usefulness when all the genes will have been identified by taking advantage of the new techniques of genomics which will be invaluable in deciphering the complex pattern of gene expression in relation to function and diseases. It is however necessary to interject a word of caution because of the permanent difficulties of raising recurrent funding and cooperation, a problem which is apparently shared by most databases, including GENATLAS. However, GENATLAS sustains its perfect congruence with the objectives of the Research Unit Inserm U781, headed by Professor Arnold Munnich where it was housed, that is to say to get a better knowledge of causes and pathogenesis of diseases, hoping that this knowledge will make their prevention easier and open new therapeutic prospects.

Community-based programme for the diagnosis and prevention of genetic disorders in Cuba: twenty years of experience

Luis Heredero-Baute

Former Director of National Centre of Medical Genetics, Higher Institute of Medical Sciences, Ministry of Public Health, Cuba.

Dear George,

I have never forgotten the time, in April 1972, when I worked for a time in your Department in Leiden as part of my training in Human Genetics. My stay opened my eyes so much about the field of medical genetics! Several years later in Havana, when I was working together with colleagues in the field of deafness to create the Cuban Society of the Deaf, I used many of your articles and books as references in our project application to set up the society, which was accepted; you were again present in my life. Your visit to Havana and my home on the occasion of the Latinamerican Congress of Genetics in 1987, was unforgettable. Now, for your 75th birthday, I want to wish you a long life, George! *Que vivas muchos años más!* I love you, my dear friend, *mi muy querido amigo*.

Thank you so much! Luis.

Abstract

The author's experience of 20 years as director of the medical genetic services programme in Cuba is presented. The setting of the infrastructure for equipment and the training of personnel for the medical genetics programme began in 1981 in the city of Havana, and was progressively extended to cover the whole country in 1988. Between 1982 and 2002, 2.8 million pregnant women were tested for sickle cell carrier status, 96,000 carriers and 4,786 couples at risk were detected and offered genetic counselling and prenatal diagnosis. In the same period, the combination of maternal serum AFP screening and foetal ultrasound led to the prenatal diagnosis of several thousand foetuses with anomalies. The accessibility to legal abortion and the autonomous decisions by the majority of couples to terminate abnormal pregnancies led to a reduction in the prevalence at birth of neural tube defects and sickle cell disease by 90% and 65% respectively by 2002. Over approximately 20 years, 22,690 pregnant women at risk received prenatal chromosomal diagnosis. Newborn screening for phenylketonuria and congenital hypothyroidism was established. Genetic counselling was offered to every detected person or family at risk for genetic conditions. A network of medical genetic services was established in the country with a very positive acceptance by the population. A very successful connection was established with the primary health care delivery.

Introduction

This paper reviews more than 20 years of experience in the implementation of a nationwide programme for the prevention and management of genetic disorders in Cuba, in the context of the recent advances of genomics, and the new support that medical genetics is receiving in the country (see also Heredero 1992, Galjaard 1994, Heredero 1997, Penchaszadeh 2002).

The first report of a programme to prevent and manage genetic diseases in Cuba was published in 1974 in two volumes of *Revista Cubana de Pediatría* which published articles and discussions of a workshop held in Havana in February 1973, entitled 'Sickle Cell Anemia in Cuba and its Prevention'. (Sickle cell disease is one of the most frequent monogenic diseases in Cuba.) In that workshop, a programme for the prevention and management of sickle cell disorders (SCD) was proposed based in the education of the population, the detection of carrier couples, genetic counselling and prenatal diagnosis, coupled with early treatment of babies born with the condition. The implementation of this proposal was made feasible by the invention of a new cost-effective method for screening sickle cell carriers using a low-cost high-speed electrophoresis screening system for haemoglobin S and other proteins (Heredero *et al.* 1974). This homemade equipment was patented and produced in Cuba, and it is still in use. In Cuba, as in other parts of the world, the control of haemoglobin disorders contributed to the development of community genetics (Modell and Kuliev 1998).

Medical Genetics became a medical specialty in 1977 included in basic sciences with a four year programme involving one year training in paediatrics, one year training in basic sciences and the rest of the time in clinical and laboratory genetics.

In the early 1980s, an inexpensive and partially automated ultramicrolitre enzyme immunoassay (ELISA) to measure maternal serum α-fetoprotein (AFP), was developed and applied to the screening for neural tube defects (NTD) in pregnancy (Korner *et al.* 1986). This system is produced in Cuba and is used for the diagnosis of a number of other conditions (*www.tecnosuma.com*). The screening of NTDs became a public health priority in Cuba after preliminary epidemiological data suggested a relatively high prevalence at birth of these anomalies (Rodríguez *et al.* 1979). A clinical protocol for the prenatal screening of congenital anomalies using maternal serum AFP determination in pregnant women was started as a pilot project by the end of 1981.

That same year, a programme known as 'Diagnosis and Prevention of Genetic Diseases' proposed by the National Centre of Medical Genetics (at that time a Department of Medical Genetics of the Medical University of Havana) was implemented under the auspices of the Maternal and Child Health Department of the Ministry of Public Health, harnessing the experience with SCD and NTD. The background, rationale and context that made the programme feasible were the following:

> The country was experiencing an increase in the proportion of genetic diseases as causes of child morbidity and mortality. (The infant mortality rate was around 20 per 1,000 at that time.)
> The Cuban population was well educated.
> Local appropriate technologies were developed for the detection of SCD carriers and the measurement of maternal serum AFP.
> The existence of a National Health Care System that covers all the population free of

charge, and a well delineated primary care level as corner-stone for the detection of genetic risks.

The programme was launched simultaneously with the development of the infrastructure and the technical qualification and training of personnel.

The approval and participation of health authorities in the organisation and implementation of the programme was secure.

The establishment of evaluation of the impact of the genetic programme on the health system and health indicators.

The harnessing of previous local experience, infrastructure, capacities and epidemiological data.

The delineation of the principles of a National Medical Genetics Organisation, the existence of Regional Medical Genetics Centres and a National Programme for Training and Education.

As described below, the initial emphasis of the genetics programme in Cuba was on expanding reproductive options for couples at risk for sickle cell disorders, neural tube defects, chromosome abnormalities and other severe foetal malformations. This programme was feasible because induced abortion is legal in Cuba. Pregnancy termination on request is permissible until ten weeks. When a foetal abnormality is demonstrated, parents (or at least the pregnant woman) can request termination of pregnancy up to 26 weeks of gestation. Since its inception, however, the programme has respected personal reproductive autonomy and freedom of choice, and has been based on non-directive genetic counselling aiming at empowering prospective parents of affected foetuses to make their own decisions regarding the fate of the pregnancy without coercion, at all stages of the process. Prenatal genetic services were coupled with the best available services for treatment and management of patients with genetic disorders, especially sickle cell disease, Down syndrome, neural tube defects, congenital hypothyroidism and phenylketonuria.

Prevention and management of sickle cell disorders (SCD)

The carrier frequency for sickle cell genes ranges from 3% in the western provinces to 7% in the east of the country, which has a higher concentration of people of African descent. The National Institute of Haematology and Immunology in Havana co-ordinates the treatment and follow-up of patients with SCD throughout the country, in conjunction with a network of haematology departments in every province and in some regional hospitals. The prevention component of SCD is under the responsibility of the National Centre of Medical Genetics (NCMG) and consisted of the detection of carrier couples at the start of gestation, genetic counselling, and the offer of prenatal diagnosis.

The programme began in Havana City in 1981, and was extended to the whole country by 1988. Detection of heterozygous pregnant women is offered at their first antenatal care visit at the primary care level, and the acceptance is around 90% (Granda *et al.* 1991). Those who turn out to be carriers are offered carrier testing for their partner (partner testing coverage is around 80%), and couples at risk receive genetic counselling and the offer of prenatal diagnosis by amniocentesis and DNA testing for SCD. The latter is performed centrally at the National Centre of Medical Genetics in Havana (Granda *et al.* 1994). From the inception of the programme until December 2002, 2,788,542 pregnant women were tested for carrier status and

95,838 of them were determined to be carriers; 4,786 couples at risk were found and prenatal diagnosis was performed in 2,874 pregnancies (the main cause for declining prenatal diagnosis was an advanced gestational age) and 563 homozygous foetuses were detected. Termination of pregnancy was requested by 70% of couples with an affected foetus. The follow-up of all couples at risk who continued their pregnancies enabled the quality control of prenatal diagnosis, the early diagnosis of affected newborns and the establishment of long term patient management beginning shortly after birth and based on prophylactic treatment with antibiotics, folic acid and health education of the parents to avoid circumstances that could trigger crises (Dorticós *et al.* 1997, Granda pers. comm.). A follow-up study of the reproductive behaviour of couples at risk for sickle cell disorders in Cuba, showed couples were well satisfied with the prevention programme (García *et al.* 1999). Given that approximately 80 babies were born annually with SCD previous to the launching of the programme (Altland and Heredero 1974), the programme effectively contributed to a reduction of the prevalence at birth of SCD by 65% by 2002 and also to better management of the disease, as the treatment started shortly after birth in close to 40% of the affected babies (Granda pers. comm.).

Prenatal screening of congenital anomalies

The first early prenatal diagnosis of an anencephalic foetus in Cuba was performed in 1980 using AFP in amniotic fluid at 16 weeks of gestation in a woman with a previous child with a neural tube defect (NTD) (Rodríguez *et al.* 1982). Population prenatal screening of NTDs by quantification of AFP in maternal serum was initiated at the end of 1981. The test is offered to all pregnant women at the primary care level by the family doctor, between 15 to 19 weeks of gestation. Those who show abnormal MSAFP levels (above 2 MoM) are referred to the secondary level of care (provincial medical genetics departments) for further evaluation (Rodríguez *et al.* 1987). After some years of training and increased expertise in foetal ultrasonography in provincial hospitals, all pregnancies are offered foetal ultrasound at 20 to 24 weeks of gestation for the detection of foetal malformations, especially cardiovascular and obstructive urological anomalies (Rodríguez *et al.* 1997, Llanusa *et al.* 1998).

By December 1999, more than 2.5 million pregnant women had had MSAFP testing. In Havana City, where one fifth of Cuba's deliveries occur, half a million pregnant women had MSAFP testing during 1982–2002. Nine percent of them had values over 2 MoM and 1,255 foetuses with severe malformations were detected. Also, more than 4,000 other obstetric conditions (multiple pregnancies, foetal deaths, etc.) were diagnosed during the clinical evaluation of elevated MSAFP and proper treatment was instituted. An additional 1,973 pregnancies with normal MSAFP were found by ultrasound to be affected with severe foetal anomalies (Rodríguez. pers. comm.). Cases detected with minor anomalies were referred for follow-up in specialised hospitals for proper treatment or follow-up after birth. Cases with cardiovascular and urological anomalies in which pregnancy is continued are referred for special management and follow-up to specialised institutions for surgery and clinical management. Periconceptional administration of folic acid to women in families with a previous child with NTD was initiated in 1984 (Vergel *et al.* 1990) and a national program of folic acid fortification of different foods is ongoing. According to the National Registry of Congenital Malformation the prevalence at birth of NTDs declined by 90% in the last 15 years.

Prenatal diagnosis of chromosomal abnormalities

Between 1984 and the end of 2001, 22,690 women all over the country received prenatal cytogenetic studies, the main indication being advanced maternal age (the age cut-up point was initially 40 years old and is currently 38 years old, varying according to the operational capacity of the six regional cytogenetic laboratories of the Cuban genetic program and the fluctuating availability of imported reagents). The aneuploidy most commonly detected was trisomy 21 (451 cases) followed by trisomy 18 and 13 respectively. Approximately 90% of the pregnant women with abnormal results opted for pregnancy termination. Detailed reports of this programme have been published elsewhere (Casana *et al.* 1986, Quintana *et al.* 1999, Quintana 1999). As the use of biochemical and ultrasound markers for foetal anomalies increased, new indications for pre-natal chromosomal studies are being followed (Solis *et al.* 2001). The Cuban Register of Congenital Malformations has documented a reduction in the prevalence at birth of auto-somal trisomies, possibly as a result of the prenatal cytogenetic program (Ferrero *et al.* 1998, World Atlas of Birth Defects 2003). Notwithstanding the existence of a free special education system for people with disabilities and the lack of genetic discrimination or stigmatisation, the most important effect of the prenatal diagnosis programme in the couples at risk is the reduction of anxiety brought about by the reassurance of not bearing a foetus with Down syndrome or other chromosome abnormality.

Newborn screening for phenylketonuria (PKU) and Congenital Hypothyroidism (CH), and detection of inborn errors of metabolism (IEM) in population at risk

Between 1984 and 2000, more than 1,831,000 newborns (90% coverage) were tested for PKU by the Guthrie test in two regional laboratories (Heredero *et al.* 1986, Barrios *et al.* 1989), with a yield of 38 cases with classic PKU (one in 48,048) and 56 with benign hyperphenylalaninaemia (one in 32,695) (Barrios *et al.* 2001, Gutiérrez *et al.* 2002). On the other hand, newborn screening for congenital hypothyroidism (CH), which also started in 1987 under the responsi-bility of the National Immunoassay Centre, determined prevalence at birth of one in 3,282 with almost total coverage. The difference in coverage for PKU and CH is explained by the fact that while for PKU a blood sample is obtained at the primary level within the first week of life, for CH cord blood is taken at birth. A change in the screening procedures is being implemented, with the use of SUMA technology (Heredero 1974), and blood sampling for PKU will soon be obtained at birth, which should improve coverage. The treatment (including replacement hormone for CH and special diet for PKU) and follow-up of all newborns with PKU and CH is provided free of charge by the Cuban national health care system.

In the same period, and using a referral system through the national network of medical genetics for the detection of inborn errors of metabolism, 8,700 patients with a clinical suspicion of these disorders were studied by clinical and laboratory methods, with the following results: lysosomal storage diseases 83 cases; mucopolysaccharidoses 48 cases; PKU 21 cases (most were adults in specialised institutions and two babies who had not been detected by newborn screening), homocystinuria nine cases; organic acidurias six cases; cystinosis six cases; alcaptonuria five cases; and biotinidase deficiency five cases (Barrios and Gutiérrez 2001). Currently, a big effort is being made to improve the capacity for detection and treatment of the IEM in our country.

Genetic counselling services and clinical genetics

This is an inherent part of the genetics programme and in increasingly closer connection with the primary health care level, through a system that has trained family doctors and nurses in genetic counselling in a M.Sc. postgraduate course which started in 1996 (Penchaszadeh *et al.* 1997) and which has recently been expanded. The goal is to have at least one family physician with genetic counselling training in every one of the 169 municipalities of the country, each one with a population of 40,000–60,000 persons. These professionals are the link between the family physicians at the primary care level and the geneticists at regional level. Clinical genetic services in provincial and regional hospitals serve about 2,500 consults per year, in addition to prenatal genetics and newborn screening. The network of medical genetics links with provincial, regional and national laboratories to which samples are referred from all corners of the Isle for special genetic diagnostic tests.

Additional public health actions that contribute to prevent congenital anomalies

As part of the general measures of the health system there are several actions that contribute to the prevention of congenital defects. The access to medications without medical indications or prescription is very low in Cuba because there is high health coverage of the population by family physicians and the health education of the population is very good. There are national environmental protection policies all over the country for pregnant women working under unsafe conditions. A national rubella immunisation program was established and not a single case of this disease has been reported since 1999 (http://www.sld.cu/anuario/indice.html). Prevention of other congenital infections, such as syphilis and HIV/AIDS, is very effective and the number of cases of these conditions is very low. Tertiary prevention is in place for birth defects, including surgical repair of heart defects in infants less than one year of age, rehabilitation procedures, early stimulation in all cases of Down Syndrome and the already mentioned free national system of special education for persons with disabilities.

Research in medical genetics

For the past 20 years, the Cuban Medical Genetics Programme engaged in research aimed at better understanding the epidemiology of genetic disorders in the country as well as devising better ways of conveying genetic information to families to make knowledge a tool in prevention.

The following are some examples of research projects accomplished.

The molecular genetics study of dominant ataxias in the north eastern part of Cuba, where the largest homogeneous group of patients in the world has been described (Orozco *et al.* 1990) and for which the localisation of the gene in the short arm of chromosome 6 was excluded, being the first evidence of genetic heterogeneity of this group of diseases (Auburger *et al.* 1990). The autosomal dominant cerebellar ataxia prevalent in Cuba was eventually assigned to a second locus (SCA2) in chromosome 12q23.24.1 (Gispert *et al.* 1993) and a CAG trinucleotide expanded repeat with CAA interruptions was finally described (Pulst *et al.* 1996). These findings are being applied in Cuba to the genetic counselling, presymptomatic diagnosis and prenatal diagnosis of the disease (Paneque *et al.* 2001).

DNA studies on cystic fibrosis demonstrated that the five most common mutations accounts for no more than 60% of cases of this disease in Cuba (Collazo *et al.* 1995).

The aetiological characterisation of severely mentally retarded patients using clinical genetics studies (Lantigua *et al.* 1999) was later expanded in a nationwide study of mental and physical disabilities in Cuba.

Molecular genetic studies in large families have led to a better clinical delineation of particular diseases such as a large family in the central part of Cuba with optic atrophy type Kjer (Lunkes *et al.* 1993), and a large family with autosomal dominant presenile dementia with a novel presenilin 1 mutation (L174) (Bertoli *et al.* 2002), which is relevant for its diagnosis in other Hispanic populations and for possible presymptomatic testing. Recently we reported suggestive linkage to chromosome 19 in a large Cuban family with late onset Parkinson's disease (Bertolí *et al.* 2003).

Discussion

There is a high expectation that the completion of the human genome project, and the development of genomics, proteomics and pharmacogenomics, will lead to major applications in the prevention, diagnosis and management of many conditions which hitherto have been impossible to control. Cuba, with its comprehensive and universal public health and education systems, with a developing biotechnology industry and with more than 20 years' experience in the implementation of clinical genetic services and the introduction of new DNA technology into clinical practice, is probably best suited to be a 'laboratory' where the potential of genomics to improve health can be tested.

Two grand challenges must be addressed if genomics is to fulfil its promise. The first is the appropriate genetic education of health professionals. It is clear that clinical geneticists will not be able to cope with the needs of medical care for all types of genetic disease, particularly in the context of the expanded concept of genetic disorders which includes common complex diseases. While it is impossible to graduate medical professionals specialised in all genetic disorders, the goal should be that all medical graduates have a functional knowledge of genetics. This will be particularly useful in primary health care, where problems are detected and prevention is implemented. However, in order to ensure that the expansion of coverage does not compromise the quality of services, some degree of centralisation and supervision from the tertiary level is necessary to guarantee the excellence of the service. In addition, Medical Centres should contain or be linked to specialists with expertise in the genetic conditions of their field, i.e. orthopaedists specialised in genetic bone changes, dermatologists specialised in genodermatoses and so on. Such specialists should be part of the medical genetics department full- or part-time (as in Cuba for the past 20 years with paediatricians and obstetricians). This will improve the diagnosis accuracy and contribute to therapy; at the same time, genetic counselling will be also improved.

The provision of sufficient physicians and ancillary staff trained in medical genetics will be a considerably challenge. Currently there are 101 clinical geneticists in Cuba, while close to 500 family physicians have received some training in medical genetics, many of them with MSc degree in genetic counselling. In addition, several dozen biologists, biochemists and molecular biologists work in the field of human genetics. The training and education in medical genetics has strong ethical foundations, where fair and just care, respect of autonomy and rejection of genetic discrimination and stigmatisation are guiding principles (Penchaszadeh *et al.* 1997).

The second grand challenge is to have a society educated in genetics, without which it will be very difficult to achieve prevention programmes at the population level. Thus, it is necessary to bring genetics closer to the understanding of common citizens. The geneticists of the new century must be knowledgeable in the new automated information services and with the proper laboratory methods and not left isolated in the community. An increase in collaboration among both developed and developing countries will be vital for the future. As Professor Galjaard stated in 2001: 'whether with old disciplines or new names does not matter', clinicians, scientists and the related industries will be able to prevent much more serious genetic diseases in the near future (Galjaard 2002).

Acknowledgements
Many persons have contributed to the results reported here. Special thanks go to all the personnel of the National Centre of Medical Genetics, and particularly the heads of the provincial departments of medical genetics. The senior staff who were responsible for each of these five aspects of the Cuban genetics programme were essential: Hilda Granda (sickle cell disorders), Lidia Rodriguez (prenatal diagnosis of congenital malformations), Jorge Quintana (prenatal cytogenetic diagnosis), Barbara Barrios (PKU screening and IEM) and Aracely Lantigua (clinical genetic services). The constructive criticisms, recommendations and support during many years from Professors Klaus Altland (Germany), Victor B. Penchaszadeh (USA) and Hans Galjaard (The Netherlands) were of great value. The support from Tania Fraga during these 20 years was invaluable. Finally, my gratitude goes to our patients, who believed in us.

The clinical legacy of Jonathan Hutchinson (1828–1913): Syndromology and dysmorphology meet genomics [1]

Victor A. McKusick

University Professor of Medical Genetics, Johns Hopkins University Baltimore, Maryland

Dedication

This paper is dedicated to George Fraser on the occasion of his 75th birthday. In 1963, he provided me with a valuable critique of a preliminary version of my catalogue of autosomal recessive phenotypes that was under preparation for *Mendelian Inheritance in Man* (1st edn, 1966). His friendship and professional advice have meant much to me for over 40 years.

At that time in 1963, in the course of surveys of genetic causes of congenital deafness and blindness, George had observed a distinctive syndrome which has carried his name since that first edition of *Mendelian Inheritance in Man*. Fraser syndrome (OMIM 219000) is an autosomal recessive, multi-system congenital disorder presenting with cryptophthalmos, syndactyly and renal defects. In recognition of one of George Fraser's major contributions, it seems appropriate to analyse how genetic syndromes are recognised, are described in their full range of phenotypic variability, and are characterised at the molecular level. To do this, I will use two genetic syndromes first described by a remarkable British surgeon of the latter half of the 19th century.

This London surgeon, Jonathan Hutchinson, was an exceptionally astute clinical observer. He was, furthermore, an assiduous and prolific recorder of observations over a wide range of fields. In his prime he did for many years all the onerous work of a surgeon at the London Hospital. He was president of the Royal College of Surgeons in 1888, and is said to have still been doing mastectomies at age 70. But he also worked at the Moorfields Eye Hospital and in 1863 published a monograph entitled *A Clinical Memoir on Certain Diseases of the Eye and Ear, consequent on Inherited Syphilis*, containing early fundus portraits prepared by his long-time artist-illustrator Edwin Burgess, scarcely a decade after the invention of the ophthalmoscope. In addition, he worked at Blackfriars Skin Hospital and was president of an international congress of dermatology (1896). He was a mentor and collaborator of pioneer neurologist Hughlings Jackson and president of the Neurological Society in 1887.

Few clinical lives are as extensively documented as that of Jonathan Hutchinson. His bibliography comprises almost 1,200 items (Kelly 1940). The core is his *Archives of Surgery*, a quarterly periodical to which he was almost the only contributor, published between 1890 and 1900. In writing on the importance of post-graduate study Osler (1900) stated:

> When anything turns up which is an anomaly or peculiar, anything upon which the textbooks are silent and the systems and cyclopaedias are dumb, I tell my students to turn to the volumes of Mr. Hutchinson's *Archives of Surgery* as, if it is not mentioned in them, it surely is something very much out of the common.

A second and related documentation of Jonathan Hutchinson's clinical career is provided by his Clinical Museum containing an extensive collection of illustrations, some photographic, but most water colours by illustrators Edwin Burgess, Mabel Green, and others. After Hutchinson's death in 1913, the collection of clinical illustrations and case reports was acquired by the Johns Hopkins Medical School through the agency of Sir William Osler and the financial generosity of Mr. W. A. Marburg of Baltimore. The collection arrived in Baltimore in 1915 in eight large crates (with British war bond posters covering the material). In that year, Osler (1915) published a small note in the *Johns Hopkins Hospital Bulletin* describing the collection:

> [It] illustrates the whole of medicine and surgery … The drawings are classified: one group comprising more than 5,000, are in large paper envelopes, the other an even larger number, in large cardboard portfolios. While they illustrate particularly the life work of the collector in syphilis and skin diseases, there is scarcely a department of medicine that has not one or two portfolios devoted to it.

In academic year 1950–51, while a senior assistant resident on the Osler Medical Service of the Johns Hopkins Hospital, I devoted all of the spare time I could to a review both of the Hutchinson collection (then reposing still in the eight crates in the Welch Medical Library) and of Hutchinson's numerous publications. On the basis of these, I produced a 25-page article entitled 'The clinical observations of Jonathan Hutchinson' illustrated with items from the Clinical Museum (McKusick 1952).

Syphilis

Mr. Hutchinson's most important observations concerned syphilis. His descriptions of the stigmata of congenital syphilis (including the Hutchinson triad) were published in 1858 when he was 30. He later wrote (Hutchinson 1913):

> I had not at that time [1858], nor, indeed, subsequently, any desire to cultivate practice in venereal diseases – rather, indeed, to avoid it – but my papers having attracted attention … private patients were sent to me, and I soon had an opportunity for observing syphilis among those whose education and intelligence enabled them to afford information far more trustworthy than that obtained in hospital practice … No week passed but some patient whom I had treated many years previously, and quite lost sight of, would turn up again and give me the history of his subsequent life and facts as to his children.

As a result, Hutchinson acquired an extraordinary grasp on the natural history of syphilis.

Venereal diseases, skin diseases in general, and eye diseases were relegated to the surgeon in British medicine of the 19th century. As Johns Hopkins medical historian Owsei Temkin pointed out to me, these 'external' disorders were considered the domain of the surgeon; the rather archaic terms *internist* and *internal medicine* date from an old dichotomy in the healing arts.

Two 'new' genetic syndromes

Although Hutchinson referred to congenital syphilis as hereditary syphilis, we are concerned here primarily with two genetic syndromes that Hutchinson described for the first time.[2] Both have recently been elucidated at a molecular (i.e., DNA) level. A discussion of these syndromes illustrates the role of syndromology and dysmorphology in clinical genetics and the way mapping and sequencing the human genome (together, let us call them genomics) have permitted identification of basic defects in genetic syndromes and 'birth defects.' The two syndromes are progeria, which Hutchinson described in 1886 (Hutchinson 1886) and which was also described by Gilford in 1904 (Gilford 1904); and the polyps-and-spots syndrome (a.k.a. Peutz-Jeghers syndrome), the pigmentary component of which Hutchinson described in a pair of identical twins in 1896 (Hutchinson 1896). Both are autosomal dominant disorders, almost all cases of progeria representing new dominant mutations.

Syndromology refers, of course, to the study of syndromes. Strictly speaking, *dysmorphology* is the study of malformations, i.e., structural birth defects. The term was invented by David W. Smith (1926–1981) as a felicitous substitute for *teratology*, which should not be used with patients or their parents. Clinicians do not like to be called teratologists; dysmorphologist is now the generally used designation.

Syndromology and dysmorphology are important parts of clinical genetics. The clinical geneticist is confronted by many individually rare genetic syndromes. These must be distinguished and characterised for proper diagnosis, prognosis, clinical management, genetic counselling, and identification of 'pure-culture' (i.e., homogeneous) groups for studies of the basic defect and pathophysiology.

Structural birth defects belong in the domain of the medical geneticist even though a genetic causation may not be established. The medical geneticist must keep both non-genetic and genetic causal factors in mind, as well as the interaction of the two.

The Hutchinson-Gilford progeria syndrome (HGPS), which is now the preferred designation for classic progeria, has an extraordinarily stereotypic phenotype and its genotypic basis turns out to be extraordinarily stereotypic as well, with change in one nucleotide out of the 3 billion in the human genome being the cause in most typical cases.

Hutchinson entitled his 1886 paper as follows: 'Case of congenital absence of hair, with atrophic condition of the skin and its appendages.' The other stereotypic characteristics are a generally senile appearance, short stature, disproportionately large head with 'pinched' facial features, lipodystrophy, stiff joints (e.g. of knees and elbows which show limitation of extension), and premature atherosclerosis.

The polyps-and-spots syndrome (Peutz-Jeghers syndrome). In 1896, Hutchinson reported identical twins with the cutaneous features of the polyps-and-spots syndrome. He emphasised the spots around the mouth and on the buccal mucosa. It is ironic that he did not recognise the association of intestinal polyps because he had written on intussusception which is a major way in which jejunal polyps are manifest.

The association of polyps with spots in Hutchinson's twins was first suggested by the statement of F. Parkes Weber (1863–1962) in a paper in 1919 (Weber 1919) that 'one of the twins died at the age of 20 years from intussusception but the other is still living and in good health and now age 35 years.' In correspondence with Weber in 1949, I learned that the family name was Howard and that the father of the twins was official ratcatcher of the Corporation of the City of London. Furthermore, Weber (although a native Brit (McKusick 1963), he always pronounced his name in the German manner 'Vayber') stated that a brother of the twins succeeded their father in the official position. A letter from that brother, R. Howard, in May 1949 stated that the second twin had died at home of breast cancer at the age of 52 (in 1936). Although in 1949 (Jeghers *et al.* 1949) we thought that the occurrence of breast cancer was coincidental, we now know that such is not the case.

An association of polyps with the spots of the lips and oral mucosa was firmly established in 1921 by Peutz (1921) who reported findings in a large Dutch family.

I was introduced to the polyps-and-spots syndrome by a teenaged patient, Harold Parker, during the final month of my Osler internship at Johns Hopkins, June 1947. The patient had the characteristic pigmentary changes. During his first decade he had chronic anaemia and frequent 'belly aches' and at the age of 11 began having episodes of intussusception requiring surgery. Jejunal polyps were discovered as the basis.

During the following year four other patients came to my attention at Johns Hopkins, three of whom were in the same family, indicating autosomal dominant inheritance. (The three subjects in one family belonged to what was subsequently called the 'Harrisburg family' (Foley *et al.* 1988).) I heard that Harold Jehgers in Boston likewise had five cases. In 1948 Jeghers moved to Washington as the first full-time professor of medicine at Georgetown University and we teamed up for a report of these ten cases in the *New England Journal of Medicine* in 1949 (Jeghers *et al.* 1949).

It was Chester S. Keefer who had called Jeghers' attention to the fact that Hutchinson had described the characteristic pigmentary changes of this syndrome. Keefer graduated in the class of 1922 from the Johns Hopkins Medical School (with two other notables, Alfred Blalock and Tinsley Harrison) and was an intern and resident on the Osler Medical Service. As chairman of medicine at Boston University School of Medicine, Keefer was later Harold Jeghers' chief when Jeghers was head of the Boston University Service at Boston City Hospital. As a student and house officer at Hopkins, Keefer must have taken Osler's recommendation concerning Hutchinson's *Archives of Surgery*, thus explaining his familiarity with the pigmentary changes. Certainly the set of *Archives of Surgery* in the Welch Medical Library at Johns Hopkins is well worn.

In 1949 (Jeghers *et al.* 1949), we concluded that the polyps in this syndrome are hamartomatous, not adenomatous in nature, showing multiple tissue types on histology, and we concluded that they are not premalignant. Subsequently, the occurrence of malignant tumours in this syndrome, particularly pancreatic cancer and characteristic types of functional testicular and ovarian tumours, was demonstrated (Lehur *et al.* 1984; Giardiello *et al.* 1987; Young *et al.* 1995). (My second patient, a teenaged female, succumbed to pancreatic cancer at the age of about 32 years.)

The designation *Peutz-Jeghers syndrome* (PJS) seems to have first been applied in a paper from the Mayo Clinic in 1954 (Bruwer *et al.* 1955).

The principle of genetic pleiotropism

I was responsible for the genetic interpretation of the polyps-and-spots syndrome in Jeghers *et al.* (1949). I was guided in that analysis by Professor Bentley Glass of Johns Hopkins University who made it clear to me that the syndrome was almost certainly an example of pleiotropism of a single mutant gene and not linkage of two separate genes, one for spots and one for polyps.

Having been schooled in the principle of pleiotropism, I could then easily recognise in my cardiologic experience in the early 1950s, the likelihood that the eye, skeletal, and aortic features of the Marfan syndrome represent the pleiotropic effect of a mutation in a single gene involved with one element of connective tissue wherever it occurred in the body. I called it a *heritable disorder of connective tissue* in my first publication on the Marfan syndrome (McKusick 1955): heritable because it was capable of being inherited but in the individual case might be the result of new mutation.

I searched for other conditions that might qualify as heritable disorders of connective tissue and came upon the Ehlers-Danlos syndrome, osteogenesis imperfecta, pseudoxanthoma elasticum, and Hurler syndrome (the prototype of the mucopolysaccharidoses) as likely examples. With the Marfan syndrome, these four constituted the five main chapters in my monograph *Heritable Disorders of Connective Tissue* (McKusick 1956, 1960, 1966, 1972; McKusick and Beighton 1993).

Medical genetics at Johns Hopkins

Medical genetics was institutionalised at Johns Hopkins on 1 July 1957 with the creation of a division of medical genetics in the Department of Medicine. The division was based in the multifaceted chronic disease clinic developed in the Johns Hopkins Hospital by J. Earle Moore (1892–1957) and after 1957 named for him. The new division of medical genetics in the Moore Clinic undertook to do for hereditary diseases what the traditional divisions such as cardiology, gastroenterology, and nephrology did for disorders in their specialty areas, namely teaching, research, and exemplary patient care.

The clinical programmes of the new medical genetics division, which extended to patients of all ages, had an immediate clientele on start-up, namely patients with heritable disorders of connective tissue. In January 1959 the extra chromosome of mongolism (later mercifully renamed Down syndrome) was discovered and in February 1959 a cytogenetic laboratory was established in the Moore Clinic by Malcolm Ferguson-Smith which attracted additional patients. Some of these patients may have been referred because of the misconception that any familial or congenital disorder should have a microscopically identifiable abnormality of the chromosomes. Patients were referred by colleagues of many specialties of medicine, and many patients and families were self-referred.

The teaching programme, engrafted on an already existing training program in chronic disease, attracted to it post-doctoral fellows from a wide area in terms of both geography and clinical specialty. At that time there was no restriction on the use of NIH training grant funds for non-citizens and a considerable number of fellows were recruited from abroad. Many of them were undifferentiated internists who were proselytised to medical genetics and subsequently pursued distinguished careers in the field, including prominent positions in paediatrics.

The research programme of the division focused on the nosology of genetic disease, which was approached in a comprehensive manner, and on gene mapping. Five sections of the Division of Medical Genetics were established: cytogenetics (with Ferguson-Smith and Borgoankar), biochemical genetics (with Boyer), immunogenetics (with Bias), statistical and population genetics (with Abbey, Murphy and Renwick), and clinical genetics. All five sections were necessary for the nosology of genetic disease and for gene mapping by genetic linkage and other methods.

The nosology of genetic disorders, i.e., the clinical delineation of distinct entities, and the description of their natural history could be studied in the Moore Clinic over the whole range of clinical specialties because of its location in a large general hospital. Many of my studies of the Old Order Amish (McKusick 1978) were nosologic in nature. Studies of these founder populations allowed delineation of the range of variability of a disorder such as the Ellis-van Creveld syndrome which was unusually frequent in the Lancaster County Amish and could be assumed to be due to precisely the same mutation in each Amish case. At the same time, it was possible to identify 'new' entities brought to light in numerous cases by inbreeding; cartilage-hair hypoplasia (CHH), for example, is a skeletal dysplasia causing dwarfism that was first described in the Amish (McKusick *et al.* 1965).

The work on dwarfism in the Amish led directly to studies of skeletal dysplasias more generally and interaction with Little People of America, Inc. (LPA), a fraternal organisation of dwarfs and midgets, or, as it now might be termed, a genetic support group. Involvement with LPA began in 1965 when I attended for the first time the annual national convention of the organisation. The skeletal dysplasias represent a highly heterogeneous category of genetic disease that has been the subject of intense nosology for the last 40 years.

There can be said to have been a heyday of syndromology and dysmorphology, and of genetic nosology in general, in the period 1955 to 1975. During that period, *Mendelian Inheritance in Man, A Catalog of Autosomal Dominant, Autosomal Recessive, and X-linked Phenotypes* (McKusick 1966 (1st cdn), 1998 (12th edn). Online version, OMIM (*www.ncbi.nlm. nih.gov/omim*) was initiated, with the first print edition in 1966. Annually for the 5 years 1968 to 1972, one-week conferences entitled the *Clinical Delineation of Birth Defects* were held at the Johns Hopkins Hospital. These conferences, at which many patients were presented at a mid-day conference, and syndromologists and dysmorphologists from various parts of the world participated, had an air of intellectual excitement. In attendance at each of these conferences were a dozen or more 'living eponyms,' persons who already had their names attached to syndromes or who, on the basis of these conferences, became eponymously honoured (Table 1). The conferences gave academic respectability to syndromology; it could no longer be considered merely rare postage stamp collecting.

Not only was pleiotropism a leading principle underlying the delineation of genetic entities and description of natural history at the week-long Birth Defects conferences, but also the principles of genetic heterogeneity (different fundamental cause of the same or nearly the same phenotype) and variability (individual differences in a phenotype that has the same genetic basis). As to genetic heterogeneity, it was pointed out that nosologists tend to be either lumpers or splitters. By and large, splitting is the usual result when a particular inherited phenotype is studied in detail. What appear to be identical disorders are found in fact to be two or more aetiopathogenetically distinct disorders. Because of uncertainties, the nosologist often ends up a 'fence straddler' however.

Table 1. 'Living Eponyms' in attendance at the five successive annual Conferences on the Clinical Delineation of Birth Defects held at Johns Hopkins, 1968–72.

Aarskog	Fraser, George	Jackson	Laron	Opitz	Smith, David
Bartter	François	Kaufman	Leroy	Potter	Spranger
Blizzard	Gardner	Keutel	Lowry	Robinow	Taybi
Char	Gorlin	Klein	Menkes	Rubinstein	Townes
Cohen	Hall	Klinefelter	Nance	Seip	Warburg
Dent	Hirschhorn	Lamy	Noonan	Shwachman	Zellweger
Emery	Gardner	Langer	Omenn	Sly	

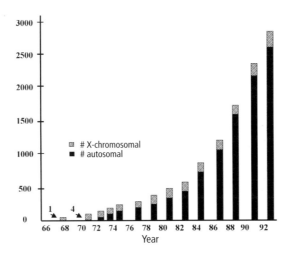

Figure 1. Progress in gene mapping: number of genes mapped to the X chromosome or a specific autosome. (Based mainly on data collated during the international Human Gene Mapping Workshops.)

Gene mapping

The period between 1970 and 1985 can be said to have been a heyday for gene mapping. The gene mapping research of the Division of Medical Genetics began with studies of the linkage of G6PD deficiency and colour-blindness (on the X chromosome) in the early 1960s, and the design of an early computer program for linkage analysis by Jim Renwick and Jane Schulze. Other research resulted in the first mapping of a specific gene to a specific autosome; in 1968 (Donahue *et al.* 1968) the Duffy blood group gene was mapped to chromosome 1 by Roger Donahue, a candidate for the PhD degree in Human Genetics at Johns Hopkins University. In that year 68 genes were known to be on the X chromosome as revealed by X-linked pedigree pattern and listed in the 1968 (second) edition of *Mendelian Inheritance in Man*, and only one, the Duffy blood group, on a specific autosome.

As shown by the curve in Figure 1, the gene mapping field took off in the 1970s using the methods of linkage analysis, somatic cell hybridisation, and, after 1980, molecular genetics. Beginning in 1973, annual or biennial conferences called Human Gene Mapping Workshops were held, at which the aficionados of gene mapping collated the information on mapping that had been acquired since the previous meeting.

By 1985 when the 'Human Genome Project' was formally proposed as an initiative to sequence the entire genome, at least 700 genes had been assigned to a specific chromosome and in most instances to specific regions of those chromosomes. Furthermore, the clinical usefulness of gene mapping (Figure 2) was becoming evident in the practice of medical genetics, in relation to Huntington disease, for example.

The Human Genome Project can be said to have started about 1970 with localisation of genes on specific chromosomes, and genomics, in the original sense for which the name was devised, can also be said to have started at that time. According to the OED, the term *genome* was first used by Winkler in 1920. He appears to have created the word by elision of the words GENes and chromosOMEs. Of course, that is what the word genome means: the complete set of chromosomes and the genes they contain. The word *genomics* is of more recent vintage, having been

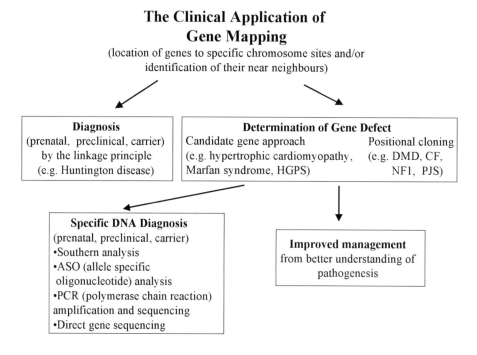

Figure 2. The clinical application of gene mapping.

suggested in 1986 by Thomas H. Roderick of The Jackson Laboratory, Bar Harbor, Maine as the title for a new journal on mapping and sequencing genomes. Thus, the word genomics in its original sense encompassed both mapping and sequencing. The inaugural editorial in *Genomics* in 1987 (McKusick and Ruddle 1987) was entitled 'A new discipline, a new name, a new journal.' Nine years later the journal's editorial was entitled 'Genomics: an established discipline, a commonly used name, a mature journal' (Kucherlapati *et al.* 1996). Furthermore, 'On mapping and sequencing the human genome' was the title of the report of the NRC/NAS committee which was commissioned in late 1986 to consider the proposed Human Genome Project (National Research Council 1988).

The Human Genome Project was formally proposed in 1985 as an initiative to sequence completely the human genome. The project was funded with an official start date of 1 October 1990, with James Watson as the first director of the NIH program. Francis Collins assumed the directorship on 1 January 1993.

The NRC/NAS committee had recommended 'Map first, sequence later.' This 'top-down' approach was recommended because in 1988 the technology was not well enough developed for efficient sequencing and the maps of marker traits and DNA clones, such as the YAC (yeast artificial chromosome) maps, would be needed as an aid in the final sequencing. The top-down approach was the one taken by the NIH programme.

In 1995 Craig Venter, Hamilton O. Smith, and their colleagues (Fleischmann *et al.* 1995) succeeded in determining the first complete sequence of the genome of a free-living organism, *Haemophilus influenzae*, the bacterium from which Smith had first isolated a restriction enzyme for which he shared the Nobel Prize. Rather than using the 'top-down' approach, they shredded the genome into multiple pieces, sequenced the pieces individually and then reassembled the sequence using an algorithm that recognised overlapping ends of the pieces. At a company called Celera founded by Venter, this 'bottom-up' approach was subsequently applied with success to many organisms including *Drosophila melanogaster*, the complete genome sequence of which was published in March 2000, and *Homo sapiens*. Meanwhile, a consortium of investigators world-wide, led by Francis Collins, undertook a collaborative enterprise for complete sequencing of the human genome. On 26 June 2000, draft sequences were announced by Venter speaking for his company Celera and by Collins speaking for the consortium. The announcement took place in the East Room of the White House and the occasion was hosted by President Clinton, with closed circuit television hook-up to 10 Downing Street, where a comparable group of British scientists was gathered. The sequence drafts were published in February 2001, that of the consortium in *Nature* and that of Celera in *Science*.

Dating 'completion' of the human genome sequencing project is a somewhat arbitrary matter; because of complexities, the sequence will always be to some extent incomplete. Thus there was justification for the 2003 timing of celebrations of the discovery of the double helix, 50 years after the eureka moment in the Cavendish Laboratory in Cambridge in February 1953, and the publication of Watson and Crick's paper in *Nature* on 25 April 1953.

With completion of the Human Genome Project, several paradigm shifts have come about, as outlined below.

Post genomic paradigm shifts

Structural genomics	→	Functional genomics
Map-based gene discovery	→	Sequence-based gene discovery
Single gene disorders	→	Complex traits (multifactorial disorders)
Genetic disease diagnosis	→	Common disease prediction (susceptibility)
Aetiology (specific mutation)	→	Pathogenesis (mechanism)
One gene at a time approach	→	Families of genes, pathways, systems
Genomics	→	Proteomics

One of them involves a shift from genomics to proteomics—from study of the genetic material to the study of the proteins encoded by the genome.

Completion of the genome sequence brought to attention what might be called the Gene-Protein Paradox. Scrutiny of the human genome sequence for identification of genes defined as coding elements reveals a much lower number than previously thought, perhaps no more than 30,000 genes. The dogma *gene → messenger RNA → protein* in its simplest form would suggest a numerical correspondence between the number of genes and the number of proteins. But clearly there are many more different proteins, perhaps at least ten times as many, than there are genes. It is generally agreed that multiple proteins are encoded by a single gene through alternative splicing, which results in a mixing and matching of the separate exons (coding segments) of a given gene, and through co- and post-translational processing. This matter of many proteins from one gene is relevant to the nature of the basic defect in the Hutchinson-Gilford progeria syndrome and a number of related conditions, as determined within the last two years.

The molecular genetics of progeria (HGPS)

Studies of the genetics of progeria over the last 30 or 35 years led to the conclusion that the disorder is due to heterozygosity for a dominant mutation, in most cases a *de novo* mutation since most cases are sporadic. Paternal age effect, an average increase in the age of fathers of sporadic cases, also pointed to new dominant mutation because of the known role of paternal age in mutation rate. The lack of increased parental consanguinity suggested that recessive inheritance is not involved. The few instances in which more than one sib was affected could be explained by gonadal mosaicism.

Map-based gene discovery was involved in the molecular characterisation of both of the genetic syndromes described by Hutchinson. The HGPS gene was initially mapped to the long arm of chromosome 1 by observations of two cases of uniparental isodisomy of 1q, and one case with a 6-Mb paternally derived interstitial deletion of one area of 1q. These findings prompted Eriksson *et al.* (2003), working with Francis Collins, to focus on genes in the 1q21 region. One of these genes, LMNA, encodes lamin A/C, intermediate filament components of the inner membrane of nuclei. Mutations in this gene had already been identified in Dunnigan-type familial partial lipodystrophy, and lipodystrophy is a feature of progeria.

Eriksson *et al.* (2003) found that in 18 of 20 patients there was a change of nucleotide 1824 in the LMNA gene from cytosine to thymine (1824C→T). This changed codon 618 from GGC to GGT, with no change in the amino acid encoded (both code for glycine), but with an alteration in splicing of the messenger RNA. The nucleotide change caused activation of a cryptic splice site within exon 11, resulting in a skipping of a stretch of coding DNA and production of a lamin A/C protein missing 50 amino acids at its COOH end.

One of 20 cases studied by Eriksson *et al.* (2003) had a different substitution within the same codon (GGC to AGC), changing the codon from glycine to serine (G608S). This mutation likewise caused activation of the cryptic splice site, resulting in the same anomalous protein missing 50 amino acids in the C terminus. (A 20th case in the original study of Eriksson and coworkers had a missense mutation, E145K; in retrospect, it was concluded that this patient instead of HGPS had mandibuloacral dysplasia (OMIM 248370), another laminopathy.)

Thus, the genotype of classic progeria is as stereotypic as the phenotype. Mutations elsewhere in the LMNA gene cause various other disorders, including at least two types of muscular dystrophy, a form of dilated cardiomyopathy, familial partial lipodystrophy, a form of Charcot-Marie-Tooth disease, and, as already mentioned, mandibuloacral dysplasia (Table 2). All of these conditions are laminopathies and show abnormalities of the nuclear membrane. Distortion of the shape of the nucleus can be demonstrated in all of the laminopathies.

The laminopathies also include a new autosomal dominant disorder described by my colleagues Daniel Judge and colleagues (Judge *et al.* personal communication) and designated cardiocutaneous progeria syndrome (CCPS). This disorder is characterised by precocious aging, calcific aortic and mitral valvular disease, and early myocardial infarction.

The fundamental defect in the Peutz-Jeghers syndrome involves a 'novel' (i.e., previously unknown) gene that was identified by positional cloning. Hemminki *et al.* (1997), in Helsinki, reasoned that the gene mutated in PJS was likely to be a tumour suppressor gene and to follow the Knudson paradigm; they assumed that in cases of familial PJS the development of polyps or cancers required a germline mutation on one chromosome and a somatic mutation on the other chromosome. They further assumed that the somatic mutation was, in many cases, a dele-

Table 2. Laminopathies resulting from different mutations in the LMNA gene which encodes the lamin A/C proteins, autosomal recessive form intermediate filament components of the inner nuclear membrane.

Emery-Dreifuss muscular dystrophy, autosomal recessive form
Dilated cardiomyopathy, one form
Dunnigan-type familial partial lipodystrophy
Limb-girdle muscular dystrophy, one form
Charcot-Marie-Tooth disease, one form
Mandibuloacral dysplasia
Hutchinson-Gilford progeria syndrome

tion. Thus, they performed comparative genome hybridisation (CGH), comparing DNA of tumours with the DNA of other non-tumour (somatic) cells of the same patient. In tumour DNA, hybridisation revealed evidence of deletion on the short arm of chromosome 19 in a number of instances. They then did conventional linkage studies with markers in Peutz-Jeghers families, focusing on the region of deletion, and in this way mapped the PJS phenotype to 19p13.3. In this area they found a previously unknown gene, a serine/threonine protein kinase, which they designated STK11. Various heterozygous point mutations in this gene were discovered in affected members of PJS families by the Helsinki group (32) and by others. Somatic mutations in the gene were also found in sporadic cases of pancreatic cancer, melanoma, and other tumours in which two somatic 'hits' were presumably necessary for tumorigenesis.

Patients with familial PJS are prone to breast cancer (the cause of death in one of Hutchinson's twins) but STK11 mutations have not been found, it seems, in sporadic breast cancer. Testicular and ovarian tumours in PJS patients are often functional with feminisation in males and isosexual precocity in females.

Bentley Glass told me in 1948–49 that even though we did not understand the mechanism of the association of polyps and spots, it must be based on pleiotropism. Even though the gene defect in PJS is now known at the level of DNA sequence, the mechanism of the spots in PJS is still unclear. Because of the finding of PJS mutations in sporadic melanoma (Guldberg *et al.* 1999), it has been suggested that the spots are small tumours. The stippled appearance of the spots in my first patient (Jeghers *et al.* 1949), and the histology of a spot on the patient's hand suggested clonality. Does repeated irritation around the mouth have any causative role in the development of spots in the genetically predisposed subject? Hutchinson claimed that in his twins the spots were not present at birth but appeared at about the age of 3 years.

A catalogue of human genes and genetic disorders

Progress in both the clinical and the molecular delineation of distinct syndromes and dysmorphological entities has been chronicled for more than 40 years in *Mendelian Inheritance in Man (MIM)* (McKusick 1966 (1st edn), 1998 (12th edn). Online version, OMIM (www.ncbi.nlm. nih.gov/omim). This catalogue has been computerised [3] since 1964; the first book edition was a pioneer in computer-based publishing. MIM has gone through 12 print editions, the first in 1966 and the most recent (in 3 volumes) in 1998. For the first 10 editions, the subtitle was 'Catalogs of Autosomal Dominant, Autosomal Recessive and X-linked Phenotypes.' Because of

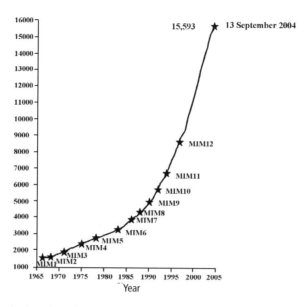

Figure 3. Growth of number of entries in the 12 print editions and in the continuously updated online version, OMIM.

the molecular information that could be added to the phenotype-based entries of earlier editions, the subtitle of MIM was changed to 'A catalog of human genes and genetic disorders' in 1992. The internet version, OMIM (online MIM), updated essentially daily, has been available world-wide since 1987. Figure 3 indicates the growth in the number of entries in MIM (and OMIM) over the years of its existence.

As of 24 May 2005, OMIM listed 1,742 genes with at least one known mutation causing a disorder or neoplasm. The total number of distinct phenotypes for which a causative mutation in one of these 1,742 disease-related genes was 2,743. The latter figure reflects the fact that several phenotypically distinct disorders can result from different mutations in a single gene, as illustrated by progeria and the other laminopathies (Table 2).

As chronicled in *Mendelian Inheritance in Man* (McKusick 1966 (1st edn), 1998 (12th edn) and its online version, most of the identification of the specific mutational basis of syndromes and dysmorphological entities has come since 1986, and much of it through map-based gene discovery, as in Peutz-Jeghers and Hutchinson-Gilford progeria syndrome.

Throughout its history, *Mendelian Inheritance in Man* has been authored and edited (curated, if you will) at Johns Hopkins where the online version was created and made generally available from the Welch Medical Library under the designation OMIM beginning in 1987. In December 1995 the World Wide Web distribution of OMIM was assumed by the National Center for Biotechnology Information at the National Library of Medicine. Enhancements instituted there and linkages with other databases and bibliographic resources have greatly increased its usefulness.

Summary

Using the work of Jonathan Hutchinson as a starting point, I have attempted to show how the description and delineation of genetic disorders goes through stages: initial description by an astute observer, refinement of the phenotype description with documentation of the range of variability and the natural history, defining of the Mendelian genetics from family studies, and through genomics (defined here as both mapping and sequencing and more), discovery of the precise molecular defect at the level of DNA. Diagnosis becomes precise when testing can be based on the actual lesion in the DNA. Deduction of the pathogenetic mechanisms by which the mutant gene gives rise to the morbid phenotype (the steps from gene to phene) may be possible. Efficacious amelioration of the disorder by means directed at these intermediate steps or at the mutant gene itself may then be devised.

Epilogue

The syndrome that bears the name of George Fraser (OMIM 219000) has now been elucidated at the level of the DNA change. Indeed, two clinically indistinguishable molecular forms have been identified: that due to mutation in the FRAS1 gene (OMIM 607830) on chromosome 4 and the other due to mutation in the FREM2 gene (OMIM 608945) on chromosome 13. Mutations in these genes were discovered after mutations had been found in the homologous genes of the mouse in two conditions thought to be models for Fraser syndrome, 'blebbed' and 'myelencephalic blebs', as discussed elsewhere in this Festschrift.

Notes

1. Presented as the Gordon Wilson Lecture at the annual meeting of the American Clinical and Climatological Association, Sea Island, GA, 23 October 2004.

2. The patients Hutchinson reported in his *Archives of Surgery* and elsewhere were referred to him by their primary physicians. Dr. J.T. Connor, for example, had previously presented the twins with perioral spots at the Aesculapian Society of London (*Lancet* 2 1169 1895). Hutchinson noted that the practitioner would give him permission to 'take a portrait,' i.e., have a watercolour illustration made by his full-time artist. There is no record that the consent of the patients was obtained. The Royal College of Physicians (London) has Hutchinson's own set of *Archives of Surgery* (1889–1901) with annotations in his hand and interleaved letters from practitioners whose patients he described.

3. In a letter to George Fraser in Jan. 1964 thanking him for review of the catalogue of autosomal recessive phenotypes, I commented that, on the advice of my computer assistant David R. Bolling, I was putting the catalogues 'on the computer.' In the days before the word processor, this avoided the introduction of errors with the periodic updates and facilitated preparation of indices and other features of the published version.

Prenatal death in Fraser syndrome

Jessica M. Comstock[1], Angelica R. Putnam[1], John M. Opitz[1,2,3,4], Theodor J. Pysher[1], and Juliana Szakacs[1,5]

[1]Departments of Pathology, [2]Pediatrics (Medical Genetics), [3]Human Genetics, [4]Obstetrics & Gynecology, University of Utah, Salt Lake City, UT, [5]Department of Pathology, Harvard University

Dedicated to George R. Fraser of Oxford, England, universally learned physician and humanist, on the occasion of his 75th birthday with gratitude for seminal contributions to human and medical genetics which shall endure for all time.

Abstract

Cryptophthalmos may be partial or complete, unilateral or bilateral, apparently non-syndromal or syndromal. A recent study of two stillborn infants at the University of Utah prompted an analysis of the developmental aspects of the syndromal form ('Fraser syndrome'). We conclude that, *per se,* cryptophthalmos is a developmental field defect on the basis of *heterogeneity* (autosomal dominant and recessive forms) and *phylogeneity* (occurrence also in the pheasant, rabbit, pigeon, dog and mouse). In humans this autosomal recessive disorder maps to 4q21, is homologous to the bleb (*bl/bl*) mouse and is due to mutations in the *Fras1* gene that codes for a 4007 amino acid protein 85% identical to the *Fras1* gene of the bleb mouse. Commonest anomalies in humans are cryptophthalmos, cutaneous syndactyly of digits, abnormal ears and genitalia, renal agenesis and congenital heart defects. Almost half of affected infants are stillborn or die in infancy and mental retardation is common. The pathogenesis evidently involves abnormal epithelial integrity during prenatal life. Older (mostly German) publications, some dating to the 19th century, allow a fascinating historical insight into the process of syndrome delineation.

Introduction

In the context of a discussion of the 'genetical load' of the human species Fraser (1962a) noted that inbreeding, specifically of first cousin mating, was associated with a significant increase in major congenital abnormalities, suggesting 'that recessive inheritance plays some part in the causation' of these and 'that the tendency was especially marked when the complex anomalies were considered' (p. 399). And then, citing Brodsky and Waddy (1940), he briefly cites his observation of two sibships in each of which two sisters were affected with various permutations of: cryptophthalmos; absent or deformed lacrimal ducts; middle and outer ear malformations; high palate; 'cleavage along the midplane of nares and tongue'; hypertelorism; laryngeal stenosis; syndactyly; wide separation of symphysis pubis; displacement of umbilicus and nipples; primitive mesentery; small bowel; maldeveloped kidneys; fusion of labia and enlargement of clitoris; bicornuate uterus and malformed Fallopian tubes. In each case one sister was stillborn, the other was viable, the latter being X-chromatin positive. 'This could well be a recessive condition ... ' To the best of our knowledge Fraser has never published details.

Reardon (1997) has pointed out that eponymy is as much a matter of luck as it is a 'haphazard process' and that George Fraser was a case in point in that the condition to which his name is attached did not reflect ' "the area of special expertise and interest" for which that clinician most deserves remembrance'. Never mind that the condition under discussion was first well-described 90 years before Fraser's paper. However, given his massive contributions to the genetics of visual impairment and of hearing loss and to the establishment of cancer genetics clinics in the UK, it would be graceless to quibble about the eponym 'Fraser syndrome' which seems to have been promoted by McKusick and other medical geneticists to refer to those cases with all manifestations except for cryptophthalmos.

We are describing two neonatally deceased infants with Fraser syndrome to honour George Fraser and as a contribution to the developmental biology of complex multiple anomaly (MCA) syndromes.

Clinical reports

Infant 1 (AU 04–84)

From a previous marriage the Hispanic mother of this infant had a healthy term infant, as she did from the first pregnancy in her present marriage. She received prenatal care; however, ultrasonography at 28 weeks in the present pregnancy showed intrauterine growth retardation (25 wk gestational size), anhydramnios, bilateral renal agenesis and a 2-vessel umbilical cord. Maternal age was 30. There was no maternal diabetes, known drug use or chemical exposure, or hypertension. The history did not indicate evidence for abruption or placenta previa. Spontaneous labour began at 36–37 wks of gestation, and the infant was delivered normally with Apgar scores of 2 and 2 but was not resuscitated and died neonatally.

Gross findings. Crown-heel length: 41.0 cm (normal 43.4 ± 5.9), crown-rump length 24.5 (nl 31.8 ± 3.9 cm); weight 1580 g (nl 2280 ± 615). Normal head size (27.8 cm). Inner- and outer canthal distances 2.3 and 6.3 cm, respectively; inter-pupillary distance was 3.5 cm, palpebral fissure lengths both 2.0 cm with upslant. Skull shape was symmetrical and scalp hair distribution was normal for a Hispanic baby; hairlines were low and back was hirsute. On the right there was a 5 mm ankyloblepharon superior. Mouth was small (2 cm), philtrum was smooth and there were virtually no vermilion borders. Right helix was 2.7, left 2.4 cm long with deficiency of the lower helix and prominence of upper helix on the right; on the left there was hypoplasia of the descending and transverse helix. External auditory meati were probe-patent but stenotic. Tip of nose was anteverted with hypoplasia of alae nasi; there was no choanal atresia. Chest circumference was 24.5 cm; chest was symmetrical and nipples, areolae and breast buds were well-developed. The 2.5 cm remnant of cord contained two vessels. Back was straight, anus imperforate and external genitalia ambiguous, not allowing clear determination of sex. An apparent vaginal orifice was probe-patent, but the vagina ended in a blind pouch. No urethral orifice was seen and the phallus was prominent. There was a bilateral varus deformity of feet with mild syndactyly of toes four and five, and dislocation of the head of the radii. On the right there was a simple palmar crease with syndactyly of digits 1–2, 2–3, and 3–4 and clinodactyly of 5. On the left all digits were syndactylous; there was hypoplasia of thumbs and thenar eminences.

Thoracic and abdominal situs was normal but the pleural cavities contained 3.5 ml of serous fluid. Mediastinum was normal but thymus small (4.9 vs 7.7 ± 5.0 g). Diaphragms were intact, but the mesenteric root was short. The heart weighed 9.8 g (vs 15 ± 5.1 g) and was normally formed; however, the semilunar valve was very thin. All vessels were normal save for absence of the left umbilical artery. The trachea branched normally but the larynx was stenotic. Pulmonary weights and location were normal. Anus was imperforate; gastrointestinal tract was normal but colon dilated; liver and pancreas were normally formed; liver weight was 60 g (vs 96.3 ± 33.7 g) and spleen weight 6.2 (vs 8.1 ± 3.1g).

There was absence of kidneys, ureters, bladder and renal veins and arteries. The uterus was bicornuate, normal in size, with cervical stenosis. Adrenal glands were normal in size and location as were ovaries, however, the right was undescended. Other endocrine organs were normal.

Skeletal radiographs showed: broad sternum, posterior homoeotic transformation of C7 into T1 (eight cervical vertebrae). Skeletal muscle mass was decreased for age.

Placenta (weight 221 g) was normally formed without amnion nodosum but with somewhat thick and opaque membranes.

Results of chromosome examination were normal. A fibroblast culture was archived for potential *FRAS1* gene analysis.

Brain meninges and cerebral vessels were essentially normally formed for age with normal brain weight.

Histological findings of placenta, viscera, central nervous system and ovaries were essentially normal.

Summary: Lethal MCA syndrome compatible with a diagnosis of Fraser syndrome comprising, in this case, intrauterine growth retardation with normal brain, single umbilical artery, unilateral ankyloblepharon, syndactyly of fingers (on the left mitten-like), hypoplastic thumbs and thenar eminences, dorsal dislocation of the heads of the radii, blind-ending vagina with bicornuate uterus and unilateral maldescent of ovary, renal-urethral-vesicular agenesis, indeterminate external genitalia, imperforate anus with dilated colon, cervical vertebral anomaly and multiple minor anomalies.

Infant 2 (PA 20–03)

This 22-year-old mother had had two normal children, a premature baby boy with 'renal agenesis' in 1999, and received adequate prenatal care during the present, fourth, pregnancy which was uneventful without exposure to drugs or chemicals until ultrasonographic detection of multiple anomalies and anhydramnios at 35 weeks. Labour was induced at 36 weeks, delivery was normal. The infant died shortly after birth.

External examination. Crown-heel length 42.5 (vs 42.3 ± 2.9 cm). Crown-rump length 30.1 (vs 30.9 ± 2.0 cm), weight 1900 g (vs 2093 ± 309 g). OFC: 31.2 cm. The forehead was narrow and the skull was dolichocephalic in shape. The left eye (OS) was normally formed; there was cryptophthalmos OD and no globe could be palpated; however, the orbital bones were well formed OU. Auricles were low-set and posteriorly angulated, external auditory meati were probe-patent. Tip of nose was flattened (part of the oligohydramnios or Potter sequence). There was no choanal atresia. Palate was highly arched with prominent central raphé without cleft and micrognathia. Umbilicus was low. Back was straight, external genitalia normally formed without identifiable penile urethra. The umbilical cord contained three vessels. Anus was patent. Hands were broad with redundant skin but without poly- or syndactyly.

Cranial Contents. Skull, meninges and cerebral vessels were normally formed; brain weight was increased (294 vs 257 ± 45 g); convolutions were compatible with a gestational age of 35 weeks.

Thoracic and abdominal organs were all normally formed and located, except that the lungs were small. Heart and great vessels were normal. *Retroperitoneally* no kidneys, or ureters were present; bladder was hypoplastic and no urethra, prostate or seminal vesicles could be identified. Adrenals and testes were normal (haemorrhagic on the left). Radiographs showed a normal skeleton. Placenta weighed only 233.1 g; amnion nodosum was noted along one margin. Chromosomes were normal (46,XY).

Histological studies showed normally developed lungs with one area of interstitial and alveolar haemorrhage; sections of skin showed hyperkeratosis, and of the penis confirmed absence of urethra. All other organs and body structures were histologically normal.

Discussion

'Oh pleiotropy, thy name is Fraser syndrome' a poetically inclined developmental pathologist might say in contemplating the enormous complexity and variability of this entity as seen in foetal and paediatric pathology.

Some 90 years before the important paper by Fraser (1962), excellent descriptions of 'Fraser syndrome' began to be published, primarily in the German literature (q.v. Chiari 1883, von Hippel 1904, 1906, the older literature cited in Thomas *et al.* 1986 and Zehender 1872). A case, or cases, may yet be discovered in the writings of J. F. Meckel the Younger, and in anatomical or pathological museum collections (q.v. the foetus from the Gordon Museum of Guy's Hospital, Elçioglu and Berry 2000). As mentioned above, the eponym that honours George Fraser represents the consensus of clinical geneticists in referring not only to these cases with cryptophthalmos of variable degree but also to those with all other manifestations but no cryptophthalmos.

The epistemological process that leads to syndrome delineation (of phenotype) and definition (elucidation of cause) is infinitely fascinating and never the same for any syndrome. We all stand on the shoulders of giants (as Newton wrote to Hooke on 2 Feb 1675), but, if they wrote in German or Latin, in 'obscure publications,' over a century ago, it is so much easier to consult the next-to-last paper or review on the subject as if history had begun yesterday.

And while the description of Pliny the Elder (born 23 or 24CE, died August 24, 79CE, Gaius Plinius Secundus) was more terse than that of Zehender (1872) it is nevertheless highly suggestive. Pliny was born in Como and served as a high imperial officer and procurator under Vespasian; and while commanding the Roman fleet at Misenum suffocated on land at Stabiae during the catastrophic eruption of Vesuvius which wiped out not only Stabiae but also Pompeii and Herculaneum. He was a gifted encyclopaedist who attempted to summarise all knowledge of natural history in his *Naturalis Historia*, available in the English translation by Philemon Holland, Doctor in Physic of 1601, online at *http://penelope.uchicago.edu/holland/index.html* by J. Eason. Lacking a copy of Pliny's work we queried Prof. Eason who noted that in Book 7.51 Pliny notes: '… *in Lepidorum gente tres, intermisso ordine, obducto membrana oculo genitos accepimu'* in the Holland translation: 'In the rase and familie of the Lepidi, it is said there were three of them (not successively one after another, but out of order after some intermission) who had everie one of them when they were borne, a little panricle or thinne skinne growing over the eye'. Ah, for the succinctness of Latin over English and, even more so, over German.

There must have been many others between Pliny and Zehender who had noticed the extraordinary phenomenon of a 'membrane growing over the eyes' in a foetus or newborn infant, and we strongly suspect that such an instance will still be found in the voluminous writings of the younger Meckel in the early 19th century.

Zehender (1872) surely was only another incisura in the history of cryptophthalmos, but without doubt a most important beginning of our modern understanding of syndromal cryptophthalmos together with Chiari and Fuchs.[1]

Zehender (1872, with collaboration of his colleagues Manz and Ackermann, sometimes cited as co-authors) described little Caroline Warner, born 5 March 1871 in Zolkow in Mecklenburg, who died at 9 months. Parents were normal and non-consanguineous; this was the first and only child. Cause: maternal impression, otherwise normal pregnancy and delivery. Bilateral cryptophthalmos with deficiency of eyebrows and lashes; globes present, with light perception OD; abdominal eventration below umbilicus; abnormal genitalia, no sphincters with total incontinence. Syndactyly of fingers and toes (except for halluces). Developed severe diarrhoea, then vomiting, died on 4 December 1871. Autopsy showed, in addition, clitoris without prepuce, no labia minora; lungs plethoric, heart enlarged, but both normally developed. First detailed description (Manz) of the eye abnormalities and the wide separation of pubic symphysis. Apparently, no abdominal or retroperitoneal examination.

Hans Chiari (1851–1916) provided a beautiful, detailed post-mortem study of a stillborn infant received on 18 October 1882 by the pathologists in the Prague Pathological-Anatomical Institute (Chiari 1883). Gestation and family history unremarkable. This baby girl (about gestation age 36 wks plus) had a complete cleft of lip and palate on the right, normal right and completely cryptophalmic left eye, small auricles, abnormal genitalia, syndactyly of all digits except for relatively free halluces. Brain normal, complete atresia of the middle portion of the larynx with pulmonary atelectasis but otherwise normally developed trachea and bronchi. Severe left renal hypoplasia (with surprisingly normal histological structure *en miniature*), small bladder, ovaries present, slightly bicornuate uterus with vaginal and distal cervical atresia.

Fuchs (1889), at a meeting of the Imperial and Royal Society of Physicians in Vienna on 29 March 1889, reported another prototypically affected infant, the ninth child of a poor peasant. Six sibs were normal, but three died. One also had an anchyloblepharon [sic]. Parents were normal and the infant was baptised a girl in spite of ambiguous genitalia (*ein männlicher Zwitter*). Tetramelic syndactylies but with thumbs and halluces relatively spared. The auricles were completely attached to the scalp and the ear canals were very narrow. The right eye was normal, but there was a complete ankyloblepharon OS.

The paper of S. Golowin (1902, Imperial University of Moscow) is a truly classical piece of work in this field from several perspectives. In the description of his first patient, the 26-year-old severely smallpox-scarred Maxim A (from the Governmental District of Kaluga) with apparently non-syndromal bilateral cryptophthalmos, he introduced a note of such poignant pathos in the care of his patient that one is inclined to suggest this paper to first-year medical students as an exercise in the medical humanities. 'The patient implored us most fervently (*inständig*) to open the skin covering his left eye in spite of our doubt that even his opened eye would be capable of vision, if only to bring about a cosmetic improvement.'

Golowin's second patient is more to the point. This 2-month-old girl with complete bilateral cryptophthalmos had, as most do, a broad bridge of nose, large fontanelles, a left parietal foramen (5.5 x 4.5 cm!), 'left fontanelle higher than the right.' This child's mother's oldest child, a girl, now 4-years-old, had the same anomaly and was blind OU. Golowin's proposita died of scarlet fever, but the older surviving, now 8-year-old sister, was severely retarded, unable to speak or walk. Another reason why this paper is important is Golowin's literature review which records in detail (q.v. his Table, pp. 183–186) a total of 13 cases (11 previously reported added to his two). It is an amazing collation.

After Zehender (1872) his case 2 is ('Hocquart 1881')[2] of a woman in her mid-thirties, brought into the anatomical theatre with bilateral cryptophthalmos but no other obvious anomalies. The cases of Chiari and Fuchs are cited above. However, Golowin's case 4 was unknown to us until recently. It refers to van Duyse (1889)[16] and his report of a 3-week-old boy with consanguineous (uncle-niece) parents with bilateral cryptophthalmos, facial asymmetry, defective development of the posterior portions of the ossa parietalia who died at age 21 months of measles. A microscopic examination was published in 1899 (yet to be analysed!). Golowin's cases 6, 8, and 10 will be mentioned below. Case 7 (Otto 1893)[16] was a newborn infant with bilateral cryptophthalmos and syndactyly of the toes of both feet. Case 9 (Kármán 1895[16], also was not studied in the original) referred to a 6-month-old child with bilateral complete cryptophthalmos; no other anomalies were mentioned. Case 11 (Blessig 1900)[16] was a 2-year-old boy with 4 normal brothers and normal parents. This was probably a syndromal instance since, in addition to typical bilateral eye involvement, the boy also had a deformed left nostril and a cleft tip of nose. Cases 6, 8 and 10 are of exceptional interest.

The first was a 1-year-old rabbit (Bach 1895) with complete cryptophthalmos OD and a partial defect OS (temporally there was a 2.5 mm long cleft). The second was a pigeon with a complete defect OD and normal OS. The last was a pheasant chick (Gillet de Grandmont 1893)[16] with complete involvement OU.

The case of von Hippel (1906, abstract 1904) was familial, the mother having had, 6 years before the birth of the present child, a stillborn infant (sex?) with severe foetal ascites as a birth impediment, bilateral cryptophthalmos, webbed fingers, bilateral genu varum and pes varus, rudimentary auricles and prominent (?) maxilla. The present infant was seen on day 10 of life with complete involvement OD and almost complete cryptophthalmos OS except for a 3 mm triangular gap at the temporal canthus. There was extensive syndactyly of fingers and toes, umbilicus abnormally low, normal scrotum and testes but tip of penis bent to the left 'like a posthorn.' He mentions that since his abstract other cases had been published (Blessig 1900, Golowin 1902, Andogsky 1900[18] – three cases, Goldzieher, 1903[18] – re-reported the Kármán case, now at age 10).

Onishi (1911) reported the first case from Japan. In the USA apparently the first two cases were published by Dr. David H. Coover in Vol 55 of *Journal of the American Medical Association*, followed by that of Eberhardt (1911) of an infant girl from Michigan City, born at term with total bilateral symblepharon and left microtia. The patient of Key (1920) had a complete defect OD, partial OS with microphthalmic globe, normal lower lid, ankyloblepharon superior, webbing of all fingers and mental retardation. The 7-line abstract of Müller (1922) refers to complete involvement of one and an upper lid coloboma of the other eye, partial syndactyly on hands and feet, attachment of the upper parts of the auricles to the scalp, atresia of the larynx and anal stenosis. Brodsky and Waddy (1940) published the first Australian case of indeterminate sex, with syndromal involvement.

The definitive review of the Fraser syndrome was published by Thomas *et al.* (1986). Of 124 cases 27 were non-syndromal (sporadic in 16, familial in 11, the familial cases occurring in three families with vertical transmission). Sib involvement in the syndromal cases with occasional parental consanguinity strongly suggested autosomal recessive inheritance. Major diagnostic criteria for cryptophthalmos syndrome were defined arbitrarily as: eye and genital involvement, syndactyly, and affected sib; minor criteria were malformation of nose, ears and larynx, cleft lip and/or palate, skeletal defects, umbilical hernia, renal agenesis and mental retardation. Nowadays, of course, one would say that Fraser syndrome is anyone with a *FRAS1* mutation and an abnormal phenotype including any or all of the Thomas *et al.* criteria. We can be grateful to Thomas *et al.* for their effort in reviewing the literature with such thoroughness and for establishing the frequency of involvement in *sibs* of index cases, as follows:

		Percentage
Cryptophthalmos	15/23	65
Syndactyly	13/20	60
Abnormal genitalia	11/15	73
Nose	7/12	58
Ear	12/14	85
Renal agenesis	8/9	89
Larynx	5/5	100
Umbilical hernia)	3/15	20
Skeletal defects	3/5	60
Mental retardation	1/3	33
Cleft lip/palate	2/20	10

In addition to their review, Thomas *et al.* reported on seven personally studied and highly informative cases. The five cases with congenital heart defects had ASD (2), univentricular heart defect (1), a 'complicated' defect (1), and one unspecified defect. Since then Hambire *et al.* (2003) reported the presence of ASD in addition to preductal coarctation of the aorta, left superior cava draining into the aortic sinus and patent ductus arteriosus.

Pathological analyses date to the 19th and early 20th centuries; however, the first group of families of Fraser syndrome studied by a pathology/clinical genetics group was that by Lurie and Cherstvoy (1984). They studied four families with a total of 14 pregnancies, including two spontaneous abortions, three normal liveborn children, and nine affected foetuses/infants. Apparently, none of the parental couples was consanguineous. Sex ratio: one unambiguous girl; three of indeterminate sex presumed to be girls with ambiguous genitalia; four males. Autopsies were performed in six, not in three of nine. Out of six, renal adysgenesis was present in five, laryngeal atresia in two; cryptophthalmos in at least eight of nine but the ninth had an upper lid coloboma OD but no 'cryptophthalmos'; four of nine had an omphalocoel.

Boyd *et al.* (1988) reported on autopsy findings in 11 foetuses with Fraser syndrome, eight dying neonatally, one being stillborn, together with two mid-trimester foetuses. Three of 11 lived as long as 60 minutes; thus, their foetuses and infants were severely affected. This underscores the great importance to be alert to the diagnosis; for if the delivering physician or midwife did not make the diagnosis then the pathologist will be the last to do so and potentially to prevent recurrence. If pregnancy losses of foetuses of unknown sex are counted as affected then there were six familial occurrences in 11 sibships. In the Boyd *et al.* study one of their foetuses had micromelia of all limbs with marked bowing of tibiae; all had eye, ear, digital, laryngeal and renal abnormalities. If suspected on the basis of gross abnormalities by ultrasonography (Serville *et al.* 1989) then confirmatory molecular testing can be performed. Schauer *et al.* (1990) made the diagnosis prenatally at 18.5 weeks of gestation (after a previous affected sib). Enlarged lungs were highly suggestive of laryngeal atresia (Stevens *et al.* 1994), but this foetus also had absence of the Eustachian tubes, connective tissue occupying the tympanic cavity and bone occluding the external acoustic meatus. In the sibship reported by Francannet *et al.* (1990) three of eight infants were affected, born to first-cousin Turkish parents. Stevens *et al.* (1994) make the instructive point that in case of renal agenesis in this syndrome there is pulmonary *hypo*plasia, in case of renal agenesis and laryngeal stenosis or atresia lung development is apparently *normal*, and without renal agenesis but laryngeal stenosis/atresia there is pulmonary *hyper*plasia. Slavotinek and Tifft (2002) reviewed the diagnostic criteria of Fraser syndrome at some length, generally accepting those of Thomas *et al.* (1986), discussed some of the complex or correlated anomalies in terms of 'phenotypic modules', by which, we think, they mean (polytopic) developmental field defects. They reviewed 117 'cases' since Thomas *et al.* (1986), conveying the impression that, a few cases of consanguinity notwithstanding, this syndrome is not at all uncommon, at least not in prenatal and perinatal pathology services.

Molecular insights

McGregor *et al.* (2003), Vrontou *et al.* (2003) and Takamiya *et al.* (2004), in work with humans and mice, have established the causal and pathogenetic nature of Fraser syndrome. Autozygosity studies mapped *FRAS1* to HSA4q21 (however, with heterogeneity). *FRAS1* encodes a member of a newly identified, phylogenetically ancient group of proteins related to the extracellular matrix (ECM) blastocoelar protein of sea urchin. The fact that the N-terminal of the FRAS1 protein contains a series of cysteine-rich repeat motifs implicated in bone morphogenetic protein (BMP) metabolism suggests that this molecule has a role in 'structure and signal propagation in the ECM' (McGregor *et al.* 2003). These authors also sequenced the *Fras1* gene of the blebbed (*bl/bl*) mouse, a presumed model of the cryptophthalmos syndrome in humans, and found a premature termination of mouse *Fras1*. The blebbed phenotype is highly variable and causally heterogeneous with at least five loci mapped to date. *FRAS1* encodes a predicted protein of 4007 amino acids, with 85% identity to *Fras1*, encoded in the human by 75 exons. In one family McGregor *et al.* (2003) found the sequence change 8620C→T resulting in premature termination of the protein product (Q2863X). They also found four additional mutations in other kindreds, each predicted to lead to loss of function. There was no correlation between genotype and phenotype.

A developmental pathology analysis of the *bl/bl* mouse with cryptophthalmos and webbed digits showed separation of the basement membrane from the dermis around Ell.5 before the development of overt blebs. Thus, the pathogenesis of the ectodermal defects in the *bl/bl* mouse is akin to a prenatal epidermolysis bullosa, however, without a collagen VII defect or continued postnatal blistering. However, this explanation does not 'work' for the renal anomalies ranging from bilateral agenesis to unilateral absence to a cystic renal dysplasia. Until day Ell.5 the mesonephros is intact. Thereafter the embryos had dilated ureteric buds with less branching than did wild-type mice; however, 20% of the embryos had relatively normal metanephric development. Thus, after initially normal mesonephric ontogeny, facultative deficiency of mesonephric/ureteric bud interaction leads to subsequent degeneration.

The work of Vrontou *et al.* (2003) confirmed that of McGregor *et al.* (2003) and showed that the FRAS1 protein is distributed in a linear fashion beneath the epidermis and basal surface of other embryonic epithelia, and that its absence leads to subepidermal hemorrhagic blisters and renal anomalies. Takamiya *et al.* (2004) extended the story by demonstrating that loss of a cytoplasmic multi-PDZ scaffolding protein, namely the glutamate receptor interacting protein 1 (GRIP1) causes the same type of subepidermal haemorrhagic blister formation as in the *bl/bl* mouse. And that GRIP1 interacts physically with Fras1 in order to localise Fras 1 to the basal side of cells. In another form of blebbed mouse, the eye-blebs (*eb*) mouse *Grip1* is disrupted by a deletion of two coding exons.

Thus, the heterogeneity of Fraser syndrome and of the blebbed mice has led to an insight of fundamental biological importance concerning the complexity of chemical and physical processes that regulate ECM and the integrity of basement membrane function during development. *FRAS1* will surely soon be followed by *FRAS2, FRAS3* …

Notes
1. But as an eponym Zehender-Chiari-Fuchs syndrome does not roll too lightly off the tongue of non-German speaking clinicians.
2. Quoted in primary sources and not referenced in bibliography.

Twenty-five years' history of Váradi-Papp syndrome (orofaciodigital syndrome VI)

Zoltán Papp

Department of Obstetrics and Gynaecology, Semmelweis University, Budapest, Hungary

I have long known George Robert Fraser because of his contributions to the literature. I had the opportunity to first meet him in person in 1984 when I spent a year in Oxford on a Wellcome fellowship, I then learned that in addition to being a great geneticist, he was also Hungarian. I have great respect for him as a clinical geneticist and appreciated him especially for his great knowledge and philosophical insight. I have also admired how the syndrome named after him (McKusick Catalog Number 219000) has become part of the curriculum in medical education (Fraser 1962a). I pay tribute to this great physician on his anniversary by presenting the 25-year follow-up of the multiple malformation syndrome described by my wife and me in 1980 (Váradi *et al.* 1980). It is of interest to note that our studies were pursued in a gypsy isolate and that George Fraser has in his career also studied rare autosomal recessive diseases in isolates. Our studies were pursued in Hungary while his most substantial studies of this type were pursued not very far away on the Croatian island of Krk (Veglia) which, until 1918, formed part of the Austro-Hungarian Empire (Fraser 1962c, 1964h).

In 1978 I studied a male gypsy child with a combination of malformations similar to those of trisomy 13 and a normal 46,XY karyotype. In this gypsy colony a further five cases were found with similar features and I asked a brilliant paediatric colleague who happened to also be my wife (Valéria Váradi MD, PhD) to examine the children and a postgraduate student of mine (László Szabó MD) to perform a pedigree analysis. The six children in the inbred gypsy isolate were published in 1980 (Váradi *et al.* 1980). The features of affected children included reduplication of the big toes, supernumerary fingers on the hands (characterised by a Y-shaped central metacarpal, 'central polydactyly' of the hands), cleft lip/palate or lingual lump, elongated skull, ocular hypertelorism, epicanthic folds, strabismus, cerebral abnormalities (absence of olfactory bulbs and tracts probably with semilobar holoprosencephaly), cryptorchidism, inguinal hernia, somatic/psychomotor retardation and occasionally congenital heart disease.

Within gypsy groups many children are born as a result of extramarital relationships, and because of the difficulties of tracing connections it was possible to draw the pedigrees for only three generations. The large number of admitted extramarital connections and the small number of surnames (only six) revealed a high degree of endogamy in these colonies. We thought that the findings in the proband and in other members of the gypsy families represented a nosological entity. Because of consanguinity in the family reported and because of the involvement of multiple siblings in this and other families, autosomal recessive inheritance was strongly suggested.

After submitting the manuscript, we confirmed a further case of this syndrome from the same gypsy colony in a sibship of 12 children (Papp and Váradi 1985). Based on these seven cases, the syndrome appeared as entry 277170 in the McKusick encyclopaedia *Mendelian Inheritance in*

Table 1. Occurrence of features of affected children with Váradi-Papp syndrome published between 1980 and 2004

Case	Reference	Sex	Age at death	Polydactyly of each hand and foot	Cleft lip	Cleft palate	Lingual nodule	Cerebral or celebellar anomaly	Crypt-orchidism	Intrauterine and/or postnatal retardation	Psycho-motor retardation
1	Váradi et al. 1980	M	3 yr	+	+	+	-	+	+	+	+
2	Váradi et al. 1980	F	6 yr	+	-	-	+	+		-	+
3	Váradi et al. 1980	M	1 dy	+	-	+	-	+	+	+	
4	Váradi et al. 1980	F	1 mth	+	-	+	-	+		+	+
5	Váradi et al. 1980	M	1 wk	+	+	-	+	+	+	+	
6	Váradi et al. 1980	F	2 wk	+	+	+	-	+		+	
7	Papp and Váradi 1985	F		+	+	+	+	+		+	+
8	Gencik and Gencikova 1983	M	?	+	+	-	+	+	?	+	+
9	Gencik and Gencikova 1983	F	?	+	+	-	+	+		+	+
10	Mattei and Ayme 1983	M		+	+	-	+	+		+	+
11	Mattei and Ayme 1983	M	1 wk	+	?	?	?	+	?	+	
12	Mattei and Ayme 1983	M	1 wk	+	?	?	?	+	?	+	
13	Howard and Young 1988	M	?	+	+	+	-	+	?	+	+
14	Marles and Chudley 1990	M	?	+	+	+	-	+	?	+	+
15	Münke et al. 1991	M	?	+	+	+	+	+	?	+	+
16	Münke et al. 1991	M	?	+	+	+	+	+	?	+	+
17	Münke et al. 1991	F	?	+	+	+	+	+	+	+	+
18	Cleper et al. 1993	M	?	+	+	+	+	+	+	+	+
19	Cleper et al. 1993	M	?	+	+	+	+	+	+	+	+
20	Ádám and Papp 1996	F	Stillborn	+	+	+	+	+	?	+	+
21	Stephan et al. 1994	M	?	+	+	+	+	+	?	+	+
22	Wey et al. 1994	F	?	+	+	+	+	+		+	+
23	Sabry et al. 1997	M	?	+	-	+	+	+		+	+
24	Doss et al. 1998	M	?	+	+	+	?	+	+	+	+
25	Doss et al. 1998	M	Stillborn	+	+	-	?	+	?	+	
26	Al-Gazali et al. 1999	F	?	+	-	+	?	+	?	+	+
27	Al-Gazali et al. 1999	M	1 dy	+	-	+	?	+	+	+	
28	Mauceri et al. 2000	M	?	+	+	+	+	+	?	+	+
29	Guven et al. 2004	M	?	+	+	+	+	+	?	+	+

Man as the Váradi-Papp syndrome (McKusick 1966–2005).

Münke *et al.* (1990) presented evidence on the basis of three unrelated patients that poly-dactyly of the hands is characterised by a Y-shaped central metacarpal, this is the most specific feature of Váradi-Papp syndrome and hypoplastic cerebellar vermis is also a consistent finding in patients with this disorder. Münke *et al.* (1990) proposed that this syndrome was the disorder present in the families reported by Gustavson *et al.* (1971), Egger *et al.* (1982), Gencik and Gencikova (1983), Haumont and Pelc (1983), Mattei and Aymé (1983) and Silengo *et al.* (1987).

Münke *et al.* (1990) provided a classification of the orofaciodigital syndrome into 7 varieties, following in part the classification of Toriello (1988). Subsequently, Váradi-Papp syndrome has been quoted as orofaciodigital syndrome VI (OFD syndrome VI, OFDS VI, oral-facial-digital syndrome, type VI) in many syndromatological databases and textbooks (McKusick 1966–2005, Cohen 1983, Gorlin *et al.* 1988, Jones 1988, POSSUM 1988–2005, Cohen 1989, Buyse 1990, Gorlin *et al.* 1990, Winter and Baraitser 1991, Pfeiffer *et al.* 1993, Donnai and Winter 1995, Twining *et al.* 2000).

Münke *et al.* (1991) described detailed studies of a foetus with clinical findings overlapping this disorder, the hydrolethalus syndrome (McK 236680), and the Pallister-Hall syndrome (McK 146510). The foetus had many manifestations in common with the twin foetuses reported by Hingorani *et al.* (1991). Al-Gazali *et al.* (1999) reported two siblings with features overlap-ping those of Váradi-Papp syndrome. Both presented at birth with dysmorphic facial features, high arched palate, dysgenesis of the cerebellar vermis, aplasia of the pituitary gland and postaxial polydactyly of both hands and one foot. Shashi *et al.* (1995) reported two brothers with findings overlapping OFD II, OFD VI and Pallister-Hall syndrome, both of whom had congen-ital absence of the pituitary gland.

Cleper *et al.* (1993) reported the cases of two male cousins, both the offspring of consan-guineous matings, with multiple congenital anomalies. They had an unusual facial appearance. Multiple bucco-alveolar frenula and notched inferior alveolar ridges were present at birth. Both had congenital heart anomalies, micropenis, and cryptorchidism. The surviving patient was men-tally retarded and had a unilateral central extra digit with partially formed metacarpal, as well as partial agenesis of the corpus callosum. Cleper *et al.* (1993) suggested that this finding overlapped with those of the Váradi-Papp syndrome and Opitz trigonocephaly syndrome (McK 211750). Bankier and Rose (1994), reviewing this paper for POSSUM (1988–2005), suggested that one of the patients had manifestations of the Váradi-Papp syndrome, whereas his first cousin only shared some of the same findings. Sabry *et al.* (1997) reported a similar case. Stephan *et al.* (1994) suggested that hypothalamic hamartoma is an occasional manifestation of Váradi-Papp syn-drome.

Doss *et al.* (1998) described the neuropathological findings in a stillborn. Autopsy findings included facial abnormalities, postaxial central polydactyly of the right hand, bilateral bifid toes, and absence of cerebellar vermis. Microscopic analysis of the cerebellum demonstrated dis-ruption or dysgenesis of the glial architecture, which may account for the cerebellar abnormal-ities in the Váradi-Papp syndrome.

Váradi-Papp syndrome has similarities to the Joubert syndrome (McK 213300) which is also an autosomal recessive disorder characterised by cerebellar (vermis) hypoplasia, hypotonia, developmental delay, and abnormal eye movements, in addition, associated malformations such as polydactyly and soft-tissue tumours of the tongue can be found rarely, in 8% and 2% of cases respectively (Chance *et al.* 1999, Haug *et al.* 2000). The Váradi-Papp syndrome also has similarities to OFD type II, III and IV (Chitayat *et al.* 1992, Smith *et al.* 1993, Haug *et al.* 2000, Yildirim *et al.* 2002).

Toriello (1993) reviewed the clinical overlap observed with the nine described types of orofaciodigital syndromes and with other entities. Metacarpal abnormalities (forked metacarpals) with central polydactyly, cerebral/cerebellar abnormalities and mental retardation distinguish Váradi-Papp syndrome from other OFD syndromes (POSSUM 1988–2005, Degner *et al.* 1999).

Table 1 shows the occurrence of most common features of 29 affected children with Váradi-Papp syndrome collected from the literature published between 1980 and 2004. The syndrome consists of *orofacial* (facial dysmorphism, cleft lip and/or palate abnormality, lingual nodule or tumour of the tongue, buccoalveolar frenula, alveolar and dental abnormalities, strabismus), *cerebral/cerebellar* (deformation of the skull, semilobar holoprosencephaly and/or absence or dysgenesis of cerebellar vermis or corpus callosum or hypothalamus or pituitary gland), *digital* (metacarpal abnormalities with central polydactyly, reduplication of the big toes) and *genital* (cryptorchidism, micropenis) anomalies. The patients are growth-retarded and when survival occurs psychomotor retardation is present. Some occasional associations include congenital heart disease and renal anomalies. Accumulation of consanguinity supports the autosomal recessive inheritance.

Proper classification of patients with variants of overlapping syndromes and distinction between them is important for the practical clinical genetics and essential for localisation and identification of mutant gene in this multiple malformation syndrome (Münke 1989, Verloes 1995, Gleeson *et al.* 2004).

Foetal Váradi-Papp syndrome using ultrasonography in the mid-trimester both in routine screening and detailed scanning can be detected (McGahan *et al.* 1990, Camera *et al.* 1994, Suresh *et al.* 1995, Ádám and Papp 1996, Guven *et al.* 2004), and termination of pregnancy can be offered to the parents.

Hirschsprung Disease:
genetic dissection of a complex disorder

Eberhard Passarge

Institut für Humangenetik, Universitätsklinikum Essen

Meinem Freund und Kollegen George Fraser mit guten Wünschen zugeeignet.

Hirschsprung disease (MIM 142623 with links to 19 related entries: McKusick *et al.* 1998) also named congenital intestinal aganglionosis or HSCR (Hirschsprung Chromosomal Region), is an aetiologically heterogeneous, surgically correctable neonatal intestinal obstruction syndrome resulting from absence of ganglion cells in variable parts of the intestines. It occurs in a syndromic form associated with other, mainly Mendelian disorders, and as a non-syndromic form restricted to manifestations in the gastrointestinal tract. The non-syndromic form has a complex genetics with inheritance patterns corresponding to a multigenic aetiology. However, in about 5–10% of patients other family members are affected in a pattern corresponding either to an autosomal dominant or an autosomal recessive mode of inheritance (Chakravarti and Lyonnet 2001, McCallion *et al.* 2003, Passarge 2003, Passarge 2007). Mutations in nine genes have been identified as contributing to this disorder, but the overall rate of detectable mutations is only about 30%. In addition to known mutated genes, other chromosomal regions and non-coding variants in a major gene are significantly associated with susceptibility for Hirschsprung disease. This major gene is the receptor tyrosine kinase *RET* (MIM 164761), which plays a pivotal role in the aetiology of Hirschsprung disease, although mutations in *RET* alone are neither necessary nor sufficient as a cause. During the past three years the genetic basis of Hirschsprung disease has been elucidated by new insights (Bolk *et al.* 2000, Gabriel *et al.* 2002, Emison *et al.* 2005). Here we review our understanding of this complex genetic disorder.

Classification

Hirschsprung disease was recognised as a clinical entity in 1888 by Harald Hirschsprung in Copenhagen (Hirschsprung 1889) almost 20 years after it was first described by Abraham Jacobi of New York, the first president of the American Pediatric Society (Jacobi 1869). In spite of good early pathological studies (Tittel 1901, Della Valle 1924), the underlying cause, absence of intestinal ganglion cells, was not recognised until 1948 (Whitehouse and Kernohan 1948, Zuelzer and Wilson 1948). The lack of the intramural ganglion cells, which are required for peristalsis, results from failure to migrate from the neural crest into the intestinal wall and to proliferate, differentiate, and establish proper function there during embryogenesis. Thus, based on its pathophysiology, Hirschsprung disease is defined as a neurocristopathy (Bolande 1973, Le Douarin and Kalcheim 1999).

For clinical and genetic reasons discussed below, Hirschsprung disease is classified into different types according to the length of the aganglionic segment. The distal border of the aganglionic segment is always present from just above the rectum in the terminal portion of the colon, whereas the upper (rostral) border is highly variable. Two types of Hirschsprung disease are distinguished: short-segment (also called *type 1* or S-HSCR), occurring in about 65–85% of patients, and long-segment (*type 2* or L-HSCR) in 15–35%. The short-segment type 1 is defined as aganglionosis between the rectum and the upper sigmoid colon, the long-segment type 2 as aganglionosis extending to the upper descending colon or beyond the splenic flexure into the transverse colon at variable sites. In a small percentage of patients (about 3–5%) the entire colon or in addition variable parts of the small intestines may be aganglionic (total colonic aganglionosis). Other rare variants are total intestinal aganglionosis and ultra-short segment aganglionosis (distal rectum until the normal aganglionic zone 2 cm above the pectinate line), and segmental aganglionosis above a normal distal segment, reported in very few cases only. The distinction between long- and short-segment aganglionosis is to some extent arbitrary, but has been useful in practice.

Clinical background

Hirschsprung disease is clinically variable, with manifestations ranging from severe neonatal or late infantile to mild manifestations with onset later in childhood (Figure 1), depending on the length of the bowels lacking intramural ganglion cells. The main clinical manifestations are chronic constipation, abdominal distension, and megacolon of variable length (see below). Secondary manifestations may be enterocolitis, malnutrition, electrolyte imbalance, and increased susceptibility to infection. The disease may present in the neonatal period as an acute, life-threatening condition due to meconium ileus or sigmoid perforation. Impaired defaecation is a leading sign in all cases. The diagnosis is confirmed by rectal biopsy demonstrating the absence of intramural intestinal ganglion cells. Surgical intervention, successful in about 75% of patients, entails removal of the aganglionic, distended part of the colon to prevent secondary manifestations. Early recognition and therapy are important.

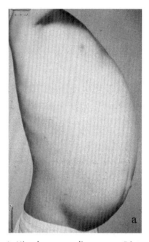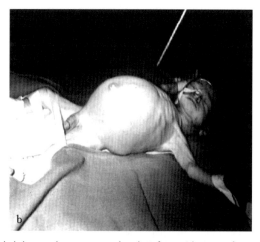

Figure 1. Hirschsprung disease. a. Distended abdomen due to megacolon. b. Infant with severe form of megacolon (Photo courtesy Dr Lester Martin, Cincinnati)

Epidemiology

With a population incidence of one in 5000 newborns in populations of European ancestry, Hirschsprung disease is the most common cause of intestinal obstruction in the neonatal period and early childhood. In Asia it occurs with a higher frequency of 1/3500. The sex ratio is about 4.5:1 male to female. Yet, the proportion of affected family members is higher in female propositae than in male propositi as discussed below. Mutations in the genes mentioned below have variable penetrance and clinical expression.

Inheritance pattern and familial risk of recurrence

About 70% of patients do not have affected family members, but an increased risk of recurrence in sibs or offspring of affected parents has to be anticipated. Hirschsprung disease was reported in sibs before its cause was known. And after successful surgical correction procedures were introduced in the 1960s it became apparent that an affected parent has an increased risk for an affected child. Three systematic population genetic studies of the familial occurrence of non-syndromic Hirschsprung disease reported from London (Bodian and Carter 1963), Copenhagen (Madsen 1964), and Cincinnati (Passarge 1967a, b), done prior to the identification of genes, indicated an overall risk of recurrence in sibs of about 4%, about 200-fold the population risk. However, the individual risk varies greatly according to the length of the aganglionic segment, the sex of the index patient, and the sex of the sib or offspring.

Most familial cases do not show a Mendelian inheritance pattern. For this reason, based on a detailed analysis taking the gender of the index patient and affected sibs into account, a multigenic (multifactorial) model has been invoked to explain the non-Mendelian inheritance pattern of Hirschsprung disease (Table 1). The empirical risk figures obtained through these studies are still a reasonable basis for genetic counselling. However, some pedigrees correspond to an autosomal dominant or an autosomal recessive mode of inheritance (Badner *et al.* 1990).

Table 1. Empirical risk for Hirschsprung Disease[1]

Sex of Index Patient		Risk to sibs (%) [2]	
		Male	Female
Male	Type I[3]	5.5	0.6
	Type II	8	3
Female	Type I	7.0	2
	Type II	18	9
Offspring of affected parent		Risk to offspring (%)[4]	
		Male	Female
Male	Type I	11	9
	Type II	18	13
Female	Type I	15	11
	Type II	28	22

[1]Monogenic forms exist (see text). Syndromic aganglionosis not included. The figures are approximations with a range.
[2]Data from Bodian and Carter (1963), Madsen (1964), Passarge (1967a, b).
[3]Length of aganglionic segment defined as type I for absence of intestinal ganglion cells caudal to the splenic flexure, and as type II for absence anywhere further rostral to this point (see text).
[4]Data from Chakravarti and Lyonnet (2001).

Molecular genetics

The era of molecular genetics of Hirschsprung disease began in 1993, when close linkage with the gene locus for *RET* (see below) at 10q11 was demonstrated in families in which the disease occurred (Angrist *et al.* 1993, Lyonnet *et al.* 1993). The first point mutations were reported soon after by Romeo *et al.* (1994). Subsequently, additional genes with functions related to the neural crest origin of intestinal ganglion cells have been recognised to contribute to the causes of this heterogeneous disorder.

Normal development of the neural crest-derived intestinal ganglion cells involves genes in three different, possibly functionally related signalling pathways: (i) the *RET* receptor tyrosine kinase pathway consisting of the genes encoding the RET receptor and its ligand, glial cell line-derived neurotrophic factor (GDNF), (ii) the endothelin type B receptor pathway with the receptor (EDNRB) and its ligand, endothelin-3 (EDN3), and (iii) the transcription factor SOX10. Mutations in their genes, as well as others (see below), may result in Hirschsprung disease. In addition to these five genes, mutations in several other genes are recognised to contribute to the causes of Hirschsprung disease (Table 2). However, in a high proportion (70%) of isolated patients with short-segment aganglionosis a causative mutation cannot be identified.

The role of the *RET* gene in Hirschsprung disease

The *RET* gene (MIM 164761), located on the long arm of chromosome 10, region 1, band 1.2, plays a pivotal role in Hirschsprung disease. *RET* was originally described as a proto-oncogene (RET, *Re*arranged during *t*ransfection, a term used when this gene was discovered in a different context). Inactivating *RET* mutations are dominant and result in aganglionosis of different types with a penetrance of 50–70% (Chakravarti and Lyonnet 2001, McCallion and Chakravarti 2004). They may occur as a new mutation or are transmitted from a parent. They account for about 50% of familial patients with Hirschsprung disease and about 35% of isolated patients. In short segment type 1, mutations in the *RET* gene are found in 32%; in long segment type 2, in 57% (Attié *et al.* 1995, Iwashita *et al.* 1996). In addition, changes in non-coding regions of *RET* contribute to the causation of Hirschsprung disease (see below).

The *RET* gene encodes a cell surface molecule, a receptor tyrosine kinase, that is expressed in derivatives of the neural crest and neuroectoderm (Chakravarti and Lyonnet 2001, McCallion and Chakravarti 2004). This receptor has functions in the development and differentiation of neural crest cell lineages, from which the intramural intestinal ganglion cells also derive. The RET receptor is a transmembrane protein of 1114 amino acids with different domains: a signal peptide; three extracellular domains consisting of the putative ligand-binding domain, a cadherin-like domain, and a cysteine-rich domain; a transmembrane domain and an intracellular catalytic domain (Figure 2). The latter is conserved in evolution and similar to other growth factor receptors. Different parts of the gene encode functionally distinct domains of the RET protein. This explains the protean nature of the gene, with mutations involved in five clinically different disorders (Hirschsprung disease, Multiple Endocrine Neoplasia type IIA (MEN2A, MIM 171400) and type IIB (MEN2B, MIM 162300), Familial Medullary Thyroid Carcinoma (FMTC, MIM 155240), and Pheochromocytoma (MIM 171300).

RET mutations identified in Hirschsprung disease are unique and are distributed throughout this gene (Chakravarti and Lyonnet 2001, McCallion and Chakravarti 2004). They appear to result in a loss-of-function of the RET receptor. Presumably insufficient expression levels of RET

Table 2. Genes involved in the causation of Hirschsprung disease [1]

Gene	Location	Main effect	Penetrance	MIM
RET	10q11.2	dominant, loss-of-function	50–72 %	164761
GDNF	5p13.1	dominant/recessive	unknown	600837
EDNRB	13q22	recessive	8–85 %	131244
EDN3	20q13	recessive	unknown	131242
SOX10	22q13	dominant/recessive[2]	>80%	602229
ECE1	1p36	dominant/recessive[2]	unknown	600423
NTN	19p13	unknown	unknown	601880
ZFHX1B (SIP1)	2q22	sporadic[2]	unknown	605802
PHOX2B	4p12	congenital hypoventilation[3]	unknown	603851

[1]Most genes appear to be interdependent and mutations usually do not segregate in a Mendelian pattern. (Data from Chakravarti and Lyonnet 2001, Passarge 2003). Predisposing chromosomal regions have been identified at 3p12 (MIM 606874), 9q31 (MIM 606875), and 19q12 (see text).
[2]Syndromic forms.
[3]Sometimes associated with Hirschsprung disease.

on the surface of neural crest cells interfere with the normal migration of ganglion cells to the colon or their differentiation *in situ* (Iwashita *et al.* 1996, McCallion and Chakravarti 2004). The majority of mutations in the *RET* gene that cause Hirschsprung disease are missense and nonsense mutations, including deletions. In contrast, mutations of the *RET* gene that result in MEN2A/FMTC and MEN2B are gain-of-function mutations that occur in MEN2A in a

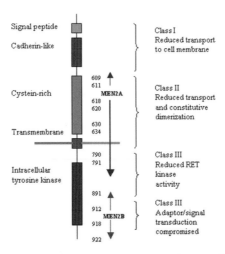

Figure 2. Schematic structure of the RET protein. Mutations in Hirschsprung disease occur throughout, whereas mutations in MEN2A/FMTC and MEN2B cluster in the domains as shown (adapted from Chakravarti and Lyonnet 2003)

cluster of six cysteine residues in exon 10 (codons 609, 611, 618, and 620) and 11 (630 and 634), and as a unique mutation in MEN2B (M918T) (Attié *et al.* 1995). Hirschsprung disease and MEN2A segregate together in some families; here they may occur in one and the same individual. In addition to mutations, polymorphic variants in the *RET* gene segregate with Hirschsprung disease in some patients (see below).

Other genes involved in Hirschsprung disease
In addition to the *RET* gene, dominant mutations have been identified in the gene of one of its ligands, *GDNF* (glial cell derived neurotrophic factor), located on 5p13.1-p12 (MIM 600837). GDNF is a TGF-β-related protein of 211 amino acids that can activate *RET* in cultured cells and is strongly expressed in developing murine gut and kidney mesenchyme (Iwashita *et al.* 1996). Mutations in the *GDNF* gene alone are rare. Usually other predisposing changes are present, such as mutations in *RET* or trisomy 21 (see below). Mutations in the other three genes (*EDNRB*, *EDN3*, and *SOX10*) contribute to about 10% of patients and often are associated with specific phenotypes.

Recessive mutations have been demonstrated in the gene for endothelin receptor type B (*EDNRB*), a G-protein-coupled receptor on 13q22 (Hirschsprung disease type 2, MIM 131244), (Baynash *et al.* 1994, Puffenberger *et al.* 1994, Amiel *et al.* 1996, Chakravarti 1996) and that of its ligand endothelin 3 (*EDN3*) on 20q13.2-q13.3 (MIM 131242), and in *SOX10* (SYR-box 10; MIM 602229) on 22q13. All mutations have a reduced penetrance and variable expression with respect to the length of the aganglionic segment. Homozygotes for a missense mutation (G to T transversion) of a highly conserved tryptophan residue at position 276 (W276C) in the fifth transmembrane domain have a risk of 74% to develop Hirschsprung disease, whereas the risk for heterozygotes is 21% (Puffenberger *et al.* 1994a, b). In some of the syndromic forms of Hirschsprung disease other cell types derived from the neural crest are also involved, such as precursors of melanocytes for pigment formation in the cranial region in Shah-Waardenburg syndrome, or cells destined to function in the sensory components of the acoustic pathway in associated deafness.

The recessive mutations at the *EDNRB* locus on 13q22 (MIM 131244) were described in an inbred Mennonite population, in which they show a reduced penetrance of 8–85% (Puffenberger *et al.* 1994b). They account for about 5% of Hirschsprung patients. Mutations in *SOX10* (MIM 602229) have been found to cause the Shah-Waardenburg syndrome, another neurocristopathy associated with intestinal aganglionosis (see below).

In addition to the five major genes mentioned above (*RET, GDNF, EDNRB, EDN3,* and *SOX10*), minor genes are involved with rare mutations: *ECE1* (endothelin-converting enzyme, MIM 600420) (Hofstra *et al.* 1999), *NTN* (Neurturin; MIM 601880) (Doray *et al.* 2001), and *ZFHX1B* (zinc finger homoeobox 1 B; MIM 605802; previously known as *SIP1* or *SMADIP1*) (Cacheux *et al.* 2001, Wakamatsu *et al.* 2001, Gregory-Evans *et al.* 2004), which occurs in a syndromic form (see below). Mutations in *PHOX2B* (Paired-like homoeobox 2 B; MIM 603851) cause life-threatening hypoventilation syndrome; this may be associated with Hirschsprung disease (Matera *et al.* 2004). Owing to reduced penetrance of the mutant genotypes, differences in the sex ratio, and variable expression, mutations of these genes can lead to different frequencies of recurrence in relatives (Chakravarti and Lyonnet 2001, Passarge 2003).

Interaction of different HSCR susceptibility loci

In general, the genotypes and the phenotypes (lengths of the aganglionic segments) in isolated Hirschsprung disease correlate poorly. The major susceptibility locus for Hirschsprung disease is *RET*. However, in addition to the genes shown to harbour causative mutations, modifier genes appear to influence penetrance and expression of aganglionosis. In particular, susceptibility loci in three chromosomal regions, at 3p12 (HSCRS2; MIM 606874), 9q31 (HSCRS3; MIM 606875), and 19q12, interact with the *RET* locus at 10q11 in the causes of short-segment Hirschsprung disease and, resulting in a 4.2, 5.0, and 8.3-fold increase of risk disease (Bolk *et al.* 2000, Gabriel *et al.* 2002, Passarge 2002). None of these susceptibility loci alone causes the disease; combination resulting from multiplicative interaction at three loci is necessary, but at the same time sufficient as a cause (Gabriel *et al.* 2002). No other loci seem to be required. These new insights represent the first genetic dissection of a multigenic, complex disorder in man (Gabriel *et al.* 2002, Passarge 2002). In addition, distorted parental transmission of susceptibility alleles at the *RET* locus was found, i.e. 21 maternal and six paternal transmissions, a ratio of 3.5:1 (Gabriel *et al.* 2002).

Furthermore, certain common polymorphic variants may contribute to the susceptibility for aganglionosis (Borrego *et al.* 1999, Fitze *et al.* 1999, Griseri *et al.* 2002, Fitze *et al.* 2003, Burzynski *et al.* 2004). Fitze *et al.* (1999, 2003) determined that expression of a haplotype defined by two variants of the *RET* promoter region -5 and -1 from the transcription start are associated with Hirschsprung disease in 68.8% of patients (110 of 160 alleles) compared with 25% in controls (60 of 240 alleles). The predisposing variants were a change from guanine (G) to adenine (A) at position -5, and from cytosine (C) to A at position -1 (Fitze *et al.* 2003). A further susceptibility locus may be that for X-chromosomal neuronal intestinal pseudo-obstruction at Xq28 (MIM 300048) (Auricchio *et al.* 1996).

SNP analysis (single nucleotide polymorphism) has identified susceptibility haplotypes in the promoter and the 5' region of the *RET* gene. Burzynski *et al.* (2004) found a strong association between six markers in the 5' region of the *RET* gene between the promoter region and HSCR (Burzynski *et al.* 2004). Homozygosity for one haplotype conferred an increased risk of developing HSCR. Presumably different susceptibility alleles contribute to differences in penetrance and expression. Iwashita *et al.* (2002, 2003) determined that genes associated with Hirschsprung disease are upregulated in rat gut neural crest stem cells compared with whole-foetus RNA. Among the genes with highest expression were *Ret, Gdnf, Sox10, Gfra1* (MIM 601496), and *Ednrb*. *Ret* was found to be necessary for neural crest stem cell migration. *Gdnf* promoted the migration of neural crest stem cells in culture but did not affect their survival or proliferation.

Recently Chakravarti and co-workers identified a common non-coding variant with enhancer function in intron 1 of the *RET* gene which was associated with a 20-fold susceptibility to Hirschsprung disease (Emison *et al.* 2005). In addition, transmission distortion in eight single nucleotide polymorphisms (SNPs) was found in a 27.6-kb segment extending from 4.2 kb 5' of the *RET* gene to the end of exon 2. The highest association was with SNP RET+ within multi-species conserved sequences (conserved in chimpanzee, baboon, cow, pig, cat, dog, rat, mouse, chicken, zebra fish, Fugu and Tetraodon) (Emison *et al.* 2005). Instead of a highly conserved cytosine (C allele) present in these species, a thymine (*RET+3:T* allele) was preferentially transmitted to affected individuals. In individuals from 51 unselected populations Emison *et al.*

(2005) found the *RET+3:T* allele to be virtually absent in Africa (frequency less than 0.01) in contrast to a high frequency of 0.45 in Asia and 0.25 in Europe (Emison *et al.* 2005). For this reason they consider the possibility of a perhaps past selective advantage of this allele.

Embryological origin of aganglionosis

Intramural intestinal ganglion cells reach the alimentary tract by migrating from the cephalic neural crest between the 6th and 12th week of embryogenesis (Hüther 1954, Okamoto and Ueda 1967, Andrew 1971). Migration follows a defined time sequence with a cranial-caudal gradient. At five weeks gestation (Carnegie stage 13), paired vagal fibres extend to the upper oesophagus, and there are a few fine fibres from the periaortic and pelvic plexuses, but ganglion cells are still absent. At six weeks (stage 15), neuroblasts are present in the oesophagus outside the circular layer and the stomach. At eight weeks (18 mm embryo, stage 20) ganglion cells are present in the small intestine and the rectum, but not the colon. At 12 weeks (70 mm) the entire plexus is innervated, presumably by further caudal ganglion cell migration. A critical period seems to be between weeks eight and 12, when most of the distal plexus develops.

The first neuroblasts to reach the alimentary tract form the myenteric plexus. The submucosal plexus is formed by neuroblasts migrating from the myenteric plexus across the circular muscle layer into the submucosa. The submucosal plexus is also formed in the caudal direction, but later, during the third and fourth months of gestation. The outer longitudinal muscle layer develops from embryonic mesenchymal tissue after the myenteric plexus has been formed in the 12th week. (Amiel and Lyonnet 2001). While vagal nerve fibres apparently play a direct role in the development of the intramural plexus, sympathetic and pelvic parasympathetic nerves are not involved.

The precursor cells of the intramural intestinal ganglion cells are derived from the neural crest. From here they migrate to the intestines and differentiate into the normal neuronal network, with all phases under the control of different genes. For example, sufficient levels of *RET* gene expression on the surface of the cells involved are thought to be required for migration and differentiation. The failure of the intestinal ganglion cells to migrate from the neural crest or to proliferate, differentiate, and survive in the intestinal wall results from several genetic causes. Since other cell types also are derived from the neural crest, intestinal aganglionosis may occur together with other neurocristopathy syndromes. These other cells are sensory ganglia, sympathetic neurons and parasympathetic neurons of the autonomous nervous system, endocrine and paraendocrine cells (thyroid C-cells, adrenal medulla, carotid cells), pigment cells and cells of the craniofacial mesectoderm as well as conotruncal heart defect (Amiel and Lyonnet 2001, Chakravarti and Lyonnet 2001, McCallion *et al.* 2003).

Genetic counselling

The risk of a sib of an index patient being affected is high (18%) if the index patient is a female with a long aganglionic segment and the sib is a male, compared with 7% if the aganglionic segment is short. In contrast, female sibs of a male index patient with type I (short) have a low risk of 0.6% and with type 2, a risk of 3% (Table 2). Thus, the risk to sibs is higher if the index patient is female rather than male. This is the opposite of the population incidence, with a ratio of four males affected to one female. This apparent paradox, first described in the 1960s by the great medical geneticist from Great Ormond Street, Cedric O. Carter, for pyloric stenosis, is called the *Carter effect*. It is interpreted as being due to a multigenic population difference between males and females. They differ in a threshold effect: it takes more contributing genetic factors for disease manifestation in females than it does in males, the latter presumably being more liable to the disease than females. The average risk to offspring of a parent affected with Hirschsprung disease is influenced also by the gender of the parent and the length of the aganglionic segment (Table 2).

Caution should be exercised when applying the empirically derived risk figures because the actual risk in a given family might come close to that of an autosomal dominant or recessive inheritance pattern. If intestinal aganglionosis occurs as a component of a systemic disorder (syndromic Hirschsprung disease), the risk of recurrence is determined by the underlying disease (see below).

Syndromic forms of Hirschsprung Disease

About 12% of patients with intestinal aganglionosis have another, additional genetic disorder (Table 3) (Amiel and Lyonnet 2001). In several of these disorders, other cell lineages derived from the neural crest are involved. In others the origin of the association is less clear. MIM lists 50 entries under the heading 'Hirschsprung disease'.

Mutations at the *RET* locus may cause both Hirschsprung disease and Multiple Endocrine

Table 3. Syndromic forms of Hirschsprung disease [1]

Disorder	OMIM
Multiple Endocrine Neoplasia type 2A	171400
Multiple Endocrine Neoplasia type 2B	162300
Familial Medullary Thyroid Carcinoma	155240
Pheochromocytoma	171300
Waardenburg syndrome type 1 and type 2	193500/10
Shah-Waardenburg syndrome	277580
Cartilage hair dysplasia	250250
Smith-Lemli-Opitz syndrome	270400
Hypoventilation syndrome	209880
Goldberg-Shprintzen syndrome	235730
McKusick-Kaufman syndrome	236700
Association with polydactyly, renal dysgenesis, and deafness	235740
Brachydactyly type D	306980
Microcephaly and iris coloboma	235730
Dysmorphic facial features and nail hypoplasia	235760
Osteopetrosis, recessive type	259700
Bardet-Biedl syndrome	209900
Piebald trait	172800
Congenital deafness, isolated cases	277580

[1]Examples from OMIM (Online Mendelian Inheritance in Man: (*http://www.ncbi.nlm.nih.gov/Omim/*).

Neoplasia type 2B (MIM 171400) in one and the same patient. In some families both disorders occur in different relatives carrying the same mutation. Other important monogenic disorders associated with congenital intestinal aganglionosis are Waardenburg syndromes type 1 (MIM 193500) and type 2 (MIM 193510), Shah-Waardenburg syndrome (MIM 277580), cartilage-hair dysplasia (MIM 250250), Smith-Lemli-Opitz syndrome (MIM 270400), and primary central hypoventilation syndrome ('Ondine's curse', MIM 209880/603851) (Amiel and Lyonnet 2001). Some mutations of *L1CAM* are also involved in Hirschsprung disease, contributing to a quantitative defect in the migration of neural crest cells to distal segments of the gut (see MIM 308840.0016).

Best understood is the association of congenital intestinal aganglionosis with defects in other cells derived from the neural crest, such as melanocytes in Waardenburg syndrome (MIM 277580), and sensory components of the acoustic pathway leading to an association with deafness (Shah *et al.* 1981). The Shah-Waardenburg syndrome (Mowat *et al.* 2003) is a disorder of the embryonic neural crest combining clinical features of Hirschsprung disease and Waardenburg syndrome (MIM 193500) with gene loci at 20q13.2-q13.3 and 22q13. This phenotype can be caused by mutations in the *SOX10* gene (MIM 602229), in the endothelin-B receptor gene (*EDNRB*; MIM 131244), or in the gene for its ligand, endothelin-3 (*EDN3*; MIM 131242). The report by Shah *et al.* (1981; see also Mowat *et al.* 2003) is based on five families with 12 infants (seven male, five female) who presented in the neonatal period with aganglionic intestinal obstruction and with a white forelock and white eyebrows and eyelashes.

Mutations in the *ZFHX1B* gene (Zinc finger homoeobox 1B gene; MIM 605802; previously called *SIP1*, Smad-interacting protein 1 (*SMADIP1* gene) cause Mowat-Wilson syndrome (MIM 235730) (Passarge 2007, Wilson *et al.* 2003, Ishihara *et al.* 2004). This is a variable, complex developmental disorder including microcephaly, agenesis of the corpus callosum, mental retardation, delayed motor development, epilepsy, dysmorphic face, congenital heart defects and other abnormalities. Wilson *et al.* (2003) found *ZFHX1B* mutations or deletions in 21 of 23 patients (Ishihara *et al.* 2004). Ishihara *et al.* (2004) identified five novel nonsense and frameshift mutations in the *ZFHX1B* gene in 27 patients with Mowat-Wilson syndrome (Passarge 1972). All of the deletions were of paternal origin. Two of their patients with large deletions (10.42 Mb and 8.83 Mb) had delayed psychomotor development. One patient also had a cleft palate and complicated heart defects, features not previously reported in this syndrome.

A striking association of trisomy 21 and congenital intestinal aganglionosis has been noted in several studies. Up to 2.5% of patients with Hirschsprung disease have Down syndrome (Passarge 1976a, Chakravarti and Lyonnet 2001, Carrasquillo *et al.* 2002, Passarge 2007). In a large Mennonite kindred with autosomal recessive Hirschsprung disease due to a mutation in the *EDNRB* gene, Puffenberger *et al.* (1994) found preliminary evidence for a genetic modifier locus on 21q22 (*HSCMR1*; MIM 600156) (Puffenberger *et al.* 1994b). It has not been established whether this is in fact responsible for the relatively frequent association of trisomy 21 and Hirschsprung disease.

About 8% of patients with Hirschsprung disease have chromosomal aberrations visible by light microscopy (Chakravarti and Lyonnet 2001, Passarge 2007). Structural aberrations include interstitial deletions. Probably in the case of the deletions 13q22 and 10q11.2 the *EDNRB* and the *RET* gene, respectively, must have become hemizygous and caused the intestinal aganglionosis.

Animal models: congenital intestinal aganglionosis in mutant mice

Neural crest-derived associations of pigmentary anomalies and intestinal aganglionosis occur in several different mutations in mice. Piebald, piebald-lethal, and lethal-spotting are recessive coat colour mutations associated with intestinal aganglionosis (Bielschowsky and Schofield 1962, Carrasquillo 2002, McCabe *et al.* 1991, Hosoda *et al.* 1994, Lane 1996). Mice homozygous for piebald-lethal and lethal-spotting also have aganglionosis. Homozygous piebald and piebald/piebald-lethal compound heterozygotes have aganglionosis in about 10% (Bielschowsky and Schofield 1962, Hosoda *et al.* 1994). These early observations indicate that mutations in different genes differ in penetrance. Knockout mice carrying mutations at several genes in the *Ret* pathway exist: *Ret, Gdnf, Ednrb, Edn, Ece1* (Omenn and McKusick 1979, McCallion *et al.* 2003).

Three differences between mice and humans carrying mutations encoding genes in the *Ret* pathway are noteworthy: (i) aganglionosis in mice occurs in homozygotes (*Ret*$^{-/-}$ and *Gdnf*$^{-/-}$) only, not in heterozygotes as in humans, (ii) penetrance is complete in mutant mice, (iii) sex bias is absent in mice but present in humans, with a two-fold higher penetrance in males than in females (Chakravari and Lyonnet 2001). Since the mouse strains studied have a homogeneous genome in contrast to the highly variable human genome, modifying genes in humans may have a greater influence on phenotypic expression. Perhaps the effect of mutations in the various genes involved in migration, differentiation, and network formation of the intramural gastrointestinal ganglion cells lies in a gene dosage gradient that eventually determines penetrance and expression (Chakravarti and Lyonnet 2001). Overo-spotted horses and the lethal white foal syndrome with aganglionosis megacolon may be an equestrian equivalent of the human Hirschprung disease (McCabe *et al.* 1990).

Conclusions

Hirschsprung disease is a neural crest disorder leading to lack of intramural intestinal ganglion cells (congenital intestinal aganglionosis) of variable parts of the colon intestines, leading to megacolon and concomitant manifestations. Hirschsprung disease is classified into short-segment aganglionosis type 1 (descending colon) in 52% and long-segment type 2 (aganglionosis of the colon rostral of the splenic flexure to the traverse and ascending colon, rarely including the small intestines) in 18%. Its population incidence is one in 5,000 with a 4.5:1 male to female ratio. In 70% Hirschsprung disease is limited to the gastrointestinal tract (non-syndromic form), in 18% it is associated with a disorder involving other organ systems and in 12% with a recognisable chromosomal abnormality, especially trisomy 21 (syndromic forms). Family studies established an empiric risk of recurrence of 4% with marked differences between short- and long-segment variety, and relatives of males and females. Transmission may correspond to an autosomal dominant and autosomal recessive mode of inheritance. Mutations in nine specific genes associated with variable, low penetrance account for about 30–50% of cases. The most important gene is the *RET* gene. Gain-of-function mutations account for 30–50% of patients. However, these mutations are rare in the population and can explain only a small percentage of cases (Emison *et al.* 2005).

The most common mutation is non-coding, with a low penetrance, and sex-dependent effects. The usually sporadic short-segment type 1 results from a combination of a mutation or a non-coding variant in the *RET* gene in combination with one other of several additional susceptibility chromosomal regions at different sites of the genome (at 3p12, 9q31, 19q12). The combination of a change in the *RET* gene and one other susceptibility region is necessary and sufficient as a cause of the disease. The presence of a non-coding variant in intron 1 of the *RET* involving an enhancer increases the susceptibility to Hirschsprung disease 20-fold over the population risk. One variant, the *RET+3:T* allele, is relatively frequent in Asia (45%) and Europe (25%) whereas it is virtually absent in Africa (1%), possibly due to an unidentified selective advantage. The interactions of gene mutations and non-coding variants recognised thus far make Hirschsprung disease arguably the currently best understood genetic dissection of a complex disorder. Additional unknown genes, other modifying effects of the genetic background, and environmental factors may also play a role in the aetiology of Hirschsprung disease.

Acknowledgement

I thank Dr Mary F. Passarge for helpful comments on the manuscript.

Familial recurrence of cot death

Michael A. Patton

Department of Medical Genetics, St George's Hospital Medical School, London SW17 0RE

George Fraser's work took him into many different areas of genetics and from the foundations he laid down our understanding of genetic disease is continuing in new ways. A major part of his work was defining the inheritance of deafness and some of the single gene disorders that contribute to this. This work is being reviewed by other contributors, and I want to illustrate how his careful and detailed clinical genetic studies have provided an important insight into the recurrence of cot death within families and the legal implications of such tragedies.

Fraser *et al.* (1964c) published their studies on families in which congenital deafness was associated with electrocardiographic abnormalities. Previously, Jervell and Lange-Nielsen had described a Norwegian family in which four of six siblings had had congenital deafness and episodes of fainting from early childhood (Jervell and Lange-Nielsen 1957). Of the four affected children three had died suddenly in childhood. The ECG abnormality reported in these children was a prolonged QT interval. Fraser *et al.* (1964i) described a further nine cases from the deafness study. The investigations confirmed the likely pattern of inheritance was autosomal recessive. Of the nine cases, eight had experienced syncopal attacks and in three of these the children had died suddenly from what was presumed to be a fatal arrhythmia. The deaths occurred at three years of age in two affected children and at 14 years of age in the remaining child.

In correspondence to *The Lancet*, Fraser and Froggatt (1966m) pointed out the importance of cardiac arrhythmia in unexplained cot death especially where it recurred in the family. They pointed out that although the sudden deaths which occurred in their previous study had all been after the age of three, the same condition could have occurred earlier in infancy, and sudden death could have preceded the recognition of deafness as there was a history of sudden unexplained death in the families they reported. There was further historical support for this as they had looked further into a case of a 10-year-old with deafness and sudden death from 'epilepsy and nervous shock' reported by Meissner in 1856. The description given is a testimony the thoroughness of their study in clinical genetics:

> It has been possible with the help of the pastor of the Evangelical-Lutheran Church of St Wenceslas in Pappendorf über Mittweida, East Germany, to obtain details of the sibship of this child. While at that period deaths in infancy were not uncommon, it is striking that four of her seven sibs died at ages ranging from 2 days to 46 months – two from 'apoplexy' and one each from 'sudden tetany' and 'cough'.

The conclusion of their letter was that a small proportion of sudden unexplained deaths in infancy was attributable to genetically determined disorders of cardiac conduction.

Unfortunately, this careful clinical observation was overlooked by those pre-eminent in the field of sudden infant death syndrome (SIDS). The recognition that patients may invent illness and falsify symptoms to gain medical attention was recognised as Münchhausen's syndrome.

Professor Sir Roy Meadow described the condition of Münchhausen by proxy to describe how some parents may harm their children in order to gain attention for themselves and how on occasion this abuse can lead the death of a child (Meadow 1977, 1982). Most paediatricians have had the experience of dealing with child abuse, often from parents who may harm their own children, and recognised the possibility of Münchhausen by proxy. However, what is an explanation for some cases of cot death became widely suspected and apparently discovered. It was particularly suspected where there were multiple cot deaths. This view was reinforced by a number of high profile cases where multiple cot deaths had occurred and the mother had subsequently confessed to suffocating the babies. One such case was Maybeth Tinning who had had nine children die in fourteen years including one adopted son (Egginton 1997). This experience was translated into the clinical aphorism 'One cot death is natural, two is suspicious and three is murder unless proved otherwise.'

In 1999, Angela Cannings was charged with the murder of two of her children. She had had four children. The first child, Gemma, died at three months of age and had been certified by the pathologist as being a case of cot death. The second child, Jason, died at six weeks of age in a similar situation. The third child, Jade, had had a near-miss cot death or acute life threatening episode, but survived and had no further problems. Her last child, Matthew, had died at four months, and, although the pathologist's diagnosis was cot death, Angela was arrested and charged with the murder of Matthew and Jason.

I was approached by Mrs Cannings's defence team and after some reflection agreed to assist by reviewing the case from a genetic point of view as it seemed that the possibility of a genetic cause for the infant deaths had to be considered as an alternative explanation to murder. It did seem as though the 'rule of three' had become established both in medical and legal belief and it seemed that the starting point was that Mrs Cannings was guilty 'unless proved otherwise'. It therefore placed an onus on the defence team to prove innocence by identifying a specific cause for the deaths. As well as considering and excluding metabolic disorders, we considered the possibility of cardiac arrhythmias and long QT in particular. An ECG had only been performed in the last child, Matthew, and although the QT interval was at the upper limit of the normal range, this was not sufficient to make a clinical diagnosis of long QT syndrome. None of the children had been diagnosed as being deaf. However by this stage there were several genes identified in the aetiology of long QT. We were able to look at the relevant exons in *SCN5A*, *KCNQ1*, *HERG*, *KCNE1* and *KCNE2*, but no mutations were found. By our estimates this had excluded about 60% of the known mutations. It did not, however, absolutely exclude the possibility of long QT as a cause for the deaths in Mrs Cannings's children.

Another aspect of the Cannings case was that there was a family history of sudden unexplained infant death in the extended family. In particular, she had had a second cousin in Ireland who had had four children two of whom had died from cot death and two of whom had had near-miss cot deaths. The evidence of the family history was presented by the defence team as indicating a possible genetic factor in the family. However, when the jury reached their verdict they found her guilty, and she was sentenced with a mandatory sentence for murder.

The subject of multiple cot deaths that George Fraser had discussed in the letter to *The Lancet* in 1966 continued to feature in the media. In November 1999, a solicitor from Manchester, Sally Clark, had been convicted in Chester Crown Court of murdering her two sons. The prosecution case had been supported by Professor Roy Meadow who had suggested the odds of two healthy babies dying in these circumstances would be one in 73 million, and therefore, an unnatural cause was more probable. Although this evidence was accepted by the court at the time, the evidence was criticised by the Royal Statistical Society after the case. They correctly argued that other possibilities such as a genetic cause had not been taken into account in the prior probability and that if, for example, it was due to an autosomal recessive disorder, the probability of two children being affected would be one in 16. The Appeal Court accepted that the statistical evidence was wrong, but it was only on a further appeal showing that the Home Office pathologist had not disclosed all the relevant microbiology that Sally Clark was freed on appeal.

A further high profile case came before Angela Cannings's appeal. A pharmacist from Reading, Trupti Patel, had had four children of whom three had died suddenly and unexpectedly in infancy. With the death of her last child, she too was charged with murder. The initial evidence had included a suggestion that rib fractures were a sign of physical assault in the last child, but in court this evidence was discredited when it was demonstrated that the rib fractures were more likely to be due to cardiopulmonary resuscitation. I was again called by the defence team and was able to obtain a full pedigree. This showed that Trupti Patel's grandmother had had 12 children of whom five had unexplained deaths in infancy. The problem was to know whether this background rate of infant death was within the normal range for rural India two generations ago. As there were no medical records available she was called to give evidence in person in court. She was 80 years of age and only able to give evidence through an interpreter. However, the history clearly emerged that three had died suddenly and unexpectedly, one had been born prematurely, and death may or may not have been associated with that, and one had died after a diarrhoeal illness. It became clear that the number of infant deaths in this family exceeded that predicted from background population statistics and most had been sudden and unexplained. After this and further evidence, the outcome was different in Trupti Patel's case. She was cleared of murder by the jury.

The tide was turning and the public media were accepting that the possibility of a genetic cause should be considered, especially in cases of multiple cot death. The 'rule of three' had been discredited and would in future fail to persuade juries that multiple cot death was equivalent to murder.

We were slowly able to gather further evidence for Angela Cannings's appeal. Through a genealogist in Ireland we were able to identify three other infant deaths in previous generations. These had not previously been recognised as the rural tradition had been if a baby boy had died, to use the same first name for the next male infant, e.g. if a boy named Michael had died the next boy would be called Michael. The family had not recognised that these infants existed but the birth and death certificates were traced and confirmed the infant deaths. When this was broadcast as part of a TV documentary an unexpected turn of events occurred. A woman in Angela's home town contacted the TV and explained that she had had a secret relationship with Angela's father and that Angela had an illegitimate half-sister who had had two children by different partners both of whom had had a near-miss cot death. With the new evidence of unexplained infant deaths and near-miss cot deaths, the evidence from the family history pointed overwhelmingly to a genetic cause rather than murder. In December 2004, Angela Cannings was freed in the Appeal Court.

In summing up, Mr Justice Judge and the Appeal Court judges recognised the difficulty in fully ascertaining the cause of death in all cases. They gave the ruling that without objective scientific evidence, it is better to err on the side of innocence rather than wrongly convict a mother for the murder of her children. In his own words:

'Unless we can be sure of guilt, the dreadful possibility always remains that a mother, already scarred by the unexplained deaths of her babies, may find herself in prison for life for killing them, when she should not be there at all. In our community, and in any civilised community, that is abhorrent.

'We recognise that justice may not be done in a small number of cases where in truth a mother has deliberately killed her baby without leaving any identifiable evidence of the crime. This is an undesirable result which however avoids the worse one.'

The challenge remains. There will be a small proportion of cases of sudden infant death in which there is a genetic cause. These are more likely to be recognised in the rare multiple cot death cases and discrediting 'the rule of three' opens the way for objective scientific research in this area. A further autosomal recessive gene for sudden infant death has been identified in the Amish (Puffenberger *et al.* 2004). The *TSPYL* gene is expressed in the midbrain, and in males will be associated with genital hypoplasia but in females will have no associated phenotypic features to identify before sudden death may occur. Another potential gene that may be involved in unexplained sudden infant death is *PHOX2B* that causes autosomal dominant chronic hypoventilation syndrome in older children, but in theory might cause sudden infant death if the first apnoeic episode occurred in the cot at home (Amiel *et al.* 2003). There are also further genes to be identified in long QT syndrome and other inherited cardiac conduction defects.

George Fraser's study on congenital deafness with syncope and sudden death is a masterpiece of clinical genetic research, showing the value of careful clinical documentation and critical thinking. In his discussion he was able to predict that the nature of the genetic defect might be a disturbance of ionic balance many years before the gene for Jervell-Lange-Neilson was identified. With his passion for ensuring that genetics is not misused for political ends, I am sure he will have taken some satisfaction from seeing the recognition of genetic factors in sudden infant death being recognised in the courts and justice being achieved for these innocent women.

The evolution of sex chromosomes and sex determination in vertebrates

Malcolm Ferguson-Smith

Centre for Veterinary Science, University of Cambridge, UK

Nature is nowhere accustomed more openly to display her secret mysteries than in cases where she shows traces of her workings apart from the beaten path ... (William Harvey, 1657).

Many of us share with George Fraser our appreciation of this famous passage from a letter written by William Harvey in 1657, which underlies the philosophy of those with a medical background who seek to understand biological systems (see page xii). Our present scant knowledge of mammalian sex determination is based on studies over 45 years ago, of Klinefelter and Turner syndromes in humans and mice that revealed the dominant Y chromosome factor in male differentiation. Studies in XX males led to the discovery of *SRY* as the testis determining factor (TDF) and almost all the other known factors in the sex determination pathway have been characterised by the careful investigation of sex reversed XY females (Table 1). The work was initiated largely by pathologists and clinicians who studied patients referred to them with disorders of infertility or sex differentiation. The solution of these problems has required the collaboration and expertise of geneticists and molecular biologists. Disorders of sex determination are not lethal and so provide an excellent model for the study of biological systems.

Some of us were introduced to this fascinating field through the study of numerical and structural abnormalities of the human sex chromosomes. In my case, this was prompted by the paradoxical nuclear sex chromatin findings that were widely thought to indicate that Turner and Klinefelter syndrome were sex reversed males and females respectively. An early chromatin-positive Klinefelter patient was observed to have XY sex bivalents and sperm in a single seminiferous tubule (Ferguson-Smith and Munro 1958) suggesting to me that he was not sex reversed, and that the answer was to be found by karyotype analysis. The 47,XXY and 45,X karyotypes were reported in 1959 (Jacobs and Strong 1959, Ford *et al.* 1959), and the location of *TDF* to the short arm of the Y was shown later (Jacobs and Ross 1966). The anomalous segregation of the Xg sex-linked blood group in families of XX males led in 1966 to the hypothesis that there had been an interchange between *Xg* on the X and *TDF* on the Y (Ferguson-Smith 1966) and to the isolation of *SRY* much later (Sinclair *et al.* 1990). In all these cases, human disorders led to an understanding of normal processes, and there are many more examples in the field of human sex determination.

Evolution of the sex chromosomes

Studies on the evolution of sex chromosome systems have been based during the last 40 years on the hypothesis of Susumu Ohno (1967) who observed the variable size of the W chromosome

251

and the invariable size of the Z in different snake families. He concluded that the W and Z must have evolved and differentiated from an ancestral pair of chromosomes.

Table 1. Genes involved in sex differentiation

SRY	Primary TDF
SOX9	Main target for SRY
DAX1	Antagonist to SRY
SF1, WT1, GATA4	Required by SRY
AMH	Activated by SOX9
WNT4, WNT7	Required by AMH
DMRT1	Required by Sertoli and germ cells
ATRX	Required for testis development
LHX9, M33	Required for gonadal development
FGF9	Required for peritubular myloid cell migration
TDA1, 2, 3	Promote ovarian development
AR (DHTR)	Required for development of male genitalia
PAX2, LIM1, EMX2	Required for Wolffian and Mullerian ducts
INSL3	Required for testis descent
DHH	Required for spermatogonial division

The same reasoning suggested that mammalian sex chromosomes must have evolved from an identical ancestral pair of homomorphic chromosomes, one of which developed a sex-deter-mining role. The pair gradually differentiated from one another, with the sex-determining region on the Y becoming isolated during paternal meiosis so that recombination with the X chromosome was prevented. Attrition of the Y and accumulation of sequences involved in male gametogenesis led to a distinctive Y chromosome in which only a small part was homologous with the X. This part is necessary for synapsis and segregation of the X and Y chromosomes during male meiosis. The X chromosomes are able to pair and to recombine with one another throughout their length in female meiosis but, in male meiosis, can only undergo recombination with the Y at the small pairing (pseudoautosomal) regions, leaving a large differential segment containing the strictly X-linked genes. The similar dosage of X-linked genes in males and females is explained satisfactorily by X-inactivation, whereby the genes in one of the two X chromosomes in the female were silenced (Lyon 1961). The conservation in form and gene content of X chromosomes in all mammalian species is explained by the two processes of meiotic isolation and X-inactivation. The fact that the extant human X and Y share a number of genes active on both their pairing and differential segments, that lead to the haploinsufficiency in Turner syndrome (Ferguson-Smith 1965), attest to their origin from an ancestral pair. Rearrangements between the X and autosomes are deleterious as they disrupt the dosage compensation mechanism; any such change would be eliminated by natural selection.

Ohno also proposed that the Z and W chromosomes in birds have evolved from an ancestral pair of chromosomes different from the mammalian ancestral sex chromosomes. Like the X among mammals, the Z chromosome has been conserved among birds. It is now known that genes on the chicken Z chromosome are homologous to genes on human chromosomes 9, 5, 8 and 18 (Nanda *et al.* 2000). Genes homologous to the mammalian X are located on chicken chromosomes 1, 4 and 12.

Nearly forty years later the basis of Ohno's hypothesis still stands, although it now requires some modification.

1. The marsupial X chromosome differs from the eutherian X chromosome in that genes homologous with the short arm of the human X are autosomal and do not require dosage compensation (Spencer *et al.* 1991).

2. The chloride channel gene (*CLCN4*) is X-linked and X-inactivated in humans and *Mus spretus* but maps to chromosome 7 in *Mus musculus* (Palmer *et al.* 1995). FISH mapping reveals copies of this gene on both sex chromosomes of *M. spretus* (unpublished). The steroid sulphatase gene (*STS*) is X-linked in humans and pseudoautosomal in mice (Keitges *et al.* 1985); the gene escapes inactivation in both. Apart from *CLCN4*, genes X-linked in one placental mammal are found to be X-linked in all.

In contrast to the mammalian X chromosome, the mammalian Y chromosome shows considerable variation between species while retaining *SRY* (which shares homology with *SOX3* on the X) and genes necessary for male fertility. Thus Y chromosome paint probes from one species seldom show much homology when cross-hybridised with the Y from other species. Within species, Y paints hybridise to the pseudoautosomal regions of the X and, in the human at least, to a chromosomal block on the long arm of the X, encompassing the X-inactivation centre.

The evolution of the human sex chromosomes has been investigated more recently by comparing the map positions of 19 X-Y gene pairs located on the differential segments of the X and Y chromosomes (Lahn and Page 1999). The X-Y gene pairs were further characterised by determining the synonymous nucleotide divergence between the X linked and Y linked coding regions, which provides a measure of the evolutionary time that has elapsed since the gene pairs started differentiating into distinct X and Y forms. At that time, recombination between the X and Y homologues would have ceased due to the meiotic isolation of the differential segments. It was found that the 19 X-Y pairs could be grouped into four 'strata' in time depending on the extent of sequence divergence, each stratum with a comparatively distinct range of values. The genes in each stratum were linked on the X chromosome map, but were somewhat scrambled on the Y map, presumably by the occurrence of inversions in the Y during evolution. The oldest group, with the greatest sequence divergence, was designated stratum 1, with the corresponding genes mapping to the long arm of the human X. Strata 2, 3 and 4 (most distal) mapped to the short arm of the X and were progressively younger in origin. Theoretically, it would be expected that the oldest stratum would have the fewest remaining Y homologues and, conversely, that the youngest would have the highest density of X genes with active Y homologues that escape X-inactivation; these expectations are all fulfilled.

It was postulated that the results indicated that X-Y recombination was suppressed regionally in four steps during evolution; each step separated by a time interval and initiated by an inversion event on the Y chromosome that led to suppression of recombination and the onset of differentiation between the X-Y pairs of genes. As the Y genes decayed and lost their function, X homologues were first up-regulated for dosage compensation in the male, and later by the evolution of X-inactivation (Reik and Lewis 2005). Lahn and Page used comparative genomics to estimate the likely timing of the origin of differentiation in each stratum. Stratum 4 (the most recent) was estimated to have arisen 30 to 50 million years ago (Mya), stratum 3 at 80–130 Mya, stratum 2 at 130–170 Mya and stratum 1 at 240–350 Mya. The differentiation of stratum 1 would have occurred shortly after the mammalian and avian lineages diverged at about 310 Mya, and stratum 2 prior to the divergence of the monotreme lineage at 210 Mya. The fourth stratum probably arose during early primate evolution at the time the simian and prosimian lineages diverged.

Independent comparative studies on vertebrate species that diverged from mammals early in evolutionary history support the idea of a stepwise evolution of the mammalian sex chromosomes. Gene mapping in marsupials and monotremes shows that genes that currently map to the short arm of the X are autosomal and escape inactivation in marsupials and, for example, are located in chromosomes 5 and 1 in the tammar wallaby (Graves and Westerman 2002). The X-linked genes on the long arm and proximal short arm of the human X are conserved in the marsupial X. The genes on the short arm of the human X and on its two pseudoautosomal regions must have been added later by translocation from autosomes, and this is confirmed by mapping mammalian X genes in the platypus as in that species these genes are also autosomal (Table 2). Hybridisation of X-chromosome-specific DNA from the tammar wallaby to human metaphases by fluorescence in situ hybridisation (FISH) confirms the location of the conserved X genes to the long arm and proximal short arm of the human X (Glas *et al.* 1999). (It is noteworthy that the marsupial X is the only marsupial chromosome that can be successfully hybridised by FISH to human chromosomes, underlying the remarkable conservation of the X as postulated by Ohno). These comparative studies strongly support the idea of X chromosome strata with the conserved X genes corresponding to stratum 1, and groups of autosomal genes corresponding to strata 2 and 3 added to the X by successive translocations. The pseudoautosomal genes seem to be later additions.

Moving further 'apart from the beaten path' of humans and mammals to birds, reptiles and fish, more is revealed about the evolution of sex chromosomes from comparative mapping in these groups. The publication of the draft genome sequence of the chicken in particular has enabled valuable conclusions to be drawn from comparisons with the map of the mammalian X. Kohn *et al.* (2004) have now made these comparisons and show that stratum 1, the X conserved region, is derived from chicken chromosome 4 with the exception of a small 1.4 Mb region, corresponding to human Xq28, that shares homology with several chicken microchromosomes including chromosome 12. The added 'X genes' are located on chicken chromosome 12 (stratum 2) and 1 (stratum 3). The findings convincingly confirm the three genome regions that are ancestral to birds and mammals and that have become incorporated into the mammalian X chromosome in a stepwise fashion at critical times of divergence in vertebrate evolution. When the comparative mapping was extended to teleost fish (zebrafish and pufferfish), Kohn *et al.* found that genes from strata 1 and 3 were together on several small chromosomes of both species, suggesting that they might have been present in a single chromosome in a teleost ancestor. Following a series of inversions and fission events the single chromosome became separated into several smaller chromosomes. When tetrapods and fish diverged 450 million years ago, strata 1, 2 and 3 may each have been on separate chromosomes, with fusion subsequently of strata 1 and 2 prior to the divergence of marsupials and monotremes estimated at 210 Mya, and fusion of stratum 3 to 1 and 2 in the eutherian X about 180 Mya. These timings differ somewhat from earlier estimates by Lahn and Page, and the authors note that proof of this mechanism must await the comparative mapping in an outgroup species, such as shark, lamprey or amphioxus.

Table 2. Location of human X-linked genes in platypus.

Human	Platy X$_1$		Platy 1	Platy 2	Platy 6
Xp	ALAS2	P	CYBB	OTC	
	GATA1	P	ZFX	SYN1	
	UBE1*	P	AMEL	POLA	
			DMD	MAOA	
			TIMP		
Xq	F8	P		BGN	ATRX
	F9	P			RBMX
	PLP	P			CDX4
	GLA	P			SLC16A2
	RCP	P			SYBL1
	AR*	D			
	GDX	D			
	MCF2	D			
	G6PD*	D			

P = present on pairing (pseudoautosomal) segment, not requiring dosage compensation.
D = present on differential segment, requiring dosage compensation.
* = gene expression equal in both sexes.

Sex determination in birds

Comparative mapping has shown that the XX/XY sex chromosome system in mammals and the ZZ/ZW system in birds have evolved independently from two different ancestral pairs of chromosomes. While SRY is the primary switch for testis differentiation in mammals, the mechanism remains unclear. SRY is absent in some mole voles such as *Ellobius lutescens* which has an XO sex chromosome complement in both sexes, and a related species *E. tancrei*, which has an XX complement in both sexes (Just *et al.* 1995). The spiny rat, *Tokudai osimensis*, also has an XO complement in both sexes, and the single X has a small region with homology to the mouse Y but without SRY (Arakawa *et al.* 2002). These rodents have evolved a system of sex differentiation that no longer needs a Y chromosome or SRY. The mechanism that has replaced SRY remains to be determined.

Apart from the above exceptions, in mammals the male (XY) is the heterogametic sex. In birds the female (ZW) is the heterogametic sex and SRY is absent. The W chromosome in most birds has undergone a similar type of attrition as the mammalian Y, although the primitive ratite birds have a Y almost equal in size to the X (Shetty *et al.* 1999). Unlike sex chromosome aneuploidy in mammals and in Drosophila, ZO and non-mosaic ZZW diploid birds are unknown (and probably embryo lethal), so that a sex-determining mechanism by a female determining gene on the W cannot be distinguished from a dosage-dependent gene for testis determination on the Z (Graves 2003). However, triploid ZZZ are viable but sterile males and triploid ZZW chickens develop as intersexes. In mammals the exceptional XO (female) and XXY (male) indi-

viduals revealed the dominant male determining role of the Y, whereas in Drosophila the observation that XXY flies are female and XO flies male led to the conclusion that X gene *dosage* determines sex in this species. Both theories of sex determination have been studied extensively in
birds.

Dosage-dependent male determination

The best candidate is the gene *DMRT1* that maps to the Z chromosome and apparently has a
role in the sex differentiation pathway in many different species. The gene has homology to the
doublesex (*dsx*) gene in Drosophila and to the *mab-3* gene in *Caenorhabditis elegans* (Raymond
et al. 1998). *DMRT1* maps to the short arm of chromosome 9 in humans and haploinsufficiency
is thought to be responsible for the sex reversal in XY females associated with deletions in the
distal end of chromosome 9p (Calvari *et al.* 2000). The gene is expressed in the gonads of chicken
embryos, with higher levels of expression in ZZ males than in ZW females prior to and during
the critical period of sex differentiation (Raymond *et al.* 1999). Thus a double dose of the
DMRT1 gene is associated with male sex and a single dose with female sex. In the ZW female
only the left gonad becomes an ovary, while the right gonad regresses. In the ZZW triploid
intersex, the right gonad becomes a testis and the left gonad develops first into an ovotestis,
during which the bird appears phenotypically female (Graves 2003). The ovarian part of the
ovotestis then degenerates and the phenotype changes to male in the adult. This suggests that any
W-linked female determinant cannot be dominant. Studies show that in the normal ZZ male
embryo both Z chromosomes express *DMRT1* and that there is no dosage compensation at this
locus. Other Z-linked genes show equal levels of expression in both sexes and are dosage com-
pensated (McQueen *et al.* 2001). As there is no sex chromatin in chicken it follows that dosage
compensation of these genes does not occur by Z inactivation.

W-linked female sex determination

The small W chromosome of birds, like the mammalian Y chromosome, has degenerated
during evolution, which suggests that it has developed a sex determining function for female differentiation. Only a few W-linked genes have been reported and several of these (including
ATP5A, *CHD* and *WPKC1/ASW*) have homologues on the Z chromosome, consistent with an
origin of Z and W from a homomorphic pair of ancestral chromosomes (Smith and Sinclair
2004). However, *DMRT1* does not have a homologue on the W. One gene, *FET 1*, may be
unique to the W and so represent a possible ovary-determinant. It is strongly expressed in the left
gonad (developing ovary) of the ZW female embryo, consistent with this role. Another W-linked
gene, *WPKC1/ASW*, has repeated copies on this chromosome and is also strongly expressed in the
gonads of ZW female embryos. A single homologue of this gene, *ZPKC1*, maps to the Z chromosome and is weakly expressed in embryos of both sexes. *WPKC1* may act by repressing
ZPKC1 action, i.e. through a dominant negative effect on testis differentiation. An alternative
theory implicates another W-linked gene (so far not characterised) which controls the expression
of a male hypermethylated (MHM) region of the Z chromosome close to *DMRT1*. This region
is undermethylated and transcribed on the single Z of ZW females, and methylated and
repressed on both Z chromosomes of ZZ males (Terinashi *et al.* 2001). It is suggested that this
mechanism serves to repress the expression of the adjacent *DMRT1* gene in ZW females. Thus,
DMRT1 dosage and W-linked factors may act in concert to determine avian sex.

Sex determination in reptiles and fish

Reptiles and fish seem to have evolved a confusing variety of mechanisms of sex determination. Snakes are like birds with female heterogamy and a ZZ/ZW system. In primitive snakes, such as the boid group, the Z and W chromosomes are homomorphic and cytologically indistinguishable, whereas in vipers and elapids the W is small and in colubrid snakes the sex chromosomes are distinguishable only by the location of the centromere (Graves and Shetty 2001). Turtles have karyotypes similar to birds with both macro- and micro-chromosomes and crocodiles have only macrochromosomes. Most turtles and all crocodiles have no sex chromosomes and sex is determined by the temperature at which the eggs incubate. In general, higher temperatures produce males in crocodiles and females in turtles, although different temperature patterns are found in some species of both groups. It has been suggested that changes in temperature affect the expression of sex determining genes in the embryo during gonadal differentiation (Deeming and Ferguson 1988). This hypothesis is supported by the observation that *DMRT1* expression in the genital ridge of turtle embryos is increased at the low temperatures that promote male rather than female differentiation (Kettlewell *et al.* 2000).

Cross-species chromosome painting with chicken Z-specific paint probes onto turtle and crocodile metaphases has revealed apparently complete homology with one homomorphic pair of chromosomes in each species, i.e. with chromosome 5 in the turtle *Chelodina longicollis* (Graves and Shetty 2001), and chromosome 6 in the red-eared slider and Nile crocodile (Kasai and Ferguson-Smith unpublished). *DMRT1* maps to chromosome 6 in both these species (Kasai and Ferguson-Smith unpublished) and to chromosome 6 in the Chinese soft-shelled turtle (Matsuda *et al.* 2005). Chromosome 6 in the crocodile thus seems to represent the ancestral pair of chromosomes from which the ZZ/ZW pair of chromosomes in birds evolved.

The karyotypes of snakes show considerable rearrangement in comparison to those of turtles, crocodiles and birds that are highly conserved. The most striking difference concerns the snake Z and W sex chromosomes that are not homologous to the avian sex chromosomes and must have evolved independently from a different pair of ancestral autosomes. Orthologues of Z genes in the chicken, including *DMRT1*, map to chromosome 2 in both the Japanese four-striped rat snake (Matsuda *et al.* 2005) and the Australian tiger snake (Kasai and Ferguson-Smith unpublished). Mapping studies indicate that the snake Z shares homology with chicken chromosome 2 and human chromosomes 7 and 10. The primary sex-determining switch and the role of *DMRT1* in sex differentiation in snakes are not yet known. These genomic differences support the recent molecular phylogenetic data that suggest that snakes are more distantly related to birds and to other reptiles than was previously thought (Kumar and Hedges 1998).

In several species of lizard, both environmental and genetic factors have been described. Temperature-dependent sex determination was first described for the lizard, *Agama agama*. In some lizards the female (ZW) is the heterogametic sex and in others it is the male (XY). Incubation temperature has an influence on sex ratio in some species with genetic sex determination. The same is found in some species of fish and in amphibians (Deeming and Ferguson 1988). In fish, sex determination in the medaka has been studied in some detail (Wittbrodt *et al.* 2002). The male (XY) is the heterogametic sex and, although the sex chromosomes are the second largest pair and morphologically indistinguishable, linkage studies reveal a sex determining cluster on a region of reduced recombination in the Y. *SRY* is absent but *DMRT1* maps to this cluster and also to a closely related autosomal homologue (Volff *et al.* 2003). It is not yet clear how the Y linked *DMRT1* gene functions in sex determination, but one possibility is that it acts, directly or by regulation, to increase autosomal *DMRT1* expression in male embryos. However, molecular analysis in other fish species does not provide evidence of a Y-linked copy of *DMRT1*, and suggests that the Y-linked gene in Medaka is a recent duplication (Volff *et al.* 2003). In zebrafish the absence of sex-linked markers has so far prevented an analysis of sex determination in this species. Only one *DMRT1* gene locus has been identified in the draft sequence of the fugu genome.

Sex determination in Monotremes

Monotremes are egg-laying mammals that have a unique mixture of mammalian, avian and reptilian features. In addition to egg-laying and absence of teeth as in birds, the duck-billed platypus has fur, mammary glands, electro-sensors in its soft bill, and a pelvis similar in structure to the reptilian pelvis. There are two other extant monotremes, namely the short-beaked and long-beaked echidna. The ancestor of these three species is believed to have diverged from the mammalian lineage about 210 Mya and some 30 million years before the divergence of marsupials and placental mammals. The monotremes have been regarded therefore as more primitive than mammalian species and are seen as forming a link between reptiles and mammals. As such, they may help to answer questions about sex determination and the origin of dosage compensation in mammals. Accurate chromosome identification is crucial for such studies, but karyotype analysis has been difficult in both platypus and echidna as both have unpaired chromosomes believed to form a multivalent chain at meiosis. Chromosome sorting and painting of platypus chromosomes have now revealed for the first time that the male platypus has a remarkable sex chromosome complement of five X chromosomes and five Y chromosomes (Rens *et al.* 2004). These form a chain of alternating X and Y chromosomes at meiosis, in which each of the ten components are linked together by pairing segments. The homology between each X and Y chromosome is revealed by chromosome-specific paint probes. The X and Y chromosomes undergo alternate segregation at meiosis so that sperm contain either five X or five Y chromosomes (Grützner *et al.* 2004). Females have five pairs of X chromosomes and do not form a chain at meiosis.

Preliminary gene mapping shows that *SRY* is absent in platypus as in birds, reptiles and fish. The large chromosome at one end of the chain, X_1, has homology with the mammalian X. Comparative mapping with the human X chromosome (Table 2) shows that many X-linked genes that map to the short arm of the human X are autosomal, as found in marsupials. However, additional X-linked genes located in the long arm of the human X are also autosomal and map to platypus chromosome 6 (Waters *et al.* 2005). These genes flank the X-inactivation centre in placental mammals and, as they also map to the marsupial X and to chicken chromosome 4, their loss from the platypus X indicates an evolutionary change that has occurred in the monotreme lineage after the monotreme-mammalian divergence. Being autosomal these genes do not require dosage compensation in monotremes, and this applies also to the X-linked genes on the pairing region of platypus X_1 which has homology with platypus Y_1. As shown in Table 2, some platypus genes homologous to genes on the long arm of the human X map to the short arm of the platypus X which is homologous to the long arm of Y_1, and thus have similar dosage in both sexes without dosage compensation. Another surprising finding is that *DMRT1* is located on X_5 in the platypus sex chromosome chain (Grützner *et al.* 2004). There is thus a double dose of *DMRT1* in the *female* platypus and a single dose in the *male*. This is in contrast to mammals where two doses seem to be required for testis determination. In order to understand this paradox it will be necessary first to investigate *DMRT1* expression in the developing gonads of platypus embryos, and then determine possible regulatory factors. It seems inconceivable that *DMRT1* is not involved. The location of *DMRT1* on one of the platypus sex chromosomes is likely to be of significance, given its proven key role in sex differentiation in so many other vertebrate and invertebrate species.

Large regions of chromosomes X_1-X_5 in platypus are without homology to Y_1-Y_5, and some form of dosage compensation is to be expected to account for the male-female difference. Several genes have been studied on platypus X_1, and the androgen receptor and *G6PD* loci that are on the differential segment of X_1 seem to have similar gene expression in both sexes (Grützner *et al.* 2003). There is some evidence that mechanisms of genome imprinting and X-inactivation are absent in monotremes (Killian *et al.* 2000) and so it seems more likely that dosage compensation is achieved by up-regulation of X-linked genes as in Drosophila.

The five Y chromosomes of the platypus deserve careful mapping and analysis to identify possible sex determinants within the substantial number of differential segments that have no homologous sequences in the female. These regions may also be expected to include loci important in male gametogenesis and fertility.

Conclusion

These observations on the evolution of sex chromosomes in vertebrates indicate that the differentiation of a testis requires a gene cascade similar in all species but differing in the primary switch mechanism. SRY is the primary switch in almost all mammals, but the mole vole among rodents and the SRY-negative XX male in humans show that a factor other than SRY must operate in these exceptional cases. Temperature regulation of the expression of *DMRT1* may be the primary switch mechanism in turtles and crocodiles, and in birds with female heterogamety the most likely mechanism is dosage of the Z-linked *DMRT1* gene. In snakes, *DMRT1* is autosomal and the ZW sex-determining system has evolved independently from the ZW system in birds, turtles and crocodiles. *DMRT1* thus has a key role in sex determination in almost all species investigated, and it is perhaps not surprising that in birds, monotremes and some fish it is found in a sex chromosome. One awaits with interest the results of further studies in vertebrates that may reveal more about the mechanism of action of this interesting gene.

This article was written in celebration of George Fraser's outstanding contributions to human genetics over many years. As a friend, I have appreciated his interest in the history of ancient science and trust that he enjoys this fragment on the evolutionary history of our sex chromosomes.

Genetics of sex differentiation: an unsettled relationship between gene and chromosome

Ursula Mittwoch

Galton Laboratory, Department of Biology, University College London, Wolfson House,
4 Stephenson Way, London, NW1 2HE, UK

'the formulation of sex determination has remained in terms of chromosomes, while the modern unit of determination is the gene … '(Bridges 1925)

George Fraser arrived in the Galton Laboratory in 1957, a recently qualified physician intent on mastering the new subject of medical genetics for the benefit of his patients. He left two years later, having written his PhD thesis on Pendred syndrome, the syndrome of sporadic goitre and congenital deafness. By visiting families of affected individuals and testing their thyroid function and hearing ability, George was able to demonstrate that the same gene was responsible for the goitre as well as the deafness and was distinct from other genes responsible for congenital deafness (Fraser *et al.* 1960). This investigation was soon followed by many others, which helped to put clinical genetics on a sound scientific basis. His review 'Our genetical load' (Fraser 1962a), besides being a masterful summary of the effects of different modes of transmission on the burden of genetically determined diseases, also describes two sibships, each containing two sisters afflicted with a multi-system syndrome, which now bears his name. A recently discovered mutation in Fraser syndrome confirms and extends the link between *FRAS1* and related genes in the mouse (Jadeja *et al.* 2005). But in spite of his formidable intellect, George remains a physician who regards helping his patients as his primary task, and remembers the fact that he was instrumental in curing a young patient of deafness as perhaps his greatest achievement.

The causes that led to my own presence in the Galton Laboratory contained a larger element of chance, as did the choice of my investigations into sex determination. Although the researches undertaken by George and myself proceeded mainly along parallel lines, our common interests – among them the desire to consult old publications to illuminate issues of the day – also led to some collaborations. His books on blindness and deafness in childhood (Fraser and Friedmann 1967, Fraser 1976) are on my shelf, and I feel privileged to have been included in the distinguished list of his friends.

The question of what causes the division of human beings into males and females has been debated for longer than that for any other human dimorphism, and a lot of effort has been expended into trying to find an answer – usually with ultimately disappointing results. The birth of genetics at the beginning of the 20th century also led to the conclusion that sex was inherited, but the basis appeared to be in a pair of chromosomes rather than in contrasting genes. Netty Stevens (1905) investigated both sexes of the common mealworm, *Tenebrio molitor*, and discovered that in males, but not in females, one chromosome was smaller than the others; and she concluded that this chromosome must be responsible for the production of males. The smaller chromosome became known as the 'Y' chromosome, its partner as the 'X' chromosome (Wilson 1909) and both as 'sex chromosomes' (Wilson 1911).

The vinegar fly, *Drosophila melanogaster*, was also found to have two X chromosomes in females and an X and a Y chromosome in males, and at the age of 22, Alfred Sturtevant (1913) published a map of sex-linked genes on the X chromosome. The X-linked genes specified body colour, eye colour and wing shape, but not female sex. The possibility that the Y chromosome might carry genes for maleness was ruled by C.B. Bridges (1916), who showed that XXY flies were fertile females, while XO flies were male, albeit sterile. In order to accommodate different data pointing to the participation of many genes on the X chromosome, having a net female-determining potency, as well as on autosomes, with net male-determining potency, in the determination of sex, Bridges (1925, 1939) used the analogy of a mechanical balance. As equal weights are added to the pan, the beam finally tips, but the last weight has no more intrinsic value than the others.

In his book *A History of Genetics*, Sturtevant (1965) wrote: 'It has sometimes been felt that the determination of sex offered the best opportunity for the study of the action of genes, and the results described here have contributed largely to our understanding; it now, however, appears that it will be more profitable to study simpler situations, and it is to these that attention is now more often turned'.

The 'simpler situations' referred to the bread mould *Neurospora crassa*, a haploid organism that could be maintained on a chemically defined medium, and which far surpassed Drosophila in the number of mutants that could be produced and detected. In 1941, George Beadle and Edward Tatum described three strains that were unable to grow on the basic medium, because each had lost the ability to synthesise an essential growth factor, either vitamin B_8, thiazole or para-amino benzoic acid; and in each case, the loss of ability to synthesise the growth factor seemed to be caused by a mutation in a single gene. The Neurospora work gave rise to the concept that the function of genes is to produce proteins, many of them enzymes.

In view of the advantages of Neurospora as an organism for genetic investigations, it seems somewhat ironical that the principles deduced from studying the mould had been discovered more than 30 years earlier by Archibald Garrod (1908) on the basis of inborn errors of metabolism in humans! Garrod wrote that the 'mode of incidence of alkaptonuria finds a ready explanation if the anomaly be regarded as a rare recessive character in the Mendelian sense', and that this character was caused by the absence of an enzyme. However, the general medical opinion of the time regarded Garrod's inborn errors as quaint curiosities rather than corner stones of a new medical science, and they received little attention.

Half a century later, Beadle and Tatum received the Nobel prize for their discovery that genes regulate chemical events, and in his Nobel lecture, Beadle (1959) paid tribute to Garrod both for initiating the concept and for formulating its essential principles. But this was more than 20 years after Garrod's death.

Threshold dichotomies, quasi-continuous variation and the simulation of Mendelism
Neurospora was the workhorse that allowed the ideas first formulated by Garrod to be tested and confirmed on a large scale. However, the fungus does not exhibit any significant cellular differentiation and does not manifest the many types of normal and pathological variation, including

sex differentiation, with which we are so familiar in our own and more closely related species. Sewall Wright pioneered the analysis of meristic traits that had a clear genetic component but failed to obey Mendelian rules.

The incidence of an extra toe in guinea pigs, *Cavia porcellus*, varies in different inbred strains and can be increased by selection. Wright (1934a, b) suggested that the formation of an extra toe is due to many genes producing a physiological threshold, one side being represented by presence and the other by absence of the extra toe. A similar phenomenon, absence of the third molar, was studied by Grüneberg (1952), who called it 'Quasi-continuous variation'.

John Edwards has drawn attention to the likelihood that this mechanism is the cause of many developmental abnormalities in humans, including cleft lip, anencephaly, spina bifida and most heart defects, 'which are probably related to the unpunctual fusion of the margins of various grooves and holes, may be conditioned by continuous variations, innate or acquired, in developmental punctuality', rather than being caused by mutations in a single gene (Edwards 1960).

The relevance of this topic to the determination of sex will be discussed in the next section but one.

Sex determination in humans

The elucidation of the mechanism that determines sex in humans had to await improvements in cytogenetic techniques in the third quarter of the last century, when the XX/XY mechanism was unambiguously established. The new techniques soon led to the discoveries of chromosome anomalies, notably XXY in male patients with Klinefelter syndrome (Jacobs and Strong 1959) and XO in female patients with Turner syndrome (Ford *et al.* 1959). These findings demonstrated that, in contrast to the situation in Drosophila, the human Y chromosome, and it soon became evident, that of other mammals, functioned as a determiner of male sex.

The last quarter of the 20th century witnessed a sustained search for the 'testis-determining' gene, named *TDF* (for 'testis-determining factor') in humans, and *Tdy* in the mouse. The choice of name was based on the idea popularised by Alfred Jost (1953) that embryos of both sexes develop in identical fashion until the hitherto undifferentiated gonad becomes a testis, whose secretions would masculinise the reproductive tract. Following the findings by Jost *et al.* (1981) that the Sertoli cell is the first identifiable cell in the testis, sex determination became identified with the process that caused uncommitted somatic gonadal cells to differentiate into Sertoli cells.

After several candidate genes had been proposed and found unsuitable, the search ended with the isolation of *SRY* from the short arm of the human Y chromosome, the cloning of a corresponding gene in the mouse and the demonstration in 1991 that the addition of *Sry* as a transgene could cause XX mouse embryos to develop into males (reviewed by Graves 1994, Mittwoch 2000). However, any idea that mammalian sex could be explained by monogenic inheritance was soon dispelled by the discovery of a growing number of other genes whose correct dosage was necessary in humans and mice for *SRY/Sry* to ensure male development (Lovell-Badge *et al.* 2002, Vilain 2002). Neither has the expectation of discovering the pathway from gene to phenotype been fulfilled (Scherer 2002). These findings may be connected with the fact that sex determination is based on a process of cell differentiation.

Sex differentiation and growth

Concurrent with advances leading to the discovery of sex-determining genes, other lines of research focused on quantitative differences between developing male and female embryos. These showed that in a variety of mammalian species, testes grow faster than ovaries and that in young embryos, a difference in size could be detected before histological differentiation was evident. These findings gave rise to the hypothesis that fast growth is a prerequisite for testis development (Mittwoch *et al.* 1969, Baker *et al.* 1990)

Furthermore, a faster growth rate is not confined to the developing testes. A recent study on the development of bovine blastocysts showed that male embryos tended to reach the blastocyst stage earlier than female embryos, supporting earlier findings in humans, mice, cattle and sheep (Nedambale *et al.* 2004). Some Y-chromosomal genes are transcribed in preimplantation embryos (Erickson 1997). Burgoyne *et al.* (1995) reported that in mice, the difference in developmental rates is due both to an accelerating effect of the Y chromosome and a retarding effect of the second X chromosome.

Genes involved in testis differentiation effecting an increase in cell proliferation

The question whether *Sry* might initiate the increased growth of the XY gonad was tackled by Schmahl *et al.* (2000) in the mouse, using BrdU (5'-bromo-2'-deoxyuridine) incorporation into dividing cells. The authors reported that an increase in cell proliferation in the coelomic epithelium was the first detectable effect of *Sry* activity.

The *SRY*-related gene, *SOX9,* is a testis-determining gene that also has important functions in the development of bones (Wagner *et al.* 1994), the heart (Akiyama *et al.* 2004) and other organs (Koopman 2005). Haploinsufficiency of the gene causes campomelic dysplasia, a severe abnormality of bone development in which most XY patients present as females. In mice, the introduction of a third copy of *Sry* as a transgene can induce male development in XX individuals (Vidal *et al.* 2001), and a human XX male patient has been reported who carried a duplication of *SOX9* (Huang *et al.* 1999). An effect of the gene on cell proliferation has been demonstrated in endocardial cushion cells of mice (Akiyama *et al.* 2004). Another related gene, *Sox8*, is also involved in testis development, but absence of the gene has relatively minor effects (Koopman 2005); in mice lacking the gene, the only detectable phenotypic effect seems to be reduced body weight.

Some genes required for testis development code for growth factors. Fibroblast growth factor 9, coded for by *Fgf9*, is required in the lineage giving rise to Sertoli cells in mice, and its absence is compatible with ovary, but not testis, development (Schmahl *et al.* 2004). Female development, associated with reduced cell proliferation, has also been reported in XY mice in which all three insulin receptors were non-functional (Nef *et al.* 2003).

Cell proliferation, energy metabolism and mitochondria

The higher rate of cell proliferation necessary for male development can be assumed to require an increased supply of cellular energy, and the same applies to other events observed in testis, but not ovary, development, i.e. cell migration from the mesonephros (Buehr *et al.* 1993) and the early establishment of a vascular supply for the XY gonad (Brennan *et al.* 2002).

I have previously suggested that the function of *SRY* and other genes necessary for male development is to amplify the amount of energy available to the cells by increasing the activity of mitochondria (Mittwoch 2004). This hypothesis, which implies that *Sry* targets mitochondria rather than nuclear genes, is supported by the recent finding of glycogen deposits in pre-Sertoli cells immediately after the onset of Sry expression in mouse embryos (Matoba *et al.* 2005). It also implies that the higher metabolic rate of the male, which is intensively documented in our own species (Lentner 1981), is at the very basis of sex differentiation.

A possible role of the Y chromosome as a stabiliser of male sex

The discovery that testis differentiation is based on multiple genes giving rise to a fast rate of cell proliferation, which may be associated with high levels of cellular energy, makes a stochastic model of the biological basis of sex determination particularly appropriate. Nevertheless, sex determination differs from other threshold dichotomies in several characteristics: The sex ratio is consistently close to 1:1, in contrast to many other threshold dichotomies that do not segregate in constant ratios. It should be remembered in this connection that reports of (usually minor) variations from a 1:1 sex ratio are concerned with the XY: XX ratio rather the incidence of sex reversal (Hardy 2002). The concordance rate for sex in monozygotic twins is virtually 100% (barring very few exceptions due to sex chromosome anomalies), compared with a markedly lower rate in threshold dichotomies. Thus, in spite of a strong genetic component, the concordance rate for cleft lip with or without cleft palate is only 40%; also, most clefts are unilateral, when there are twice as many clefts on the left as on the right side (Cohen 2002).

By contrast, in sexual development unilateral manifestations are confined to the rare abnormality of true hermaphroditism (TH). Large surveys of patients (van Niekerk and Retief 1981, Krob *et al.* 1994), comprising 403 and 234 cases respectively in which the position of both gonads was recorded, revealed that TH exhibits bilateral asymmetry, with the majority of ovaries situated on the left side, and most testes and ovotestes on the right (Mittwoch 1996). The rare failure of the sex-determining process reveals its stochastic basis, modulated by the left/right asymmetry of embryonic development (Levin 2005).

It is tempting to postulate that the apparently close resemblance of sex to a monogenic condition is connected with the location of *SRY* on the Y chromosome, which minimises the spectrum of variation in testis development seen in XX male patients carrying *SRY* on one of their X chromosomes (Boucekkine *et al.* 1992). In mice there is evidence that the masculinising effect of *Sry* in XX transgenic animals is dependent on genetic background (Hacker *et al.* 1995), an effect that is not normally seen in the presence of a Y chromosome. We may assume that at the very least the location of *SRY* on the Y chromosome protects the gene from neighbouring genetic elements that threaten to interfere with its activity. But might there still be other genes on the Y chromosome, in addition to *SRY* and genes required for spermatogenesis (Okabe *et al.* 1998) that promote male development?

The Y chromosome got a bad press in recent years, having been regarded as little more than a decayed X chromosome and a redundant parasite among the rest of the chromosomes, but this view may be changing. A major collaborative study (Skaletsky *et al.* 2003) has resulted in a new image of the human Y chromosome. The male-specific region, in which there is no crossing over with the X chromosome, comprises 95% of the chromosome's length and contains at least 156 transcription units, including 78 protein-coding genes; and a more recent study points to the likelihood of some additional genes (Kirsch *et al.* 2005).

Evidently, the Y chromosome contains more genes than had been expected, and an understanding of their function will be of primary importance. In view of known and suggested functions of Y-chromosomal genes, including stature (Kirsch *et al.* 2002) and endurance running (Moran *et al.* 2004), it would not be surprising if additional genes were to increase cell proliferation and/or metabolic rate, or, indeed, that any of them might be active during testicular differentiation; but this remains to be established.

Sex and Down syndrome: a brief comparison

Down syndrome resembles the sexual dimorphism in being a cytogenetically defined entity with a high concordance rate in monozygotic twins (Penrose and Smith 1966). But how does trisomy of chromosome 21 exert its phenotypic effects? Attempts to discover which genes on this chromosome when over-expressed contribute to specific phenotypic features have proved a difficult task, as is evident from a recent report of a workshop on the biology of chromosome 21, convened to tackle this question (Gardiner 2005).

In a section on mitochondria, Jorge Busciglio reported that mitochondrial morphology and function are altered in the cells of Down syndrome patients, suggesting that chronic energy deficits may contribute to their pathology. Pursuing this approach in trisomy 21 and those of other chromosomes could provide a fruitful approach towards an understanding of the relationships between karyotypes and phenotypes.

Conclusion and outlook

It has to be admitted that, though we may have reached the post-genomic age, our knowledge of 'cytogenomics' is still in its infancy. True, during the three quarters of a century that have elapsed since Bridges (1925) made the remark quoted at the beginning of this essay, *SRY* and many other genes involved in the differentiation of the mammalian male gonad have been discovered, but the function of these genes remains unknown, as does that of the Y chromosome; and the causes of abnormal sexual development in many patients remains unexplained.

Could it be that different chromosomes carry arrays of genes that control the levels of energy available to cells, and that the Y chromosome, as suggested by Laurence Hurst (1994), has been particularly successful in attracting growth factors from other chromosomes for the benefit of male development? Could it be that, in the mammalian male, genes that increase the levels of cellular energy are the equivalent – mutatis mutandis for relevant differences in male and female growth rates between mammals and Drosophila – of the male- or female-producing genes in the balance theory of Bridges (1939)? Since effective techniques for investigating energy metabolism in individual cells are now available, their application to the study of sex differentiation could bring the subject a significant step forward.

Finally, it may be pertinent to mention that the notion of male development in humans being based on an increased energy supply is not so different from the idea held by the philosophers of Ancient Greece that males are hotter than females. I hope that George will appreciate the comparison.

Towards the responsible human being

Jean Dausset

Association française pour le Mouvement Universel de la Responsibilité Scientifique MURS France,
Hôpital Saint-Louis, Paris

Dedication

Of course, I recollect with emotion our meeting in Mexico in 1994. I would ask you to transmit all my friendship to Maria and I wish you a Happy Birthday on 3 March.

I congratulate you on your whole career, both as a geneticist and as a humanist. Since long ago, you have been an honorary member of the Mouvement Universel de la Responsabilité Scientifique (Worldwide Movement for Scientific Responsibility) in view of the fact that you contributed to the International Conference *Biology and the Future of Man* in 1974 when the MURS was founded, and that you helped the Rector of the Sorbonne, Robert Mallet, for several months in 1975 in the office of the MURS in the Boulevard St Germain. You are therefore a 'mursian' of greater seniority than I am. Thus, Robert Mallet was the first president from 1974 to 1983, then I from 1983 to 2002, then Gérard Mégie for two years until his death, and now I am president again.

Each human being is unique, life is unique. Each human being is unique, life also is unique. How can we reconcile this apparent paradox?

It is this unique character which confers on Man his dignity and therefore the respect which is owed to him. It is on this dogma that all philosophies are based, in particular our Western philosophies, born of Christianity.

Until now, to think that each human being is unique was only an intuition, rooted very deeply in the minds of the first men, since they attributed an eternal life to the souls of the dead.

But no scientific proof came to the support of this 'evidence'. This proof was provided in 1958 by the discovery of the human tissue groups, those whose identity between donor and recipient we must respect, in the case of an organ transplant. Each one of us is, in fact, the carrier of a unique combination of the genes of the HLA system (with the exception of identical twins). This is a veritable identity card or an individual bar code.

As for life, let us note straightaway that it has been designated throughout time and by all peoples by one sole word: LIFE, felt by all to be an indivisible entity.

The demonstration of the unique nature of life has been long suspected by good minds in the light of the obvious similarities among the members of each line of living beings. But the concept of evolution of one species into another was only formulated very late on. We remember the ideological battle, not so long ago, in about 1830, between Cuvier, fixist/creationist, that is to say champion of a divine immutable creation of all species, and Bernard de Saint Hilaire, pupil of Lamarck, author of the first theory of evolution by progressive maintenance of acquired traits. It was necessary to wait for Darwin to understand that this maintenance is due to the natural selection of the fittest.

But the mystery of life stayed complete until the discovery of DNA, this chain composed of four, and only four, chemical compounds (the nucleotides). And the true secret was only unveiled when Watson and Crick showed that this chain was double (the double helix) and that in uncoiling, it made the transmission of hereditary traits possible. We still had to understand how to decipher this message. The genetical code was discovered. It is formed of four nucleotides and corresponds to the 23 amino-acids of which proteins are composed.

All living beings possess a chain formed out of the same nucleotides, and the code is universal.

And this is the demonstration of the unique nature of life. Marvellous adventure that it is, life, which began 3.5 billion years ago, became progressively diversified through 'a game of chance and of need' (Jacques Monod). We have a deceptive impression of a great harmony of nature. But we pretend not to know that, in fact, it is a horrible jungle. The animals eat each other and, in addition, only survive thanks to the plants which, alone, are able to transform solar energy into biological matter. And the plants themselves struggle with each other for their share of sunlight!!! And all this without taking into account parasites of every kind.

And whatever is Man doing here? He is a predator just like all the other animals.

However, Man, in the course of millennia, has known how to sublimate his instincts in order to become a rational being on whom we pride ourselves. How has this miracle come about just with the assistance of the gentle modifying influences provided by survival of the fittest?

It is a miracle attributable to the upright position which has been its good fortune and its motor. Hands, once liberated, allow many experiences and apprenticeships, and much memorising and reflection, all to be transmitted to the young. The descent of the larynx and the formation of the resonating cavity of the pharynx have allowed speech to develop. Then, the volume of Man's brain increased progressively.

And it is here that we must find perhaps the most important consequence of the upright position, which has long gone unrecognised: it is the downward inclination of the front of the pelvis. As a result, the newborn, with an ever-increasing brain size, has a more difficult passage through the birth canal. To counteract this, the human infant is born immature, unable like a chicken, for example, to scurry around immediately after delivery.

This immaturity of the newborn will be put to good use to fashion the network of his neurones of which many are available at this stage for any experiment. The infant adapts to his environment, listens to, and soon learns, the language of his parents (we know that wolf-children cannot learn a language). The extreme importance of this period of life, and the essential role of the mother and of teachers, becomes understandable.

In the course of millennia, the material ingenuity of Man has developed. Soon, he became aware of himself. His instincts were sublimated into abstract notions. For example, the sexual instinct which encourages the partners to choose each other, according to the criteria of the norms of the species. The male showing the greatest conformity is systematically chosen, and it is the same for the female: we would say that they are the most beautiful. The sense of the beautiful is universal, even among the most primitive peoples.

In the same way, benevolence doubtless flows from gregarious instincts which imply devotion, abnegation and even sacrifice.

Can we see traces of this evolution of the human spirit during the course of the past 10–12 millennia on which we have some information, first archaeological, then historical?

A very long time before the historical period, human beings had the idea of the beyond. They buried their dead. They deified the forces of nature which they had to appease by sacrifice. The notion of evil (which is not possible except within society) appeared. The soul of the evil-doer had to pay eternally in the beyond, or had to be reincarnated in a being recognised as inferior. The definition of evil-doing is very variable between cultures, but crimes of blood represent a constant component.

Respect for the other culminated in the monotheistic religions. The notion of justice is very ancient, but that of love of the neighbour was particularly exalted in Christianity. The notion of equality between human beings is very recent. It dates from the Enlightenment. We need to take account of the fact that until then, in the absence of any power other than that of muscle (horse or Man), society had need of common labourers, first slaves and then subordinates – even Jesus Christ did not condemn slavery. It was the European philosophers who were the pioneers in this respect. The French Revolution established the relevant social principles.

At that time, equality between human beings was proclaimed, and, therefore, between men and women, slavery was abolished, the rights of the oldest child with respect to inheritance were abolished, the liberty of thought and of expression was proclaimed. And, as a result, fraternity. A beautiful programme which has been, and still is, systematically flouted.

Condorcet 'outlined the progress of the human spirit' since the beginning of time, passing by the dazzling Greek civilisation, the wise men of Rome, Saint Augustine, etc …

Condorcet went further. His prophetic views considered the social aspects deliberately forgotten by the revolutionary parliamentarians. He foresaw the Social Security programme, the Right to Housing, etc …

Despite these beautiful promises, the industrial revolution brought into being growing injustice: the odious child slavery of the first factories in England and France, so well described by Flora Tristan, the grandmother of Gauguin. An unimaginable exploitation of wretched poverty; seven hours of sleep for the workers in the spinning-mills of Lyon.

This contempt for human life continued for a long time in the West. The deaths of numerous illiterate soldiers counted for little. Obligatory education represented a great victory.

It was necessary for the First World War to happen for the hecatomb to bring about reflection, if not remorse, in the international conscience. A second significant advance of the human spirit: the League of Nations, with its headquarters in Geneva, makes pronouncements on great principles such as the status of prisoners of war, the prohibition of torture, protection of civilians in wartime … Democracy, the vote for women, etc …

Principles which were quickly forgotten in the face of the rise of fascism and of communism. A Second World War brings its share of suffering and of horrors never before imagined – the Holocaust and the Gulag – bringing about a fresh upsurge of the collective conscience. A veritable international legislation is established within the framework of the United Nations with its many agencies (WHO, UNESCO, FAO, UNICEF, the refugees (UNHCR), the atomic agency (IAEA), and others). The Universal Declaration of Human Rights of 1948 brings great principles to the fore, in particular the abolition of the death penalty. In this rush, other advances are secured such as the right of interference in the affairs of other states, the status and protection of charitable works, the prohibition of anti-personnel mines, equality between men and women.

In the face of lightning-like technical progress, the first ethical committees are created.

One can remain sceptical, especially in the face of the present unfurling of violence, but one needs to do justice to the remarkable effort made during the course of the past century to try to establish international rules which can be qualified as representing progress. But is it really a question of 'progress of the human spirit' or of simple rules of the game? I maintain that it is a question of victories of reason, which are due to responsible individuals, perhaps Utopian but also visionaries.

Let us understand well: it is not during the course of a few centuries – a mere drop of time on the geological scale – that the human brain has been modified; the brains of the Greeks of the 4th century before our era were probably identical to ours. But collective thought has made considerable progress in the direction of respect for every human being, whoever he is, and of peace among men. We can, therefore, vouch for a 'progress of the collective spirit of Man' but we may doubt whether any profound anatomical modification of the human brain has occurred in such a short time, a modification which, in any case, would only have involved a very small proportion of the inhabitants of our planet.

It seems to us that important consequences for the human spirit should develop in the centuries to come from the generalisation of technologies, including those of which we know nothing as yet and which may turn out to be as astonishing as those which we have experienced during barely a century, in particular the most recent ones: information science, the mobile telephone, GPS, nanotechnology, etc … It is enough to see the skill with which children adapt to them in their play. In one generation, the modalities of communication, and even of human relationships, have evolved. An abolition of time and space.

Who can foresee the influence of this new situation on the human spirit? Being useful and attractive, these technologies will rapidly become generally available to the totality of humankind, unless catastrophes occur which are unforeseeable or, unfortunately, provoked by Man himself: change of climate, atomic bombs, or others.

It is probable that this new environment, this new way of life, will give rise to new neuronal connections which, if they are transmitted, will favour the more fit. One can also conceive that scientific knowledge of cerebral mechanisms will reach a stage where Man will not be able to hold himself back from intervening. But such interventions would only involve a small number of individuals and not the species. By analogy with reproductive cloning, these interventions, necessarily limited in their scope, will not be incorporated into the genome.

So, is there reason to be optimistic or pessimistic?

Certainly, the present world situation can hardly be regarded as being encouraging. Violence persists and grows. Despite international principles and laws, poverty and wretched misery have not been conquered. Many signals show red: over-population, climate change, deforestation, desertification, drastic reduction of biological diversity, scarcity of fresh water, etc … Has humanity launched itself into a fatal cycle which threatens to escape its control, so great has its power become and so small will its planet, this small pebble, remain? There is no doubt that this alarming situation is in large part due to the prodigious resources which scientists have bequeathed to the human beings of this century. Is science at fault?

The answer is certainly NO, certainly NOT in having provided new knowledge, this being the honour of Man, but perhaps scientists are indeed at fault in having allowed this knowledge to be applied in an undiscerning and anarchic manner.

But this situation is perhaps only a crisis of adaptation, or at least we may hope so, doubtless extending over a very long period because it can only develop according to the slow rhythm of the succession of generations. The duty of scientists is, therefore, to accelerate this adaptation in emphasising to the public the benefits to be expected from the new technologies and in alerting those who make decisions to the risks inherent in their premature and inappropriate utilisation.

If we take no notice of these years of adaptation and consider what could happen in the longer term, if science continues its fabulous ascent, applying itself judiciously to the benefit of humanity, then we become resolutely optimistic.

The human spirit will have mastered its new way of life. The progress of the neurosciences may perhaps be able to favour evolution in the direction of what we have elsewhere called 'sublimated' Man, that is to say one who has known how to sublimate, extend, develop his innate abilities, thus able to establish a superior order, dominated – albeit in a material world – by the incorporeal.

Will humanity then find the wisdom to overcome its rivalries and conflicts, and to expand the treasure represented, even today, by the universal morality created around the profound need of the human soul to rise even more above its animal nature? There exist universal values recognised and venerated by all human beings whatever their culture, whatever their religion, whatever their philosophy, culminating in the essential: the end of contempt, that is to say, a true tolerance without disdain, accepting the other with all his differences, without ulterior thoughts of some kind of superiority.

I wish for the alliance of the right brain and the left brain, that is to say of poetry and of reason as well as the alliance of all brains united by information science, an alliance which could permit the attainment of a new evolutionary stage of the sublimated human being, responsible for himself and for his species.

Man has a need to give a meaning to his life; it seems that there is more and more of a need of spirituality. Would not the responsible man have need of it? Freed from all myths and superstitions, he also has a need to dream, to sublimate (yet again!) his rationality. I believe that this need can be satisfied by the cult of beauty, physical as well as moral, and by a tenderness towards humanity (some might say compassion, a term to which I take exception because it carries a heavy historical load); tenderness suits me: it evokes the indulgent regard of the mother.

This contribution is a translation from French by G.R. Fraser of a paper in *Cahiers du MURS*, number 43, (Second Semester 2004), pages 43–48, entitled *Vers l'homme responsable*.

Is modern genetics the new eugenics? [1]

Charles J. Epstein

Department of Pediatrics, RH584B, University of California, San Francisco, San Francisco, CA 94143–2911

On 25 April 2003 we celebrated the fiftieth anniversary of the elucidation of the structure of DNA by Watson and Crick, an accomplishment of which we are all extremely proud – and justifiably so. A vast amount is being written and said about the significance of this discovery and about where genetics in general, and human genetics in particular, are now and where they are going.

Along with the celebration there has been much speculation about what the future will bring. For example, the next fifty years will bring, according to *Time's* consultants, food without fat, cellular therapies and new organs, the $100 personal genome, slowing of the ageing process, germ-line genetic engineering, and reverse engineering of the human brain (Anonymous 2003a). As might be expected, speculations such as these have engendered considerable concern – how much, how fast? (Kolata 2003) What are the ethics and morality of making new and better babies? Is this a return to the eugenics thinking of the past?

In 1990, in a book describing his view of human genetics, Berkeley sociologist Troy Duster spoke about a *back door to eugenics*, one that is made up of 'screens, treatments, and therapies,' and in 1997, Arthur Dyck, a professor of ethics at Harvard, wrote that

'Science, medicine and law at present willingly provide the information, rationale, and technical know-how for current eugenic practices in the United States, some of them quite coercive and arguably unethical … Eugenics is not simply a matter of history. Eugenics is practiced today … [and] the very ideas and concepts that informed and motivated German physicians and the Nazi state are in place.'

Dyck and Duster were not alone in telling us that eugenics is actively being pursued in the practice of human and medical genetics. For example, Dorothy Wertz (1998) said it outright: 'Eugenics is alive and well.' The basis for her assertion is that whereas only some geneticists regard what they are doing as being eugenics, counselling for prenatal diagnosis is 'pessimistically biased' or 'slanted' and counsellors have a 'pessimistic view of persons with disabilities,' – perhaps not so much in the English-speaking countries, but certainly in the rest of the world. Similarly, Holden (2003), in reporting a survey on cloning, tells us that 'Eugenics in gaining broader acceptance overall,' in this instance equating eugenics with prenatal diagnosis for desirable traits and the use of genetic engineering to produce these traits. And, in a comment cited in an article reporting that the governor of Virginia recently apologised for Virginia's 1924 law authorising involuntary sterilisation for eugenic purposes, Barbara Bieseker is quoted to the effect that prenatal diagnosis may be operating in a 'milieu of personal eugenics' (Fox 2002).

The worst accusation that can be levelled against modern human genetics and medical genetics is that they are eugenic – if not a literal return to the eugenics of the past, at least a reincarnation of that eugenics in a new guise. The mere use of the word 'eugenics' brings forth very visceral responses. Dawkins (1999), of *The Selfish Gene* fame, tells us that, 'If cannibalism is our greatest taboo, positive eugenics ... is a candidate for the second ... In our time, the word [eugenics] has a chilling ring. If a policy is described as "eugenic," that is enough for most people to rule it out at once ... ' And, according to Diane Paul (1998), 'the term is wielded like a club. To label a policy "eugenics" is to say, in effect, that it is not just bad but beyond the pale.'

How should we respond?

I must confess that I was quite taken aback when I first heard human genetics being equated with eugenics. Indeed, when in 1998 the American Society of Human Genetics was discussing the 1994 Chinese eugenics law, it did not really occur to me that anyone would construe what human and medical geneticists were doing in the United States as being eugenic. My first response was to deny the eugenics association out of hand, thinking that it might be more an issue of semantics than of substance. Then, when I began to look further into the matter, I went through a period of self-doubt and disbelief, perhaps even guilt, and I found myself beginning to identify with the critics and to question everything that I had taken for granted.

However, with the passage of time, I regained my sense of balance and, while acknowledging that the critics have raised many important issues, came to the conclusion that the human genetic enterprise is basically sound. I also came to the realisation that if human and medical geneticists are to make an effective response to the critics, they – and they are we – must first listen carefully and non-defensively to what they have to say. I think that we should try to avoid the stance that some of our most prominent scientists took when confronted with public concern about recombinant DNA research (Turney 1998). Their approach was to attribute the public's reservations to a combination of anti-science and ignorance. There's nothing wrong with us – it's the rest of the world that's wrong! There may well be kernels of truth in what they said, but I do not think that making the anti-science claim is going to be a productive way to deal with the issues and it certainly will not resolve them. A better response would have been be to ask whether the accusation is a meaningful one, whether – in our case – it is really true that we are setting goals and engaging in practices that could – perhaps 'should' would be a better word – be considered as eugenic.

What is eugenics?

The term 'eugenic', originally suggested about 1883 by Galton, literally means 'well-born', and if we were dealing with only this definition, there would perhaps be little to talk about. Much of what is done in life has just this goal in mind – to have healthy children free of disease and with prospects of leading productive and fulfilling lives. If looked at this way, most human societies could be regarded as being eugenic, with their members taking a large number of steps to increase the likelihood that their children will indeed be wellborn, even though none of these activities has historically been regarded as being eugenic.

But the definition of eugenics goes beyond wellborn and introduces the notion of process as well as intention, of what we are willing to do to ensure that the children will be wellborn. Thus, even in Galton's original usage, the concepts of 'cultivation of race' and the production of 'men of a high type' were present. (*Oxford English Dictionary* 1970) In modern dictionaries, 'eugenic' is defined as 'relating to the production of good offspring,' and 'eugenics' as 'a science that deals with the improvement (as by the control of human mating) of hereditary qualities of a race or breed.' (*Webster's* 1981) Thus, the emphasis is on the control of the genetic properties of future offspring. Furthermore, it is implicit in all definitions that the traits that are the subjects of concern are indeed under genetic control and that they can be altered by making genetic changes and by improving the quality of the gene pool. Social and other environmental factors were thought to have relatively little effect.

The early concepts of eugenics as developed principally in England were derived from the belief that the upper (Anglo-Saxon) social classes were in danger of being diluted by the expansion, because of their higher birth rates, of the lower social classes and the allegedly inferior races. Galton's proposed solution to the problem was along the lines of positive eugenics – the encouragement of the breeding of those among the upper classes who possessed the desirable characteristics. However, he also allowed for more than that – for, as he put it, 'whatever tends to give the more suitable races or strains of blood a better chance of prevailing over the less suitable than they otherwise would have had.' (Buchanan *et al.* 2000) Thus, even if the original notion was of positive eugenics, the actual implementation of eugenic principles very quickly began to run along negative eugenic lines. Rather than permit the Darwinian survival of the fittest to control the gene pool, the object was to ensure the non-survival of those considered to be unfit.

The principal means of implementing negative eugenics were by discouraging or preventing the reproduction of the unfit by prevention of marriage and of racial mixing, institutionalisation, sterilisation, and sometimes castration. Added to this were quotas on the immigration of the supposed unfit and a general stigmatisation of and discrimination against them. And, if this was not sufficient, abortion was employed to prevent their birth. Although all of this was theoretically voluntary, it rapidly became compulsory in many countries abroad and in many states in the United States.

However, it is, of course, in Nazi Germany that eugenics reached its ultimate and most terrible application – the Holocaust, with its wholesale extermination of millions of Jews, gypsies, homosexuals, mentally retarded, epileptics, mentally ill, and others deemed unworthy of life and certainly of reproduction. This was to be the ultimate cleansing of the gene pool. No longer was breeding of the undesirable to be controlled – rather, the breeders who were thought to carry the undesirable genes were to be eliminated altogether.

As this main-line eugenics was flourishing in Nazi Germany, its popularity in other countries was decreasing. It was replaced by a so-called 'reform eugenics' that did not include the Nazi approach and called into question the unscientific basis of much of the original eugenics and rejected the blatant racism and antifeminism that characterised it. However, this was not a rejection of the basic goal of eugenics but only of the excesses and unscientific attitudes that characterised it. So, despite the change in attitudes, laws providing for voluntary (although very often coerced) and compulsory sterilisation were passed in more than half of the American states and in several countries of northern Europe, and sterilisations under these laws were carried out into the 1960s – right here in San Diego [the city in which the address from which this contribution was taken was given]! (Abate 2003).

In the latest iteration of reform eugenics, it was argued that what was required was a combined positive and negative eugenic approach – increasing the proportion of children born to those who achieve most and decreasing the proportion of children born to those with least achievement, all of this to be voluntary, of course (Osborn 1968). In addition there was to be a voluntary reduction of the incidence of deleterious genes by means of the heredity clinic (which is equated to genetic counselling), institutional care, and the expanded use of contraception. And, lest there be any misunderstanding, modern genetic counselling is described as being 'in a special sense a eugenic activity ... a form of negative eugenics, in that it attempts to prevent the conception or birth of individuals with most serious forms of maldevelopment ... '(Osborn and Robinson 1987)

Eugenics as practised was intrinsically wrong

Given its history, there is no question that eugenics as it was actually practised was wrong. Whether positive or negative, it violated fundamental human rights in numerous ways. It was coercive, discriminatory, and racist. It substituted control by the state for personal choice and autonomy. It denigrated and stigmatised the disabled and others deemed, for whatever reasons, as being unfit. It deprived people of their ability to reproduce, of their freedom, and, in the extreme, of their lives. The effects of the various eugenics movements were widespread, and they ranged from terrible to horrific.

Eugenics as conceived was intrinsically wrong

Not only was eugenics flawed in its applications, it was flawed in its basic genetic assumptions. The first was the belief in genetic determinism, the belief that all of the undesirable diseases and traits that the eugenicists were anxious to eliminate were genetically predetermined and, as a result, not alterable by environmental alteration. This notion is a major conceptual foundation for eugenics and, like eugenics, is a target for attack in its own right. The second flaw was the assumption that there were simple genetic explanations – in the extreme, recessive inheritance – for many, if not all of the traits of greatest interest to the eugenicists – criminality, alcoholism, psychiatric illness, and mental retardation. Indeed, as eminent a geneticist as R. A. Fisher argued that the incidence of 'feeblemindedness' could be reduced by as much as 17 to 36% in one generation by 'segregation or sterilisation' of affected individuals (Fisher 1924).

In addition to being genetically wrong, eugenics as originally conceived was also morally wrong. This was the conclusion reached by Buchanan, Brock, Daniels and Wikler who conducted an interesting *ethical autopsy of eugenics* (Buchanan *et al.* 2000). In this 'autopsy' they considered many issues of importance to ethicists, the most important to them being the issue of justice, and this is what they had to say:

> The eugenics movements of 1870–1950 insisted ... that humankind ... stood to gain a large benefit (more able, fit people) if humans would submit to the kind of breeding programs that had been used to improve plants and livestock. But who would benefit, and at whose expense? The internal logic of eugenics provides the answer. The 'underclass' is simultaneously the group of people whose genes were not wanted and the people who, through involuntary sexual segregation, stigmatisation and denigration, sterilisation, and even murder, paid the price. The injustice of this distribution of benefits and burdens is evident.

From this and the considerations discussed earlier, I think that it is fair to conclude that eugenics both as actually practised and as originally conceived was intrinsically wrong and now want to turn to the question posed in the title of this address: is modern genetics the new eugenics? This is indeed a very broad question, and I shall restrict myself to consideration of just one aspect of modern medical genetics – prenatal diagnosis. The questions then are whether prenatal diagnosis is eugenics and, more particularly, whether it manifests the attributes that made eugenics intrinsically wrong.

Is prenatal diagnosis eugenics?

In the strictest sense, the question of whether prenatal diagnosis is eugenics translates into another question – does prenatal diagnosis seek to ensure the birth of well-born children? – and the answer is quite straight forward. *Yes*, it does indeed seek to ensure the birth of well-born children! However, if we go to the broader conception of eugenics as seeking to improve the genetic qualities of populations, the answer is clearly *no*! Populations and gene pools are not the point at issue.

So much for definitions! We can then ask whether prenatal diagnosis is based on the basic assumptions underlying the eugenics of the past – unproven genetic mechanisms and genetic determinism. As currently practised, the answer again is *no*.

However, these are not really the most important questions. The real issue is whether prenatal diagnosis as currently practised exhibits the attributes that made the practice of eugenics wrong in the past: coercion and abrogation of reproductive freedom, state control, discrimination and racism, denigration and stigmatisation of the disabled, deprivation of life. Let's consider each of these in turn, starting with the matter of coercion.

Several commentators have suggested that prenatal diagnosis and selective abortion are subject to a variety of forces that are coercive in nature. Their concerns can be summarised in five statements:

What is routine is, by its very nature, coercive.
The very fact that prenatal diagnosis and screening are widely available may be considered as implying that they *should* be used (Green and Statham 1996).

The context in which they are operating prevents counsellors from being truly even-handed and unbiased when laying out possible options.
The existence of a test may be regarded as implying not only that it should be used, but also that the condition being screened or tested for is one that should be avoided. Therefore, even with the best of intentions, it is claimed, counsellors cannot truly act in a non-directive manner when it comes to helping the potential parents to decide whether testing should be done at all or, if an adverse test outcome is obtained, whether the pregnancy should be terminated (Duster 1990, Harper and Clarke 1997, Appleyard 1998).

Pressure to be tested is exerted by those who offer prenatal testing, often for their own purposes.
This is based on the assertion that the rationale for offering prenatal diagnosis and screening is not only or even really medical, but also legal – to ward off potential malpractice suits (Dyck 1997).

When foetal abnormality is detected, there may be medical or social pressure to abort the pregnancy.
Although this pressure may sometimes be overt, for most pregnant women with abnormal foetuses, any external pressure to abort is more likely to be the result of perceived societal pressure against having abnormal babies (Green and Statham 1996, Hume 1996, Hubbard and Wald 1999).

Even in the absence of external pressure, the process of prenatal diagnosis has an internal momentum that drives toward abortion.

Given what has already been said about what the existence and performance of prenatal testing imply, the contention that 'it is very difficult to get off the roller coaster once embarked' can readily be understood. With minor exceptions, the process is usually undertaken with the expectation – implicit if not explicit – that abortion will be the likely outcome if foetal abnormality is diagnosed (Green and Statham 1996).

In sum, then, although there is no claim that women are truly coerced into undergoing prenatal diagnosis, I do not think that we can avoid the fact that there are indeed forces at work, some subtle and others not so subtle, that do exert a coercive force toward utilisation of prenatal diagnosis and termination of pregnancy if an abnormal foetus – however that is defined – is detected.

As you can see from this list, the genetic counsellor is perceived as playing a major role in the prenatal diagnosis process and his or her ability to be neutral or impartial, as current dogma requires, has been called into question. Just what the counsellor should be doing is, of course, a very major question.

I now turn to the second negative attribute of eugenics practice and ask the following question: what is at the role of the government in prenatal diagnosis? With the possible exception of China and a few other countries, governments do not appear to be mandating prenatal diagnosis and selective abortion. Nevertheless, governments are very much in the business of prenatal diagnosis, and it is necessary to consider why this is. Duster (1990) encapsulates the issue very neatly.

When the state pays for genetic screening, counselling, and treatment, it becomes the 'third party' in the transaction. Here is the final inherent contradiction: the state cannot both (a) insist that such genetics disorder control programs are designed to assist the individual, whatever his or her choice, and then (b) argue cost effectiveness of service utilisation on the grounds that such services as utilised in the aggregate cost the public less money.

This contradiction that has troubled me quite a bit, and the paragraph above a table from a report on the California triple marker screening program is an example of what about it has troubled me. 'It is useful to reflect on the missed opportunities for the avoidance of birth defects ...' (Cunningham and Tompkinson 1999). And what missed opportunities are being spoken about? The table make this clear – they are the numbers of women who did *not* elect to terminate a pregnancy after the detection of an abnormal foetus – about 50% overall with a chromosomal abnormality and 30% with a neural tube defect. However, Cunningham and Tompkinson (1999) are probably correct in asserting that cost-benefit analyses of genetic services are probably here to stay, and the New England Regional Genetics Group Social and Ethical Concerns Committee (1999) agrees. Therefore, it would augur well if the cautions of this group could be heeded. They point out that cost-benefit analyses attempt to convert seemingly incommensurable units (costs and benefits to patients, families, and society) solely to monetary terms. In doing so, they have several ethical limitations, in particular the frequent overlooking or inadequate weighting of non-monetary costs and benefits and the omission of considerations of equity and fairness because costs and benefits are aggregated across all individuals concerned. The latter results in benefits to individuals being compared with benefits to society, a situation that may not be the most ethical one if the interests of society are regarded as paramount. Genetics, which deals with analysis of risk and variable outcomes, is considered to be 'particularly recalcitrant to accurate and value-neutral' analysis.

Nevertheless, perhaps we need to look past the rhetoric and consider what might be accomplished. If we can agree that it is good to give prospective mothers the opportunity to have their pregnancies tested – if that is what they truly wish, if their choice to do so is truly informed and voluntary, and if they fully understand the implications of being tested – then perhaps we can look beyond the cost-benefit analyses and accept the state-run screening programs on that basis. If such analyses are what are needed to satisfy the requirements of governmental due diligence, so be it, but let's not believe that they are what truly matters.

I am going to skip over the issue of discrimination since I do not see it as a major problem except possibly in a reverse context. If discrimination exists it is when certain ethnic groups are being *denied* access to prenatal diagnostic services rather than being forced to undergo it. However, the next question is of paramount importance. Does prenatal diagnosis denigrate and stigmatise the disabled? Prenatal diagnosis programs, however they may be justified, clearly do have as their goal the prevention of the birth of children with various types of disability. Similarly, even if not targeted against specific racial or ethnic groups, negative eugenic policies were certainly aimed at persons with disabilities, particularly those associated with mental retardation and illnesses believed to be hereditary. It is, therefore, not difficult to understand how prenatal diagnosis has been viewed as allying itself with the eugenic denigration of the disabled and as implying (rather than stating openly, as did many of the eugenic programs of the past) several negative things about the disabled: that they, rather than their disability or their treatment by society, are the problem (Parens and Asch 1999); that they are less desirable members of society and thus have no place in our society (Headings 1998); that disability is inherently bad and that people with disabilities lead blighted, tragic lives (Wikler and Palmer 1992); and that their lives are not worth living and, therefore, that they should not exist (Nelkin and Lindee 1995, Appleyard 1998).

Therefore, the critics believe, the new genetics and old eugenics are not very different in that they both promote exclusion of the disabled rather than inclusion (Buchanan *et al.* 2000). Accusations against prenatal diagnosis and genetic counselling based on these types of thinking have been made by several groups of persons with congenital conditions – [I had to catch myself when I was writing this. I originally wrote 'defects', a clearly prejudicial term in the present context.] – particularly by persons with dwarfism, spina bifida, and deafness, as well as by the broader disability rights community. To get a real sense of how a group of disabled person feels, we should look at the *Position Statement on Genetic Discoveries in Dwarfism* posted by the Little People of America (2003) on their website:

> What will be the impact of the identification of the genes causing dwarfism, not only on our personal lives but on how society views us as individuals? … Some members [of LPA] were excited about the developments that led to the understanding of the cause of their conditions, along with the possibility of not having to endure a pregnancy resulting in the infant's death [from homozygous achondroplasia]. Others reacted with fear that the knowledge from genetic tests such as these will be used to terminate affected pregnancies and therefore take the opportunity for life away from ourselves and our children.

The Little People of America have had a long and positive history of interactions with medical genetics. Therefore, one cannot read such a statement without coming away with the feeling that persons with disabilities are indeed deeply threatened and affected by the rhetoric of the proponents of prenatal diagnosis. However, this is not all that the statement says:

> The common thread throughout the discussions was that we as short statured individuals are productive members of society who must inform the world that, though we face challenges, most of them are environmental (as with people with other disabilities), and we value the opportunity to contribute a unique perspective to the diversity of our society.

If there is any question in your mind that this is an important and highly emotional issue, the cover from the *New York Times Magazine* of 16 February 2003 with the headline, 'Should I Have Been Killed at Birth? The case for my life', should convince you otherwise. Although the article itself is concerned with a debate between Princeton philosopher Peter Singer and the author, Harriet McBryde Johnson, whose picture in a wheelchair is shown on the cover, over the euthanasia of newborns with disabilities, it might just as well have been about prenatal diagnosis and abortion.

I have thought a lot about the arguments that are encompassed by the disability rights critique (see Appleyard 1998 for summary). This thinking has been particularly focused by my involvement with the National Down Syndrome Society which brought me into extensive contact with many persons with Down syndrome and with their parents and friends – not in the formal role of physician and teacher and research scientist, but as their advocate and friend. In the latter capacity I was forced to look at the world from their point of view and to ponder why it was so uncomfortable for me to speak about prenatal diagnosis in their presence. I shall not go into the arguments on both sides of the disability rights critique. Nevertheless, I do have to say that while I do *not* accept the arguments of the disability rights critique as they apply to prenatal diagnosis, I am nevertheless greatly troubled by what the critique says about how human and medical geneticists are perceived and about what they – about what we – are doing. Gillam (1999) said it very well:

> For people with disabilities, to have someone else look at their lives from the outside, and make judgments about how fulfilling and how happy they are, must be deeply offensive … However … the fact that a practice is offensive to some section of the community does not make it morally wrong to engage in it, or make it to be discriminatory to the minority … This does not mean that offence or psychological distress caused to people with disabilities does not matter. It certainly does, and there is good moral reason to try to avoid or at least to minimise such effects.

I come now to the final question regarding prenatal diagnosis and eugenics – does prenatal diagnosis involve deprivation of life? The answer, in real terms, is certainly *yes*. Whatever the theory might be with regard to prenatal diagnosis as merely providing information, prenatal diagnosis and abortion are inextricably linked. Without entering into the arguments pro and con, I have made my peace with the right of parents to terminate pregnancies that are unwanted. Nevertheless, I must confess that I agree with Bryan Appleyard that 'whatever one's feelings about abortion, everybody must agree that this does not amount to a very positive achievement for this new form of medicine' (Appleyard 1998).

So, is prenatal diagnosis eugenics?

Does it seek to ensure the birth of well-born children? YES, it certainly does!

Does it seek to improve the genetic qualities of populations? NO, it does not!

Does it adhere to the basic assumptions underlying the eugenics of the past – to unproven genetic mechanisms or to genetic determinism? Again, the answer is NO.

Does prenatal diagnosis exhibit the attributes that made the practice of eugenics intrinsically wrong?

Is there coercion and abrogation of reproductive freedom? I would say NO, but, as I mentioned earlier, there may well be coercive elements at play.

Is there state control? Again, the answer is NO, but the state – in the literal sense, the states – do seem to have a vested interest in the outcome.

Is there discrimination and racism? Again, NO. These do not appear to be major issues except perhaps from the point of view of lack of access to services rather than over-utilisation.

Is there denigration of the disabled? I would say NO, but there are many who feel very strongly that it does occur.

Finally, is prenatal diagnosis murder? Prenatal diagnosis may lead to abortion and deprivation of fetal life, but NO, I do not equate this with either euthanasia or murder.

So, by the strictest definition – 'eugenic' being equated with 'well-born', prenatal diagnosis is eugenics. But, it is not concerned with populations, it does not adhere to the basic assumptions underlying eugenics of the past, and, as currently practised, it does not truly exhibit attributes that made the practice of eugenics intrinsically wrong. Therefore, from my point of view, prenatal diagnosis is, at worst, eugenics in name only. BUT!!! ... But, I would hope that we all agree that there is still much for us to think about with regard to what we are now doing, and even more to think about with regard to what we might be doing in the future – about carrier screening before marriage, presymptomatic genetic testing, pre-implantation diagnosis, gene therapy, the search for genetic components of behaviour and intelligence.

My view of how the genetics community should address the public concerns about a resurgence of eugenics is philosophically quite close to the view expressed by Harper and Clarke in the introduction to their very insightful and valuable volume, *Genetics and Society and Clinical Practice*, which appeared in 1997, and I shall close by quoting their comments on the subject.

... a questioning and critical attitude, both to new developments and to established concepts, is important and necessary in a field like medical genetics, which inevitably impinges on so many controversial areas ... There is all the more reason to maintain a critical attitude in the light of the disastrous abuses that have already been carried out in the name of genetics by professionals and by an entire social system in the past ...

We need to keep alive our awareness of these past abuses and to maintain our vigilance that new developments are not misused in the future ... We feel that it is much healthier for this questioning to come from within the genetics community, rather than for those in the field to 'close ranks' against external criticism.

Dedication

I am pleased to contribute this article on eugenics to the Festschrift for George Fraser, a colleague for many years who has greatly advanced the scientific, clinical, and ethical basis of modern human and medical genetics.

Acknowledgement

The research for this presentation was carried out in part in 2000 while I was a Scholar in Residence at the Rockefeller Foundation Bellagio Study & Conference Center, Bellagio, Italy.

Note

1. Presidential address: Delivered at the 2003 Annual Clinical Genetics Meeting of the American College of Medical Genetics, San Diego, CA (Epstein 2003)

The genetic counsellor: a personal view

F. Clarke Fraser

Professor Emeritus, Departments of Human Genetics and Pediatrics, McGill University, Montreal, PQ, Canada*

It gives me great satisfaction to contribute to this Festschrift and *liber amicorum* for my old friend, George Fraser. In my contribution, I will trace the emergence of the category of genetic counsellors, mainly from a Canadian perspective, and muse upon some of the complexities that counsellors meet that test the professional guidelines set up for their direction. George will be familiar with these complexities in that throughout his career, he participated actively in genetic counselling. In fact, such activities became his full-time occupation during the eight years before his retirement when he created and directed a Cancer Genetic Clinic in Oxford, one of the first such clinics in the United Kingdom, between 1990 and 1997 (Fraser 1999).

Genetic counsellors are at the interface between the knowledge provided by genetical research and its application to the problems of particular families. The original genetic counsellors were geneticists, usually in biology departments, who were approached by physicians for an opinion as to whether such and such a disease would happen again in a particular family. The geneticist would look up the mode of inheritance (if known) and use the Mendelian laws to come up with a probability for that family situation. Simple!

In the late 1940s a few forward-looking hospitals in Canada (Toronto, Montreal), and elsewhere, decided it would be nice to have a geneticist on site, who would collect data on modes of inheritance, and talk to the families about their problems, what the risks were, and what to do about them. There wasn't much they *could* do. Decide not to have any more babies, which required abstinence, sterilisation, or unreliable methods of birth control. Or, if a pregnancy were already under way, decide whether to apply for an abortion. Abortions were hard to get. It required applying to a hospital committee, to whom the geneticist wrote a letter stating the risk and burden of the disorder, and that in their opinion these would justify the procedure. Permission was not always granted; I can recall a number of women who almost lost their lives having an illegal abortion after refusal, and a woman who, following repeated refusals or delay in processing her applications, ended up with three haemophilic sons (Fraser 1968).

Things have certainly changed. Since the 1950s, we have had treatable inborn errors of metabolism, newborn screening programs, the subspecialty of biochemical genetics, chromosomal disorders, and cytogenetics. Since the early 1970s we have had prenatal diagnosis. Since the early 1980s we have had molecular genetics, for the diagnosis of a rapidly increasing number of Mendelian disorders, both pre- and postnatally.

Obviously the PhD geneticist of the 1940s was not up to this. Genetic counselling became more and more a team affair, depending on various laboratory resources (cytogenetics, DNA analysis, biochemistry), clinical diagnosticians and, of course, geneticists, who had to keep up with a bewildering and rapidly increasing array of conditions about which they were asked to provide information. Genetic diagnosis and counselling were increasingly provided in centres,

---

* Now residing at 81 Chute Road, Bear River, NS, Canada, B0S 1B0.

usually affiliated with medical schools, where the necessary skills and resources were concentrated. Genetic counsellors needed a lot more skills than in the days of yore.

In the early 1970s it became embarrassingly clear that a few of those who were acting as genetic consultants, not associated with centres, were providing advice that was inappropriate, inadequate, or downright erroneous. There was a need for quality control, and the Canadian geneticists responded by setting up an accrediting body for those providing genetic services, the Canadian College of Medical Geneticists (CCMG). (At that time, the Royal College of Physicians and Surgeons of Canada did not recognise medical genetics as a specialty and, in any case, they would not accept PhDs, who needed to be accredited too.) A small steering committee was chosen, who reviewed the qualifications of those practicing medical genetics, and decided who would be admitted to the College as grandparents, without examination, in the initial phase. The College took responsibility for approving the training of applicants, and set examinations through which Fellowship in the CCMG could be gained. They also evaluated for approval the centres providing the services, and the training of candidates for Fellowship, and set up quality control programs for the associated laboratories. The CCMG was incorporated in 1976, and the Americans formed a College, the ACMG, a few years later. The Royal College of Physicians and Surgeons of Canada recognised the specialty of medical genetics in 1989, and the two Colleges recognise each other's programmes.

In response to the growing demand for genetic services, a new class of counsellors emerged. These are professionals at the MSc or RN level, who provide valuable support in the form of family history-taking, record searching, literature review, counselling, and post-counselling follow up of families. They are an integral and invaluable component of genetic services. They too have formed a professional association, the Canadian Association of Genetic Counsellors, and have developed accrediting procedures. There are close interactions with the CCMG. For a while, this kind of genetic counsellor was called a Genetic Associate, since many medical geneticists considered themselves genetic counsellors, but the Masters level counsellor now seems to have expropriated the term.

As one might gather from the rigorous training required, genetic counselling is a formidable job (Clarke Fraser 1974). Counsellors must keep up with an enormous number of conditions that they may be consulted about. There are thousands of such conditions, many of them so rare that most people have never heard of them, and the number is constantly growing. Counsellors must be able to find their way around this massive and complex database. They must know about how these conditions are diagnosed, how they are inherited, how variable they may be, if and how they can be treated, and if and how they can be diagnosed prenatally. They must be able to calculate risks of recurrence, which can sometimes be quite a complex procedure. They must be able to impart this information to the couple, sense whether they have grasped it, and help them decide what to do about it without *telling* them what to do about it. This requires tact and empathy.

Counselling sessions are usually scheduled to last about an hour, and several sessions may be required. The information the counsellor provides to the couple (the consultands) is rarely simple. It usually involves a probability of being affected with a disorder, and the burden of that disorder. The couple then has to decide, whether they want a baby badly enough to take the given risk of having a child with the given burden. Nowadays, more and more often, prenatal testing turns the probability into a certainty, but there may still be uncertainty as to whether the disorder will be mild or severe. And when prenatal diagnosis detects a disorder, there may not be enough time to make a well-considered decision about abortion. No wonder that, even with the best possible counselling, couples may feel frustrated and angry that they are not getting the answers they want.

To illustrate these complexities, here are some examples of the kinds of information that genetic counsellors can give, and the kinds of uncertainties that may result.

Firstly, as we have said, the answers can often be stated only as probabilities. There is one chance in two that your next baby will have the same thing. Or one in four, or one in 20, or whatever. One in four may be formidable to most couples, and one in 100 reassuring, but there are exceptions. 'Gee, I thought it was 100%. If it's only one in four, that's pretty good.' Or, you may say that it's one in 20, or one in 100, or one in 1000, but that one never goes away. Or 'either it will happen or it won't, and that's 50:50.' Or 'you say the odds are only one in 1000, but it happened to *us*, didn't it?' (Lippman-Hand and Fraser 1979) The problem is that probabilities, no matter how precise, are unsatisfying. What couples want is a simple yes or no answer – which the counsellor cannot provide.

But even when predictive testing does give a yes or no answer, that may not be simple either: Yes, your unborn son has inherited your gene for neurofibromatosis, but we can't say whether that means that he will have just a few brown spots on his skin, or will be mentally retarded and develop cancers – we can only give you the probabilities. Or, yes, your unborn daughter has an opening in her spine, but we can't say whether she will have just a mild club foot or be paralysed from the waist down and have hydrocephalus with mental retardation. Or, your unborn son will have a serious anaemia that may cause his death in his early teens, but it's quite possible that a treatment will soon be found that will make it a lot less severe. These uncertainties are not the counsellor's fault, they are facts of life, but they make it a lot harder for the couple to decide whether to have an abortion. And for this they may feel angry at the counsellor.

There are other sources of complexity in decision making too. Couples vary not only in their perception of risk, but of burden. Polydactyly may be a trivial anomaly to one couple and a disgrace to another. The specific family situation may be crucial. If a couple already has a child with cystic fibrosis, or spina bifida, who is a loved member of the family, the thought of aborting a foetus who has it may be seen as tantamount to aborting their loved one, and therefore unacceptable. On the other hand, a couple at risk for having a child with cystic fibrosis or spina bifida may say 'there's no way I want to bring a child into the world that would have to suffer like my sister.' A woman who has worked with children with Down syndrome may be much more anxious about her pregnancy than one who has not. A person who has polycystic kidney disease that was entirely asymptomatic and only discovered on routine examination at the age of 30 may be much more willing to take the 50% chance of passing it on to their child than someone who already has a kidney transplant.

In short, people vary greatly in their perception of risk and burden (Fraser 1956). Every family is different. The counsellor wants to relate to the couple emotionally, as well as intellectually, and yet to be objective, which requires a certain detachment.

It is a tenet of genetic counsellors not to be directive. Being nondirective is not easy. One may feel an aversion to the thought of a consultand taking a high risk of bringing yet another child with a severe disease into the world, but one should not assume that the decision to take such a chance is irresponsible. If the couple makes a carefully considered decision to take the chance, counsellors are expected to respect that decision and try to avoid any sign of disagreement or disapproval, hard though it may be.

It is even harder to be non-directive when the decision seems (to the counsellor) ill-considered. For example, a child has a severe immune deficiency causing repeated infections and many hospital admissions. The mother, who bears the brunt of her care, does not want to take the one in four chance – or indeed *any* chance – that the next baby will have it too. The father thinks one in four isn't bad; after all, that is three to one in their favour. It is sometimes hard for counsellors not to take sides, but they must not.

It is sometimes assumed that one of the aims of genetic counsellors is to promote eugenics – to improve the human gene pool. Genetic counsellors do not do what they do with the *aim* of improving the gene pool. As a matter of fact, some genetic screening and prenatal diagnosis programs may have consequences that are actually *dys*genic (Fraser 1972f, 1974e; see also his contributions to the present volume).

Occasionally a consultand will bring up the question of what eugenic consequences their course of action might have. For example, 'If I decide to take my one in four chance of having a child with cystic fibrosis, and get away with it, I may still be passing this deleterious gene on to the next generation. Isn't that morally wrong?' A counsellor who was dedicated to eugenics might say 'Yes it is, and you should take that into consideration when making your decision.' But a more appropriate answer would be 'We know so little about what makes genes "good" or "bad" and we are doing so many things that are changing our gene frequencies in ways we know nothing of (the pill, the baby bonus, wars, treating genetic disorders) that the consequences of your action are not only highly insignificant but impossible to evaluate.' So to repeat, the *effects* of genetic counselling and prenatal diagnosis may or may not have eugenic (or dysgenic) consequences, but the *aims* of the counsellor should not include eugenic motives.

Counsellors do considerable soul-searching about the ethics of what they do, but often have trouble fitting the nice clear-cut ethical principles to the complex and muddy situations they become involved with. They may invoke 'situational ethics', trying to be as fair as possible in trying to resolve the complexities of the conflicting rights that the various parties may have.

The most troublesome problem is the question of how severe a disorder should be in order to justify aborting a foetus and terminating what is unquestionably a life. There are disorders so dreadful that no counsellor would question the decision to abort, apart from those who would oppose abortion on any grounds. At the other end of the spectrum is a healthy foetus of the unwanted sex. The idea of aborting a healthy foetus is rebarbative, but in practice it is not easy, or simple, to refuse such a request from a woman whose life may be shattered if she has yet another baby of the 'wrong' sex. The counsellor may take refuge behind the professional guidelines that say non-medical conditions are not an indication for prenatal diagnosis. But what about a condition that is not life-threatening, but potentially serious – for example the 20% chance of ovarian failure in a fragile-X carrier female? Would this justify an abortion? My view is that if the woman thinks it is serious enough to be willing to undergo the stress of an abortion, it *is* serious enough. Women are seldom, if ever, referred for trivial conditions. But there are concerns that this may change, as diagnosis becomes possible at progressively earlier stages of pregnancy.

Among the other problems that counsellors worry about are: the conflict between the rights of those who need to know certain genetic facts about a relative in order to make informed reproductive decisions, and the right to privacy of the relative who does not wish to reveal these facts;

the balance between not providing enough information and providing so much that the consultands are confused or unduly alarmed; the rights of the two parents when they disagree. One cannot draw up categorical rules that will resolve such quandaries – each situation is unique. One can only try to be fair, and try to see that others are being fair.

Even when counsellors try their best to be non-directive, they may give away their biases by subtle inflections, glances, or other body language. And the consultand may perceive directiveness even when it is not there. One consultand, when asked by a researcher whether I had been directive, said 'no, but if Dr. Fraser had thought I shouldn't have any more babies he would have said so, wouldn't he?' And sometimes consultands seem to *want* direction (Lippman-Hand and Fraser 1979a). Or at least they want to know what the norms are, what other people do. When counsellors are asked what they would do in the consultand's situation, they usually say 'It's not for me to say, it's your decision.' But when, towards the end of a long session, the question comes up for the third or fourth time, they might say 'I don't know what you would do, because I am not you, and I am not even sure what I would do, but putting myself in your situation as best I can, I think I might … And I do know of couples who did that, and had a good outcome.' Some counsellors consider even this as too directive.

The problem of nondirectiveness may also involve informed consent and informed choice. Everyone agrees that the consultands must understand what the problem is in order to give informed consent to any procedures agreed upon, and what the options are for its resolution, so there can be informed choice. But sometimes it is not that simple. Counsellors often begin the interview by asking the consultands why they are here, what they see as the problem. Sometimes the consultands will say they have no idea, the doctor told them to come so here they are. The counsellor may have to suggest what the problem may be, and it may turn out that the couple does not see this as a problem, and that is the end of it. Or the counsellor may feel it necessary to acquaint the couple with the fact that they have a problem, a task requiring sensitivity and tact. For example, consider a very poor, uneducated, South American couple; the woman, after having one child, became infertile because of endometriosis. The endometriosis disappeared, and they thought they would like another child. She told her doctor about her three retarded brothers, and he sent them for genetic counselling. The counsellor establishes that it is not possible to reach the brothers for diagnostic testing. He asks if they are concerned that the brothers' condition might appear in their future child. No, they are not. Is that because they did not know it was possible, or if it did happen it would not be a problem for them, or that they just didn't think it would happen to them? The last. The counsellor brings up the idea that some kinds of mental retardation 'run in the family', that this might well be the case for her brothers, and that there is indeed a chance that it could happen in their children, with a risk as high as one in four if it were a son. This is not the time to explain about genes, dominance and recessiveness, X-linkage, and the fragile-X syndrome. The possibility of prenatal diagnosis, with abortion if the foetus is a male, is brought up. Eventually the husband says he thinks that they will just take their chances, and that God will not let this happen to them. The counsellor feels that they don't really appreciate the problem.

Does the counsellor try to go over the explanation again? Does he make an appointment for the couple to come in again after they have had time to think about it, even though the couple does not seem interested in more discussion? Wouldn't either of these options be directive? If he does not, wouldn't he be failing to achieve informed choice? The counsellor settles for writing them a letter, spelling out what he has told them, and writing to the referring doctor. But he is not comfortable about it.

This illustrates one of the problems with informed consent. In this case the couple's lack of education prevented them from understanding fully the basis for their being at risk. Is it enough for them just to appreciate that they *are* at increased risk, without following all the complexities involved in the estimate? In some cases the situation is so complex that even well educated consultands may have trouble understanding the choices, and their merits. They may turn to the counsellor for direction, so there is not only informed choice but non-directiveness to worry about.

Women offered amniocentesis after a positive screening test are another example. There are those who claim that the counsellor should provide a detailed discussion of Down syndrome, including pictures, the burden of raising a retarded child, the blessings of caring for a retarded child, the support services available (or not) – and all this for a woman who took the test just to rule out a small risk about which she was not anxious in the first place. They argue that not to do so means the woman is not giving informed consent to the testing. Others would say that, at this stage, the benefits of doing this do not justify the anxiety aroused in women who see their risk as low – the place for that counsel is in the minority of cases where a disorder is detected. Who is right? Should there be rules, or should it be left up to the judgment of the counsellor? Or should there be periodic reviews by an ethics committee, with lay representation?

The demand for counsellors' time often exceeds the supply, leaving them feeling dissatisfied. They may also sometimes feel inadequately prepared to meet the psychodynamic complexities of the family problems they meet. For example, the Italian parents of a child with a progressively crippling muscular dystrophy, recessively inherited, are told that they could reduce the chance of recurrence from one in four to about one in 600 by donor insemination. The mother, who very much wants another child, thinks this is a great way out. The father says that one in 600 is unacceptably high. Surprise, but the counsellor tries not to show it, and probes a little deeper. A sordid story of childhood abuse slowly unfolds, that has led the father to want no children at all. He consented to the first one as a concession to her, he says, and look what happened! Now it's her turn to concede. By the end of the two-hour session the couple are in tears, and the counsellor is having a hard time not to weep himself. He feels out of his depth, and suggests that the parents might want to get some help from a psychiatrist or marriage counsellor, but the father doesn't want that. They leave, with the situation unresolved, and the counsellor (and presumably the parents) feeling frustrated at his inadequacy. The mother calls a few weeks later, to apologise (!) for giving the counsellor such a hard time. The counsellor says don't worry, that's what we are here for – and would they like to come back for another session, or to have a referral for marriage or psychiatric counselling? Thanks, but my husband isn't ready for that just yet. And that is the last the counsellor sees of them, though he thinks of them sometimes, and wonders how he might have done it better.

What is it like to be a genetic counsellor? Most of them go into it because they like genetics, and like to help people, not for personal aggrandisement or financial reward. Genetic counselling is not a good way to get rich, but it is a rich and rewarding vocation. When I was a genetic counsellor, I used to look forward each day to a new opportunity to get to know a wide variety of mostly nice people, listen to their problems, try to answer, and sometimes help formulate, their questions, and to the satisfaction of feeling that I may sometimes be of some help.

Medical knowledge in the service of informed parental reproductive decision and choice: absence of implications for eugenic ideology

G.R. Fraser

Sometime Professor of Human Genetics, University of Leiden, Netherlands
Present address: 1 Woodstock Close, Oxford OX2 8DB, UK

Article 1 – All human beings are born free and equal in dignity and rights. They are endowed with reason and conscience and should act towards one another in a spirit of brotherhood.

From the Universal Declaration of Human Rights adopted and proclaimed by the General Assembly of the United Nations on 10 December 1948.

This paper is based on the conviction, which I believe to have been shared by Professor Lionel Penrose, my late teacher, mentor and friend, to whose memory this paper is dedicated, that there are no genetically superior human beings and no genetically inferior human beings; there are only human beings who are all fellow-members of one species. There is no place for notions of superiority and inferiority when applied to the infinite variety which characterises the individual genetical endowments of the members of our species. In this context, there are no objective criteria whereby superiority or inferiority can be measured, and there are no judges who possess either the wisdom or the knowledge to enable them to deliver valid judgments in this matter. Unfortunately, the gross ethical aberrations which have occurred in the name of eugenics, such as those which gave rise to unimaginable suffering during the first half of the 20th century, were based on the false premise that the genetical endowments not only of individuals but of entire ethnic groups are superior to those of other individuals and of other ethnic groups who, on these grounds, were stigmatised as being undesirable.

Since this stigmatisation has no validity, either in the case of ethnic groups or in the case of individuals, certain inalienable rights inhere in each member of our species, among them being the right to satisfy his or her needs with respect to food, clothing, and shelter, the right to health care and to education, and the right to procreate. The very large, and increasing, inequalities which obtain today with respect to economic status, and therefore with respect to the fundamental right of adequate access to food, clothing, shelter, health care and education, should be a source of concern and of shame to the more fortunate members of our species. The problem of reducing these inequalities is one of enormous difficulty, involving complex considerations in many dimensions, social, economic, and political, and cannot be treated in this paper which will be confined to the question of the right to procreate, even though, of course, this question cannot be regarded in isolation from the inequity which prevails universally with respect to the determinants of economic status, leading to unacceptable extremes of wealth and of poverty.

The desire to procreate represents the most deeply seated instinct of human beings, assuring the very survival of the species. All human beings who do not suffer from a condition which renders them infertile have the right to procreate. This right is enshrined in the United Nations Universal Declaration of Human Rights (*Article 16-1 – Men and women of full age, without any limitation due to race, nationality or religion have the right to marry and have a family.*) and in the International Covenant on Civil and Political Rights (*Article 23-2 – The right of men and women of marriageable age to marry and found a family shall be recognised.*). No human being should have the power to interdict the exercise of this right by another human being. The role of genetical counselling is to give advice to individual couples in their exercise of this right. Thus, the medical knowledge of today should be applied to the facilitation of informed parental decision and choice with respect to reproduction. There is no proper role for medical knowledge in defining groups of individuals who are fitter or less fit to procreate because their genetical endowments are superior or inferior, concepts which are inappropriate in this context.

There has been much discussion of the supposed social duty of parents who happen to be carriers of mutant alleles which are deleterious or potentially deleterious, to refrain from allegedly polluting the gene pool of future generations by procreating and passing on these mutant alleles. In my view, parents owe a duty to their children and to their grandchildren; they cannot be expected to owe a duty to future generations in the abstract. In the same way, the medical geneticist and the genetical counsellor owe a duty to the family members who have asked for advice, not to future generations in the abstract and not to the human species as a whole. Nor does it come within the remit of such professionals to modify their advice in order to spare the society within which they work a potential economic and financial burden, as long as their advice is not based on access to therapeutic modalities which are in fact totally unavailable to that society.

In any event, the human genetical endowment is so complex that no exact analysis can be made of the total effects on the gene pool of reproduction by a particular couple or individual. The simplest situation involves the many gene loci at which a mutant allele in single dose can give rise to serious disease in heterozygotes. A good example of such autosomal dominant inheritance is afforded by retinoblastoma, an eye cancer of childhood. An affected individual will pass on the mutant allele and the disease to half of his or her offspring. Such transmission did not take place in the past because the cancerous process spreads from the eyes and causes death in infancy or childhood when treatment is not available. It was only in the late 19th century that surgical treatment enabled a proportion of sufferers to survive to adult life and to procreate. For many years, survival was accompanied by blindness, the disease in its hereditary form usually being bilateral, but more recently it has become possible to preserve sight in a proportion of affected individuals by treating the cancer by radiotherapy rather than by surgical excision of both eyes.

When survival beyond childhood of persons with this disease enabled them to procreate, half of their offspring were similarly affected. Thus, the incidence of the condition at birth, which had previously been maintained only by the occurrence of new mutations, has increased. On the assumption that the initial fertility of affected individuals is zero while their fertility, following the widespread introduction of treatment, is 0.5, it can be shown that the incidence of retinoblastoma doubles in the population over a number of generations – hardly a disastrous increase in the incidence of a rare condition from about 1 in 20,000 births to 1 in 10,000 births, especially when this increase is spread over such a long period.

Similar calculations can be made in the case of other rare Mendelian conditions, whether inherited in a dominant or a recessive manner and whether X-linked or autosomal. Medical treatment of persons with such conditions could be regarded as representing a *dysgenic* intervention, in the sense that the frequency of deleterious genes is increased. These increases are so small in absolute terms, however, and take place over such an extended period encompassing many generations, especially in the case of an autosomal recessive disease such as cystic fibrosis, that they cannot be considered to be a significant burden to society. Thus, there is no valid reason why influence or, *a fortiori,* coercion should be brought to bear on such individuals to induce them to refrain from their right to procreate. Such arguments can readily be extended to common diseases which are not of monogenic Mendelian causation but are determined by complex interactions of hereditary and environmental factors.

We should resist the temptation, moreover, to view this problem within the limitations of the knowledge and technology available to us today. Undoubtedly, future generations will see farther than our generation with respect to the preservation of our biological heritage, as long as we refrain from making use of our newly acquired power widely to pollute, and otherwise jeopardise, the environment, thereby giving rise to so major a degree of degradation of the physical legacy which we bequeath to our descendants that they will become unable to continue the process of building on the scientific foundations which have previously been laid down, and even unable to continue the pursuit of science as a whole.

Thus, our primary task is to ensure the careful conservation of the environment, and hence the transmission of an intact and undiminished physical heritage to our descendants. These descendants will then be able to ensure the careful conservation of an intact gene pool and the continuing transmission of an undiminished biological heritage, using means whose nature we cannot predict, or even imagine, since they will be based on discoveries which are yet to be made.

Of course, the central theme of molecular biology today, within the framework of the human genome project, is the mapping and, eventually, sequencing of the human genome. Thus, in the case of retinoblastoma, the birth of normal children to an affected parent can now usually be assured by techniques of prenatal diagnosis, followed by abortion of foetuses detected to be carrying a deleterious allele which would give rise to retinoblastoma. Such procedures will be acceptable to many couples, but by no means to all. Acceptability of abortion will vary not only between individual couples but also between ethnic and religious groups. Pre-implantation diagnosis by selection of embryos following *in vitro* fertilisation is being developed at a rapid pace, and may, in some circumstances, provide a more acceptable substitute for prenatal diagnosis followed by selective abortion.

In addition, gametic selection may become possible in the future, although at the present time even the separation of sperm into X-chromosome- and Y-chromosome-carrying groups, in order to determine the sex of the foetus by means of artificial insemination, cannot be reliably achieved, despite several decades of experimentation. Such a technique would be useful, for example, in families where the mother is a heterozygote for a deleterious allele giving rise to Duchenne muscular dystrophy, in order to enable her to give birth only to daughters, thus avoiding the 50% risk that any son would be affected by this tragic disease.

Whatever advice is appropriate, there should in no case be any question of direction or coercion, even implicitly by communication, whether overt or subliminal, of the disingenuous message that society cannot carry the financial burden of the care and treatment of a handicapped child. With some re-ordering of priorities, economically developed states are sufficiently prosperous to provide treatment and care for a small additional number of cases of retinoblastoma and of other rare Mendelian diseases. By eliminating directive advice and, of course, the possibility of coercion, I believe that the practice of medical genetics and of genetical counselling can be exonerated from accusations of propagating the discredited principles and outdated doctrines of negative eugenics with respect to reproduction.

Thus far, I have discussed individual couples. With the increase of knowledge to be expected within the framework of the human genome project, it will become increasingly apparent that wide divergences exist in the genetical endowments of different ethnic groups. Recognition of the gene as the determinant of heredity is very recent, and discussion of the topic of differential reproduction by genotype is essentially restricted to the 20th century, although it was broached to some extent in the late 19th century. The 20th century can be divided in this respect into two halves, the periods before and after 1945. In order to avoid the abominations perpetrated in the name of eugenics which marked the first of these periods, it is incumbent on us to take a global view of the human genetical endowment. While it is self-evident that humankind is ethnically heterogeneous from this point of view, the concepts of superiority and inferiority, of goodness and badness, of desirability and undesirability, are just as inapplicable to the ethnic components of genetical variability as they are to the individual components.

It may well be that with increasing knowledge of the variability of the human genome, some ethnic groups will be identified where the frequencies of certain alleles, or combinations of alleles, which may be regarded as conditionally, or even unconditionally, deleterious, are unusually high. It would be entirely inappropriate, however, to stigmatise the group in question as being genetically inferior, or even genetically compromised, on the basis of such a finding. For one thing, the high frequencies of the alleles under discussion could well be balanced by correspondingly higher than average frequencies of alleles which could be regarded as being advantageous; alleles of this second kind will always remain more difficult to detect than alleles of the first kind. Not only, as I have already emphasised repeatedly, are the concepts of inferiority and superiority entirely out of place in this context, but the organisation of the human genome is so complex that it is simply impossible, with the partial exception of rare anomalies such as retinoblastoma, to define the result of the interactions of its component genes, whether in an individual or in an ethnic group.

I see no reason to take the fatalistically pessimistic view that transgressions are bound to occur which will take humanity back to the ethical enormities of the pre-1945 period. It should be remembered, however, that, quite apart from the abominations perpetrated in the name of eugenics and of 'race-hygiene' within the framework of the Nazi 'New Order' in Europe, these discredited practices played a shameful role in motivating programmes of enforced sterilisation in many countries both before and after 1945; in some countries, unfortunately, such programmes have continued until the present day.

Paradoxically, my relative optimism about the future stems from the triviality of the problems associated with the application of medical knowledge to reproductive practices, which is under discussion here, in comparison to the vast difficulties in the social, economic, and political spheres, which humanity must overcome in the near future in order that our species may survive in a recognisable form through the third millennium of our era. I can claim no priority for the statement that it is the nature of our recent technological discoveries, leading to mastery of physical modalities of potential self-destruction, rather than the possible misuse of molecular biology and of genetics that represents the real danger to our civilisation and to the survival of our species.

Thus, humankind has reached a watershed in its evolutionary journey in that the human community now faces a situation where there is a clear choice between the self-destruction for which the technological modalities exist for the first time, and progress towards new levels of human creativity and dignity. There is in my view no third choice. If we do not resolve the unstable equilibrium in which we now find ourselves by learning to share the resources of our planet more fairly between all the members of our species, and do so, moreover, in the near future, then it is likely that prolonged social, economic, political, ethnic, and religious tensions will eventually lead to our planet being rendered uninhabitable by the misuse of one or other of the technological innovations which have recently come into our hands. The prospect of such a catastrophe would, of course, render all discussions relating to eugenics, and, indeed, to medicine and to science in general, irrelevant. On the other hand, there can be no doubt that if we are able to attain these new levels of human creativity and dignity, the issues with respect to eugenics under discussion here will be resolved by the recognition of the essential equality, in particular relating to genetical endowment, of the individual members of our species and of the various ethnic groups into which it is subdivided.

I have perhaps allowed myself to stray too far from the strict confines of the theme which I am addressing. I should like to return by referring to a concrete manifestation of the disquiet which has been expressed about the possibility of a recrudescence of eugenic policies promulgated by the state. I refer to reactions to the Law of the People's Republic of China on Maternal and Infant Child Care which became effective on 1 June 1995. It is claimed by some that this law contains clauses promoting eugenic ideology. It is not my intention to enter this debate. I cannot read Chinese at all, and I am not able to make a judgment about the subtleties of interpretation to which the English translation may not have done justice, nor can I claim to understand the cultural differences between Chinese and European attitudes to these problems.

Instead of entering this debate, I wish to use it to illustrate a point which I made earlier. I have said that the question of the right to procreate cannot be regarded in isolation from the determinants of economic status. It is, of course, a well-known fact in this context that China is a developing country with a small gross domestic product per head. Because of advances in average living standards accompanying political changes in recent times in China, there has been a sharp decrease in mortality with a corresponding increase in life expectancy. As a result, explosive population growth has occurred. It is also a well-known fact that the government of the People's Republic of China has adopted a policy of encouraging individuals in most regions of the country to restrict their family to one child. It is against this unique background that the provisions of the Law on Maternal and Infant Child Care should be assessed.

It is important to stress that these attitudes towards the quantitative and qualitative aspects of reproduction and their interaction, which I have discussed briefly in this Chinese context with which, unfortunately, I am not well acquainted, need to be extrapolated to a worldwide context. I have said that humanity faces a clear choice between self-destruction and progress towards new levels of human creativity and dignity. In other words – and I stress that these are ideas which are far from original – we stand at a watershed between oblivion and utopia. One of the main elements which will determine the direction in which we proceed from this watershed is whether we will succeed in controlling the quantitative aspects of human reproduction. This problem is of far greater importance than control of the qualitative aspects of human reproduction which is the theme of this colloquium. As I have indicated, the effects of what can be done from the point of view of controlling quality are of marginal significance to the welfare of humanity as a whole, whereas successful control of quantity with respect to reproduction is of central importance to the survival of the species.

Just as I have indicated my belief that the practice of medical genetics and of genetical counselling in developed countries can be exonerated from accusations of propagating the discredited principles and outdated doctrines of eugenics, so it is my belief that the numerical restrictions in family size which must occur if our civilisation is to survive, will not involve eugenic trends. The only effective means which we have at our disposal in order to achieve this end is worldwide education. These means must be applied universally without favouring preferential reproduction by any ethnically or socio-economically defined group. Positive eugenics is as irrelevant to reaching this milestone in the progress of humanity towards new levels of creativity and dignity as is negative eugenics with its extremely small effects.

Thus, the proper practice of medical genetics and of genetical counselling does not involve eugenic principles in dealing with individual patients and their families. The proper practice of family planning is far more important from the point of view of the future of the human species, and again does not involve eugenic principles in dealing either with individual patients or with sub-groups of our species defined on the basis of ethnic, religious, or socio-economic criteria.

In relation to the title of this paper, I have discussed only a small part of the work of the profession of medical genetics and genetical counselling. I have taken the extreme case of persons affected with diseases which are genetically determined wholly or in part, and I have noted the opposition which prevails within this profession to the introduction of principles of negative eugenics in the shape of discouraging such individuals from procreating. The spectrum of diseases under consideration is vast, ranging from rare monogenically determined conditions such as retinoblastoma, through to common diseases such as adult-onset, or non-insulin dependent, diabetes mellitus where a major degree of genetical determination is mediated by the interaction of several alleles. In general, it is the view in this profession that even in the case of diseases which rendered procreation impossible before the introduction of medical treatment, part of our therapeutic role is to discuss with such individuals the optimal conditions under which they can undertake reproduction, rather than to advise them not to do so on the basis of some ill-formulated concept related to negative eugenics, to the effect that they should be discouraged from contaminating the human gene pool by adding to the number of deleterious alleles. In passing, artificial insemination by a donor is a simple way to bypass this problem in many situations, and may be acceptable to some couples.

The weakness of a position with regard to reproduction, which is based on the ideology of negative eugenics, is well illustrated by considering diabetes mellitus of adult-onset. Doubtless, with increasing knowledge of the human genome, the genetical basis of this common disease will be elucidated. This will lead to accurate definition of the number and nature of the alleles involved. Some of these are very common. If we are to deny reproduction to persons carrying any one of these alleles in a misguided attempt to 'purify' the gene pool, we would be denying reproduction to a very large proportion of humanity. I shall have occasion to revert to this aspect of the refutation of the doctrines of negative eugenics.

Of course, the role of the medical geneticist and of the genetical counsellor extends beyond giving advice on reproduction to persons with a disease-producing genotype. In fact, in the realm of Mendelian conditions, we are far more often called upon, for example, to advise individuals who are healthy heterozygous carriers of alleles which can cause autosomal recessive disease in homozygous offspring when both parents are heterozygotes (such as cystic fibrosis) or X-linked recessive disease in male offspring (such as Duchenne muscular dystrophy) when the mother is a heterozygote.

Our role in this respect is developing very rapidly. Whereas previously we were restricted to retrospective counselling in that it was only the birth of an affected child which revealed that a couple was at risk, we can now provide prospective counselling because of the introduction of genetical testing and of genetical screening for heterozygotes for alleles which in homozygous form can give rise to a large number of such conditions, including cystic fibrosis, thalassaemia, sickle-cell anaemia, and Tay-Sachs disease. I have chosen these examples – and the number of such conditions is increasing rapidly as new discoveries are made – because each has a characteristic ethnic distribution.

I have referred to the fact that a large proportion of the human population are carriers of alleles which can contribute to the causation of diabetes mellitus of adult onset. This is true also of almost all other common diseases. In addition, probably every single individual carries one or more alleles which can cause autosomal recessive disease in homozygous form. It should be remembered that if we assume that a rare autosomal recessive disease with a birth frequency of one in 40,000 is homogeneously distributed throughout the human population (although very few such diseases, if any, would in fact fulfil this criterion of homogeneous distribution), then 1% of individuals would be heterozygotes. Since there are many hundreds of such diseases, it is readily apparent that almost every one of us is carrying at least one such potentially deleterious allele.

Thus, discussion of the extension of the principles of negative eugenics to discouraging the spread of such alleles by reproduction would represent a *reductio ad absurdum*, since all of us carry alleles which are potentially deleterious. In the case of common diseases such as diabetes mellitus of adult onset, this deleterious effect will occur in the individual when the allele in question is combined with certain others, while in the case of alleles which can cause rare autosomal recessive diseases in homozygous form, this deleterious effect will occur in a proportion of the offspring, if the marriage partner carries the same allele.

It should also be noted that the concept of desirable and undesirable, advantageous and disadvantageous, alleles, is, with a few exceptions as in the case of those giving rise to retinoblastoma, relative and conditional rather than absolute and unconditional. Thus, an allele which gives rise to sickle-cell (S) haemoglobin rather than normal haemoglobin, is undesirable and disadvantageous in that if two healthy heterozygotes marry, a quarter of their offspring will suffer from a very serious disease, sickle-cell anaemia. The same allele, however, has been desirable and advantageous throughout long periods of human evolution, in that heterozygotes have been more resistant to the effects of the parasite, *Plasmodium falciparum,* in causing malaria than normal homozygotes (see D.J. Weatherall in this volume).

In the case of diabetes mellitus of adult onset, the hypothesis has been put forward (Neel 1962) that the genotypes responsible have been desirable and advantageous in the past in the sense that they have determined thriftiness with respect to carbohydrate metabolism during long millennia of limited food supply, and that they have only become undesirable and disadvantageous very recently in the context of the grossly excessive and unbalanced diets which characterise our modern civilisation.

In general, we are not yet in a position to provide informative genetical counselling in common diseases such as diabetes mellitus of adult onset, because the genotype which is responsible is represented by several unidentified alleles which interact to cause the disease. A partial exception to this rule occurs in the case of some common cancers, largely involving the breast, the ovary and the colorectum. A small minority of cases of such cancers have been shown over the last two decades to be due to Mendelian transmission of mutant alleles leading to dominant monogenic inheritance of a major degree of cancer susceptibility. These cancers are very common, however, and the involvement of such alleles even in a small minority of cases, perhaps five percent, would mean that as much as one percent of the human population may be harbouring one of these deleterious alleles.

This population incidence, if confirmed, is vastly greater than the incidences of other types of monogenically determined Mendelian diseases. As a result, this situation has created a new and very large problem in medical genetics and genetical counselling. It is of interest to note, however, that despite the fact that these susceptibility alleles cause a great deal of suffering and premature death because of the occurrence of cancers at inappropriately early ages, sometimes even in the twenties and thirties, there has been very little discussion of reproductive or eugenic issues in this context. Individuals who carry such alleles think in terms of early diagnosis and of the possibility of successful treatment of such cancers. They procreate and do not seem unduly worried by the question of whether their children have inherited the abnormal allele, trusting that by the time that they become adults, better therapeutic modalities of dealing with the problem will have been developed. Possibly, this situation may serve as a paradigm, in that treatment rather than abortion may be increasingly emphasised in the future even in the case of hereditary diseases which manifest themselves long before adult life.

To return to the case of the many rare autosomal recessive diseases which, in general, manifest their effects in childhood rather than in adult life, following pre-reproductive detection of a couple at risk, there are a number of measures at our disposal. Advice not to marry is acceptable in certain Ashkenazi Jewish communities in connection with Tay-Sachs disease. In other cases, prenatal diagnosis followed by selective abortion of affected foetuses is acceptable, even though this is somewhat of an anomaly in relation to the principles underlying medical practice in the past. Thus, we can reduce the birth frequency of cystic fibrosis in this way, or even eliminate thalassaemia altogether in certain communities such as those living in the Greek part of Cyprus where the Orthodox Church encourages premarital testing and permits the abortion of abnormally homozygous foetuses. In the past, however, we would not have spoken of preventing or eliminating tuberculosis by killing the patient. We can hope that future developments in foetal medicine will permit us, in some cases at least, to treat the abnormally homozygous foetus, rather than proposing abortion. I have already mentioned the possibility that selection of embryos after *in vitro* fertilisation and pre-implantation diagnosis may be expected to play an increasing role in dealing with the problems under discussion without the necessity for abortion.

I shall not discuss here the various aspects of genetical engineering and germ line gene therapy, for I regard these as futuristic strategies with uncertain prospects.

I do not believe that these applications of medical knowledge to the facilitation of informed parental decision and choice with respect to reproduction, involving avoidance of marriage, gene testing and screening, prenatal and pre-implantation diagnosis, artificial insemination and *in vitro* fertilisation, have any but very trivial effects on the gene pool. Thus, although these interventions will increase the frequencies of abnormal alleles in many circumstances, the changes will be very small and will be spread over a very long period of time measured in terms of many generations.

In fact, exactly the same considerations apply as have previously been discussed in connection with the treatment of hereditary disease followed by the reproduction of individuals who would not otherwise have been able to reproduce. There is, therefore, no appreciable implication of these measures from the point of view of eugenic doctrine, and we are not leaving any intractable problems for our descendants from the point of view of contamination of our gene pool. As I have already stated, providing these descendants with an intact physical legacy is immeasurably more significant from the point of view of the continued survival of our species.

This is all the more true since at present these medical techniques are only being applied to a tiny minority of privileged families living in economically advanced countries. This inequality in access to medical care is, of course, a symptom of the gross inequalities, in the distribution of material resources in general, which disgrace our civilisation. It could be argued that these gross inequalities in familial incomes are of more significance from the point of view of eugenic ideology than inequalities in the familial distribution of genes and genomes. Thus, Plato's proposal in his *Laws* (v, 744), of a society in which no man would possess more than four times that level of wealth which is sufficient to ensure a decent existence, and in which no man would fail to reach this level, is of far more importance in the context of avoiding childhood handicap, and in improving the health of future generations, than all the minor genetical interventions which I have been discussing put together.

I shall summarise these interventions briefly in relation to the transmission of mutant alleles which are deleterious or potentially deleterious. In order to obviate ideological implications, this summary refers to *increases* or *decreases* of the population frequencies of such alleles in preference to the use of the terms *dysgenic* and *eugenic*. In addition to their inappropriate ideological connotations, the use of these terms leads to logical inconsistencies, in that the avoidance of the birth of a handicapped child, which may be regarded as a nominally *eugenic* intervention, is very often accompanied by a nominally *dysgenic* increase in the population frequency of the mutant allele which gives rise to the handicap in question. These points are discussed *in extenso* in Fraser (1972f).

A Treatment of monogenically determined Mendelian diseases (and of common diseases of complex and partially hereditary determination) *increases* the population frequencies of mutant alleles which are deleterious or potentially deleterious, to the extent that reproduction and the transmission of such alleles would have been impossible in the absence of treatment.

B Avoidance of marriages between heterozygotes by pre-marital screening in the case of autosomal recessive Mendelian diseases (for example, Tay-Sachs disease in Ashkenazi Jews) *increases* the population frequencies of mutant alleles which are deleterious or potentially deleterious, because their elimination in homozygotes no longer occurs.

C Artificial insemination by a donor *decreases* the population frequencies of mutant alleles which are deleterious or potentially deleterious and which cause autosomal dominant or X-linked recessive Mendelian diseases, if the husband carries such an allele which, as a result of the artificial insemination by a donor, is not transmitted.

D Prenatal diagnosis and selective abortion of affected foetuses *decreases* the population frequencies of mutant alleles which are deleterious or potentially deleterious in the case of autosomal dominant disease, but *increases* these frequencies in the case of recessive disease, whether autosomal or X-linked, because elimination of these alleles in homozygotes or hemizygotes no longer occurs and reproductive compensation will lead to increased transmission of such alleles.

E Selection of embryos after *in vitro* fertilisation and pre-implantation diagnosis may turn out to be a more acceptable alternative to D, when these techniques are developed further. In addition to enhanced acceptability from the medical and moral points of view, the *increases* in the population frequencies of mutant alleles which are deleterious or potentially deleterious in the case of recessive disease, whether autosomal or X-linked, which occur in connection with D, may be avoided by selecting only homozygous normal embryos for implantation. Thus, the mutant alleles involved are not transmitted and their population frequencies *decrease*.

This list can be greatly extended, but this would not affect the conclusion that the interventions in question have small and indeterminate effects on the gene pool. In fact, these effects are dwarfed by concomitant changes in reproductive patterns which are occurring quite independently of the application of medical knowledge. For example, the break-up of isolates and the resultant increase in outbreeding in the modern era lead to *increases* in the population frequencies of mutant alleles which are deleterious or potentially deleterious and give rise to autosomal recessive Mendelian diseases, because of reduced elimination of the mutant alleles in homozygotes. The spread of chemical agents and radiation as accompaniments of modern civilisation also leads to *increases* in the population frequencies of mutant alleles which are deleterious or potentially deleterious, because of the direct effects of such agents in increasing mutation rates.

On the other hand, improvements in living standards connected with medical and other advances, may lead to *decreases* in the population frequencies of mutant alleles which are deleterious or potentially deleterious and which have been maintained by the type of mechanism which has been postulated in the case of the mutant allele giving rise to the production of haemoglobin S, involving heterozygote advantage through increased resistance to the effects of the malaria parasite.

Probably, the most important change in reproductive behaviour which is occurring is intimately related to the point which I have made with respect to quantitative control of reproduction. If our civilisation is to survive the hazards imposed by gross over-population of the planet, then the pattern of limitation of reproduction which we already see in developed countries will have to be replicated on a worldwide scale through economic development accompanied by education. Typically, a woman will have one or two or three children in her lifetime. The emphasis will be on quality rather than quantity. In this context, every pregnancy will be prepared with the same care and attention to detail as a patient would exercise when faced with the prospect of major elective surgery. Proper importance will be attached both to adequate spacing between births and to the optimal parental age range for reproduction. Fortunately, such changes will tend to *decrease* the population frequencies of mutant alleles which are deleterious or potentially deleterious, in that a reduction in births to older parents will lead to a reduction in mutation rates. Thus, there is a well-known association between raised maternal age and an increase in the birth frequency of one of the most common chromosomal anomalies, trisomy 21 or the Down syndrome, which could in this context be regarded as being due to the effect of mutation within the wider meaning of the term. More conventional types of mutations leading to the formation of deleterious or potentially deleterious alleles will also occur less frequently with a reduction in average parental, especially paternal, ages at birth.

A quarter of a century ago, a noted geneticist (Glass 1972) seemed overcome by gloom as he contemplated the man of tomorrow, the result of failing to implement the doctrines of negative eugenics for several generations, 'beginning his day by adjusting his spectacles and his hearing aid, inserting his false teeth, taking an allergy injection in one arm and an insulin injection in the other, and topping off his preparations for life by taking a tranquilising pill'.

I have made it abundantly clear that I do not share this apocalyptic vision of the effects of the application of medical knowledge to reproductive practices. I believe that we can continue to develop these applications along the lines that have already been initiated in order to facilitate informed parental decision and choice with respect to reproduction, without fearing consequences of the type presented by Glass (1972). I do not believe that these modest interventions have, or will have, any but trivial effects from the point of view of the doctrines of negative eugenics.

I have repeatedly stated my conviction that a *sine qua non* with respect to advances in this context in the coming century and in the coming millennium must be the preservation of our civilisation and of our planet from major catastrophes resulting from our newly acquired mastery of the technological means of self-destruction. Provided that this will be so, I am confident that our descendants will be dealing with the problems which I have adumbrated in an effective and totally different manner. I have no idea wherein these differences will lie, but I believe that the present emphasis on prenatal diagnosis followed by selective abortion represents only a temporary phase which will be superseded by methods which will be more generally acceptable.

I can only speak for the applications of medical knowledge, particularly in the areas of medical genetics and genetical counselling, to the question of facilitating parental decision and choice with regard to reproduction. Physicians, however, will only be able to function as a small group within a large team in this context. Thus, especially with respect to quantitative aspects of limitation of reproduction, input will be required not only from physicians but also from scientists in many disciplines, from educators, from sociologists, from economists, from ethicists, from lawyers, from theologians, from politicians, and from many others. After the physical survival of our planet, the quality of human life is the most important element of our future. Granted the physical survival of our planet, I do not believe that the discredited principles and outdated doctrines of eugenics represent a danger in this context.

I may, of course, be entirely wrong in this assessment. Müller-Hill (1997, 1998) has written of the 'spectre of kakogenics', as the result of an unholy alliance between politicians and geneticists (see also B Müller-Hill in this volume), and, in his view, we may yet see attempts to re-introduce at least some the practices which culminated in the abominations perpetrated in the name of eugenics and of 'race-hygiene' in the first half of our century. Such a retrograde step would represent a far greater danger now because of the stark choice between utopia and oblivion which lies before us. Thus, the recrudescence of these horrors would betoken such a debasement of our science and such a decline of our civilisation that the inevitable accompaniment of social unrest and war would, in the altered technological circumstances of today, imperil the very survival of our species.

We are united in our hope that such an Armageddon will not come to pass, and I should like to close on an optimistic note in quoting a reflection of Xenophanes, a pre-Socratic philosopher of the 6th century BC. This is one of the few fragments of his writings on the limitations of human knowledge which survive.

οὔτοι ἀπ᾽ ἀρχῆς πάντα θεοὶ θνητοῖσ᾽ ὑπέδειξαν;
ἀλλὰ χρόνῳ ζητοῦντες ἐφευρίσκουσιν ἄμεινον.

The gods did not reveal all things to mortals in the beginning:
but in long searching man finds that which is better.

We have been searching for knowledge about the human genome for a very short time; in this context, it is salutary to reflect that it is only 140 years since Mendel's discovery laid the basis for the entire science of genetics. Although even our entire third millennium still represents only a short time for this difficult search, we shall slowly but surely find that which is better not only with respect to knowledge but, more importantly, with respect to the wisdom which will guide our use of this knowledge.

Acknowledgements

This paper is an expanded version of a lecture presented during an International Colloquium: L'eugénisme après 1945: formes nouvelles d'une doctrine périmée (Eugenics after 1945: new forms of an outdated doctrine) held on 22–23 September 1997 in Dijon, France. A translation into French of the full text of the paper has been published as 'Le savoir médical au service de choix reproductifs informés: une pratique sans implication eugénique'. In *L'éternel retour de l'eugénisme* (editors Gayon J and Jacobi D with the collaboration of Lorne M-C) pp. 7–27 Presses Universitaires de France, 2006.

I should like to express my sincere gratitude to Professors Jean Gayon and Daniel Jacobi for having invited me to Dijon in September 1997 and for their permission, as well as that of Les Presses Universitaires de France, to reproduce a slightly amended version of the text of my lecture in this Festschrift and *liber amicorum*. I should also like to thank them, as well as Dr Marie-Claude Lorne, for affording me the opportunity to play a small part in the preparation for publication of the excellent translation into French of which the initial version was made by Professor Martine Semblat to whom also it is a pleasure to acknowledge my appreciation.

Human genetics today: hopes and risks

G R Fraser

1 Woodstock Close, Oxford, OX2 8DB England

It is in this very same city of Rome that I gave my first talk at an international congress, 44 years ago in 1961, on 'The pool of harmful genes in human populations' (Fraser 1962b), a topic subsumed within that of the present subject, 'Human genetics today: hopes and risks', a vast panorama which retraces the entire history of the development of our human species as documented in the common heritage of our human genome in all its diverse, and yet harmonious, complexity. In this short talk, in conformity with the written presentation of the aims of this conference, I shall confine myself mainly to the perspective of health, leaving aside many other aspects of human genetics; this leads to the corollary that the term human genetics will be largely interchangeable with that of genetic (or genomic) medicine. And I shall speak of the hopes and of the risks attendant on the process of integrating genetic medicine into the science and the art of medicine and medical practice as a whole, a process which is being pursued intensively at the present time.

As part of this perspective of the application of human genetics to health, the recent description of the sequence and topography of the human genome (the Human Genome Project) represents an important advance which has led to the accumulation of a vast and rapidly increasing body of knowledge about our hereditary material. A vast body of knowledge, however, is insufficient in itself to make a major impact on health; the wisdom to make use of the knowledge is also indispensable. In the acquisition of this wisdom, humility rather than arrogance must be our guiding principle. Thus, the pre-Socratic philosopher, Xenophanes of Colophon, wrote, among the few fragments which have come down to us of his thoughts about the limitations of human knowledge:

> *The gods did not reveal all things to mortals in the beginning;*
> *but in long searching man finds that which is better.*

When we consider our human genetics of today, while we may justifiably claim that we have made progress along the road towards that which is better, we should not delude ourselves into believing that in our searching, we have taken more than the first few steps along this road; nor should we flatter ourselves that these few steps have all been taken in the modern era. For example, among the ancient Greeks from the sixth century BC onwards, Anaximander, Hippocrates, Aristotle, and Plato, among others, wrote about the mechanisms of inheritance as well as about those of evolution. To introduce a very striking example which says much about one of the major risks of human genetics today in the context of unjustified and unjustifiable attributions to individuals of hereditary superiority and inferiority, Plato applied such ideas extensively 2500 years ago to what we would now call eugenic principles, especially in the fifth book of the Republic, a chronicle of the regulations which would govern the functioning of the hypothetical city where justice would reign supreme. In this quotation, Plato passes from considering the practices involved in the breeding of hunting dogs in order to maximise their skills, to the mating patterns which should be applied to the Rulers or Guardians of the city.

It is necessary, given our agreed premises, that it is the best men must mate with the best women, and that as frequently as possible, while it is the worst men who must mate with the worst women, and that as infrequently as possible; and we should rear the offspring of the former, but not of the latter, if our flock is to be of the highest quality.

To move fast forward two and one half millennia to the 19th century, in the words of Theodosius Dobzhansky – 'Genetics, an important branch of biological science, has grown out of the humble peas planted by Mendel in a monastery garden.' The achievements of Abbot Gregor Johann Mendel can without any hint of hyperbole be described as unique in the annals of science and we have built on these achievements over the past 140 years to accumulate the vast body of knowledge of which I have spoken.

Until very recent times, the role of the human geneticist in medicine has been largely confined to giving advice and counselling about the role of heredity in the causation of disease and handicap among those who consult him, and their offspring, especially with respect to the evaluation of rare monogenic Mendelian disorders, congenital malformation syndromes, and chromosomal anomalies. His role has definitely not been in following the eugenic precepts of Plato in encouraging reproduction in one group of the population and discouraging reproduction among others. One advantage which has accrued to us with the accumulation of knowledge is that we now know that, in addition to the lack of any moral basis for such selection of the parents of the next generation which was being practised until recently, there can also be no rational basis for any such selection from among the virtually infinite variety of our genetical endowment.

Thus, in 1966, my late mentor, teacher, and friend, Professor Lionel Penrose wrote: 'The social and biological values of hereditary differences are continually altering as the environment changes. We cannot be sure that any gene will be bad in all circumstances and much less sure that any gene is always good. At the moment we are only scratching the surface of this great science and our knowledge of human genes and their action is still so slight that it is presumptuous and foolish to lay down positive principles for human breeding. Rather each person can marvel at the prodigious diversity of the hereditary characters of Man and respect those who differ from him genetically. We all take part in the same gigantic experiment in natural selection' (Penrose 1966).

Human genetics today is in no sense to be regarded as a surrogate for the discredited eugenic ideas of the past; it is rather to be regarded as an integral part of medicine with the role of incorporating the knowledge which is being gained about the human genome into diagnostic, preventive, and therapeutic activities, whether directed towards the foetus, the infant, the child, the adolescent, or the adult, within the context of the family unit, there being no discontinuity between these phases of human existence. In this context, the practice of selective abortion of foetuses diagnosed *in utero* as being affected by a genetically determined disease or malformation is contrary to the tenets of our medical tradition with its emphasis on the preservation of life, and this may best be regarded as a transient phase in the development of genetic medicine, which will be superseded in time by methods which will be more generally acceptable.

Pre-implantation diagnosis has been put forward as an alternative, and is being practised in some centres, but this has given rise to ethical controversies connected with the moral status of the embryo with respect to personhood. Turning from selection at the level of the foetus and of the embryo to selection at the level of the gamete, reliable separation of the husband's sperm into X- and Y-chromosome-bearing fractions, followed by fertilisation of the wife using only the X-chromosome-bearing sperm in order to avoid the births of males when the wife is a carrier of an X-linked disorder such as Duchenne muscular dystrophy, may become possible in the near future and may be less open to objection on ethical grounds. It would be foolhardy, however, to forecast the attitudes and practices of future generations with respect to these problems in general, especially since they will have access to technologies whose nature we cannot now foresee.

There are experts assembled here who will be telling you over the next three days in detail about each and every aspect of this integration of human genetics into the practice of medicine. Of course, the present important role of the medical geneticist as a nondirective counsellor will always continue to be needed, and will expand in parallel with the expansion of knowledge about the role of heredity in the aetiology and pathogenesis of disease. A remarkable example of such expansion has been the relatively recent creation of a very large subspecialty of genetical counselling in the field of cancer, as a result of the discovery of gene loci harbouring alleles at a substantial frequency which give rise to a strong predisposition to various forms of this common disease, often at a tragically young age.

In fact, it has always been evident that heredity plays a role in the aetiology and pathogenesis of virtually every disease state, but we have not been able previously to define the mechanisms of this involvement. We are now in a position to begin this task, and, as a result, to start thinking about therapeutic modalities of correcting inherited defects, diseases, and disabilities. There is a perception that genetic profiling is going to be the cornerstone of this endeavour, in determining the genetic complement of individuals at large numbers of chromosomal loci and thus defining their specific susceptibilities to common diseases, indicating possibilities of prevention and treatment, and also, in the same manner, predicting the variations to be expected in the efficacy or toxicity of therapeutic agents, providing pointers to tailoring therapy to the patient (pharmacogenomics). This genetic profiling will supplement the many types of genetic testing and screening for individual genes, which is currently being performed and which will continue into the future.

Apart from the prospects opened up for advances in health care by such genome-wide extensions of determining the genetic complement of the individual, genetic medicine has been enriched by a new armamentarium of methods involving gene therapy – the treatment of genetic disorders by introducing specific engineered genes into the cells of the patient, which began in 1989 with the successful treatment of a rare recessive immune disorder, adenosine deaminase (ADA) deficiency, also known as severe combined immunodeficiency (SCID) syndrome. The possibilities of such gene therapy in replacing the function of defective genes will increase with the increased use of cultured and modified pluripotent stem cells which have retained plasticity in that they can be integrated into multiple tissues. It may become possible to derive such pluripotent cells from adult tissues in the future, thus obviating the ethical problems of using embryonic stem cells.

I should emphasise that this brief summary of possible applications of advances in our knowledge of the human genome to the treatment of disease represents hopes, or prospects, for the future, rather than present-day realities. We are, in fact, at the starting point of the long searching for that which is better which I mentioned at the beginning of this talk in connection with the sentiment of Xenophanes. We can just begin now to entertain hopes for improvements in the treatment of common diseases of adult life such as diabetes mellitus of adult onset (type 2), cancer, atherosclerosis, hypertension, and others. Such diseases have multifactorial causation mediated by complex interactions of environmental factors with multiple genes determining susceptibility, which are now being intensively investigated by the establishment of the haplotype map, or HapMap, of the small proportion of gene loci where the DNA sequence varies between individuals (SNPs or single nucleotide polymorphisms). It should be noted that many agents which have been implicated as environmental triggers of such diseases in genetically susceptible individuals, are so deeply ingrained in our way of life that they are hard to avoid. These contributory factors include smoking, unbalanced or excessive diets, a sedentary existence, and exposure to stress, primarily among inhabitants of wealthy nations where excess rather than scarcity is the hallmark of society, and where life expectancy has increased so that the large majority of the population survive into adulthood. Unfortunately, the abolition of the use of tobacco and its products, and the adoption of a healthy diet, are projects which, while very simple in their conception, are difficult in the extreme in their execution. In this context, fields of genetic medicine are being developed today specifically in the area of studying the effects on the individual of pollutants and toxins such as tobacco (toxogenomics or toxicogenomics) and of diet (nutrigenomics).

In the case of cancer, a group of diseases in which mutations in somatic rather than germ cells usually play a primary role, pharmacogenomics can already contribute to the better care of some types. For example, if testing reveals that the genetic complement of a newly diagnosed breast cancer results in the presence of human epidermal growth receptor 2 (HER-2) on the surface of the cell, as is the case in 25–30% of all such patients, then adding a drug such as herceptin (trastuzamab) to the therapeutic regime, is thought to lead to a remarkable improvement in prognosis. Much is also expected of gene therapy for the improvement of prognosis in various other forms of malignant disease. Improved knowledge of the genetic mechanisms involved in pathogenesis can also contribute to earlier detection of cancers when the malignant cell mass is much smaller and the prospects of successful treatment much more promising.

It should be noted that the approaches to therapy by defining aberrations of cell function as a preliminary to correction, which have been discussed, must be accompanied by major advances in computational biology. It is not sufficient to have a catalogue or database of the gene complement of an individual. At least an elementary understanding is needed of the patterns of the virtually infinite number of first, second, and higher order interactions between the myriads of genes (genomics), mRNA gene transcripts (transcriptomics), proteins (proteomics), glycans (glycomics), and metabolites (metabolomics) which underlie the phenotype and the physiology of a single cell. And the functioning, healthy or diseased, of a body organ cannot be defined merely by making an inventory of the functioning of its constituent single cells.

Because of the complexity of multifactorial causation depending on the intricate interactions of environmental factors with multiple genes determining susceptibility, the development of prevention and of treatment for common diseases, including those which have been mentioned, will always be very difficult. But these difficulties will have their compensations. For example, in the case of cancer, the discovery of gene loci (*BRCA1* and *BRCA2* responsible for breast and ovarian cancer, and others), harbouring, at a substantial frequency, alleles which give rise to a strong predisposition to various forms of this common disease, often at a tragically young age, has already been mentioned. All common diseases are heterogeneous in their aetiology and pathogenesis, and there are unusual rare forms subsumed within the generality of each common disease, which are inherited in a simple monogenic manner as in the case already cited of breast and ovarian cancer due to mutations at the *BRCA1* and *BRCA2* gene loci. Intensive study of these monogenically determined subgroups of cancer, diabetes mellitus of adult onset (type 2), atherosclerosis, hypertension and other common diseases, and the uncovering of the genetical abnormalities involved in rare and unusual families of this type, will throw light on the aetiology and pathogenesis of these diseases in general, in part by suggesting leads to the discovery of genes involved in susceptibility to the more common forms.

During these voyages of discovery, as in so many others, we shall be following the precepts contained in a prescient and much-quoted extract of great beauty and power from a letter written by William Harvey in 1657, six weeks before his death, to John Vlackveld, a physician of Harlem in the Netherlands, in reply to an enquiry about a patient. Harvey wrote his letter in Latin, the lingua franca in use within the medical profession at that time.

> Nature is nowhere accustomed more openly to display her secret mysteries than in cases where she shows traces of her workings apart from the beaten path; nor is there any better way to advance the proper practice of medicine than to give our minds to the discovery of the usual law of Nature by careful investigation of rarer forms of disease. For it has been found in almost all things, that what they contain of useful or applicable nature is hardly perceived unless we are deprived of them, or they become deranged in some way.

(Translation from the original Latin, given on p. **xii** into English by R. Willis (1847))

Thus, careful investigation of these rare monogenically determined forms of heterogeneous common diseases cannot fail to throw a great deal of light on the mechanisms of pathogenesis of the corresponding common disease in general.

To turn now to rare genetically determined malformation syndromes, to which, of course, this perspicacious precept of Harvey applies *a fortiori*, I should like to say a word about the disease with which the career of my late colleague and friend, Professor Jérôme Lejeune, was mainly associated, the Down syndrome due to a chromosomal aberration resulting in a trisomy of chromosome 21 which he himself discovered in 1959.

In 1981, Professor Lejeune wrote: *Que savons-nous de la trisomie 21, après vingt ans de recherche? Quel sujet de méditation et d'inquiétude aussi! Certes, nous avons appris bien des choses, et même à reconnaître la maladie chez des enfants très jeunes, encore au ventre de leur mère. Mais si ce pouvoir nouveau a suscité chez certains la tentation d'éliminer les malades extrêmement jeunes, cette connaissance n'a fait en aucun cas régresser la maladie.*

Et c'est pourtant la maladie qu'il faut vaincre, et les patients qu'il faut guérir! (Lejeune 1981)

'What do we know of trisomy 21, after twenty years of research? What a topic this is for meditation and also for deep concern! It is true that we have learned many things, even to the point of recognising the disease in very young children, still carried within the body of their mother. But if this newly acquired power has given rise in some to the temptation to eliminate these extremely young patients, this knowledge has not given rise in a single case to any regression of the disease.

'And yet it is the disease which must be vanquished and the patients who must be cured!'

Professor Lejeune always remained convinced that these patients could be treated and cured. He thought of the manifestations of the Down syndrome as symptoms of a disease to be vanquished, and he totally disagreed with many of his medical colleagues who thought of the condition, following antenatal diagnosis, as a symptom of death, thereby perverting the traditional goal of medicine from a cure to an assault on the patient. He said that he looked forward to the day when a patient with the Down syndrome, treated successfully, becomes a successful geneticist. Possibly, in the new era of functional genomics, some way will be discovered of silencing the supernumerary copy of chromosome 21, perhaps by learning from the mechanism of inactivation of one of the X chromosomes in females.

The question and statement by Professor Lejeune quoted above, echoes another reflection written by William Harvey, extracted from the dedication of his remarkable book *Exercitatio Anatomica de Motu Cordis et Sanguinis in Animalibus* (1628), in which he describes his discovery of the circulation of the blood. The intellectual humility manifested in this extract, despite its exalted provenance from a classic text occupying a fundamental position in the annals of scientific progress, is as relevant today as it was when it was written in the 17th century. It serves as a reminder, if one were needed, of the vast extent of our present ignorance, and as a cautionary antidote to any misplaced conceit with respect to the limited extent of our present knowledge in the field of human genetics, as in other fields of scientific endeavour.

Nec tam angusti animi ut credant quamuis artem aut scientiam adeo omnibus numeris absolutam et perfectam a veteribus traditam, ut aliorum industriae, et diligentiae nihil sit reliquum: cum profiteantur plurimi, maximam partem eorum quae scimus, eorum quae ignoramus minimam esse.

William Harvey

Nor are they so narrow-minded as to believe that any art or any science was ever so absolutely or perfectly taught in all points by the Ancients, that there is nothing remaining to the industry and diligence of others, for there are indeed a great many who openly confess that the greatest part of those things which we do know, is the least of the things which we know not.

(Translation from the original Latin into English by G. Whitteridge 1976)

I should like briefly to mention one more rare genetically determined malformation syndrome, the autosomal recessive multiple malformation syndrome of cryptophthalmos to which my name has been attached as an eponym on the basis of a paper which I wrote more than four decades ago (Fraser 1962a). Over the past three years, two gene loci have been identified where mutant alleles responsible for this heterogeneous entity reside, and the mutant alleles themselves and the proteins to which they give rise have been characterised. Thus, the cryptophthalmos syndrome has entered the modern age of genomics with its hopes for eventual prevention and treatment.

The disabilities associated with the fully expressed form of this condition are even more serious than those associated with the Down syndrome, and yet there are families who are bringing up such children in an atmosphere of harmony and happiness. In a recent case in Germany, antenatal diagnosis of a fully affected female was followed by major foetal surgery on her malformations to ensure her survival, rather than by abortion – not gene therapy *in utero* as yet, but at least conventional treatment in foetal life of a fatal Mendelian disease (Kohl *et al.* 2006).

One of the girls in my 1962 paper was eight years old at that time. I had first met her at the age of five years when she was deaf and blind; she was behaving like a wild animal, reacting with frightened screaming to any attempted approach. I noticed that she liked to press loudly ticking clocks to her skull and, suspecting that her deafness was largely conductive, I took her to an otorhinolaryngologist of my acquaintance. He reconstructed her malformed outer and middle ears, and, at the age of six years, she heard speech for the first time; 45 years later, she lives a life which has been more tolerable than it would have been without any hearing. Perhaps this small contribution to the improvement in the quality of the life of this girl represents a greater achievement than that of my name becoming attached to a syndrome, and even, in abbreviated form as *FRAS1*, to a gene. (See page 187 for a recent photograph.)

There is no doubt that the bringing up of such a disabled child can be an enriching experience for the family if society is prepared to provide some relief from the economic burden involved. Moreover, society as a whole can benefit from such an investment in terms of rewarding employment and spiritual enlightenment, as a result of co-operation in the support and education of such children.

As long ago as the 16th century, in his essay, Of a Monstrous Child, Michel Eyquem de Montaigne (1533–1592) wrote: 'Those whom we call monsters are not so with God, who in the immensity of his work sees the infinity of the forms therein contained'. Four hundred years later, Professor Lionel Penrose who shared with Professor Lejeune his love for patients with the Down syndrome, wrote in 1971, the year before he died – 'The object of medical science in civilised communities is to keep people alive. This principle has no exceptions and it applies also to low-grade defectives of all kinds … … Not only are these low-grade defectives harmless, they are not responsible for their own condition; they can be happy and they can stimulate human feelings and parental love. By all canons of civilised society, they have a right to demand care and comfort even if they are unable to give adequate returns. The ability of a community to make satisfactory provision for its defectives is an index of its own health and progressive development; the desire for their euthanasia is a sign of involution and decay of human standards.' (Penrose 1972)

I should like now to return to the quotation from Plato with which I began this lecture, containing ideas associated with positive eugenics, or improvement of human qualities, to be realised by controlled assortative matings between individuals with superior qualities, however their superiority is defined. The application of human genetics to medicine should be thought of in terms of the individual and of the family, as in the case of all other branches of medicine, and not in the context of future generations. It should also be thought of primarily in terms of combating disease and not in the context of enhancement of the potential of the child. Genetic engineering directed towards the transfer of genes determining higher levels of intelligence or of musical ability, or modifying behaviour in socially desirable directions may never be realised; it is not even going to be possible to find universally acceptable definitions of the quality of the 'enhancement' which could be obtained by such methods if and when such genes are individually identified. We have neither the knowledge nor the wisdom to pursue the eugenic illusions of the past with respect to any hypothetical improvement of our species, and Plato's analogy drawn from breeding hunting dogs is as inappropriate and as irrelevant as analogies drawn from present-day cattle-breeding in order to maximise milk production. While human beings are not candidates for genetical engineering of 'improvements' by such means, they are able to benefit greatly from improvements in education, both qualitative and quantitative in the sense of extending opportunities for the best education to wider segments of society.

With respect to the goal of negative eugenics, also reflected in Plato's writings, of reducing or eliminating the reproduction of the inferior and unfit, not only do we lack the knowledge and wisdom to define the inferior and unfit, but, in addition, all human beings who do not suffer from a condition which renders them infertile have the right to procreate. This right is enshrined in the United Nations Universal Declaration of Human Rights (*Article 16–1 – Men and women of full age, without any limitation due to race, nationality or religion have the right to marry and have a family.*). No human being has the power to interdict the exercise of this right by another human being. And if genetic medicine allows reproductive transmission of hereditary traits which precluded reproduction before treatment became available, any increase in the population frequency of the genes determining such traits will represent a totally insignificant modification of the gene pool.

The primary task of our society is to ensure, entirely independently of the introduction of genetic medicine, the careful conservation of the environment, and hence the transmission of an intact and undiminished physical heritage to our descendants. This is a *sine qua non* with respect to the preservation of our civilisation and of our planet from major catastrophes resulting from our newly acquired mastery of the technological means of self-destruction. Provided that this will be so, these descendants will then be able to ensure the careful conservation of an intact gene pool and the continuing transmission of an undiminished biological heritage, using means whose nature we cannot predict, or even imagine, since they will be based on discoveries which are yet to be made. In the words of Francis Bacon (1561–1626) – 'Men must pursue things which are just in present, and leave the future to the Divine Providence.'

What then are the hopes and the risks of this modern transition of human genetics to illuminate and inform a much broader spectrum of medical activity—human genetic medicine? The main hope is a simple one and it will undoubtedly be realised – an improvement in the health of the population. But the realisation of this improvement will carry within itself its own risk – the accentuation of inequality and injustice. Thus, these new medical activities are both very expensive and very difficult to implement, and their benefits will not be equally available to all members of the human family.

Unfortunately, moreover, the main current sources of morbidity and mortality in the materially disadvantaged, or, in many cases, dispossessed, majority of the population of our planet, are derived from scourges such as, for example, famine, infectious diseases, and lack of access to clean water, where the predominant role of the environment in the determination of the associated adverse effects overwhelms the role of genetical variation. In striving to mitigate the effects of such scourges, we must also ensure that the potential benefits of genetic medicine will not be confined to a few individual members of prosperous societies. Inequalities in access to the benefits of genetic medicine are particularly anomalous precisely because the human genome which is the topic of this conference, is the joint heritage of the world-wide family represented by the members of our species, and constitutes both the pledge and the indelible hallmark of its intrinsic and immanent unity.

Indeed, progress in the direction of equity in access to health care in general is not an optional matter; rather, it is an essential requirement for the long-term stability of our society. Thus, for example, in 1999, women in Japan had a life expectancy of 84.3 years while in Sierra Leone women had a life expectancy of 42% of this figure at 35.4 years. Pope John Paul II wrote in the Evangelium Vitae in 1995 – 'Life is always a good. This is an instinctive perception and a fact of experience, and man is called to grasp the profound reason why this is so.' Such an inequitable distribution of this good is a disgrace to our civilisation as well as a major source of social and political unrest.

To return to our hopes for genetic medicine, I shall not be mentioning possible applications and uses in the fields of reproductive cloning, germ-line gene therapy and postponement of ageing; these topics lie outside the main stream of this short talk. In passing, it would seem presumptuous to embark on major projects involving germ-line gene therapy with its potential effects on future generations, at a time when somatic cell gene therapy has only reached an early experimental stage.

As I have already indicated, some of the major avenues which will lead to therapeutic advances will be in the field of pharmacogenomics, a science which will lead to the development of small-molecule drugs to modulate disease-related pathways in the desired direction, and will refine our knowledge of the variation in the reaction to drugs already in use, helping to avoid the occurrence of side effects which can often be serious. The genetic basis of such reactions to the large range of antipsychotic drugs which are being so widely prescribed in our society, would seem to be a potentially fruitful field for study. I have also mentioned gene therapy as a therapeutic modality which already shows promise in the treatment of diseases such as severe combined immunodeficiency (SCID) syndrome by replacing the function of a defective gene by adding a normal one, even though progress is slow and attended by setbacks representing undesirable results of the procedures used to introduce the normal gene. An alternative strategy which shows promise of avoiding detrimental effects is to replace only the mutated sequences of the abnormal gene, to repair its function rather than to replace it.

Predictive genetic tests for predisposition for common disease will become available, allowing tailoring of protective measures involving changes in life styles and drug administration, to individual genetical susceptibilities. In this field, particularly great care will have to be taken to avoid discrimination with respect to employment and insurance. The mantra of genetic medicine is that genetic profiling will permit the tailoring of health care, preventive strategies, treatments, and interventions to the individual, and will therefore make personalised medicine possible. There are major ethical, legal and social implications of this genetic profiling information particularly in areas such as privacy, insurance, employment, and education. All these implications are connected with risks, and within the Human Genome Project, 3–5% of all funding has been devoted to the study of such ethical, legal, and social issues (ELSI). In addition, I have already mentioned that the expense of genetic medicine can be considered as a risk involving discrimination on economic grounds in that inequitable distribution of health benefits will heighten the already considerable tensions besetting our society.

There is a risk, moreover, that the vast accretion of knowledge about their own genetic complement and the potential harmfulness of some of its constituents may be psychologically deleterious for many individuals and cause them severe anxiety rather than influence them to adapt their life style to their genetic constitution. The remedy is to increase the awareness and expertise in these respects of members of the health care profession so that they can play an enhanced educational role.

We must also consider the risks attendant on the commercialisation of every aspect of human genetics and genetic medicine. This involves the introduction of unjustified testing on a large scale, the patenting of DNA sequences so that royalties become payable on laboratory procedures, greatly increasing their cost, the reluctance to develop drugs for the treatment of rare 'orphan' diseases, and the development and patenting of drugs on the basis of false premises. For example, the Federal Drug Administration (FDA) in the USA has very recently approved a drug called BiDil to treat heart failure in self-identified African Americans and only in self-identified African Americans on the basis of a trial (A-HeFT or African American Heart Failure Trial) which only enrolled such individuals. This episode certainly does not represent an advance in pharmacogenomics or in personalised medicine, and, in fact, has some very grave implications from the point of view of racial discrimination which have been discussed by Kahn (2005).

Thus, while the distribution of genes at the very small proportion of loci where variation occurs, may be different in different 'races', 'race' in itself is not a genetic trait. In this connection, in 1966, Professor Penrose wrote: 'The exact description of the hereditary polymorphisms in our species, which overrun the boundaries of antiquated ideas of racial groups. helps us to comprehend, rather than to deplore, each other's inborn peculiarities.' (Penrose 1966). Nevertheless, discrimination on the basis of supposed and even real differences in the frequency of various genes between different ethnic groups represents a substantial risk arising from the development of human genetics. There are various very large research projects collecting blood and other samples to provide genetic information about gene distributions in large groups of individuals. There is the Human Genome Diversity Project of HUGO (Human Genome Organisation) and there are many Biobanks of various sorts containing large quantities of blood and other samples. While such projects have value in studying such topics as human evolution and migration, as well as differential susceptibility to disease, they also present major risks of misuse and abuse in the realms of political, social, educational, medical, and economic discrimination. Privacy and confidentiality must be primary considerations affecting storage of the data. Such risks of misuse and abuse are much intensified when attempts are made to identify genes responsible for intelligence and for normal and abnormal, including criminal, behaviour, with a view to studying their differential ethnic distribution. So far, no genes have been identified, which have a direct effect on non-pathological variation in intelligence and behaviour, not associated with disease; it may never be possible to do so.

As I have indicated, the hopes attendant on human genetics today can be summarised as the promise of better health in the long term, inseparably connected with the promise of a greater respect for human life in association with the increasing realisation that all the members of our species are inextricably bound together by the possession of the common heritage of our human genome. These hopes can only be achieved against a background of improved education of the public and especially of the healthcare profession, and of a concerted attack on the global inequality which shames our civilisation.

Many of the risks of human genetics today involve geographical, ethnic, and class discrimination leading to restrictions in the beneficial use of the information which is being so rapidly accumulated. There is also the risk that we shall not only fail to use this new information for the benefit of all members of our human species, but actually misuse it to obscure and justify existing inequalities, thereby entrenching the privileges of a small self-selected minority of members of our society who consider themselves to be superior, to the extent of denying the right to reproduction and of life itself to the majority of the weak and vulnerable, whether the weakness and vulnerability are based on poverty or ill-health, or on unjustifiable attributions of lack of fitness on the grounds of physical or intellectual inferiority. A comment by Professor Penrose is relevant in this context – 'New discoveries may take everyone by surprise and scientists have to be continually on guard against misuse of their discoveries by those whose knowledge is incomplete' (Penrose 1966).

We have in part uncovered the nature of the human genome; within the boundaries of this unitary human genome, we cannot define individual overall genetic complements which are qualitatively superior or inferior. Unfortunately, however, such misguided notions of superiority and inferiority, based as they are on incomplete knowledge, are deeply entrenched in our collective psyche and are expressed explicitly even by such revered thinkers as Plato. The disastrous corollaries of these notions in the form of exclusion and of murder have been put into practice repeatedly, culminating in the vast scale of the exclusions and murders which have blighted many regions of the world during the troubled century just past. As we face the dangers of the century to come, we must find the wisdom to prevent the great scientific achievements which have led to the human genetics of today, and which can be regarded as the antithesis of these reverses to which our society has been subject during this period, being used to support such divisive practices, thereby increasing the risk that social and political discord and dissension will jeopardise the future of our civilisation.

In general, we must guard the integrity of our human genome, the joint heritage of the world-wide family represented by the members of our species, entirely independently of class and of ethnic origin, by putting an end to our ecological depredations and by applying ourselves sedulously and meticulously to the protection of the environment from further damage, thereby ensuring the transmission of an intact and undiminished physical heritage to all our descendants so that they are enabled to foster their biological heritage in an appropriate manner in order to underpin improvements in their physical and social circumstances. In this context, I conclude by quoting the last two sentences of the talk which I gave in Rome 44 years ago and which I mentioned at the beginning of this article (Fraser 1962a). 'The complexity of the genetical dynamics of a population can be compared to the complexity of the molecular organisation of a cell. Yet just as a cell functions as a beautifully integrated whole so does the total genetical constitution of a population; and just as a cell is sensitive to a variety of insults, so is the hereditary material of our species to any uncontrolled changes in its environment.'

Acknowledgements

This paper is an expanded version of one presented at the Vatican at the XX International Conference of the Pontifical Council for Health Pastoral Care: The Human Genome – Biological, Medical and Ethical Prospects, held from 17 to 19 November 2005 in the New Synod Hall. The full text has been published in the English, French, Italian and Spanish editions of *Dolentium Hominum*, Number 61, 2006. The bibliographical reference to the English edition is Fraser GR 2006 Human genetics today: hopes and risks. *Dolentium Hominum* **21** (1) 16–23.

I should like to express my sincere gratitude to His Eminence Cardinal Javier Lozano Barragán, to His Excellence Bishop José Luis Redrado, and to the Reverend Professor Angelo Serra for having invited me to Rome to give this talk, and for their permission to reproduce the text in this Festschrift and *liber amicorum*. I should also like to thank them and their staff in the Pontifical Council for Health Pastoral Care for the gracious hospitality which they extended to me, as well as to my wife Maria, during our very pleasant stay in the Holy City.

Is there a biological concept of race?

Jean Gayon

University Paris 1-Panthéon Sorbonne Philosophy Department, 17 rue de La Sorbonne, F-75005 Paris

I am glad to dedicate this philosophical meditation to George Fraser. Both as a geneticist and as humanist thinker, he has thought a lot on the kind of subjects that I have dealt with in this paper.

Introduction

Does our current biological knowledge allow us to construct the category of race such that it describes objectively-existing units in nature? The question is philosophical rather than historical.

In practice, contemporary biologists have generally abandoned the use of the term in scientific discourse, and in particular with regard to humans. Other words are used to categorise intra-specific taxonomic diversity: sub-species, variety, strain, local population, deme, etc. These words are ideologically more neutral than 'race'. Nevertheless, biologists find it difficult when they hold discussions with a public that continues to use the vocabulary of 'race'. For example, when a biologist says 'races do not exist', the exact meaning is generally unclear. Does he or she mean that the notion of race is confused; or that the term does have a precise meaning, but that this does not exist, either in nature in general, or in humans? Pseudo-concept or fiction? This is the question I will dissect in this paper.

First I will study what the category of race could mean for modern biologists as a whole, then I will examine those aspects that specifically relate to humans.

Race and species: a general approach

The modern view of races among biologists is dominated by their understanding of the category of the species. The work by Mayr that is the most incisive in terms of the definition of the species is entitled 'Species concepts and definitions'. It forms the introductory chapter to a collective work *The Species Problem*, edited by Mayr (1957). In this text, Mayr argues that there are three fundamental species concepts that form general 'philosophical' frameworks underlying the various empirical species definitions which include his own 'biological species' concept, the 'morphological' concept, the 'genetic' concept, the concept based on inter-sterility and the 'practical' concept.

The first theoretical or 'primary' species concept is the typological concept, based on the notions of similarity and difference. The biologist using this approach considers the species as a group of individuals that share certain essential properties. The important aspect of the species concept is that the class of individuals that is identified as a species differs from other individuals of the same rank by its 'degree of difference'.

The second primary concept is the 'non-dimensional' concept: 'The essence of this concept is the relationship between two coexisting populations in a non-dimensional system, that is, at a single locality at the same time (sympatric and synchronous)' (Mayr 1957). The relation in question is that of interfertility. Two groups are described as being two species if there is reproductive isolation. This species concept is relational in the same way that, for example, 'being the brother of' is. 'Being a brother' is not inherently relational. Similarly, the nondimensional species concept implies that it cannot be applied to a group of organisms independently of any other group. The nondimensional concept is centred on a distinction, and not a difference.

The third theoretical species concept is the 'multidimensional' concept, which considers that the species is a group of interfertile populations distributed in space and time. This 'collective' species concept thus spreads the notion of reproductive community through space and time. It is thus by definition 'dimensional' because it specifies the spatio-temporal extension of the species. But unlike the nondimensional concept, it does not always lead to clear distinctions.

Mayr's three theoretical species concepts can be applied through various empirical criteria: morphological, genetic, reproductive, ecological or other kinds. These theoretical concepts can help us to clarify the meaning of the concept of biological races.

1. Just as there is a typological species concept, there is a typological concept of race. However, in both cases, it is difficult to apply the concept in the light of modern evolutionary biology: it would require constructing a class of individuals that possess a set of essential properties that other individuals do not possess.

2. There is no 'non-dimensional' concept of race, because there is no reproductive discontinuity between two sympatric and synchronous races. The most operational theoretical species concept thus has no equivalent concept of race.

3. On the other hand, the multidimensional concept can at least partially be applied to races. What biologists and anthropologists call or have called 'races' can be described as populations or groups of populations, situated in space and in time and forming a real or potential reproductive community. This concept of race does not describe the most inclusive or most strictly delimited reproductive community, but it most clearly does describe such a community.

This closes our attempt at definition. If we wish to have a biological concept of race, it will be an entity that oscillates between a type and an evolving reproductive community. The biologist may try to link the two aspects, or to dissociate them. The former approach would be that of the racial anthropology of the beginning of the 19th or 20th centuries, in which the notion of race is a mixture that includes, in a contradictory manner, aspects of an objective classificatory category and of an evolutionary category. On the other hand, the second attitude, adopted by the biologists who developed the evolutionary synthesis, involved dissociating the two classificatory and evolutionary aspects of the concept of race. Dobzhansky, in his 1937 book *Genetics and the Origin of Species*, which founded the evolutionary synthesis, argued that, for the biologist, races as classificatory groups were mere convention, whereas the process of 'raciation', that is of the differentiation of partially reproductively-isolated local populations, is objective. I am not convinced that this terminological approach is in fact coherent. The term 'race' is retained merely for its ease of use. A more radical approach – which is today widely adopted and which can be traced back to the 1930s – would be to stop talking of races for evolving populations, and, noting the lack of any real objective basis to the notion of racial type in natural species, and in particular in human beings, to banish the term completely from biology. Ashley Montagu, although he was not the first person to put forward the idea, persistently argued this point from the beginning of the 1940s onwards (Ashley Montagu 1942).

Bacterial races and species

These debates situate the question of race, and more generally that of infra–specific taxa, in the context of the species concept. Such debates generally take place with regard to sexually-reproducing organisms, in which the reproductive cohesion of the species is clear. Furthermore, such debates are generally motivated by the desire to clarify the meaning of the term 'race' for the species for which it has been most often used, the human species. Before dealing with this point, I would like to provide a counterpoint to the previous discussion, by examining a group of organisms for which the concept of race might be clearer than that of species: bacteria.

Bacteria generally reproduce asexually, by simple division. However, some bacterial processes allow an exchange of genetic material. This exchange takes on a very different form from that shown by eukaryotic interfertile individuals (I rely here on Cowan 1996).

1. In most groups of bacteria, genetic recombination is rare. For example, its frequency is 10^{-8} per genic segment and per generation in *Escherichia coli*.

2. Bacteria rarely exchange their genes, but they can do so with species that are relatively distant in terms of their DNA sequences. Homologous recombination can be seen in species that may diverge for up to 25% of the sequence of their homologous genes. This figure is amazingly high. As a comparison, even in the most primitive animals the level of divergence is not greater than 2%. Furthermore, genes are also exchanged in plasmids, with even more divergent forms.

3. The quantity of genetic material thus exchanged is small compared with the amount exchanged by eukaryotes during a cycle of sexual reproduction. Whereas a eukaryote receives about as equal amounts of DNA from each of its parents, a bacterium receives only a small fraction of the genome of another individual (a few kilobases, less than a thousandth of the bacterial genome).

These aspects of bacterial genetic exchange produce evolutionary scenarios that are very different from those found in the eukaryotic world. In sexually-reproducing animals and plants, natural selection has only a limited effect on genetic diversity. The extremely high level of recombination (which is strictly speaking maximal because, at each generation, two entire genomes are associated in the same organism) means that an advantageous mutation is diffused in a population without bringing with it the whole genome of the mutant individual. In bacteria, given that recombination is rare, the selection of an advantageous mutation carries with it the whole genome of the mutant individual. Any natural selection event therefore leads to a 'purge of diversity' in the local clone in which it occurs. Therefore genetic exchange does not preserve genetic diversity among bacterial populations. Furthermore, various models show that, even if two neighbouring populations are not absolutely isolated one from each other from a reproductive point of view (that is, if the intensity of genetic exchange is the same within and among neighbouring populations), the rate of recombination is too small for preventing the ecological divergence of the populations.

In such a biological situation, the traditional category of race is much more plausible than in fully sexually reproducing organisms. A local population, ecologically adapted, can be characterised both as a robust type (that is, as a class of homogeneous individuals) and as an evolutionary line with a limited life-span. Similarly, the notion of species is much more problematic than in the case of eukaryotes, where the species is a closed space of genetic exchange. Each eukaryotic species thus has to develop its fresh adaptations *de novo*. For example, if several insect species become resistant to a given insecticide, this will have taken place on the basis of their own specific mutations, and perhaps through very different physiological mechanisms. In bacteria, however, adaptation to local conditions can often pass from one species to another. This often takes place, for example, with factors associated with resistance to antibiotics. The existence of this kind of exchange makes it difficult to apply the category of species to bacteria. In the biological sense of the term, bacterial species are biological species, that is, cohesive and isolated reproductive communities, only for a faction of their genome – the fraction that is not involved in genetic exchange, and that is transmitted in a strictly vertical manner (that is, through the asexual process of reproduction by division).

Bacteria therefore present us with a set of organisms in which the notion of race is easier to deal with than the notion of species. Locally adapted populations of bacteria are good candidates to be considered as races, because they possess the very properties that make this status so problematic in the case of sexually-reproducing plants and animals. They are entities than can be described both as types (transitory types, but types nevertheless), as temporary but coherent reproductive communities, and furthermore as reproductive communities that are not isolated from other similar groups.

I have dealt with this example in some details in order to show that the category of race is not in and of itself undermined by such conceptual problems that it cannot be applied to any real biological situation. Of course, the naturalists and anthropologists who have used the category of race (or those that criticised it) since the 18th century did not know the details of bacterial genetic exchange. They hardly developed the concept to explain the particularities of the life-cycle of these organisms. Nevertheless, the example of bacteria shows that the pertinence of the category of biological race is not only a conceptual question. It also has empirical aspects. In turn, this implies that there well may be good reasons to reject the use of the concept in some biological situations that we find particularly interesting, in particular in the case of humans.

Human races

I will distinguish two aspects of the question of the meaning of the category of race in the case of humans: its possible biological pertinence, and its social meaning.

Are there human races, biologically speaking?

If biologists declare that human races do or do not exist, what concept of race are they using in arriving at their position?

Since the Second World War, the main position has been to say that, biologically speaking, there are no human races. The general historical context of suspicion with regard to racial and/or racist ideologies is no doubt largely responsible for this evolution on the part of biologists, but this should not lead us to ignore the scientific argument that underlies it. This argument is linked to an important change in the methods used to describe human biological diversity.

At the beginning of the 20th century, physical anthropology was based on obvious morphological characters: height, height-weight ratio, skull shape, jaw shape, skin and hair colour, hair texture, and other similar characters. Such characters are highly heritable, and generally correspond to those that form popular opinions of racial differences. However, these characters posed an enormous problem for biologists who wanted to find objective criteria to separate races. In reality, the geographical distributions of most of the anthropological indices show little or no overlap. A way round this difficulty was found by developing composite indices, thus presenting human races as mean types, 'omelette statistics' as Montagu put it.

In the 1937 book cited above, Dobzhansky strongly criticised this method. He pointed out that, from a genetical point of view, the only things that are transmitted from generation to generation are genes. To compare populations from the point of view of their hereditary attributes, it would be necessary to measure gene frequencies rather than the means of quantitative characters (height, weight, etc.). When this kind of study is carried out on animals, populations are rarely differentiated by alleles that are present in all the members of one population, but absent in all the members of another. In the case of humans, Dobzhansky gave the example of the only genetic system which was reasonably well understood at the time in human populations, the genetic system that underlines the ABO blood types. The ABO system does not reveal the existence of clear-cut races, but only of clines, that is of continuous gradients of gene frequencies. After the war, this result was found time and time again with a wide variety of genetic systems, with the added finding that the distributions of these gene systems did not overlap any more than did those of anthropometric characters.

This modern biologist's rejection of the word 'race' is based on a typological concept of race or, to put it another way, a concept of race that has been constructed with the aim of classification. Some biologists, like Dobzhansky, have argued that it is impossible to construct human types on the basis of genetic criteria by saying that racial classifications, although they may sometimes be useful from a practical point of view, are nevertheless constructions of convenience. However, Dobzhansky argued that it would be better not to ban the use of the term race, because he considered that racial hatred could also feed on other terms. It would therefore be better to retain a term which, despite being scientifically confused, is nevertheless inevitable in all discussions between scientists and the public. Other scientists, such as Ashley Montagu – one of Dobzhansky's close friends – argued in favour of the complete abandonment of racial vocabulary, at least in the case of humans (for more details of this story, see Gayon 2003).

A new situation developed around 1970, when population geneticists like Luca Cavalli-Sforza began to develop statistical methods that made it possible to use the genetic makeup of current human populations in order to infer the history of the differentiation of modern humans (Cavalli-Sforza *et al.* 1994). These methods involved studying as many genetic systems as possible (over 200 are known today), all of them selectively neutral, and calculating an index of the genetic distance between populations. By relating palaeontological and prehistoric indices, it is possible to construct a plausible scenario of the spread of human beings over the planet from a single hypothetical small population, which might have lived in Africa or the Middle East more than 100,000 years ago. This scenario, which had a substantial impact on the popular imagination, implies that populations of modern humans became sufficiently reproductively isolated as they spread across the face of the globe to become differentiated. This view of the origins of modern human diversity replaced the adaptationist interpretations that dominated from the 1940s to around 1980 (Dobzhansky was an important representative of this position). Natural selection is generally invoked today only for the most obvious morphological characters (size, body-shape, skin-colour, etc.). These characters, although they are of major importance in the social perception of the physical differences between human populations, correspond to a mere fraction of the actual genetic diversity.

Cavalli-Sforza has always rejected any use of the concept of human races. However, an American philosopher, Robin Andreasen, has recently argued that the modern interpretation of the origins of human genetic diversity in fact renders the application of the notion of race to humans plausible, in the following sense. Andreasen (1998) defines the race, or sub-species, as a monophyletic unit. The idea, which is in fact relatively straightforward, is that genetic anthropology is able to make phylogenetic inferences about the origin of human diversity and to construct phylogenetic trees because it has shown (or, in some contexts, postulates) that, for a certain length of time, the populations of modern humans that migrated from Africa were reproductively isolated. According to Andreasen, this kind of analysis involves a concept of race that has the following characteristics: it is a diachronic concept, which does not suggest that races exist today, but that they have existed, not as types, but as reproductively isolated evolutionary lines. Andreasen calls this concept of race, which she imputes to genetic anthropology, a 'cladistic' conception. This is based on the idea that races, if they exist, should be considered as clades (monophyletic units), and not as grades (degrees on a progressive evolutionary scale). The 'cladistic conception of race' in fact simply signifies that races are conceived as monophyletic groups, or at least as sufficiently isolated to be approximately treated as such. In fact, Andreasen emphasises that there is no *a priori* reason that the concept should apply to the whole history of a species, and to that of humans in particular. In other words, the human species may well have been composed of races (as in clades) in the past, even though it can no longer be characterised as such today.

To summarise, the debates on human races raised by genetic anthropology confirm the bipolarity of this notion: sometimes it refers to the notion of type (a classificatory entity), sometime to a spatio-temporal entity (a reproductive community).

Races as social signifiers

I will close by relativising the biological approaches to the notion of race to which I have devoted most of this communication. In a remarkable book published around 30 years ago, the sociologist Colette Guillaumin argued that we should distinguish two levels of discussion in the question of human races: the 'concrete' level which, she argues, is that of biological research, and the 'symbolic' level, which relates to the function of the signifier 'race' in modern societies. Guillaumin insisted that the question of race as social signifier is separate from that of the result of scientific debates on races as natural objects.

This thesis was associated to two others. Firstly, Guillaumin did not consider that race as a social operator was necessarily a notion with a conceptual nature. Rather, she argued it was a fetish-notion, the point of which was not to know whether it existed or not, but to know what it produced in practice. '*That* does not exist. That leads to death. It is a murder machine, a technical murder machine. Of proven efficacy. It is a way of rationalising and organising the murderous violence and the domination of some social groups over other social groups that have been rendered powerless' (Guillaumin 1981).

Second, the symbolic race, although it might not have any objective biological meaning, nevertheless maintains a close relation with biological signifiers. Races are social categories covered in a biological veneer. It does not really matter whether this veneer is real, fictitious or artificial (as for example with circumcision amongst Jews), the important point is that the social space that is the race is marked by a *physical* trait, an inscription on the body. The function of this 'biologification' is to radicalise differences by presenting them as unchanging and irreversible.

Guillaumin's sociological analysis can shed light on the profound causes of the historical situation that I mentioned at the beginning: naturalists have used the term 'race' much more often in reference to humans than to other organisms. It should also be added that this term has always been widely, if not systematically, used with regard to varieties of domestic plants and animals, in other words, with regard to organisms that are marked by human constraints. This preferential presence of the racist vocabulary in the strictly human habitat, in our *domus*, can be understood if 'race' is above all a social signifier. If we look at the question from this point of view, it becomes difficult to consider neutrally the use of the word 'race' by scientists.

Smugness compounded

Robert J. Gorlin*

Oral Pathology and Genetics, University of Minnesota, Minneapolis

I have known George Fraser for a very long time. The most substantial thing which we have in common professionally is that in 1976 two books were published – B.W. Konigsmark and R.J.Gorlin, *Genetic and Metabolic Deafness* W.B. Saunders, Philadelphia, 1976 and G.R. Fraser, *The Causes of Profound Deafness in Childhood: A Study of 3,535 Individuals with Severe Hearing Loss Present at Birth or of Childhood Onset* (with foreword by V.A. McKusick) Johns Hopkins University Press, Baltimore and London, 1976.

Victor McKusick played an important role in the production of both books. In the case of the first, he had encouraged Bruce Konigsmark, a neuropathologist by training, to turn his attention to the genetics of deafness before his tragically early death a few months before our book was published. In the case of the second book, he had encouraged the Johns Hopkins University Press to publish George's manuscript, and he had written a foreword.

In 1977, I was asked to review George's book. In the penultimate sentence of my review (*American Journal of Medical Genetics* 1 119–121, 1977), I wrote: 'In summary, I liked this text.' Now almost thirty years later, I hope that George will like my text for his Festschrift and *liber amicorum*.

One of the problems of putting on years is that we often tend to become smug. A few of us start out that way but much of the smugness which we older ones acquire is based on the assumption that years of experience give us a special cachet. Well, let me tell you of two examples that show how flawed that logic is.

My first instance occurred almost ten years ago – when I was barely 72 – a patient with naevoid basal cell carcinoma syndrome told me about her 'scratchy eyes'. I was a bit nonplussed! Did she mean 'itchy eyes'? Casual inspection revealed no incipient conjunctivitis. Was she allergic? She revealed no history of stuffed or runny nose. My initial reaction was to shrug my shoulders and refer her to a friend in ophthalmology, to assure her that she had 'nothing'. Inviting me to observe, he everted her tarsal plates and showed me several superficial palpebral cysts. They, fortunately, were evanescent and needed no therapy. Although at that time I had seen several hundred patients with the syndrome, I had never heard of this complaint. Now that my curiosity was aroused, I needed to know how often they occurred and how many patients were aware of their existence. At National Support Group meetings, on asking those attending, I found that about 40% were cognisant of their presence but had never complained about them.

Another example is more recent. Raoul Hennekam e-mailed me about a random hair-

---

* Sadly, Robert Gorlin died on 29 August 2006.

swatch in patients with the syndrome. Good Lord, I had seen about 500 patients in these 50 years in the field. Surely, Raoul, this is a chance finding! Within hours of posting the enquiry on the web, at least half a dozen patients told me about extra patches of hair – one on the wrist which was covered by a large watch, another above the knee which was shaved daily, still others on the lateral aspect of the neck, below the scapula, and in sundry other sites. What a shock! Smugness strikes again – and still again!

Cancer, genetics, ethics and genetic counselling

Ernest B. Hook

School of Public Health, University of California, Berkeley CA 94720–7360

I have known George Fraser in many capacities. I first became aware of his work in reading an important paper on the human genetic load which examined aspects of human variation. In passing, this presented information on a syndrome, later to be named 'Fraser syndrome' (now also known as 'Cryptophthalmos syndrome'), which provided an interesting example pertinent to human 'load' and the issues involved in analysing it (Fraser 1962a). I found it extraordinary that an illustrative case described in an article on population genetics should have later taken on a life of its own, so to speak, in clinical genetics.

Later I became aware of other major contributions to medical genetics, for instance important monographs by him on blindness (Fraser and Friedmann 1967) and deafness (Fraser 1976). Clinical geneticists and those working in the specialties covered found them of immense value. And they served as testament to his immense ability and energy in pursuing and synthesising relevant material. Further indications of his contributions to clinical genetics appear in the index of the 10th edition of *Mendelian Inheritance in Man* which contains 29 references to his work in this field.

I did not observe the professional activities in which he had been involved before and after we overlapped at the medical genetics program at the University of Washington, Seattle. I knew he had worked at various times in statistical genetics, had held a major chair in clinical genetics in Newfoundland, and undertaken important administrative roles bearing on human and medical genetics both in Ottawa with the Canadian National Agency concerned with such matters, and then later at National Institute of Health (NIH) in Bethesda, Maryland in the USA. Whenever I have encountered him, he has projected an image of a person with complete and absolute dedication to his work at the time and to the responsibilities of his then professional role.

In the 1990s he returned to the United Kingdom, the country in which he began his first serious exposure to genetics as an undergraduate student of R.A. Fisher, and later, in which he achieved all his higher degrees. Here, when I visited him in 1995 and then 1998 at the Oxford University Clinical Genetics Unit, I had for the first time an opportunity to observe his activities at close hand.

He then ran the genetic counselling section of the cancer genetics part of the unit. He gave this the complete dedication he had given to his previous professional activities. He had no previous extensive involvement in genetic counselling for cancer. And, in many clinical genetic units, such work often gets short shrift from physicians, who delegate it to nurses or master's degree trained genetic counsellors. But while he had no such assistants and practically no support or facilities except small office space, he threw himself totally into this work.

In his counselling sessions he took immense efforts. He explained matters carefully and patiently to clients, then he counselled them, then he reviewed it all again. He followed this up with sympathetic, supportive and informative letters which he typed then printed and sent himself. He composed his letters with lengthy painstaking care. On occasion I would find him at night in his small office preparing them.

Frequently, in his counselling sessions, he could not resolve the unresolvable dilemmas in which many clients found themselves as to whether she/he or his/her children would eventually manifest cancer. He demonstrated patience, sympathy and ability to listen and to explain the situation and what meagre steps might be available. Judging by their responses, his clients felt great gratitude for his efforts. And he did not give false hope to those to whom such could not ethically be offered. His clients respected him for his integrity in this. Nevertheless, he enabled them to deal better with the despair of uncertainty. All of his clients recognised their experience with him as different from that which they had with previous physicians. His obviously carefully thought out extensive follow up letter led many to reply, writing of their gratitude for their experience with him.

To George Fraser, the expressed gratitude of his clients when they left their sessions with him, and the letters which they took the effort to write afterwards in appreciation of his work, provided great satisfaction, happiness and a sense of fulfilment. They testified directly to his contribution to the welfare of his fellow human beings. I think he regarded these responses as the ultimate tangible reward for the immense time and labour he put into his efforts, indeed efforts far and above that called for by his part time position.

This contribution to humanity which he took on in this work was consistent with the attitude to society he demonstrated as long as I have known him: a deep concern for his fellow human beings and for ethical aspects of behaviour. He thought about these issues considerably and tried to live his life, to the extent anyone can, in balancing dilemmas about the choices one makes. Sadistic or callous behaviour by others – whether on large scale or small – troubled him greatly, especially when they did not appear of concern to those aware of them.

His knowledge of classical literature enabled him to share annually with his friends extracts from a pre-Socratic philosopher Xenophanes (*fl.* 570 BCE). These extracts led me to investigate the writings of Xenophanes, and commentaries upon them. Xenophanes' precepts on ethics, human behaviour, and the role of society obviously had influenced George Fraser.

Xenophanes we would regard today as one who tried to reform religion from an ethical perspective. He opposed the anthropomorphic polytheism of Homer and Hesiod. He attacked them as directly responsible for the moral corruption of the time for 'Such things of the Gods that are related by Homer and Hesiod as would be shame and abiding disgrace to any of mankind; promises broken, and thefts, and the one deceiving the other' and as well adulteries (cited in Lewes 1875, p. 38). And if the Gods can do this with impunity what guidance then is their behaviour to that of humanity? Xenophanes did not reject religion but felt that a divine 'being' must provide ethical goals for humanity. A human society which achieved these goals would cherish all of its citizens. Whatever tragedies occurred would not originate knowingly from human behaviour.

These views of course may have been expressed earlier by others whose works did not survive even to later classical writers. Indeed one wonders to what extent a predecessor, of whom we are unaware, influenced him, and to what extent these ideas originated with him. They certainly resonate with those of many later philosophers and indeed of ourselves today. In any event, the writings of Xenophanes, to my knowledge, provide their first such explicit surviving expression. And I suspect they provided George Fraser since an early age when he first encountered them a guide to conduct that has stayed with him all of his days.

Blood transfusion, HIV and 'race': the case of South Africa

Trefor Jenkins

Professor Emeritus, Department of Human Genetics, National Health Laboratory Service,
and Division of Bioethics, University of the Witwatersrand, Johannesburg

Many countries in sub-Saharan Africa have human immunodeficiency virus (HIV)-positive prevalence rates in adults of up to 30%; in most of them blood transfusion services have collapsed. In those countries still boasting a blood transfusion service, it is estimated that 5–10% of HIV-positive individuals have acquired the infection from blood transfusion; this is not surprising because at least 25% of blood transfused in Africa is not screened for HIV (WHO 2001). Ironically, the main indication for blood transfusion in Botswana is AIDS-induced anaemia.

Although many African countries do not offer population screening for HIV, even in those that do, like South Africa, the uptake of the test is, regrettably, still extremely low. Fear of stigmatisation, denial of the problem, and a fatalistic view of life are all factors contributing to this.

After centuries of racial discrimination, and over 40 years of institutionalised racism, called *apartheid*, which only ended in 1994, the black population of South Africa still suffers the consequences of living under these unjust systems. Until a year ago, the South African National Blood Service (SANBS) employed a policy of assigning blood donors to risk categories, using 'race' as a surrogate for high risk behaviour (Heyns *et al.* 2006). Those categorised as being 'high risk' were largely Black African donors. However, 'race' is associated with many variables, including culture, health beliefs (the majority of Africans, when sick, are said to consult traditional healers as well as practitioners of allopathic medicine), language, and, particularly, socioeconomic status. This practice was discriminatory and may be illegal in terms of the South African constitution, unless the discrimination were to be considered by the Constitutional Court as not unfair. Census statistics still use 'race' as a category with the objective of using these data to monitor and to redress the past discrimination.

Following the publicity surrounding the exposé of the alleged 'racist' practices of the SANBS, in not using the whole blood of black donors for transfusion, the Minister of Health ordered the SANBS to introduce a new practice model, ensuring the safety of the blood supply by (a) treating the blood donated by Black blood donors in the same way as they treat blood from other donors; (b) testing all donated blood units for HIV by the sensitive nucleic acid technology (NAT), which will, it is predicted, reduce significantly the 'window period' of infectivity (when the blood is infective even though the conventional antibody test for HIV is negative) from weeks, or even months, to 5–11 days; and (c) placing greater emphasis on encouraging donors to give regular donations, a strategy known to reduce the proportion of donors who are HIV-positive, or in the 'window period'. Bekker and Wood (2006) have claimed that the introduction of this policy illustrates how planners of public health medicine tend to ignore the societal roots of poor health in favour of interventions which operate 'further along the pathway'. In the present case, it seems to be easier to introduce more sophisticated screening technologies, at high cost, rather than tackle the underlying social inequalities which, it is argued, are responsible for the high HIV infection rate in the majority Black African population group.

Titmuss (1970), in his classic book on blood transfusion, extolled the virtues of voluntary, non-remunerated blood donors (the only donors in the UK – then, as now!), contrasting the situation in the UK with that in the USA where, at that time, only about 10% of blood donations came from 'the voluntary community donor', 90% coming from the down-and-outs, often selling their blood to pay for their next 'fix'.

Blood can be lethal to the patient receiving it – not only from incompetent cross-matching or mislabelling or careless misidentification of the patient, but from the risks of disease transmission: first it was syphilis; then hepatitis (B and C) were added and, since 1982, HIV/AIDS, at present still the most feared of blood transfusion risks. If the prospective donor has not acknowledged his/her recent 'high risk' behaviour (on the form which s/he is obliged to complete before the blood donation is actually collected), or is not aware of a partner's HIV status or 'high risk' behaviour, the donated blood can be issued to some needy patient with disastrous consequences. The donor may be genuinely unaware of the 'high risk': her husband, for example, may have acquired HIV from a sexual contact outside the marriage. Upon the honesty of the donor (or his or her sexual partner) depends the life of the recipient of his or her blood. How can a blood transfusion service encourage maximum truthfulness on the part of the prospective donor? In a society like South Africa, at the present time, the alarmingly high HIV prevalence rate is accompanied by denial which does not encourage individuals to seek counselling and/or testing.

Unlike the situation that existed, as late as 1968, in Louisiana and Alabama, where blood was labelled with the race of the donor and transfusions limited to patients/recipients of the same race, in South Africa, during the apartheid era, there was no such restriction. In fact, over 90% of donors were white, sufficient to supply the needs of most patients; white blood donors were not deterred from donating blood because of knowing that their blood was very likely to be used to treat a black fellow countryman.

When hepatitis B became a major transfusion hazard in the 1970s, and testing for its presence was not very efficient, Dr M Shapiro (1976), one of the founders, and the first Director, of the South African Blood Transfusion Service, from 1940 to 1990, justified giving blood from black donors only to black patients, claiming that, as a population, black patients would have a higher proportion of patients who possessed antibodies to the virus! Shapiro was criticised for this practice of giving 'black' blood only to black patients, and did not seem to acknowledge that the practice was racist. When in December 2004, the SANBS was accused of applying a racist policy with respect to donors (i.e. not using the blood collected from African donors), it needed to be pointed out that neither White nor Black patients were being transfused with the whole blood obtained from Black donors – i.e. at this level of implementation of blood transfusion policy, it was not racist. It was a case of the safest possible blood for all patients, irrespective of 'race'; it represented major progress from the practice in South Africa in the 1970s!

Blood transfusion has become an essential component of modern medical practice. The collection of blood, and its distribution and transfusion into patients, raise a number of ethical and legal issues:

Informed consent and the privacy of prospective donors
Donors can certainly expect to be informed of the details of the procedures involved in donating;

they will be warned of the mild pain experienced, and told of the confidential nature of the interview prior to acceptance as a donor – all are agreed that the mere completion of a questionnaire by the donor is not adequate and that an experienced interviewer (in fact, a 'counsellor'), ought to elicit an accurate history of possible high risk behaviour, residence in countries where tropical diseases are common, etc. Such a thorough counselling session is not standard practice. The potential donor should understand that tests for HIV, hepatitis B and C (employing the new NAT-technology), as well as for other pathogens, will be carried out on the blood and the results conveyed to him/her, emphasising the need for post-test counselling. This procedure of pre- and post-test counselling, coupled with voluntary HIV testing, has become an integral part of blood transfusion services in developing countries in Africa, where high HIV prevalence rates have virtually destroyed blood transfusion services. However, at least 25% of blood transfused in Africa is not screened for HIV (WHO 2001), and there is an urgent need for community-based national blood services to be established and for them to recruit panels of voluntary non-remunerated donors. Some success in this regard has been achieved in Botswana and Nigeria, where the US-based *Safe Blood for Africa Foundation* has assisted these and other sub-Saharan African countries, in setting up such services (*www.safebloodinternational.org*) (Field 2004).

Privacy of the blood donor is a very important ethical issue which tends to be neglected or compromised. Blood collection clinics are, it is hoped, busy events, when the staff are motivated to collect as many units of blood as possible. They are held in public places like church and community halls and even at shopping centres. Space is often at a premium and it is difficult to provide sufficient private rooms in which to interview the potential donors. A *Self-exclusion Questionnaire* is also completed; there are fewer questions, but they certainly need to be asked by a skilled interviewer in soundproof rooms. Using the conventional procedure in 2004, where potential donors completed the standard form, an analysis of the nearly 523,000 donations, collected by one region of the SANBS, showed that 351 were found to be HIV-positive; when broken down into the different population groups, gender and risk categories, it was found that the HIV-positive rate (expressed as number positive per 100,000) was over 4,000 in new Black female donors (i.e. 4%) and around 3000 in Black male donors (i.e. 3%); the HIV-positive rate dropped as one progressed through the other categories, with the 'safest' donors, from the point of view of HIV risk, being 'repeat' female Asian donors. (See Heyns *et al.* 2006 for a more detailed analysis of larger data sets over a longer period of time.)

Do all individuals have the right to donate blood, irrespective of their sexual orientation or of their population's 'risk profile'? The issue with respect to gay men has been discussed by Brooks (2004), with particular reference to the situation in the USA. He concluded that changing the rule and allowing gay men to donate blood for transfusion would increase the risk of HIV transmission. Such a dilemma is part of the broader issue of the responsibilities of blood services to donors and recipients, and Brooks (2004) concluded: 'Blood services should base decisions regarding donor suitability on science rather than on their donors' desires. Blood Services must recognise that the rights of blood recipients should supersede asserted rights of blood donors.'

Schuklenk (2006) posed the question: 'Do men who have sex with men in South Africa really have a higher prevalence of HIV infection than other groups of people in the country?' The data probably do not exist – and may be very difficult to ascertain – but the main issue he highlights is that some categories of heterosexual women and men have a very high prevalence of HIV (approaching, or even exceeding 30% in many areas of the country), and yet they are not at present excluded from donating in terms of the new policy. If the laboratory testing of donated blood has been refined to the extent projected, then most donors in the window period of HIV and Hepatitis B and C infections will be detected and yet whole blood and certain blood products from these donors would not be used for transfusion purposes. Schuklenk suggests that the current policies in South Africa, with respect to gay male donors, may not be equitable.

Informed consent of patients in need of blood transfusion

Medical colleagues are extremely lax about the ethical requirement (which, incidentally, is also a legal requirement!) of obtaining the informed consent of a patient before administering a blood transfusion. In fact, there seems to be some confusion about who is responsible for giving the transfusion! The prescribing doctor, one would have thought, would discuss with the patient the need or indication for the transfusion and the consequences of not consenting to it. It is also important that the patient clearly understands the necessity, or merely the possible or probable benefit of the treatment (if it is a life-saving procedure there will be no discussion!), as well as the risks of acquiring an infection from the blood transfused; the patient may want to know the probabilities in his or her country of residence.

There is a need to emphasise that the patient's signature on an admission to hospital from 'consenting' to everything that might be done to her in the hospital, does not constitute 'informed consent'; in fact, a signature at the bottom of a form does not prove 'valid consent'.

Allocation of scarce resources

The new model adopted by the SANBS in 2005 to use the blood donated by all population groups ('races') was facilitated, to a large extent, by NAT-testing every unit of blood donated for transfusion. The cost of health care has, as a result, been increased. The ethical principle of distributive justice is the rubric under which this issue must be discussed. There never seems to be enough money available for the health care which needs to be provided by the State for the majority (about 80%) of its people in South Africa who cannot afford private health care and the State's health care service is experiencing difficulties in coping with the demands placed on it. Many nurses, doctors, pharmacists, etc move to the private health care sector (for better pay and better working conditions) or leave the country.

The introduction of NAT-technology by the SANBS and by the Western Province Blood Service (WPBS) has resulted in a 20% increase in the cost of blood and blood products, the total cost being R120–150 million per year. About half of the blood is used in the private sector and, as a consequence, members of medical insurance schemes will have to pay more for 'cover'; but the public sector's health budget will also have to be increased or economies made elsewhere. This raises ethical issues around the allocation of scarce resources or distributive justice. Does this modification of the practice of blood transfusion medicine justify the 20% increase in cost? Could the former model have been retained and the money be put to better use?

This is another example which illustrates the concern of Becker and Wood (2006), mentioned above, that expensive treatments may take the place of prevention because the latter, which requires addressing underlying social inequities, is too difficult.

There is no doubt that, as portrayed in the press in December 2004 and in the months following the 'exposé', this is an extremely emotional issue, viewed as a form of racism, unacceptable in the new South Africa. The practice of collecting blood from all volunteers who presented and complied with the stated criteria, but then not *using* the donation for the stated purpose because the donor belonged to a 'high risk category' was deceitful and ethically unacceptable.

Consequences of the new blood transfusion policy: dealing with any problems
It is expected by the SANBS (and the WPBS) that the new policy governing blood donation will ensure a safe, adequate supply of blood for the country. The new NAT-technology holds great promise but we cannot ignore the phenomenally high prevalence of HIV in the country and the estimate that 500,000 individuals are being infected with HIV each year – there are no signs of the epidemic abating!

If there were an increase in the number of patients infected with HIV, due to the failure of NAT-technology to detect individuals who had only very recently been infected with HIV, or to the other changes in policy being ineffective, the State ought, it may be argued, accept responsibility. The SANBS will be unlikely to secure insurance cover for such an eventuality – as it did prior to the switch to the new policy. Would the health care services give priority to such patients for enrolment in their anti-retroviral (ARV) programme which is reaching a small proportion of AIDS patients?

Directed or designated donor blood transfusion
Some private hospitals in the USA have their own blood banks and these are well placed to deal directly with patients who may choose to have directed blood donors supply their transfusion needs. I have found a number of medical colleagues, fearful of a decline in the quality of the blood transfusion service because of the new policy, who say they would elect to have directed blood donations if the need arose.

It seems unlikely that a private hospital-based, directed blood donation service would be licensed in South Africa; it would be at variance with the stated policy of having only one licensed national blood donation service.

If the new model of donor safety which was implemented in October 2005 turns out to be less safe than the former practice of the SANBS, it is possible that the private health care service, which caters for the needs of about 20% of the South African population, including a disproportionately large proportion of white citizens, might be tempted to set up its own blood transfusion service, claiming that it uses directed blood donors. Such a dichotomy would widen the gap between the private and the public health sectors even further, the poor and disadvantaged members of society being the main losers.

Population 'profiling' and public health risk
The racial profiling used by the SANBS for several years prior to the introduction of its new policy in October 2005, was widely criticised in South Africa by the Minister of Health, the Human Rights Commission, the president of the South African Medical Association, Human Rights groups, and many other individuals and social groups.

It was considered to be racist and reminiscent of the worst practices used by the South African government, and its agencies, during the *apartheid* era. 'Race classification' was a tool used by the exclusively White government to subjugate the black population and the relatively small groups of 'Coloureds' (people of mixed ancestry), and 'Indians or Asians'. The scientific basis for such a classification is not clear and the genetic profiles of these populations are not clearly demarcated from each other, as the 'race classification boards', and the courts which sat in judgment during the apartheid era, found out.

The identity of these major population groups is not clear-cut – the overlap in gene frequencies between the groups reflects the long history of intermarriage and 'mixing' between them. Nevertheless, the names given to the four groups were entrenched in laws, passed by the apartheid government, and give members of these groups identities which are in common usage to this day in census returns and for gathering health statistics. Blood transfusion services in South Africa were required by law to *collect* the blood of members belonging to these different race/ethnic groups in separate facilities; the 'race' of the donor had to be stated on each unit of blood. The variable race/ethnicity categories are associated with a range of other variables, causally and incidentally and can sometimes serve as a proxy for other variables, one of which is HIV-status, although, in the view of some people, the use of race/ethnicity for the profiling of blood donors is considered to stigmatise black donors. But HIV-status has nothing to do with genetics which determines race/ethnicity. HIV status is a reflection, rather, of 'the role that stereotyping, discrimination and related "structural violence" (Farmer 2003) have played in creating these associations. In the case of HIV in South Africa, this would involve emphasising the role that apartheid played in the differential spread of HIV and in the legacy of inequalities in education, income, health and access to health care that continue to influence the impact of HIV/AIDS' (Ellison 2005).

Dedication

It is a pleasure and a privilege to contribute to this Festschrift and *liber amicorum* in honour of George Fraser on the occasion of his 75th birthday.

George is one of the doyens of our discipline, internationally acclaimed for his classic volumes on the causes of blindness in childhood (1967) and the causes of profound deafness at birth or of childhood onset (1976).

Some of George's early research efforts were concerned with the (red cell antigen) blood groups; this is hardly surprising when one considers that his undergraduate teachers at Cambridge were R. A. Fisher, who contributed so much to our understanding of blood-group genetics including that of the Rhesus system, and Robin Coombs, who made seminal contributions to the techniques employed in the laboratory demonstration of 'incomplete' antibodies, in particular those which were necessary for elucidating the genetics of the Rhesus and other blood-group systems. As an example of this interest in blood groups, George, together with Oliver Mayo, one of the editors of this Festschrift and *liber amicorum*, and with George Stamatoyannopoulos and Steve Wiesenfeld, discovered the fascinating association between the ABO blood groups and serum cholesterol (Mayo *et al.* 1969c, 1971i), a finding which needs to be explored further at the molecular level. (See Edwards's contribution in this volume.)

George Fraser will long be remembered, with gratitude, for his translation of Benno Müller-Hill's book *Murderous Science* (1988), which provided non-German-speaking human geneticists with insights into how and why German human genetics descended into the perverted science of the Nazi era.

The major concern which George so clearly shows for people with congenital abnormalities and defects of sensory perception, extending to a passionate advocacy for the rights of the disabled, was kindled by his early association with Lionel Penrose, whose faithful disciple he has remained for the first 50 years of his professional life.

Birthday Greetings, George, and many Happy Returns!

In friendship,
Trefor Jenkins

Concerns

Benno Müller-Hill

Institute of Genetics, University of Cologne, Zülpicher Str. 47, 50674 Cologne

I did not know George Fraser, when he informed me that he would like to translate my book *Tödliche Wissenschaft* (Müller-Hill 1984) for Oxford University Press. So we corresponded about the translation (Müller-Hill 1988) and not about his work in Human Genetics. We talked mainly about the German past, for example about Mengele and colleagues. I only later came to appreciate his deep knowledge in human genetics, but slowly, after he gave me copies of his books.

My book begins with a chronicle of the crimes of human genetics in Germany, and ends: '25 April 1953: Watson and Crick define the three-dimensional structure of DNA … Rapid, almost explosive, advances in the science of genetics begin … Has anything been learned from the outbreak of barbarism in Germany or will it be repeated on a worldwide scale in a yet more dreadful form and to a yet more dreadful degree?' We discussed my concerns. They have not disappeared. It is the proper occasion to revisit them.

The Human Genome Project

The Human Genome Project has been an excellent scientific enterprise. Its goals, detailed physical maps of the 24 chromosomes of the human genome and a catalogue of all DNA sequences, were exciting, and the results are most useful. The same can be said about related genome efforts of various model organisms: yeast, Arabidopsis, Drosophila, Caenorhabditis, zebra fish, Fugu, the mouse and the rat. All these results enable human geneticists who concentrate on particular genes, which are connected with particular diseases or behaviour, to locate and to analyse the genes they are interested in. Finally, it has become clear among the interested scientists how little we actually know about the action of the human genome (or even the mouse genome). And knowledge is better than no knowledge. Only knowledge enables us to make rational and wise decisions.

To write here about concerns where most other scientists write about success seems out of place. Genetics is one of the most successful young branches of present day science. Why worry about success? In Germany, most media are biased against *all* genetics. They see possible risks and dangers everywhere. And so, the German public knows much more about the possible dangers of genetics than about actual facts. There are intellectuals in Germany who seriously talk of a long and perhaps even never-ending moratorium on all genetic research. I certainly do not want to strengthen their positions.

As a large scale project, the Human Genome Project is linked to politics, an uncomfortable arrangement, because science differs from politics. In politics one can say 'Everything is possible', as Bill Clinton said some years ago in Berlin. And everything implies war and peace, justice and injustice, wealth and poverty. In science proper – and this is of course also true for genetics – not everything, but only very special events are possible. Experiments which will be repeated by some, and logic which has to be understood by many, define the area of science. Unlike politics, science has a *logical* structure, which is determined by the internal logic of the physical world. This is sometimes forgotten.

So what are my concerns? I am concerned that:

1. some geneticists make promises to politicians and the people which cannot be fulfilled;
2. some geneticists have lost the proper perspective of what genetics as a science is all about;
3. some politicians may misuse genetics; and
4. the media, particularly the German media, inform the public and politicians badly about genetics. There is – particularly in Germany – a general climate of misunderstanding genetics.

I will first list individual cases. Then I will try to generalise.

False promises

The temptation to inflate the importance of a particular discovery is so common that I barely dare to mention the phenomenon as dangerous. When Nixon declared war against cancer more than thirty years ago, most scientists were jubilant. They expected more money for their research in molecular medicine and said that their experiments would help to win this war. I immediately thought they would never win it. Ironically, Nixon died from cancer, and the mortality figures for cancer today are somewhat better than thirty years ago, but they are not that much better. Such grandiose promises scare me: final victory against cancer when? A vaccine against AIDS or malaria next year? Cheap and safe somatic gene-therapy for major diseases? I am concerned that most of these promises cannot be kept and that the public will eventually also distrust scientists when they should be trusted.

To sum up: Science delivers facts, but some scientists prefer to deliver hope and promises. Hope is a spiritual drug which can not replace reality. I think such promises have to be paid for, if not by the present generation of scientists, then by their students and successors. Thus I warn against making such promises. Let us tell the public what we are trying to do and what we have done. The facts of science are interesting enough to be reported.

Loss of history and perspective

Genetics has a history and the history of a science proper is logical in contrast to the history of politics. So let's have a brief look at the history of genetics. The history of genetics is short. It begins in 1900 with the discovery of Mendel's papers. It has its first high points with Drosophila and maize genetics between 1910 and 1940. It has its lowest point in human genetics in Nazi-Germany 1933–45. I think it is fair to say that present day students of molecular genetics know almost nothing about this history. When I examined a student from the Max-Delbrück-Laboratory in Cologne in a doctoral exam, he knew a lot about transgenic mice, which were the topic of his thesis, but he could not tell me the experiment which made the patron of his institute, Nobel Prize winner Max Delbrück, famous.

Some would argue that such knowledge is not necessary. In fact, I know molecular biologists who state that to study genetics before 1953, i.e. before the structure of DNA was known, is a waste of time. This clever move gets rid of the crimes of the German human geneticists who formed a coalition with the Nazis in Germany. Modern geneticists can thereby claim that these events do not belong to the history of DNA genetics anymore, and so it becomes impudent to compare any writings of present day human geneticists with the writings of their German predecessors.

Illusions

On the other hand, the illusion that genetics will solve everything in the future is widespread. Let me report a revealing anecdote. Some years ago, I met a close colleague at Frankfurt airport. We had come from different meetings and were going to different places. I asked him what was new. He told me about a conversation he just had with Juri Ovchinnikov, then the most important man of Soviet molecular biology and now deceased. Juri had told him that he was about to solve the problem of human language and thought. His team had cloned the cDNAs for the most important ion channels from human and from monkey brain. Since these channels were intimately involved in the production of language, the sequence difference would show the essence of human existence: language and thought. I thought at first that my colleague, who is respected internationally, was joking. But he was utterly in earnest. This view which sees culture as a phenotype of human genes is a self-deception of the scientists who indulge in this attitude, and a deception of the public about the role of human genes.

Genetic difference and the stigmatisation of the different

The premise of all genetics is that there are genetic differences in all living matter. We all enjoy genetic difference among us humans. Blue eyes, brown eyes, black hair, red hair; today, we see nothing wrong with it. But people with black skin have been particularly mistreated as slaves by those with white skin. There is an endless bloody history of stigmatisation of the different.

Human genetics may tell today whether a particular person who looks and is quite healthy will become severely ill years later with a particular affliction – Huntington's chorea, Alzheimer disease or breast cancer. If something can be done about it, it is good to know beforehand. But in Huntington's chorea or Alzheimer disease, nothing can be done today to prevent the onset of the disease. Insurance companies and employers may see an advantage in denying insurance or employment to those carrying such mutant genes. All geneticists should be aware of this and defend the common rights of their patients and of all possible patients in general. In Europe, and particularly in Germany, with its still functioning general health insurance systems, this seems less of a problem than in the USA.

Behavioural genetics and genetic determination

Human behavioural genetics was in the past a curious hybrid and it is still today a curious hybrid. In human behavioural genetics, the phenotypes are actually outside the realm of genetics and science. They belong to psychology and psychiatry which, I argue, are not authentic sciences. One may say that behavioural phenotypes are part of culture and civilisation, and there is an interface between genetics and culture. But where is this border exactly and how is it defined? For me the answer is straightforward. Genotypes may set the limits of intellectual achievement, but they will not determine the content of thoughts. Genotypes may set the limits of emotions, but they do not determine the distinctive reactions to the joys and horrors of this world. Genotypes may limit the capability to produce music, paintings or books, but they will not determine form or content of concerts, paintings or books. The form and contents of concerts, paintings or books are outside the realm of genetics. To sum up: genes define the borders within which free will is active.

But these days a different view exists among some geneticists, which I will summarise in the following manner: culture cannot be understood properly as culture in its own terms, but only by understanding the human brain in its molecular terms. Religion, literature, painting, music, all this will only be *properly* understood by those who understand the molecular structure of the human brain. It is therefore essential not to waste a minute on the epiphenomena of culture but to leave them for the time being, grudgingly, to those who play these games without understanding them. Culture is a *Fata Morgana* which will sooner or later be replaced by science. The active geneticist, who has very, very little spare time outside of his science, is enchanted on hearing this message. It tells him that it is a waste of time to listen to what is outside of science. This attitude is a form of determinism, of Calvinistic predestination even. What is determinism? The determinist claims that free will is an illusion and that human behaviour is to a large extent genetically determined. It is their genes and not their culture which drive people to do what they do in every detail. This attitude has consequences I will now discuss.

The inheritance of intelligence

The genetic inheritance of intelligence has been discussed by many people from the beginning of the 20th century. Family and twin studies have been cited repeatedly as proof that intelligence (for example measured as IQ) is genetically inherited i.e. that there are variant forms of genes which determine intelligence. Racial differences in intelligence have been claimed, again and again between Caucasians and Blacks. Several things can be stated on this matter where most would agree:

1. So far there is no evidence on the DNA level for this claim.
2. Most of the proponents of the claim are determinists. They see attempts of increased education and learning as failed and as a waste of money. They propose measures which their opponents call racist.
3. The Human Genome Project has produced the detailed chromosome maps which will allow scientists to look for and possibly to find the genes and their variant forms which determine in one way or the other intelligence, i.e. our cognitive capabilities. The validity of this notion remains to be tested.

It is wise to anticipate both possible outcomes of this enterprise:

1. Indeed, one or several genes can be isolated which limit intelligence. The sequence of the genes must allow accurate predictions about the type and level of intelligence.
2. No such gene can be isolated in the next twenty years. The proponents of the hypothesis will then argue that intelligence is determined by many genes. This would indeed make the isolation of one of them difficult if not impossible. Then the belief cannot be falsified on the DNA level with the techniques available. Then one can forget about it.

The conclusion, how to deal with either result, is not a matter of science or genetics. It is a matter of politics. Most of us live in democracies, where all citizens have equal rights. If a majority of the citizens decides that the proper response to being able to genetically measure intelligence is the increased and better education of those who are genetically disadvantaged, this would be a great step forward.

However, if the predominating viewpoint is that social help and education of the stupid does not help anyhow and should be abolished because of its high costs, then indeed we are going to repeat some of the experiences we had in Germany between the two wars 1914 and 1945. Human geneticists then claimed what will now again be claimed, that those with low intelligence breed faster than those with high intelligence. It was predicted then, and will be predicted now, that a cultural disaster will destroy the USA and Europe if this process is not stopped. In Germany the Nazis were the only political party who seriously listened to and repeated these claims. So the German human geneticists formed a coalition with the Nazi party which could only be conquered by a major war.

The worst example of stigmatisation by genetics: crime as phenotype

One may think the scenario I discussed so far is rather bad, but there exists an even worse one. More than ten years ago an article appeared in *Science* (Brunner *et al.* 1993) which presented evidence that some male members of a Dutch family carry a nonsense mutation in the X-linked monoamine oxidase (MAOA) gene. It was already known that the absence of the enzyme monoamine oxidase causes mood changes if they eat certain food. They better avoid eating chocolate and Chinese food. Now the afflicted persons could learn which type of food to avoid. So far the discovery is nice from a medical-genetic point of view, but somewhat boring from a media point of view. However the abstract of the *Science* paper states that these 'males are affected by a syndrome of borderline mental retardation and abnormal behaviour. The types of behaviour that occurred include impulsive aggression, arson, attempted rape and exhibitionism'. If one reads the text of the paper, one realises a discrepancy: the science of the genotype is excellent, but the description of the phenotype, i.e. the behaviour, is sloppy. If it were science, one would ask for a table listing the frequency of the typical aggressive acts and all comparable acts of their MAOA wild type brothers. The text reveals that only one of these aggressive acts had been tried by a court. The other acts are only documented by family lore. A table would hurt the family members, who could be easily identified. On the other hand, the absence of a table for such reasons does not ennoble the phenotypes as being scientifically established.

But worse, the aggressive acts can be summarised as *crime*. This was not done by the authors, presumably under the assumption that crime is indeed outside of the scope of science. The authors assume that if one calls attempted murder and rape aggression but not crime, then it is science. Can there be a gene whose phenotype is crime, an action which is deeply connected with the notion of value? Can there be the genetics of good deeds and bad deeds? I think it can be done but it should not be done. The moment it is done genetics ceases to be just science. It turns partly into a religion. I think this has happened here, but it should be avoided. A genotype like MAOA may determine impulsiveness or spontaneity, but to claim it determines aggressive acts, i.e. criminal behaviour, stigmatises its carriers. This has social consequences.

There will be lawyers who will like the idea of crime genes. Maybe their client who committed murder carries such a gene? Would this fact not make the accused not guilty? Psychiatrists have always liked to psychiatrise justice. More *MAOA* mutants would give them strong support. More then ten years have passed since the discovery of the Dutch family where several males had the same nonsense mutation in their *MAOA* gene. No other nonsense or missense mutants in the

MAOA gene have been discovered among the large population of criminals since then. However a promoter difference in the upstream region of this gene has been found in about half the Caucasian population: its influence on behaviour is marginal. The less active promoter has a weak effect favouring aggression when its carrier has been severely mistreated as a child (Caspi *et al.* 2002).

In a recent report of the 73rd Annual Meeting of the American Association of Physical Anthropology (Gibbons 2004) the *MAOA* gene is called a 'warrior' gene. The less it is expressed the more aggressive are monkeys. Thus its evolutionary history of monkeys and man can be followed. One may ask what about the French Foreign Legion or the German SS? Is this real or higher nonsense?

Stigmatisation of ethnic groups

Ethnic groups have been stigmatised again and again before and after the advent of genetics. I recall anti-Semitism. However, genetic knowledge of various ethnic groups is rather limited at present. How are they related to each other, how do they differ from each other? To answer these questions a large attempt is being made to collect DNA from all possible ethnic groups. Understandably, the reaction of those whose DNA is to be analysed is rather mixed. I give just an example: after two centuries of oppression the Australian aborigines are adamant in their opposition against such a study. And this has to be accepted as long as genetic determinism is going strong in the minds of some geneticists and of many citizens in the Western world. It seems to me that this enterprise will not so much demonstrate the quality of the Human Genome Project, but rather the present moral state of Western civilisation.

Misuse of genetics by politics

In politics everything is possible, to quote Bill Clinton again. Genetics has indeed been terribly misused. The worst example is Nazi Germany, but other countries like the US or the Scandinavian countries also had sterilisation programs for genetic reasons, which were utterly unjust. It is important to ensure that this history is not going to be repeated, not even in part. Stigmatisation of genetically characterised groups is such a beginning. If these groups are ethnic groups this is racism. Wherever they arise, stigmatisation of ethnic groups or racism must be adamantly resisted.

The view of the public and the role of the German media

The ordinary German knows that genetics (called gene technology in Germany) is dangerous because DNA itself is dangerous. The public has been fed with constant warnings of the indefinable dangers of gene cloning by the media. The media, particularly *Der Spiegel* and the public TV stations, have sold again and again the message of *Angst*: those media said: working with DNA, in particular plant DNA, is dangerous and should be stopped. Thus the German chemical industry quietly left Germany when it invested in either research or production of gene technology. This is bad for the German students of these fields. They have to leave Germany if they want to find a job in industry. It would be good if the media spoke out for knowledge and genetics and against stigmatisation and determinism.

Summary

To sum up: I think some of the concerns from within genetics are as real as some of the concerns from without. I am optimistic that the community of geneticists will provide a tremendous amount of useful knowledge. I see problems at the interface between genetics and the media and politics. The political decisions of how we should deal with genetic stigmatisation touch science, but they will also determine the character of our future society.

The eugenic prospects of technically assisted reproduction: The pre-implantation genetic diagnosis

Angelo Serra

Professor Emeritus of Human Genetics, 'A. Gemelli' Medical School,
Catholic University of the Sacred Heart, Rome

A premise

The prospect of 'eugenic selection' has emerged with full force with the growth in the advances in the field of genetics. The operational plan was launched at the Third International Conference on Human Genetics by the Nobel Prize winner Herman Muller (1967) when he invited the two thousand or so participants 'to engage in a strong offensive for the control of human evolution'. And he gave the reasons for this: 'Modern culture by maximal saving of lives and fertility, unaccompanied by a conscious planning which takes the genetic effects of this policy in account, must protect mutations detrimental to bodily vigour, intelligence or social predisposition … If genetic defects and shortcomings were to be allowed to accumulate to an unlimited extent among us, as seems to be happening now, the condition would eventually be reached in which each person likewise would present an immense, yet in his case distinctive, complex of problems of diagnosis and treatment'. He himself outlined the lines of this offensive. The first – *germinal selection* – was to lead to the 'production' of a human subject of the 'desired quality'; the second – *genotypic selection* – was to involve, after an early diagnosis during pregnancy, the elimination, through 'abortion' either on demand or imposition of a subject who ran the risk of manifesting a serious illness; the third – *gene selection* – was to lead to the *improvement* of the human species, as soon as the advances in knowledge about the *human genome* had opened up the pathway to its realisation.

The first objective, the 'production of a human subject', has been in part achieved with the creation of human embryos *in vitro* (Serra 1999). However, serious technical problems, which still persist after twenty-five years since the birth of the first baby conceived *in vitro*, do not allow us to foresee an easy achievement of a given 'desired quality', or even if this would be possible.

The second objective, 'genotypic selection', proceeds with dizzy exponential speed. Through major scientific advances in the field of cytogenetics and genomics, there has been indeed a major spread of the prenatal diagnosis (PND) of syndromes caused by alterations in the genetic information which are observable at the chromosomal level (chromosomal syndromes) or analysable at a molecular level (monogenic or polyfactorial genetic diseases). Unfortunately, however, because in the majority of these illnesses there is no possibility of prevention or cure, a strong social pressure has developed – which has by now become a cultural fact – not to accept the responsibility of keeping alive a subject with a 'quality of life' that is held to be unworthy of the human person. Hence the orientation towards 'selective abortion', which has by now become a legally recognised and often recommended practice, and which can be extended in some countries to the third month of pregnancy as well, and even to birth.

Handyside *et al.* (1990) published the first report of the birth of twins whose sex had been identified through cells taken from the embryos before implantation: a new technique of 'genetic selection', known as preimplantation genetic diagnosis (PGD), was introduced. This has by now become established not only as a precautionary measure in the medical practice of fertilisation *in vitro* but also an effective negative eugenic measure to be applied in all families where there is a risk of having children afflicted by serious illnesses because of chromosomal or gene alterations present in the parents. From January 1999 to the end of August 2003, Medline assessed 578 scientific works directed both to the improvement of the technique in itself and to its evaluation as a safe guarantee for a 'healthy child' through *in vitro* fertilisation.

A stimulating but reductive vision of this new technology and its future was presented by Kuliev and Verlinski (2002) who had been working in this field for years at the Reproductive Genetics Institute of Chicago. They wrote as follows: 'More than 4000 preimplantation genetic diagnosis (PGD) cycles have been performed, suggesting that PGD may no longer be considered a research activity. The important present feature of PGD is its expansion to a variety of conditions, which have never been considered as an indication for prenatal diagnosis ... PGD has also become a useful tool for the improvement of the effectiveness of IVF, through avoiding the transfer of chromosomally abnormal embryos, representing more than half of the embryos routinely transferred in IVF patients of advanced maternal age and other poor prognosis patients. PGD is of particular hope for the carriers of balanced chromosomal translocations, as it allows accurate pre-selection of a few balanced or normal embryos ... PGD may soon be performed for both chromosomal and single gene disorders using the same biopsied polar body or blastomere. The available clinical outcome data of more than 3000 PGD embryo transfers further suggest an acceptable pregnancy rate and safety of the procedure, as demonstrated by the follow-up information available for more than 500 children born from these PGD transfers'.

In the face of such optimism, an examination of the features and results of this new technology is necessary to assess and evaluate what it really involves.

Pre-implantation Genetic Diagnosis (PGD)

PGD involves the genetic analysis of one or two cells taken from the embryo in order to detect the existence or otherwise of chromosomal aberration and gene mutations that, obviously enough, would impede normal development.

The general protocol can be summarised through an identification of the following stages: 1) *ovarian stimulation* followed by aspiration of oocytes; their *in vitro fertilisation* through either the ordinary process or intra cellular spermatozoon injection (ICSI); *culture* setting; 2) 72 hours after fertilisation – 'the best moment in humans' (de Vos and van Steirteghem 2001) – *embryo biopsy* for the removal of one or two blastomeres (out of 7–8) at cleavage-stage, either by direct puncture or partial mechanical dissection of the zona pellucida, or through acidic tyrode chemical zona drilling, or laser-assisted zona opening; 3) *karyotyping* or suspected *gene* search; 4) *in utero transfer* of 'healthy' embryos. Two observations require especial emphasis.

The first observation concerns the *number of cells taken from the embryo* in order to obtain a reliable diagnosis of its normality and thus of its capacity for *in utero* transfer. A careful study of *188 cycles*, in which only embryos from which respectively one or two or three blastomeres had been taken, and which on the basis of the examination should have been thought to be 'normal', led the authors to advise an analysis of *two cells* of embryos of seven or more cells so as to make the diagnosis more accurate and reliable (van de Velde *et al.* 2000). This indication was confirmed by a mathematical model developed to find new strategies by which to increase the accuracy of this technique (Lewis *et al.* 2001). It also emerges that notwithstanding the manipulation that the embryos undergo during the process of PGD, the levels of pregnancy achieved appear to be comparable with those obtained in ordinary *in vitro* fertilisation (IVF). The results presented in the work indicate, in fact, a rate of pregnancies begun for each cycle of 29.1% (55); an implantation rate of 18.6% (35); and a birth rate of 14.2% (27). However, De Vos and Van Steirteghem (2001) emphasise with great clarity: 'More data are needed in order to reassure that none of the biopsy procedures applied clinically interferes with implantation rates on ongoing pregnancy rates, allowing the birth of healthy children'.

The second observation relates to the *methods of diagnosis*. There are essentially two goals to be achieved. The first is to define the presence in the cells that have been removed of *chromosomal aberrations* – aneuploidies, deletions, inversions and translocations – through the application of the FISH (Fluorescence *In Situ* Hybridisation) technique, which, with all the advances that have now been made, allows a definition in an individual cell of the numerical and structural anomalies of the chromosomes (Ferguson-Smith 2002, Pearson 2002). The second is to define the presence of *gene mutations* through the process called Polymerase Chain Reaction (PCR), which allows, starting with the DNA of an individual cell, an efficient and rapid amplification of the fragment affected in a given illness and an accurate definition of its alteration (Findlay *et al.* 1999, Blake *et al.* 2001, Harper *et al.* 2002, Fiorentino *et al.* 2003). Obviously enough, errors are not absent due to a notable extent to allele-specific amplification failure or *allele dropout* (ADO) (Hussey *et al.* 2002, Bermudez *et al.* 2003). It has been estimated that for recessive illnesses two genotypings of two separate blastomeres are needed to ensure a minimum risk (< 1%) of transferring an affected embryo *in utero* (Lewis *et al.* 2001).

Another possible analysis is use of the *polar bodies* (Strom *et al.* 2000, De Vos and Van Steirteghem 2001). This technique has the major limitation that it can give information only on the genetic contribution of the mother. The possibilities are two in number: an examination of the first *polar body* only and an examination of the two *polar bodies*. In the first case, the mature oocyte can be used for fertilisation if the polar body certainly carries the expected chromosomal or gene alteration, given that the soundness of the information that remains in the oocyte is then certain. Because of the possibility of ADO, any doubt can be settled through an examination of one or two more of the mature oocytes. In the second case the extraction of the two *polar bodies* can take place only after the fusion of the gametes has occurred. In the view of Rechitsky *et al.* (1999) this is absolutely necessary in the case of the diagnosis of single-gene disorders in order to avoid the notable difficulties that are encountered in the analysis of the DNA of a single cell, amongst which may be listed *DNA* contamination, undetected ADO, and *preferential amplification*, which can all lead to *a misdiagnosis*.

The results of PGD applied to man

In the face of this new advance, welcomed by medicine as a further instrument by which to reduce the number of children born with serious or grave pathologies, the question immediately arises as to what the results of the application of this new technology have been in the fourteen years or so since it was first used.

We have little data on the use and results of PGD through the technique of *sequential polar body removal* (PBR). From the work of Rechitsky *et al.* (1999), we learn that of 529 oocytes in 48 clinical cycles of 26 patients, only 106 embryos had been transferred in 44 clinical cycles, which were followed by 17 (10%) unaffected pregnancies. And Strom *et al.* (2000), when presenting their results, especially on the state of health at birth and during the first six months of the first 109 children who were born following use of this technique for the diagnosis of Mendelian disorders and aneuploidies, concluded as follows: 'The data presented here demonstrate that PGD by PBR is a safe and accurate technique for couples at high genetic risk to avoid having children with genetic abnormalities, without the anxiety of awaiting prenatal diagnosis and the potential of having to terminate affected foetuses'.

A broader and more complete answer emerges from the analysis of a by now notable number of data obtained using the technique of *blastomere biopsy*. Table 1 presents the most representative data published since 1999.

Table 1. Frequency of abnormal embryos found by PGD in assisted reproduction techniques (s: singleton, tw: twin, tr: triplet).

Risk	Cycles (couples)	Biopsied Embryos			Born
		Total	Abnormal	Transferred	
European sample (Nygren and Andersen 1999)	258,460	none		638,508	87,347 (13.7%)
Single gene defects and chromosome aberrations (Vandervors *et al.* 2000)	183 (92)	1079		301	34 (11.6%) 23 s, 5 tw, 1 tr
Aneuploidy					
FISH 6 chrom.	71	406	247 (61%)	159	7 (4.4%)
FISH 9 chrom. (Gianaroli *et al.* 1999)	45	236	146 (62%)	90	4 (4.4%)
ICSI (Kahraman *et al.* 2000)	(72)	329	136 (41%)	193	6 (3.1%)
ICSI (Frydman *et al.* 2002)	71 (59)	312	185 (59%)	127	18 (5.8%) 7 s, 4 tw, 1 tr
Translocations (Iwarsson *et al.* 2000)	11 (7)	64	47 (73%) (mosaics)	14	4 (6.25%) 3 s, 1 tw
Sex determination (Hanson *et al.* 2001)	30 (13)			18	3 (10.3%)
Genetic disorders (Pickering *et al.* 2003)	100 (60)	473			24 (5.1%)

The first observation from the analysis of all these data is the enormous number of embryos – human subjects at the beginning of their lives – which are sacrificed, that is to say literally *killed*. Table 2 presents the sum of data of the five studies in Table 1, which presented respectively: 1) the *total number* of biopsied embryos; 2) the *number of abnormal* embryos because of the presence of chromosomal aberrations – which are the most frequent errors – all of which were rejected, that is to say directly *killed*; 3) the number of embryos that were *transferred in utero*; and 4) the number that were *born*.

Table 2. Outcomes of biopsy

| Total | Biopsied Embryos | | |
	Abnormal	Transferred	Born
1,347	761	583	39
	56.5%	43.3%	6.7%

The number of embryos that were born is, obviously, to be evaluated as a fraction both of the number of *biopsied* embryos, that is to say the *total* number of embryos that were produced and used, and of the number of embryos that were *transferred*. It is clear that 97.1% of the embryos that were produced were lost and 56.6% were directly *killed* because they had a chromosomal abnormality and 40.5% were consciously *exposed to foreseen and willed death*.

This situation is confirmed by a comparison between the number of *transferred embryos* and *those born* in two groups as presented in Table 3: group 'A', which contains the data of another three studies – which are also presented in Table 1 – in which only the number of *transferred* embryos and the number of born embryos is reported; and group 'B', which presents the corresponding data of the sample examined above.

Table 3. Survival rates to birth of non-biopsied (Group A) and biopsied (Group B) embryos

| | Biopsied Embryos | |
	Transferred	Born
Group A	638,508	87,347 (13.7%)
Group B	583	39 (6.7%)

$\chi_1^2 = 3.103$ P > 0.05

It is evident, from the value of χ^2, that the difference between the two groups is not significant. Hence, given the notable information that comes from the enormous European sample, the frequency of embryos born – despite the very high selection obtained through PGD – is markedly lower than that obtained in the ordinary processes of Fertilisation *In Vitro* and Embryo Transfer (FIVET) and Intracytoplasmic Sperm Injection (ICSI), in which selection by PGD is not carried out. This difference could be ascribed to various causes. Whatever the case, the embryos, even if apparently selected following PGD, are in the same situation of high precariousness as – indeed they are perhaps in an even worse situation than – the embryos produced and used in the ordinary processes in which selection occurs spontaneously.

In the face of these results, collected in a serious way by those who wanted, and want, to make a contribution to human comfort in so many situations of suffering and pain, but which indicate also a lack of understanding of the true reality of the human embryo which is reduced instead to a pure technological instrument, a recent statement of a pioneer and protagonist in this field, Winston, seems to me very correct and of great resonance (Winston and Hardy 2002). He concluded his analysis of the state of the technologies of technically assisted reproduction with the following statements: 'Patient desperation, medical hubris and commercial pressures should not be allowed to be the key determining features in this generation of humans. Bringing a child into the world is the most serious human responsibility. We cannot ignore the clouds lowering over these valuable therapies. To do so could have a profound influence on the progress of medical science, not only in this high-profile field, but in others too' (Winston and Hardy 2002).

The ethical prospects of PGD in the medical field

After this consideration of the techniques of preimplantation genetic diagnosis and analysis of the effects and the results of the application of this recently new technology, it is not only useful but also incumbent on us to reflect on the reasons that led to this new step in medical diagnosis not only in the field of *assisted reproductive technology* but also in that of *genetic pathology* and others which are now opening up. Certain statements, gathered from the writings of researchers in this field and from people working in public health care, allow us to understand the principles that are held to justify preimplantation genetic diagnosis – principles that have by now become widely accepted both in the scientific and medical fields and within society.

Savulescu (2001) observes: 'Eugenic selection of embryos is now possible by employing in vitro fertilisation (IVF) and preimplantation genetic diagnosis (PGD) ... I will defend a principle which I call Procreative Beneficence: couples (or single producers) should select the child, of the possible children they could have, who is expected to have the best life, or at least as good a life as the others'. Cameron and Williamson (2003) argue that 'PGD and implantation of an unaffected embryo is a more acceptable choice ethically than prenatal diagnosis (PND) followed by abortion for the following reasons: Choice after PGD is seen as ethically neutral because a positive result (a healthy pregnancy) balances a negative result (the destruction of the affected embryo) simultaneously'. Robertson (2003), in a detailed discussion of the ethics of the use of PGD, states: 'While recognising the strong objections of some people to PGD, ... the following discussion assumes that the use of PGD to screen for aneuploidy and serious Mendelian disorders is ethically and legally acceptable'. And after discussing new uses of PGD for screening embryos for susceptibility to cancer, for late-onset diseases, for HLA-matching for existing children, and for sex, he concludes: 'Except for sex selection of the first child, most current extensions of PGD are ethically acceptable and provides a framework for evaluating future extensions for nonmedical purposes that are still speculative'. The concerns felt by Robertson in respect of sex selection were overcome by Dahl (2003) who argued: 'After considering five potential objections, I conclude that parents should be permitted to use PGD to choose the sexual orientation of their children'. A similar openness to the justification of PGD has been ascertained in a social context by an empirical survey carried out by Katz *et al.* (2002) by questionnaires given to 121 subjects after a previous accurate consultation. Of these: 41 had presented themselves for a PGD because of *gene disorders*; 48 for *aneuploidy screening*; and 32 were about to commence their *first IVF cycle* as a control group. The authors concluded: 'All groups found PGD to be a highly acceptable treatment. They expressed little concern about its extension to testing for non-disease states such as sex, and they were strongly in favour of a shared decision-making model in which couples have considerable autonomy over decisions about the embryo(s) to transfer'. However, these authors also emphasised that 'Whilst our society supports reproductive autonomy there is also concern about the impact of genetic manipulation and genetic enhancement of embryos. There may not be the same community support if the move was towards embryo enhancement, eugenics and even HLA matching'.

In relation to these positions, which are characteristic of a *negative eugenic approach* that is today prevalent and strongly sustained, objections and forms of resistance, however, are not absent (Gianaroli *et al.* 2001, Scriven *et al.* 2001, Lavery *et al.* 2002). The first, and the strongest, relates to the grave abuse of the human embryo, which is reduced to a mere technological instrument. This objection was formulated in 1984 by three members of the Warnock Committee and their opinion was included in the final report in the form of an 'expression of dissent'. It reads as follows: 'It is in our view wrong to create something with the potential for becoming a human person and then deliberately to destroy it. We therefore recommend that nothing should be done that would reduce the chance of successful implantation of the embryo'. This position was openly recognised and emphasised at point n. 17 of the report of the Donaldson committee, which had been established in 1999 by the British government for the regulation of research on embryonic stem cells, where it is stated that 'A significant minority of people believe that the use of any embryo for research purposes is unethical and unacceptable' (Carriline 1984). The second objection, emphasised by Robertson himself, 'arises from the fact of selection itself, and the risks of greatly expanded future selection of embryos and children … Any form of selection or manipulation turns the child into a "manufacture" and thus impairs human flourishing … Increasing the frequency and scope of genetic screening of prospective children will move us toward a eugenic world in which children are valued more for their genotype than for their inherent characteristics, eventually ushering in a world of "designer" children in which genetic engineering of offspring becomes routine'.

From these brief notes the contrast between the two ethical positions is clear: one position is fully in favour of the use of PGD not only for treatment involving *in vitro* fertilisation but also in any case in which a serious possibility of pre- or post-natal pathology exists for a wanted child; the other position is decidedly opposed. It is opposed not out of some whim but for the simple and clear reason that through such a procedure one seeks a 'good', albeit justly wanted, through an action that involves a 'grave wrong' – the intentional killing, even in a single case, of one or more human embryos who have begun their lives. Whoever recognises the scientific truth of the human embryo as a real human subject cannot but recognise the moral value and the correctness of this position.

Habermas (2000), the famous philosopher of the Frankfurt School, dwells at length upon this subject. 'For years the discussion about genetic research and engineering has continued to centre uselessly round the question of the moral status of prepersonal human life. Thus I will adopt the perspective of an imaginary present, projected into the future, beginning from which the practices presently under discussion could retrospectively appear to us as a sliding into a form of liberal genetics, that is to say genetics governed by the law of supply and demand. Research on embryos and the preimplantation diagnosis preoccupy spirits above all because they exemplify the dangers evoked by the metaphor of "*selective eugenics*" in relation to the human race'. Later on he makes clear his thought on the matter: 'Let us suppose that the experimental use of embryos generalises a practice by which the defence of prepersonal human life is seen as being of secondary importance in relation to other possible ends (including the to be wished-for development of noble "collective goods", for example new methods of treatment). The widespread acceptance of this practice would render our vision of human nature less sensitive and would open the door to a form of liberal genetics. In this we can now see what in the future will appear to us as a *fait accompli* of the past, to which the proponents of liberal genetics will appeal as a Rubicon that we have already actually crossed'.

One must admit that in reality, with this new step of preimplantation genetic diagnosis, the very heights have been reached of the overbearing arrogance of scientists, who have wanted not to acknowledge the true reality of the human embryo, degrading it for the first fifteen days of its existence to a '*pre-embryo*' (McLaren 1986): a mass of cells without any law synthesising them into an *organised whole*, a cumulus of disposable cells for any kind of scientific or technological use. Faced with this situation, Testart, the technical father of the first child conceived *in vitro* in France, with evident worry wrote: 'What is in the making is a veritable revolution in ethics, transcending the frontiers of any given country' (Testart and Séle 1995). And he, with a sense of responsibility, concluded: 'Beyond technical performance, individual interest and naïve desire, the problems are more complex than we are led to believe. We ought to approach these problems with a concerned effort and determined humility to uphold the ethical dimensions of human life'.

One must honestly recognise that the great expectations that the progress of science and medicine seemed to have opened up in the vital field of procreation are being transformed into a serious threat to society, in which '*values*' and '*ethical aspects*' are losing their meaning. The reason appears clear: in the prevalent scientific-technological system the value of the constant 'man' – which is indispensable in maintaining the equilibrium of the whole system – has been seriously altered, if not completely annulled. We urgently need to return to the recognition of his *real value*, and thus his *dignity* and his *rights*. However, the methodologies of science and technology cannot calculate or estimate the value of this constant. It is necessary for scientists and technologists, who today have a notable power in directing and effecting social development, not to remain closed within their axiomatic reductive system but to become open to, albeit respecting their own prerogatives, and to welcome, the stimuli of a 'sapiential' system that reflects thought and light that come from the deepest part of ourselves, critically explored, examined and assimilated. Only from this research can one obtain the value of the constant '*man*' and, as a result, rediscover a sense of limits and deduce from this what our responsibility towards him really is. It is man in his integrated reality that must dictate, from his interior being, the set of rules to apply to his action, the basis of every form of responsible behaviour. What is required is that it should be sought for and that there should be a will not to reject it.

It is necessary to transform the *closed* scientific-technological system, which today prevails, into an *open* system in which the real value of 'man' is recognised, and thus his *dignity* and his *rights* but also his *responsibilities* and his *duties*. Only in this way can science and technology – and medicine in particular – find how to meet the needs of every human person, deciding when and in what forms *this or that behaviour* is ethically correct, and thus create real social progress.

John Paul II (2003), when addressing the members of the Pontifical Academy of Sciences, laid stress upon this aspect: 'We must not allow ourselves to be beguiled by the myth of progress, as though the possibility of conducting research or of applying a technique would immediately qualify them as morally good. The moral goodness of all progress is measured by its genuine benefit to man, considered in relation to his twofold corporeal and spiritual dimension; as a result, justice is done to what man is; if the good were not linked to man, who must be its beneficiary, it might be feared that humanity were heading for its own destruction. The scientific community is ceaselessly called to keep the factors in order, situating scientific aspects within the framework of an integral humanism; in this way it will take into account the metaphysical, ethical, social and juridical questions that conscience faces and which the principles of reason can clarify' (n.5).

Dedication

This paper, dedicated to Professor George Fraser on the occasion of his 75th birthday, is intended to be a very modest expression of my high appreciation and admiration for his great scientific contribution to Medical Genetics, clearly inspired by the Hippocratic principles.

Reproduced by kind permission from
The Dignity of Human Procreation and Reproductive Technologies: anthropological and ethical aspects Proceedings of the Tenth Assembly of the Pontifical Academy for Life Vatican City, 20–22 February 2004 Libreria Editrice Vaticana 2005, pp 127–140.

Similar growth rates of monophyletic surnames in Europe

I. Barrai, A. Rodriguez-Larralde[1], C. Scapoli and E. Mamolini

Department of Biology, University of Ferrara, Via L. Borsari 46, I-44100 Ferrara, Italy
[1]Center of Experimental Medicine, Laboratory of Human Genetics, IVIC, Caracas, Venezuela.

Abstract

The present use of family names in Europe dates back to the Middle Ages. In most European countries the insurgence of surnames was gradual, and may have lasted a few centuries. Most surnames were assigned on the basis of the physical characters of persons, of their trade, of their place of origin. Therefore, many surnames may have a polyphyletic origin. Surnames are polyphyletic between countries, as seen from surnames indicating the same trade in different languages, such as Müller and Miller and Molinari. Within the same country or language, some indication that the same surname is polyphyletic comes from its high frequency in different places. Most of the frequent trade names, and the trait names, have similar geographical distributions. It is therefore questionable whether the observed levels of isonymy obtained from surnames validly indicate inbreeding levels.

We analysed 1.7 million surnames in Switzerland in 26 Cantons, 5 million surnames in Germany in 106 towns and 5 million in Italy in 123 towns. We identified for each town the 30 most frequent surnames (20 for the Cantons), and among these the one which was most specific. We defined as more specific among those most frequent, the surname whose ratio town/country was highest. The specific surname so identified is probably monophyletic, and from its frequency we may obtain estimates of inbreeding. We observed that the ratio Frequency/Sample size for surnames so identified tends to be constant in Switzerland, Italy, and Germany. The estimate of the number of generations from the inception of surnames to the present so obtained is somewhat consistent with the historical data for the three countries studied.

Introduction

Family names in Europe appeared gradually in the Middle Ages. It is believed that the first family names were those of feudal lords, based on membership and pride for ancestors (Capeto, Plantagenet, Absburg, Dandolo, Diaz). From castles, the process continued in towns where bourgeois families acquired a name (Luther, Medici, Villon, Becket, Romero) and spread to the villages and the country, where the common people followed the new identification system (Forgeron, Smith, Schmidt, Ferrari, Herrera). Although the process was already general at the turn of the millennium, it was probably completed in continental Europe only at the end of the Middle Ages.

The new system, based on two identifiers instead of three as in the Roman system, required a large number of surnames. There was no fixed rule for assignment. Many surnames were given on the basis of the physical characters of persons, of their trade, or of their place of origin. Therefore, many may be polyphyletic. In Europe, the most common surnames in most countries are trade names having the same meaning in different languages (Müller, Miller, Moulin,

Molinari, Molinas; Schmidt, Smith, Forgeron, Ferrari, Herrera; Schneider, Taylor, Tailleur, Sartori, Sastre). Since mills, ironsmiths and tailors were present in almost every village, it is very likely that persons living in different areas and sharing the same and frequent trade surname were not related. It seems reasonable to assume that names derived from the same trade are polyphyletic between countries, and may be largely polyphyletic also within country. For these surnames, it cannot be proposed that they are clonal copies of the same ancestral surname.

In 1965, Crow and Mange proposed the use of surnames as Y-linked genetic markers for the assessment of the level of random inbreeding in human populations. They described the basic relation existing between isonymy (the probability of two surnames being equal by descent) and inbreeding (the probability of two alleles being equal by descent). Under monophyletism, isonymy is exactly four times inbreeding. The Crow and Mange relation, under the neutral evolution theory (Kimura 1983), permitted considerable progress in the application of population genetics models in the past four decades (Yasuda *et al* 1974, Lasker 1983, Zei *et al*. 1983, 1984, Mascie-Taylor and Lasker 1984, 1985, Rodriguez Larralde *et al*. 1998a,b, 2003). Since population genetics studies are usually marred by the limited availability of genetic data, surnames were a widely used substitute. In the past decade, we used surnames for testing minor hypotheses in population genetics, and our results have been based on very large samples, of the order of millions of individuals (Barrai *et al*. 1996, 1997, 2000, 2004). However, surnames are only recent markers, and their main utility of surnames lies in the possibility of assessing recent migration and approximate levels of inbreeding. Practically, no sample size limit exists for surnames, and the surname distributions of entire nations have been analysed (Dipierri 2005). The underlying hypothesis in these studies is that surnames are monophyletic, which is a very severe restriction.

Purpose of the present work

Since many surnames have a polyphyletic origin, it would be of considerable advantage for population genetics analysis to be able to identify those which are indeed monophyletic, or that have a high probability of being so. We have studied surnames, in Switzerland, Germany and Italy, among others. All these countries have undergone recent demographic expansion.

We have observed that the most frequent surnames are the same in different places (Barrai *et al*. 2000). This does not mean that these must necessarily be polyphyletic; however, their extremely high frequencies are incompatible with genetic drift, even assuming that some of them originated earlier than others. These high frequencies cannot be due to bottlenecks followed by exponential growth, since in such case the frequent surnames would be different in different places of the same country or language. It is therefore questionable whether the observed levels of isonymy obtained from a mixture of mono- and polyphyletic surnames are reliable indicators of inbreeding levels (Cavalli-Sforza and Bodmer 1971, Crow 1980, Rogers 1991). Differences in the parameters obtained from isonymy were observed when mono- and polyphyletic surnames were used (Tay and Yip 1984, Rojas-Alvarado and Garza-Chapa 1994, Garza-Chapa and Rojas-Alvarado 1996).

We have available surnames from Swiss Cantons, and from German and Italian towns. We selected for each town the 30 most frequent surnames, and among these the one which was more specific. We define as most specific among the 30 most frequent in a town, the surname whose ratio town/country is the highest. The specific surname so identified may be monophyletic, and from it we may obtain estimates of equality by descent for surnames.

In Switzerland our procedure was the same, except that we studied the 20 most frequent surnames in each Canton. We ordered the 20 most frequent surnames inside each Canton, according to the ratio of the frequency Canton/Confederation. In some cases, more than 50% of the persons having the same surname were found in the same Canton, and the rest in the other 25 Cantons. This confers high specificity to a surname. For example, of the 308 persons with surname Crameri we found in Switzerland, 217 are in the Grisons. It seems reasonable to believe that the surname is specific to this area, and that the Crameri did not emigrate in large numbers to the rest of the Confederation. This however is no evidence that Crameri is monophyletic, unless the present number of the persons having the name is consistent with the population growth expected in the area since the origin of surnames, which may be up to one thousand years ago. So, population growth and the large sample size of each Canton are complicating the picture, and simultaneous occurrence of more than one Crameri in the Grisons at the time of origin is by no means excluded.

Some of the surnames are ubiquitous, in the sense that they are present in high absolute frequency in almost all Cantons. For example, Müller is found in such high frequency all over the Confederation that it cannot be accepted that genetic drift is the only factor which has produced the increase in frequency in the growing population. Then, if we compare the distribution of Crameri and Müller in Switzerland, we can decide that Müller is probably polyphyletic, and Crameri probably is monophyletic. In so doing, we assume that if a surname appeared once in a location and drifted toward high number and diffused slowly from the area of origin, it is expected to be frequent in that area and not outside it. The criteria for monophyletism deriving from these considerations are then that (1) the surname must belong to the class of most frequent surnames in an area and (2) it must be rare outside the area. At this time we do not worry about correcting for the sample size inside and outside the area, as we should have done for the Cantons of Switzerland. In fact, in Germany and Italy we compare towns with the rest of the nation so the correction may be unnecessary. We are well aware that in reducing the diffusive process of the surname to two points, the town and the rest of the country, our data do not satisfy the conditions for the study of enhanced transport induced by population growth (Vlad *et al.* 2004), for which the geographical distribution of a surname should be available (Figs 4 and 5). *Faute de mieux*, we assign the original position of the surname to the town where it is most specific and frequent. Using these criteria, we can identify specific surnames in an area and consider them as monophyletic for hypothesis testing. The criteria will be considered valid if, when we have identified the different surnames which are presumably monophyletic each in a different area, they will give the similar results when one estimates time of origin under genetic drift. The underlying (and weak) hypothesis is that all presumably monophyletic surnames present today must have approximately the same age.

Materials and methods

The data
We studied 1.7 million surnames in Switzerland, distributed in 26 Cantons (Barrai *et al.* 1996), 5 million in Germany, distributed in 106 towns (Rodriguez-Larralde *et al.* 1998b), and 5 million in Italy distributed in 123 towns (Barrai *et al.* 1999). The samples were obtained from the CD-ROMs of telephone users, from where the private users were extracted. Details of the

data and of the procedures are given elsewhere (Barrai *et al.* 1996, Rodriguez Larralde *et al.* 1998a,b).

Estimates of equality by descent

The ratio between the frequency of most specific surname in town I, S_i, and the sample size N_i for the same town was calculated. This ratio is an estimate of ϕ_i, the present level of equality by descent due to drift for that surname. We observe then that, setting

$$k = \frac{\log(S_i)}{\log(N_i)}$$

then

$$\phi_i = N_i^{(k_i - 1)}$$

For each specific surname, there is a value of k_i. There is a strong linear correlation between $\log(S_i)$ and $\log(N_i)$ (Figs 1–3), and we observed that the value of k tends to be a constant, which would mean that under exponential growth S_i and N_i undergo the same rate of increase. In Figure 1 the regression of $\log(S_i)$ over $\log(N_i)$ for the German towns is given, that for Italy in Figure 2 and for Swiss Cantons is given in Figure 3. The expected values of k are 0.44 in Germany, 0.52 in Italy and 0.50 in Switzerland. In Germany, the value of k varies from 0.33 to 0.55, in Italy from 0.22 to 0.66, and in Switzerland from 0.38 to 0.62, indicating that we may deal here with a parameter which is expected to be constant by country. For further analysis, we used k averaged by country to calculate the expected value of ϕ_i.

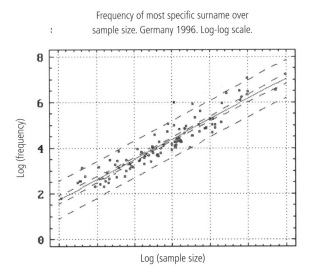

Frequency of most specific surname over sample size. Germany 1996. Log-log scale.

Figure 1. Variation of $\log(S_i)$ on $\log(N_i)$ in Germany, showing 106 towns thoughout the country. The rate of growth of monophyletic surnames appears to be similar for all towns.

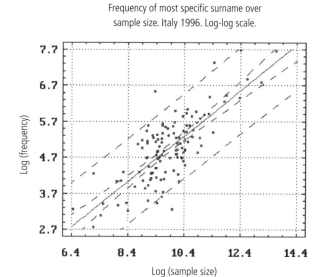

Figure 2. Variation of $\log(S_i)$ on $\log(N_i)$ in Italy, showing 123 towns throughout the country.

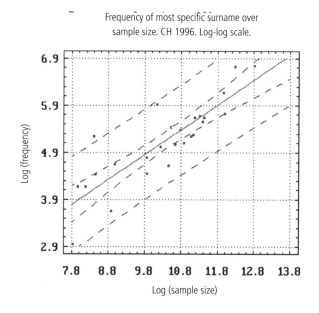

Figure 3. Variation of $\log(S_i)$ on $\log(N_i)$ in Switzerland over 26 Cantons.

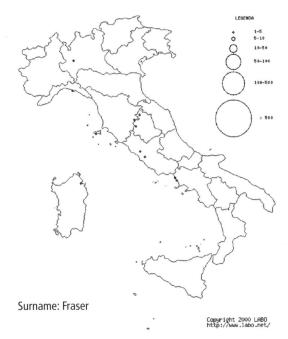

Surname: Fraser

Figure 4. Geographical distribution of one probably polyphyletic surname in Italy. These surnames are probably of immigrants who have selected areas of high geographical interest.

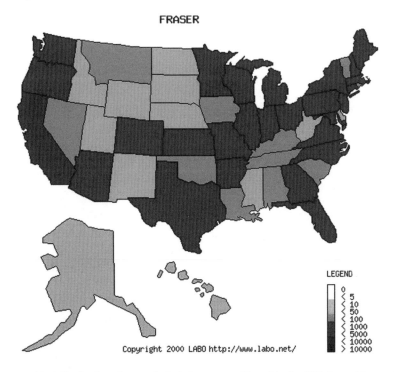

Figure 5. Geographical distribution of one polyphyletic surname (Fraser) in the USA. It would seem that the preferred immigration settlements were in California.

Time to origin of surnames

For the discrete model, under genetic drift and isolation with $\phi_0=0$ the equivalent of the random mating fraction of the population of constant effective size N at generation n is

$$(1 - \phi_n) = (1 - {}^1/_{2N})^n$$

which for Y linked markers becomes

$$(1 - \phi_n) = (1 - {}^1/_N)^n$$

from which

$$n = \frac{\log(1 - \phi_n)}{\log(1 - {}^1/_N)}$$

We can use this crude approximation to estimate n with our data since we have an estimate of ϕ_n and several options for the value of the effective population number.

Historically, values of n for monophyletic surnames in different areas should be of the order of 30–40 generations, namely, if one average human generation has been in the last millennium about 30 years, the time estimated should be of the order of 1000 years. If such will be the case, it will indicate that the surnames used in the calculation may indeed be monophyletic. For ϕ_n, we use the value obtained by averaging k over one area, and for N, three different values for each town, namely (1) the observed sample size, (2) the observed number of surnames and (3) the value today of α, namely the effective number of surnames which, present once, would always result in the same isonymy as observed today.

Results

Most specific surnames

The most specific names in 26 Swiss cantons, in 106 German towns, and in 123 Italian towns are listed in Tables 1, 2, and 3 respectively. Although these surnames are invariably among the 30 most frequent surnames in a town, rarely those which are most represented are also the most specific. In Germany the surname Schmitz, possibly a graphic variant of Schmidt, is specific in Düsseldorf and in Wuppertal, and the Scandinavian Petersen is specific in Hamburg. In most other towns there is a truly specific surname, which emerges with an immunofluorescence effect from the ubiquitous Müllers, Schmidts, Bauers, Fischers, Hubers and so on. The same can be said also for Italy, where few surnames are at the same time the most specific and the most represented. In Rome, the Proietti, in Genoa the Parodi, and in Venice the Vianello are specific and most frequent. These few Italian surnames represent a problem, in so far they all have the same meaning. Proietti are those which were *projectos sub urbem*, ejected, limited to the suburban area of the Town. Parodi (as Caruso in Naples) are those living in the alleys, *para-odos* in Greek. So are the Vianello, those from the paths. So, although these surnames are classified as specific according to our criteria, they may be polyphyletic, and were used at the insurgence of surnames to indicate suburban origin. Unfortunately we cannot construe this reasoning for Germany and Switzerland, since our knowledge of early German is practically nonexistent.

Table 1. The most specific and the most frequent surnames in 26 Swiss Cantons. The most specific surname (1) belongs to the 20 most frequent in the Canton (2) has the highest ratio Canton/Confederation.

Canton	N	S	Specific	Canton	Switzrl	Most freq.	Freq.
Appenzelle	7907	2915	Inauen	105	319	Manser	133
Appenzelli	2853	801	Preisig	65	295	Frischknecht	94
Argovie	81366	18703	Lüscher	290	1008	Müller	1159
Baselland	67748	16993	Thommen	195	639	Müller	812
Baselstadt	88606	23006	Schaub	257	995	Müller	931
Bern	215747	29921	Wenger	830	1922	Müller	1884
Fribourg	27741	8160	Baeriswyl	1511	328	Baeriswyl	1511
Geneva	155647	46185	Muller	477	1092	Muller	477
Glarus	2451	1178	Freuler	19	110	Leuzinger	43
Grisons	47667	12808	Crameri	217	308	Schmid	369
Jura	7190	2718	Chèvre	39	73	Fleury	54
Luzern	71492	13764	Lustenberger	282	537	Müller	977
Neuchatel	37023	11335	Huguenin	230	463	Dubois	234
Nidwald	4667	1666	Lussi	84	203	Odermatt	199
Oberwald	4503	1768	Burch	189	357	Burch	189
Sangallen	92454	19298	Bischof	282	771	Müller	977
Schaffause	18980	6710	Bührer	86	289	Müller	320
Schwyz	19282	5107	Auf der Maur	122	232	Kölin	571
Solothurn	41480	10661	von Arx	160	641	Müller	469
Thurgovie	34384	10568	Stäheli	101	427	Müller	447
Ticino	63707	20989	Rezzonico	189	248	Bernasconi	531
Uri	3499	1122	Herger	65	247	Gisler	230
Valais	52328	12651	Zenklusen	165	230	Rey	303
Vaud	163913	37543	Pasche	308	449	Martin	635
Zug	25008	7424	Iten	371	871	Müller	367
Zurich	364644	58701	Rüegg	842	1626	Müller	4194

Table 2. The most specific and the most frequent surnames in 106 German towns. The most specific surname (1) belongs to the 30 most frequent in town (2) has the highest ratio Town/Germany.

Town	N	S	Specific	Town	Germany	Most freq.	Freq.
Augsburg	81420	27652	Wiedemann	194	1234	Müller	765
Bamberg	21160	8535	Burgis	71	80	Müller	256
Bayreuth	23764	8788	Nützel	63	111	Schmidt	306
Bielefeld	108560	32893	Diekmann	171	819	Schmidt	674
Bonn	93114	35546	Breuer	154	1067	Müller	885
Brandenburg	15421	7027	Thiele	42	2315	Schmidt	153
Braunschweig	87256	27629	Brandes	194	940	Müller	863
Bremen	185230	48751	Behrens	517	2928	Meyer	1873
Bremerhaven	37743	14626	Döscher	71	174	Müller	321
Brilon	4176	2076	Klaholz	24	36	Becker	82
Coburg	12196	5499	Forkel	31	77	Müller	189
Cottbus	35101	12031	Buder	73	194	Lehmann	355
Cuxhaven	15099	7359	Tiedemann	41	841	Schmidt	126
Darmstadt	43235	18198	Schmitt	142	6249	Müller	449
Deggendorf	8053	3755	Bielmeier	26	119	Müller	68
Donaueschingen	5690	3026	Hauger	32	64	Müller	68
Dresden	110318	23597	Böhme	385	1619	Müller	1416
Duisburg	33305	14918	Maas	62	717	Müller	213
Düsseldorf	213646	66990	Schmitz	1162	6766	Müller	1907
Eberswalde	8506	4355	Otto	16	3546	Müller	110
Elsterwerda	2559	1343	Manig	13	55	Müller	49
Emden	16347	5591	Bakker	53	130	Janssen	282
Eschwege	7224	3386	Zeuch	48	113	Müller	77
Essen	219899	55985	Schmitz	652	6766	Müller	1546
Flensburg	26434	10158	Lorenzen	160	770	Hansen	589
Frankfurt aM	214704	72306	Schmitt	569	6249	Müller	1789
Frankfurt aO	20148	8098	Noack	42	1268	Schulz	281
Freiburg in Br	54777	21093	Maier	206	4734	Müller	576
Fulda	18595	7236	Wehner	99	837	Müller	277
Giessen	21933	10664	Kreiling	54	77	Müller	262
Gorlitz	13017	4759	Heinze	46	1512	Schmidt	147
Göttingen	33485	14812	Ahlborn	71	264	Müller	287
Greifswald	12476	5676	Dinse	32	194	Schmidt	157
Gustrow	9705	4406	Ohde	27	151	Schmidt	86
Halle	60258	17737	Voigt	152	2717	Müller	695
Hamburg	587912	114190	Petersen	1392	3675	Schmidt	4290
Hannover	178422	55359	Krueger	446	9079	Müller	1543
Heidelberg	37327	16997	Mohr	76	2076	Müller	391

Heilbronn	31930	13698	Hagner	61	163	Müller	345
Hildesheim	33753	13531	Engelke	82	633	Müller	309
Hofsaal	17506	6114	Köppel	59	179	Müller	235
Holzminden	6534	3450	Kumlehn	26	45	Müller	80
Husum	6933	3177	Feddersen	31	210	Hansen	237
Itzehoe	10014	5015	Lohse	25	717	Schmidt	88
K'lauten	36281	13476	Christmann	123	531	Müller	478
Karlsruhe	70771	25671	Kiefer	131	1143	Müller	737
Kassel	62658	22424	Siebert	271	1720	Schmidt	628
Kempten	20728	9117	Prestel	39	121	Müller	232
Kiel	73494	24647	Hamann	133	1204	Schmidt	591
Koblenz	32631	12848	Dötsch	56	159	Müller	444
Konstanz	21764	10438	Sauter	44	633	Müller	255
Landshut	16771	7652	Weinzierl	30	331	Huber	149
Limbach-Oberfro	2734	1403	Landgraf	23	437	Müller	54
Limburg	5263	3149	Kremer	19	1320	Müller	74
Lorrach	9776	5276	Sütterlin	22	97	Müller	92
Lubbenau	5000	2721	Heinze	13	1512	Lehmann	80
Lübeck	60910	20485	Wulf	112	1087	Schmidt	474
Lüneburg	19782	9594	Burmester	40	672	Meyer	216
Mainz	51527	20548	Schmitt	378	6249	Müller	637
Mannheim	93819	31762	Seitz	133	1519	Müller	1011
Marburg	13994	7782	Naumann	36	1486	Schmidt	155
Meppen	8701	3632	Altmeppen	33	42	Schulte	110
Minden	26052	10013	Riechmann	113	165	Müller	196
Monchen-Gbch	85192	24037	Kamphausen	155	218	Schmitz	1053
München	455904	111138	Wimmer	689	1326	Müller	3285
Münster	86425	30902	Brüggemann	108	920	Müller	452
Neubrandenburg	23085	8953	Voß	61	2250	Schmidt	246
Nienburg Weser	10468	5138	Beermann	24	296	Schmidt	114
Nordlingen	4307	2344	Schöppel	14	54	Müller	35
Oldenburg	47496	17110	Röben	56	116	Müller	442
Oranienburg	5399	3173	Henning	12	1574	Schulz	60
Osnabruck	50961	18948	Hörnschemeyer	96	112	Meyer	426
Paderborn	41904	15994	Schulte	123	2678	Müller	302
Passau	14345	6574	Seidl	39	887	Bauer	129
Penzberg	4343	2537	Knoblach	11	56	Wagner	27
Prum	1558	942	Raskopp	6	11	Schmitz	25
Puttgarden	2590	1547	Lafrenz	13	51	Schmidt	24
Ravensburg	8710	4762	Steinhauser	20	217	Müller	110
Regensburg	41455	14864	Brandl	127	841	Müller	291
Rosenheim Bayer	14248	7661	Lechner	27	830	Huber	125
Rostock	50750	17105	Rohde	96	1878	Schmidt	501
Rothenburg odT	3794	2107	Kandert	10	15	Schmidt	61

Saarbrucken	35912	12961	Simon	84	3360	Müller	513
Salzgitter	29236	12145	Fricke	64	1990	Müller	281
Salzwedel	5552	2869	Busse	14	1088	Schulz	147
Sassnitz	2947	1782	Köning	16	23	Müller	36
Schweinfurt	17169	7201	Pfister	48	551	Müller	235
Schwerin	31030	11824	Kröger	146	1927	Schmidt	262
Siegen	37078	12045	Röcher	73	79	Schneider	608
Simbach a Inn	2411	1446	Pinzl	12	24	Huber	34
Soltau	6821	3509	Eggersglüß	15	25	Müller	81
Stralsund	16048	6656	Ewert	40	682	Schmidt	176
Stuttgart	175542	53920	Schmid	672	5530	Müller	1540
Travemünde	4871	3169	Freitag	10	1698	Müller	34
Trier	31920	11042	Görgen	71	213	Müller	445
Tübingen	17851	9355	Kehrer	45	210	Schmid	161
Ulm	29587	13210	Unseld	50	88	Müller	259
Weiden i.d.Opf	11708	4361	Kick	53	178	Bauer	104
Weissburg i Bay	4626	2401	Satzinger	14	42	Schmidt	43
Wiesbaden	71284	27378	Jung	127	4111	Müller	747
Wilhelmshaven	31242	11872	Janßen	404	1580	Janßen	404
Wittemberg	6368	3312	Köhn	18	799	Schulz	77
Wolfsburg	37843	15273	Fricke	71	1990	Müller	471
Wuppertal	124578	37472	Schmitz	603	6766	Müller	1033
Würzburg	40979	15260	Endres	106	667	Müller	510
Zeitz	7232	3409	Rothe	23	1074	Müller	80

Table 3. The most specific and the most frequent surnames in 123 towns in Italy. The most specific surname (1) belongs to the 30 most frequent in the Town (2) has the highest ratio Town/Italy.

Town	N	S	Specific	Town	Italy	Most freq.	Freq.
Alessandria	30240	9420	Bocchio	143	235	Rossi	202
Ancona	31306	7585	Burattini	125	230	Rossi	151
Aosta	11878	5382	Fazari	43	90	Fazari	43
Arezzo	25462	5441	Gallorini	138	209	Rossi	476
Ascol Piceno i	13003	2779	Celani	106	345	Angelini	194
Assisi	4265	1520	Tardioli	31	118	Rossi	53
Asti	23983	7171	Cerrato	133	497	Musso	215
Avellino	11790	3055	Iannaccone	169	419	Iannaccone	169
Avezzano	9296	2556	Cipollone	85	383	Fracassi	97
Bari	79914	11545	Lorusso	594	1719	Lorusso	594
Belluno	10590	3438	Darold	200	250	Darold	200
Benevento	10440	2545	Zollo	71	201	Deluca	137
Bergamo	36298	9770	Rota	419	1605	Rota	419
Biella	15174	6353	Ramella	215	487	Ramella	215
Bologna	148306	25387	Venturi	613	2109	Rossi	792
Bolzano	29383	10236	Pichler	110	139	Pichler	110
Brescia	59773	13917	Bonometti	151	222	Ferrari	459
Brindisi	22086	4655	Guadalupi	262	357	Guadalupi	262
Campobasso	12122	3232	Mignogna	81	224	Palladino	153
Canosa di Pugl.	7435	1375	Dinunno	179	312	Dinunno	179
Carrara	16954	4110	Menconi	258	364	Menconi	258
Casale Monf.	12397	4599	Patrucco	127	238	Patrucco	127
Caserta	10570	3617	Giaquinto	71	315	Fusco	120
Castellamare St	14990	2576	Schettino	265	522	Esposito	662
Castrignano C.	1333	293	Schirinzi	70	192	Schirinzi	70
Catanzaro	18608	3936	Rotundo	203	335	Rotundo	203
Cesena	23731	3665	Casadei	400	1866	Casadei	400
Chieti	14482	3952	Iezzi	167	449	Iezzi	167
Chioggia	11692	1842	Boscolo	681	1503	Boscolo	681
Chivasso	7401	3749	Cena	143	280	Cena	143
Civitavecchia	14814	5060	Biferali	37	48	Larosa	62
Comacchio	9044	3418	Cavalieri	267	1308	Cavalieri	267
Como	25904	9108	Tettamanti	161	238	Bianchi	347
Cosenza	13371	3418	Filice	148	220	Filice	148
Cremona	24100	6020	Bodini	119	361	Ferrari	332
Crotone	7533	2139	Vrenna	58	80	Vrenna	58
Cuneo	14879	4782	Dutto	276	359	Dutto	276
Dobbiaco	629	337	Taschler	26	33	Taschler	26

Fabriano	8840	2282	Stroppa	111	258	Stroppa	111
Fasano	8867	1351	Cofano	172	259	Cofano	172
Ferrara	43534	7569	Balboni	192	550	Rossi	277
Firenze	136235	25922	Innocenti	523	1689	Rossi	715
Foggia	32258	6087	Dellicarri	226	232	Russo	414
Foligno	14127	3554	Trabalza	59	126	Angeli	75
Forli	31926	5849	Ravaioli	231	501	Fabbri	438
Formia	6308	2309	Demeo	160	566	Demeo	160
Frosinone	10849	3148	Turriziani	187	248	Spaziani	243
Genova	243430	39002	Parodi	2131	2654	Parodi	2131
Gioia Tauro	3140	1043	Cedro	29	98	Romeo	41
Gorizia	12302	5331	Bregant	63	86	Bressan	75
Grosseto	22460	6685	Corridori	77	141	Rossi	249
Imperia	13502	5500	Amoretti	110	241	Amoretti	110
Isernia	3540	1319	Antenucci	58	229	Antenucci	58
Ivrea	8152	4354	Gillio	47	97	Gillio	47
La Spezia	35019	10341	Cozzani	266	341	Cozzani	266
Lamezia T	8842	1794	Mastroianni	164	575	Mastroianni	164
Latina	21079	8053	Pietrosanti	26	134	Rossi	77
Lecce	22322	4751	Quarta	214	566	Greco	256
Lecco	14590	4056	Valsecchi	184	587	Colombo	309
Livorno	52869	11556	Falleni	126	150	Rossi	216
Lucca	14659	3872	Lucchesi	117	646	Bianchi	130
–	12688	4078	Ianni	132	494	Ianni	132
Macerata	11874	3428	Giustozzi	81	136	Moretti	89
Mantova	15519	5086	Benedini	42	222	Ferrari	113
Massa Car.	18168	4529	Mosti	246	325	Ricci	316
Matera	12572	2486	Montemurro	207	457	Montemurro	207
Milano	513552	75523	Brambilla	889	1596	Rossi	2330
Modena	56147	11388	Malagoli	321	506	Ferrari	814
Monfalcone	9815	4636	Miniussi	59	103	Visintin	67
Monza	39187	12643	Galbiati	269	701	Villa	477
Napoli	235971	26905	Scognamiglio	574	971	Esposito	5247
Novara	31989	11510	Brustia	99	136	Rossi	155
Novi Lig.	10115	3922	Fasciolo	62	194	Repetto	154
Orvieto	5694	1958	Prosperini	49	147	Prosperini	49
Ostia Lido	26739	12243	Proietti	56	3079	Rossi	119
Padova	66167	16100	Rampazzo	359	514	Schiavon	594
Parma	56791	11394	Campanini	189	509	Ferrari	779
Pavia	26595	8962	Rognoni	69	349	Sacchi	212
Perugia	37464	10249	Alunni	169	344	Rossi	334
Pesaro	25790	5570	Marchionni	111	571	Cecchini	212
Pescara	33149	8472	Camplone	133	153	Digirolamo	155
Piacenza	32775	7772	Tagliaferri	175	783	Rossi	455

Piombino	12402	3731	Paini	31	323	Rossi	86
Pisa	28632	9617	Sbrana	163	262	Rossi	174
Pordenone	15696	5990	Santarossa	130	196	Santarossa	130
Potenza	22711	4393	Telesca	290	433	Telesca	290
Prato	49295	10857	Gori	395	1899	Gori	395
Ravenna	44622	9536	Mazzotti	234	700	Montanari	413
Reggio C.	35534	5545	Morabito	374	1006	Romeo	582
Reggio E.	42181	8330	Davoli	343	653	Ferrari	566
Rieti	12229	3118	Festuccia	117	209	Angelucci	127
Rimini	34738	7899	Semprini	225	363	Fabbri	484
Roma	858318	84583	Proietti	2125	3079	Rossi	4082
Rossano C.	3820	1048	Sapia	73	162	Graziano	75
Rovigo	14035	3612	Osti	100	368	Ferrari	117
Salerno	33620	6534	Memoli	180	318	Santoro	222
S.Benedetto T	12888	3318	Capriotti	181	468	Capriotti	181
Sapri	1324	731	Scarpitta	16	50	Scarpitta	16
Savona	24273	8371	Briano	190	225	Briano	190
Siena	19614	5385	Pianigiani	108	221	Rossi	154
Sondrio	6630	2557	Marveggio	74	84	Bordoni	87
Susa	1847	1125	Favro	30	92	Favro	30
Taranto	51291	9442	Albano	291	1276	Russo	364
Tarvisio	1721	1112	Vuerich	22	80	Vuerich	22
Teramo	9194	2907	Pompilii	33	41	Difrancesco	87
Terni	33194	7365	Massarelli	89	219	Proietti	302
Tolmezzo	2983	1081	Iob	57	101	Iob	57
Torino	308090	51235	Ferrero	966	1749	Ferrero	966
Trento	30773	7367	Tomasi	344	1135	Tomasi	344
Treviso	24802	7925	Pozzobon	119	205	Pavan	173
Trieste	90055	22529	Vascotto	219	256	Furlan	273
Udine	30054	10202	Degano	108	209	Rossi	123
Urbino	4417	1650	Ligi	25	104	Rossi	52
Varese	27041	10571	Macchi	189	720	Bianchi	245
Vasto	8639	2509	Marchesani	109	218	Marchesani	109
Venezia	92754	17574	Vianello	1535	2063	Vianello	1535
Ventimiglia	9078	4361	Ballestra	51	91	Lorenzi	129
Verbania	10208	4614	Caretti	70	171	Caretti	70
Vercelli	15787	6068	Francese	61	295	Ferraris	148
Verona	77966	16669	Avesani	368	392	Ferrari	396
Vicenza	32922	9223	Bedin	136	333	Rossi	167
Viterbo	16825	5225	Morucci	74	212	Rossi	114
V.Veneto	8590	2393	Piccin	230	366	Piccin	230

Table 4. Estimates of the number of generations from insurgence of surnames in three nations, obtained using the most specific surnames and three estimators of the population effective size.

Nation	Number of towns	Generations from origin of the specific surname			
		k	α	No. surnames	N
Switzerland (Cantons)	26	0.50	31.1 ± 2.7	104.6 ± 10.8	209.2 ± 26.0
Germany	106	0.44	19.7 ± 0.3	45.0 ± 1.9	66.9 ± 3.5
Italy	123	0.52	33.4 ± 1.2	80.8 ± 3.8	164.0 ± 11.0

Time from origin

If our hypothesis is correct, we expect the estimate of the time from the time of origin of the use of specific surnames to be consistent with history, namely it should place the origin of surnames around the turn of the first millennium. The results are given in Table 4. There, we give the average values of k for the three nations studied for specific surnames, and the number of generations from their inception using three different values for the effective number.

The time estimates differ significantly by nation and by estimator of constant population effective size. They are inconsistent when sample size and present number of surnames are used, but are surprisingly consistent with historical data when Fisher's Æ is used as an estimator of constant effective size. For the Swiss Cantons, which however are not homogeneous with the towns of Italy and Germany, we obtain 31 generations, about 930 years. For Italy, the estimated number of generations is 33.4, or 1002 years, again assuming that the length of one generation is 30 years. For Germany, we obtain 19.7 generations, or 591 years, which seems somewhat short compared to historical data, but still within a plausible range, it puts the origin around 1300 AD. The estimates for the Swiss Cantons and for Italy are very consistent with history, for the origin of surnames is located in the period 1000–1100 AD.

When we use the most frequent, possibly polyphyletic surnames for the same estimates, the results are grossly inconsistent, since the origin of surnames is placed more than one hundred generations in the past.

Conclusion

We propose then that, under exponential population growth and simultaneous drift, it is possible to assess whether a surname is either mono- or polyphyletic, if large samples are available in a large number of groups, such as in the towns of an entire nation. Our basic reasoning is that, if the surname is very frequent only in one limited area and is rare elsewhere, then it may be monophyletic and can be classified as such. On the other hand, if the surname is frequent in all groups, then it may be polyphyletic and can be so classified.

HLA-DQB1 genotype variation between Cretans and Athenians with type 1 diabetes mellitus

Christos S. Bartsocas[1][*], Peristera Paschou[1], Evangelos Bozas[1], Maria Dokopoulou[1], Christina Giannakopoulou[2], Dimitris Mamoulakis[2] and Andriani Gerasimidou-Vazeou[1]

[1]Department of Paediatrics, Faculty of Nursing, University of Athens, Greece
[2]Department of Paediatrics, University of Crete, Greece
[*]Present address: Mitera Paediatric Hospital, 6 Erythrou Stavrou St., GR-15123 Marousi, Greece

This article is written to honour George Fraser for a long-lasting friendship.

Introduction

Type 1 diabetes mellitus (T1DM) is on the increase in most countries around the globe (Onkano *et al.* 1999, EURODIAB ACE Study Group 2000). It is a multifactorial, multigenic disorder associated with severe complications and significant costs. T1DM is caused by an autoimmune destruction of the insulin-secreting β-cells of the pancreatic islets, eventually resulting in a total exogenous insulin dependency (Bach 1995, Akerblom *et al.* 1997).

Genetic susceptibility to the disease is complex, consisting of an unknown number of inter-acting loci. Several genome scans (Davies *et al.* 1994, Mein *et al.* 1998, Concannon *et al.* 2005) have indicated the existence of more than 20 loci possibly linked to T1DM. Definite linkage to the disease, however, has only been confirmed for two of these loci; the HLA class II region and a VNTR locus at the 5'-end of the insulin gene (Retwein *et al.* 1986, Todd *et al.* 1987, Ronningen and Thorsby 1993). The HLA region is the major genetic determinant of T1DM accounting for about 42% of the familial clustering of the disease (Davies *et al.* 1994), a relative risk (RR) greater by far than the attributed to any other single locus. HLA-DQ heterodimer mol-ecules encoded by HLA-DQA1 and -DQB1 genes are believed to play the most important role, while the DR molecules and -DRB1 genes may have an independent effect and might in some cases modify the risk for the disease. HLA-DQB1 genes *0201 and *0302, encoding for a non-Asp57 DQβ-chain, increase the risk for the development of T1DM, while alleles *0301, *0602 and *0603 have a protective effect (Todd *et al.* 1987, Ronningen and Thorsby 1993).

The strong linkage disequilibrium observed between the HLA-DQA1, -DBQ1 and -DRB1 loci simplifies in most cases the complicated analysis of HLA genotypes. One can often deduce the alleles of other loci based on the determination of the alleles in one locus, so that the whole haplotype is known. It has been shown that the susceptibility allele HLA-DQB1*0602 is almost always found with the protective DQA1*0102, while DQBQ1*0301 is always protective regard-less of the DQA1 allele in the haplotype (Ilonen *et al.* 2002). However, typing for HLA-DQA1 becomes very informative for individuals carrying DQB1*0201. When associated with DQA1 alleles 0301 or 0501 they confer a strong predisposition to T1DM, although they are protective or neutral when found with DQA1*0201(Ronningen and Keiding 2001). See Table 1 below.

Table 1. Comparison of HLA-DQB1 genotype incidences between controls in Athens and Crete.

| GENOTYPE DQB1* | Athens | | Crete | | |
	N	%	N	%	P
0201,0302	**36**	**2.0**	**16**	**4.0**	**0.05**
0302,y	97	5.4	20	5.0	NS
0301,0302	93	5.2	24	6.0	NS
0201,0301	193	10.8	36	9.0	NS
0201,x	245	13.7	46	11.5	NS
0302,0602–0603	13	0.7	3	0.8	NS
X	119	6.7	21	5.3	NS
0301,z	634	35.5	156	39.0	NS
0201,0602–0603	44	2.5	7	1.8	NS
0301,0602–0603	109	6.1	18	4.5	NS
0602–0603,w	152	8.5	28	7.0	NS
Total	1784		396		

x = other than HLA-DQB1 *0302, *0301, *0602, *0603, *0604
y = other than HLA-DQB1 *0201, *0301, *0602, *0603, *0604
z = other than HLA-DQB1 *0201, *0302, *0602, *0603, *0604
w = other than HLA-DQB1 *0201, *0301, *0302, *0604
NS = Non significant

Although the role of genes in T1DM is essential, the penetrance of the predisposition or protection alleles is determined by yet unidentified environmental factors. The remarkable north to south gradient of the incidence of the disease in Europe is considered an indication of the important role of the environment in the pathogenesis of T1DM, although it may also be due to the distribution of genetic factors (Ronningen and Keiding 2001). The incidence of T1DM in Europe (expressed as new cases/100,000 children under the age of 14 years) is highest in Finland (50/100,000/year), and it decreases moving south (Levy-Marchal *et al.* 1995). Variation in the incidence of the disease may also be observed within countries. In Greece, there is an impressive clustering of new cases of T1DM observed in the Athens metropolitan area, where the incidence of the disease is over 10/100,000/year, compared with 7.5/100,000/year in Crete (Dacou-Voutekakis *et al.* 1995, Bartsocas *et al.* 1998).

Methods

Blood samples were collected from 130 Athenians and 66 Cretans with T1DM. Reference samples for the Athenian population were 1748 cord bloods of consecutive births in a large maternity hospital in Athens. Cord blood samples were also collected from 396 healthy newborns in Crete. Analysis was performed directly on a drop of blood dried on filter paper.

All samples were typed for five HLA-DQB1 alleles, proven to be associated with increased risk for the development of T1DM (*0201, *0302) or protection from it (*0301, *0602, *0603). The method used is based on time-resolved fluorometry (TRF) and has been described in detail previously (Sjöroos *et al.* 1995, Ilonen *et al.* 1996). In brief, 158 base pairs of the second exon of the DQB1 gene from each sample were amplified by polymerase chain reaction (PCR) using a primer pair with a biotinylated 3' primer (5'-GCATGTGCTACTTCACCAACG and 3'-CCTTCTGGCTGTTCCAGTACT) (Tables 2 and 3).

Table 2. Oligonucleotide primers used for multiplication HLA-DQB1.

Locus	
HLA-DQB1	5' GCA TGT GCT ACT TCA CCA ACG 3'
	5' CCT TCT GGC TGT TCC AGT ACT 3' biotin

Table 3. Oligonucleotide primers used for allelomorphs of HLA-DQB1.

Allelomorphs HLA-DQB1*	Oligonucleotide sequence of hybridisation	
0602–3, 0607–8, 0610–11, 0613–15	5' GTA CCG CGC GGT 3'	Europium
0302, 0304–5, 0307–8	5' GCC GCC TGC CG 3'	Europium
0301, 0304	5' ACG TGG AGG TGT AC 3'	Samarium
0603–4, 0607–8, 06014	5' TTG TAA CCA GAC ACA 3'	Samarium
0201–3	5' GAA GAG ATC GTG CG 3'	Terbium
All allelomorphs controls	5' CTT CGA CAG CGA CG 3'	Terbium

The amplification product was bound to streptavidin coated microtitration plates and denatured with NaOH. After washing, bound DNA was hybridised with allele specific oligonucleotides (ASOs), each labelled with one of the three fluorescent lanthanides, europium, samarium or terbium (Hurskainen *et al.* 1991). The unique properties of the lanthanides allow simultaneous typing of all five detected alleles, using a combination of just two different hybridisation solutions. The probe hybridisation is measured using TRF (EG and G Wallac). The different emission wavelengths and delay times are used to distinguish the signals of each lanthanide label (Dahlen *et al.* 1994). The results were analysed with the Multicalc software.

Statistical evaluation was performed using the χ^2 test. RR was estimated according to the formula

$$RR = \frac{a(b+d)}{b(a+c)}$$

in which a and b are the number of patients who were positive and negative for the marker, respectively, and c and d the respective numbers of unaffected subjects.

Results

HLA-DQB1 is strongly associated with both the development of T1DM and protection from it, in both populations studied (Tables 4 and 5). HLA-DQB1*0201, *0302, y and *0201, x are average risk factors in Cretans, while HLA-DQB1*0301, z, *0301,0602–0603 and *0602–0603 are protective for T1DM. Comparing the two control populations it seems that Cretans present a statistically significant difference in the HLA-DQB1*0201,0302 genotype, which is double in Cretans that in the mainland Greeks. HLA-DQB1*0201,0302 confers similar degrees of risk for

T1DM in both Athenians and Cretans, while *0301 is very frequent in both reference populations and protects strongly against the disease. Furthermore, DQB1*0302 increases the risk for T1DM in mainland Athenians as well as Cretans.

Table 4. Incidences of HLA-DQB1 allelic genes in Athenians with T1DM, controls and relative risk for development of T1DM.

| GENOTYPE DQB1* | T1DM | | Control Population | | | |
	N	%	N	%	RR	P
0201,0302	20	15.5	36	2.0	6.22	<0,001
0302,y	15	11.63	97	5.4	2.17	<0.01
0301,0302	5	3,88	93	5,2	0,76	NS
0201,0301	4	3.10	193	10.8	0.29	<0.01
0201,x	62	48.08	245	13.7	4.94	<0.001
0302,0602–0603	0	0.00	13	0.7	0.00	NS
x	8	6.20	119	6.7	0.95	NS
0301,z	5	3.88	634	35.5	0.08	<0.001
0201,0602–0603	1	0.77	44	2.5	0.33	NS
0301,0602–0603	0	0.00	109	6.1	0.00	<0.01
0602–0603,w	2	1.55	152	8.5	0.18	<0.01
Total	129		1784			

x = other than HLA-DQB1 *0302, *0301, *0602, *0603, *0604
y = other than HLA-DQB1 *0201, *0301, *0602, *0603, *0604
z = other than HLA-DQB1 *0201, *0302, *0602, *0603, *0604
w = other than HLA-DQB1 *0201, *0301, *0302, *0604
NS = Non significant
RR = Relative Risk

Discussion

It is still unclear whether the striking regional variations in the incidence of T1DM in Europe are due to environmental factors, genetic factors, or a combination of both. The effort to disentangle the relative contribution of genes and environment to the pathogenesis for T1DM is a difficult task, mainly because the environmental factors involved, as well as because many of the genetic factors, remain obscure. The present study provides information on the T1DM association of one of the major genetic determinants of the disease (HLA-DQB1) in populations of South-Eastern Europe with very low incidence (Paschou et al. 2004). Studies in low-risk populations can serve as a reference when compared to the high-risk populations of Northern Europe (Herman et al. 2004). Furthermore, our study provides data on the molecular typing of patients with T1DM in Crete for the first time.

As expected there were no remarkable differences in HLA alleles between Athenians and Cretans. The results strongly suggest genetic homogeneity for these markers. It should be noted that significant differences were observed between Greeks and Finns, which possibly partly explains the extreme differences in the incidence of T1DM in the two populations.

These results demonstrate that the same HLA-DQB1 genotypes confer risk for the develop-

Table 5. Incidences of HLA-DQB1 genotypes in Crete (T1DM patients, controls) and RR for development of T1DM.

GENOTYPE DQB1*	T1DM		Control Population			
	N	%	N	%	RR	P
0201,0302	17	25.8	16	4.0	4.51	<0.001
0302,y	11	16.7	20	5.0	2.78	<0.001
0301,0302	2	3.0	24	6.0	0.52	NS
0201,0301	3	4.5	36	9.0	0.52	NS
0201,x	27	40.9	46	11.5	3.69	<0.001
0302,0602–0603	0	0.0	3	0.8	0.00	NS
x	4	6.1	21	5.3	1.13	NS
0301,z	1	1.5	156	39.0	0.03	<0.001
0201,0602–0603	1	1.5	7	1.8	0.87	NS
0301,0602–0603	0	0.0	18	4.5	0.00	<0.05
0602–0603,w	0	0.0	28	7.0	0.00	<0.05
Total	66		396			

x = other than HLA-DQB1 *0302, *0301, *0602, *0603, *0604
y = other than HLA-DQB1 *0201, *0301, *0602, *0603, *0604
z = other than HLA-DQB1 *0201, *0302, *0602, *0603, *0604
w = other than HLA-DQB1 *0201, *0301, *0302, *0604
NS = Non significant
RR = Relative Risk

ment of T1DM on the people of the Greek mainland and the Cretans. However, the HLA-DQB1 *0201, 0302 genotype although more common in the Cretan population has a lower estimated relative risk (4.51), as compared with a RR of 6.22 in the mainland population. This may be related to the fact that T1DM has a higher prevalence in the cities than in the rural of semi-rural areas, as in Crete. This could be because certain environmental factors related to T1DM are more prevalent in the cities. Among these factors, viruses, dietary products and stress should be mentioned. In conclusion, it is evident that these two populations of the study differ genetically as concerns the HLA-DQB1*0201, 0302 genotype.

Further comparisons between low-disease incidence and high-risk populations will shed more light into the complex imbalance between environmental triggers and genetic predisposition than leads to the onset of T1DM. A better understanding of the disease will ultimately result in the development of prediction and prevention strategies.

Effect of the prevention of genetic diseases on gene frequencies

Bernardo Beiguelman and Henrique Krieger

Laboratory of Genetic Epidemiology, Department of Parasitology, Instituto de Ciências Biomédicas, Universidade de São Paulo, Brasil.

This article was written in homage to our good friend George Fraser, well-known scientist, for his lasting contributions to human genetics. His collaboration with his Brazilian colleagues will never be forgotten. In writing this paper our intention was primarily to give theoretical support to students with little mathematical experience in a subject in which George was a pioneer: providing population dynamics answers to modern medical practices.

Introduction

Although the statement may seem paradoxical, genetic counselling differs from any program of prevention of genetic diseases since they have different aims. In fact, genetic counselling intends to help individuals or families who are, or suppose to be, at risk of occurrence or recurrence of genetic defects to take rational decisions about procreation. Moreover, according to the bio-ethical principle of autonomy, genetic counselling should not be directive, and should provide individuals who look for genetic advice with both a wide comprehension of their diseases or genetic risks and the options offered by present medical resources for therapy and/or prophylaxis. Therefore the goal of genetic counselling will be fulfilled either if the couple, who is at risk of giving birth to an affected child, decides to avoid the birth of this child or, in contrast, decides to continue the pregnancy of the child who exhibits a genetic disease. Otherwise stated, genetic counsellors are not worried about the eugenic or dysgenic consequences of their patients' reproductive options, since the individuals concerned make their choices freely after receiving sufficient information.

Eugenic principles fundamentally intend to protect interests of society. Instead, genetic counselling intends to preserve the well-being of individuals and their families, helping them to solve problems related to genetic diseases, clearing up their doubts, and diminishing or avoiding suffering.

Consider, for instance, that genetic counselling was given to two young couples including in each case one member with an early manifested dominant autosomal form of inherited blindness that cannot presently be corrected. After being conscious that they have a 50% risk of transmitting the same inherited disease to their offspring, one couple takes the decision not to have biological children, as the spouses have concluded that such behaviour will avoid suffering of their offspring and themselves. The other couple decides instead to have biological children, since the spouse who exhibits the complete visual deficiency considers him/herself a happy and useful person, being his/her opinion shared with the spouse with normal vision. This couple disdains the high risk of giving rise to a child who will be blind, since they consider that this child could be as happy and useful as the spouse with the inherited visual deficiency. Thus, in spite of the first couple having decided for an eugenic option while the second for a dysgenic procedure, genetic counselling has reached its goal in both cases, since both couples have demonstrated a wide comprehension of all the implications of the genetic defect under discussion, and have taken rational decisions with respect to procreation.

In contrast to genetic counselling, the programs of prevention of inherited diseases have a directive hallmark, varying from paternal advice to coercion, as in the Law on Maternal and Infant Health Care, promulgated on 1 June 1995 in the People's Republic of China. Two articles (10 and 16) of this Law violate the bioethical principles of autonomy, privacy, and justice, since their tenor is as follows:

Art.10 – Physicians shall, after performing the pre-marital physical check-up, explain and give medical advice to both the male and the female who have been diagnosed with certain genetic disease of a serious nature which is considered to be inappropriate for child-bearing from a medical point of view. The two may be married only if both sides agree to take long-term contraceptive measures or to be submitted to tubal ligation operation for sterility.

Art. 16 – If a physician detects or suspects that a married couple in their child-bearing age suffer from genetic disease of a serious nature, the physician shall give medical advice to the couple, and the couple in their child-bearing age shall take measures in accordance with the physician's medical advice.

Obviously, all conscious geneticists become deeply concerned with:

1. the mandatory childlessness for those with certain genetic disease of *a serious nature,* since no two human geneticists would agree on which disorders are included under such a classification of diseases;
2. the possibility of both mandatory prenatal diagnosis testing and mandatory termination of pregnancy when a foetal abnormality is identified or *suspected;*
3. the possibility of mandatory subsequent childlessness for couples who have given birth to a child with *a serious* birth defect, which may not in fact have a genetic basis.

All preventive programs have similar goals since they intend to obtain an eugenic effect, but, instead of this, the result may be dysgenic. We will try to demonstrate this phenomenon in following pages.

Autosomal diseases

Let us consider a pair of autosome alleles *A,a* for which the homozygous *aa* genotype determines a recessive disease that causes its bearers to die before reaching reproductive age. As a consequence, such homozygous individuals will be engendered only by heterozygous couples (*Aa* × *Aa*), while in the population there will be only three types of couples concerning the autosomal *A,a* pair, that is to say, *AA* × *AA*, *AA* × *Aa* and *Aa* × *Aa* couples.

Let us also accept both that the *Aa* × *Aa* couples may be detected in prenuptial examinations and that prenatal diagnosis of the *aa* homozygous foetuses is possible. If *Aa* × *Aa* couples, who have planned to have a certain number of children, always terminate pregnancy when the foetus is diagnosed as *aa*, allowing the birth of a foetus with *AA* or *Aa* genotype, a reproductive compensation will result, since the *aa* children who are lost are replaced by individuals who have *AA* or *Aa* genotype. This preventive measure will provoke a dysgenic effect, since it is easy to demonstrate that the frequency of the *a* allele in the population will decrease less rapidly than in the absence of preventive measures.

In fact, when $Aa \times Aa$ couples do not end the gestation of aa foetuses, they may generate children with AA, Aa or aa genotypes with probabilities $1/4$, $1/2$, and $1/4$ respectively, with the aa individuals genetically inactive, since they do not reach reproductive age. However, when $Aa \times Aa$ couples decide to end the pregnancy of aa foetuses, but allow the generation of AA or Aa individuals, they change the probabilities of these genotypes, which become $1/3$ for AA and $2/3$ for Aa. According to Tables 1 and 2, $Aa \times Aa$ couples that decide to terminate pregnancies of aa foetuses and adopt reproductive compensation, by allowing the generation of AA and Aa children, will transmit a larger amount of a alleles to the following generation than couples which do not adopt such preventive measures. For instance, an $Aa \times Aa$ couple with three children that

Table 1. Number of a alleles transmitted by $Aa \times Aa$ couples to their children when the pregnant women do not opt for prenatal diagnosis.

No. of sibs	Genotypes	Probability	a alleles	Transmissible a alleles
1	AA	1/4	-	-
	Aa	1/2	1	1/2
	aa	1/4	-	-
	Total			1/2 = 0.5
2	AA-AA	1/16	-	-
	AA-Aa	4/16	1	4/16
	Aa-Aa	4/16	2	8/16
	AA-aa	2/16	-	-
	Aa-aa	4/16	1	4/16
	aa-aa	1/16	-	-
	Total			16/16 = 1
3	AA-AA-AA	1/64	-	-
	AA-AA-Aa	6/64	1	6/64
	AA-AA-aa	3/64	-	-
	AA-Aa-Aa	12/64	2	24/64
	AA-Aa-aa	12/64	1	12/64
	Aa-Aa-Aa	8/64	3	24/64
	Aa-Aa-aa	12/64	2	24/64
	AA-aa-aa	3/64	-	-
	Aa-aa-aa	6/64	1	6/64
	aa-aa-aa	1/64	-	-
	Total			96/64 = 1.5

Table 2. Number of *a* alleles transmitted by *Aa* × *Aa* couples to their children when there is reproductive compensation following abortion of *aa* foetuses prenatally diagnosed.

No. of sibs	Genotypes	Probability	*a* alleles	Transmissible *a* alleles
1	*AA*	1/3	-	-
	Aa	2/3	1	2/3
	Total			2/3 = 0.67
	AA-AA	1/9	-	-
2	*AA-Aa*	4/9	1	4/9
	Aa-Aa	4/9	2	8/9
	Total			12/9 = 1.33
	AA-AA-AA	1/27	-	-
	AA-AA-Aa	6/27	1	6/27
3	*AA-Aa-Aa*	12/27	2	24/27
	Aa-Aa-Aa	8/27	3	24/27
	Total			54/27 = 2

adopted reproductive compensation after terminating pregnancy of *aa* foetuses prenatally diagnosed will transmit two *a* alleles to the next generation (Table 2) instead of 1.5 alleles transmitted by an *Aa* × *Aa* couple who do not adopt the preventive measures mentioned above (Table 1). Tables 1 and 2 also show that the number of *a* alleles transmitted to the next generation increases as the number of children rises.

Therefore, when the three types of couples who are genetically active (*AA* × *AA*, *AA* × *Aa*, and *Aa* × *Aa*) are taken into account, it may be foreseen that if they would have, on the average, the same number of children, the *a* allele will decrease more slowly in a population that adopts preventive measures than in a population that does not. In fact, let as consider that, in the first generation, the frequency of individuals with the homozygous dominant genotype *AA* is D, and that H is the frequency of the heterozygous *Aa* individuals. In this case, it seems clear that D^2 will be the frequency of *AA* × *AA* couples, 2DH the frequency of *AA* × *Aa* couples, and H^2 the frequency of *Aa* × *Aa* couples.

Table 3 shows that if no preventive measures are taken, the frequency of the *a* allele in the generation resulting from the *AA* × *AA*, *AA* × *Aa* and *Aa* × *Aa* couples will be estimated as

$$H\left[\frac{D}{2} + \frac{H}{4}\right]$$

However, if reproductive compensation is adopted by *Aa* × *Aa* couples after terminating the pregnancy of *aa* foetuses prenatally diagnosed, the frequency of the *a* allele will be estimated as

$$H\left[\frac{D}{2} + \frac{H}{3}\right]$$

Table 3. Distribution of families in a theoretical population in which a pair of autosomal alleles (A,a) is segregating, and one of the genotypes (aa) is completely selected against. The frequency of the a allele is estimated when preventive measures are taken and when there is reproductive compensation following abortion of aa foetuses.

	Children					
Couples	No preventive measures			Preventive Measures		
	AA	Aa	aa	AA	Aa	aa
$AA \times AA$ (D²)	D²	-	-	D²	-	-
$AA \times Aa$ (2DH)	DH	DH	-	DH	DH	-
$Aa \times Aa$ (H²)	$\dfrac{H^2}{4}$	$\dfrac{H^2}{2}$	$\dfrac{H^2}{4}$	$\dfrac{H^2}{3}$	$\dfrac{2H^2}{3}$	-
Frequency of the a allele in the offspring generation	$\dfrac{DH}{2} + \dfrac{H^2}{4} = H\left(\dfrac{D}{2} + \dfrac{H}{4}\right)$			$\dfrac{DH}{2} + \dfrac{1}{2}\dfrac{2H^2}{4} = H\left(\dfrac{D}{2} + \dfrac{H}{3}\right)$		

Figure 1 shows graphically the difference between the intensities of elimination of an autosomal allele with a homozygous deleterious effect in successive generations if no preventive measures are taken and when they consist of reproductive compensation after termination of the aa pregnancy.

Since the number of a alleles transmitted to the next generation increases as the number of children born to $Aa \times Aa$ couples rises (Tables 1 and 2) it may also be foreseen that the frequency of this allele in successive generations will be higher than predicted in Table 3 if there is excess reproductive compensation, i.e., if the replaced non-aa children are more numerous than the lost aa foetuses, resulting in more children born to $Aa \times Aa$ than to other couples.

Obviously, the dysgenic effect of the prevention of recessive autosomal diseases will be smaller if the identification of the $Aa \times Aa$ couples will only be possible after they bear an aa child. Table 4 illustrates this situation in $Aa \times Aa$ couples who, after bearing an aa homozygous child whose inherited disease is detected at birth, decide to terminate the pregnancy of subsequent aa foetuses and to undertake reproductive compensation. The total number of transmissible a alleles in Tables 4.1 and 4.2, which take into account $Aa \times Aa$ couples which planned to have two and three children, respectively, should be compared to the total number of this allele in Table 2. The dysgenic effect of this type of prevention is, of course, less marked when the disease determined by aa genotype is not manifested at birth, and stronger when the inherited disease is manifested shortly after birth.

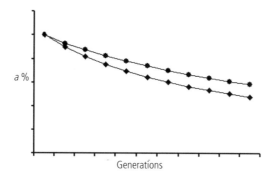

a %

Generations

Figure 1. Rates of elimination of an autosomal allele with a homozygous deleterious effect (*aa*) when no preventive measures are taken (♦), and when reproductive compensation is adopted after abortion of *aa* foetuses(●).

Sex-linked diseases

Let as consider a sex-linked pair of alleles *A,a*, in which the *a* allele determines a severe disease in hemizygous males that does not allow them to reach reproductive age. This also means that homozygous $X^a X^a$ females are not found in the population, the female genotypes being only $X^A X^A$ or $X^A X^a$. Couples which include $X^A X^a$ women may generate $X^A X^A$ or $X^A X^a$ daughters and $X^A Y$ or $X^a Y$ sons, each of these genotypes having $1/4$ probability of occurrence. The termination of the pregnancy of undesired male foetuses detected by prenatal diagnosis may also have a dysgenic effect. The intensity of this effect will depend on the possibility of recognising prenatally the hemizygous $X^a Y$ foetuses, and on the existence of reproductive compensation.

Now, let us consider couples that include $X^A X^a$ women who intend to have two children. In this case, the daughters of $X^A X^a$ women will transmit, on average, 0.5 *a* sex-linked allele to the next generation, as is pointed out in Table 5. The other 0.5 *a* sex-linked allele will be lost through the $X^a Y$ sons who cannot reach the reproductive age. Nevertheless, the number of transmitted *a* sex-linked alleles will be higher if all $X^A X^a$ women could be identified prenuptially, and decided to terminate the pregnancy of $X^a Y$ foetuses detected by prenatal diagnosis followed by reproductive compensation in favour of female children ($X^A X^A$ or $X^A X^a$ genotypes) or male children with $X^A Y$ genotype. As is shown in Table 6, the heterozygous $X^A X^a$ daughters will transmit, on average, 0.667 of *a* sex-linked allele to the next generation, provoking, therefore, a dysgenic effect.

Table 4. Mean number of *a* alleles transmitted to sibships with two or three individuals born to *Aa* × *Aa* couples who, after bearing an *aa* homozygous child, decide to terminate the pregnancy of subsequent *aa* foetuses and to opt for reproductive compensation.

a. Sibships with two individuals

Sibship	Probability	*a* alleles of heterozygous individuals	Transmissible *a* alleles
AA - AA	(1/4)(1/4) = 1/16	-	-
AA - Aa	2(1/4)(2/4) = 2/16	1	2/16
Aa - AA	(2/4)(1/4) = 2/16	1	2/16
AA - aa	(1/4)(1/4) = 1/16	-	-
aa - AA	(1/4)(1/3) = 1/12	-	-
Aa - Aa	(2/4)(2/4) = 4/16	2	8/16
Aa - aa	(2/4)(1/4) = 2/16	1	2/16
aa - Aa	(1/4)(2/3) = 2/12	1	2/12
Total			1.04

b. Sibships with three individuals

Sibships	Probability	*a* alleles of heterozygous individuals	Transmissible *a* alleles
AA-AA-AA	(1/4)(1/4)(1/4) = 1/64	-	-
AA-AA-Aa	3(1/4)(1/4)(2/4) = 6/64	1	6/64
AA-Aa-Aa	3(1/4)(2/4)(2/4) = 12/64	2	24/64
Aa-Aa-Aa	(2/4)(2/4)(2/4) = 8/64	3	24/64
AA-Aa-aa	(1/4)(2/4)(1/4) = 2/64	1	2/64
AA-aa-Aa	(1/4)(1/4)(2/3) = 2/48	1	2/48
aa-AA-Aa	(1/4)(1/3)(2/3) = 2/36	1	2/36
Aa-AA-aa	(2/4)(1/4)(1/4) = 2/64	1	2/64
Aa-aa-AA	(2/4)(1/4)(1/3) = 2/48	1	2/48
aa-Aa-AA	(1/4)(2/3)(1/3) = 2/36	1	2/36
AA-AA-aa	(1/4)(1/4)(1/4) = 1/64	-	-
AA-aa-AA	(1/4)(1/4)(1/3) = 1/48	-	-
aa-AA-AA	(1/4)(1/3)(1/3) = 1/36	-	-
Aa-Aa-aa	(2/4)(2/4)(1/4) = 4/64	2	8/64
Aa-aa-Aa	(2/4)(1/4)(2/3) = 4/48	2	8/48
aa-Aa-Aa	(1/4)(2/3)(2/3) = 4/36	2	8/36
Total			1.61

Table 5. Sib pairs who may be engendered by women who are heterozygous for a sex-linked a allele ($X^A X^a$) which impedes hemizygous $X^a Y$ men to reach reproductive age. Average number of a alleles transmitted through the heterozygous daughters when preventive measures are not adopted.

Sib pairs	Probability	a alleles of heterozygous daughters	Transmissible a alleles
$X^A X^A - X^A X^A$	$(1/4)(1/4) = 1/16$	-	-
$X^A X^A - X^A X^a$	$2(1/4)(1/4) = 2/16$	1	2/16
$X^A X^a - X^A X^a$	$(1/4)(1/4) = 1/16$	2	2/16
$X^A X^A - X^A Y$	$2(1/4)(1/4) = 2/16$	-	-
$X^A X^A - X^a Y$	$2(1/4)(1/4) = 2/16$	-	-
$X^A X^a - X^A Y$	$2(1/4)(1/4) = 2/16$	1	2/16
$X^A X^a - X^a Y$	$2(1/4)(1/4) = 2/16$	1	2/16
$X^A Y - X^A Y$	$(1/4)(1/4) = 1/16$	-	-
$X^A Y - X^a Y$	$2(1/4)(1/4) = 2/16$	-	-
$X^a Y - X^a Y$	$(1/4)(1/4) = 1/16$	-	
Total			8/16 = 0.5

Table 6. Sib pairs who may be engendered by women who are heterozygous for a sex-linked a allele ($X^A X^a$) which impedes hemizygous $X^a Y$ men from reaching reproductive age. Average number of a alleles transmitted through the heterozygous daughters when reproductive compensation is adopted after terminating the pregnancy of $X^a Y$ foetuses.

Sib pairs	Probability	a alleles of heterozygous daughters	Transmissible a alleles
$X^A X^A - X^A X^A$	$(1/3)(1/3) = 1/9$	-	-
$X^A X^A - X^A X^a$	$2(1/3)(1/3) = 2/9$	1	2/9
$X^A X^a - X^A X^a$	$(1/3)(1/3) = 1/9$	2	2/9
$X^A X^A - X^A Y$	$2(1/3)(1/3) = 2/9$	-	-
$X^A X^a - X^A Y$	$2(1/3)(1/3) = 2/9$	1	2/9
$X^A Y - X^A Y$	$(1/3)(1/3) = 1/9$	-	
Total			6/9 = 0.667

Let us now consider the case in which the hemizygous $X^a Y$ foetuses cannot be recognised, and couples including an $X^A X^a$ women that, wishing to avoid the birth of an affected male, decide to terminate pregnancy when a male foetus is detected *in utero*. In this case, the mean number of transmitted a alleles will be doubled, since these couples will generate only $X^A X^A$ and $X^A X^a$ daughters with the same probability ($1/2$), as pointed out in Table 7.

Concerning the sex-linked alleles A,a, two types of couples are observed in populations, i.e. $X^A X^A \times X^A Y$ and $X^A X^a \times X^A Y$. If both types of couples have, on average, the same number of children, it may be foreseen that the allele with a deleterious effect (a) will decrease less rapidly in a population which obeys preventive measures than in others in which such measures are not adopted.

Table 7. Sib pairs who may be engendered by women who are heterozygous for a sex-linked a allele ($X^A X^a$) which impedes hemizygous $X^a Y$ men from reaching reproductive age. Average number of a alleles transmitted through the heterozygous daughters when reproductive compensation is adopted after terminating the pregnancy of male foetuses.

Sib pairs	Probability	a alleles of heterozygous daughters	Transmissible a alleles
$X^A X^A - X^A X^A$	$(1/2)(1/2) = 1/4$	-	-
$X^A X^A - X^A X^a$	$2(1/2)(1/2) = 2/4$	1	2/4
$X^A X^a - X^A X^a$	$(1/2)(1/2) = 1/4$	2	2/4
Total			$4/4 = 1$

In fact, if in the parental generation D and H are the frequencies of the homozygous $X^A X^A$ and heterozygous $X^A X^a$ women, respectively, it is clear that the frequencies of the $X^A X^A \times X^A Y$ and $X^A X^a \times X^A Y$ couples will be D and H respectively, since all genetically active men will be $X^A Y$. Table 8 reveals that, if no preventive measures are observed, $H/6$ will estimate the frequency of the sex-linked a allele in the next generation. However, if the $X^A X^a \times X^A Y$ couples, after terminating the pregnancy of $X^a Y$ foetuses, adopt reproductive compensation for the generation of a $X^A Y$ son or a daughter, then

$$\frac{2H}{3(3 + H)}$$

will estimate the frequency of the sex-linked a allele in the next generation (Table 9). Finally, if $X^A X^a \times X^A Y$ couples terminate all pregnancies of male foetuses and adopt reproductive compensation in favour of daughters, the frequency of the sex-linked a allele will be estimated by means of

$$\frac{H}{3(1 + H)}$$

Table 8. Distribution of families and frequency of a sex-linked a allele in a theoretical population in which no preventive measures are taken and hemizygous $X^a Y$ males cannot reach reproductive age, being therefore completely selected against.

Couples	Daughters		Sons	
	$X^A X^A$	$X^A X^a$	$X^A Y$	$X^a Y$
$X^A X^A \times X^A Y$ (D)	D/2	-	D/2	-
$X^A X^a \times X^A Y$ (H)	H/4	H/4	H/4	H/4
Total at birth	D/2 + H/4	H/4	D/2 + H/4	H/4
Total at marriage, according to sex	D + H/2	H/2	1	—
Frequency of the a allele in the offspring generation	$1/3 \times H/2 = H/6$			

Table 9. Distribution of families and frequency of a sex-linked *a* allele in a theoretical population in which hemizygous X^{a}Y males are completely selected against. Reproductive compensation is adopted after terminating the pregnancy of X^{a}Y foetuses.

	Daughters		Sons
Couples	X^{A}X^A	X^{A}X^a	X^{A}Y
X^{A}X^A × X^{A}Y (D)	D/2	-	D/2
X^{A}X^a × X^{A}Y (H)	H/3	H/3	H/3
Total at birth	D/2 + H/3	H/3	D/2 + H/3
Total at marriage, according to sex	$\dfrac{3D + 2H}{3D + 4H}$	$\dfrac{2H}{3D + 4H}$	1
Frequency of the *a* allele in the offspring generation	$\dfrac{1}{3} \times \dfrac{2H}{3D + 4H}$	$= \dfrac{2H}{3(3 + H)}$	

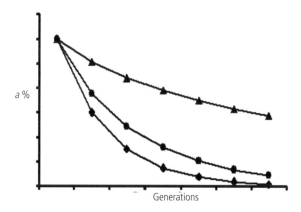

a %

Generations

Figure 2. Intensity of elimination of a sex-linked allele which has a hemizygous deleterious effect (X^{a}Y) when no preventive measures are taken (◆) and when reproductive compensation is adopted after terminating the pregnancy of X^{a}Y foetuses (●) or of male foetuses (▲).

(Table 10). Figure 2 gives a graphical notion of the different intensities of elimination of a sex-linked allele, which has a deleterious effect in hemizygosis, in populations under different conditions of preventive measures.

Comments

The decrease of the intensity of elimination of autosomal or sex-linked alleles that determine recessive monogenic diseases in populations adopting preventive measures against these alleles would, of course, not occur if the termination of pregnancy were extended to heterozygous foetuses. Nevertheless, since the heterozygous individuals are expected to be healthy and every human being is, with a probability near to certainty, heterozygous for a significant number of deleterious alleles, this practice is logically absurd, and indefensible from the ethical and scientific point of view.

Table 10. Distribution of families and frequency of a sex-linked *a* allele in a theoretical population in which hemizygous X^aY males are completely selected against. Reproductive compensation is adopted after terminating the pregnancy of male foetuses.

Couples	Daughters		Sons
	X^AX^A	X^AX^a	X^AY
$X^AX^A \times X^AY$ (D)	D/2	-	D/2
$X^AX^a \times X^AY$ (H)	H/2	H/2	-
Total at birth	(D + H)/2	H/2	D/2
Total at marriage, according to sex	1/(1 + H)	H/(1 + H)	1
Frequency of the *a* allele in the offspring generation		H/3(1 + H)	

Thus, the elimination of a foetus which is heterozygous for a certain *a* allele may result, after reproductive compensation, in the generation of another individual who, in spite of not being heterozygous for the *a* allele, may be heterozygous for genes belonging to other loci, which may provoke effects much more severe than those of the *a* allele.

The explanations given above may have created the impression that the authors are concerned about the dysgenic effects that can result from preventive programs directed to genetic diseases. This is not true. The authors are fully aware that present medicine acts in opposition to natural selection, and that the dysgenic effect resulting from programs for the prevention of monogenic recessive diseases is similar to that observed when a clinical or surgical treatment can resolve them.

Let us consider again an autosomal pair of alleles *A,a* with frequencies *p* and $q = 1 - p$, respectively. Let us also suppose that the homozygous *aa* individuals do not reach the reproductive age. In this case, if mutations are not taken into account, and supposing that the heterozygous *Aa* do not have reproductive advantage, it is easy to demonstrate that after *n* generations the frequency of the *a* allele (*q*) will decrease to

$$q_n = \frac{q}{1 + nq}$$

Let us now admit that, as a result of a new treatment or due to a special diet, the selective process against the homozygous *aa* is discontinued, enabling the individuals with *aa* genotype to have the same reproductive possibilities as the individuals with genotypes *AA* or *Aa*. Such treatment would result in a dysgenic effect since the *a* allele would not be eliminated while more *a* alleles would be introduced in the population by mutation, reaching a new equilibrium value, solely dependent on mutation rates. Thus, the frequency of the *a* allele would increase. The new value will depend on the mutation rate μ of *A* to *a* allele, that is to say μp, as well as to the *A* alleles generated by the reverse mutation rate ν that is to say νq.

In the case of dominant monogenic diseases, the dysgenic effect due to possible clinical or surgical treatment of the affected individuals will be as striking as the difference between the selection coefficients after and before the introduction of these treatments becomes larger. For instance, let us consider a dominant disease caused by an *A* allele, that determines genetic transmission incapacity either by early death or by suppressing the reproduction of the affected individuals. In this case, all affected individuals who arise in the population exhibit the *A* mutation, which is completely selected against (*s* = 1, being *s* the selective coefficient). The persistence (*i* = 1/*s*) of the *A* mutation, i.e. the mean number of generations that this mutation will be maintained in the population will be only one generation, since *i* = 1 when *s* = 1. Otherwise stated, the frequency of the *A* allele will be equal to the mutation rate.

If the selection against the *A* allele decreases (*s* < 1) as a consequence of the introduction of a treatment for the affected individuals, the persistence of the *A* mutation will inversely increase. The *A* mutation may even persist indefinitely in the population, this extreme situation occurring when all bearers of the *A* allele will have the same fitness as the non-bearers, *i.e.* when the selection coefficient is null. In fact, *i* approaches ∞ when *s* = 0, since *i* = 1/0, i.e. ∞.

The authors are, as noted earlier, aware of the anti-selective role played by modern medicine, and, of course, their intention is not to write against the dysgenic effect resulting from prevention of inherited recessive diseases, since they do not judge the decisions which result in dysgenic effects when they are taken by couples after non-directive genetic counselling. Moreover, the authors believe that this effect should not cause concern, since the spectacular development of molecular biology gives us hope that in the near future the replacement of genes with deleterious effects by normal ones will be an ordinary tool included in the clinical therapeutic arsenal.

We address the subject differently from Fraser (1972) but the conclusions are very similar. The intention of the present authors is to emphasise that the supposed eugenic effect used as an argument for carrying out preventive programmes for inherited diseases is a great mistake, especially when they are coercive. No arguments exist to justify a violation of the principle of autonomy. Nobody should violate the individual's right to take conscious decisions about her or his health condition and reproduction. The health authorities have the responsibility to provide suitable information to achieve such consciousness. Nobody should interfere with somebody's right to refuse genetic tests or the detection of pre-symptomatic diseases, even those that may cause early death. No result of genetic tests should be used to influence the termination of pregnancy. People undertaking a genetic test may refuse to know the result of this test or to restrict the use of it. In the sense used here, nothing can justify the violation of the principle of privacy.

Summary
The differences between the aim of genetic counselling and the goal intended by programs of prevention of genetic diseases are stressed. It is demonstrated that preventive programs result in a dysgenic effect, since the frequency of autosomal or sex-linked alleles decreases less rapidly when preventive measures are taken.

Down's Syndrome: maternal age and inheritance

Howard Cuckle[1] and Svetlana Arbuzova[2]

[1]Reproductive Epidemiology, Leeds Screening Centre, University of Leeds,
Gemini Park, Sheepscar Way, Leeds LS7 3JB, UK
[2]Interregional Medico-Genetic Center, Central Hospital, Clinic number 1,
57 Artem Street, 83000 Donetsk, Ukraine

George Fraser is a man who takes seriously the concept of a mentor. His teachers included two great figures of 20th century biology, R. A. Fisher and Lionel Penrose, and at the 1998 mini-symposium held to mark the centenary of Penrose's birth, George gave a talk (Fraser 1998) entitled 'Lionel Penrose as Scientist and Mentor: Recollections and Lifelong Legacies'. Consequently we thought that it would be appropriate in the current celebration of George to revisit Penrose's work on Down syndrome, principally maternal age and inheritance (Smith and Berg 1976), to see how much has been learnt subsequently.

Penrose's contribution

Penrose's major research resource, the 'Colchester Survey', consisted of the 1280 residents of the Royal Eastern Counties Institution in Colchester, England, and their families (Penrose 1938). His belief that mental abnormality had a biological rather than social aetiology was confirmed by the survey. It yielded clear evidence to support a number of salient features of mental abnormality: large excess of males, heterogeneity of expression, continuum between normal and intellectual impairment, environmental causes and familial aggregation. This seminal work led to more focussed studies by Penrose on tuberous sclerosis, phenylketonuria and Down syndrome.

In the Colchester Survey, there were 63 residents with Down syndrome and Penrose observed an association with increased maternal and paternal ages. In a more detailed study in a larger population he was able to show, using regression analysis, that the primary effect was maternal (Penrose 1933, and see Oliver Penrose's contribution to this volume). The mean paternal age was not significantly different from the value expected given the maternal age distribution, whereas there was a highly significant difference in the mean maternal age from that expected given the paternal ages. He similarly showed that gravidity, parity, and between-pregnancy interval were not significant after maternal age had been taken into account.

Penrose also contributed to the investigation of Down syndrome inheritance. He was the first to report familial transmission of chromosome 21 translocations (Penrose et al. 1960) and subsequently showed that there was no maternal age effect in the inheritance of such translocations. He also suggested that in families with recurrent standard trisomy 21 and apparently normal parental chromosomes, there may be mosaicism. Through his study of parental dermatoglyphics he concluded that about 10% of mothers of single children with Down syndrome, and perhaps 50% of those with more than one affected child, were mosaic for trisomy 21 (Barnicot et al. 1963).

Maternal age – current knowledge

Penrose studied the association between maternal age and Down syndrome in children and adults. Subsequently it has been possible to determine single-year maternal age-specific rates of the disorder, at birth and in early pregnancy, and to study the effect of maternal age on aneuploidy in miscarriages, embryos and gametes.

In recent years, the widespread use of antenatal screening, invasive prenatal diagnosis and selective termination of pregnancy has enabled rates in early pregnancy to be determined. At the same time this change in practice has reduced the ability to estimate birth prevalence accurately and of itself introduces potentially strong selection bias. This arises because of the high intrauterine lethality of Down syndrome, such that a large proportion of prenatally diagnosed pregnancies with the disorder is not viable. It is possible that screening detects cases with reduced viability on average, and termination of prenatally diagnosed cases leads to over-diagnosis since some will be non-viable.

Advances in assisted reproduction technology such as *in vitro* fertilisation (IVF), intracytoplasmic sperm injection (ICSI) and pre-implantation genetic diagnosis (PGD) have provided an insight into the earliest stages of human aneuploidy. Whilst the data need to be interpreted cautiously, as the populations studied may not represent the reproductively fit and the material may be sub-optimal (for example, oocytes that have failed to be fertilised or morphologically abnormal embryos), this should not bias any associations with maternal age.

Birth prevalence

The best available estimate of age-specific rates is obtained by meta-analysis, where data from several series are combined and a regression equation fitted. Four meta-analyses have been published based on eleven series studied before prenatal diagnosis became common and including a total of more than 5000 Down syndrome and five million unaffected births (Cuckle *et al.* 1987, Hecht and Hook 1994, 1996, Bray *et al.* 1998).

Over the age range 15–24 there is virtually no change in the observed age-specific rates for all series combined: 0.00%, 0.06%, 0.07%, 0.06%, 0.06%, 0.06%, 0.07%, 0.06%, 0.07% and 0.08%. Hence the claim that prevalence is relatively high at extremely young ages (Erickson 1978) is not sustained. Error in recording maternal age is the most likely explanation for anomalous results in some studies as under-recording will result in a J-shaped prevalence curve.

Across the 25–45 age range all the regression curves are similar, yielding an additive-exponential pattern of increase. The estimated rates from one curve are 0.07%, 0.11%, 0.26%, 0.89% and 3.5% at 25, 30, 35, 40 and 45 respectively (Cuckle *et al.* 1987). After age 45 the regression curves do differ, so that at 50 the estimated rates range widely from 6% to 20%, and in the National Down Syndrome Cytogenetic Register for England and Wales, established in the post-screening era, the estimate at 50 is just 3.8% (Morris *et al.* 2003), although this could be due to selection biases (Cuckle 2002). When the Register data were supplemented by three previously unpublished series there was a decline in prevalence from age 47 to 52: 4.2%, 3.1%, 2.2%, 2.9%, 1.5% and 1.2% respectively (Morris *et al.* 2005).

It is possible to estimate the overall birth prevalence rate for a specific country, in the absence of prenatal diagnosis, by the average age-specific rate weighted by the proportion of maternities at each completed year of age. For England and Wales in 2002 the calculated rate is

0.19% (Cuckle 2005); since the rate is 0.06–0.07% under age 25 it means that about two thirds of cases today are maternal-age related.

Early pregnancy

Several studies have reported Down syndrome rates at mid-trimester amniocentesis or first trimester chorionic villus sampling (CVS) in single years of age but they are all restricted to women aged over 35. In a meta-analysis of nine such studies, there were 1159 Down syndrome cases diagnosed at amniocentesis and 341 at CVS (Bray and Wright 1998). The observed rates at the two stages of pregnancy for women aged 35 were 0.3% and 0.5%; at 40, 1.4% and 1.8%; at 45, 4.4%, 6.1%.

In the absence of comparable data on younger women it is usual to estimate rates from the birth prevalence curves after adjustment for Down syndrome foetal loss between the stages of pregnancy. The loss rate can be estimated from the average difference in incidence compared with birth in the prenatal diagnosis series, and from follow-up series of women who have refused termination following prenatal diagnosis, who again are largely older (Hook *et al.* 1989, Morris *et al.* 1999). This shows about one half of Down syndrome pregnancies are lost after CVS and one quarter after amniocentesis (Cuckle 1999).

It is also possible to calculate risks at individual weeks of early pregnancy by applying gestation-specific foetal loss rates. These can be derived using a regression formula obtained from a series of over 57,000 prenatal diagnoses at 9–16 weeks in women aged 35–45 (Snijders *et al.* 1999). The data were first grouped into three gestational periods and the observed number of cases was compared with that expected from age-specific birth prevalence. However, it would be more accurate to do the regression on single gestational week comparisons, constrained to 100% at 40 weeks. When this is done, and the term risks are those of Cuckle *et al.* (1987), the loss rates are 46%, 44%, 43%, 40%, 38%, 36%, 34% and 32% for each week from 9 to 16.

These calculations of age- and gestation-specific rates assume that foetal loss rates do not vary with maternal age. However, the overall loss rates are largely based on women aged over 35, so this cannot be readily examined. Since the foetal loss rate in general increases markedly with maternal age (Nybo Andersen *et al.* 2000), it is likely that this will also happen in Down syndrome pregnancies.

There have been two large studies, in New York and Hawaii, in which consecutive miscarriages were karyotyped (Hassold *et al.* 1984). These showed that the incidence of Down syndrome increases with maternal age, so that in a total of 92 cases from the combined results the mean age was 31 years, compared with 27 years in the chromosomally normal miscarriages. But the studies are not large enough to judge whether the magnitude of the increase in Down syndrome incidence with maternal age among foetal losses is the same as that among births. Pregnancy losses after IVF and subsequent demonstration of foetal cardiac activity have also been karyotyped. In one study the foetal loss rate was 12% (233/2014), and there was aneuploidy in 82% of losses among women aged over 40 years compared with 65% in younger women (Spandorfer *et al.* 2004).

Cleavage-stage embryos and gametes

There is a strong relationship between chromosome abnormalities in early embryos and maternal age. In one study of 1255 embryos, the aneuploidy rate increased from 3.1% in women aged

20–34 years to 17% in those 40 or older (Marquez *et al.* 2000). In a similar study of 2058 embryos, many karyotyped using PGD, an increase with age was specifically seen for trisomy 21 (Munne *et al.* 2004).

The largest study of oocytes that failed to fertilise after IVF included over 1000 karyotypes and the aneuploidy rate was strongly related to maternal age: 5.6%, 7.2%, 7.4%, 15%, 34% and 65% in those aged under 27, 27–30, 31–34, 35–38, 39–42 and 43–46 (Pellestor *et al.* 2003). A similar effect was seen in both single chromatid and whole chromosome non-disjunction. The frequency of diploid sperm also increases with age but this is not due to disomy 21 (Bosch *et al.* 2001).

Twins

There are no studies of age-specific Down syndrome rates in twins, although predicted rates have been derived from singleton rates on the assumption that in dizygous twins the probability of the second being affected is independent of the first and in monozygous twins there will be concordance for Down syndrome (Meyers *et al.* 1997). However, the overall observed prevalence of Down syndrome in twin pregnancies is considerably less than these theoretical predictions.

A meta-analysis of five cohort studies (four cited in Wald and Cuckle (1987) and Doyle *et al.* (1990)) includes a total of 106 affected twins and the overall prevalence was only 3% greater than for singletons. In a study of prenatal diagnoses, there were no Down syndrome cases in 512 samples from 278 twin pregnancies (Jamar *et al.* 2003). This was a statistically significant deficit compared with the 0.9% rate in samples from over 19,000 singletons.

Maternal age standardisation

Since maternal age is such a strong risk factor it can confound the effect of other variables. To overcome this it is necessary either to stratify the data and make comparisons within maternal age groups or carry out an analysis of variance. When there is a high degree of correlation between age and a variable, the former may not be sufficient unless the strata are small since residual confounding could remain within strata.

Paternal age

Maternal and paternal ages are highly correlated. Moreover, there is relatively little variability in the age difference between the parents so an extremely large number of couples need to be investigated in order to discern any independent paternal age effect. Some studies of couples have reported a small effect in births (Stene *et al.* 1987) and miscarriages (Hatch *et al.* 1990), but many others, like Penrose, found none. A recent study demonstrated that residual confounding could produce a spurious paternal age effect if 5-year strata were used (Kazaura and Lie 2002).

One convincing piece of epidemiological evidence for an independent paternal age effect comes from a study of French donor insemination centres (Lansac *et al.* 1997). Since it is simply a matter of chance whether a young woman receives sperm from an older man, maternal age confounding can be discounted. However, there was a much larger age difference between donors and recipients than would be seen in spontaneous pregnancies.

Another possibility, which might account for some of the conflicting results between studies is a synergistic effect whereby paternal age is not important unless there is advanced maternal age. This was suggested by statistical modelling applied to a large cohort of 3419 Down syndrome births in New York State (Fisch *et al.* 2003).

Parity

Maternal age and parity are also highly correlated. Only three studies have reported on parity after stratifying by single year of age (Kallen 1997, Chan *et al.* 1998, Doria-Rose *et al.* 2003). Two of them reported a significant positive association but this could have resulted from bias as they excluded terminations of pregnancy and the acceptability of prenatal diagnosis tends to decline with parity. When one of them was reanalysed after excluding pregnancies where the birth certificate (in the US) reported that amniocentesis had been performed, the effect was reduced and no longer statistically significant (Doria-Rose *et al.* 2003). There is gross under-reporting of amniocentesis on US birth certificates so the real effect is likely to have been even smaller.

Reproductive ability

Fertility declines with age so an association between Down syndrome prevalence and infertility is to be expected. However, there is some evidence for reproductive aging independent of age.

In two studies of women with Down syndrome births mean menopausal age was 2.6 and 0.7 years earlier than controls (Freeman *et al.* 2000, Bartmann *et al.* 2005) and in a study of trisomic miscarriages it was 1.0 years earlier (Kline *et al.* 2000). Furthermore, women with a previous Down syndrome pregnancy have been reported to have elevated serum follicle stimulating hormone (FSH) levels (van Montfrans *et al.* 1999), on average, indicating impending ovarian failure, as have women undergoing early abortions for social reasons where karyotyping revealed foetal aneuploidy (Nasseri *et al.* 1999). But there are contradictory data from a study of trisomic miscarriages (Kline *et al.* 2004) on the number of antral follicles and serum dimeric inhibin B, as well as FSH.

Surgical removal or congenital absence of one ovary, which impairs reproduction, is associated with a ninefold increase in Down syndrome risk (Phillips *et al.* 1995). Women with Turner syndrome have low fertility and from reports in the literature (Tarani *et al.* 1998, Birkebaek *et al.* 2002) they too have a very high Down syndrome risk of 1.8%.

Down syndrome risk does not appear to be greater in pregnancies achieved by assisted reproduction technology than in naturally conceived pregnancies but the total number of cases is small. In four age-matched or age standardised IVF series (Bergh *et al.* 1999, Westergaard *et al.* 1999, Ericson and Kallen 2001, Koivurova *et al.* 2002) combined the prevalence was 0.23% (32 cases) compared with the weighted average of 0.21% in the controls, and in two ICSI series (Bonduelle *et al.* 1998, Loft *et al.* 1999) it was 0.32% (seven cases) compared with 0.24% expected from the average maternal age and gestation of diagnosis.

Ethnic origin

The series included in the meta-analyses of age-specific birth prevalence are almost entirely based on women of European origin. A large number of studies have been published from other populations in 5-year age groups and since the maternal age distribution varies between populations it is necessary to make comparisons in terms of age-standardised rates. A meta-analysis of age-standardised rates has been carried out which compared 49 populations with sufficiently detailed and reliable age information (Carothers *et al.* 1999). Two ethnic groups had age-standardised rates greater than Europeans: Hispanics were 19% and 30% higher in two studies, and non-Ashkenazi Jews were 27% higher. Groups with markedly reduced rates were likely to be due to incomplete ascertainment.

Smoking

Several early studies reported that maternal smoking was less common in Down syndrome pregnancies, but the latest meta-analysis of 17 published studies failed to find a significant association (Rudnicka *et al.* 2002). Smoking habits are subject to strong birth cohort effects so it is important to take full account of maternal age. Some of the early studies either did not do so or stratified the data using too broad age groups. This was demonstrated in one study which found a relative risk of 0.87 with broad grouping, 0.89 adjusting for additional variables and 1.00 when age adjustment, together with the additional variables, was in single years (Chen *et al.* 1999).

Inheritance – current knowledge

Investigators today, particularly with the availability of molecular methods, are able to obtain a more detailed understanding of non-disjunction than in the past.

Non-disjunction

Cytogenetic analysis of peripheral blood from individuals with the Down syndrome phenotype shows that 95% have a non-disjunction of chromosome 21, 4% a Robertsonian translocation, mostly t(14;21) or t(21;21), and 1% are mosaic (Mutton *et al.* 1995). DNA analysis of peri-centromeric polymorphisms in non-disjunction cases and their parents is used to determine which parent was the source of the additional chromosome. In the largest series, including 724 cases, 89% were maternal, 9% paternal and 2% had post-zygotic mitotic non-disjunction (Hassold and Sherman 2000). A similar distribution was found in a series of prenatally diagnosed cases (Muller *et al.* 2000).

The same method also provides information about the stage in meiosis at which the error occurred. When the chromosomes are heterozygous it is inferred that it was in MI, and this happens in about three quarters of the maternal and half of the paternal cases. However, two alternative interpretations of these data have been put forward which would suggest that the supposed MII cases are either the result of MI errors or not essentially MII.

The first follows from the observation that trisomy 21 non-disjunction is associated with altered recombination (Warren *et al.* 1987, Lamb *et al.* 1996, 1997, 2005). In MI cases the genetic linkage map is markedly shorter than the normal female map, showing reduced recombination primarily in the proximal region of 21q. In MII cases the map is increased especially near the centromere suggesting increased proximal recombination in MI.

The second follows from studies of aneuploid oocytes where there are considerably more additional free chromatids than extra whole chromosomes (Angell 1997). If bivalent coherence is gradually lost during the long ovarian sojourn time, at completion of MI they will become four single chromatids held together only by chiasmata. At metaphase, stable orientation along the spindle is achieved by tension between the kinetochores and univalent pairs will be rotated until there is a stable reorientation. Distal chiasmata will require 90° rotation leading to heterozygous free chromatids, whereas univalent pairs with proximal chiasmata will orientate normally and produce homozygous chromatids (Wolstenholme and Angell 2000).

Not surprisingly the mean maternal age is greater in maternally derived cases of Down syndrome than those in which there is a paternal error: for example, 32 and 28 years in the study of Hassold and Sherman (2000). But the mean maternal age does not differ according to the meiotic stage of the maternal or the paternal error.

Reduced recombination in MI cases could render chromosome 21 susceptible to non-disjunction, since the absence of chiasmata or distal sites would be less stable during meiosis. If so, instability must increase with maternal age since the susceptible configurations become less frequent: in a total of 400 cases 78% among those aged less than 29 years, 34% 29–34 years and 19% older women (Lamb *et al.* 2005).

Recurrence risk

The risk of recurrence is dependent on the parental karyotype and the detailed pedigree. In most cases the parents are normal but a small number have a structural chromosome rearrangement, the most frequent being a heterozygous Robertsonian balanced translocation. Female carriers have a very high recurrence risk whilst males do not; in one amniocenteses study of 255 couples, 15% of carrier mothers had foetuses with a translocation whilst there were none to male carriers (Boué and Gallano 1984).

The recurrence risk for woman with a previous Down syndrome but a normal karyotype has been estimated at three points in pregnancy. In an unpublished study of more than 2,500 women who had first trimester CVS for this indication, the Down syndrome incidence was 0.75% higher than that expected from the maternal-age distribution (Kypros Nicolaides, personal communication). In a meta-analysis of four second trimester amniocentesis series to such women totalling 4,953 pregnancies the excess was 0.54% (Arbuzova *et al.* 2001) and in a meta-analysis of 433 live-births the excess was 0.52% (Hook 1992). The weighted average of these excess risks, allowing for foetal losses is 0.77% in the first trimester, 0.54% in the second and 0.42% at term. Assuming that the excess risk is additive the recurrence risk is relatively large for young women but by the age of about 40 it is not materially different from the normal age-specific risk. There is also an increased risk of other types of aneuploidy (Arbuzova *et al.* 2001, Warburton *et al.* 2004) and neural tube defects (Barkai *et al.* 2003). Moreover, the recurrence risk appears to be similar whether the index case was diagnosed in a spontaneous abortion or at birth (Warburton *et al.* 2004).

Inheritance of genes or polymorphisms

Alzheimer disease has been reported to be more common in the mothers of children with Down syndrome (Schupf *et al.* 1994). This study has recently been extended and, as well as confirming the original observation, suggests that most of the association is in women who were under 35 at the time of the affected pregnancy (Schupf *et al.* 2001). The series now includes the parents of 200 Down syndrome probands and 252 controls; the relative risk of Alzheimer disease was 4.8 in the younger mothers, compared with no statistically significant difference for older mothers (although the relative risk was 1.8) or fathers of any age. The ε4 allele of the apolipoprotein E gene has been investigated in parents of Down syndrome individuals. In a series of 188 families the allele distribution was not altered overall compared to a control population but there was significantly increased frequency in young MII mothers (Avramopoulos *et al.* 1996).

Polymorphisms of genes involved in folate metabolism have been investigated in mothers of Down syndrome individuals since abnormal folate and methyl metabolism can lead to DNA hypomethylation and abnormal segregation. Some studies have found an excess of the 677C→T polymorphism in the 5,10-methylene-tetrahydrofolate reductase gene (James *et al.* 1999), but this is not a consistent finding. The two studies investigating the 66A→G polymorphism in methionine synthase reductase found an excess and also reported that a combination of the two mutations conferred a higher risk than either alone (Hobbs *et al.* 2000, O'Leary *et al.* 2002).

The complete mitochondrial (mt) DNA was sequenced in a peripheral blood sample from the mother of a Down syndrome child who was the originator of the additional chromosome 21 (Arbuzova 1998). There were four point mutations not previously described, each of which is likely to disrupt mitochondrial function.

Explaining the maternal age effect

Oocyte selection

The 'production-line' hypothesis proposes that the order in which oocytes are selected for maturation and ovulation from the 'resting pool' is determined by the order in which they were produced *in utero* (Henderson and Edwards 1968). Since those formed in late foetal life have fewer chiasmata and more univalents they might be susceptible to non-disjunction. The idea has been tested in animal models using various experimental methods with no clear and consistent supportive evidence (e.g. Meredith and Doolin 1997).

Zheng *et al.* (2000) have constructed a mathematical model in which monthly selection favours euploidy so that the proportion of actual or potential aneuploid oocytes in the resting pool increases with age. Although this fits prevalence rates there is no direct evidence for it. Depletion of the pool by accelerated atresia could have a similar effect (Kline and Levin 1992). When inbred CBA mice, which have a small pool that is completely depleted when ovulation ceases, were given a unilateral oophorectomy they had increased aneuploid rates and earlier onset of irregular cycles (Brook *et al.* 1984). The human evidence on reproductive aging cited above, although inconsistent, also tends to favour this idea.

Oocyte damage

Over the many years' sojourn time when oocytes await MI completion just before ovulation, there are many potential opportunities for damage. The frequency of persistent nucleoli might be increased, leading to mal-segregation of acrocentric chromosomes where nucleolar fusion holds together the short arms. And generally declining sister chromatid cohesion would lead to non-disjunction of other chromosomes (Wolstenholme and Angell 2000). Spindle damage might occur due to intrinsic factors or environmental insults such as irradiation – radio-sensitivity increases with age (Tease and Fisher 1991) – and heavy metal ions affecting free-radical production or oxidative effects. Hormonal imbalance and reduced micro-vasculature around the ovarian follicle could lead to a cascade of events: increased carbon dioxide and lactic acid inside the follicle, decreased pH in the oocyte, reduced mitotic spindle size and spindle displacement (Gaulden 1992).

The frequency of mtDNA mutations increases with age (Keefe *et al.* 1995), and incidentally mtDNA is almost entirely of maternal origin. These mutations lead to a decline in ATP level and increased production of free-radicals, which could affect the spindle, accelerate telomere shortening and alter recombination (Arbuzova 1998). In a mouse model, it has been shown that mtDNA mutations can modulate the expression of an inheritable MI error in oocytes (Beerman *et al.* 1988).

The secondary oocyte remains in MII metaphase in the Fallopian tube until fertilised and a delay could lead to spindle defects. Since there is decreased frequency of coitus with advancing age (German 1968) the chance of delay will increase. Some animal experiments seem to support this although there are difficulties of interpretation; for a review see Martin-DeLeon and Williams (1987). Infrequent coitus could also contribute to a paternal age effect through delayed utilisation allowing immature diploid sperm to mature.

Relaxed selection

Aymé and Lippman-Hand (1982) hypothesised that the propensity to selectively miscarry affected embryos decreases with advancing maternal age. But this it is inconsistent with IVF results using young donor oocytes in older recipients (Eichenlaub-Ritter 1998). If there was relaxed selection the mean maternal age in Down syndrome would be lower among miscarriages than births but the evidence for this is inconsistent (Hook 1983, Hassold *et al.* 1984), and mean maternal age would be increased in Robertsonian translocation cases too, and it is not (Mutton *et al.* 1996).

Explaining the inheritance pattern

Recurrence in older women may be due to chance alone but in the young it is more likely to have a genetic cause. Gonadal mosaicism is one possibility but is technically difficult to study directly and most data relates to peripheral blood or fibroblasts. In one study five of 13 families with recurrent Down syndrome had demonstrable parental mosaicism (Pangalos *et al.* 1992) and in another study low level mosaicism was demonstrated using more sensitive molecular techniques in two couples aged under 35 whereas whilst no genetic cause was found for the recurrence in two older couples (James *et al.* 1998). A third study, using molecular techniques in couples aged under 35, demonstrated mosaicism in two out of three compared with none of the five control couples who had normal offspring (Frias *et al.* 2002). They also investigated 22 couples who so far only had one Down syndrome child and two had mosaicism. Nevertheless, even gonadal mosaicism requires explanation as it may not be the primary cause but rather a consequence of factors which also lead to meiotic non-disjunction.

There are 14 case reports in the literature of families with either two Down syndrome cases or one Down and another aneuploidy in which there were different reproductive partners in the parental or grand-parental generation (Arbuzova *et al.* 2001). In 13, recurrence was on the maternal side, suggesting the inheritance of a cytoplasmic factor.

Conclusion

Penrose's observations on maternal age and inheritance in Down syndrome have been generally confirmed and there is now a more detailed description of both. Nevertheless, a satisfactory explanation of these phenomena remains elusive.

Genome scans and the 'old genetics'

J.H. Edwards

Biochemistry Department, South Parks Road, Oxford OX3 7LP

Abstract

The term 'new genetics' was used by Julian Huxley to cover Lysenko's imposition of what may be termed 'genetics without the nucleus'. In 1980, following the 'Southern Blot', and the possibility of standard techniques to type very large numbers of variant loci, it was again invoked to relate to the associating of genotypes to phenotypes directly without help from either prior expectations or intervening mechanisms: 'genetics without the cytoplasm'. The ultimate DNA genotype was to resolve the causes of phenotypic variation in health and disease by objective methods freed from the vagueness of informed expectation and the distractions of previous work. This unleashed technology-driven strategies, replacing the problem-oriented and curiosity-driven approach of the past with analytical methods of great complexity based on assumptions of disarming simplicity.

A decade earlier Motulsky, Fraser and their colleagues advanced substantial evidence for the location of a locus influential in a group of severe related disorders. They compared genotypes inferred from typing the red cells in blood and a phenotype, the level of a substance in the surrounding plasma that contributed to the furring of arteries, an event known to impose increased risks of high blood pressure, heart attacks and strokes (Fraser 1969c).

At that time no restrictions were imposed on investigations involving minor procedures. When blood was taken, or rather given, for research purposes the donors expected it to be used efficiently within the resources of the investigator – a simple if informal ethical imperative. They also expected any abnormal and treatable finding to be handled efficiently and confidentially. The main disorder in my experience was anaemia, noted on separating cells, sometimes severe and usually simple to treat.

Introduction

The technological imperative imposed by the simultaneous typing of a million loci, at less than a thousandth the cost per locus of typing multiallelic loci, or variations in size or charge of products of transcription, transformed the strategies for both collecting and analysing data. But at a cost. The technical procedures unleashed made possible the 'HapMap Project' (The International HapMap Consortium 2005) with the scheduled release (in November 2005) of provisional analyses of data from 279 individuals typed at over a million loci.

This formidable feat of organisation aimed to provide both data and analyses with a synoptic view of the average genome, as revealed by common base pair variants. This condition of funding allowed those expert in small segments of the genome, or in related phenotypes, or in comparative mapping, or merely curious, to make detailed 'ground level' analyses by simpler methods, an occupation better suited to small departments.

This advance, based on the power to generate and marshal such massive data, is comparable to the recent mapping of the world by satellite photography, a project also necessarily undisturbed by the fragmentary maps in various formats, dominated by centres of human activity, offering information whose interpretation is beyond the resolution of the electronic eye and creating a similar need for complementary earthbound investigation.

While these studies provide an essential foundation through which changes and other unexpected findings can be surveyed from 'from the ground', their efficient exploitation requires matching funds to support those with appropriate expertise and resources, or the ability to develop new techniques or methods of analysis.

The HapMap data included ten half-megabase segments, one near the ABO locus, each a six thousandth of the genome, in which, it is claimed, 'essentially all information about common DNA variation has been extracted' (HapMap 2005). A claim hardly justified after exclusion of over 10% of the loci on statistical grounds when the technique used confounds homozygotes and hemizygotes. These segments are short enough – with a little more than a mere 70 or so recombinant events, shared unequally between the parents, and known to be distributed very unevenly, for the chance of a recent recombinant event to be very low. But neither recombination nor mutation offers a simple explanation of the events underlying HapMap Figure 7.

The intention of this paper was to consider how far one could go by using only procedures in established use in the sixties, most well established in the first half of the last century. However this has not been possible on account of unexpected and as yet unresolved problems in the raw data downloaded, many loci having only a single allele or an unduly rare second homozygote in over 10% of individuals (see Appendix Table for an example). This work will be submitted elsewhere when resolved. The problems of ascertainment have been discussed by Clark *et al.* (2005). However, I hope something relevant can be learned from the much neglected past and the strategy underlying Fraser's 'Greek village work'.

Table 1. χ^2, θ and R^2 and map distance (cM) relating a pair of loci in one hundred individuals with allele frequencies a,b in first locus and c,d in second where b and d refer to the rarer allele. For locus 1, $a = b = 50$. The expected value of Pearson's r conforms to his value for a bivariate normal surface.

c	d	$b-d$	$\dfrac{b-d}{N}$	χ^2	R^2	r	Expected Pearson's r	cM	θ
50	50	0	0	0	0.00	0.00	0.0	0.0	0.00
54	46	4	0.04	0.32	0.00	0.04	0.1	4.1	0.04
58	42	8	0.08	1.29	0.01	0.08	0.2	8.4	0.08
62	38	12	0.12	2.92	0.01	0.11	0.2	12.9	0.11
66	34	16	0.16	5.25	0.03	0.15	0.3	17.6	0.15
70	30	20	0.20	8.33	0.04	0.18	0.4	22.7	0.18
74	26	24	0.24	12.22	0.06	0.22	0.5		
78	22	28	0.28	17.01	0.09	0.25	0.5		
82	18	32	0.32	22.82	0.11	0.28	0.6		
86	14	36	0.36	29.78	0.15	0.31	0.7		
90	10	40	0.40	38.10	0.19	0.34	0.8		

Some basic problems of genome trawls

The HapMap paper, both in its text and its references, showed little respect for the past, and little consideration of major evolutionary strategies established in all well studied animals and plants large enough to see. There is no mention of order variants, or substantial duplications or deficiencies, both obvious in humans on optical microscopy and likely to be common and influential in a rapidly evolving species in which most deaths after conception occur before birth. Nor is there any mention of the limitations in assuming these variants play no major role in evolution, with its necessary scars from major environmental changes and migration in the past and mechanisms for limiting life expectation: the latter particularly necessary to renew leadership in tribal species and highly relevant to common disorders. Over a hundred references, all since 1980, excepting two pages from Sewall Wright, hardly reflect the golden age of population genetics in the first half of the 20th century.

Even allelic association, on which the HapMap project is based, quotes papers of this century as 'early information'. But allelic association is a necessary feature on whose existence the concept of linkage analysis was based once a linear order with recombination had been demonstrated (Sturtevant 1913, Jennings 1917). The statistical consequences of a new allele with dominant expression were documented by Robbins as 'linkage inequality' (Robbins 1918; see this reference for Robbins's earlier work) and used by Bernstein to define the first human linkage, connecting the A and B loci to form the triallelic ABO locus almost a century ago (Bernstein 1924 1925). The later contributions of Fisher, Wright, Hogben, Waddington and Penrose on the nature of multifactorial disorders hardly deserve downgrading to 'recent experience bears out the hypothesis that common variants have an important role in disease' on the first page of HapMap (HapMap 2005).

Apart from the coalescent, reviewed by its author (Kingman 2000 but see also Kingman 1982) and the development of the Monte Carlo Markov Chain (MCMC) approach, usually attributed to Heath (1997) it is difficult to find any substantive concept formation, as opposed to those secondary to such technical advances as forensic typing, in situ mapping, comparative mapping and RNA typing since the first edition of Falconer in 1960: five years after Morton's tabulation of lod scores for linkage analysis (Morton 1955).

The major contributions of Morton to mapping by allelic association were also excluded, including recent papers highly relevant to HapMap (Maniatis *et al.* 2002, Zhang *et al.* 2002).

The statistical basis of inferring haplotypes in HapMap invokes the square of Pearson's correlation coefficient r, derived indirectly via inferred haplotypes, as the main estimator and determinant of the informative grids displayed in colourful graphics. A correlation is primarily a quantitative measure summarising a pair of variates or an attribute and a variate. Pearson sought to extend it to pairs of attributes by assuming the underlying mechanism could be represented by a bivariate normal surface with volumes in the quadrants cut by vertical and horizontal slices equal to the observed frequencies *a, b, c, d* in the underlying 2 × 2 table (see Figure 1).

Later Pearson and Elderton (in 1914; see Pearson and Elderton 1923 for references) extended it to cases with unequal marginal totals by formidable feats of numerical integration, using cogwheel calculators, to tabulate the tetrachoric coefficients. These, assuming normality an adequate approximation, provide an exact solution. Pearson never used r^2. Hill and Robertson later showed D^2 was simple to derive as χ^2/N (it is equal to r^2) and to have powerful properties in evolutionary genetics (Hill and Robertson 1968, Hill 1974, Weir *et al.* 2005). The relationships

Principal Components Analysis

The London, Edinburgh and Dublin Philosophical Magazine and Journal, Volume 6, Issue 2, 1901.

p. 566

X = UΛU' where U'U = UU' = I$_n$

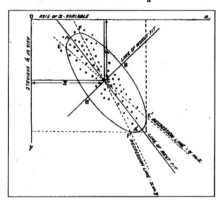

Figure 1. This is taken from Pearson's two contributions to this journal, the second showing a diagram of a horizontal section through volume bounded by a bivariate normal surface as advanced by Galton, but with the addition of orthogonal lines through major and minor axes that on rotation would have represented product moment correlations from 0 to 1. (See Pearson 1901)

among these metrics are shown in Table 1.

Almost all QTL approaches now follow the multifactorial model of Wright introduced in the 1930s (see Wright 1968 for references) and Penrose, also known as the quasi-continuous model of Grüneberg (1963), although Penrose preferred the term quasi-discontinuous. The matter was discussed by Hill and Robertson following a paper by Dempster and Lerner in 1950 and later applied by Falconer to human disorders coded by presence or absence of disease (Dempster and Lerner 1950, Falconer 1965). This model was used in the Greek paper, where raw quantitative data were available and did not need to be derived from qualitative data.

The problem of inferring *r*, Pearson's product moment correlation coefficient, was discussed by Pearson in 1901, shortly after he had established the convention of *a, b, c, d* for the numbers in a 2 × 2 table, and also introduced *r*, χ^2, δ, from which Robbins derived the capital form Δ, now usually written as *D*. His formula was used by Brownlee in a key paper on the efficacy of vaccination against smallpox (Brownlee 1905). It is given in detail and later illustrated by a diagram (Pearson 1901) which showed he was already working on Galton's model of a bivariate normal surface, adding orthogonal lines through the centre covering correlations from zero to one on rotation: this explains the redundant trigonometric multiplier below. See Appendix.

His formulae are:

$$k^2 = \frac{4abcdN^2}{(ad - bc)^2(a + d)(b + c)} \qquad \text{where } N = a + b + c + d$$

$$r = \frac{\sin(\pi/2)}{(1 + k)^2} \qquad (\sin(\pi/2) \text{ is equal to unity})$$

$$\frac{1}{(1 + k)^2}$$

The standard formula for χ^2 viz.

$$\chi^2 = \frac{(ad - bc)^2 N}{(a + b)(c + d)(a + c)(c + d)}$$

is simpler and covers unequal allele proportions, as well as giving different answers: the latter is too well established and studied in depth by Yule and Fisher to be in error.

Hill and Robertson (1968) showed χ^2 to offer a simple estimator of D^2, which is equal to r^2, with powerful properties in the analysis of evolution.

Robbins's coefficient Δ, usually written D, relates to the decay with time of the complete association of an allele with a phenotype manifest after mutation at a nearby locus. The problem presented by HapMap is the detection of similarity in neighbouring loci, and the removal from analysis, by excluding loci that are so similar that only one in a set of similars need be typed. An estimate of similarity is needed. In the 'vertical' analysis used to compare differences at the same locus in different sets of individuals, differences being proportional to the time required for a single recombinant to have occurred. In the 'horizontal' problem of different loci, the time required for a single recombinant is proportional to the product of their genetic distance, usually measured in centimorgans (cM), and the time required for a single recombinant at a unit distance, such as the mean interlocus distance. If any pair of loci is considered in a set of individuals, the first locus having alleles a and b, and the second c and d, where b and d are the less common allele, then the same analysis can be used but interpretation will differ: r will be a measure of distance, as will the simpler estimate, the absolute value of $(b - d)/N$, if both loci are diallelic and have an equal number typed. If h designates a heterozygote then, if the pair of loci has alleles ac or ah or hc or bd or bh or hd, they will probably be from a common ancestor as the prior probability will be high with closely spaced loci: the direct probability can be computed from the allele proportions. If they are ad or bc then a recombinant is likely to have occurred. A mutation has a similar expectation but is unlikely to back-mutate to the same allele. As a single recombination has the same effect as many, and with closely linked loci multiple breaks are unlikely to be in the same individual, the number of individuals differing due to recombinants between several loci will be an approximate measure of the recombination fraction from which the distance in cM may be calculated.

Until the late seventies the number of loci available was so limited, with few assigned to chromosomes, that the null hypothesis was based on sound expectations. Now that analysis is usually restricted to single chromosomes, within which even the most distant loci will show some allelic association. There are no null hypotheses and no justification for assuming allelic association absent when it is merely non-significant. With numerous loci serially ordered by physically defined positions, the expectation is that most adjacent loci will show either complete association, with $D = r = r^2 = 1$, as clearly illustrated in Figure 1 and in a recent paper using HapMap data to examine population structure, with its expected distribution clearly documented (Weir *et al.* 2005). But apparently not as used in HapMap, when the elegant graphics based on r^2 estimated from inferred haplotypes, as explained in two papers unfortunately neither published at the same time as HapMap nor freely available (see Minchin *et al.* 2006). An estimate based on inferred haplotypes selected for $r^2 > 0.85$ averaged over all loci in a segment of genotypes, determined by inferred overlapping haplotypes, would appear to involve some circularity. The problem is not simple. It is not only said to be highly correlated with D, as is D', but when $D' = 1$ 'despite great simplicity of haplotype structure r^2 values display a complex pattern, varying from 1 to 0.0003, with no relationship to physical distance. This makes sense ... ' (HapMap p 1305). This at least suggests problems in estimating r^2 from inferred haplotypes.

Prior information relevant to allelic association

The value of prior information in interpreting whole genome data, however sparse, was discussed by Morton in advancing his 'rule of three' (Morton 1955), the value of a lod score for a single test that was comparable to the standard (1/20) level of significance, now widely used and often mis-quoted and misapplied: later Renwick (1969) considered other aspects of this problem. Prior expectations based on Malécot's work were used in the papers of Maniatis *et al.* (2002) and Zhang *et al.* (2002) and related papers from the 'Southampton group' (Malécot 1966, 1969).

With only a single degree of freedom for pairs of loci, supplementary information from other loci is needed to provide evidence distinguishing mutation from recombination. A single differ-ence between two otherwise identical and substantial haplotypes strongly implies mutation: such haplotypes might be termed 'peppered haplotypes'. The mapping of high recombination rate seg-ments based on the coalescent has been successful with total genetic distances consistent with chiasma counts. The discrepancies with direct observations are not more than might be expected from the necessary if unrealistic assumptions about population size and breeding systems demanded by the coalescent. This should be regarded as a powerful 'prior' so that it can be used as an option. Direct evidence of recombination based on these data would be expected to differ from that of the coalescent – but it is not obvious that the high precision the coalescent offers compensates for the necessary bias.

While such prior information can advance various estimates in inferring haplotypes it cannot help in the deduction of haplotypes. These are simple to deduce from adjacent sequences of homozygotes: it is also possible to deduce pairs of haplotypes from stretches of homozygous genotypes bracketed by two or more adjacent heterozygotes separated by a single heterozygote. The proportion of heterozygotes, varying from a half in the most informative loci to zero in single-allele loci, which should be excluded as uninformative in specific data sets even if selected for overall heterogeneity in other data.

The HapMap set of data restricts analysis to loci conforming to the simple genetic back-ground imposed by rejecting genotypes inconsistent with Mendel's first law and consistent with what Stern termed the Hardy-Weinberg law (Hardy 1908, Weinberg 1908, Stern 1943). The rejects – the golden dross for the recognition of recessive lethals – are not discussed in detail although they account for over 10% of loci even though rejection was based on very high levels of significance (P < 0.001).

The algorithms recently advanced for the analysis of law-abiding loci include the assumption that most matings are between randomly selected members of very large populations, with a con-sequent absence of inbreeding. This allows the luxury of the coalescent in the definition of seg-ments within a chromosome separated by other segments prone to far higher levels of recombination. A more serious assumption is the use of algorithms that minimise the number of haplotypes consistent with the genotypes. With true haplotypes of substantial length the recent common ancestors will rarely be so recent that their number will be much less than the number of parents, or twice the number of individuals.

The estimation-maximisation cycle, termed the EM algorithm in a paper given in London, distinguished by its authorship from Boston, its generosity of acknowledgement, and the inclusion of an extensive discussion (Dempster *et al.* 1977). In this discussion, C.A.B. Smith advanced a more powerful algorithm than the EM for that class of circumstances in which the likelihood surface peaked between, and not at, the limits of its possible range: then any pair of values would define the underlying near-parabolic surface sufficiently closely to define the peak with adequate resolution. Smith had already pointed out the restriction of such algorithms to likelihood distributions with a single non-terminal likelihood peak in an earlier paper when he introduced this class of algorithms to genetics: he expressed it more forcibly in a second paper (Ceppellini *et al.* 1955, Smith 1957). Even so the number of steps necessary, at least 2^h where h is the number of heterozygotes, limits the procedure to small segments of the genome, with the exclusion of true haplotypes spanning any designated segments.

These 'inferred haplotypes', usually abbreviated to 'haplotypes', denote deduced segments of the paternal gamete, after about 30 recombinations, and the maternal after about 40, present in the offspring but enjoying mutual concealment through the impossibility of deducing the identity of phase between any pair of heterozygotes, excepting when one heterozygote has immediate homozygous neighbours.

The word haplotype was introduced by Ceppellini *et al.* (1967) to mean 'a particular combination of genes along a chromosome', and, until recently, was used consistently in this context especially in relation to the HLA phenotypes. Much confusion has been generated by the use of 'haplotype' both as true or classical haplotypes (Ceppellini's term) and as 'inferred haplotypes'. A recent definition, in a glossary accompanying a definitive article states 'Linkage disequilibrium (LD): The non-random association of alleles at tightly linked markers. Tight linkage can induce strong correlation between the genetic histories of neighbouring polymorphisms and, when LD is very high, alleles of linked markers can sometimes be used as surrogates for the state of nearby loci.' (Cardon and Abecesis 2003). But this definition is part tautology and consolidates the use for LD, originally defined as a measure of difference in single loci with time, to become a measure of similarity, or 'Locus Equality' (LE) in pairs of loci. Following Robbins's 'Linkage Inequality' and Δ, now LD and D, LD became established as a measure of difference between observed and expected allele proportions at one locus over time or space. It is not easily interpreted as a measure of identity between two loci. LE might be a better acronym. It is possible to change the standard allele counts of the relevant 2x2 table from *a, b* and *c, d* to *b, a – 2b* and *d, d – 2c* to give a correlation coefficient of unity for pairs of loci with identical allele proportions, using a measure of distance to define similarity and maintaining consistency with the use of r^2 or D^2, which should be equal. However there are also advantages in conforming to simplicity by entering allele counts directly and expressing the result as measure of closeness with a relevant meaning. The simplest, in this context, is the recombination fraction, which will be zero between loci with identical allele counts, and increase with distance between loci. If not zero it may be confounded with mutation, but at least a zero value will always define identity and absence of either recombination between the loci or a mutation at either locus. Mutation will usually be distinguished from recombination by its purely local effect in disrupting a monotonic increase of recombinations with distance and usually obvious on visual inspection of the alleles involved.

If loci are linked then alleles are associated. However, if alleles are associated loci are not necessarily linked. Allelic association is an observation that may be due to inbreeding, to different allele numbers at both loci, or to the loci being physically linked by a chromosome when they will always show some degree of allelic association in the absence of very unusual meiotic activity. There is no null hypothesis and absence of significant association cannot imply absence of association. The term allelic association defines an observation in need of further interpretation (Edwards 1980).

The main aim of the HapMap project is to infer haplotypes with very different qualities from haplotypes, as originally defined, that maintain 'a high average value of r^2', and thereby reduce the number of units of inheritance for analysis. Although, with the present procedure for typing, not the number of loci typed. In true haplotypes $D = r = r^2 = 1$ between all included loci. A haplotype cannot survive a single recombinant event, which will produce two shorter haplotypes, but reduce the average value of r^2 between any pair of loci from each new haplotype to that of the average between any randomly selected pair. These inferred haplotypes, if they had high similarity values for all included loci, would have the advantage of reducing the number necessary for analysis by a factor equal to the number of loci in an inferred haplotype, at a cost of reduced power to detect influential loci.

This benefit from minimising the number of inferred haplotypes by including incompletely correlated loci is achieved at a cost of the inferred haplotypes having unusual properties inconsistent with the true haplotypes they aim to represent. The haplotypes inferred by apparently similar, if not identical, techniques in analysing chromosome 22 (Dawson *et al.* 2002) inferred an unduly small number of very long haplotypes with the unexpected mirror-image or 'yin-yang' property leading to structures that could not evolve, survive or coexist within the restraints of what is known of meiosis and must be regarded as algorithmic artefacts. Such inferences are to be expected from algorithms that include minimising the number of necessary inferred haplotypes that are only required to satisfy the demands of most loci on most genotypes. The number of haplotypes of substantial length will rarely be less than the number of parents.

The assumed equivalent size of the breeding population

The elegant assistance of the coalescent requires a large randomly mating population. What, we may ask, can be assumed of the past? We are related to the larger apes and, in the period between splitting from contemporary hominids several million years ago until substantial urbanisation a few centuries ago, most individuals would have had little chance of meeting more than a dozen or so adults, excluding sibs, of appropriate age and sex. For any given ancestor the chromosomes of any descendant could be 'banded' for any segments of genotypic identity beyond a defined length. An imaginary subset of such segments was used by Fisher in his book on inbreeding (Fisher 1949). Any ancestor's parental haplotypes that could be inferred would be biased by the inclusion of 'shoulders' of allelic identity from other ancestors. In the presence of inbreeding these could be substantial. In the case of the smallest chromosomes, 21 and 22, the whole chromosome would be transmitted intact in over a third of meioses. Increasing distance from any common ancestor would increase the number, and reduce the length, of these common segments as well as increasing the proportion of fortuitously identical alleles. For any chromosome most of the bands of identity from a defined ancestor of would be on other chromosomes, a fact usually excluded from study by the technical and computational advantages of analysing chromosomes one at a time.

Diallelic loci and linkage disequilibrium

Given the advantages and limitations of strategies based on economy per locus, the analytical simplicity of diallelic loci, and the sheer magnitude of typing a million loci in parallel has required the development of mathematical approaches which combine extreme simplicity of assumptions with extensive exclusion of statistically unlikely observations. The complexity of the algorithms involved is inevitably beyond the understanding of most experienced observers, and most, and sometimes all, of the authors of most many-author papers. A situation not helped by the lack of an explicit list of assumptions that might inhibit some authors expert on phenotypes or genetics from accepting authorship rather than making an independent contribution to the more exacting problem of generating reliable data.

These approaches claim to explore the genetic background adequately, but the genomic terrain is too rugged to be mapped on simple assumptions with suppression of the unexpected. The assumptions are not consistent with a highly polymorphic and rapidly evolving species, the only mammal known in which most deaths after conception occur before birth. Deficiencies lethal in the homozygote cannot be detected by methods economically viable as most hemizygotes can only be distinguished on statistical grounds, leading to the exclusion of related loci from analysis and, at present, as an option or downloading as an inclusive set, not allowing the recipient to distinguish the causes of rejection. The claim made in the introduction that 'essentially all information about common DNA variation has been extracted' made in the abstract needs to be qualified by 'most of the information about common variants not related to point mutations has been excluded'.

Empirically direct analysis of the genotype, using algorithms based on Malécot (1969) and on Hill and Robertson (1968), has been performed by the Southampton group (e. g. Maniatis *et al.* 2002, Zhang *et al.* 2002). The elegance and prior support from Malécot and the ability to extract phase information from heterozygotes (Hill and Robertson 1968) are not necessary for a simpler, if less efficient, solution. The greater freedom from unacceptable assumptions and a more direct approach should offer advantages over the intermediary of inferred haplotypes. Direct genotypic analysis using the MCMC approach (Heath 1997, Daw *et al.* 2003) also offer solutions without the intermediary of inferred haplotypes.

The ABO and cholesterol problem (The Greek paper)

In 1969 Fraser and his colleagues were responsible for one of the earliest linkages between markers on red blood cells and a quantitative phenotype (Fraser 1969c). They found an association between alleles at the locus of the ABO blood group and cholesterol level by comparing the qualities of five blood group loci with quantitative estimates in serum from the same sample.

Cholesterol was one of the few substances in plasma commonly estimated in hospital laboratories in the sixties: high levels were known to be directly related to the furring of arteries predisposing to hypertension, heart attacks and strokes. Their analysis based on single samples of blood revealed strong evidence of linkage between the ABO locus and a locus influential in cholesterol level.

In the HapMap project, plasma, which occupies half the volume of blood, and is termed serum after separation, was neither analysed nor preserved: in consequence no quantitative phenotypic data were available. Had they been, and had this strong suggestion of a linkage been confirmed, the project would have had adequate data both to locate this and other influential loci and sequence the susceptibility and resistance alleles responsible for a range of common disorders dominating causes of both restricted life and early death.

Their Greek paper is available as a PDF file on the present author's website. Its conclusions are presented after an exacting, efficient and fairly simple analysis, notwithstanding a surprising variation in both allele proportions and cholesterol level in neighbouring villages (Fraser 1969c).

The absence, in these data, of any correlation between the cholesterol levels of spouses is strong evidence that the obvious nutritional effects are, firstly, not primarily qualitative and, secondly, strongly dependent on the genotype. As the clinical evidence shows an obvious relationship with obesity the difference must be on quantity of food. An interpretation consistent with recent data from Boston on 49,000 women studied for eight years showing no effect on intake of fats (Freeman 2006).

The data are summarised in Table 2: Figure 2 shows a graphical display. The data for the rarer ABO genotypes do not suffice to clarify their influence, and recent work defining the 'O' as a true null, as suggested by Bernstein's terminology, suggests it is likely to be the only outlier. In any case there is evidence of a substantial difference between some of the various susceptibility and resistance alleles at this locus, the exact nature of which is irrelevant to the need for clarification by comparing similar data at the higher resolution possible by allelic association.

This strongly suggestive evidence of association between five segments of the genome, with only one including an influential locus, has the support of none of the other four blood groups showing evidence of association. This supports the absence of secondary correlations not resolved by the somewhat complicated method of standardisation used. If such differences can prevail in neighbouring villages not obviously distinguished by ancestry or lifestyle the merging of unstratified data between even apparently similar, but more distant, populations needs to be handled with extreme caution. The insular and continental Saxons, contributing to the UK and Utah populations used in HapMap, and analysed together, were well separated by time and distance: the UK groups either walked to England before it was separated from Europe over 5000 years ago or came by sea between 400 and 1000 AD: those reaching Utah in the 18th and 19th centuries were mainly Saxons from Germany. The fairly equal Norse and Irish genetic contributions to Iceland in the ninth and tenth centuries impose both difficulties and advantages to analysis although as yet not exploited (Bjarnason *et al.* 1973). The Welsh and English differ substantially in ABO distribution. There is no reason to assume other loci are less variable, or other large populations would not benefit from weighted estimators based on name, language, ancestral language or apparent ancestral origin, all individually unreliable but each an advance on no attempt to reduce the effects of obvious admixture.

It is convenient to discuss the subset of the data from the Greek paper shown in Table 2 (available as a PDF file from the present author).

All four other blood-group loci typed lacked significant associations, providing strong supporting evidence of the reliability of the data, and of the procedures used to correct for the remarkable differences in both allele proportions and cholesterol level in the two neighbouring villages. The simpler but possibly less efficient method of weighted estimators for each population of Woolf (1955) could have been used by those of us who find incorporating linear algebras within QTL approaches difficult. The key data are shown graphically in Figure 2.

If as is consistent with both expectation and the data, the standard errors are equal, we can

Table 2. Greek data on cholesterol levels (modified from Mayo *et al.* 1969, Fraser 1969c).

Relationship	Number of sibships	Number of sibs	Correlation
Sibs	167	424	0.351
brothers	129	230	0.579
sisters	113	194	0.230
		Number	Correlation
father-son		163	0.261
father-daughter		151	0.322
mother-son		190	0.322
mother-daughter		150	0.179
Spouses		120	−0.030

Mean corrected cholesterol scores

ABO	O	A	AB	B
	-6.15	2.71	5.68	1.75
SD	1.82	3.04	3.66	8.83
N	259	182	17	74
*	0.00	8.86	11.83	7.90

*line added to standardise against 'O'

infer the underlying difference in terms of the means in units of standard deviation of the presumed underlying normal curves.

Simple weighting, assuming the standard deviations are equal to their weighted average, gives a value for difference in means in units of the standard deviation of the cholesterol score of

$$2.40 = \frac{(1.82^2 \times 259 + 3.04^2 \times 182)}{\sqrt{(259 + 182)}}$$

and a value of χ^2 of 13.6. This is a very high estimate. In the simplest and most easily visualised comparison, height and sex in adults, men are 1.7 standard deviations taller, or at least were in the recent past in the UK. The first ABO disease associations were between the ABO blood groups and cancer of the stomach and ulcers of both stomach and duodenum (see, e.g. Mayo 1978 for a review of the early studies and Edwards 1965 for details of the ABO-duodenal ulcer data and their interpretation).

Clearly this ABO-cholesterol effect is substantial and made even more remarkable by the absence of correlation between spouses, suggesting quantity, rather than variety, of food may dominate the association making preventive measures simpler, if less likely to be effective.

Figure 3 shows the distribution of a pair of curves of equal variance differing by only one unit of standard deviation, and, since the overlap is too small to display the nature of the data, a simpler case with unit difference.

If the 'threshold' beyond which a qualitative phenotype, such as 'high cholesterol level', is

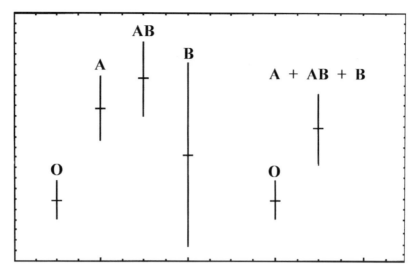

Figure 2. Showing the relative level of association and its standard error. The key O:A ratio is based on the largest pair of genotypes: it shows very substantial difference in overlap and is fully consistent with the 'O' and 'not O' difference.

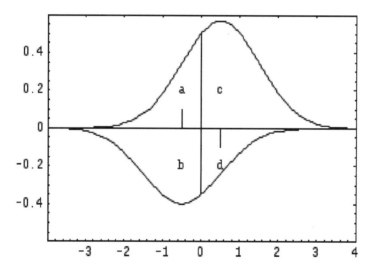

Figure 3. The relationship of the distribution of variates against two attributes.

considered 'abnormal', a somewhat vague concept, the set of data can be split into two groups. In practice 'abnormal' means a level at which most experienced clinicians would wish to modify their behaviour, including taking pills. An exasperatingly vague concept to most biometricians, unless concerned by their own level. Fortunately, due to the similarity of the logistic curve and the normal curve, this threshold, which leads to a standard ratio of the cross-product ratio (ad/bc) for the logistic, is not greatly influenced if the distribution is normal. There are advantages in working with the median as extreme values, sometimes assumed to offer more efficient methods of analysis (e.g. Lander and Kruglyak 1995, Risch and Zhang 1995), are known to be due, in part, to Mendelian variants with different loci causing some of the extreme levels. A feature present in the levels of many enzymes and other proteins relevant to most common disorders, and obvious in simpler phenotypes in which outliers are often related to increased proportions of Mendelian variants at different loci. In height there are numerous Mendelian varieties of dwarfism and a few of gigantism, none related to the same locus. Intellect offers numerous Mendelian forms at the lower end but, so far as is known, none at the upper end: nor are any expected as the obvious concentration of high intellect in some families rarely extends beyond a few generations. In hair and skin, Mendelian variants include albinism and various syndromes associated with pallor or white patches but there are no Mendelian forms of blackness known in humans. In sheep, albinism is very rare but blackness is a common recessive. Alleles conferring blackness are common and obvious variants in mice, rabbits, cats, pumas, dogs, cattle, and gibbons.

The false-positive problem

In the Greek paper there is no problem of false positives: the formally obvious statement that since it was one of five tests then the required level of significance should be increased, or the level reduced by the equally obvious and simpler alternative of multiplying the threshold of statistical significance by five, or more exactly by subtracting from unity the probability that five successive tests would fail to show a significant result.

While this is a relatively minor problem in small studies or those family studies giving very high levels of significance, or association studies with defined quantitative phenotypes, it is crucial when the number of comparisons is large and the determinants likely to be only weakly associated. Bonferroni is usually quoted as responsible for the corrections considered above, but they would seem obvious to the earliest gamblers: what seems clear from informed references, which seem few, is that what Bonferroni pointed out was not what had long been obvious, but firstly that what seemed obvious cloaked a subtle inequality: a triumph of formal statistics, but one too slight to be of much practical significance. A more important but less obvious feature to which he drew attention is that when associations under test are correlated the computation of significance is far from simple. In linkage any analyses involving appreciably correlated measures of closeness with neighbouring loci within the same 'likelihood hump' will invalidate any procedures that ignore it. There are problems in access to his two much quoted papers. A search of all common sources by an experienced librarian located only one copy in the UK, held for the British Library in a distant warehouse for books rarely read. At over fifty pages available at two pounds a page access is difficult. There are of course other simple approaches to the false positive problem, such as Fisher's probability correction (e.g. Fisher and Yates 1963).

Several difficult papers on the problem have recently appeared on the medical weeklies in the UK, but none deals in the simple currency of the expected number of false positives, or simplifies the problem by dealing with the controls and affected populations separately. In the HapMap data these sets are combined, to the extent that the controls are those in the non-affected category, and no common disorder manifest as such will exceed 10%. Some of the more intransigent, such as schizophrenia are a mere 1%.

Various articles on the power of association mapping in which affected individuals are formally ascertained and paired, or at least accompanied by comparable numbers of controls, both in the thousands, have limited relevance to HapMap.

Association mapping, as currently planned, with examples of efficient execution and the delivery of results, is very different from that with which HapMap must contend (Wang *et al.* 2005).

If the prevalence is p the affected: control ratio will be $p:1-p$. The ascertainment strategy in association studies usually ensures similar numbers of affected and controls, with the option of reducing irrelevant causes of an increased variance, by stratifying by sex and ancestry, and within these obvious categories reduce the environmental effects by pairing affected and control individuals by such methods as common sense and local knowledge can offer. An additional and major practical advantage of association studies is that the ascertainment procedures will usually ensure both higher and more homogeneous standards of diagnosis.

Some simple matters seem clear. The number of loci sufficiently influential to show up is limited, a feature obvious from the failures of the extensive data now collected from affected parented sib-pair analyses which have yielded little information relevant to the disorders studied, but a larger number of associations common to unrelated disorders and presumably related to events before ascertainment, usually before birth (Edwards 2003). The inferences are consistent with what is known of relevant metabolic pathways, the complexities of metabolic maps and the even greater complexities of the various transcription factors acting on overlapping groups of loci. It seems unreasonable to expect that more than four or five loci would be sufficiently influential to be detected by exclusively DNA related strategies with data of realistic size would reach even the lowest level of significance without the helping hand of chance.

We can have limited expectation of more than a few influential loci, some, related to sequential enzymatic actions with limited alternative routes, may be unearthed from DNA data alone by the cyclic Bayesian procedures discussed earlier. These offer hope from methods of great complexity with which simple procedures cannot compete, and statistical significance can be interpreted, or quarantined, by those expert in the metabolic pathways unearthed, and offer other loci relevant to neighbours on the metabolic map. But the problems of contamination by background noise', often both familial and environmental, and the need for homogeneous populations impose difficulties that, in principle could be resolved by these methods, already achieving success yet to be consolidated by homology with metabolic maps or known associations. At least these might be surmounted at a cost of even more elaborate algorithms, and the results could be validated or quarantined by relevant experts on cytoplasmic mechanisms or the phenotype, after 'mapping' a linear map of influential loci that reflected closeness of related enzymes to the multidimensional metabolic map. But with the association studies now under development HapMap is unlikely to offer competitive data to find influential loci, as opposed to the essential provision of detailed mapping of segments whose relevance is suggested by other studies.

In practice it is difficult to envisage that any realistic method of 'casting out' similar loci by either direct methods or those through the intermediary of inferred haplotypes could reduce the number of elements to less than a thousand over the whole genome before Bonferroni's warning of the bias of correlated elements would be applicable. Worse still, the expected level of association would be such that the loci exhibiting the highest levels of significance would strongly imply the hand of chance – highly significant levels would have to be rejected as biologically implausible and lower levels as statistically inadequate. Even if it is a genuine effect, but aided by the hand of chance, as is likely, the intensity of association will be exaggerated by amounts that defy estimation, making it unwise to assume the apparent degree of association, a figure on which the high cost of further action must depend. As with drug trials, where the effect is relatively minor, the results must be either no effect or the drugs is less effective than claimed.

We have the problem that in DNA restricted genome-studies false positives would greatly outnumber true positives in the absence of prior information based on previous work, or supplemented by studies on the intermediary mechanisms connecting the genotype to the genotype, for which serum provides the ideal resource: it is difficult to see how the present strategy could provide information directly relevant to human disease. At least the data now available, and increasing in volume, quality and, one hopes, variety – as yet no 'warts and all' downloads with all varieties of 'QCs', no grouping of trios into families, and in the names identification of sex, can justify the present availability of data, but the question remains as to when 'enough is enough'.

At present any apparent success with a disorder will leave a gross excess of false positives compared to true positives, with no means of defining which is which, except by conducting similar studies of similar magnitude in other populations or accepting the need to integrate the DNA analyses with simultaneous studies of the proteins that mediate between the DNA and the phenotype. Is history repeating itself with the approach of Karl Pearson? His formidable attempts to define the constituent peaks of a summation of normal distributions, based on the action of the very numerous loci assumed, some sufficiently influential to cause detectable peaks, eventually succumbed to an algebraic exuberance devoid of substantive discoveries, but not of formal impossibilities. A recent major precedent, equally divisive on grounds of prior assumptions, and not including open data, was the diallel-cross project in the UK: now little known, in part due to the recent synonymous use of the hybrid term biallelic for diallelic.

Appendix

HapMap raw data (HapMap 2005)

Downloaded from ENr131.2q37.1_CEU.txt.gz February 2006

rs# SNPalleles chrom pos strand genome_build center protLSID assayLSID panelLSID

QC_code NA06985 NA06991 NA06993 NA06994 NA07000 NA07019 NA07022 NA07029 NA07034 NA07048 NA07055 NA07056 NA07345 NA07348 NA07357 NA10830 NA10831 NA10835 NA10838 NA10839 NA10846 NA10847 NA10851 NA10854 NA10855 NA10856 NA10857 NA10859 NA10860 NA10861 NA10863 NA11829 NA11830 NA11831 NA11832 NA11839 NA11840 NA11881 NA11882 NA11992 NA11993 NA11994 NA11995 NA12003 NA12004 NA12005 NA12006 NA12043 NA12044 NA12056 NA12057 NA12144 NA12145 NA12146 NA12154 NA12155 NA12156 NA12234 NA12236 NA12239 NA12248 NA12249 NA12264 NA12707 NA12716 NA12717 NA12740 NA12750 NA12751 NA12752 NA12753 NA12760 NA12761 NA12762 NA12763 NA12801 NA12802 NA12812 NA12813 NA12814 NA12815 NA12864 NA12865 NA12872 NA12873 NA12874 NA12875 NA12878 NA12891 NA12892

rs7425424 A/G Chr2 51633336 + ncbi_b34 broad urn:lsid:wicgr.hapmap.org:Protocol:genotype_protocol_1:1 urn:lsid:wicgr.hapmap.org:Assay:HapMap_EncodeCleanup-rs7425424_116.4:1 urn:lsid:dcc.hapmap.org:Panel:CEPH-30-trios:1 QC+

GG GG GG AG NN GG GG AG AG AG GG GG GG GG GG GG GG GG GG GG GG GG GG GG GG GG GG GN NN AG GG AG GG GG GG GG GG GG AG GG GG AG GG GG GG GG GG GG GG GG GG AG AG NN GG AG GG AG GG GG AG GG GG AG GG AG GG AG AG AG GG GG GG GG GG GG GG GG AG AA GG AG AA AG AG NN GG GG AG GG GG GG

rs1852777 C/G Chr2 51633397 − ncbi_b34 mcgill-gqic urn:LSID:illumina.hapmap.org:Protocol:Golden_Gate_1.1.0:1 urn:LSID:mcgill-gqic.hapmap.org:Assay:1019265:1 urn:lsid:dcc.hapmap.org:Panel:CEPH-30-trios:1 QC+ CC CC CC CC CC CC CC CC CC CC CC CC CC CC CC CC CC CC CC CC CC CC CC CC CC CC CC CC CC CC CC CC CC CC CC CC CC CC CC CC CC CC CC CC CC CC CC CC CC CC CC CC CC CC CC CC CC CC CC CC CC CC CC CC CC CC CC CC CC CC CC CC CC CC CC CC CC CC CC CC CC CC CC CC CC CC CC CC CC CC

rs7349348 A/T Chr2 51635678 + ncbi_b34 mcgill-gqic urn:LSID:illumina.hapmap.org:Protocol:Golden_Gate_1.0.0:1 urn:LSID:mcgill-gqic.hapmap.org:Assay:555464:1 urn:lsid:dcc.hapmap.org:Panel:CEPH-30-trios:1 QC+ AA AA AA AA AA AA AA AA AA AA AA AA AA AA AA AT TT AT AA AT AT AA TT AT AT AA AA AT AT TT AA AT AA AT AT AT AT AT AT TT AA AT TT AT AT TT AT AT AA AT AT AA AT AT AA AT AT AT AT AA AA AT AA TT AT AT AT AT AT TT AT AT TT AT AT TT AA AT AA AA AT AA AA AA AT AA AA AT AT AT AT

rs7349275 C/T Chr2 51635706 + ncbi_b34 mcgill-gqic urn:LSID:illumina.hapmap.org:Protocol:Golden_Gate_1.0.0:1 urn:LSID:mcgill-gqic.hapmap.org:Assay:307169:1 urn:lsid:dcc.hapmap.org:Panel:CEPH-30-trios:1 QC+ TT TT TT NN TT CT TT NN TT TT TT CT TT TT TT CT CC CT TT CT CT TT CC CT CT TT TT CT CT CC NN CT TT CT CT CT CT CT CT CC TT CT CC CT CT CC CT CT TT CT CT TT CT CT TT CT CT CT CT TT TT CT TT CC CT CT CT CT CC CT CC CT CT CC NN CT NN TT CT TT NN TT NN TT TT CT CT CT CT

rs13013045 A/G Chr2 51635787 + ncbi_b34 mcgill-gqic urn:LSID:illumina.hapmap.org:Protocol:Golden_Gate_1.1.0:1 urn:LSID:mcgill-gqic.hapmap.org:Assay:1018878:1 urn:lsid:dcc.hapmap.org:Panel:CEPH-30-trios:1 QC+ AA AA AA AA AA AA AA AA AA AA AA AA AA AA AA AA AA AA AA AG AA AA AA AA AA AA AA AA AA AA AA AA AA AA AA AA AA AA AA AA AA AA AA AA AA AG AA AA AA AA AA AA AA AA AA AA AA AA AA AA AA AA AA AA AA AA AA AA AA AA AA AA AA AA AA AA AA AA AA AA AA AA AA AA AA AA AA AA AA AA

AA AA AA AA AA AA

rs13396266 C/G Chr2 51636251 + ncbi_b34 mcgill-gqic

urn:LSID:illumina.hapmap.org:Protocol:Golden_Gate_1.1.0:1 urn:LSID:mcgill-gqic.hapmap.org:Assay:1018706:1

urn:lsid:dcc.hapmap.org:Panel:CEPH-30-trios:1 QC+

CC CC CC CC CC CC CC CC CG CG CC CC CC CC CC CC CC CC CC CC CC CC CC CC CC CC CC CC CC CC CG CC CC

CC CC CC CC CC CC CC CC CC CC CC CC CC CC CC CC CC CC CC CC CC CC CC CC CC CC CC CC CC CG CC CC CC

CC CC CC CC CC CC CC CC CC CC CC CC CC CC CC CG CC CC CC CC CC CC CC CC CC

rs1206413 A/G Chr2 51637950 + ncbi_b34 mcgill-gqic

urn:LSID:illumina.hapmap.org:Protocol:Golden_Gate_1.0.0:1 urn:LSID:mcgill-gqic.hapmap.org:Assay:307171:1

urn:lsid:dcc.hapmap.org:Panel:CEPH-30-trios:1 QC+

AG AA AA AG AG AA AA AG AG AA AA AA AA AA AA AA AA AA GG AG AA AG AA AA AG GG AA AA AA AA AA AG

GG AG AG AA AA AA AG AA AA AG AA AG AG AA AG AA AA AG AA AA AA AA AA AG AA AA AA AG AG AA AG AA

AA AA AA AA AA AA AG AA AA AG AA AA AA AA AA AA AA AG AA AA GG AA AA AA AA AA

rs10184263 C/T Chr2 51638297 + ncbi_b34 mcgill-gqic

urn:LSID:illumina.hapmap.org:Protocol:Golden_Gate_1.1.0:1 urn:LSID:mcgill-gqic.hapmap.org:Assay:1019567:1

urn:lsid:dcc.hapmap.org:Panel:CEPH-30-trios:1 QC+

CC CC CC CT CT CC CC CT CT CT CC CC CC CC CC CC CC CC CC CC CC CC CC CC CC CC CC CT CT CC TT CC CC

CC CC CC CT CT CC CC CT CC CC CC CC CC CC CC CC CC CT CT CT CC CT CC CT CT CC CT CC CC CT CC CT CC CT

CT CT CC CC CC CC CC CC NN CT NN CC CT TT CT CT CT CC CC CT CC CC CC

rs17868116 C/T Chr2 51638297 + ncbi_b34 mcgill-gqic

urn:LSID:illumina.hapmap.org:Protocol:Golden_Gate_1.0.0:1 urn:LSID:mcgill-gqic.hapmap.org:Assay:555467:1

CC CC CC CT CT CC CC CT CT CT CC CC CC CC CC CC CC CC CC CC CC CC CC CC CC CC CC CT CT CC TT CC CC

CC CC CC CT CT CC CC CT CC CC CC CC CC CC CC CC CC CT CT CT CC CT CC CT CT CC CT CC CC CT CC CT CC CT

CT CT CC CC CC CC CC CC CC CT TT CC CT TT CT CT CT CC CC CT CC CC CC

rs17874912 C/T Chr2 51638454 + ncbi_b34 mcgill-gqic

urn:LSID:illumina.hapmap.org:Protocol:Golden_Gate_1.0.0:1 urn:LSID:mcgill-gqic.hapmap.org:Assay:555468:1

urn:lsid:dcc.hapmap.org:Panel:CEPH-30-trios:1 QC+

CC CC CC CC CC CC CC CC CC CC CC CC CC CC CC CC CC CC CC CC CC CC CT CC CC CC CC CC CC CC CC CC CC

CC CC CC CC CC CC CC CC CC CC CC CC CC CT CC CC CC CC CT CC CC CC CC CC CC CC CC CC CC CC CC CC CC

CC CC CC CC CC CC CC CC CC CC CC CC CC CC CC CC CC CC CC CC CC CC CC CC CC

rs1206415 G/T Chr2 51638838 + ncbi_b34 broad urn:lsid:wicgr.hapmap.org:Protocol:genotype_protocol_1:1

urn:lsid:wicgr.hapmap.org:Assay:HapMap_EncodeCleanup-rs1206415_121.2:1

urn:lsid:dcc.hapmap.org:Panel:CEPH-30-trios:1 QC+ GT TT TT GG GG GT TT GG GG GT TT NN TT TT TT GT GG GT

GG GG GT GT GG GT GG GG TT GG GG GG GG GG GG GG GG GT GG GG GG GG GT GG GG GG GG GG GG GT TT

GG GG GT GG GT GT GG GG GG GT GG GT GT GG GG GG GT GG GG GG GG GG GT GG GG TT GG GG TT GG

GG GG GT GG GG TT GG GT GT GT

rs1919416 A/G Chr2 51638865 − ncbi_b34 mcgill-gqic

urn:LSID:illumina.hapmap.org:Protocol:Golden_Gate_1.0.0:1 urn:LSID:mcgill-gqic.hapmap.org:Assay:555469:1

urn:lsid:dcc.hapmap.org:Panel:CEPH-30-trios:1 QC+

GG GG GG AG AG GG GG AG AG AG GG GG GG GG GG AG AA AG GG AG AG GG AA AG AG GG GG AA AA AA AA

AG GG AG AG AG AA AA AG AA AG AG AA AG AG AA AG GG GG AG AA AG AA AG AG AA AA AG AG GG AG AG

AA AA AG AA AA AA AA AG AA AG AG AA GG AA NN GG AA AA AG AG AA GG GG AA AG AG AG

Fraser and the genetic load

Warren J. Ewens

Department of Biology, University of Pennsylvania, Philadelphia PA, 19104, USA

Summary

The concept of the genetic load has led to perhaps more controversy and argument than any other in population genetics theory, rivalled only by the still continuing debate between the followers of Fisher and those of Wright on various aspects of evolutionary theory. In this note a summary of various load definitions and calculations will be given, with special reference to the place of the work of George Fraser on the load concept. The overall conclusion of this note is that whereas Fraser made many enlightened comments about the mutational load, other authors made misleading calculations in using this and other load concepts in other areas, especially those connected with the neutral theory of evolution.

The mutational load

In his various papers on the load concept, Fraser concentrated on what is now called the mutational load. As a physician, he was mainly interested in that load deriving either from naturally occurring deleterious mutations or, as in the case of atomic radiation, from deleterious mutations deriving from the activities of mankind. In particular he assessed, in some detail, the long-term consequences of the treatment of persons having genetically caused diseases, a topic of increasing relevance today. In this note I summarise some of his views, and then contrast these with the misinterpretations that I believe have been made by others as a result of mathematical analyses of the load concept.

To my knowledge the first mathematical calculation that was made about what later became called the mutational load was that of Haldane (1937), who did not use the word 'load' in that paper. Haldane's calculations concerned a situation where a favourable allele has a stable equilibrium frequency slightly less than unity, the deficit from unity being due to mutations from this allele to the less favoured allele. Haldane compared the mean fitness of the population that would arise if these mutations were not to occur to the mean fitness when they do, and reached an initially surprising conclusion, namely that this difference (later called the mutational load) is essentially independent of the selective difference between the favoured homozygote and the heterozygote genotypes.

This independence is easily explained in common-sense terms. The larger this selective difference the smaller is the frequency of the less favoured allele. These two opposite trends more or less exactly compensate for each other, leading to the conclusion that Haldane observed. Haldane used his calculations to estimate the selective disadvantage of various disease-related mutations in humans, although here he had to make the debatable assumption that the deleterious genotypes had reached their stationary frequencies in the human population.

The purely mathematical analysis is straightforward, and is outlined here for the simple case of a non-recessive disease. Assuming that homozygotes for the disease allele essentially never arise, all that is needed for the appropriate calculations is the mutation rate u from the normal allele to the disease allele and the fitness of the heterozygote genotype *Aa* (with *A* being normal allele and *a* the disease allele) relative to that of the normal homozygote *AA*. We take the latter fitness to be 1 (an assumption discussed in some detail below) and that of *Aa* to be $1 - s$, where *s* is positive. Then to a sufficiently close approximation the equilibrium frequency *x* of *Aa* is

$$x = u/s, \tag{1}$$

and the frequency of the normal homozygote is $1 - u/s$. It is easy to calculate, from this, that the population mean fitness is essentially $1 - 2u$, independent (as mentioned above) of *s*.

The next major contribution to the topic was made by Muller (1950), who perhaps is responsible for the introduction of the word 'load'. Muller viewed the deficit calculated by Haldane as a burden 'felt in terms of death, sterility, illness, pain and frustration'. To quote Fraser and Mayo (1974d), it is important to investigate the 'nature, extent, maintenance and importance' of such a burden, or load, and, one might add, to find the most appropriate mathematical definition of it. A similar attempt is needed to find the most appropriate mathematical definition of the substitutional load and the segregational loads introduced below.

Fraser and Mayo analysed the three definitions of load offered by Crow (1970). The first of these was the fraction by which the population mean fitness differs from that of some reference genotype. This definition obviously raises the question of which existing genotype is to be taken as the reference, or whether some idealised but non-existent genotype is to be taken as the reference. Crow's second definition also concerned the population mean fitness. The load relating to any factor affecting fitness is defined, under this definition, as the extent to which the population mean fitness is decreased in the presence of this factor compared to the mean fitness in its absence. Crow's third definition appears close to that used in subsequent analyses of the substitutional and the segregational load, namely the rate of reproductive excess to maintain a population at a stable size. Relations between these definitions requires some connection between population mean fitness and population size, and in the literature this is generally taken as the requirement that a population mean fitness of unity implies constant population size. We adopt this relationship in the following sections.

In a paper preceding Crow's, Fraser (1962a) had already used a different definition of the mutational load, one which however had also been proposed by Crow (1958). This definition focuses on a single gene locus, and stated that the load at that locus is the proportion by which the mean fitness for the genotypes at that locus falls short of that of the optimal genotype. For the case leading to the calculation in [1], this load is thus, for all practical purposes, $2u$. We note that under this definition the reference genotype, referred to in the preceding paragraph, becomes the optimal genotype. Fraser attacked this definition on two grounds. The first is that the optimal genotype might be hard to characterise. The second is that since under this definition genetical variation always produce a load, the 'loaded' property of this word has the unsatisfactory consequence of implying that genetic variation is harmful, contrary to the standard view that genetic variation, in particular additive genetic variation, is seen as essential for evolution. I myself do not go the whole way with Fraser's criticism, since to me the essential factor is the purpose for which the load definition, whatever it might be, is used, and whether it is used appropriately for this purpose.

Fraser's (1962a) discussion of genetic load started with, and to some extent focused on, a consideration of the mutational load and its implication for various genetic diseases. He used estimated mutation rates to the deleterious allele for each disease and estimated frequencies of the heterozygotes, in conjunction with equations such as [1], to estimate the fitness deficit s of the typical heterozygote. He also pointed out that [1] is an equilibrium formula and that in man equilibrium might not obtain for some of the diseases that he investigated, so that these estimates might be inaccurate.

He then went on to discuss the so-called segregational load (described in more detail below). His discussion was carried out again largely in the context of genetic diseases, in particular those diseases where heterozygous carriers of this disease gene seem to be at a selective advantage over the normal homozygotes (the sickle cell anaemia case springs to mind). Here of course the disease allele occurs at a much higher frequency that that given in [1]. In doing so he considered one disease (and one disease locus) at a time, although he did refer to Morton's calculation that the typical human carries some three to five genes in heterozygotes each of which, had they occurred in homozygous form, would have been sufficient to have caused premature death of their carrier.

Fraser's analysis of the segregational load was quite different from that carried out a few years later in the context of the neutral theory. This latter analysis focuses on the load arising when a large number of gene loci are segregating. The motivation for this analysis came largely from the discoveries of Harris (1966) and Lewontin and Hubby (1966), in humans and Drosophila respectively, of large numbers of loci with two (or more) alleles segregating, both at appreciably frequencies. It is interesting to speculate how Fraser would have modified his 1962a paper had these data been available to him.

Kimura (1968) used load arguments, together with the observations of Harris (1966) and Lewontin and Hubby (1966), to argue that much of the variation seen by these authors must be selectively neutral. The discussion was of a different type from Fraser's. There was little focus on disease loci, and the calculations, either explicitly or implicitly, referred to genetic variation at many loci simultaneously. The loads calculated by Kimura were often huge, and this was seen by him and others as a prime argument in favour of the neutral theory for which, by definition, the load was calculated as zero. In later sections these load calculations are briefly reviewed (and criticised). I have given a more extensive discussion of these calculations in Ewens (1991, 2004).

It is noteworthy that Fraser did not refer to the two papers by Haldane (1957, 1961) introducing the so-called 'cost of natural selection'. This 'cost' later became called the 'substitutional load', an unfortunate change in terminology since the difference in viewpoint between a cost and a load is an important one, as discussed later. This brings us to Haldane's cost calculations.

The 'substitutional load'

The initial term given by Haldane to the substitutional load was the 'cost of natural selection', and in this section I use the word 'cost' for this quantity. This word is far more indicative of what is involved than is the word 'load'.

Haldane's point of departure was the observation that in the classic industrial melanism example, possibly half the conspicuous phenotypes of *Biston betularia* might be eaten in one day. He then observed that if ten independent characters were subjected to the same level of selection, only about one in, or roughly one in a thousand, of the original genotype would survive, so that the species would probably have become extinct. On the other hand, he claimed that the species might well survive ten such individual selective episodes if they occurred at widely spaced intervals. Thus 'we see, then, that natural selection must not be too intense'. His 1957 paper was aimed at finding the intensity of natural selection that can be tolerated, given the reproductive capacity of the species.

It is easiest to carry out the calculations in terms of haploids, and then make a more or less subjective adjustment for the value to be found in the more interesting diploid case. Suppose then that there are two haploid genotypes, A and a, and that '1 − s of a survive for every one of A'. (We have changed Haldane's notation k to s to conform with current usage and with that adopted above.) If in generation n the frequency of A is p_n and that of a is $q_n = 1 − p_n$, then according to Haldane the fraction of selective deaths is sq_n. It is easy to set up a recurrence relation giving the value q_{n+1} of the frequency of a in generation $n+1$, and this can then be used to find D, the 'total of the fractions of selective deaths', as

$$D = \sum_{n=0}^{\infty} sq_n \qquad [2]$$

A simple approximation to this sum by using the recurrence relation for the q_n and a continuous-time calculation yields the value −log p_0, where p_0 is the initial frequency of A. This is independent of s, and the reason for this is essentially the same as the reason why the mutational load is independent of selective differences – a larger value of s leads to a larger amount of 'selective death' per generation, but over a smaller number of generations. Taking the approximate value $p_0 = 10^{-4}$, this gives a total of about 9.2 for the total number of selective deaths over all the generations in the gene substitution procedure. In diploids the value appears to be somewhat higher, and Haldane took 30 as a representative value.

Haldane next turned to the calculation appropriate when gene substitutions are occurring at a number of loci simultaneously, with a new substitution starting every n generations. His calculations indicated that the mean fitness, compared to the case where all individuals had a fitness of 1, would be about $e^{-30/n}$. Assuming that the population could not tolerate a decrease of more than about 10%, this would imply that the population could not tolerate selective substitutions at more than one starting every 300 generations. This value and this calculation then became enshrined in the load literature. Haldane later (1961) made slightly more precise calculations on this problem, but the essence of the conclusion remains the same. In particular the mean fitness calculation just given remains the same.

It is relevant to observe that Fraser's 'reference' or 'optimal' genotype in the above argument is that multilocus genotype in which the favoured homozygote arises at all the loci undergoing the substitution. It is, however, important to look at this calculation from a different point of view. If the mean fitness of a population of constant size is to be taken as 1, then the calculations imply that the fitness of the optimal reference genotype has to be $e^{+30/n}$. It is in effect this calculation which is the relevant one for an examination of load arguments as used in support of the neutral theory. We now turn to these arguments.

From the data available to him at the time, Kimura (1968) estimated the value of n to be $1/6$ (so that six new substitution processes start in every generation). From this and the fitness requirement of the optimal genotype just given, he and Ohta (Kimura and Ohta 1971) concluded that 'to carry out mutant substitutions at the above rate, each parent must leave $e^{180} \approx 10^{78}$ offspring for only one of the offspring to survive'. This calculation was then stated as the main reason why the neutral theory, claiming that the substitutions must on the whole be selectively neutral, was proposed.

What does this calculation really mean? It is clear that the argument shows that it is the fitness of Fraser's 'reference genotype', not that of every parent, is required to be 10^{78}. If all fitness differentials related to fertility differences, the fitness requirement of the optimal genotype is irrelevant to all but individuals of that genotype. It is only when fitness differentials relate to differential mortality that this calculation has any relevance to each individual in the population.

The calculation leading to the value is, however, based on several dubious assumptions. It is easy to see that the calculation 10^{78} implicitly assumes that fitnesses are multiplicative over loci. That is, it is implicitly assumed that a single-locus fitness for any individual is found, and that his overall fitness is then the product of these single-locus fitnesses, the product being taken over all the loci currently substituting. This reductionist approach is surely unacceptable. As genomic data increasingly indicate, epistasis clearly occurs, and the unreasonableness of the multiplicative assumption was stressed as long ago as Wright (1930), and has been stressed by many others since then.

Even allowing the multiplicative assumption, the reference genotype for which the calculation is implicitly made is nothing much more than a figment of the imagination. A straightforward calculation (Ewens 2004) shows that in the situation considered above, the probability that any individual taken at random from the population is homozygous for the favourable allele at each locus substituting, that is who is of this optimal genotype, is about $10^{-23,200}$. Any theory centred on such an individual, whose very occurrence is so unlikely, must be reconsidered. This point was also stressed by Wright (1977, page 481) – I suspect in response to load arguments – and subsequently by other authors.

Suppose then that one centres one's calculation around the fitness of an individual likely to actually arise in a population. Even making the unreasonable multiplicative assumption referred to above, it can be shown (Ewens 1970) that in the case considered above, if the mean fitness of the population is scaled to the value 1, the variance in fitness of a randomly chosen individual is s/n. In the representative case s = 0.01 and with $n = {}^1/_6$, this is 0.06, yielding a standard deviation in fitness of about 0.25. Even in a population of size many millions, it is unlikely that one observes a value more than about five standard deviations above the mean, or in this case 2.25. Doubling this value as the requirement of a mating couple, we reach a number (four or five) of children which is well within the capacity of the most fit type in a population. (Indeed in the human population a major problem is to stop individuals reproducing at this rate.) With a model more reasonable than the multiplicative one, the requirement is less. Thus in a population of any reasonable finite size, no problems arise from load arguments, and thus no support is given to the neutral theory from these arguments.

There are many other considerations that have to be taken into account, for example ecological aspects, whether selection is through viability, fertility differences, or both, the effects of linkage disequilibrium, the possibility of frequency-dependent selection, and so on, one is hard pressed to find strong support for the neutral theory based on substitutional load arguments.

There are two further points to be made. The first is this. Crow and Kimura (1970, page 252) calculate a 'cost in terms of variance [in fitness]'. The result of their calculation is that, in the example considered above, where the less favoured allele has fitness $1 - s$ and the favoured allele a fitness of 1, one requires a genic fitness variance of s/m in each generation to change the frequency of the favoured allele at each substituting locus from a small value to a value close to 1 in m generations. An essentially identical calculation arises for the diploid case if one replaces 'genic variance' by 'additive genetic variance'. This is a very small variance if we take, for example, $s = 0.01$ and $m = 300$, and is surely well within the realms of possibility. The small value arises because it is very unlikely that any given individual has a numbers of favoured and unfavoured alleles very far from the average, and it can be shown that this calculation and that made two paragraphs above are just two different ways of presenting the same result. It has always puzzled me that Kimura published this result, in effect implying a tolerable genetic load, at much the same time as he elsewhere claimed a quite unbearable load.

The second point is that load calculations post-Haldane are based on assumed fitness values. In this they follow Haldane's original 'cost' calculations, which he identified as a reproductive excess, were also based on these values. Thus the Haldane 'reproductive excess' was in effect seen as being driven by selective differences. However, if one allele is to replace another in a population, then a reproductive excess is required whether the reason for this replacement is selection or pure random chance in a selectively neutral case. This argument, which I first heard (from Lewontin) in 1971, has from that time at least has formed part of the unspoken doubts about load theory. It certainly casts doubts on the load calculations as made above, depending as they do on selective differences, as an argument for the neutral theory of evolution.

The 'segregational load'

The observation of substantial genetic variation in a population, as noted by Harris (1966) and Lewontin and Hubby (1966) led to further calculations concerning segregational load. Suppose that the observed variation is due to the classic case of heterozygote selective advantage at all the loci at which genetic variation is indeed observed. As the simplest possible case, suppose that at each of m such loci, both forms of homozygote have fitness $1 - s/2$ and the heterozygote has fitness $1 + s/2$, and that both alleles have frequency $1/2$. These values lead to a population mean fitness of unity, taken (as mentioned above) as describing a population of constant size. Adopting the (unreasonable) multiplicative fitness model, an individual heterozygous at all loci has fitness $(1 + s/2)^m$ and when m is large this can be a very large quantity – for example, if $m = 10,000$ and $s = 0.01$, this is about 10^{22}. As with the substitutional load calculations given above, taking this load calculation at face value unthinkingly would lead one to question whether the observed polymorphisms are arising as a result of heterozygote selective advantage.

However, again it is necessary to examine the assumptions, implicit and explicit, in these calculations. As for the substitutional load, the multiplicative assumption is an unreasonable one. Arguing again as for the substitutional load, one has to question whether one is at all likely to see an individual of the optimal genotype, here an individual heterozygous at all loci segregating. Using the (inappropriate) multiplicative model, the probability that an individual taken at random is of this genotype is found, for the case considered above, to be,

$1/2^{10,000} \approx 10^{-3,010}$ and again it seems inappropriate to base load calculations on such a non-existent individual. As with the substitutional load, it seems more appropriate to consider the variance in fitness of the individuals in the population. In this case an individual is a heterozygote at k of the 10,000 loci segregating with probability

$$\begin{bmatrix} 10,000 \\ k \end{bmatrix} (1/2)^{10,000}$$

and such an individual has fitness $(1 + s/2)^k (1 - s/2)^{10,000-k}$. The population variance in fitness is thus

$$\sum_{k=0}^{k=10,000} (1/2)^{10,000} (1 + s/2)^{2k} (1 - s/2)^{10,000-2k}$$

and this calculation reduces to $(1 + s^2/4)^{10,000} - 1$ or about 0.28 if $s = 1$. This quite small variance, as with the similar calculation for the substitution load, is simply a reflection of the fact that most individuals will not be heterozygous at a number of loci greatly exceeding, or greatly falling short of, 5,000. From this point of view it is easy for the population to carry the 10,000 segregating loci, and it would be even easier if a more realistic model than the multiplicative fitness model were chosen.

Altogether the conclusions of this and the previous section is that load arguments have been used in an ill-advised way in support of the neutral theory. That theory well might be correct, and might be supported by observational data, but there seems to be little theoretical support for it from load arguments.

Summary

Fraser considered load arguments in much of his work on diseases relevant to man. These centred mainly on the mutational load, but he also considered segregational load arguments. In doing so he did not fall into the errors made by others after him in using segregational load arguments in support of the neutral theory. Nor did he fall into errors deriving from substitutional load, of the 'cost of natural selection', calculations, despite the fact that Haldane had put forward these arguments just a few years before Fraser's main work on genetic loads.

Bloom's syndrome. XXII. Numerous founder mutations bear witness to the persistence of mutant alleles in different human populations

James German[1] and Maureen M. Sanz[1,2] and Nathan A. Ellis[3]

[1]Weill Medical College of Cornell University, New York, NY 10021
[2]Molloy College, Rockville Centre, NY 11571
[3] Department of Medicine, Gastroenterology Section, University of Chicago, Chicago, IL 60637

The genetically determined trait known as Bloom's syndrome (BS) (German 1993) was described half a century ago by a dermatologist in private practice in New York City who named it 'congenital telangiectatic erythema resembling lupus erythematosus in dwarfs' (Bloom 1954). Although additional clinical features were recognised and the skin lesion turned out not to be congenital, a sun-sensitive erythematous skin lesion limited mainly to the face in an unusually small person remain the constant features of this very rare disorder. A genetic aetiology for BS was not mentioned in the original report, but later it was shown to be inherited in autosomal recessive fashion.

Although BS is a clinical entity barely mentioned in standard paediatrics and medicine textbooks (presumably because it is rare), it became interesting biologically when cells from affected persons were found to exhibit microscopically visible 'chromosome breakage.' Further investigation of this cytological feature revealed that BS cells are the most hypermutable and hyper-recombinable known. A programme of surveillance of affected persons referred to as the Bloom's Syndrome Registry (German and Passarge 1989) now in its fifth decade, the purpose of which was to learn the biological significance of the 'breakage,' has shown that two of man's serious age-dependent diseases, diabetes mellitus and malignant neoplasia, arise not only exceptionally commonly in persons with BS but at exceptionally young ages.

A decade ago the gene that when mutated is responsible for the trait BS was named *BLM*, mapped to chromosome band 15q26.1, and isolated (Ellis *et al.* 1995). *BLM* encodes a protein named BLM which is a DNA helicase of the highly conserved RecQ subfamily of helicases, a group of nuclear proteins important in the maintenance of genomic stability.

A search for Bloom's syndrome-causing mutations

With the BS gene in hand, a molecular search for BS-causing mutations was initiated. The Registry's policy had been to obtain blood and skin biopsies from registered persons whenever possible. Cultured cell lines had been developed from these, and the lines as well as the uncultured cells had been cryopreserved. To define the *BLM* mutations at the molecular level, DNA was extracted from these cell sources. From 1995 through 1999 the DNAs that had been accumulated from persons in the Registry were analysed (summarised in German and Ellis 2002). In families in which more than one sib had BS the DNA from only one of those affected was examined. In most families, DNA had been obtained from parents of the affecteds, making it possible to determine the parental origins of the different mutations, to compare haplotypes, and to rule out *de novo* mutation and uniparental disomy. In a few families in which DNA was unavailable from the person with BS, analysis of parental DNA samples defined the BS-causing mutations. Our purpose here, rather than to discuss clinical and cellular aspects of BS, is to re-examine the mutational data that had been generated and now organised as displayed in Tables 1 and 2. To accomplish this, the molecular work was in each case correlated with the information held in the Registry, about the affected person and family. In the Registry a file is maintained on each of the registered persons, most of which contain extensive clinical and genetic information. The pedigrees extend back several generations and record places of birth and residence as well as health information.

DNA was successfully analysed from a person with BS from each of 134 families, and in 125 of them at least one mutation at *BLM* could be defined in molecular terms (Tables 1 and 2, column (col.) 1): Both mutant alleles were identified in 117 of the 125 persons, whereas in 8 persons, one mutant allele only could be identified (those 8 being identifiable by having an entry in neither col. 2 nor col. 3 of Tables 1 and 2). Finding persons with but a single identifiable mutation at *BLM*, despite thorough searching for mutations (including the performance of Southern analysis to look for large deletions) of genomic DNAs and cellular mRNAs where available, is taken as evidence for the existence of additional as-yet-unidentified mutations. Finally, in nine persons, no BS-causing mutations at *BLM* could be detected (Table 3), leaving open the possibility that a locus other than *BLM* can when mutated result in the clinical trait BS; however, because of the eight persons just mentioned, technical limitation is considered a better explanation for these nine.

Seventy five of the affected persons proved to be homozygous for their BS-causing mutation, 50 to be genetic compounds. The 64 mutations consisted of the following: 29 single base substitutions in exons ($C{\rightarrow}T$ or $A{\rightarrow}G$ being 12 of those), 10 of which were missense mutations; 17 deletions (13 of those being of but one or two bases but four extending over one or more exons); nine single base substitutions in intervening sequences; eight base insertions (seven being of but a single base); and, one deletional/insertional mutation, this last to be referred to as *blm^{Ash}* (see footnote 3b to Table 1). In addition, 17 DNA changes were identified in the analysis that for various reasons were classified as not causing BS.

Dedication; and a historical note

The subject we have chosen for a paper with which to honour geneticist George Fraser on his 75th birthday is one particular aspect of BS's extensive genetic heterogeneity that became evident as the mutation data we had accumulated in the Registry were analysed. That is the existence in BS of numerous groups of persons who inherited their mutated *BLM* gene identical by descent from a recent common ancestor. And, George, a new word in your honour – *foundred*.

A paper on Bloom's syndrome is appropriate for this particular volume because, as George and I realise now, it was in the same year, 1960, that he began his adventure with the cryptophthalmos syndrome with which his name is associated (Fraser syndrome) and co-author J. G. saw his first patient with what was to become known as Bloom's syndrome, and set up a culture of her cells for cytogenetic analysis. Those were heady days for those of us whose business is *finding out* and whose field is human genetics, for that field finally was shedding the pall that had been cast over it by the grievous misuse of genetics by the Nazis; much 'new' now could come to light.

A guide to the tables

The mutations that were identified in *BLM* that are considered to be BS-causing are listed in the first columns of Tables 1 and 2. The *BLM* mutation on the second No. 15 chromosome by which a given person became either a homozygote or a compound heterozygote is shown in the second and third columns of Tables 1 and 2. Table 1 pertains to the *BLM* mutations that were present in multiple individuals with BS, Table 2 to those mutations present only in single individuals. Table 3 lists persons with BS in whom, although a successful molecular search was carried out, a mutation at *BLM* could not be identified. The persons with BS whose DNAs were analysed are identified in cols 4 of Tables 1 and 2 and col. 1 of Table 3; they are identified as they have been in all publications from the Bloom's Syndrome Registry since 1969 (German 1969). Those whose parents knew themselves to be related are indicated in cols 5 of Tables 1 and 2 and col. 2 of Table 3. The last two columns of the three tables show the geographic and ancestral origins of the persons analysed. The reduction of this large amount of data to three tables made the 15 footnotes to the tables necessary, to which the reader's attention is directed.

Founders and foundreds

In the case of approximately a third of the mutations detected (19 of the 64), two or more of the persons with BS turned out to have inherited the same mutation (Table 1, col. 1). Although in principle some of the recurring mutations listed in col. 1 of Table 1 could have arisen independently, for present purposes they are assumed to be founder mutations, and the persons having inherited the same mutation identical by descent from a common ancestor are considered to be members of a population descended from that one person, the progenitor of a progeny that we refer to as a *foundred* (foun.). The mutations identified in 102 of the 134 successfully analysed individuals with BS allowed the assignment of each of those 102 persons to one of 20 foundreds; furthermore, for approximately a third of them (39 of the 102 persons), assignment to 2 different foundreds was possible. (The second foundreds for those 39 are indicated in Table 1, col. 3.) Two foundreds are constituted by one and the same mutation, the reasons for which will be explained below.

Forty-five of the 64 mutations in a total of 42 persons were found not in foundreds but, each, in a single individual with BS (Table 2); i.e., each of the 45 is unique within the present survey, and in this paper they will be referred to as such. Eighteen of these 45 unique mutations were homozygous in the affected persons (Table 2, cols. 1 and 2); the other 27 were present in compound heterozygosity with a different *BLM* mutation (Table 2, cols. 1 and 3). In 19 of 27 persons the second mutation was one of the 19 founder mutations listed in Table 1 (the relevant foundreds identified in Table 2, col. 3); in three persons the second mutation was unique (in 65(AnPa), 98(RoMo), and 112(NaSch)); and, in two persons a second mutation could not be identified (in 205(JoRob) and 223(GiZa)).

The 20 foundreds are comprised of different numbers of persons: the most populous, Foun. 1, is comprised of 35 persons, 11 foundreds of from three to 19 persons, and eight of two persons. (Note: to facilitate counting, those individuals whose mutations assign them to two foundreds are identified by square brackets at their second appearance in the last four columns of Table 1.)

An argument could be made for there being 27 rather than 20 foundreds in our material, seven of them, seemingly contradictory in terms, having been identified not in a population of persons with BS but each through a single person! These persons are the seven in Table 2 each of whose parents, although not known to be related, carried identical *BLM* mutations, which indicates that each pair of parents shared a common distant ancestor – 40(DoRoe), 104(NaSp), 123(FaCar), 186(MoIs), 211(NaMac), 215(AnDonas), and 223(GiZa). In these seven families each pair of *parents* would have constituted a foundred, rather than multiple propositi/ae as in the 20 foundreds of Table 1.

Parental consanguinity in persons in foundreds

The Bloom's Syndrome Registry is comprised of 238 persons with BS of whom 151 are non-Jewish. The parental consanguinity rate is not greatly increased over that in the general population in Ashkenazi Jews with BS because of the approximately 1% frequency of *blm^{Ash}* in the latter (Li *et al.* 1998). In 51 of the 151 non-Jewish, however, the parents are closely related (29.1%), 'closely' here meaning at least as second-cousins. In the present report, which deals only with those persons with BS in the Registry whose *BLM* locus was successfully searched for mutations, of those whose mutations assigned them to a foundred (all those listed in Table 1), the parents were related in 16 of the 102 (only 15.7%). In contrast, for the persons with BS who had inherited at least one unique BS-causing mutation (those listed in Table 2), 11 of the 42 had closely related parents (26.2%). And in even greater contrast, in the 23 persons in Table 2 neither of whose mutations were represented in a foundred (i.e., in persons both of whose mutations were unique), 11 had consanguineous parents (47.7%). It is interesting (but not explained), that those with BS in whom no *BLM* mutations at all could be identified, those comprising Table 3, the parents were closely related in seven and probably eight of the nine (see the table's footnote 2). These very different incidences of parental consanguinity agree with the intuition that although all of the mutant alleles with which we are dealing are very rare in the general non-Jewish population, those that turn up in multiple affecteds – here in foundreds – are relatively less rare than those that proved to be unique.

The Registry has been informed that persons with BS are being diagnosed with a fair regularity in Turkey and in Tunisia. Nine Turkish individuals have been accessioned to the Registry, four in three families living in Turkey proper and five in three families that have emigrated and now live in Germany or Belgium. Mutation analysis was carried out on the index cases in the three Turkish families in Germany and Belgium (identified in col. 6 of Table 2); in all three, the parents were closely related, and all three of the affecteds were found to be homozygous for their BS-causing mutations, which were different in each case. The Tunisian with BS analysed (Table 2), also born to consanguineous parents, also was homozygous, and, again, for a unique mutation.

In the survey, two of the individuals with BS have parents who are French-speaking and consider themselves *Canadien-Français*. The parents of both individuals are closely related. One individual, who lives in Ontario, is homozygous for a unique mutation and so appears in Table 2 (81(MaGrou)); the other, from Quebec, also is homozygous for a mutation, but one that assigns her to a foundred, Foun. 9 in Table 1 (176(CoGui)).

Persons who share two founders as ancestors

Table 1 shows that three pairs of the persons in foundreds share not one but two distant ancestors, and specifically, had as ancestors two persons who, each, had become the progenitor of a foundred! The three pairs are the following: (i) 91(GeCro) and 118(MaIn), whose recent ancestors have lived, respectively, in The Netherlands and Germany. Together they comprise Foun. 16; but, they are members also of Foun. 7, so they both can claim as distant ancestors the founder of Foun. 16 and the founder of Foun. 7; (ii) 177(KeSol) and 183(ZaRoge), both of whom are in Foun. 7 as well as Foun. 8, live, respectively, in Michigan and Utah. The parents of 177(KeSol) are descendants of immigrants to Canada and the United States from Germany and France, those of 183(ZaRoge) of immigrants from England, Denmark, and Germany. Both of these sets of parents, it now can be said, had as distant ancestors the founders of not only Foun. 7 but also Foun. 8; (iii) 141(YaFo) and 237(YiRen) who together comprise Foun. 2, both Ashkenazi Jews, one today living in Israel and the other in the United States, share as distant ancestors not only the founder of Foun. 1 but also him/her of Foun. 2. It is the father of 141(YaFo) who carries insT2407; his ancestors lived in Berlin, Bukovina, and (the Polish) Galicia. It is the mother of 237(YiRen) who carries it; this mother's paternal grandparents lived in Poland and Czechoslovakia, her maternal grandparents in Germany. The genetic histories provide no suggestions as to the dwelling place of the founder of Foun. 2, and too few chromosomes have been identified to allow dating of the mutation by haplotype analysis.

Brief comments on the several foundreds identified

Foun. 1 is comprised of persons who carry the complex mutation referred to as *blm*[Ash]. Each is a descendant of a Jewish man or woman who migrated to Eastern Europe sometime after the early thirteenth century and became the ancestor of approximately one percent of those called in Hebrew *Ashkenazim* (Li *et al.* 1998). Foun. 2 is comprised of only two of the Ashkenazi Jews we analysed; they inherited, along with *blm*[Ash], the insertional mutation insT2407. *blm*[Ash] and insT2407 may be referred to as the *Ashkenazi mutations* that determine the trait BS when either homozygous or present in compound heterozygosity with another BS-causing mutation. (That a second example of founder effect exists in Ashkenazi Jews is of obvious significance in the genetic counselling so widely practised in that community with such impressive results.)

Foun. 3 is comprised of non-Jewish persons, all Spanish-speaking Christians, who also have inherited *blm*[Ash] (Ellis *et al.* 1998)! Here *blm*[Ash] is the same mutation at *BLM*, and on the same haplotype, as that segregating in Ashkenazi Jews, but the founder in Eastern Europe and the founder in Spain's 16th century *Nuevo Mundo* were two different people. In the population that originated in *El Nuevo Mundo*, *blm*[Ash] may be referred to as the *Central American/United States Southwest mutation*.

Foun. 4 is comprised of two Spanish-speaking individuals, one born in Mexico and one in New Mexico. Their mutation, delAG2506–7, may be referred to as the *Mexican mutation*.

All persons with BS having Portuguese or Brazilian ancestry fall into Founs. 5 and 6. Their mutations, delExons20–22 and delG3587, may be referred to as the *Portuguese/Brazilian mutations*.

All those in foundreds who are of Italian ancestry fall into Foun. 17 and 18, in company with many North Americans who reported no Italian ancestry during the recording of their genetic histories. Their mutations, C2098T and G3164C, may be referred to as *Italian mutations*. (Seven other mutations in Italians, each unique, are listed in Table 2.)

Founs. 19 and 20 are comprised, with one exception, of Japanese individuals the exception being an American of mixed Western African, American Indian, and Western European ancestry.

The two mutations delCAA557–559 and insA1544 may be referred to as the *Japanese mutations*; they are and will be responsible for most of the BS in Japan.

Founs. 7–16 are comprised of approximately half of the persons with BS analysed in this survey. All but one of these persons are non-Italian Europeans, North Americans, or Australian, the exceptional person being an emigrant from El Salvador to North America. (The two exceptional individuals mentioned here and in Founs. 19–20 are most easily explained by independent mutational events rather than founder effect.) Founs. 7–16's ten mutations may be referred to as *Euro-American mutations*. (Other so-far unique Euro-American mutations are listed in Table 2.)

Thus, many different BS-causing mutations are segregating today in different peoples and in people inhabiting different parts of the earth. Nevertheless, because the mutations recognised thus far all, when present in two copies, result in the absence of BLM protein's DNA helicase activity, the clinical trait 'Bloom's syndrome' is similar enough among the affecteds to allow the astute physician to obtain cytogenetic analysis by which the diagnosis will be unequivocally confirmed or excluded.

A closer look at just one of the foundreds

In the section above, brief comments were made on the several foundreds identified in the present survey. But consideration of many of the individual foundreds in a historical context along with the information obtained from the families during the genetic history-taking also can be interesting. Space constraints in the present report permit us to give but a single example of this, as follows:

Mutation C2098T, carried by the founder of Foun. 17, is present today in the genomes of two Italians with BS, both of Sicilian parents – 102(GiBuc) and 103(SeDip). Yet the same mutation also is present in North America, geographically widely dispersed in persons who neither know themselves to have a common distant ancestor nor to have Italian ancestry. How might this be explained? The ancestors of the two individuals lived about 80 miles apart, those of 102(GiBuc) in Canicatti, a city centrally located on the island, those of 103(SeDip) near Barcellona Pozzo di Gotto not far from the Strait of Messina (in Porto Salvo and Contra da Comicia, villages about 45 miles apart). Both of these young men with BS are genetic compounds, but they do share mutant allele C2098T. (Each also inherited one of the unique mutations listed in Table 2.)

The six non-Italians with BS who also are in Foun. 17 live (as mentioned) in widely separated parts of the United States, indicating that C2098T arrived on the North American continent early as it was being populated by Europeans. Yet, the significant immigration to the United States of Italians, including Sicilians, took place only between the late 1880s and the early 1920s. Therefore, a hypothesis to explain Foun. 17's today being comprised of Sicilians and inhabitants of the United States is the following: (i) The mutant allele C2098T was introduced into Sicily by a Norman, one of the 'Northmen' (evolving into the English *Norsemen*) who in the 7th and 8th centuries invaded and settled in present-day Normandy and who in the 11th century conquered and controlled both Sicily and England; (ii) one of that Norman's relatives introduced C2098T into England following the conquest of 1066; and, (iii) sometime during the 16th and 17th centuries an English emigrant of Norman descent – one of that 11th century Norman's descendants – brought it to England's colonies in North America, thence its wide dissemination throughout the present-day United States. (Note: C2098T is the only one of the ten 'Italian mutations' that is found in persons lacking known Italian ancestry, specifically in the six in Foun. 17.)

The data as organised in Table 1 permit such conjectures when integrated with historical events and migrations of peoples.

Discussion

Employing Bloom's syndrome as a genomoscope

In the present work, we have taken advantage of our familiarity with a large proportion of the world's BS population to peer, so to speak, directly into the human genome and in doing so to obtain some understanding of the extent to which mutation affected the germ lineages of our ancestors. Finding so many different mutations to have arisen along the length of just this one gene, a gene that encodes a protein already selected for very early in evolution and thereafter conserved (witness the orthologue of *BLM* in the budding yeast, *sgs1* (Gangloff *et al.* 1994)), is perhaps not surprising as they are apparently all neutral in single dose. However, the analysis of *BLM* mutations has brought to our attention two interwoven themes that reflect a recurring process in human populations: (i) the *BLM* mutations recur in different persons, having been inherited identical by descent from a common ancestral individual – a founder – and (ii) these founders occur in essentially all the different human populations studied and not just those populations that so often are associated with founder mutations, such as the Ashkenazi Jews. Foundreds 1 and 3, which are constituted by the mutation *blm^Ash* relatively common in the Ashkenazi Jewish population, no doubt owe their origins to that form of genetic drift known as *founder effect*, wherein a new population is established from small numbers of individuals who come from a larger population and who marry exclusively amongst themselves. The history of the Ashkenazi Jews in Europe and the Sephardi Jews in Spain's *Nuevo Mundo*, in which Founs 1 and 3 have arisen, suggest founder effect because the migrations are well-documented. For the other foundreds, it is not clear what roles migration and small population size played in the widespread recurrence of *BLM* mutations. Although each of the recurring mutations is restricted to particular populations, in none of them is the frequency of *BLM* mutation uncommonly elevated as it is in the Ashkenazi Jewish population, which would be expected if founder effect had played a role. An alternative explanation may be that exponential population growth experienced by the human species over the last three centuries facilitates a persistence of recessive mutations, because the mutant allele frequency remains low even as the absolute number of mutant alleles is ever increasing by mutation and by a greater likelihood of transmission through large families. Indeed, the generally small size of most of the non-Jewish foundreds suggests that none of the mutations therein are particularly old.

BS has permitted us to have a view of what has gone on and continues to go on in the genome of a sexually reproducing species that has survived a very long time and proven itself capable of inhabiting in large numbers much of the earth's surface. What we perceive that has gone on at *BLM*, and still is, may be assumed to have gone on and be going on at all coding loci – myriad mutations along the lengths of the genes, many of which will be retained to segregate in the general population.

None of this is unexpected, of course, but this particular view into the human genome through BS provides sound supporting evidence for what is generally assumed to have taken place. It again confirms that, though *stability* is one of the most awe-inspiring features of the genetic material, which remains largely unchanged from cell division to cell division in both somatic and germ lineages and in the germ lineages of all continuing species, an equally impressive feature is its *ability to mutate* – and thereby to provide the raw material for continuing evolution.

These mutational data in BS also constitute new and possibly useful information for those interested in estimating the number of ancestors of the present-day population of *Homo sapiens* (as Harding and McVean, 2004).

The importance of organising the data

The organisation of the mutation data into Tables 1 and 2 gives special value to what otherwise was little more than an accumulation of clinical, genetic, cytogenetic, and molecular genetic information that had been gathered carefully and reliably during the course of the long-term surveillance of a small population with one rare genetically determined disorder coupled with the application of technological advances as they had become available. Lacking this re-organisation, we had accomplished little more than confirm BS's recessive transmission. With the re-organisation of the data, the several foundreds emerged as distinct entities. Then their consideration inevitably fused genetics and history in fascinating ways, one example of which was given above.

But more importantly, by viewing the data as set out in Tables 1 and 2 certain possible, and serious, implications for human health become apparent, and raise the following question: Are there not untoward consequences for man the species of the accumulation of a very large number of mutations at coding loci, even though as far as is known they have no developmental consequence in single dose?

The price being paid

We propose that the accumulation of mutations in this young species indeed is responsible for certain human health problems that affect many people, i.e., that a pretty price is being paid by the species for its having survived for so long. The first of three health problems to be mentioned is not in question; the other two we offer as reasonable possibilities:

First, some relatively small proportion of the mutations that have arisen at coding loci are responsible for the live birth of persons with recessively inherited traits that qualify as disease. Although most of the conditions are rare, collectively they are not; they in fact are responsible for much of paediatric disease.

Second, we propose also that the mutations that have accumulated in the human genome are responsible for an unknown but significant proportion of intellect-deficient and often variously dysmorphic individuals, specifically the proportion not due to chromosome imbalance. Given the high quality of medicine and the social attitudes of many of the advanced societies, thousands of such people today survive into adulthood.

Third, the accumulation of mutations, most of them silent in single dose, may be responsible for a significant, very possibly the predominant, proportion of what is often referred to as pregnancy wastage, which in the human is enormous. Approximately a fifth of pregnancies of which women become aware are eliminated spontaneously, and a much larger proportion of conceptuses are silently lost even earlier. Aneuploidy explains approximately a fifth of spontaneously aborted pregnancies of which women are aware, and teratogenic chemicals and infectious agents, a few of which have been identified, can be assumed to be responsible for a small proportion; however, the vast majority of naturally (i.e., spontaneously) eliminated conceptuses go unexplained. Our proposal is that homozygosity and compound heterozygosity for some, very possibly many, of the myriad mainly-to-remain-unidentified mutant alleles in the genomes of humans is the basis for a major part of the loss before birth of human conceptuses.

The knowledge that the genome we each inherit is peppered with recessive mutations that have been accumulating over the long period we have been a species – as our view into the genome per Bloom's syndrome has shown us to be the case – may make it easier to accept and pay the price, it being the price just *to be* here! (See the contributions on 'load' in this volume.)

Acknowledgements

The collection of clinical and genetic information and of biological specimens which began in 1960, the preparation of DNA samples, the isolation of *BLM*, and the identification and early molecular characterisation of many of the mutations of *BLM* were carried out with the support of grant Nos. HD-00635, HD-04134, AG00051, CA-38036, HD 19146, CA37327, and CA-50897 from the National Institutes of Health; No. VC-104 from the American Cancer Society, and No. GB1868 from the National Science Foundation. The molecular characterisation of the *BLM* mutations listed in Tables 1 and 2 of the present report, which was under the supervision of Nathan A Ellis, PhD, was carried out by Susan Patton Ciocci, David Lennon (now PhD), and Tian Yi.

Table 1. Foundreds[1] in the Bloom's Syndrome Registry

The mutations at *BLM*			The individuals with Bloom's syndrome comprising the foundreds[2]			
The mutations carried by the founders – the founder mutations[3]	The mutations at *BLM* on the homologous chromosomes[4]		Identification[5]	Parental consanguinity[6]	Noteworthy parental ancestry[7]	Places of birth and/or principal residence
	Identical (i.e., homozygous)	Different (i.e., heterozygous)				
Jews						
Foundred 1 (n = 35)						
blm^4sh			2(SuBu)		AA	New York
blm^4sh			3(HoCo)		AA	New York
blm^4sh			9(EmSh)		AA	Israel
blm^4sh			14(LeSi)		AA	New York
blm^4sh			15(MaRo)		AA	California
blm^4sh			16(EtFi)		AA	New York
blm^4sh	delT3261 (Foun. 10)		26(SaTi)		A-	Indiana
blm^4sh			27(LySe)		AA	Ohio
blm^4sh			32(MiKo)		AA	New Jersey
blm^4sh			34(AlSti)		AA	Illinois
blm^4sh			42(RaFr)		AA	Israel
blm^4sh			44(AbRu)		AA	Israel
blm^4sh			45(ZvSha)		AA	Israel
blm^4sh			47(ArSmi)		AA	New York
blm^4sh	delExons20–22 (Foun. 5)		50(JeBl)		A-	Ontario
blm^4sh			53(StAs)		AA	New Jersey
blm^4sh	G2672A		54(AlTu)		A-	New York
blm^4sh			56(JoGr)		AA	Ohio
blm^4sh			57(AmEl)		AA	Israel
blm^4sh			79(MeDer)	+	AA	Connecticut
blm^4sh	T3163G		87(AlFra)		A-	Michigan

Allele 1	Allele 2	+	Patient	Ethnicity	Ancestry	Origin
blm^Ash	blm^Ash		106(JaHe)	AA		Florida
blm^Ash	blm^Ash		107(MyAsa)	JA		New York
blm^Ash	blm^Ash		119(AbHal)	AA		New York
blm^Ash	blm^Ash		121(NiRos)	AA		Israel
blm^Ash	IVS2+1(G→T)		126(BrNa)	AJ		Brazil
blm^Ash	insT2407 (Foun. 2)		141(YaFo)	AA		Israel
blm^Ash	blm^Ash		142(MaMar)	AA		Israel
blm^Ash	blm^Ash		171(CrSpe)	AA		District of Columbia
blm^Ash	blm^Ash		172(AkGroun)	AA		Michigan
blm^Ash	blm^Ash		178(SaNu)	AA		New York
blm^Ash	blm^Ash		182(JuLi)	AA		Belgium
blm^Ash	T3510A		225(RaRose)	A-		Minnesota
blm^Ash	insT2407 (Foun. 2)		237(YiRen)	AA		New York
blm^Ash	blm^Ash		238(MiGur)	AA		Israel
Founded 2 (n = 2)						
insT2407	blm^Ash (Foun. 1)		[141(YaFo)]	[AA]		[Israel]
insT2407	blm^Ash (Foun. 1)		[237(YiRen)]	[AA]		[New York]
Mexicans and Americans of Spanish Ancestry						
Founded 3 (n = 5)						
blm^Ash	delAG2506–7 (Foun. 4)	+	127(TaLu)		Spanish via Mexico	Colorado
blm^Ash			130(ChVa)		Mexican, Texan, New Mexican	Utah
blm^Ash	G3197A		179(AtPon)			Mexico
blm^Ash			197(BrPar)		El Salvadoran and Ecuadoran	Texas
blm^Ash			200(JeJa)		El Salvadoran	Maryland
Founded 4 (n = 2)						
delAG2506–7	blm^Ash		[130(ChVa)]		[Mexican, Texan, New Mexican]	[Utah]
delAG2506–7			173(RaLem)			Mexico
Portuguese and Brazilians						
Founded 5 (n = 4)						
delExons20–22	blm^Ash (Foun. 1)		[50(JeBl)]	[A-]		[Vancouver]
delExons20–22	delExon20–22		181(ElPi)		Portuguese	Switzerland

delExons20–22	delExon20–22	195(BrCarv)		Brazil
delExons20–22	delG3587 (Foun. 6)	216(LeCoe)		Brazil
Founded 6 (n = 2)				
delG3587	delExons20–22 (Foun. 5)	[216(LeCoe)]	+	[Brazil]
delG3587	delG3587	217(ArC0s)		Brazil
Non-Italian Europeans and North Americans				
Founded 7 (n = 18)				
C1933T		11(IaTh)		Ontario
C1933T	C1933T	61(DoHop)	+	California
C1933T	C1933T	64(CrFe)		Ohio
C1933T	insAAAT2250–2253 (Foun. 15)	67(SuSc)		The Netherlands
C1933T	G1284A (Foun. 16)	91(GeCro)		The Netherlands
C1933T	C2725T	105(ShWo)		Texas
C1933T	insA3223	114(BeNeu)		Germany
C1933T	C2098T (Foun. 17)	115(AsHol)		Florida
C1933T	G1284A (Foun. 16)	118(MaIn)		Germany
C1933T		133(JaBer)		Germany
C1933T	delC2923 (Foun. 9)	147(BjSchr)		Belgium
C1933T	C1933T	164(DaGo)		New Brunswick
C1933T	delT3261 (Foun. 10)	169(ThGe)		Ontario
C1933T	C2695T (Foun. 8)	170(ErBor)		New York
C1933T	C2695T (Foun. 8)	177(KeSol)		Michigan
C1933T		183(ZaRoge)		Utah
C1933T	C1933T	188(YvMul)	+	Belgium
C1933T	C1933T	189(KaReg)		California
Founded 8 (n = 9)				
C2695T	C2695T	21(RaRe)	+	California
C2695T	delG3028 (Foun. 11)	30(MaKa)[6]	+	Germany
C2695T	C3415T	109(BeGra)		New York
C2695T	delEx11+12 (7812bp)	111(JaKir)		Massachusetts

Allele 1	Allele 2	Index (family)		Ethnic origin	Geographic origin
C2695T	C1933T (Foun. 7)	136(JoNae)			Maryland
C2695T	C1933T (Foun. 7)	[177(KeSol)]			[Michigan]
C2695T		[183(ZaRoge)]			[Utah]
C2695T	C2695T	185(HeRa)		El Salvadoran	Ontario
C2695T	C2695T	191(LaHam)			South Carolina
Founder 9 (n = 7)					
delC2923	delExon15 (1583bp)	6(DeSou)			Maine
delC2923		20(ViShr)			Maryland
delC2923		69(AnSl)			Washington
delC2923	C2098T (Foun. 17)	137(KrMu)			Australia
delC2923		146(HoNo)			Minnesota
delC2923	C1933T (Foun. 7)	[164(DaGo)]			[New Brunswick]
delC2923	delC2923	176(CoGui)	+	French Canadian	Quebec
Founder 10 (n = 4)					
delT3261	delT3261	22(ElHa)	+		Illinois
delT3261	blm^{Ash} (Foun. 1)	[26(SaTi)]		[A-]	[Indiana]
delT3261	C1933T (Foun. 7)	[170(ErBor)]			[New York]
delT3261	insAAAT2250–2253 (Foun. 15)	194(RaTe)			California
Founder 11 (n = 3)					
delG3028	C2695T (Foun. 8)	[30(MaKa)][7]	[+]		[Germany]
delG3028	delAAAGA991–995 (Foun. 14)	52(PaDü)			Germany
delG3028	delAAAGA991–995 (Foun. 14)	207(KiPet)			Kentucky
Founder 12 (n = 2)					
C1642T	IVS7+5(G→A)	167(RoTha)		Adopted (American)	Virginia
C1642T		202(JuCy)		Adopted (German)	Germany
Founder 13 (n = 2)					
A2015G	IVS5–2(A→G)	31(CaDe)			Belgium
A2015G	T3163C	139(ViKre)			Illinois
Founder 14 (n = 2)					
delAAAGA991–995	C1642T (Foun. 12)	[167(RoTha)]		[Adopted (American)]	[Virginia]

Allele 1	Allele 2	Founder		Ethnicity	Location
delAAAGA991–995	delG3028 (Foun. 11)	[207(KiPet)]			[Kentucky]
Foundred 15 (n = 2)					
insAAAT2250–2253	C1933T (Foun. 7)	[64(CrFe)]			[Ohio]
insAAAT2250–2253	delT3261 (Foun. 10)	[194(RaTe)]			[California]
Foundred 16 (n = 2)					
G1284A	C1933T (Foun. 7)	[91(GeCro)]			[The Netherlands]
G1284A	C1933T (Foun. 7)	[118(MaIn)]			[Germany]
Italians and North Americans					
Foundred 17 (n = 8)					
C2098T	C2098T	5(JaOa)	+		Colorado
C2098T	C2098T	51(KeMc)	+		Kentucky
C2098T		59(FrFit)			Connecticut
C2098T	A3191T	102(GiBuc)			Italy (Sicily)
C2098T	delTT3475–6	103(SeDip)		Italian (Sicilian)	Switzerland
C2098T	C1933T (Foun. 7)	[115(AsHol)]			[Florida]
C2098T	IVS16+3(A→T)	134(ViBai)			Missouri
C2098T	delC2923 (Foun. 9)	[146(HoNo)]			[Minnesota]
Foundred 18 (n = 3)					
G3164C	C311A	80(ErPal)		Italian	Massachusetts
G3164C	G3164C	113(DaDem)	+	Italian	New York
G3164C	G3164C	222(ElGen)			Italy
Japanese					
Foundred 19 (n = 5)					
delCAA557–559	IVS8+1(G>T)	78(AkSak)			Japan
delCAA557–559	delCAA557–559	97(AsOk)	+		Japan
delCAA557–559	insA1544 (Foun. 20)	100(YuMat)			Japan
delCAA557–559	delCAA557–559	110(MaKur)	+		Japan
delCAA557–559	delCAA557–559	128(YaWa)			Japan
Foundred 20 (n = 5)					
insA1544	C2254T	71(HaEn)		West African	Pennsylvania

insA1544	insA1544	86(NoKi)		Japan
insA1544	insA1544	93(YoYa)		Japan
insA1544	delCAA557–559 (Foun. 19)	[100(YuMar)]		[Japan]
insA1544	insA1544	129(MaWat)	+	Japan

[1] The term founded (foun.) refers to a group of persons with BS each of whom has a particular BS-causing mutation, all of whom, therefore, are descended from a person who carried that mutation – the founder. n, the number of persons comprising the founded.

Note. Dividing the 20 foundeds broadly into 6 groups as attempted in this table – Jews, Non-Italian Europeans and North Americans, Italians and North Americans, etc. – is arbitrary and imperfect, exceptions readily noted being the following: In Foun. 1, four non-Jewish parents, identifiable by 'A-' in Col 6; in Foun. 5, a person with an Ashkenazi Jewish father and a non-Jewish North American mother (A-' in Col. 6), whereas all others in Founs. 5 + 6 are Portuguese or Brazilian; in Foun. 8, an El Salvadoran homozygous for a mutation limited otherwise to North Americans and Europeans; in Foun. 9, an Australian; and, in Foun. 20, a North American of West African ancestry living in the United States, all others in Founs. 19 + 20 being Japanese living in Japan.

Note. Only two of the 153 families represented in the survey knew of any 'blood' relationship. The mothers of 9(EmSh) and 44(AbRu), two young men with BS in Foun. 1, shared two of their great grandparents, i.e., were 2nd cousins twice removed. Both families lived in Israel.

[2] Their mutations assign 20 persons not to just 1 but to 2 foundeds. To facilitate counting, square brackets are used in Cols. 4–7 to identify such persons (and those data that pertain to them) who had been listed earlier in the table.

[3a] The nomenclature employed for the mutations is that in use at the time the survey reported here was completed. The reasons for employing it still are, first, that they are clear in meaning; second, that the 'official' nomenclature seems still to be evolving; and, most important, to avoid introducing errors as the 'names' of the many mutations we report here would be altered.

[3b] The mutation referred to as blm^{tib} is a 6-bp deletion and 7-bp insertion at nucleotide 2281 described as 2207–2212delATCTGAinsTAGATTC. For reasons given in the text, blm^{tib} appears in this table as the founding mutation in both Foun. 1 and Foun. 3.

[4a] A blank in both Cols. 2 and 3 for a given individual indicates that although DNA was available and successfully analysed, only a single BLM mutation could be identified. The six such individuals are in Founs. 7, 8, 9, and 17.

[4b] For a mutation that is in another founded also, that other founded is indicated in parenthesis in Col. 3.

[4c] Re the possibility of identifying mutational hot spots, the only instances in either Table 1 or 2 of the same nucleotide's being mutated in more than one way are in individuals 87(AlFra) in Foun. 1 in whom a T becomes a G and 139(ViKre) in Foun. 13 and Table 2 in whom the same T becomes a C.

[5] Individuals with BS are identified as in the Bloom's Syndrome Registry.

Note. The number that has been assigned to each of the 238 persons in the Bloom's Syndrome Registry signify nothing more than the order in which they were accessioned. In this table, in each founded, persons are listed in numerical order (Col. 4).

[6] A second cousin or closer relationship existed in all those indicated by a plus sign except in 30(MaKa) whose parents are fourth cousins. Re 30(MaKa): his parents did not themselves know that they were related, but they were found to be fourth cousins in an exhaustive search of church records in the valley in Germany's Sauerland where their families long had dwelt (Passarge, 1991); 30(MaKa) is the only persons with parental consanguinity who is not homozygous at BLM.

[7a] 'Noteworthy' in the sense that the ancestry differs from that expected from the individual's place of birth or principal dwelling place shown in Col. 7.

[7b] For those who were adopted, the ancestry of their biological parents is shown in parenthesis.

[7c] A, an Ashkenazi Jewish parent; J, a Jewish parent but not Ashkenazi (see (d) below); -, a non-Jewish parent in a union in which the other parent is Jewish; the father's designation appears before the mother's, so that A- indicates that the father was Ashkenazi Jewish and the mother non-Jewish.

[7d] The non-Ashkenazi but Jewish father of 107(MyAsa) in Foun. 1 was Bulgarian, and he emphasised that he was Sephardi; he, therefore, provides evidence that blm^{tib} is segregating in the Sephardi Jewish population of Eastern Europe. The Jewish but non-Ashkenazi mother of 126(BrNa) in Foun. 1, who transmitted to 126(BrNa) the unique mutation IVS2+1(T→G), was born to a Sephardi Jewish father and an Egyptian Jewish, but non-Sephardi, mother.

Table 2. The 45 unique mutations at *BLM* identified in the Bloom's Syndrome Registry[1]

The unique mutations	The mutations at *BLM* on the homologous chromosomes[2]		The individuals with Bloom's syndrome			
	Identical (i.e., homozygous)(i.e., heterozygous)[3]	Different	Parental consanguinity	Noteworthy parental ancestry[4]	Identification[4]	Places of birth and/or principal residence
C3847T	C3847T		+	Italian	7(RoTa)	New York
C3118T	C3118T		+		10(GrSt)	Manitoba
insG1968	insG1968		+		17(ChSm)	Pennsylvania
delExon15 (1583bp)		delC2923 (Foun. 9)			20(ViShr)	Maryland
IVS5−2(A→G)		A2015G (Foun. 13)			31(CaDe)	Belgium
G2702A	G2702A				40(DoRoe)	Alberta
G2672A		*blm*^Ash (Foun. 1)		A-	54(AlTu)	New York
insA2488		delA3681			65(AnPa)	The Netherlands
delA3681		insA2488				
C2254T		insA1544 (Foun. 20)		West African	71(HaEn)	Pennsylvania
delC1346	delC1346		+	Turkish	74(OmAy)	Germany
IVS8+1(G→T)		delCAA557−559 (Foun. 19)			78(AkSak)	Japan
C311A		G3164C (Foun. 18)		Italian	80(ErPal)	Massachusetts
C1784A	C1784A		+	French Canadian	81(MaGrou)	Ontario
T3163G[5]		*blm*^Ash (Foun. 1)		A-	87(AlFra)[5]	Michigan
delEx11+12(6126bp)	delEx11+12(6126bp)		+	Italian	92(VaBia)	New York
G1701A		A1090T			98(RoMo)	West Virginia
A1090T		G1701A				
A3191T		C2098T (Foun. 17)		Italian	102(GiBuc)	Italy
delTT3475−6		C2098T (Foun. 17)			103(SeDip)	Switzerland
insA3727	insA3727				104(NaSp)	Germany
C2725T		C1933T (Foun. 7)			105(ShWo)	Texas

Col. 1	Col. 2	Col. 3	Individual	+	Descent	Geographic origin
C3415T		C2695T (Foun. 8)	109(BeGra)			New York
delExons11+12(7812bp)		C2695T (Foun. 8)	111(JaKir)			Massachusetts
IVS6+3(A→G)	A814T		112(NaSch)			Wisconsin
A814T	IVS6+3(A→G)[8]		114(BeNeu)			Germany
insA3223		C1933T (Foun.7)	123(FaCar)		Mestiza	Mexico
C2887T	C2887T		126(BrNa)[6]		AJ[6]	Brazil
IVS2+1(G→T)[6]		blm^{Ab} (Foun. 1)	134(ViBai)			Missouri
IVS16+3(A→T)		C2098T (Foun. 17)	139(ViKre)[5]			Illinois
T3163C[5]		A2015G (Foun. 13)	149(SeSaf)	+	Turkish	Belgium
G2643A	G2643A		166(BrDje)	+		Tunisia
insT3255	insT3255		186(MoIs)		Indian (Gujurat)	England
delA275	delA275		192(DuAk)	+	Turkish	Germany
T1628A	T1628A		197(BrPar)		Ecuadoran & El Salvadoran	Texas
G3197A		blm^{Ab} (Foun. 3)	198(SaBar)	+	Italian	Germany
IVS18+1(G→A)	IVS18+1(G→A)		202(JuCy)		Adopted (German)	Germany
IVS7+5(G→A)		C1642T (Foun. 12)	205(JoRob)			Wales
IVS9+2(T→G)	delTC771–772		211(NaMac)		West African	New York
delTC771–772	C3278G		215(AnDonas)			Brazil
C3278G			218(AnArm)			Portugal
IVS11+2(T→G)	G2855T		223(GiZa)			Italy
G2855T		blm^{Ab} (Foun. 1)	225(RaRose)		A-	Minnesota
T3510A	delT582		232(JePauc)	+	Ecuadoran and El Salvadoran	Minnesota

[1] Mutations that are referred to as unique are those that were detected in only one of the persons analysed in the survey.

[2] A blank in both Col. 2 and Col. 3 for a given individual means that only one mutation at *BLM* could be identified by the techniques employed in this survey – in 205(JoRob) and 218(AnArm). Three of the 42 individuals with BS represented in this table had inherited not one but two unique mutations – 65(AnPa), 98(RoMo), and 112(NaSch); each of their mutations is listed in Col. 1, but not necessarily also in Col. 3.

[3] The founders (Table 1) in which mutations that appear in this column were the founder mutation are indicated in parentheses.

[4] Footnotes 5 and 7 to Table 1 apply. In this table (Table 2), individuals are listed in Col. 4 in the numerical order with which they had been accessioned to the Registry.

[5] Footnote 4c to Table 1 applies.

[6] Footnotes 7c and 7d to Table 1 apply.

Table 3. Persons with Bloom's syndrome whose DNAs were successfully analysed but in whom Bloom's syndrome-causing mutations could not be identified.

Identification[1]	Parental consanguinity[2]	Noteworthy parental ancestry[3]	Places of birth and/or, principal residence
60(AnDav)	+	West African	California
63(KeKr)			Germany
70(DiYus)	+		Lebanon
122(RoHer)	+	West African	New Jersey
140(DrKas)	+	Persian Jewish	Israel
144(NiIa)	+		Italy
161(MaBeni)[2]			Argentina
206(SiKak)	+	West African	New York
219(AyAks)	+		Turkey

[1]Individuals with BS are identified as in the Bloom's Syndrome Registry.
[2]Parental consanguinity is believed to be a good possibility in a eighth of the nine individuals listed here: 61(MaBeni)'s purported parents and grandparents lived in an underdeveloped area of northeastern Argentina where family structure, particularly paternity assignment, regularly is undefinable. Furthermore, in this isolated population a rare recessive disease is unusually common, namely the Ellis-van Creveld syndrome (Castilla and Sod 1990).
[3] 'Noteworthy' in the sense that the ancestry differs from that expected from the individual's place of birth or principal dwelling place shown in Col. 4.

Environment and complex disease

Oliver Mayo

CSIRO Livestock Industries, PO Box 10041 Adelaide BC, South Australia 5000

Introduction

'Many agents which have been implicated as triggers of [the common diseases of adult life] are so deeply ingrained in our way of life that they are hard to avoid. These include unbalanced or excessive diets, a sedentary existence, and exposure to stress, as important contributory factors to such common diseases as diabetes mellitus, hypertension and atherosclerosis. At the moment, our understanding of the nature of the underlying genetical predisposition is insufficient to define individuals especially at risk but presence of the disease in a close relative should always be regarded as an indication of this type. Perhaps, later on, our understanding both of the exact nature of the environmental factors and of the underlying predisposition will become sufficiently refined to be able to give rather more specific advice of a useful kind in this situation.'

So wrote George Fraser in 1975. At that time, in the same book, (the now sadly late) Charlie Smith provided simple tools for estimating quantitative genetical parameters such as heritability and for predicting risk from incidence in relatives as against population incidence. Considering the question later on, after the triumph of the Human Genome Project, can we give specific, indeed useful advice?

Dissection of the interaction of genotype and environment

The traditional path to such dissection has been through quantitative genetics, and this has progressed to the stage where second-order statistics (variances, covariances, correlations, regressions, path coefficients) allow the patterns of association to be correctly estimated and clearly displayed. Table 1, taken from Whitfield *et al.* (2002), shows the relationship of γ-glutamyl transferase (GGT) activity and a range of risk factors for cardiovascular disease and other disorders. The level of GGT in the blood is itself a risk predictor for cardiovascular disease, stroke and diabetes. Its heritability was estimated by Whitfield *et al.* (2002) at 0.52. At the very least, these results show that, even as empirical risk prediction through the association of a risk factor with the incidence of a disorder improves, the nature of the causation of the disorder is not necessarily simplified.

The likelihood of the discovery of 'genes for' the disorder, that is, point mutations or small DNA changes in one or a few exons, is not raised by a strong association of a biochemical kind, even when the risk factor, like GGT, is a gene product. Indeed, there can be substantial genetically determined variation in a gene product that is the product of a defective, disease-causing gene, as was demonstrated many years ago for sickle haemoglobin, by Nance and Grove (1972). Hence, major 'genes for' complex diseases may lie rather in complex interactions, except perhaps for those discussed by Fearnhead, Whinney and Bodmer (p. 110). In the case of cardiovascular disease, a rare disorder, familial hypercholesterolaemia, can be caused by such a major gene, and its ascertainment leads to counselling issues of the kind raised by Clarke Fraser (p. 281 and see Newsome and Humphries 2005).

Table 1. Genetic and environmental correlations between log GGT and other variables (ALT alanine aminotransferase, AST aspartate amino transferase, BMI body mass index = body-weight (kg)/height^2 (m^2), apo apolipoprotein, LDL-C low density lipoprotein cholesterol, HDL high density lipoprotein, BP blood pressure) (from Whitfield *et al.* 2002)

Variable	Genetic correlation	Environmental correlation
g ALT	0.35	0.34
log AST	0.42	0.49
BMI	0.34	0.23
log triglycerides	0.45	0.29
apo B	0.37	0.17
apo AI	-0.04	-0.03
apo AII	0.22	0.19
apo E	0.35	0.19
LDL-C	0.25	0.16
HDL-C	-0.25	-0.15
log apo(a)	-0.08	0.00
glucose	0.20	0.11
log insulin	0.25	0.22
uric acid	0.22	0.28
diastolic BP	0.27	0.16
systolic BP	0.25	0.15
iron	0.05	-0.12
transferrin	0.10	0.07
log ferritin	0.24	0.24

Mapping quantitative trait loci

The report of Whitfield *et al.* (2002) constitutes a traditional quantitative genetical analysis of complex disorders together with risk factors that can be the quantitative level of the product of a single structural gene or a variable that is itself a complex trait, e.g. BMI. A subsequent step has been the combination of quantitative and Mendelian genetics, where alleles of a single gene are associated with different relative risks, e.g. apolipoproteins (see e.g. Miller 1987, Tiret *et al.* 1993). In such a case, it appears likely that the different alleles are associated functionally, that is causally, with the disease, even though they do not account for more than a modest proportion of the variance in the trait.

In another association study, Freeman *et al.* (2003) found that a polymorphism in the structural gene for cholesteryl ester transfer protein was associated strongly with risk of cardio-vascular disease in non-smokers but not in smokers. As with other genes showing associations with disease, not all polymorphic alleles were associated with increased or deceased risk. The association was not fully mediated through HDL or LDL, and it was present whether or not subjects were taking a cholesterol-lowering drug (pravastatin).

The results just described arose in the course of the genetical analysis of a trait complex with known risk factors (see e.g. Mayo 1994 for an analysis of the possible approaches). A quite different single-gene methodology is now available since the sequencing of the human genome has been completed. This is the genome scan using polymorphic DNA markers.

A genome scan begins with the assay of a set of DNA polymorphisms spaced out across the entire genome in a set of affected individuals (and possibly their families) and in an unrelated control set of unaffected individuals. The sets can then be compared by a range of standard statistical techniques to determine which alleles of which polymorphic genes/sites differ in their frequencies in the two groups. At the same time, or subsequently, family studies show whether or to what extent the elevated-frequency alleles segregate with the trait. Consequently chromosomal sites can be mapped which contribute to the risk of the trait. These are termed QTL (quantitative trait loci), though of course the trait, if it is a disease, is discrete insofar as the diagnosis of the disease is concerned.

A successful genome scan may be combined with other knowledge. For example, Zhu *et al.* (2004) conducted a genome scan for eye colour (coded as (blue/grey, intermediate, brown) in a large Australian twin sample of predominantly European ancestry, the use of MZ and DZ families giving the study good statistical power. They found a QTL on chromosome 15 that absorbed 74% of eye colour variance. The QTL was mapped close to the gene responsible for Type II oculocutaneous albinism. This gene has been suggested to be the eye colour gene that had previously been mapped to chromosome 15q (Eiberg and Mohr 1996). Thus, over 1200 persons were scanned for about 400 DNA markers, with further scans of subsets of the original 502 families, to confirm previous results. These families are under study for many other conditions, so the scan's demands are spread over many traits.

Genome scans for individual genes or regions can be followed by genome scans for interactions, though, as discussed by Mayo (2004), these have to date been undertaken largely in experimental organisms. Further developments along this path are occurring rapidly (e.g. Van Driessche *et al.* 2005), but have yet to be applied to humans.

Environmental factors and disease prevention

As noted earlier, and illustrated in Table 1, environmental influences are very important in the causation of complex diseases. Obesity, assessed through the metric BMI, is seen in many studies to influence the predisposition to many complex diseases. In Table 1, its interconnection with other risk factors is clear. Prevention of obesity-related diseases can be addressed by individual therapy (e.g. prescription of cholesterol-lowering drugs), by advice through the family or by population-wide measures, such as provision of nutritional advice.

Death rates for cardiovascular diseases have shown remarkable secular variation over the last century, as may be illustrated by Australian values (Australian Institute of Health and Welfare (AIHW) 2001). These diseases accounted in 2000 for 49,700 or 39% of all deaths, yet the rates have changed astonishingly. For men, the age-standardised rate per 100,000 rose from 376 to 843 between 1907 and 1968, after which it fell steadily to 256 in 2000, while for women the values for 1907, 1952 and 2000 were 328, 583 and 173. The rate for women, always lower than for men, peaked earlier but declined steadily over the years 1968–2000. Stroke peaked later for women than other cardiovascular diseases. Morbidity in both sexes for all of these diseases has increased over the past 30–40 years. It has been suggested that some of these secular changes are causally related to changing diet (less meat and animal fats, more olive and other plant oils, etc.), changing exercise patterns and the introduction and prescription on a wide scale of cholesterol-lowering drugs. If these suggestions are correct, then there has been a substantial success for preventive medicine through advice and example. However, we should at the same time note that the prevalence of smoking, the most important single environmental cause of cardiovascular diseases (e.g. Doll *et al.* 2004), declined at a lower rate than did the death rates, and is still over 20% in adults (AIHW 2005). Predisposition to smoke is strongly influenced by heredity and environment (heritability of smoking habit 30–60%, common environment absorbing about 15–40% of the phenotypic variance, e.g. Heath *et al.* 1993). Furthermore, overweight and obesity have increased in prevalence, that for obesity being 15–17% in 2001, and the increases appear to be as great in young children as in adults, or even greater (Booth *et al.* 2003, AIHW 2005). These risk factors are negatively associated with socio-economic status. It is clear that people at risk through genetical predisposition will only obtain effective advice if they are highly aware of current knowledge and interested in their own health. Screening followed by advice, even if feasible, appears unlikely to be fully effective, given the countervailing pressures in society.

Where screening appears likely to be helpful, is it being used? Consider the case of cyclo-oxygenase-2 (COX-2) inhibitors used for pain-relief in sufferers from rheumatoid arthritis. One, Vioxx®, has been withdrawn because of an increased incidence of myocardial infarction in persons taking it (Anonymous 2004c). This drug has also been claimed to have fewer gastrointestinal side-effects than other COX-2 inhibitors and earlier non-steroidal anti-inflammatory drugs. Here, surely, is a case for a genome scan, of tissue from those who died while using these drugs, those suffering illness while using these drugs, and those who suffered no side-effects but experienced appropriate pain-relief. Differential responses to drugs have been dealt with under the rubric pharmacogenetics, now relabelled pharmacogenomics, for half a century (e.g. Mayo and Brock 1978), and subgroups of people who can safely take a drug as against those who cannot should constitute an important market, if nothing of more substantial value (Shastry 2006). Should the different drug responses result from complex interactions among many genes, then such markets might not exist, but at least our understanding of our own lack of knowledge and power (in Francis Bacon's sense) would have been advanced.

Thirty years on from Fraser's enunciation of his reasonable hopes, Hunter (2005) has written:

> A common model for future preventative health care proposes that, initially, physicians will test their patients for hundreds or thousands of genetic variants, and that ultimately we will all have our entire genome sequence on a card or chip. Advice on disease prevention will be based on this information, implying that the relevant gene-environment interactions will have been proposed, replicated and validated, so that this advice is evidence-based. We face the prospect that affordable individual genome sequencing will be the easy part; developing a credible database on replicable gene-environment interactions will be the challenge.
>
> In addition, the concept of 'personalised prevention' might also seem to conflict with the view … that population-wide interventions are usually more effective in reducing the incidence of common diseases than interventions that target high-risk individuals.

See Epstein in this volume and Epstein (2004) for thoughtful development of the case mentioned in the second quoted paragraph. That Hunter could make his sensible observations as if they were new in 2005 suggests that Fraser's expectations have not yet been met, his lessons not yet learnt.

Genetic loads half a century on

N.E. Morton

Human Genetics Division, University of Southampton, United Kingdom

During most of the last century population genetics had much theory but little data. Deleterious mutations were known from research on Drosophila and the mouse to have a large recessive component, but clinical and epidemiological branches of human genetics were in their infancy. The urgency of mutation research was apparent in efforts to assess hazards of ionising radiation and other mutagenic agents. Even if a 'doubling dose' were estimated from another species, how could its human effect be anticipated without knowing our spontaneous mutation rate? Children of consanguineous marriages provide one avenue, which was exploited by a paper that estimated the deleterious mutation rate for quasi-recessive genes as 0.06–0.15 per gamete per generation (Morton *et al.* 1956). This corresponds to $6 – 15 \times 10^{-6}$ such deleterious mutations per contributory locus, assuming 10^4 of the latter. With all the advances during the last half century, we have no better estimate today.

However, its validity has been disputed on epidemiological and genetic grounds. The *epidemiological* argument is that estimates of the number of lethal equivalents are variable among studies, depending on sampling error, past inbreeding, definition and levels of mortality and morbidity, accuracy of diagnosis, and possible ascertainment bias or confounding with environmental variables that differ among populations and over time. Ideally non-consanguineous marriages of siblings would be used as controls, but today the joint requirements of good medical information and a conspicuous frequency of consanguineous marriage limit these studies to populations in which consanguinity is traditionally preferred, and therefore varies in frequency among social classes. Genetic load theory predicts survival or non-affection as $S = e^{-(A+BF)}$, where F is the inbreeding coefficient. A significant departure from this prediction, especially between F = 0 and F > 0, has been observed only in samples where no attempt has been made to detect or control covariance with environmental variables.

The *genetic* argument is directed against the assumption that B/A >> 1 is pathognomonic of quasi-recessive deleterious mutations. The main competitor is heterozygote advantage (overdominance), which can give a high value of B/A if all homozygotes at a highly polymorphic locus are rare and deleterious, all heterozygotes are favoured, and the deleterious effects are included in the mortality or morbidity being assessed. To my knowledge the only known example of such a system is death of male Hymenoptera homozygous at the sex-allele locus, which make a large contribution to B (Kerr 1975). Genetic load theory discriminates this locus from the significant mutation load (Morton 1975). Among the millions of human polymorphisms, the few known examples of heterozygote advantage are functionally diallelic and decrease B/A. Evidence from Drosophila has helped to undermine belief in ubiquitous overdominance (Crow 1987).

The most interesting response to consanguinity studies was to overlook the evidence they provided while defending other issues in population genetics. Bruce Wallace (1970) wrote a book of 116 pages on *genetic load* in which the last chapter of three pages was devoted to *uncovering genetic load by inbreeding*, giving an imaginative example of *soft selection* but no evidence. Haldane and Jayakar (1965) wrote 'polymorphism is found in all human groups; and no explanation, other than the increased fitness of heterozygotes, is available'. They probably would not make that argument today when confronted with more than 10 million polymorphisms, most of them present in all human groups. The wonder is that the number of polymorphisms is so much less than the 3000 million nucleotides in our genome. Sewall Wright (1960) proposed that 'as a first step in the problem of appraising genetic damage, it is desirable to attempt some classification of human phenotypes with respect to social value'. Such a balance sheet should never be attempted, but its appeal was strong enough to be included seventeen years later in volume 3 of his great opus on population genetics.

These and many other commentaries were remarkable in attributing *genetic loads* to the Morton *et al.* (1956) paper, in which that expression was never used. We preferred *mutational damage*, a more precisely defined term unconnected to the earlier paper on *our load of mutations* (Muller 1950), which was brilliant but tarred by the brush of eugenics. Theodosius Dobzhansky (1955) called overdominance 'heterosis' and coined the term *mutational load*, including both deleterious and overdominant mutations, a usage that has not survived. The term *mutation load* is preferable when there is an apparently successful intent to exclude overdominance. His evidence from Drosophila confounded chromosomes with loci and was the basis for his dictum that 'all genes are heterotic, but some genes are more heterotic than others'. Wright found this complexity congenial to his shifting balance theory of evolution and tended to side with Dobzhansky against simpler interpretations of the mutation load, giving little weight to evidence from inbreeding and depressed viability of heterozygotes for mutants detected as lethals in homozygotes (Stern *et al.* 1952, Falk 1961, Crow and Temin 1964).

The main extension of research on the mutation load *sensu strictu* has been to detrimental traits not measured as mortality (Morton 1960, 1981). The B/A ratio tends to be greater for some types of abnormality than for mortality, arguing for a larger recessive component in the former. Extending the analysis to retrospective studies of rarer disease with more precise clinical definition, there was good agreement between inbreeding and segregation estimates of the proportion of sporadic cases. Omitting sporadics, some inferences can be made about the number of contributing loci and the mutation rate per locus. I shall return to implications of the fact that subsequent research has shown that the minimal estimates of numbers of loci for quasi-recessive detrimentals were too low and the mutation rate per locus correspondingly too high, but genes causing post-reproductive disability and quasi-additive detrimentals neglected in classical theory are far from negligible.

In the last phase of genetic load theory, James Crow (1970) sought to reconcile different schools of evolutionary genetics in a paper dedicated to Dobzhansky. He defined *genetic load* in terms of deviation from an optimal genotype, which is rarely feasible. Separating mutation from heterosis, he considered eight types of load. One of these was recombination between haplotypes that conform to the dictum of E. B. Ford (1940) that 'genetic polymorphism is the occurrence together in the same locality of two or more discontinuous forms of a species in such proportions that the rarest of them cannot be maintained by recurrent mutation', a fiat to which many evolutionists subscribed in the days when few polymorphisms were known (Lewontin 1964). Like most of the other loads, balanced haplotypes have sometimes been postulated, rarely proven, and never assessed at a genomic level. The theory for the genetic load due to mother-child incompatibility was developed and applied to data in the ABO blood groups (Crow and Morton 1960), but the large effects estimated from the data available then have not been confirmed.

Genetic load theory led to some unanticipated results. One was recognition that population genetics is an amalgam of two disciplines that diverge most conspicuously in their approach to genetic loads. *Evolutionary genetics* was lineage-oriented, neutrality-biased and highly theoretical, and practised differential-equation biology. *Genetic epidemiology* was family-oriented, disease-biased and more empirical, and practised statistical biology (Morton *et al.* 1967). Today the two disciplines remain distinct but are being redefined under the unifying effect of analytic genomics.

A less predictable effect of genetic load theory was its influence on Motoo Kimura and through him on evolutionary genetics. Haldane (1937) showed that the effect of recurrent deleterious mutation on fitness expressed through reproduction is measured by the mutation rate, independently of gene frequency and selection coefficient, a result anticipated by Danforth (1923) but not widely known until cited by Muller (1948) and subsequently. Haldane (1957) returned to this in a paper on the *cost of natural selection* that assumed a constant population size and tried 'to estimate the effect of natural selection in depressing the fitness of a species'. His model was industrial melanism, which was favoured by smoke pollution over the lighter type that had been selected against a background of pale lichens. As the lichens were killed by pollution, birds decimated light moths and population size plummeted. Obviously the melanic successor raised species fitness, and its increase in the new environment was a boon rather than a burden. The same logic applies to other selective sweeps, and so few scientists found merit in Haldane's model. One of them was Kimura (1960), who used load as a synonym for cost. His 'principle of minimum genetic load' supposes that evolution tends to minimise the sum of the mutation load and Haldane's cost. He did not explain why evolution should act to minimise the sum of a differential and an integral, but from his principle he was able to deduce a relationship among the total mutation rate, the total mutational load, the degree of dominance, the rate of allelic substitution, and the selection coefficient against homozygotes. Using other estimates of the first three parameters (Morton *et al.* 1956), he found remarkably good agreement. However, his estimate of the mean selection coefficient against homozygotes constituting the inbred load was less than one percent, leading to a high ratio of detrimental to lethal inbred loads, in contradiction to the data on quasi-recessive mutation. In the face of criticism (Sved 1968, Fraser and Mayo 1974), Kimura soon abandoned the theory of minimal genetic load, but it stimulated his development of neutral mutation theory as a credible explanation of much of the unexpectedly high frequency of genetic variation that was beginning to be revealed at the molecular level. Meanwhile productive research in genetic loads was diminishing with the rate of consanguineous marriage, despite improved medical services in populations where such marriages are preferred (Bundy and Alam 1993). Mutation research was being directed to effects of mutagens, and molecular genetics was producing data more testable than the speculative issues summarised above.

For many of us, the attraction of consanguinity studies has been their relevance to genetic risks and population structure, and the evidence that inbreeding is 'a knife that will cut out the mutation load and will allow examination of it' (Morton 1963). This has been relatively successful for major detrimental genes that are much more severe in homozygotes than in heterozygotes, although some broad classes of disability have a large and uncertain environmental component that inflates the panmictic load A and precludes estimation of the number n of contributory loci. However, the contributions of completely penetrant quasi-recessive genes can be isolated to give $A = \Sigma q^2$, $A + B = \Sigma q$ where Σq is the sum of gene frequencies over n loci (Morton, 1960). Then the mutation rate per gamete is $U = s[A + \alpha(A + B)]$, where α is the inbreeding coefficient under which the population reached equilibrium and s is the selection frequency (assumed constant) under which the population reached equilibrium. Estimates of U pose no problem, and the mean mutation rate per contributory locus is u = U/n, where n is the number of such loci (assuming that their effects are independent, which is not restrictive because all the genes in question are rare). Likewise the mean deleterious allele frequency is Q = (A + B)/n. But what is n? We are not concerned with the unknown values of n in an isolate or in our species, but in some poorly specified population. Because q is variable, it was shown that $n \geq (A + B)^2/A$ and $Q \leq A/(A + B)$. Using current estimates and numbers of recessive loci, n increases while Q and u decrease. As more causal loci are identified, this trend can exaggerate n (and thereby underestimate Q and u) by inclusion of loci monomorphic even in large populations, without data accurate enough to correct for this. It is the rule rather than the exception for deleterious mutations within one locus to affect multiple phenotypes, more easily recognisable through medical genetics in humans than in experimental organisms. These complexities raise unsolved problems in estimation of u.

Moreover, mutation parameters for genes of small effect and more nearly additive effects cannot be estimated from consanguinity studies. Sperm may provide in vitro estimates of mutation for particular nucleotides, but with uncertain reliability. Extension to a sufficiently large sample of ova would be more difficult. If these meiotic studies become feasible and reliable, they have the additional problem that all n' causal sites (nucleotides or longer sequences) contribute to the deleterious mutation rate per locus, which is n'u', where u' is the mean deleterious mutation rate per causal site. Calculation of u' is daunting for several reasons. First, most sites that would produce deleterious mutations in an infinite population are highly unlikely to have such mutations in a small population. Secondly, a mutation may be deleterious in one population at one time, but not universally. Thirdly, 'quasi-neutral' deleterious genes may by drift or close linkage to an advantageous mutant become polymorphic, and therefore not amenable to consanguinity analysis. Even highly deleterious recessives pose a serious problem because some compound heterozygotes may have the phenotype of an affected homozygote, but not all such alleles may give affected homozygotes, a situation that simulates a negative inbreeding coefficient that is difficult to identify unless normal heterozygotes are typed.

To illustrate how our understanding of genetic architecture has changed from the 'one gene, one enzyme' dictum of George Beadle, consider two of George Fraser's favourite genes. Bardet-Biedl syndrome has such a distinct phenotype with four primary features (some shared by heterozygotes) that in the pre-molecular era it was assumed to be monogenic. Although some cases are caused by one nucleotide, there are at least eight loci with evidence that expression sometimes requires three particular alleles at two of these loci (one homozygote and one heterozygote). In contrast, Pendred syndrome with no greater clinical distinctness is caused by a single gene, some alleles of which also produce non-syndromal deafness. In both cases several deleterious alleles have been observed. Cystic fibrosis is perhaps the most extensively studied single-gene disease. A deletion allele is the most common cause, but a great many other alleles act in the same way in compound heterozygotes. Pure homozygotes have been observed in some consanguineous families, but the list of deleterious alleles and their interactions is far from complete.

Diseases of post-reproductive onset (and therefore more deleterious in a clinical than Darwinian sense) are at the opposite extreme, with little counter-selection, unknown allele frequencies, and a significant recessive component that is seldom measured (Schull and Neel 1965, Conterio and Barrai 1966, Morton 1991).

Analytic genomics has revealed a further complication. Many of the mutations of current interest do not conform to the quasi-recessive model of classical load theory, which assumed a low penetrance in heterozygotes. Instead, the 'common disease/common variants' hypothesis urges us not to neglect genes of small effect in homozygotes (Reich and Lander 2001). On the contrary, Drosophila experiments show that such genes typically approach additivity as their homozygous effects decrease (Morton *et al.* 1968), a result anticipated by Maclaurin's theorem. Despite their significance for disease, such genes currently defy the global consanguinity analysis that has been relatively successful for rare genes that are quasi-recessive. Using estimates of non-synonymous mutations in 46 orthologous gene pairs from human and chimpanzee (and assuming 60,000 genes in the human genome, an average length of protein coding sequence of 1.52 kb, divergence time of 6 My, average human generation time of 25 years, and deleterious mutations represented exclusively by all non-synonymous substitutions), the rate was estimated to be 0.8 ± 0.4 deleterious mutations per haploid genome per generation (Eyre-Walker and Keightley 1999). Assuming only 20,000 loci, this becomes 0.27 ± 0.13 deleterious mutations/generation, which is almost twice the upper limit from consanguinity studies, suggesting that rare recessives make up about half of the mutation load, the remainder perhaps corresponding to the 'common disease/common variants' hypothesis and consistent with mutation accumulation in Drosophila studies (Crow 1997). The uncertainty of this inference is obvious, suggesting that the last fifty years have added significantly but quite modestly to information about the human mutation load. New ways to augment this information must be sought. (See the next contribution.)

Patients not of European origin can lead to previously unrecognised causal loci and alleles. Consanguineous families are valuable to enrich rare homozygotes for every locus with deleterious recessive alleles. From the Levant to the Bay of Bengal lies a swathe of populations with preferential consanguineous marriage. The concurrence of non-European origin with high consanguinity makes these populations of exceptional interest for studies that combine clinical and molecular research with genetic epidemiology, given the necessary motivation, training, support, and peace. This would be an apotheosis both of mutation loads (Fraser 1962a) and of the population studies that George Fraser initiated, now fused with analytic genomics into the human genetics of the future.

Genetic risks of consanguineous marriages

P.A. Otto, O. Frota-Pessoa and J. Vieira-Filho

Department of Genetics and Evolutionary Biology (former Department of Biology), Instituto de Biociências, Universidade de São Paulo, PO Box 11461, 05422–970 São Paulo SP, Brazil

This paper is dedicated to George Fraser on the occasion of his 75th birthday.

Introduction

The estimation of genetic risks to the offspring of consanguineous marriages, especially those of first cousins, has been exhaustively studied in the literature. Even so the subject still remains controversial, as shown by the results of a recent inquiry performed among medical geneticists and genetic counsellors in the United States by Bennett *et al.* (1999). In fact, analysing the answers they received from 309 professionals, these authors showed that the risks of genetic disease given by these professionals to first-cousin couples ranged from 0.5% to 20% and that there was no consensus as to the types of defects and diseases they included in the calculation of these risks.

Empirical risk estimates

Most estimates of consanguinity risks stem from surveys comparing the frequencies of defects in samples of children born to first cousins and in children born to matched controls (unrelated couples). Four important, well designed surveys were performed in the fifties and sixties in France (Sutter and Tabah 1952), Sweden (Böök 1956), the United States (Slatis *et al.* 1958), and Japan (Schull and Neel 1965). The overall frequencies of diseases and defects, including mental retardation, varied in these four surveys from 3.5% to 9.8% (average value of about 6.5%) among controls and from 11.7% to 16.2% (average value of about 14%) among children born to first-cousin couples. The defects and diseases had therefore frequencies of 3.2% to 12.0% (average value of about 7.5%) higher among the offspring of first cousin couples than in the offspring of unrelated people. As Stern (1973) points out, the values obtained in the four surveys commented above (as well as in others in the literature) differ from one another because 'the results are obtained by different investigators, at different times and places and with different methods.' The criteria used for inclusion of defects and the procedures to recognise them obviously also vary from sample to sample. For instance, tuberculosis was included in three out of the four studies mentioned above and the research conducted with Japanese children considered only major congenital malformations. In any event, the frequency of affected individuals showed to be in every one of these and other studies consistently higher among the inbred offspring than among children born to unrelated parents. Some researchers also argue that the results of such studies are biased, because the offspring of related persons tend to be more carefully studied than the children of non-consanguineous pairs.

Methods for estimating the risk to the offspring of consanguineous couples

In the lines which follow we describe some simple operational methods which we developed for dealing with the question of risk estimation for the offspring of consanguineous couples. Their general framework has already been delineated by Frota-Pessoa *et al.* (1975), Otto and Frota-Pessoa (1979), and Otto (1989); some of the material presented here was originally expanded in the doctoral thesis of Vieira-Filho (1992), under the guidance of the senior author.

A first simplified method

The simplest method assumes that the average frequency of genes determining autosomal recessive diseases is somewhere between $q = 0.001$ and $q = 0.01$. Using the average figure of about $q = 0.005$, this should imply that on average there exists an increase of the relative risk $RR = (q^2 + pqF)/q^2 = 1 + pF/q$, with $p = 1 - q$ and $F = 1/16$, of about 13 times in the offspring of first-cousins, when compared to that of unrelated couples. Since the overall frequency of recessive diseases among newborn children is roughly of the order of 0.5% in the general population, this implies that the risk of any autosomal recessive disease in the offspring of first-cousin couples will be about 6.5%. On this argument, the frequency of children affected by recessive diseases and defects would be about 6% higher among the offspring of first cousins than among children born to unrelated parents.

Method based on the distribution of frequencies of autosomal recessive genes determining diseases

Let us suppose that the total of autosomal recessive diseases is $N = 10$, and that these diseases are determined by genes with the following frequencies: 0.005 (three diseases), 0.010 (five diseases) and 0.015 (two diseases). Obviously, the risk R_1 of autosomal recessive disease in the offspring of first cousins ($F = 1/16$) would be

q_i	$f(q_i)$	$Nf(q_i)$	$q_i^2 + q_i(1 - q_i)/16$	$Nf(q_i)[q_i^2 + q_i(1 - q_i)/16]$
0.005	0.3	3	0.000336	0.001008
0.010	0.5	5	0.000719	0.003594
0.015	0.2	2	0.001148	0.002297
-	1.0	10	-	$0.006899 = R_1$

whereas in the offspring of unrelated couples ($F = 0$) the corresponding risk R_2 of recessive disease would be given by

q_i	$f(q_i)$	$Nf(q_i)$	q_i^2	$Nf(q_i)q_i^2$
0.005	0.3	3	0.000025	0.000075
0.010	0.5	5	0.000100	0.000500
0.015	0.2	2	0.000225	0.000450
-	1.0	10	-	$0.001025 = R_2$

As a direct result of the application of the method to the data above we get the estimates $R_1 = 0.007$ and $R_2 = 0.001$ respectively for the risks of autosomal recessive disease in the offspring of first cousins and in that of unrelated couples.

Formally, the method can be described as follows: let q_i be the average (midclass) point of gene frequency and $f(q_i)$ the density function of this class of gene frequency, such that $\Sigma f(q_i) = 1$. In the present case, $f(q_i)$ is to be inferred from reliable data from the literature and then adjusted to families of functions (Bronstein *et al.* 2002) by using interactive computer programs of non-linear regression analysis such as NLREG (Sherrod 1997). The overall frequency of affected children with inbreeding coefficient F is approximately

$$R(F) = N\{\Sigma f(q_i)[q_i^2 + F\, q_i\, (1 - q_i)]\}$$
$$= N\{\Sigma f(q_i)[q_i\, F + q_i^2\, (1\text{-}F)]\} ,$$

which simplifies to

$$R(0) = N[\Sigma f(q_i)\, q_i^2]$$

in the corresponding case of the offspring of unrelated couples.

Consulting books and treatises on human and medical genetics, genetic data banks and specialised articles in the literature of medical and clinical genetics, we found non-biased gene frequency estimates for 125 different autosomal recessive conditions relatively well studied in Caucasoid populations, all exhibiting narrow geographic frequency variation, including mental retardation (number of loci rounded to 100) and non-syndromic deafness (estimated in 35 loci variants), adding up to a total of 258 loci. These data enabled us to estimate roughly the distribution of frequencies of autosomal recessive genes causing disease, which is shown in the graph below.

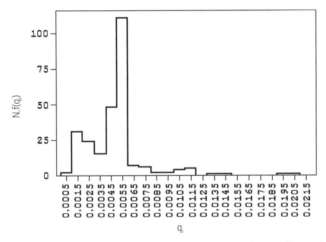

Figure 1. Distribution of gene frequency estimates based on data of 125 different deleterious autosomal recessive diseases and defects.

Using arbitrarily the gene frequency interval midpoints of 1/1000 for 123 conditions (excluding mental retardation and non-syndromic deafness) we adjusted the absolute observed frequencies of each class of gene frequency to the generalised function $y = ax^b e^{cx}$ (Bronstein *et al.* 1973). The function $y = ax^b e^{cx}$, where $x = 10^4 q$, that best fits the observed data is that with a = 0.708, b = 1.775, and c = −0.085 ($F_{(2,18)} = 74.08$; P = 0.00001).

The graph below shows, together with the absolute observed numbers of classes of gene frequencies, the above mentioned function.

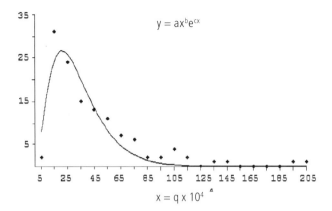

Figure 2. Function $y = ax^b e^{cx}$ ($x = 10^4 q$) that best fits gene frequency mid-points observed for 123 different autosomal recessive conditions; these were obtained from the histogram shown in Figure 1 excluding mental retardation and non-syndromic deafness cases.

Dividing a by 123 and each f(qi) by $\Sigma f(qi)$ we obtain the density function $y' = a'x^b e^{cx}$, where a' takes now the value a' = 0.0065 and both coefficients b and c do not change in relation to the values obtained for the first distribution. The average gene frequency obtained by using the rough data from 123 conditions is q = 0.0044; when the function $y' = a'x^b e^{cx}$ is used instead, the average mean frequency turns out to be

$$q = \Sigma f(q_i)q_i = 0.0032.$$

The two groups of heterogeneous conditions (isolated mental retardation and non-syndromic deafness) were not included in the above calculations because their respective gene frequency distributions are unknown. However, both conditions have a mean gene frequency (0.0050 and 0.0450 estimated respectively for 100 different mental retardation loci and for 35 different non-syndromic deafness loci) of the same order of magnitude as the one obtained from the empirical distribution excluding them. Assuming that the distribution of the frequencies of the genes responsible for both conditions can be represented by the same function above, all we have to do to obtain the risk estimates for an offspring with a fixed value of F is to multiply $\{\Sigma f(q_i) [q_i^2 + Fq_i(1-q_i)]\}$ by N = 258 instead of 123.

Using either the empirical observed distribution (corrected for N = 258) with individual gene frequencies grouped and the distribution given by the function $y' = a'x^b e^{cx}$, we estimated the overall risk of autosomal genetic disease, including mental retardation and non-syndromic deafness, in the offspring of non-consanguineous couples (F = 0), of first cousins (F = $^1/_{16}$) as well of couples with inbreeding coefficients taking values of $^1/_4$, $^1/_8$, $^1/_{32}$ and (respectively for the offspring of brother-sister and uncle-niece or aunt-nephew unions, and marriages between first cousins once removed).

Table 1. Estimated risks of autosomal recessive genetic disease in the offspring of consanguineous marriages (F = $^1/_4$, $^1/_8$, $^1/_{16}$, $^1/_{32}$) and of unrelated couples (F = 0), based on the gene frequency distribution of deleterious autosomal recessive genes.

F	$^1/_4$	$^1/_8$	$^1/_{16}$	$^1/_{32}$	0
risk[1]	0.2885	0.1483	**0.0782**	0.0431	**0.0081**
risk[2]	0.2128	0.1083	**0.0560**	0.0299	**0.0037**

[1]Estimates using the absolute observed class frequencies adjusted to a total of N = 258 autosomal recessive loci.
[2]Estimates using the function $y' = a'x^b e^{cx}$.

We obtained the risks of at least R(F) = 0.056 and R(0) = 0.004 respectively for the offspring of first cousins and unrelated couples. These figures refer only to diseases or defects produced by autosomal recessive mechanism. Since about 0.04 = 4% of all newborns present or will present a physical defect or mental retardation, we should add to R(F) the risk corresponding to other types of disease, which is of the order of 0.04 – 0.004 = 0.036. Thus we get to the global risk of any disease in the offspring of first cousins: R'(F) = 0.056 + 0.036 = 0.092 or about 9%, a risk 5% higher than the one for the offspring of unrelated people. If we use the less conservative estimates 0.078 and 0.008, the global risk for the offspring of first cousins increases to 0.078 + 0.032 = 0.11 or 11%. On average, considering all types of disease, there is an overall increase of the order of 6% in relation to the risk for non-inbred children.

The OMIM data bank of genetic diseases organised by Dr. McKusick at Johns Hopkins University Medical School lists more than 3,000 autosomal recessive conditions, over half of them still unconfirmed. Were we to reduce drastically to 500 the number of conditions with this type of mechanism not included in the list we used for our calculations, and also assume that all of them occur with extremely low frequencies, say of the order of 1/500,000 to 1/1,000,000 in the general population, we expect a total of more 500 × 1/500,000 = 0.001 (about 0.1%) to 500 × 1/1,000,000 = 0.0005 (about 0.05%) children affected in the offspring of non-consanguineous couples (therefore an increase that doesn't change meaningfully the previous estimate of about 4%) and a total of about 500 × 45.1/500,000 = 0.045 (about 4.5%) to 500 × 63.4/1,000,000 = 0.03 (about 3%) in the offspring of first cousins. If we take into account this in our previous calculations, the risk of recessive affection in the offspring of first cousins increases to about 9% to 10% (and the global risk for any disease increases to about 13% to 14%). In any case, as shown by the calculations presented before, this risk seems not to be inferior to 9%, thus indicating that the reproduction of couples of first cousins and other near relatives should not be encouraged.

Indirect method based on the average number of pathological recessive genes carried per person

Variations of this method are found in some textbooks (e.g. Murphy and Chase 1975, Stevenson and Davison 1976) but only Penrose (1956) and Stern (1973) treated the problem with an appropriate probabilistic model.

The method is in principle very simple. To obtain the risk of recessive disease in the offspring of consanguineous couples it uses the probability of any individual being a heterozygote as to a deleterious recessive gene having a population frequency q. This probability has value $P(Aa) = 2pq = 2q(1-q) \approx 2q$. If we let $m = 2Npq \approx 2Nq$ be the average number of deleterious recessive genes in heterozygous state per individual, the probability of a person having x of these genes at different loci can be expressed by formula (density function of the Poisson distribution) $f_1(x) = m^x e^{-m}/x!$, since N (the number of different pathological recessive genes) is large, q is small and $m = 2Nq$ is a finite quantity. The numerical value of m can be estimated directly from the distribution of gene frequencies just derived in the previous section. The chance of a pathological recessive gene present in heterozygous state in an individual of being present also in his or her consanguineous mate is r (relationship coefficient, which equals two times the value of the inbreeding coefficient F of the couple's offspring if no ancestors of the couple are inbred). In the case both individuals share the same pathological gene, the probability that they transmit it to their offspring is $1/4$. The joint probability taking into account both events is $r/4 = F/2$. If we assume that each individual has on average one of these genes in heterozygous state, a child born to a couple having a coefficient of relationship r has a risk $r/2 = F$ of being affected by a recessive disease, since if the first spouse is a carrier, on average, of one pathological recessive gene, it is obvious that his or her mate is also, on average, a carrier of one pathological recessive gene at another locus. If a given individual (1), married to a relative, has x_1 of these genes, the probability of transmission, by both spouses, of at least one out of these x_1 genes is

$$f_2(x_1) = 1 - \left[1 - \frac{r}{4}\right]^{x_1} = 1 - \left[1 - \frac{F}{2}\right]^{x_1}$$

If his or her consanguineous mate (2) has x_2 of these genes, the probability of transmission, by both spouses, of at least one out of these x_2 genes is

$$f_2(x_2) = 1 - \left[1 - \frac{r}{4}\right]^{x_2} = 1 - \left[1 - \frac{F}{2}\right]^{x_2}$$

The probability of homozygosis of the child born to this couple as to at least one out of the x_1 genes present in the first parent or at least one out of the x_2 genes present in the second parent is clearly given by

$$f_2(x_1,x_2) = 1 - [1 - f_2(x_1)][1 - f_2(x_2)]$$
$$= 1 - \left[1 - \frac{r}{4}\right]^{x_1}\left[1 - \frac{r}{4}\right]^{x_2}$$
$$= 1 - \left[1 - \frac{r}{4}\right]^{x_1 + x_2}$$

If $f_1(x)$ or $f_2(x)$ is the Poisson probability of an individual having x = 0, 1, 2, 3 … of these pathological genes in heterozygous state, evidently the risk of recessive disease for the offspring of a consanguineous couple with a coefficient of relationship r is the double summation (from $x_1 = 0$ to ∞ and $x_2 = 0$ to ∞)

$$f_3(x) = \Sigma\Sigma[f_1(x_1).f_1(x_2).f_2(x_1,x_2)]$$

$$= \Sigma\Sigma(m^{x_1}e^{-m}/x_1!)(m^{x_2}e^{-m}/x_2!)\left[\,1 - \left(1 - \frac{r}{4}\right)^{x_1 + x_2}\,\right]$$

$$= 1 - e^{-2m}\Sigma[(m - mr)^{x_1}/x_1!]\,\Sigma[(m - mr)^{x_2}/x_2!]$$

$$= 1 - e^{-2m}e^{(2m - mr/2)} = 1 - e^{-mr/2} = 1 - e^{-mF}$$

$$= mF - m^2F^2/2 + m^3F^3/6 - \ldots$$

Since our estimate of the average gene frequency q = 0.0032 was based on data from 258 loci, m takes the value 2Nq = 1.65. The table below shows the values of the risk R = $f_3(x)$ = 1 − e^{-mF} for selected values of m and for the values of F = $^1/_4$, $^1/_8$, $^1/_{16}$, and $^1/_{32}$.

Table 2. Estimated risks of autosomal recessive genetic disease in the offspring of consanguineous marriages (F = $^1/_4$, $^1/_8$, $^1/_{16}$, $^1/_{32}$) and of unrelated couples (F = 0), based on the average number m of deleterious autosomal recessive genes in heterozygous state per individual. R = 1 − e^{-mF}

m	F = $^1/_4$	F = $^1/_8$	F = $^1/_{16}$	F = $^1/_{32}$
0.0	0.0000	0.0000	0.0000	0.0000
0.5	0.1175	0.0606	0.0308	0.0155
1.0	0.2212	0.1175	0.0606	0.0308
1.5	0.3127	0.1710	0.0895	0.0458
1.65	**0.3380**	**0.1864**	**0.0980**	**0.0503**
2.0	0.3935	0.2212	0.1175	0.0606
2.5	0.4647	0.2684	0.1447	0.0752
3.0	0.5276	0.3127	0.1710	0.0895

The risks shown in this table are larger than the ones we estimated using the frequency distribution of pathological autosomal recessive genes. As we pointed out, many diseases were not included in our list of 258 loci. In the method we just used, however, whatever the frequencies of the non-included genes might be, the final quantity m = 2Nq should not be less than the figure already obtained, m = 2Nq = 1.65. If we add (as we did at the end of the section dealing with the distribution of gene frequencies) the expected effect of 500 more diseases, all produced by genes with very low frequencies, we get a risk figure of about 9% to 10% of recessive diseases in the offspring of first cousins. Then both estimates agree fairly well. Other less conservative estimates for the risks discussed above might be even higher. Anyway, the increase in risk due to consanguinity (marriages between first cousins or nearer relatives) seems not to be as negligible as some papers suggest.

A beautiful method of analysis

Oliver Penrose

Department of Mathematics, Heriot-Watt University, Riccarton, Edinburgh EH14 4AS, Scotland, UK

Introduction

It is a pleasure to dedicate this article to my friend George Fraser. I have a curious symmetrical relationship with George: neither of us knew the other until recently, but each of us knew the other's father. I knew George's father through playing chess with him in various tournaments and club matches. He was a strong player and won the British Chess Championship in 1957; the only way I ever found of beating him was to tempt him occasionally with an opportunity for an over-optimistic piece sacrifice. George and my own father, the human geneticist L. S. Penrose, knew each other very well, since my father was George's thesis supervisor (and, as a keen chess player himself, he also knew George's father). Nevertheless it was only about five years ago that George Fraser and I first met, when he spoke at a conference held at University College London in 1998 to commemorate the 100th anniversary of my father's birth.

My own introduction to human genetics came at a very early age. When I was about four years old I would sometimes venture into my father's study, to find him doing what I described at the time as 'red and blue busywork'. I like to think that the work he was doing with his red and blue pencils was connected with the data collection and analysis for his 1933 paper *The relative effects of paternal and maternal age in mongolism* (Penrose 1933) which he was working on around that time. This paper was a milestone in the use of quantitative methods in human genetics. Until that time, according to Harris (1973) nothing was understood about the causation of the condition now known as Down syndrome (DS). It was known that the affected children were more often born to elderly parents and often came late in the order of birth, but the relative importance of father's age, mother's age, and birth rank was a mystery. (At that time the possibility that the explanation was a chromosomal abnormality, suggested in a remarkably prescient passage by Waardenburg (1932)[1] was not taken seriously (Penrose 1966). My father set out to disentangle these effects, and the 1933 paper showed convincingly that the mother's age, but not the father's, was an important causative factor.

When I got a little older, the object of greatest interest in my father's study was a handle-powered desk calculating machine called the Brunsviga. This was the latest thing in computing technology at the time and it made a very satisfying crunching noise when you turned the handle to do a big multiplication or division sum. A lot of the number crunching he did around that time must surely have been the analysis of the data for the 1933 paper. The collection of these data was a tour de force of dedicated field work by my father and his assistants Miss M. Newlyn and Dr M. Gunther, working from the Royal Eastern Counties Institution in Colchester. They took family histories of 150 families each of which included at least one child with DS, a total of 727 children, recording, among other things, the age of each parent at the time of the child's birth. But it was my father on his own who invented the method of analysis that induced this confusing jumble of data to give up its secret.

Despite the interest generated by these colourful early experiences, it was not until 65 years later that I took the trouble to find out exactly what my father's 'red and blue busywork' might have consisted of, when I looked up the 1933 paper as part of my preparation for a lecture (Penrose 1998) about my father and his family, given at Wisbech in 1998. What I found when I did look at the paper was a seemingly simple statistical argument, but the more I thought about it the more fascinated I became by the subtle understanding that underlay the apparent simplicity. The object of this note is to reveal some of these subtleties. I hope that the historical importance of that paper, and the interesting statistical method used in it, make it worth while to give it this reconsideration, even after a lapse of over 70 years.

The 1933 paper begins with a preliminary investigation of the relative importance of the father's and the mother's ages, based on the method of partial correlations. The partial correlation of the father's age with the occurrence of DS in the child, after elimination of the effect of the mother's age, turned out to be very small, only –0.01, whereas the partial correlation of the mother's age with DS in the child, after elimination of the effect of father's age, was much larger, 0.22, with an estimated statistical error of order 0.04 in both numbers. Thus the partial correlation method provided strong *prima facie* evidence that the mother's age was an important factor and that the father's age was not. However there was a difficulty with this method, namely that the standard tests for deciding whether or not the observed correlation is statistically significant were not applicable. These tests depend on the assumption that the random variables in the problem obey a multivariate Gaussian probability distribution. But in the present case this assumption cannot be used. Although the two age variables (father's age and mother's age) are capable of a continuous range of values and therefore might without too much violence to the facts be assumed to obey a Gaussian distribution, the third variable is nothing like Gaussian because it can take only two values, depending on whether the child does or does not have DS.

My father came up with a beautiful method of analysis which enabled him to obtain reliable deductions from the data he and his assistants had collected. The method avoided the manifestly false assumption that all the random variables were Gaussian, while at the same time making it unnecessary to work out *ab initio* the corresponding theory of statistical tests for partial correlations when one of the variables can take only two values. Like all the best ideas, the method looks very simple – once you have been shown how to do it.

The idea is to test two competing hypotheses against the data. One of them, which I shall call **M**, is the hypothesis that, of the ages of the two parents at the time of birth of the baby, only the mother's age is relevant to whether the baby will have DS. This hypothesis is compared with a 'control', namely a hypothesis **F** that only the father's age is relevant. Because of the symmetry between the two hypotheses, they are easily compared. The test used is a prediction of the age of the parent whose age does not matter, based on the age of the parent whose age does matter and on whether or not the child has DS. The results of the test are shown in Table 1 (a simplified version of Table II of Penrose 1933). The top half of the table summarises the test of hypothesis **M**. On average, the prediction of the age of the 'irrelevant' parent (in this case the father) based on the age of the 'relevant' parent is very good: although the prediction in any individual case would of course be very inaccurate, the average of the predicted ages of the 'irrelevant' parents is in error by only a few weeks. For the competing hypothesis **F**, on the other hand, the predictions are much worse, the error in the predicted average being more than ten times as large. Thus the evidence strongly favours hypothesis **M** over hypothesis **F**, and the conclusion drawn in the paper was that 'paternal age is not a significant factor, while maternal age is to be regarded as very important.'

Table 1. Summary of the method and the results.

Test of hypothesis **M**, that only the mother's age matters

Average age of fathers	Observed	Predicted	Error
of affected babies	39.38	39.47	0.09
of unaffected babies	33.83	33.8	−0.03

Test of hypothesis **F**, that only the father's age matters

Average age of mothers	Observed	Predicted	Error
of affected babies	37.25	35.71	−1.54
of unaffected babies	31.25	31.68	0.43

Two things about this analysis are particularly intriguing. One is that it depends entirely on the one set of data. There was no separate survey of the population as a whole. The necessary information about the general population was gleaned from the group of parents studied – even though that group is far from typical, consisting entirely of parents of DS children. The other intriguing aspect is the relation between the two hypotheses **F** and **M**. Most statistical tests use just one hypothesis, but this one uses two. In the basic theory they are treated completely symmetrically. However, the data reveal an asymmetry between them, and so in the end the same data serve two distinct purposes at the same time: analysed according to one hypothesis, they provide information in support of that hypothesis, analysed according to the other, they provide a foil against which the performance of the first hypothesis can be evaluated. In the rest of this paper the method and the results will be examined more closely, to see better how it has all been achieved. Readers who are allergic to mathematics may skip to the beginning of the fourth section.

The probability model and the two competing hypotheses

The underlying probability model of the 1933 paper can be set out in the following way. A child is described in the model by just three variables: the mother's age m (measured in years) at the time of birth, the father's age f at the time of birth, and a non-numerical variable c describing the clinical situation. The variable c is capable of just two values: A if the baby is Affected with DS and U if the baby is Unaffected. (Thus, $c = U$ means 'the child is unaffected'). We can call the triple (f, m, c) the 'state' of the child. Other characteristics of the child, such as its sex, its date of birth, and the number of brothers and sisters, are ignored.

We consider some relevant large population, which might be (but in fact is not) all the infants born in Great Britain during a particular year. For each possible state (f, m, c) we denote the number of children in the chosen population having that state by $N(f, m, c)$. The probability that an infant randomly chosen from that population would have had the state (f, m, c) is then $N(f, m, c)/N$, where N is the total number of children in the population.

A couple expecting a baby will naturally be interested in knowing whether their child is likely to be affected with DS. The statistician cannot predict the future, but if he knows their ages f, m and the probability distribution function $N(f,m,c)/N$ he can tell them the fraction of couples of their age whose babies were affected with DS in the past. This fraction, which I will denote $p(A|fm)$, is the conditional probability of the child's being in the state A, given the ages f,m of the two parents at the time of its birth. As a formula, it is given by setting $c = A$ in the formula

$$p(c|fm) = N(f,m,c)/N(f,m)$$

$$c = \text{either } A \text{ or } U \tag{1}$$

where

$$N(f,m) = \Sigma_{c'} =\ _{A,U} N(f,m,c') \tag{2}$$

denotes the total number of children in the population whose fathers and mothers at the time of the child's birth were aged f,m.

The analysis that follows will use other conditional probabilities besides $p(c|fm)$; for example, the conditional probability that the child's state is c and that in addition the father's age (at the time of birth) is f, given that the mother's age at that time is m, is defined by

$$p(cf|m) = p(fc|m) = N(f,m,c)/N(m) \tag{3}$$

where

$$N(m) = \Sigma_{c'} \Sigma_{f'} N(f',m,c') \tag{4}$$

is the total number of children in the population who were born to mothers aged m; and the conditional probability that the father's age is f, given that the mother's age is m, and independent of the state of the child, is

$$p(f|m) = N(f,m)/N(m) \tag{5}$$

The following identity is a direct consequence of the definitions (1), (3), (5):

$$p(cf|m) = p(c|fm)p(f|m) \text{ for all } f,m,c \tag{6}$$

In principle, the probability that the baby born to a particular couple with ages f,m will turn out to have DS can depend on both parents' ages, i.e. the conditional probability $p(A|fm)$ as defined in (1) may depend on both the variables f,m. However, each of the two hypotheses to be tested can be phrased as a statement that $p(c|fm)$ depends on only one of these variables.

Hypothesis **M**: *only the mother's age affects the newborn baby's chance of having DS; the father's age is irrelevant* (i.e. $p(c|fm)$ is independent of f).

This statement of the hypothesis is equivalent[2] to the formula

$$p(c|fm) = p(c|m) \text{ for all } f,m,c \tag{7}$$

Two alternative mathematical statements of this hypothesis are[3]

$$p(fc|m) = p(f|m)p(c|m) \tag{8}$$

$$p(f|cm) = p(f|m) \quad \text{provided that} \quad p(c|m) > 0 \tag{9}$$

Hypothesis **F** *only the father's age affects the newborn baby's chance of having DS; the mother's age is irrelevant.* As a formula, this is

$$p(c|fm) = p(c|f) \qquad \text{for all } f, m, c \tag{10}$$

Two alternative formulations of hypothesis **F**, analogous to (8) and (9), can be obtained by interchanging the symbols f and m in (8) and (9).

A prediction

The statistical test summarised in Table 1 is based on the idea of using the hypothesis **M** or **F** to predict the average ages of the 'irrelevant' parents at the birth of affected children and also of unaffected children. Thus, one of the two tests of hypothesis **M** is to use it to predict the average age of the fathers of the affected children, given the ages of the mothers of those children. This average age, corresponding to the first entry in Table 1, can be written in terms of conditional probabilities as $E(f|A)$, where

$$E(f|c) = \Sigma_f f p(f|c) \qquad c = A \text{ or } U \tag{11}$$

denotes the conditional expectation of f in the sample, conditional on the given value of c. To use hypothesis **M** we express the conditional probability in the above formula as a sum over maternal ages:

$$p(f|c) = \Sigma_m p(fm|c) \qquad \text{by equations analogous to (5) and (3)}$$

$$= \Sigma_m p(f|mc) p(m|c) \qquad \text{by an analogue of (6)}$$

$$= \Sigma_m p(f|m) p(m|c) \qquad \text{by hypothesis } \mathbf{M} \text{ (equation (9))} \tag{12}$$

so that equation (11) becomes (after interchanging the two summations)

$$E(f|c) = \Sigma_m E(f|m) p(m|c) \qquad \text{under hypothesis } \mathbf{M} \tag{13}$$

where we have defined $E(f|m) := \Sigma_f f p(f|m)$, which is the expectation (average) of the father's age, given the age of the mother.

To evaluate the right side of (13) we make a standard simplifying assumption used in statistics, the assumption of *linear regression*. This assumption is that $E(f|m)$ depends linearly on m, i.e. that a relation of the form

$$E(f|m) = Am + B \qquad \text{for all } m \tag{14}$$

holds, where A, B are constants, which can be estimated by the least-squares method. Putting (14) into (13), we obtain a formula for the predicted value of $E(f|c)$ under hypothesis **M**, in terms of $\Sigma_m m p(m|c)$ which is the same as $E(m|c)$, the expectation of the mother's age conditional upon the state c of the baby. Since the baby may be either affected or unaffected, there are in fact two predictions, one for each of the two possible values of c, which can be tested against the actual data. Written out explicitly these predictions are

$$E(f|A) = A \, E(m|A) + B$$

$E(f \mid U) = A \, E(m \mid U) + B$ both under hypothesis $\mathbf{M}$ (15)

where $E(m \mid c)$ means $\Sigma_m \, m p(m \mid c)$, the conditional expectation of the mother's age for given condition of the child, in analogy with equation (11).

The sample

The question that now arises is how to estimate the numbers on the left and right sides of equations such as (15). The 1933 paper treated this as a perfectly straightforward matter. The regression coefficients A, B were estimated by the least squares method from the data from the 150 families, and the expectations $E(f \mid A)$, etc. were estimated by taking appropriate averages of these data. But is it really quite so straightforward?

A standard method of estimating such expectations would be to take a random sample of children from the general population, and to treat the sample as being typical of the general population. In the 1933 paper, however, the sample was not taken at random from the general population, but consisted of a very particular class of children, namely those families of children with DS whom the researchers had access. Because of this method of selection, the sample is far from being typical of the general population; for example, since the parents of children with DS tend to be older than parents in the general population, we might expect the parents of the children in the sample to be older, on average, than parents in the general population. Even worse, the families in the sample all contain at least one child affected with DS, so one would expect the frequency of DS in the sample to be much higher than in the general population – and indeed, the sample contained 153 cases of DS among 727 children, a frequency of about 1 in 5, whereas the frequency of DS in the general population is about 1 in 600.

Because of this bias, we cannot be sure of getting reliable results from the usual assumption of sampling theory that the members of the sample were drawn at random from the general population. For example, we have no reason to believe that the average age of the fathers of the unaffected children in the sample is even approximately the same as $E(f \mid U)$, the average age of the fathers of such children in the general population. A more reliable assumption would be that the sample was drawn at random from what might be called the special population, consisting of those families in the general population containing at least one child with DS. But analysing this assumption properly would be a complicated task, and moreover it would require statistical information not supplied in the 1933 paper, about things like the sizes and age structures of families. The following analysis is instead based on a plausible simplifying assumption which eliminates any need for additional information of this kind.

To formulate this simplifying assumption, let us extend the model described in section 2 by including one further variable into the description of the 'state' of a child. In addition to the mother's age, the father's age and the clinical state of the child, we include a fourth variable s capable of two non-numerical values which will be represented by symbols as follows: $s = {}^*$ if the child is included in the sample and $s = {}^\wedge$ if it is not.[4] In the standard random sampling procedure each child has the same probability of being included in the sample, regardless of its clinical state and the ages of its parents; that is to say, the conditional probability $p({}^* \mid fmc)$ is independent of f, m and c. By a mathematical argument very similar to the one leading from (5) to (7), this statement about independence implies

$p(fmc \mid {}^*) = p(fmc)$ for standard random sampling (16)

if we exclude the possibility that $p(*|fmc)$, the probability of going into the sample, is zero. Equation (16) says that the expected relative frequency of state (f,m,c) in the sample is the same as in the general population.

For the special sampling method used in (Penrose 1933), however, not all children in the general population have the same probability of going into the sample; that is to say, $p(*|fmc)$ depends on the values of some or all of f, m and c. For example $p(*|fmc)$ will be larger for $c = A$ than for $c = U$ since the intention of the researchers was to include as many affected children into their sample as possible. In general it will depend on both f and m as well; for it is quite likely that the other children in the family are born within a few years before or after the child being considered, and so one would expect that an unaffected child of older parents is more likely to have a sibling with DS, and therefore more likely to be in the sample, than a child of younger parents. Working out the actual dependence would require some fairly complicated theory and would also require additional factual information about things like the distribution of children's ages in families with more than one child. However there is a simple way of evading all this, if one of the following plausible extensions of the hypotheses **M** and **F** is accepted:

Hypothesis **M'** (to be used with hypothesis **M**): *only the mother's age affects whether or not the baby has (or will have) a sibling with DS, so that $p(*|cmf)$ depends only on c and m, but not on f.*

Hypothesis **F'** (to be used with hypothesis **F**): *only the father's age affects whether or not the baby has (or will have) a sibling with DS, so that $p(*|cmf)$ depends only on c and f, but not on m.*

By manipulations analogous to those used in deriving (7), (9) we can state the new hypothesis **M'** in either of the following ways:

$$p(*|cmf) = p(*|cm) \qquad \text{for all } c,m,f \tag{17}$$

$$p(f|cm*) = p(f|cm) \qquad \text{for all } c,m,f \tag{18}$$

provided, in equation (18), that $p(*|cm) > 0$ Equation (18) tells us that, under this new hypothesis **M'**, the conditional probability distribution of f at given c,m is the same in the sample as it is in the general population.

Hypothesis **M'** is certainly not a truism, nor is it a logical consequence of its close relative **M**. Moreover, it is not strictly biological, but contains a sociological component as well. If it were the case, for example, that old fathers in our society tended to have smaller families, then an unaffected child born to an old father would be less likely than one with a young father to enter the sample later on as a result of the subsequent birth of an affected child (the mother's age being the same in both cases). In that case, unaffected children with old fathers would be under-represented in the sample. Nevertheless **M'** is a useful working hypothesis. It can be combined with **M** to give the following composite hypothesis:

Hypothesis **M***: *both **M** and **M'** are true; i.e. only the mother's age matters, both as to whether the child will be affected and as to whether any of its siblings are, or will be, affected.*

For a mathematical statement of hypothesis **M***, we combine **M** in the form (9) with **M'** in the form (18), to obtain

$$p(f|cm*) = p(f|m) \qquad \text{under hypothesis } \mathbf{M*} \tag{19}$$

Hypothesis $\mathbf{M}^*$ also implies

$$p(f|m^*) = \Sigma_c \, p(fc|m^*) \qquad \text{by analogues of (5) and (3)}$$

$$= \Sigma_c \, p(f|cm^*) \, p(c|m^*) \qquad \text{by an analogue of (6)}$$

$$= \Sigma_c \, p(f|m) \, p(c|m^*) \qquad \text{by (19)}$$
$$= p(f|m) \qquad \text{since } \Sigma_c \, p(c|m^*) = 1. \qquad (20)$$

Equations (19) and (20) can be combined with (9) to show that hypothesis $\mathbf{M}^*$ implies the following analogue of (9):

$$p(f|cm^*) = p(f|m^*) \qquad (21)$$

This is just like (9), but the asterisks show that it is a statement about the probabilities in the special rather than the general population.

By a calculation just like the one that led to (15), but with stars inserted everywhere to make all the probabilities refer to the special population, we can now derive the following prediction, which, unlike its analogue (15), involves only things that can be estimated from the sample:

$$E(f|A^*) = A^* \, E(m|A^*) + B^*$$

$$E(f|U^*) = A^* \, E(m|U^*) + B^* \text{ both under hypothesis } \mathbf{M}^* \qquad (22)$$

where A^*, B^* are the coefficients in the linear regression formula

$$E(f|m^*) = A^* m + B^* \qquad \text{for all } m \qquad (23)$$

The quantities in (22) are the ones that were estimated from the data in (Penrose 1933) the regression coefficients were estimated by the least squares method to be $A^* = 0.944$, $B^* = 4.304$, and the various conditional expectations were estimated by the conditional averages in column 2 of Table I. As noted already, the agreement with equation (22) was good: the data are consistent with hypothesis $\mathbf{M}^*$.

The hypothesis to be compared with $\mathbf{M}^*$ is the one obtained by combining $\mathbf{F}$ and $\mathbf{F}'$:

Hypothesis $\mathbf{F}^*$: *only the father's age matters, both as to whether the child will be affected and as to whether any of its siblings are, or will be, affected.*

For this hypothesis a calculation just like the one that gave (22), but with m and f interchanged throughout, leads to the following predictions:

$$E(m|A^*) = A^{*\prime} \, E(f|A^*) + B^{*\prime\prime}$$
$$E(m|U^*) = A^{*\prime} \, E(f|U^*) + B^{*\prime} \qquad \text{both under hypothesis } \mathbf{F}^* \qquad (24)$$

where $A^{*\prime}, B^{*\prime}$ are the coefficients in the linear regression formula

$$E(m|f^*) = A^{*\prime} f + B^{*\prime} \qquad \text{for all } m \qquad (25)$$

The least-squares estimates of the coefficients $A^{*\prime}$, $B^{*\prime}$ given in (Penrose 1933) are $A^{*\prime} = 0.726$, $B^{*\prime} = 7.120$. The results in Table II show that (24) does not agree with the data, so that at either the biological component $\mathbf{F}$ or the sociological component $\mathbf{F}'$ of the composite hypothesis $\mathbf{F}^*$ (or both) should be rejected.

Conclusion

In Penrose 1933 it was concluded from the data that paternal age is not a significant factor in determining whether the child would have DS, and that maternal age is very important. The statistical argument used was highly ingenious and original, but could be criticised on the grounds that it does not make any explicit allowance for the bias in the sample of children studied. The present paper suggests a way of allowing for that bias, and the conclusion about the biological importance of maternal age is the same as before, although the strength of the conclusion in respect of paternal age is weaker because it may only be the sociological hypothesis **F'** that has been falsified by the data rather than the biological hypothesis **F**.

Moral: Originality, ingenuity, and tireless observation are more important than flawless statistics when it comes to doing good science.

Acknowledgements

I am grateful to George Fraser for providing references Harris (1973) and Waardenburg (1932) and for obtaining the translation in my first footnote, and to Ursula Mittwoch for doing the translation. I am also grateful to Mathew Penrose and Shirley Hodgson for informative discussions.

Notes

1. On pages 47–8 of this reference, Waardenburg says 'In view of the persistent uncertainty of its genetic basis, I may have given too much space to this anomaly. On the other hand, the unfailing recurrence of a whole series of symptoms in mongoloid patients affords a fascinating problem. I would like to persuade the cytologists to investigate the possibility that we may be dealing with a particular chromosome aberration in man. If would surely not be surprising if such conditions could occasionally occur in man and that, if the effect is not lethal, they would be the cause of a far-reaching constitutional anomaly. Investigations should be carried out to see whether mongolism is associated with "chromosomal deficiency" caused by "nondisjunction", or on the contrary, we might be dealing with "chromosomal duplication". It is of course also possible that the cause could be due to an anomaly of only parts of chromosomes (chromomeres): a "sectional deficiency" caused by a "translocation", or a "sectional duplication": this would be more difficult to demonstrate cytologically. [These terms were introduced by Morgan, Bridges and Sturtevant, The Genetics of Drosophila, *Bibliographia Genetica* 1925 **II** 1–262]. My hypothesis has the advantage that it is testable, and it might also be able to explain the influence of maternal age, whereas in men, because of the very large number of meiotic divisions, the chance of chromosome aberrations would be expected to be increased even without any special influences due to age. If my hypothesis were shown to be correct, it would result in an important insight into problems of human constitution and the manifestations of syndromes' (English translation by Ursula Mittwoch)

2. To prove (7), sum the identity (6) over f and use **M** which says that $p(c|fm)$ has a common value independent of f; then use the sum rules $\Sigma\, p(cf\,|m) = p(c|m)$ and $\Sigma\, p(f\,|m) = 1$ to obtain $p(c|m) =$ (common value of $p(c|fm)$.

3. To prove (8), put (7) into the identity (6) and use the symmetry exhibited on the left side of (3). To prove (9), interchange f and c in (6) to get $p(f\,|cm) = p(fc|m)/p(c|m)$; then use (8)in the right side of this last formula.

4. A curious feature of the variable s is that its value may not be known at the time of birth: unless the infant already has an affected older brother or sister the value of s is not known until either an affected sibling is born or it becomes clear that no more children from that family will be included in the sample. Moreover, its definition depends on when the sample is chosen: an unaffected only child born in 1932 with an affected sister born in 1934 would have $s=^{*}$ for a survey done in 1933, but $s=^{\wedge}$ for one done in 1935.

Genetic variation: the quest for meaning

Francisco Mauro Salzano

Departamento de Genética, Instituto de Biociências, Universidade Federal do Rio Grande do Sul,
Caixa Postal 15053, 91501–970 Porto Alegre, RS, Brazil

Reminiscences

The year was 1961. I had received a Rockefeller Foundation grant to attend the Second International Congress of Human Genetics, to be held in Rome, and afterwards to stay one month at the Medical Research Council (MRC) Population Genetics Research Unit in Oxford, to acquaint myself with the techniques of human chromosome research. These were exciting years for human genetics, which was emerging from a simple catalogue of rare inherited diseases to an intellectually demanding discipline based on cytogenetic and biochemical techniques.

The trip did not start well. I had been previously invited to attend a meeting in São Paulo, and from there I would travel to Rome. In the meantime, however, a serious national political crisis started, and at its climax I received a phone call that the younger of my two sons had an accident and was severely ill. All the flights between São Paulo and Porto Alegre had been cancelled, and I had to travel by ground for 12 hours, in a state of high anxiety, just to verify that the boy had entirely recovered from the accident and was then quite well.

The political crisis also delayed my trip to Rome, but finally I managed to arrive there in good shape and to meet persons as important as Sir Ronald Fisher, J. B. S. Haldane, H. J. Muller, J. V. Neel, C. E. Ford, W. C. Boyd, N. E. Morton, R. R. Race, A. S. Wiener, J. Lejeune, and many others. The congress closing ceremony was unforgettable. Hundreds of pairs of twins greeted us at the entrance bridge of Castel Sant'Angelo, where a banquet was held.

After these remarkable events I finally arrived at the MRC Unit in Oxford. It was situated on the outskirts of the town and had been founded by Alan C. Stevenson, a well-recognised clinical geneticist. There I was going to work under the guidance of Marco Fraccaro, who was conducting important human cytogenetic research at the time. But all the staff (which included Pat Stewart, Alan's secretary, and Mary Glen-Bott, who helped in the medical questions) turned to be very amicable and receptive. It was then that I first met a young, bright, somewhat eccentric person named George Robert Fraser, who was a Scientific Officer there. He had also been at the congress in Rome, where he presented a paper on the pool of harmful genes in human populations (Fraser 1962b), but we had not met there.

The stay finished and I came back to Porto Alegre. In the following year a paper presented one of our findings there (Fraccaro *et al* 1962). My relationship with George was strengthened when he spent five months in 1970 as a Visiting Professor at the Biology Department of São Paulo's University, since we could meet several times during this period.

George began his courses in English but soon switched to Spanish, a language which he knew well and which was more easily understandable to his audience because of its similarity to Portuguese, a language of which George knew nothing. However, as his experience of Portuguese grew with time, the language of his lectures gradually changed to '*Portunhol*' (a mixture of

Portuguese and 'Espanhol', or Spanish), often spoken by Brazilians and by Latin Americans in general, when they have some knowledge of the other language in addition to their mother tongue). As a result his lectures became more understandable to his audience with every week that passed.

An amusing episode occurred three years later. Both George and I were attending the XIII International Congress of Genetics in Berkeley, and he invited Ursula Mittwoch (who has also written a contribution to this Festschrift) and me to dinner with him in San Francisco; after a relatively long highway drive we arrived there to find out that for some unknown reason all restaurants we looked for were closed. We finished eating miserable hot dogs at the margin of the highway, and returned in low spirits to Berkeley.

Afterwards I have memories of George's visits to Brazil in 1974, 1992 (10th Latin American Congress of Genetics), and 1996 (IXth International Congress of Human Genetics). On each occasion, he gave some lectures including one in my own department in Porto Alegre in 1996, followed by an address to the meeting of the Brazilian Society of Genetics in Caxambu. Over the years, he had cultivated his Portunhol to such an extent that for stretches of these last two lectures in our country, his language became almost pure unadulterated Brazilian Portuguese.

During all these years we also exchanged much correspondence, a good proportion of it in Portunhol on George's part, and I am proud to have two of his books (Fraser and Friedmann 1967, Fraser 1976), as well as his translation of Müller Hill's (1988) book, autographed by him. His lively mood and erudition were and are much appreciated, and in more recent times we also enjoyed the agreeable personality of his wife Maria.

The number of persons devoted to science worldwide is increasing every day, but few would have the persistence, intelligence, and acuity for dealing with complex problems as George. I am much gratified for being able to honour him in this book.

From genetics to genomics

The progress in genetic knowledge that took place in the past half century has no adequate adjectives for its description. When George was starting his scientific career there were probably just one or two cases of genetic linkage firmly established in humans. Today we have the entire human genome at our disposal. In the field of population genetics I was able to adapt to at least two paradigm changes. In the fifties the only good markers available were the blood groups, investigated through immunological techniques. Then the protein electrophoresis era became dominant, to be substituted now by molecular techniques which directly investigate the genetic material.

Just to exemplify the types of research that are being conducted worldwide, I will give some details of an investigation we have recently conducted (Fagundes *et al.* 2005). The subject of the study was a genetic region responsible for the formation of a cell surface glycoprotein, the low density lipoprotein receptor (LDLR). This substance plays a key role in the maintenance of plasma cholesterol levels, mediating the intermembrane flux of LDL and other cholesterol-carrying substances. The *LDLR* gene, located on chromosome 19p13.1–3, consists of 18 exons, spanning 45 kilobases (kb) of DNA. At its 3'-untranslated region (3'-UTR) an area of high density of *Alu* insertions occurs. We investigated the variability of the middle part of this 3'-UTR, containing two complete and one partial *Alu* insertions. According to their location one complete and the partial insertions were labelled upstream (*Alu* U), and the remaining insertion downstream (*Alu* D). The study covered about 800 base pairs (bp) of 222 chromosomes from African,

Asian, European, and Amerindian samples.

Twenty-one polymorphic sites were found, which could be grouped into 16 arrangements or haplotypes; these, in turn, could be subdivided into three clusters. All haplotypes present in more than 20 chromosomes (with one exception) were found in all continental groups, indicating that most of this variability predated the intercontinental migrations. Interestingly, while the non-*Alu* U site showed an evolutionary rate (0.166% per million year or My) close to the average *Alu* insertion evolutionary rate, for *Alu* U the rate was much higher (0.632% per My).

To investigate the possible reasons for this high rate (which is of course reflected in the amount of nucleotide diversity), a series of simulations of different demographic scenarios were performed. For both the whole sample and for each continent separately neutrality was strongly rejected. Moreover, the low intercontinental variation, and the absence of any haplotype substructure, suggested a kind of homogeneous balancing selection in all the different places. Coalescent analysis indicated an effective population number (the number that really matters in evolutionary studies) of 11,600 individuals and a time for the most recent common ancestor of 1.155 My ago (Mya). These figures are similar to those obtained using other nuclear loci, and attest to the power of these techniques to unravel details about the remote past of our species.

In which ways may an untranslated genetic region be important for survival? One of the proposed models involves the association of messenger RNA with cytoskeletal elements through this 3'-UTR, with the *Alu*-rich region being proposed as a cytoskeleton binding domain (Wilson *et al.* 1998). Knouff *et al.* (2001), on the other hand, demonstrated that deletion of this region doubled the expression of the homologue of this gene in mice.

The Amerindian microcosm

We have been working with Amerindians for a long time now (reviews in Salzano and Callegari-Jacques 1988, Crawford 1998, Salzano 2002). They are a remarkable group from many points of view. Distributed all over the continent, they faced a wide variety of habitats to which they had to adapt, developing an equally large array of socioeconomic and cultural devices.

The first question that may be asked is about their origins, as well as the time and the processes involved in the colonisation of the Americas. I have recently reviewed the geological, archaeological, palaeoanthropological, morphological, linguistic, and genetic evidences (Salzano 2007). It is impossible to examine all this material here, but at present the most likely picture that emerges is of a single major migration without significant discontinuities in time. The First Americans should have entered the continent at least 15 thousand years before the present, probably using the Pacific coast route, and the main source of these migrants should have been the Altai Mountains of southern Siberia.

At the time of the European Conquest there should have been about 43 millions of natives in what is now Latin America, living in distinct stages of socioeconomic development, from small, simple bands of hunter-gatherers to sophisticated empires. Today this number is estimated as 54 million, a somewhat higher figure. But actually the genetic contribution of these first inhabitants is much higher, since their genes occur in a significant way in the 221 million people of mixed ancestry who now live in this region (Salzano and Bortolini 2002).

Extending our interest in the *Alu* insertions, we investigated 12 of them distributed all over the chromosomes in 179 individuals belonging to South American Native, Siberian, and Mongolian populations. These data were then integrated with those from 488 other persons, to ascertain the relationships between Asian, Northern Arctic, and Amerindian populations. A decreasing trend for heterozygosity and gene flow was observed in the three sets, in the order indicated above. However, no clear structure could be observed within South American Natives, indicating the importance of dispersive (genetic drift, founder effects) factors in their differentiation (Battilana *et al.* 2006).

A recurrent theme in discussions of human population variability is whether groups from different continents present similar or distinct levels of heterozygosities. More specifically in relation to Amerindians, the presence or absence of a significant population reduction ('bottleneck') in the continent's colonisation has been considered several times. To address this question we selected information concerning 404 microsatellites (or short tandem repeats) and 2–9 site haplotypes for 17 loci, all of them widely distributed over the human chromosomes and retrieved from two large data banks.

Seven Amerindian populations plus control populations from Asia, Europe and Africa were examined. Basically the two sets of data pointed in the same direction. The interpopulation variabilities were almost identical, the heterozygosity distributions had the same gradation, and the interpopulation relationships were also very similar. The African Yoruba clearly separated from the other groups, and the Asian Han were closest to the Amerindians, Europeans occupying an intermediate position. North and Central American Natives also clearly separated from the South American groups.

As for the question asked, colonisation of the Americas may have led to some loss of genetic variability, but the range of differences found among the Amerindian populations was two times higher than those observed between the most variable Amerindian (Maya) and the Yoruba (Salzano and Callegari-Jacques 2006).

The meaning of genetic variation

D'où venons nous? Que sommes nous? Où allons nous? (From where did we come? Who are we? Where are we going?) These three questions were inscribed by Paul Gauguin (1848–1903) in a painting made in Tahiti in 1897, when he was facing poverty and ill health, and exemplify not only problems of a philosophical nature, but also the objectives of the study of organic evolution.

Genomic evolution can be viewed as just a part of the universe's evolution, made possible by the origin of life. The whole process can be envisaged as a permanent struggle between the dialectical agents of change and order. Disordered chaos prevailed in the beginning of life, but the building of structures and channelling of processes soon established limits to variation. Natural selection was the primary agent responsible for the evolution of a genetic region or organism in one or another direction, through mechanisms of positive (emphasising novelty) or negative (protecting the *status quo*) selection. Mutation itself was influenced by this factor, which determined rates of change and forbidden options.

Life arrived early; there are indications of photo-synthesising cyanobacteria 3,500 Mya, and the history of our solar system is thought to have started 4,500 Mya. But the human species is a late comer, since anatomical and archaeological evidences of a human nature dates to just 0.13 Mya.

Goodman (1999) once said that 'we humans are only slightly remodeled apes'. This is true in relation to most of our biology, but of course there is a basic difference between us and the other living beings, and that is the possession of culture. Its development and result, technology, enabled *Homo sapiens* to become partially independent of the vagaries of environmental change. But at the same time this situation created many ethical challenges. We are a very dominant species, and the realisation that most of our DNA is shared with other organisms has had the result that the concept of universal brotherhood has changed from an abstract principle to a cold reality.

Science, anti-science, and ethics

Concomitantly with the vertiginous development of science there is a growing anti-science movement appearing everywhere that should be carefully watched and opposed. Striking aspects of this movement are the anti-evolution position of several religious groups, especially in the USA, and the world campaigns against the use of genetically modified organisms and stem cell research. Activists are also making genetic studies among humans increasingly difficult. The infamous book by Tierney (2000), besides throwing mud at respected innocent scholars, resulted in the practically complete halt of all field genetic studies among Amerindians. This situation is unfortunate because we are in this way being unable to test important aspects of the history of humanity. Actively preventing studies that comply with all ethical principles is clearly unethical (Salzano and Hurtado 2004).

Another unfortunate recent development is the increasing relationship between genetic research and business. The number of companies that are exploring different aspects of the genetic material is enormous, especially in the USA, Europe, and Japan, and one of its most controversial aspects is gene or genome patenting. Despite many views against it (for instance, Andrews 2002), more than 25,000 DNA-based patents had been issued in the USA by the end of 2000 (Cook-Deegan and McCormack (2001). The argument for patent laws is to create incentives for technological innovation. By excluding others from making, using, or selling his or her invention for, say, 20 years, a considerable financial return is expected in relation to the investment made in the development of a given product.

It happens that genes are not products. For instance, DNA sequences are not simply molecules, they are also information; and the granting of patents on parts of genes or different alleles creates a tangle of rights that prevents innovation. An extreme form of this legal procedure is whole-genome patenting. O'Malley *et al.* (2005) listed five claim-specific and five contextual whole-genome patents granted in the USA. Where are we going? Not even God (granted its existence) should have exclusive rights on life.

An SN-haplo-map in isolates of common ancestral origin: the Sardinian challenge

Marcello Siniscalco

455 East 86th New York, New York 10028, USA

Introductory notes on today's latest goals of human population genetics: an appropriate salute (or an invitation?) from one allegedly retired geneticist to another.

A personal prologue for G.R. Fraser on the occasion of his alleged retirement

Dear George, if you want to know why I chose this type of salute to honour your long-term great contribution to Human Genetics, here is the answer: I think that minds like yours have never been so necessary for the organisation and understanding of the enormous body of parcelled data that today can be piled up with a speed never known before. Thus, the best wish I can add to this highly deserved recognition of your past contribution to human genetics is for you to join the efforts of the many self-appointed experts of genomic statistics – myself included – to scan for the right answers amidst the overwhelming amount of detailed information.

Potentials of a European HAP-MAP project

There is no question that gathering information on the distribution of single nucleotide polymorphisms (SNPs) at a worldwide level promises to be the most powerful tool for the analysis of genome diversity within and between species. Soon after their discovery it became apparent that among the most important attributes of SN-variants are their average occurrence at every thousand nucleotide sites, and their apparent stability throughout long generation intervals. Thus, in spite of their individual bi-allelic nature, they offer the possibility of being classified in terms of clusters of multi-allelic haplotypes (haplotype blocks) just like the closely linked major histocompatibility genes of the murine and human genomes which, thus far, have been the only efficient tools available to search for linkage between multi-allelic haplotype combinations and specific complex phenotypes.

The latter connotation has evidently motivated the launching of the International HAP-MAP Project (IHMP) whose immediate goal has been that of classifying SNPs through the search of common patterns of stable DNA sequence variants (haplotypes) at specific chromosomal sites within and between populations. Accordingly, the first step of the IHMP has been the arrangement of 'an international consortium' meant to detect 'common patterns of DNA sequences in the human genome, by characterising sequence variants, their frequencies and correlations between them in DNA samples from populations with ancestry from parts of Africa, Asia and Europe' (Editorial 2003).

It is generally acknowledged that the availability of SN-haplotype blocks along each human chromosome (a reality of the very near future) will provide the chromosomal address for every yet undiscovered single gene mutation leading to a specific mono-factorial trait or disease. However, the HapMap application to the search for multiple genetic and environmental factors underlying the expression of multifactorial phenotypes meets the same difficulties thus far encountered in both family and population studies carried out in large heterogeneous populations or in too small highly inbred isolates. The reasons for this failure have been extensively analysed (Daly *et al.* 2001, Gabriel *et al.* 2002b, Collins *et al.* 2003) through elegant reconstruction experiments, each suggesting the application of a different remedial strategy (Lander 1996, Terwilliger and Weiss 1998, Eaves *et al.* 2000, Zack *et al.* 2001) to maximise the chance of detecting the expected difference (linkage disequilibrium or LD) in the distribution of suitable genetic markers of one or more SN-haplotype blocks in groups of carriers of the same complex phenotype versus that observed among the relevant normal controls. The present proposal is essentially based on the same rationale but for the novelty of recommending the opportunity to search for *associations* between a given complex phenotype and a specific set of markers in a sort of iterative manner by searching for it in a series of sub-populations sharing the same ancestral origin (thus almost certainly sharing the multiple mutants necessary for the expression of the same complex phenotype) and yet diversified from one another throughout their millennarian isolation. A recurrent association between a given complex phenotype and one or more SN-markers in all the Mendelian Breeding Units (MBUs) could obviously be best explained by LD. But recurrent associations limited to only a fraction of the patients within or between separate MBU would necessarily suggest more complex interactions of genetical and/or environmental nature.

The Sardinian challenge

The dramatic increase of the number of genetic markers now available has caused population genetics to become the almost exclusive approach with which genetic analysis of complex phenotypes can be approached. The island of Sardinia has attracted the attention of geneticists for decades on account of the monophyletic origin of its population (nearly 1.7 million inhabitants) distributed in a few large towns and over 300 isolated villages with many centuries of existence and, often, very diverse ecology. In the middle 1950s, these circumstances permitted the substantiation of J. B. S. Haldane's hypothesis (see Weatherall's contribution in this volume) that malignant malaria might have been one of the strongest ecological factors responsible for the maintenance of some lethal and sub-lethal mutations in the lowland villages of the island (Ceppellini 1958, Siniscalco *et al.* 1961). The marked genetic diversity of the individual Sardinian isolates in spite of their derivation from a common founder group of the late Neolithic was substantiated in the pre-genomic days by the elegant studies of Cavalli-Sforza and colleagues (Cavalli-Sforza *et al.* 1994 and Fig. 1).

This is to say that – with the exception of a few major towns whose present large populations have been the result of the *one*-way *urbanisation flow* of the last fifty years – the majority of Sardinians are still subdivided into about 300 isolated villages, stemming from the same ancestral founder group of the late Neolithic, yet diversified from one another throughout the millennarian impact of well-known triggers of genomic diversity such as founder effect, non-random 'assortative' mating, adaptive selection and pure chance. These circumstances, enriched by the availability of precious genealogical data (from governmental archives since 1860 and from ecclesiastical records for several centuries) for each of the now living individuals and their fam-

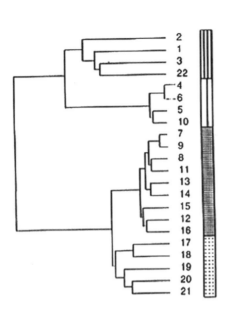
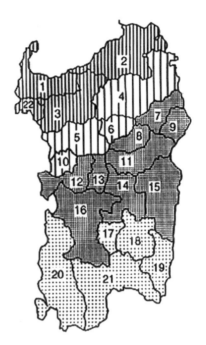

Figure 1. The monophyletic origin of Sardinians from a single Neolithic founder group is supported by the study of Cavalli Sforza and colleagues showing the most likely genealogy of the 22 sub-populations into which the inhabitants of Sardinia have been grouped on the basis of gene frequency distributions for blood groups, HLA system, serum protein and red cell enzyme polymorphisms (Cavalli-Sforza *et al.* 1994)

ilies, offer unique potentials for analysing the major factors of genome variation between Sardinian isolates which have existed for many centuries, and for exploiting their existence in the search for meaningful associations between genomic or environmental factors and specific complex traits or diseases.

Preliminary studies carried out within an ongoing research project ('Genome Diversity in Sardinia', funded by the Italian Ministry of Education, MIUR) have thus far permitted our group to single out twenty-two Sardinian isolates of nearly 8,000–10,000 inhabitants (MBUs) distributed in critical areas of the island with diverse ecological backgrounds (Fig. 2). These studies have proven that quick and reliable estimates of the genomic profiles of the single MBU can be accurately derived even from small samplings of forty individuals per village, if they are proven descendants of ancestors born in the same villages in the early 19th century and had no ancestors in common for the last four to five generations (Siniscalco *et al.* 1999, Heath *et al.* 2001).

The choice of the individual subjects fulfilling the above requirements has been possible through consultation of the municipal (and sometimes ecclesiastical) archives with the official authorisation of the governmental office of the 'privacy guarantor' under the condition that: (i) information collected would be utilised exclusively for the selection of individuals of either sex (only one per genealogical tree) fulfilling the requirements needed for a voluntary and anonymous participation in the study with the donation of a small blood sample (ca 15 ml); (ii) the vacuum tubes for blood collection would carry a label indicating the village, sex and numerical age but not the identity of the donor; (iii) the study should have the written approval of the

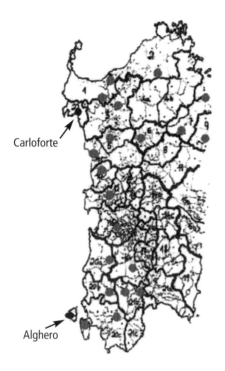

Carloforte

Alghero

Figure 2. The drawing reports the distribution of 22 long term isolated villages whose populations can be considered true MBUs possibly diversified from one another by the classical agents of genetic diversity such as founder effect, long term consanguineous and assortative mating, differential fitness in strongly diverse environments and pure chance. The two MBUs marked are the villages of Carloforte, founded by a group of Genovese fishermen in 1738, and Alghero whose present population is thought to have a Catalan ancestry. All the other MBUs are historically proven to have been derived from small founder groups of the same ancestral origin.

regional and municipal authorities in addition to the 'informed consent' of the individual voluntary participants.

In conclusion, we maintain that, apart from its primary value for studies on genomic variation, a detailed characterisation of SN-HAP blocks on DNAs pertaining to multiple Sardinian MBUs of common ancestral origin could be of special significance for searching for genetic and environmental factors involved in the expression of complex traits and diseases as well as for answering the following basic questions of SN-biology:

> 'Are single SN- polymorphisms stable enough to serve as "bona fide" suitable markers for evolutionary studies?'
> 'Is the stability of SN-polymorphism affected by their site of occurrence such as in the exons versus introns?'
> 'Should the description of SN-haplotypes at critical sites of each chromosome become the necessary preamble to identify the role of genetic and environmental factors underlying the expression of complex traits and diseases?'

Deleterious mutation and the genetic load

John Sved

School of Biological Sciences A12, The University of Sydney, NSW 2006

Abstract

The overall rate of deleterious mutation can be estimated from evolutionary considerations, by calculating how much of the genome has been substituted at less than the substitution rate for neutral mutations. In humans, it has been estimated in this way that more than one deleterious mutation occurs per diploid genome. Such a high mutation rate to deleterious alleles would have important consequences for population fitness under Haldane's principle, in which mildly deleterious mutations have the same overall effect on fitness as severely deleterious mutations. Haldane's principle is based on the assumption that deleterious mutations act independently in their effects on fitness. It is argued here that this assumption is unrealistic for most mildly deleterious mutations, and that a high overall deleterious mutation rate need not have dramatic consequences for population fitness.

The Haldane/Muller principle

Haldane (1937) enunciated a simple principle, describing how the fitness of a population is affected by recurrent deleterious mutation. He showed that slightly deleterious mutations are expected to have the same effect on population fitness as are highly deleterious mutations. Haldane's arguments were amplified by Muller (1950), who introduced the 'genetic load' terminology.

The notion that mutations of small effect can have the same effect on population fitness as mutations of large effect seems counterintuitive. The result comes from the fact that equilibrium is generally to be expected between the rate of production of deleterious alleles via mutation and their loss via selection. Slightly deleterious mutations are expected to come to higher equilibrium frequencies than are severely detrimental mutations. Overall, the lowering of fitness is the same for the two classes.

An intuitive explanation for the result comes from the fact that, eventually, one 'genetic death' is needed to remove a deleterious mutation, no matter how small the effect of that mutation. The situation is, however, expected to be different for dominant and recessive mutations. For recessive mutations, the 'genetic load' attributable to mutation at a locus is μ, the rate of mutation to deleterious alleles. For dominant mutations, the load is equal to 2μ. The difference of a factor of two is due to the fact that for recessive mutations, each genetic death removes two deleterious alleles. For a complete discussion of the calculations, including intermediate dominance and inbreeding, see Crow and Kimura (1970, pp. 299–303). For a more general discussion of different types of genetic load in man see Fraser and Mayo (1974d).

The genetic load for highly deleterious mutations seems straightforward and unambiguous. Lethal genes, by definition, cause death irrespective of the genotype at other loci. For mutations of small effect, on the other hand, the situation is more complicated. At equilibrium, each individual can be expected to have some, and possibly many, mutations of small effect. In this situation, it is unclear how effects at different loci should be combined.

As a first approximation, Haldane assumed that the probability of surviving each mutation is independent of the effect of others, so that fitness values at each locus can be multiplied. Working in terms of log fitnesses, the fitnesses can be added. Thus an overall mutation rate can be given, and the mutational load is equal to this overall mutation rate.

The idea that fitness values at individual loci can be multiplied has been challenged by a number of authors. The arguments on this point will be detailed later in this paper.

Measuring the rate of mutation

Spontaneous mutation rates are generally quite low. Therefore the difficulties of measuring mutation rates, particularly of mutations of small effect, would appear to be insuperable. Remarkably, however, a method that relies on evolutionary considerations has been developed by Kondrashov and Crow (1993), following on the neutral mutation theory of Kimura (1983).

The method is based on the expectation that the mutation rate per locus for neutral mutations is equal to the rate at which such mutations are substituted in the population over evolutionary periods of time (Kimura 1968, King and Jukes 1969). Assuming that the neutral substitution rate can be estimated from inter-species comparisons for non-functional substitutions, e.g. synonymous substitutions, pseudogenes, introns, the rate of mutation can be estimated. Assuming that the same rate of mutation applies to deleterious mutations, which are not substituted, the overall rate of mutation can be calculated by summing over all gene regions that have a substitution rate significantly reduced from the neutral rate.

The method does not allow for any distinction between mildly deleterious and severely deleterious mutations. The only criterion is that the selective disadvantage be sufficient to prevent the mutation being fixed by chance. If, however, mildly and severely deleterious mutations have the same effect on overall fitness, as discussed above, this distinction is not important. The overall mutation rate, as measured by these evolutionary considerations is exactly the rate that is critical in determining the population fitness at any given point in time.

Two studies have applied these principles in detail to estimate the overall deleterious mutation rate. Eyre-Walker and Keightley (1999) used the synonymous substitution rate in a range of functional genes to estimate the rate of production of deleterious mutations per diploid genome per generation as 1.6. They argued that from several points of view this was probably an underestimate. This estimate used the information from the human lineage following the divergence between chimpanzees and humans. Nachman and Crowell (2000) used rates of divergence in pseudogenes between humans and chimpanzees to estimate the neutral mutation rate, and used this to give an estimate of the genome deleterious mutation rate as 3.0.

These estimates of genome-wide mutation were obtained assuming that the number of protein-coding genes is 60,000 – 70,000. The actual number has now been revised downwards to less than 25,000 (see e.g. Pennisi 2003), suggesting that these genome mutation values are overestimates. However this number of genes would account for only 1% of the genome, and there are suggestions of regions of high conservation in the remaining 99% (e.g. Woolfe *et al.* 2005). Potentially, therefore, these estimates will need to be increased rather than reduced. A tentative estimate of 2.0 mutations per genome has been used for calculations below.

These estimates of genome mutation rates clearly lead to a quandary from the point of view of Haldane's principle. If it were true that each mutation needed one 'genetic death' for its elimination, the result would imply a high death rate, or more generally lowering of fitness, in the population.

The main point of this paper is to examine the manner in which Haldane's principle assumes that mutations act independently in their effects on fitness. Superficially this independence seems to be a sensible first approximation, given that most mutations will probably affect different, unrelated developmental or physical characters in an organism. The important point that needs to be examined, however, is whether independence of action of mutations in development translates into independence of action on the scale of fitness. The arguments given below show that this may be far from the truth.

A simple numerical model
The key concept that needs to be examined is the relationship between an individual's genotype, as measured by the number of deleterious mutations it contains, and its fitness, as measured by zygote-to-adult survival and fertility. For mutations of large effect, e.g. early acting developmental mutations, presence of at least one such deleterious dominant mutation assures a fitness of zero. However the situation is much less prescribed if an individual contains a number of deleterious genes of small effect.

The situation can best be understood using a simple numerical model. It will be assumed that all deleterious mutations are of small and equal effect. Lack of dominance will be assumed, but the situation is similar for deleterious recessives. Accepting an overall rate of production of 2.0 mutations per diploid genome, and a total gene number of 25,000, the mutation rate per locus, μ would be equal to 4.0×10^{-5}, somewhat higher than past estimates. This higher value may reflect the inability to detect mutations of small effects in conventional mutation detection experiments (Nachman and Crowell 2000).

At equilibrium, the frequency of deleterious alleles per locus is approximately $2\mu/s$ (Crow and Kimura 1970, p. 300), where $s/2$ is the selective disadvantage of mutations in heterozygous condition. To illustrate the calculation for mutations of low effect, we assume a value for s of 1%, making $s/2 = 0.005$. The notion of assigning a fixed selective disadvantage will be discussed below.

Under these conditions, the equilibrium frequency of deleterious mutations at each locus is $2 \times 4.0 \times 10^{-5} / 0.01 = 0.008$. The average individual will contain $25,000 \times 2 \times 0.008 = 400$ deleterious mutations. Each mutation carries a selective disadvantage of 0.01. The overall genetic load would thus be $400 \times 0.005 = 2.0$, in agreement with Haldane's principle. The addition of loads implies a logarithmic measure. In non-logarithmic terms, the probability of survival of an individual carrying 400 mutations would be $(1 - 0.005)^{400}$, which is approximately $e^{-2} = 0.135$.

The addition of loads in the above example implies that the mean individual in the population has a survival rate of around e^{-2}. This is, in fact, the rate of survival of an average individual compared to the survival rate of an individual containing no deleterious mutations. It is therefore relevant to consider the likelihood of an optimal genotype, or, in general, the population distribution of the number of deleterious mutations.

The average individual in the example contains 400 deleterious mutations. In the absence of systematic linkage disequilibrium, the number of mutations per individual is binomially distributed, with mean 400 and variance also equal to approximately 400. Assuming that the normal distribution adequately approximates the binomial, it is seen that 99.9% of individuals in the population are expected to have between 334 and 466 deleterious mutations. The probability of an individual having half the average number of deleterious mutations, 200, let alone zero mutations, is vanishingly small.

The relationship between genotype and fitness

It is convenient at this stage to introduce the more general notion of a fitness function (cf. Sved 1976). This function assumes that the genotype can be subsumed in terms of the number of deleterious mutations carried (x). It requires that all have the same effect. Similar functions have been used by other authors, including Milkman (1978) and Crow and Kimura (1979).

Figure 1 shows examples of fitness functions for the numerical example introduced above. The population frequency, $f(x)$ is shown (dashed lines), with the 99.9% range outlined in dotted lines. Four examples of fitness function are shown, plotted on a logarithmic scale. The first of these functions, $w_1(x)$, is the function for multiplicative interaction, which is linear on a logarithmic scale.

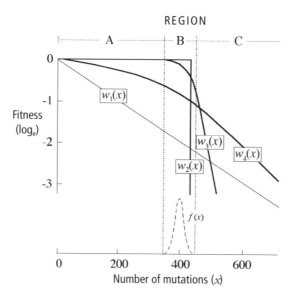

Figure 1. The relationship between genotype and fitness for multiplicative selection, $w_1(x)$, and three truncation models, $w_2(x)$, $w_3(x)$ and $w_4(x)$. The population distribution is shown at the bottom.

The other functions are calculated assuming a truncation or threshold model (King 1967). In the model introduced by King, it is assumed that there is a linear scale of 'fitness potential', to which genotype contributes, as well as environment and chance. Individuals above the threshold then are given a fitness of one, while those below the threshold have a fitness of zero.

There are two parameters that specify this model: (a) the contribution of the genotype (equivalent to a heritability), and (b) the position of the threshold. Function $w_2(x)$ shows the case of pure truncation selection, in which genotype entirely determines the fitness, and the heritability is 100%. A truncation rate of only 5% is needed in this case. The other functions, $w_3(x)$ and $w_4(x)$, show respectively a heritability of 0.1, combined with a truncation rate of 17%, and a heritability of 0.01 combined with a higher truncation rate (62%). In each case, the combination of threshold and heritability has been chosen to give a selective disadvantage of each mutation of approximately 0.005.

The selective disadvantage at each locus may be calculated in the following way. In a genotype with x deleterious mutations, the selective disadvantage of an extra deleterious mutation is equal to $[w(x+1) - w(x)] / w(x)$. Taking into account the total frequency with which genotypes of x deleterious mutations are expected to occur, $f(x)$, the total selective disadvantage of an extra mutation is equal to:

$$\sum_{x=0}^{n} f(x) \frac{w(x+1) - w(x)}{w(x)}$$

It is convenient, as in Figure 1, to depict the fitness functions as continuous functions. Passing to a continuous function, the selective disadvantage may be written as:

$$s = \int_{0}^{n} f(x) \frac{w'(x)}{w(x)}\, dx$$

or as:

$$s = \int_{0}^{n} f(x)\, \frac{d}{dx}\, [\ln(w)]\, dx \qquad (1)$$

Thus if fitness is plotted on a logarithmic scale, as in Figure 1, the selective disadvantage at individual loci is equal to the weighted slope of the fitness curve. Essentially it is the slope of the curve in the vicinity of the population that determines the selection at individual loci.

Kimura and Crow (1978) have given a formula very similar to (1):

$$s = \frac{1}{\bar{w}} \int_{0}^{n} f(x)\, w'(x)\, dx \qquad (2)$$

In practice, there seems to be little difference between (1) and (2). However, (1) is more convenient for comparison against a multiplicative model.

Equations (1) and (2) cannot be used for a pure truncation model with no chance or environmental contribution, function $w_2(x)$. In this case, the slope is infinitely large over an infinitesimally small frequency range, so that the right hand sides of equations (1) and (2) become undefined. For this case, the marginal selective value can be calculated from the effect of a single extra mutation around the population mean (see King 1967, Milkman 1967).

The shape of the fitness function

The fitness function may roughly be divided into three regions, as in Figure 1. Region B contains most of the population, while regions A and C lie to the left and right respectively.

As pointed out above, the intensity of selection at individual loci is determined by the slope of the fitness function in region B. It seems appropriate when comparing different fitness functions to adjust the models so that they have the same selective values, i.e. the same slope in region B.

The key difference between the models is therefore the slope in region A. This slope determines the mean fitness in the population compared to the optimum, or the 'genetic load'. The multiplicative model predicts a straight line relationship, in terms of log fitness, in this region of the curve. It is important to note that this relationship cannot be tested, since, in general, it is not possible to obtain genotypes that lie in region A. The multiplicative model predicts, nonetheless, that the fitness function defining these non-existent genotypes would have the same slope in region A as in region B.

The underlying rationale for such a prediction is that each deleterious mutation reduces fitness by a fixed amount, independently of how many other deleterious mutations the individual contains. Wallace (see e g. Wallace 1970) describes this as 'hard selection'. Many examples can be given of hard selection, involving genotypes with severely deleterious effects, where it seems irrelevant what the genotypes are at other loci. When it comes to considering deleterious mutations of small effect, however, it is difficult to find examples that can be unambiguously described as hard selection.

The opposite model is the soft selection model. In this model, it is assumed that all individuals are capable of surviving and reproducing, given optimal conditions. In conditions of resource limitation, however, individuals that have higher numbers of deleterious mutations are less likely to survive and reproduce. This form of selection therefore naturally explains why the slope of the fitness function should be heavily influenced by the position of the population rather than independent of it.

Soft selection and the truncation model

The terms soft selection and hard selection introduced by Wallace do not readily translate into calculations on genetic load, although Sved *et al.* (1967) attempted a general argument of this type. However, the truncation model introduced by King (1967) and Milkman (1967) is a specific attempt to calculate selective values consistent with competitive selection as envisaged under the soft selection model. The terms truncation selection, rank-order selection (Wills 1979), and soft selection have often been used interchangeably in discussions of natural selection.

Crow and Kimura (1979) have argued against the applicability of truncation selection. However, their argument is based on the extreme form of the model in which there is a sharp threshold based on genotype below which all individual survive and above which none survive. Although not a realistic description of the action of natural selection, this extreme form of truncation selection is important in providing an upper limit to the amount of selection that can be achieved at individual loci for a given death rate or genetic load in the population (Maynard-Smith 1968).

The model introduced by King (1967), which leads to the fitness functions of Figure 1, introduces non-genetic factors into the calculation. Sharp truncation occurs, but on a scale to which the genotype contributes only a small proportion of the variance. As argued above, the shape of the fitness function given by the model in regions A and B is consistent with expectation under the soft selection model. King argues that the truncation model is a statistical model rather than an attempt to accurately model the action of natural selection.

One weakness of the truncation model can be seen in the prediction for region C. The model predicts a sharp drop in region C, leading to a fitness of zero on a linear scale or $-\infty$ on a logarithmic scale. As with region A, this is not a realistic region of the curve for deleterious mutations, although in discussions on heterozygote advantage the occurrence of inbreeding does lead to genotypes in this region of the curve.

A realistic interpretation of the soft selection model would probably not lead to the zero prediction. Under soft selection, some individuals will, by chance, have sufficient resources to survive and reproduce irrespective of genotype. A lower limit of fitness greater than zero is expected.

The truncation model could be modified by adding an extra parameter that specifies a lower limit of fitness independent of genotype. Perhaps more satisfactorily, soft selection could be more realistically modelled by a resource limitation/competition model. The circumstances of natural selection are, however, different for different organisms, and usually differ between females and males, so that a single model could not satisfactorily deal with all soft selection.

Conclusion

Current estimates of the deleterious mutation load put the number of new mutations at more than one per individual, an estimate that could, as argued above, increase if more conserved non-coding regions of the genome are discovered. The main argument of this paper is that these estimates of deleterious mutation do not necessarily imply a high genetic load in the population. Theoretically, the overall genetic load could be as low as 5%, although realistically it needs to be somewhat higher.

These calculations are made on the soft selection model, which assumes that there is no necessary death or lowering of fitness associated with any deleterious mutation. Although the mutation rates are calculated for humans, the selection calculations are such as would apply to any organism, rather than specifically to humans in historic times. They have little relevance to current human populations. Such calculations also leave unanswered the question of whether, in human populations, deleterious mutations are consequential for the population. Crow (1997) in his insightful discussion of deleterious mutation, asks 'How many headaches, stomach upsets, depressed periods, and such small things that don't reduce viability or fertility, would be eliminated if our mutation rate had been lower? I suspect that the number is substantial'.

The frequency of deleterious mutations in the population has presumably been increasing, and will continue to increase, following the reduction in the traditional sources of natural selection. Will this have consequences for species survival? Crow's conclusion, similar to those of the present paper, is that the deleterious mutation rate is not a problem for species survival, although it may be a problem for health. The problems of overpopulation, climate change, environmental degeneration etc. are vastly more urgent than any attributable to mutation.

Coda

The completion of this book

Carolyn R. Leach

School of Molecular and Biomedical Sciences (Genetics), The University of Adelaide, Adelaide, SA 5005

As the time to send final manuscript of this book to the publisher fast approaches I have been reminded by my co-editor that all last/final/absolutely-final/we-really-mean-it-this-time deadlines have passed me by and my contribution has still not been submitted. I can but reply in the words of the Greek historian Xenophon (BC 431–350):

'Fast is fine, but accuracy is everything'.

In this case the term 'accuracy' is more relevant to the approach we have both tried to apply in the preparation of this volume but 'fast' has certainly not been possible. It has been a great pleasure and privilege to take part in this project though little did I know what I was agreeing to when Oliver asked me whether I would like to share the task with him. I was only slightly personally acquainted with George before the project began but by now I feel I have known him well and for a very long time. Oliver and I were both in the Department of Genetics in the University of Adelaide when George arrived in 1966 to take up his appointment. Oliver has maintained a close association with George from that time onwards and I have had a close association with Oliver since we were students together in the early 1960s. I have met George only once again after his time in Adelaide, when all three of us donned fetching raincoats for a trip under Niagara Falls.

I have been involved in teaching Genetics at the University of Adelaide since 1964 and in this capacity I have seen/used/collected articles and texts written by many of the contributors to George's Festschrift so once again I feel an affinity with the overall project. And it seems to me that while George may be about to celebrate his 75th birthday for other contributors that birthday is in the past, adding weight to the assertion of Aristotle that 'Education is the best provision for old age.' and perhaps too, to 'Clarke's First Law' that 'If an elderly but distinguished scientist says that something is possible, he is almost certainly right; but if he says that it is impossible, he is very probably wrong'. (Clarke 1962)

I will close with a quotation from Samuel Johnson who said 'There is no part of history so generally useful as that which relates the progress of the human mind, the gradual improvement of reason, the successive advances of science, the vicissitudes of learning and ignorance, which are the light and darkness of thinking beings, the extinction and resuscitation of arts, and the revolutions of the intellectual world.'(Johnson 1759) It is quite clear from the contributors to this Festschrift and the quality and diversity of their contributions that they and George have all made a substantial contribution to history and I am proud to have played a part in its being shared with the rest of the world.

George Robert Fraser

Curriculum vitae

Name	George Robert Fraser
Date of birth	3 March 1932
Home address	1 Woodstock Close
	Oxford OX2 8DB
	England
Telephone:	44–1865–515745
email:	F.Fraser.50@cantab.net

Education

1944–49	Open Scholar, Winchester College
1949–50	Sorbonne, Paris
1950–53	Open Scholar in Mathematics
	Trinity College
	University of Cambridge
1953–56	Clinical Training in Medicine, London Hospital Medical College,
	University of London
1956–57	Pre-registration House Officer, London Hospital
	House Physician to Professorial Medical Unit
	House Surgeon to Emergency Room
1957–59	Medical Research Council Scholar for Training in Research Methods,
	Galton Laboratory, University College, University of London

Degrees and Diplomas

1950	Diplomas in French, Russian, and Hungarian languages, University of Paris
1952	Part I Natural Sciences Tripos, Class 2 Division 1
	Anatomy, Physiology, Biochemistry, Mathematics
1953	Part II Natural Sciences Tripos, Class 1
	Genetics with Serology (Professors R. A. Fisher and R. R. A. Coombs),
	University of Cambridge
1953	BA University of Cambridge
1956	BChir University of Cambridge
1957	MB University of Cambridge
1960	MA University of Cambridge
1960	PhD University of London
	Human Genetics, Faculty of Science (Supervisor, Professor L. S. Penrose)
	Title of Thesis: *Deafness with Goitre (Syndrome of Pendred) and some Related Aspects of Thyroid Disease*
1960	Diploma in Medical Statistics
	Association of Incorporated Statisticians, London

1966	MD University of Cambridge
	Title of Thesis: *The Causes of Blindness in Childhood:*
	a Study of 776 Children in Special Schools
1978	DSc University of London
	Awarded on the basis of examination of publications
	in the field of Human Genetics

Honours and Awards

1944	Open Scholarship, Winchester College
1948	Open Scholarship in Mathematics
	Trinity College, University of Cambridge
1953	Senior Scholarship
	Trinity College, University of Cambridge
	Awarded on the basis of results in BA degree examination (Class I)
1956	George Riddoch Prize in Neurology
	Prize in Clinical Obstetrics and Gynaecology
	London Hospital Medical College, University of London
1966	Raymond Horton-Smith Prize
	Awarded by University of Cambridge for best MD thesis of academic year
	Title of Thesis: *The Causes of Blindness in Childhood:*
	a Study of 776 Children in Special Schools
	A revised version of this thesis was published as a book in 1967 by
	The Johns Hopkins Press under the title:
	The Causes of Blindness in Childhood: a Study of 776 Children with
	Severe Visual Handicaps

Postgraduate Fellowships and Diplomas

1975	Fellowship of the Royal College of Physicians and Surgeons of Canada
	in the Division of Medicine
1976	Fellowship of the Canadian College of Medical Geneticists as
	Clinical Geneticist
1984	Diploma by examination of the American Board of Medical Genetics as
	Clinical Geneticist
1990	Affiliate Membership of the Royal College of Physicians of London
1992	Recognition of the American Board of Medical Genetics by the American
	Board of Medical Specialties leading to Board Certification
1993	Corresponding Founding Fellowship of the American College of
	Medical Genetics
1995	Fellowship of the Royal College of Physicians of London
1996	Inclusion as Clinical Geneticist in the Specialist Register established in the
	United Kingdom in 1996 under the European Specialist Medical
	Qualifications Order (1995)

Career

1959–61	Scientific Officer, Medical Research Council, Population Genetics Research Unit, Oxford, England
1961–63	Research Fellow, Division of Medical Genetics, Department of Medicine, University of Washington, Seattle, Washington, USA
1963–66	Lecturer, Godfrey Robinson Unit (Royal National Institute for the Blind), Department of Research in Ophthalmology, Royal College of Surgeons, London, England
1966–68	Reader, Department of Genetics, University of Adelaide Associate Physician in Clinical Genetics, Adelaide Children's Hospital, Adelaide, South Australia
1968–71	Associate Professor, Division of Medical Genetics, Department of Medicine Associate Professor, Department of Preventive Medicine Consultant in Clinical Genetics, University and Children's Orthopedic Hospitals, University of Washington, Seattle, Washington, USA
1971–73	Professor of Human Genetics, Faculty of Medicine Director of Institute of Medical Genetics Consultant in Clinical Genetics University Hospital, University of Leiden, The Netherlands
1973–76	Professor of Medical Genetics Consultant in Clinical Genetics, University Hospitals, Memorial University of Newfoundland, St John's, Newfoundland, Canada
1976–79	Chief, Department of Congenital Anomalies and Inherited Diseases Department of National Health and Welfare, Federal Government of Canada Consultant in Clinical Genetics, Children's Hospital of Eastern Ontario, Ottawa, Ontario, Canada
1979–80	Associate Professor, Departments of Epidemiology and Otolaryngology and Centre for Human Genetics, McGill University, Montreal, Quebec, Canada
1980–84	Special Expert (Medical Officer) in Human Genetics, Lister Hill National Center for Biomedical Communications, National Library of Medicine Attached to Medical Genetics Clinic, National Institutes of Health, Bethesda, Maryland, USA Attached to Moore Clinic for Medical Genetics (as Lecturer in 1983–84) Department of Medicine, Johns Hopkins Hospital Johns Hopkins University, Baltimore, Maryland, USA
1984–1997 (mandatory retirement)	Senior Clinical Research Fellow Imperial Cancer Research Fund Cancer Genetic Clinic, Churchill Hospital, Oxford, England Honorary Consultant in Clinical Genetics, (Oxfordshire Health Authority) Department of Clinical Genetics, Churchill Hospital, Oxford Honorary Visiting Geneticist, St Bartholomew's Hospital London (until 1991)

Editorships

| 1963–66 | *Journal of Medical Genetics* |
| 1968–71 | *American Journal of Human Genetics* |

Visiting Professorships

| 1970 (June–October) | Division of Human Genetics Department of Biology, University of São Paulo, Brazil |
| 1988–89 (June–February) | Department of Immunogenetics Université Paul Sabatier, Toulouse, France |

Languages and Translations

Variable reading, writing, and speaking knowledge of Czech, Dutch, French, German, Greek, Hungarian, Italian, Portuguese, Russian, Serbo-Croat, Spanish

Translation from German

Müller-Hill, Benno *Tödliche Wissenschaft: Die Aussonderung von Juden, Zigeunern und Geisteskranken 1933–1945.* Rowohlt Taschenbuch Verlag GmBH, Reinbek bei Hamburg, 1984.

Murderous Science: Elimination by Scientific selection of Jews, Gypsies, and Others Germany 1933–1945. Oxford University Press, 1988.

A revised and expanded version with an afterword by J D Watson was published in 1998.

Murderous Science: Elimination by Scientific selection of Jews, Gypsies, and Others Germany 1933–1945. Cold Spring Harbor Laboratory Press, 1998.

Clinical and Research Experience

1957–59	General human genetics, population genetics, and clinical genetics as PhD student at University College London with Professor L S Penrose. Title of Thesis: Deafness with Goitre (Syndrome of Pendred) and some Related Aspects of Thyroid Disease.
1959–61	Oxford, England. Clinical and population genetics, especially relating to the causes of profound hearing impairment in childhood and to thyroid disease. Field work in the British Isles and Yugoslavia.
1961–63	Seattle, USA. Clinical and population genetics specifically relating to the medical and epidemiological aspects of the spread and persistence of pathological alleles (especially those responsible for haematological disorders such as the thalassaemias, haemoglobin S and G6PD deficiency) in human populations. Field work in Greece and Yugoslavia.
1963–66	Royal College of Surgeons, London, England. Clinical and population genetics specifically relating to the causes of severe visual handicap among 776 children in special schools in England and Wales. This work was presented as an MD thesis to the University of Cambridge and was awarded the Raymond Horton-Smith Prize for the best MD thesis of the academic year

1965–66. Title of Thesis: *The Causes of Blindness in Childhood: a Study of 776 Children in Special Schools.* A revised version of this thesis was published in 1967 as a book under the title: *The Causes of Blindness in Childhood: a Study of 776 Children with Severe Visual Handicaps* by The Johns Hopkins Press.

1966–68   Adelaide, South Australia. Clinical and population genetics specifically relating to the causes of severe hearing impairment and of visual handicap in childhood in South Australia. Theoretical studies of linkage and of the parental origin of X-chromosomal aneuploidies. Established and directed a Medical Genetics Clinic at the Adelaide Children's Hospital.

1968–71   Seattle, USA. Continuation of work commenced in 1961–63 on spread and persistence of pathological alleles, especially those responsible for haemato-logical disorders. Theoretical and clinical studies of colour blindness. Conducted Medical Genetics Clinics in University Hospitals. Theoretical studies of potential long-term effects of changes in the practice of genetical counselling on the genetical constitution of Man.

1971–73   Leiden, The Netherlands. Continuation of theoretical studies mentioned in preceding sentence. Established and directed a Medical Genetics Clinic at the University Hospital of Leiden. Studies of the distribution of genetical polymorphisms in the population of the Netherlands, and of possible inter-actions with physiological variables (physical, anthropological, biochemical, haematological).

1973–76   St John's, Newfoundland, Canada. Established and directed a Medical Genetics Clinic at the Memorial University Hospitals. Set up a register of genetically determined and other childhood handicaps in the population. Studies of genetical and environmental factors in the determination of a high incidence of Hodgkin's disease and related conditions in an isolated commu-nity living on the west coast of Newfoundland.

1976–79   Ottawa, Ontario, Canada. Directed register of congenital malformations and birth defects in six Canadian provinces as part of an international study. Conducted Medical Genetics Clinics at the Children's Hospital of Eastern Ontario, University of Ottawa, and set up register of genetically determined diseases in the local population.

1979–84   National Library of Medicine, National Institutes of Health, Bethesda, Maryland, USA. Director of Human Genetics. Knowledge Base participating in a group effort, using *Mendelian Inheritance in Man: Catalogs of Autosomal Dominant, Autosomal Recessive, and X-linked Phenotypes* (V.A. McKusick) to explore application of computer techniques to problems of interactive com-munication of biomedical knowledge to the medical profession. Participated in Medical Genetics Clinics at National Institutes of Health and at the Moore Clinic, Department of Medicine, Johns Hopkins Hospital, Johns Hopkins University, Baltimore, Maryland, USA.

1984–97   Oxford, England. Associated with the creation of a local Registry of Familial Cancer, comprising families with unusual aggregations of common cancers.

Families from this registry were referred to appropriate colleagues whose laboratories were engaged in linkage studies and mutation testing to localise and define genes involved in the determination of a major susceptibility to common cancer. In the main, such families contained persons with breast, with ovarian, and with colorectal cancer. In some cases, the pattern of aggregation is site-specific, while in others two or more sites may be involved in the familial aggregation. Instrumental in developing a Cancer Genetic Clinic where individuals who were concerned about their own risks in view of the occurrence of common cancers among their relatives, were counselled. In addition to counselling, long-term periodical screening programmes were arranged through the clinic in appropriate cases. These programmes involved breast screening, ovarian screening, and colorectal screening. In some instances, the same individual entered two of these programmes or even all three. These activities involved a close collaboration in the Department of Clinical Genetics, with the Oxford Breast Diseases Group and with other colleagues in Oxford in the Departments of Gynaecology, Oncology, Radiology, Radiotherapy, and Surgery. Collaborated in the National Study of Familial Ovarian Cancer, at the local level in connection with the Cancer Genetic Clinic, at the regional level as a representative, and at the national level as a member of the steering committee. Held general Medical Genetics Clinics at St Bartholomew's Hospital, London, in connection with an honorary appointment as a Visiting Clinical Geneticist.

Licences to Practise Medicine

1957 to present	United Kingdom – Full Licence
1966–67	South Australia – Full Licence
1968–1971	Seattle, Washington, USA – University Hospitals Practice Specialist Licence
1971–73	Leiden, The Netherlands – University Hospital Practice Specialist Licence
1973–75	Newfoundland, Canada – Full Specialist Licence
1976–79	Ontario, Canada – Hospital Practice Specialist Licence
1979–80	Quebec, Canada – Hospital Practice Specialist Licence

George Robert Fraser

Bibliography

Books

Fraser GR and Friedmann AI 1967 *The Causes of Blindness in Childhood. A Study of 776 Children with Severe Visual Handicaps* (with preface by LS Penrose). pp 245 Johns Hopkins Press, Baltimore. This book is a revised version of a thesis presented by GR Fraser which was accepted for the Degree of MD of Cambridge University, and awarded the Raymond Horton-Smith Prize for the best thesis of the academic year 1965–6.

Fraser GR and Mayo O eds 1975 *Textbook of Human Genetics.* pp 524 Blackwell Scientific Publications, Oxford.

Fraser GR 1976 *The Causes of Profound Deafness in Childhood. A Study of 3,535 Individuals with Severe Hearing Loss Present at Birth or of Childhood Onset* (with foreword by VA McKusick). pp 410 Johns Hopkins University Press, Baltimore and London.

Book Translated from German

Müller-Hill Benno 1984 *Tödliche Wissenschaft: Die Aussonderung von Juden, Zigeunern und Geisteskranken 1933–1945.* Rowohlt Taschenbuch Verlag GmBH, Reinbek bei Hamburg.

Murderous Science: Elimination by Scientific selection of Jews, Gypsies, and Others Germany 1933–1945. Oxford University Press, 1988.

A revised and expanded version with an afterword by J D Watson was published in 1998.

Murderous Science: Elimination by Scientific selection of Jews, Gypsies, and Others Germany 1933–1945. Cold Spring Harbor Laboratory Press, 1998.

Other Publications

Fraser GR, Harris H and Robson EB 1959a A new genetically determined plasma protein in man. *The Lancet* 1 1023–1024.

Fraser GR 1959b Four cases of sporadic goitre and congenital deafness. *Proceedings of the Royal Society of Medicine* 52 1039.

Fraser GR 1959c Retinal degeneration and nerve deafness. *British Medical Journal* 2 1404.

Fraser GR, Morgans ME and Trotter WR 1960 The syndrome of sporadic goitre and congenital deafness. *Quarterly Journal of Medicine* 29 279–295.

Fraser GR 1961a Cretinism and taste sensitivity to phenylthiocarbamide. *The Lancet* 1 964–965.

Fraser GR, Morgans ME and Trotter WR 1961b Sporadic goitre with congenital deafness (Pendred's syndrome). In: *Advances in Thyroid Research* (editor Pitt-Rivers R) (Transactions of the Fourth International Goitre Conference). pp. 19–21 Pergamon Press, Oxford.

Fraser GR and Calnan JS 1961c Cleft lip and palate: seasonal incidence, birth weight, birth rank, sex, site, associated malformations and parental age. *Archives of Disease in Childhood* 36 420–423.

Fraser GR 1961d Dosage for the newborn. *British Medical Journal* 2 1361.

Fraser GR 1962a Our genetical 'load'. A review of some aspects of genetical variation. *Annals of Human Genetics* 25 387–414.

Fraser GR 1962b The pool of harmful genes in human populations. *Acta Geneticae Medicae et Gemellologiae* 11 283–287.

Also in: Gedda, L, ed: Proceedings of the Second International Congress of Human Genetics. volume 1, pp. 52–56: Edizioni dell'Istituto 'Gregorio Mendel', Roma, 1963.

Fraser GR 1962c Genetic studies in isolates. In: *The Genetics of Migrant and Isolate Populations* (editor Goldschmidt E) (Proceedings of a Conference on Human Population Genetics, Jerusalem, 1961) pp. 201–203. Williams and Wilkins, Baltimore.

Fraser GR 1962d Some genetical determinants in the pathogenesis of goiter. *Journal de Génétique Humaine* **11** 240–250.

Fraser GR 1962e Severe deafness in childhood. *British Medical Journal* **1** 1274–1275.

Fraser GR 1962f Streptomycin as a cause of congenital deafness. *Developmental Medicine and Child Neurology* **4** 219.

Fraser GR 1963a Parental origin of the sex chromosomes in the XO and XXY karyotypes in man. *Annals of Human Genetics* **26** 297–304.

Addendum: 1966 *Annals of Human Genetics* **29** 323.

Fraser GR 1963b A genetical study of goitre. *Annals of Human Genetics* **26** 335–346.

Fraser GR 1963c Goitre in Klinefelter's syndrome. *British Medical Journal* **1**: 1284.

Fraser GR 1964a Genetical aspects of thyroid disease. In: *The Thyroid Gland.* (editors Pitt-Rivers R and Trotter WR) pp. 271–297 Butterworth, London.

Fraser GR and Trotter WR 1964b Goitre par défaut de fixation de l'iode. *Exposés Annuels de Biochimie Médicale* **25** 205–212.

Fraser GR, Froggatt P and James TN 1964c Congenital deafness associated with electrocardiographic abnormalities, fainting attacks and sudden death. A recessive syndrome. *Quarterly Journal of Medicine* **33** 361–385.

Fraser GR Defaranas B, Kattamis, CA Race RR, Sanger R and Stamatoyannopoulos G 1964d Glucose-6-phosphate dehydrogenase, colour vision, and Xg blood groups in Greece: linkage and population data. *Annals of Human Genetics* **27** 395–403.

Sorsby, A and Fraser GR 1964e A statistical note on the components of ocular refraction in twins. *Journal of Medical Genetics* **1** 47–49.

Also in: *Journal of Pediatric Ophthalmology* **3** 33–35, 1966 .

Fraser GR 1964f The association of sporadic goitre with congenital deafness. In: *Proceedings of the Second International Congress of Human Genetics.* (editor Gedda L) volume 2, pp. 1899–1907: Edizioni dell'Istituto 'Gregorio Mendel', Roma.

Fraser GR 1964g A study of causes of deafness amongst 2,355 children in special schools. In: *Research into Deafness in Children.* (editor Fisch L) pp. 10–13 Blackwell Scientific Publications, Oxford.

Fraser GR 1964h Studies in isolates. *Journal de Génétique Humaine* **13**: 32–46.

Fraser GR, Froggatt P and Murphy T 1964i Genetical aspects of the cardio-auditory syndrome of Jervell and Lange-Nielsen (congenital deafness with electrocardiographic abnormalities). *Annals of Human Genetics* **28** 133–157.

Fraser GR, Stamatoyannopoulos G, Kattamis C, Loukopoulos D, Defaranas B, Kitsos C, Zannos-Mariolea L, Choremis C, Fessas P, and Motulsky AG 1964j Thalassemias, abnormal hemoglobins and glucose-6-phosphate dehydrogenase deficiency in the Arta area of Greece: diagnostic and genetic aspects of complete village studies. *Annals of the New York Academy of Sciences,* **119** 415–435.

Motulsky AG, Stransky E and Fraser GR 1964k Glucose-6-phosphate dehydrogenase (G6PD) deficiency, thalassaemia, and abnormal haemoglobins in the Philippines. *Journal of Medical Genetics* **1** 102–106.

Fraser GR Giblett ER, Stransky E and Motulsky AG 1964l Blood and serum groups in the Philippines. *Journal of Medical Genetics* **1** 107–109.

Fraser GR 1964m Profound childhood deafness. *Journal of Medical Genetics* **1** 118–151.

Reprinted by the National Foundation, New York as RS-126 in Birth Defects Reprint Series.

Fraser GR 1964n Partial monosomy 18. *The Lancet* **1** 664.

Fraser GR 1964o Heredity in juvenile diabetes. *British Medical Journal* **1** 433.

Fraser GR and Froggatt P 1964p Congenital cardiac arrhythmia. *The Lancet* **2** 648.

Fraser GR 1964q Diabetes mellitus and insipidus. *British Medical Journal* **2** 1329.

Fraser GR 1965a Association of congenital deafness with goitre (syndrome of Pendred). A study of 207 families. *Annals of Human Genetics* **28** 201–249.

Motulsky AG, Lee T-C and Fraser GR 1965b Glucose-6-phosphate dehydrogenase (G6PD) deficiency, thalassaemia, and abnormal haemoglobins in Taiwan. *Journal of Medical Genetics* **2** 18–20.

Fraser GR Giblett ER, Lee T-C, and Motulsky AG 1965c Blood and serum groups in Taiwan. *Journal of Medical Genetics* **2** 21–23.

Fraser GR 1965d Sex-linked recessive congenital deafness and the excess of males in profound childhood deafness. *Annals of Human Genetics* **29** 171–196.

Fraser GR 1965e Profound deafness in the pediatric age group. *Pediatrics Digest* **7** 37–46.

Fraser GR 1965f Phenylthiocarbamide-tasting in cretins. *The Lancet* **2** 134.

Fraser GR, Grunwald P and Stamatoyannopoulos G 1966a Glucose-6-phosphate dehydrogenase (G6PD) deficiency, thalassaemia, and abnormal haemoglobins in Yugoslavia. *Journal of Medical Genetics* **3** 35–41.

Fraser GR 1966b The role of Mendelian inheritance in the causation of childhood deafness and blindness. In *Mutation in Population* (*Proceedings of the Symposium on the Mutational Process*, Prague, 1965) (editor Hončariv R) pp. 129–138 Academia, Praha.

Fraser GR 1966c Genetics and medicine. *British Medical Journal* **2** 345–347, 397–399, and 453–455. Translated into Spanish
Fraser GR 1967 Genética y medicina. *Día Médico* (Buenos Aires) **39** 297–301, 327–329, and 335–338.

Friedmann I, Fraser GR and Froggatt P 1966d The pathology of the ear in the cardio-auditory syndrome of Jervell and Lange-Nielsen (deafness with electrocardiographic abnormalities). *Journal of Laryngology and Otology* **80** 451–470.

Rose FC, Fraser GR, Friedmann AI and Kohner EM 1966e The association of juvenile diabetes mellitus and optic atrophy: clinical and genetical aspects. *Quarterly Journal of Medicine* **35** 385–405.

Stamatoyannopoulos G, Fraser GR Motulsky AG, Fessas P, Akrivakis A and Papayannopoulou T 1966f On the familial predisposition to favism. *American Journal of Human Genetics* **18** 253–263.

Motulsky AG, Vandepitte J and Fraser GR 1966g Population genetic studies in the Congo. I. Glucose-6-phosphate dehydrogenase deficiency, hemoglobin S, and malaria. *American Journal of Human Genetics* **18** 514–537.

Fraser GR 1966h Population genetic studies in the Congo. II. Effect of hemoglobin S and glucose-6-phosphate dehydrogenase deficiency on mortality and fertility. *American Journal of Human Genetics* **18** 538–545.

Fraser GR, Giblett ER and Motulsky AG 1966i Population genetic studies in the Congo. III. Blood groups (ABO, MNSs, Rh, Js^a). *American Journal of Human Genetics* **18** 538–545.

Giblett ER, Motulsky AG and Fraser GR 1966j Population genetic studies in the Congo. IV. Haptoglobin and transferrin groups in the Congo and in other African populations. *American Journal of Human Genetics* **18** 546–552.

Sorsby A, Leary GA and Fraser GR 1966k Family studies of ocular refraction and its components. *Journal of Medical Genetics* **3** 269–273.

Fraser GR 1966l XX chromosomes and renal agenesis. *The Lancet* **1** 1427.

Fraser GR and Froggatt P 1966m Unexpected cot deaths. *The Lancet* **2** 56–57.

Fraser GR and Froggatt P 1967 Sudden death in infancy. *Pediatrics* **40** 140.

Fraser GR 1968a Some complications of the use of the three-generation method in the estimation of linkage relationships on the X chromosome in man. *Annals of Human Genetics* **32** 65–79.

Fraser GR 1968b Causes of severe visual handicap among schoolchildren in South Australia. *Medical Journal of Australia* **1** 615–620.

Fraser GR 1968c The spectrum of causation of profound deafness in childhood. *Journal of the Oto-Laryngological Society of Australia* **2** 25–33.
Also in: *Proceedings of the International Conference on Oral Education of the Deaf*, Clarke School, Northampton, Massachusetts. 1967 volume 1, pp. 182–205 Alexander Graham Bell Association for the Deaf, Washington, 1967.

Fraser GR and Friedmann AI 1968d Choroideremia in a female. *British Medical Journal* **2** 732–734.

Stamatoyannopoulos G, Sofroniadou C, Akrivakis A and Fraser GR 1968e Further data from Greece on recombination between the Xg blood group and glucose-6-phosphate dehydrogenase deficiency. *American Journal of Human Genetics* **20** 528–533.

Fraser GR and Mayo O 1968f A comparison of the two-generation and three-generation methods of estimation of estimating linkage values on the X chromosome in man with special reference to the loci determining the Xg blood group and glucose-6-phosphate dehydrogenase deficiency. *American Journal of Human Genetics* **20** 543–548.

Fraser GR, Friedmann AI, Patton VM, Wade DN and Woolf LI 1968g Iminoglycinuria-a 'harmless' inborn error of metabolism? *Humangenetik* **6** 362–367.

Friedmann I, Fraser GR and Froggatt P 1968h Pathology of the ear in the cardio-auditory syndrome of Jervell and Lange-Nielsen: report of a third case with an appendix on possible linkage with the Rh blood group locus. *Journal of Laryngology and Otology* **82** 883–896.

Fraser GR 1968i Children of phenylketonuric women. *Pediatrics* **41** 155.

Fraser GR, Steinberg AG, Defaranas B, Mayo O, Stamatoyannopoulos, G, and Motulsky AG 1969a Gene frequencies at loci determining blood-group and serum-protein polymorphisms in two villages of Northwestern Greece. *American Journal of Human Genetics* **21** 46–60.

Fraser GR Grunwald P, Kitchin FD, and Steinberg AG 1969b Serum polymorphisms in Yugoslavia. *Human Heredity* **19** 7–64.

Mayo O, Fraser GR and Stamatoyannopoulos G 1969c Genetic influences on serum cholesterol in two Greek villages. *Human Heredity* **19** 86–99.

Fraser GR 1969d The genetics of thyroid disease. *Progress in Medical Genetics* **6** 89–115. Reprinted by the National Foundation, New York as RS-145 in Birth Defects Reprint Series.

Fraser GR 1969e Malformation syndromes with eye or ear involvement. In: *Proceedings of the First Conference on Clinical Delineation of Birth Defects. Part II, Malformation Syndromes.* (editor Bergsma D) The National Foundation, New York. *Birth Defects Original Articles Series* **5(ii)** 130–137.

Fraser GR, Friedmann AI, Maroteaux P, Glen-Bott AM and Mittwoch U 1969f Dysplasia spondyloepiphysaria congenita and related generalized skeletal dysplasias among children with severe visual handicaps. *Archives of Disease in Childhood* **44** 490–498.

Thuline HC, Hodgkin WE, Fraser GR and Motulsky AG 1969g Genetics of protan and the deutan color-vision anomalies: an instructive family. *American Journal of Human Genetics* **21** 581–592.

Fraser GR 1969h Estimation of the recombination fraction between the protan and the deutan loci. *American Journal of Human Genetics* **21** 593–599. Addendum: *American Journal of Human Genetics* **21** 694.

Fraser GR Friedmann, AI Delhanty JDA, Edwards JH, Glen-Bott AM, Insley J, Lele KP, Mittwoch U and Mutton D 1970a Karyotype studies among children with severe visual handicap. *British Journal of Ophthalmology* **54** 79–89.

Fraser GR 1970b Genetical aspects of severe visual impairment in childhood. *Journal of Medical Genetics* **7** 257–267.

Ikkos DG, Fraser GR, Matsouki-Gavra EN and Petrochiles M 1970c Association of juvenile diabetes mellitus, primary optic atrophy, and perceptive hearing loss in three sibs with additional idiopathic diabetes insipidus in one case. *Acta Endocrinologica (København)* **65** 95–102.

Fraser GR 1970d The causes of profound deafness in childhood. In: *Ciba Foundation Symposium on Sensorineural Hearing Loss* (editors Wolstenholme GEW and Knight J) pp. 5–40 Churchill, London.

Fraser GR 1970e Some aetiological aspects of non-endemic childhood hypothyroidism. *Journal de Génétique Humaine* **18** 169–189.

Fraser GR 1970f Clonal origin for individual Burkitt tumours. *The Lancet* **1** 948.

Fraser GR 1970g The handless and footless families of Brazil. *The Lancet* **1** 1171.

Fraser GR 1970h The linkage between the protan and the deutan loci. A*merican Journal of Human Genetics* **22** 691–693.

Fraser GR 1970i Book Review *An Inquiry concerning Growth, Diseases and Ageing.* By Burch PRJ, University of Toronto Press, Toronto, 1969. *American Journal of Human Genetics* **22** 489–490.

Fraser GR 1971a The genetics of congenital deafness. *Otolaryngologic Clinics of North America* **4** 227–247.

Barrai I, Bodmer WF, Boyo AE, Cavalli-Sforza LL, Edwards JH, Fraser GR, Frota-Pessoa O, Jacquard A, Morton NE, Nazarov KN and Schull WJ 1971b Methodology for family studies of genetic factors. *World Health Organization Technical Report Series* **466** 1–37.

Fraser GR 1971c The role of genetic factors in the causation of human deafness. *Audiology* **10** 212–221.

Fraser GR 1971d Genetic approaches to the nosology of deafness. In: *Proceedings of the Second Conference on Clinical Delineation of Birth Defects.* (editor Bergsma D) Part IX. Ear The National Foundation, New York. *Birth Defects Original Articles Series* 7(**i**) 52–63.

Motulsky AG, Fraser GR, and Felsenstein J 1971e Public health and long-term genetic implications of intrauterine diagnosis and selective abortion. In: *Symposium on Intrauterine Diagnosis.* (editor Bergsma, D) The National Foundation, New York. *Birth Defects Original Articles Series* 7(**v**) 22–32.

Fraser GR 1971f Profound Deafness in Childhood. A Study in Human Biology. (Inaugural address delivered on his entrance into office as Professor of Human Genetics at the University of Leiden on October 29th 1971) pp 15: Universitaire Pers, Leiden.

Fraser GR 1971g More on renal iminoglycinuria. *Journal of Pediatrics* **79** 174.

Fraser GR 1971h Congenital internal carotid-internal jugular fistula. *Journal of Pediatrics* **79** 343.

Mayo O, Wiesenfeld SL, Stamatoyannopoulos G, and Fraser GR 1971i Genetical influences on serum-cholesterol level. *The Lancet* **2** 554–555.

Fraser GR 1972a The effects on society of the treatment of hereditary disturbances of metabolism. *Pharmaceutisch Weekblad* **107** 285–291.

Fraser GR 1972b The short-term reduction in birth incidence of recessive diseases as a result of genetic counselling after the birth of an affected child. *Human Heredity* **22** 1–6.

Fraser GR 1972c Selective abortion, gametic selection, and the X chromosome. *American Journal of Human Genetics* **24** 359–370.

Veltkamp JJ, Mayo O, Motulsky AG and Fraser GR 1972d Blood coagulation factors I, II, V, VII, VIII, IX, X, XI, and XII in twins. *Human Heredity* **22** 102–117.

Fraser GR 1972e Unsolved Mendelian disease. In: *The Biochemical Genetics of Man.* (editors Brock DJH and Mayo O) pp. 639–652: Academic Press, London. Second edition pp 727–742: Academic Press, London, 1978.

Fraser GR 1972f The implications of prevention and treatment of inherited disease for the genetic future of mankind. *Journal de Génétique Humaine* **20** 185–205.

Also in: *Recent Progress in Biology and Medicine – Its Social and Ethical Implications* (editor Btesh S) (7th CIOMS Round Table Conference). pp. 105–126 Council for International Organizations of Medical Sciences, Geneva, 1972.

Also in: *Eugenics: then and now. Benchmark Papers in Genetics* (editor Bajema, CJ) (Volume 5): Dowden, Hutchinson, and Ross, New York, 1976.

Fraser GR 1973a Syndromes. In: *Clinical Genetics* (editor Sorsby A) (2nd edition). pp. 590–616 Butterworth, London.

Fraser GR 1973b Genetical implications of antenatal diagnosis. *Annales de Génétique* **16** 5–10.

Fraser GR 1973c Eigen keus en toekomst van de mensheid. *Cahiers Bio-wetenschappen en Maatschappij* (Utrecht) **1**(**iii**) 53–59.

Fraser GR 1973d Diabetes as a risk of parents of children with rare recessive disorders. *The Lancet* **2** 457–458.

Fraser GR 1974a Epidemiology of profound childhood deafness. *Audiology* **13** 335–341.

Fraser GR 1974b Severe visual handicap in childhood. In: *Genetic and Metabolic Eye Disease* (editor Goldberg MF) pp. 19–35 Little Brown, Boston.

Doeglas HMG, Bernini LF, Fraser GR, van Loghem E, Meera Khan P, Nijenhuis LE, and Pearson PL 1974c A kindred with familial cold urticaria: linkage analysis. *Journal of Medical Genetics* **11** 31–34.

Fraser GR and Mayo O 1974d Genetical load in man. *Humangenetik* **23** 83–110.

Fraser GR 1974e The long-term genetical effects of recent advances in the treatment and prevention of inherited disease. *British Journal of Psychiatry* **125** 521–528.

Also in: *Proceedings of the Third Congress of the International Association for the Scientific Study of Mental Deficiency* The Hague, The Netherlands, 4–12 September 1973. pp. 82–90.

Fraser GR, Volkers WS, Bernini LF, de Greve WB, van Loghem E, Meera Khan P, Nijenhuis LE, Veltkamp JJ, Vogel GP and Went LN 1974f A search for associations between genetical polymorphic systems and physical, biochemical, and haematological variables. *Human Heredity* **24** 424–434.

Fraser GR Volkers WS, Bernini LF, van Loghem E, Meera Khan P and Nijenhuis LE 1974g Gene frequencies in a Dutch population. *Human Heredity* 24 435–448.

Buehler SK, Fodor G, Marshall WH, Firme F, Fraser GR and Vaze P 1975a Common variable immunodeficiency, Hodgkin's disease, and other malignancies in a Newfoundland family. *The Lancet* 1 195.

Fraser GR and van de Kamp JJP 1975b Populatie-genetica. *Nederlands Tijdschrift voor Geneeskunde* 119 1502–1506.

Fraser GR 1975c Heredity, early identification of hearing loss, and the risk register. In: *Early Identification of Hearing Loss* (editor Mencher GT) pp. 23–32 Karger, Basel.

Carter ND, Auton JA, Welch SG, Marshall WH and Fraser GR 1976a Superoxide dismutase variants in Newfoundland – a gene from Scandinavia? *Human Heredity* 26 4–7.

Fraser GR 1976b Genetics. In: *Medical Ophthalmology* (editor Rose FC) pp. 521–534 Chapman and Hall, London.

Fraser GR 1976c Contributions to round table on Genetic Variability of Man; Effect of Behaviour. In: *Proceedings of the International Conference – Biology and the Future of Man.* (editor Galpérine C) pp. 328–332. Universities of Paris.

Fraser GR 1976d Genetical basis for otolaryngological disorders. In *Scientific Foundations of Otolaryngology.* (editors Hinchcliffe R and Harrison D) pp. 121–130 Heinemann, London.

Fraser GR 1977 Genetic counselling in the case of non-familial deafness. In: *Medical Genetics* (editors Szabó G and Papp Z) (Proceedings of the Symposium at Debrecen-Hajdúszoboszló, Hungary). pp. 739–740 Akadémiai Kiadó, Budapest, and Excerpta Medica, Amsterdam.

Fraser GR 1979a Šereševskij-Turner's syndrome or Turner's syndrome? *Human Genetics* 46 353.

Fraser GR 1979b Long-term genetic consequences of advances in treatment and prevention of hereditary disease. In: *Service and Education in Medical Genetics.* (editors Porter IH and Hook EB) pp. 21–28 Academic Press, New York.

Fraser GR 1979c Genetic counselling in hearing impairment. *Volta Review* 81 291–298.

Marshall WH, Buehler SK, Crumley J, Salmon D, Landre M-F and Fraser GR 1979d A familial aggregate of common variable immunodeficiency, Hodgkin disease and other malignancies in Newfoundland. I. Clinical features. *Clinical and Investigative Medicine* 2 153–159.

Salmon D, Landre M-F, Fraser GR, Buehler SK, Crumley J and Marshall WH 1979e A familial aggregate of common variable immunodeficiency, Hodgkin disease and other malignancies in Newfoundland. II. Genealogical analysis and conclusions regarding hereditary determinants. *Clinical and Investigative Medicine* 2 175–181.

Motulsky AG and Fraser GR 1980 Effects of antenatal diagnosis and selective abortion on frequencies of genetic disorders. *Clinics in Obstetrics and Gynaecology* 7 121–134.

Siegel ER, Sneiderman CA, Falkenberg GCW and Fraser GR 1981a Quality filtration of the published biomedical literature: some methods and strategies. In: *Proceedings of the Symposium on Information Transfer in Toxicology.* pp. 99–109 US Government Publications,

Fraser GR 1981b Genetically determined hearing defects. In: *Audiology and Audiological Medicine* (editor Beagley HA) pp. 302–316 Oxford University Press.

Fraser GR 1985 Health registers: theory and practice. In: *Familial Cancer* (editors Müller H and Weber W) pp. 258–259 Karger, Basel.

Fraser GR 1987a The genetics of deafness. In: *Scott-Brown's Otolaryngology. Paediatric Otolaryngology* (editor Evans JNG) (5th edition). pp 26–34 Butterworth, London.

Fraser GR 1987b Causes of deafness and blindness in childhood. In: *Past Present and Future of Human Genetics* (editors van Omenn GJB, Meera Khan P, Klasen EC, Millington Ward A and Pearson PL) pp. 51–64 University of Leiden.

Fraser GR 1992a Familial forms of common types of cancer. *Revista Brasileira de Genética* 15 (supplement 1) 305–310.

Fraser GR 1992b Familial forms of common cancers in a Cancer Genetic Clinic. *Anticancer Research* 12 1813–1814.

Fraser GR and Huson SM 1994 The Oxford Cancer Genetic Clinic; experience of the first 4 1/2 years. *European Cancer News* 7 9–15.

Fraser GR 1995 Plato on social inequality. *British Medical Journal* **1** 528.

Gausden E, Coyle B, Armour JAL, Coffey R, Grossman A, Fraser GR Winter RM Pembrey ME, Kendall-Taylor P Stephens D, Luxon LM, Phelps PD, Reardon W and Trembath R 1997a Pendred syndrome: evidence for genetic homogeneity and further refinement of linkage. *Journal of Medical Genetics* **34** 126–129.

Fraser GR and Huson S M 1997b The Oxford Cancer Genetic Clinic; experience of the first 5 years. In: *Hereditary Cancer 2* (editors Haefliger J-M, Müller H, Scott RJ and Weber W) pp. 57–71 Ligue Suisse contre le Cancer, Berne.

Fraser GR 1998 Lionel Penrose as scientist and mentor: recollections and lifelong legacies. In: *Penrose: Pioneer in Human Genetics* (Report on a symposium held on 12th and 13th March, 1998, to celebrate the centenary of the birth of Lionel Penrose) pp. 35–39 Centre for Human Genetics at University College London.

Fraser GR 1999 The Oxford Cancer Genetic Clinic; experience of the first seven years. In: *Familial Cancer and Prevention – Molecular Epidemiology: A New Strategy toward Cancer Control* (editors Utsunomiya J, Mulvihill JJ and Weber W) pp. 51–58: Wiley-Liss, New York.

Fraser GR 2001 Mendel, Gregor. In: *Encyclopedia of Genetics* (editors Brenner S and Miller JH) pp 1168–1171 Academic Press, NewYork.

Fraser, GR 2006a Medical knowledge in the service of informed parental reproductive decision and choice: absence of implications for eugenic ideology. Lecture delivered to the International Colloquium: *L'eugénisme après 1945: formes nouvelles d'une doctrine périmée* (Eugenics after 1945: new forms of an out-dated doctrine). Dijon, France, 22-23 September 1997. The full text has been published in a translation into French as Le savoir médical au service de choix reproductifs informés: une pratique sans implication eugénique. In *L'éternel retour de l'eugénisme* (editors Gayon J and Jacobi D with the collaboration of Lorne M-C) pp 7–27 Presses Universitaires de France, 2006.

Fraser, GR 2006b Human Genetics Today: Hopes and Risks. Lecture delivered to the XXth International Conference of the Pontifical Council for Health Pastoral Care: The Human Genome-Biological, Medical and Ethical Prospects, held from 17 to 19 November, 2005 in the Vatican. The full text has been published in the English, French, Italian and Spanish editions of *Dolentium Hominum*, Number 61, 2006. The bibliographical reference to the English edition is: Fraser GR 2006 Human genetics today: hopes and risks. *Dolentium Hominum* 21(1) 16-23.

References

Aase JM 1990 *Diagnostic Dysmorphology.* Plenum Medical Book Company, New York.

Abate T 2003 California's role in Nazis' goal of purification. *San Francisco Chronicle*, (10 March 2003) E1.

Abushaban L, Uthaman B, Kumar AR and Selvan J 2003 Familial truncus arteriosus: a possible autosomal-recessive trait. *Pediatric Cardiology* 24 64–66.

Ádám Zs and Papp Z 1996 Prenatal diagnosis of orofaciodigital syndrome Váradi-Papp type. *Journal of Ultrasound Medicine* 15 714.

Adams JA 1814 *A Treatise on the Supposed Hereditary Properties of Diseases* Calow, London.

Addison PK, Berry V, Ionides AC, Francis PJ, Bhattacharya SS and Moore AT 2005 Posterior polar cataract is the predominant consequence of a recurrent mutation in the PITX3 gene. *British Journal of Ophthalmology* 89 138–141.

Addy JH 1990 Mendelian inheritance of propranolol responsive hypertension in an extended Ghanaian family. *Ghana Medical Journal* 24 164–169.

Addy JH 1992 Genetics of hypertension. *The Lancet* 340 377–380.

Akerblom HK, Knip M, Hyöty H, Reijonen H, Virtanen S, Savilahti E and Ilonen J 1997 Interaction of genetic and environmental factors in the pathogenesis of IDDM. *Clinica Chimica Acta* 257 143–156.

Akiyama H, Chaboissier M-C, Behringer RR, Rowitch DH, Schedl A, Epstein JA and de Crombrugghe B 2004 Essential role of Sox9 in the pathway that controls formation of cardiac valves and septa. *Proceedings of the National Academy of Sciences USA* 101 6502–6507.

Al Hajj M, Wicha MS, Benito-Hernandez A, Morrison SJ and Clarke MF 2003 Prospective identification of tumorigenic breast cancer cells. *Proceedings of the National Academy of Sciences USA* 100 3983–3988.

Alberdi F, Allison AC, Blumberg BS, Ikin EW and Mourant AE 1957 The blood groups of the Spanish Basques. *Journal of the Royal Anthropological Institute* 87 217–221.

Alcântara VM, Chautard-Freire-Maia EA, Picheth G and Vieira MM 1991 A method for serum cholinesterase phenotyping. *Revista Brasileira de Genética* 14 841–846.

Alcântara VM, Chautard-Freire-Maia EA, Scartezini M, Cerci MSJ, Braun-Prado K and Picheth G 2002 Butyrylcholinesterase activity and risk factors for coronary artery disease. *Scandinavian Journal of Clinical Laboratory Investigation* 62 399–404.

Alcântara VM, Lourenço MAC, Salzano FM, Petzl-Erler ML, Coimbra Jr. CEA, Santos RV and Chautard-Freire-Maia EA 1995 Butyrylcholinesterase polymorphisms (*BCHE* and *CHE2* loci) in Brazilian Indian and admixed populations. *Human Biology* 67 717–726.

Alcântara VM, Oliveira LC, Réa RR, Suplicy HL and Chautard-Freire-Maia EA 2003a Butyrylcholinesterase and obesity in individuals with the CHE2 C5+ and CHE2 C5- phenotypes. *International Journal of Obesity* 27 1557–1564.

Alcântara VM, Rodrigues LC, Oliveira LC and Chautard-Freire-Maia EA 2001 Association of the *CHE2* locus with body mass index and butyrylcholinesterase activity. *Human Biology* 73 587–595.

Alcântara VM, Rodrigues LC, Oliveira LC and Chautard-Freire-Maia EA 2003b The variable expression of the $C_{4/5}$ complex of human butyrylcholinesterase and body mass index. *Human Biology* 75 47–55.

Aleck KA and Bartley DL 1997 Multiple malformation syndrome following Fluconazole use during pregnancy: report of an additional patient. *American Journal of Medical Genetics* 72 253–256.

Al-Gazali LI, Sztriha L, Punnose J, Shather W and Nork M 1999 Absent pituitary gland and hypoplasia of the cerebellar vermis associated with partial ophthalmoplegia and postaxial polydactyly: a variant of orofaciodigital syndrome VI or a new syndrome? *Journal of Medical Genetics* 36 161–166.

Aligianis IA, Johnson CA, Gissen P, Chen D, Hampshire D, Hoffman K, Maina EN, Morgan NV, Tee L, Morton J, Ainsworth JR, Horn D, Rosser E, Cole TR, Stolte-Dijkstra I, Fieggan K, Clayton- Smith J, Megabarbane A, Shield JP, Newbury-Ecob R, Dobyns WB, Graham JM Jr, Kjaer KW, Warburg M, Bond J, Trembath RC, Harris LW, Takai Y, Mundlos S, Tannahill D, Woods CG and Maher ER 2005 Mutations of the catalytic subunit of RAB3GAP cause Warburg Micro syndrome. *Nature Genetics* 37 221–223.

Allen J, O'Donnell A, Alexander NDE, Alpers MP, Peto TEA, Clegg JB and Weatherall DJ 1997 α⁺-thalassaemia protects children against disease caused by other infections as well as malaria. *Proceedings of the National Academy of Sciences USA* **94** 14736–14741.

Alles GA and Hawes RC 1940 Cholinesterases in the blood of man. *Journal of Biological Chemistry* **133** 375-390.

Allison AC 1964 Polymorphism and natural selection in human populations. *Cold Spring Harbor Symposium on Quantitative Biology* **29** 137–149.

Allison AC and Blumberg BS 1961 An isoprecipitation reaction distinguishing human serum protein types. *The Lancet* **I** 634–637.

Allison AC, Blumberg BS and Gartler SM 1959 Urinary excretion of beta-aminoisobutyric acid in Eskimo and Indian populations of Alaska. *Nature* **183** 118–119.

Allison AC, Blumberg BS and Rees A 1958 Haptoglobin types in British, Spanish Basque and Nigerian African populations *Nature* **181** 824–825.

Alter HJ and Blumberg BS 1966 Studies on a 'new' human isoprecipitin system (Australia antigen). *Blood* **27** 297–309.

Altland K and Heredero L 1974 Diagnóstico prenatal de la anemia a hematies falciformes. *Revista Cubana de Pediatría* **46** 237–244.

Altshuler D, Brooks LD, Chakravarti A, Collins FS, Daly MJ, Donnelly P and International HapMap Consortium 2005 A haplotype map of the human genome. *Nature* **437** 1299–1320.

Alves-Silva J, da Silva Santos M, Guimarães PE, Ferreira AC Bandelt HJ, Pena SDJ and Prado VF 2000 The ancestry of Brazilian mtDNA lineages. *American Journal of Human Genetics* **67** 444–461.

Amedofu K, Brobby GW and Ocansey G 1999 Congenital non-syndromal deafness at Adamarobe, an isolated Ghanaian village. Prevalence, incidence and audiometric characteristics of deafness in the village (Part 1). *Journal of the Ghana Science Association* **1** 63–69.

American Society of Human Genetics 1998 *http://www.faseb.org/genetics/ashg/pubs/policy/pol-30.htm*

Amiel J and Lyonnet S 2001 Hirschsprung disease, associated syndromes, and genetics: a review. *Journal of Medical Genetics* **38** 729–739.

Amiel J, Attie T, Jan D, Pelet A, Edery P, Bidaud C, Lancombe D, Tam PKH, Simeoni J, Flori E, Nihoul-Fekete C, Munnich A and Lyonnet S 1996 Heterozygous endothelin receptor B (EDNRB) mutations in isolated Hirschsprung disease. *Human Molecular Genetics* **5** 355–357.

Amiel J, Attiee-Bitach T, Marianowski R, Cormier-Daire V, Abadie V, Bonnet D, Gonxales M, Chemouny S, Brunelle E Munnich A, Manach Y and Lyonnet S 2001 Temporal bone anomaly proposed as a major criteria for diagnosis of CHARGE syndrome. *American Journal of Medical Genetics* **99** 124–127.

Amiel J, Laudier B, Attie-Bitach T, Trang H, de Pontual L, Gener B, Trochet D, Etchevers H, Ray P, Simonneau M, Vekemans M, Munnich A, Gaultier C and Lyonnet S 2003 Polyalanine expansion and frameshift mutations of the paired-like homoeobox gene PHOX2 in congenital hypoventilation syndrome. *Nature Genetics* **33** 459–461.

Amiel J, Watkin PM, Tassabehji M, Read AP and Winter RM 1998 Mutation of the MITF gene in albinism-deafness syndrome (Tietz syndrome). *Clinical Dysmorphology* **7** 17–20.

Andreasen R 1998 A new perspective on the race debate. *British Journal of Philosophy of Science* **49** 199–225.

Andreotti F, Porto I., Crea F and Maseri A 2002 Inflammatory gene polymorphisms and ischaemic heart disease: Review of population association studies. *Heart* **87** 107–112.

Andrew A 1971 The origin of intramural ganglia. IV. The origin of enteric ganglia: a critical review and discussion of the present state of the problem. *Journal of Anatomy* **108** 169–184.

Andrews LB 2002 Genes and patent policy: rethinking intellectual property rights. *Nature Reviews Genetics* **3** 803–808.

Andrikopoulos K, Liu X, Keene DR, Jaenisch R and Ramirez F 1995 Targeted mutation in the col5a2 gene reveals a regulatory role for type V collagen during matrix assembly. *Nature Genetics* **9** 31–36.

Angell RR 1997 First meiotic division disjunction in human oocytes. *American Journal of Human Genetics* **61** 23–32.

Anglian Breast Cancer Study Group 2000 Prevalence of BRCA1 and BRCA2 mutations in a large population based series of breast cancer cases. *British Journal of Cancer* **83** 1301–1308.

Angrist M, Kauffman E, Slaugenhaupt SA, Matise TC, Puffenberger EG, Washington SS, Lipson A, Cass DT, Reyna T, Weeks DE, Sieber W and Chakravarti A 1993 A gene for Hirschsprung disease (megacolon) in the pericentromeric region of human chromosome 10. *Nature Genetics* **4** 351–356.

Anonymous 1891 Book Review: *The Right Honourable Arthur MacMurrough Kavanagh: a Biography.* By his cousin, Sarah L Steele, Macmillan and Co., London and New York, 1891 *Lancet* **i** 608.

Anonymous 2003a Future visions. *Time* **161**(7) (17 Feb. 2003) 60–61.

Anonymous 2003b The International HapMap Project. *Nature* **426** 789–796.

Anonymous 2004a Finishing the euchromatic sequence of the human genome. *Nature* **431** 931–945.

Anonymous 2004b The ENCODE (ENCyclopedia Of DNA Elements) Project. *Science* **306** 636–640.

Anonymous 2004c Withdrawal syndrome. (Editorial) *Nature Medicine* **10** 1143.

Ansquer A, Gautier C, Fourquet A, Asselian B and Stoppa-Lyonnet D 1998 Survival in early-onset BRCA1 breast-cancer patients. *The Lancet* **352** 541.

Antley RM and Bixler D 1975 Trapezoidocephaly, midfacial hypoplasia and cartilage abnormalities with multiple synostoses and skeletal fractures. *Birth Defects* **11** 397–401.

Antonarakis SE and McKusick VA 2000 OMIM passes the 1,000-disease-gene mark. *Nature Genetics* **25** 11.

Antoniou A, Pharoah PD, Narod S, Risch HA, Eyfjord JE, Hopper JL, Loman N, Olsson H, Johannsson O, Borg A, Pasini B, Radice P, Manoukian S, Eccles DM, Tang N, Olah E, Anton-Culver H, Warner E, Lubinski J, Gronwald J, Gorski B, Tulinius H, Thorlacius S, Eerola H, Nevanlinna H, Syrjakoski K, Kallioniemi OP, Thompson D, Evans C, Peto J, Lalloo F, Evans DG and Easton DF 2003 Average risks of breast and ovarian cancer associated with BRCA1 and BRCA2 mutations detected in case series unselected for family history: a combined analysis of 22 studies. *American Journal of Human Genetics* **72** 1117–1130.

Anuario estadístico de salud 2000. Ministerio de Salud Publica de la Republica de Cuba. *www.sld.cu/anuario/indice.html*

Appleyard B 1998 *Brave New Worlds. Staying Human in the Genetic Future.* Viking, New York, pp. 129, 133, 137.

Arakawa Y, Nishida-Umehara C, Matsuda Y, Sutou S and Suzuki H 2002 X-chromosomal localization of mammalian Y-linked genes in two XO species of the Ryukyu spiny rat. *Cytogenetics and Genome Research* **99** 303–309.

Arbuzova S 1998 Why it is necessary to study the role of mitochondrial genome in trisomy 21 pathogenesis? *Down Syndrome Research and Practice* **5** 126–130.

Arbuzova S, Cuckle H, Mueller R and Sehmi I 2001 Familial Down syndrome: evidence supporting cytoplasmic inheritance. *Clinical Genetics* **60** 456–462.

Arias S 1971 Genetic heterogeneity in the Waardenburg syndrome. *Birth Defects Original Article Series* **7** 87–101.

Aristotle 1912 *De Generatione Animalium.* (English translation by A Platt). Clarendon Press, Oxford.

Aristotle 1970 *Historia Animalium.* (English translation by AL Peck). William Heinemann, London.

Aristotle *Diogenes Laertius, Lives of Eminent Philosophers* scanned for Peithô's Web from The Lives and Opinions of Eminent Philosophers translated by CD Yonge. http://www.classicpersuasion.org/pw/diogenes/index.htm

Arnes JB, Brunet JS, Stefansson I, Begin LR, Wong N, Chappuis PO Asklen LA and Foulkes WD 2005 Placental cadherin and the basal epithelial phenotype of BRCA1-related breast cancer. *Clinical Cancer Research* **11** 4003–4011.

Arpagaus M, Kott M, Vatsis KP, Bartels CF, La Du BN and Lockridge O 1990 Structure of the gene for human butyrylcholinesterase. Evidence for a single copy. *Biochemistry* **29** 124–131.

Arteaga CL and Baselga J 2004 Tyrosine kinase inhibitors: why does the current process of clinical development not apply to them? *Cancer Cell* **5** 525–531.

Ashburner M, Ball CA, Blake JA, Botstein D, Butler H, Cherry JM, Davis AP, Dolinski K, Dwight SS, Eppig JT, Harris MA, Hill DP, Issel-Tarver L, Kasarskis A, Lewis S, Matese JC, Richardson JE, Ringwald M, Rubin GM and Sherlock G 2000 Gene ontology. Tool for unification of biology. The Gene Ontology Consortium. *Nature Genetics* **25** 25–29.

Asher RA, Morgenstern DA, Properzi F, Nishiyama A, Levine JM and Fawcett JW 2005 Two separate metalloproteinase activities are responsible for the shedding and processing of the NG2 proteoglycan in vitro. *Molecular and Cellular Neuroscience* **29** 82–96.

Ashley Montagu MF 1942 *Man's Most Dangerous Myth: The Fallacy of Race*, Harper and Brothers, New York.

Ashworth A 2003 Institute of Cancer Research, Personal communication.

Atherton JC, Cao P, Peek RM Jr, Tummuru, MK, Blaser, MJ, Cover TL 1995 Mosaicism in vacuolating cytotoxin alleles of *Helicobacter pylori*. *Journal of Biological Chemistry* **270** 17771–17777.

Atherton JC, Cover TL, Twells RJ, Morales MR, Hawkey CJ and Blaser MJ 1999 Simple and accurate PCR-based system for typing vacuolating cytotoxin alleles of *Helicobacter pylori*. *Journal of Biological Chemistry* **37** 2979-2982.

Atkin WS, Morson BC and Cuzick J 1992 Long-term risk of colorectal cancer after excision of rectosigmoid adenomas. *New England Journal of Medicine* **326** 658–662.

Attié T, Pelet A, Edery P, Eng C, Mulligan LM, Amiel J, Boutrand L, Beldjord C, Nihoul-Fekete C, Munnich A, Ponder BAJ and Lyonnet S 1995 Diversity of RET proto-oncogene mutations in familial and sporadic Hirschsprung disease. *Human Molecular Genetics* **4** 1381–1386.

Auburger G, Orozco G, Ferreira R, Gispert S, Paradoa M, Estrada M, García M, Farral M, Williamson R, Chamberlain S and Heredero L 1990 Autosomal dominant ataxia. Evidence for locus heterogeneity from a Cuban founder-effect population. *American Journal of Human Genetics* **46** 1163–1177.

Auricchio A, Brancolini V, Casari G, Milla PJ, Smith VV, Devoto M and Ballabio A 1996 The locus for a novel syndromic form of neuronal intestinal pseudoobstruction maps to Xq28. *American Journal of Human Genetics* **58** 743–748.

Australian Institute of Health and Welfare 2001 *Heart, Stroke and Vascular Diseases: Australian Facts 2001*. AIHW CVD14. Canberra, AIHW.

Australian Institute of Health and Welfare 2005 *Living Dangerously: Australians with Multiple Risk Factors for Cardiovascular Disease*. AIHW Bulletin No. 24. Canberra, AIHW.

Auwerx J, Avner P, Baldock R, Ballabio A, Balling R, Barbacid M, Berns A, Bradley A, Brown S, Carmeliet P, Chambon P, Cox R, Davidson D, Davies K, Duboule D, Forejt J, Granucci F, Hastie N, Hrabé de Angelis M, Jackson I, Kioussis D, Kollias G, Lathrop M, Lendahl U, Malumbres M, von Melchner H, Müller W, Partanen J, Ricciardi-Castagnoli P, Rigby P, Rosen B, Rosenthal N, Skarnes B, Stewart AF, Thornton J, Tocchini-Valentini G, Wagner E, Wahli W and Wurst W 2004 The European dimension for the mouse genome mutagenesis program. *Nature Genetics* **36** 925–927.

Avise JC 2000 *Phylogeography: The History and Formation of Species*. Harvard University Press, Boston.

Avramopoulos D, Mikkelsen M, Vassilopoulos D, Grigoriadou M and Petersen MB 1996 Apolipoprotein E allele distribution in parents of Down's syndrome children. *The Lancet* **347** 862–865.

Aymé S and Lippman-Hand A 1982 Maternal-age effect in aneuploidy: does altered embryonic selection play role? *American Journal of Human Genetics* **59**: 558–565.

Aymé S and Philip N 1995 Possible homozygous Waardenburg syndrome in a fetus with exencephaly. *American Journal of Medical Genetics* **59** 263–265.

Bach JF 1995 IDDM is a β-cell targeted disease of immunoregulation. *Journal of Autoimmunity* **8** 439–464.

Badner JA, Sieber WK, Garver KL and Chakravarti A 1990 A genetic study of Hirschsprung disease. *American Journal of Human Genetics* **46** 568–580.

Baehr W and Chen CK 2001 RP11 and RP13: unexpected gene loci. *Trends in Molecular Medicine* **7** 484–486.

Baker CG 2004 An administrative history of NCI's viruses and cancer programs, 1950–1972. http://history.nih.gov/articles/SpecialVirusCaPrgm.pdf

Baker PJ, Moore HDM, Penfold LM, Burgess AMC and Mittwoch U 1990 Gonadal sex differentiation in the neonatal marsupial, *Monodelphis domestica*. *Development* **109** 699–704.

Balasubramanian AS and Bhanumathy CD 1993 Noncholinergic functions of cholinesterases. *FASEB Journal* 7 1354–1358.

Balci B, Uvanik G, Dincer P, Gross C, Willer T, Talim B, Haliloglu G, Kale G, Hehr U, Winkler J and Topaloglu H 2005 An autosomal recessive limb girdle dystrophy (LGMD2) with mild mental retardation is allelic to Walker-Warburg syndrome (WWS) caused by a mutation in the POMT1 gene. *Neuromuscular Disorders* 15 271–275.

Baldini A 2004 DiGeorge syndrome: an update. *Current Opinion in Cardiology* 19 201–204.

Baldwin CT, Hoth CF, Amos JA, da-Silva EO and Milunsky A 1992 An exonic mutation in the HuP2 paired domain gene causes Waardenburg's syndrome. *Nature* 355 637–638.

Balmain A, Gray J and Ponder B 2003 The genetics and genomics of cancer. *Nature Genetics* 33 Suppl 238–244.

Banerjee-Basu S and Baxevani AD 1999 Threading analysis of the Pitx2 homeodomain: predicted structural effects of mutations causing Rieger syndrome and iridogoniodysgenesis. *Human Mutation* 14 312–319.

Bankier A and Rose CM 1994 Varadi syndrome or Opitz trigonocephaly: overlapping manifestation in two cousins. *American Journal of Medical Genetics* 53 85–86.

Barkai G, Arbuzova S, Berkenstadt M, Heifetz S and Cuckle H 2003 Frequency of Down's syndrome and neural-tube defects in the same family. *The Lancet* 361 1331–1335.

Barnicot NA, Ellis JR and Penrose LS 1963 Translocations and trisomic mongol sibs. *Annals of Human Genetics* 26 279–285.

Barrai I, Rodriguez-Larralde A, Mamolini E and Scapoli C 1999 Isonymy and isolation by distance in Italy. *Human Biology* 71 947–961.

Barrai I, Rodriguez-Larralde A, Mamolini E, Manni F and C. Scapoli 2000 Elements of the surname structure of Austria. *Annals of Human Biology* 27 607–622.

Barrai I, Rodriguez-Larralde A, Manni F, Ruggiero V, Tartari D and Scapoli C 2004 Isolation by language and distance in Belgium. *Annals of Human Genetics* 68 1–16.

Barrai I, Scapoli M, Beretta C, Nesti C, Mamolini E and Rodriguez-Larralde A 1996 Isonymy and the genetic structure of Switzerland. I. The distributions of surnames. *Annals of Human Biology* 23 431–455.

Barrai I, Scapoli M, Beretta C, Nesti C, Mamolini E and Rodriguez-Larralde A 1997 Isolation by distance in Germany. *Human Genetics* 100 684.

Barrios B and Gutiérrez E 2001 4 años de Experiencia en el Diagnóstico de Errores Innatos del Metabolismo en Población con riesgo. *III Jornada Provincial de Ciencias Médicas 'Girón 2001'*. Libro de resúmenes.

Barrios B, Gutiérrez E, Cuadras Y and Damián A 2001 Programa Nacional de Prevención de Hiperfenilalaninemias en Cuba. *III Jornada Provincial de Ciencias Básicas Médicas 'Girón 2001'*. Libro de resúmenes.

Barrios B, Heredero L, Damiani A, Gutiérrez E, Cobet G and Machill G 1989 Estudio de Hiperfenilalaninemias en el Occidente de Cuba. *Revista Cubana de Pediatría* 61 99–106.

Bartels CF, James K and La Du BN 1992a DNA mutations associated with the human butyryl-cholinesterase J-variant. *American Journal of Human Genetics* 50 1104–1114.

Bartels CF, Jensen FS, Lockridge O, van der Spek AFL, Rubinstein HM, Lubrano T and La Du BN 1992b DNA mutation associated with the human butyrylcholinesterase K-variant and its linkage to the atypical variant mutation and other polymorphic sites. *American Journal of Human Genetics* 50 1086–1103.

Bartels CF, van der Spek AFL and La Du BN 1990 Two polymorphisms in the non coding regions of the BCHE gene. *Nucleic Acids Research* 18 6171.

Bartmann AK, Araujo FM, Iannetta O, Paneto JC, Martelli L and Ramos ES 2005 Down syndrome and precocious menopause. *Journal of Assisted Reproduction and Genetics* 22 129–131.

Bartsocas CS, Dacou-Voutetakis C, Damianaki D, Karayanni CH, Kassiou C, Quadreh A, Theodoridis CH, Tsoka H and Green A 1998 Epidemiology of childhood IDDM in Athens: trends in incidence for the years 1989–1995. EURODIAB ACE G1 Group. *Diabetologia* 41 245–248.

Bashir R, Britton S, Strachan T, Keers S, Vafiadaki E, Lako M, Richard I, Marchand S, Bourg N, Argov Z, Sadeh M, Mahjneh I, Marconi G, Passos-Bueno MR, de S Moreira E, Zatz M, Beckmann JS and Bushby K 1998 A gene related to *Caenorhabditis elegans* spermatogenesis factor fer-1 is mutated in limb-girdle muscular dystrophy type 2B. *Nature Genetics* **20** 37–42.

Bastos-Rodrigues L, Pimenta JR and Pena SDJ 2006 The genetic structure of human populations studied through short insertion-deletion polymorphisms. *Annals of Human Genetics* **70** 310–317.

Battilana J, Fagundes NJR, Heller AH, Goldani A, Freitas LB, Tarazona-Santos E, Munkhbat B, Munkhtuvshin N, Tsuji K, Krylov M, Benevolenskaia L, Arnett FC, Batzer MA, Deininger PL, Salzano FM and Bonatto SL 2006 *Alu* insertion polymorphisms in Native Americans and related Asian populations. *Annals of Human Biology* **33** 142–160.

Bayer ME, Blumberg BS and Werner B 1968 Particles associated with Australia antigen in the sera of patients with leukemia, Down's syndrome and hepatitis. *Nature* **218** 1057–1059.

Bayes T 1731 *Divine Benevolence.* John Noon, London, p. 46. (reprinted in AI Dale 2003. *Most honourable remembrance: the life and work of Thomas Bayes*, Springer-Verlag, Berlin p. 125)

Baynash AG, Hosoda K, Giaid A, Richardson JA, Emoto N, Hammer RE and Yanagisawa M 1994 Interaction of endothelin-3 with endothelin-B receptor is essential for development of epidermal melanocytes and enteric neurons. *Cell* **79** 1277–1285.

Beadle GW 1959 Genes and chemical reactions in Neurospora. *Science* **129** 1715–1719.

Beadle GW and Tatum EL 1941 Genetic control of biochemical reactions in Drosophila. *Proceedings of the National Academy of Sciences USA* **27** 499–506.

Beales PL 2005 Lifting the lid on Pandora's box: the Bardet-Biedl syndrome. *Current Opinions in Genetics and Development* **15** 315–323.

Beckmann JS 1996 The Reunion paradox and the digenic model. *American Journal of Human Genetics* **59** 1400–1402.

Beckwith JB 1963 Extreme cytomegaly of the adrenal fetal cortex, omphalocele, hyperplasia of kidneys and pancreas, and Leydig-cell hyperplasia: another syndrome? Presented at 11th Annual Meeting of Western Society for Pediatric Research, Los Angeles, November 11, 12, 1963. No 20.

Beckwith JB, Kiviat NB and Bonadio J 1990 Nephrogenic rests, nephroblastomatosis, and the pathogenesis of Wilms' tumor. *Pediatric Pathololology* **10** 1–36.

Beerman F, Hummler E, Franke U and Hansmann I 1988 Maternal modulation of the inheritable meiosis I error Dipl I in mouse oocytes is associated with the type of mitochondrial DNA. *Human Genetics* **79** 338–340.

Ben Yoau R, Muchir A, Armura T, Massart C, Demay L, Richard P and Bonne G 2005 Genetics of laminopathies. *Novartis Foundation Symposia* **264**, 81–90; discussion 90–97, 227–30.

Bennett RL, Hudgins L, Smith CO and Motulsky AG 1999 Inconsistencies in genetic counseling and screening for consanguineous couples and their offspring: the need for practice guidelines. *Genetics in Medicine* **6** 286–292.

Benson DA, Karsch-Mizrachi I, Lipman DJ, Ostell J and Wheeler DL 2004 GenBank: update. *Nucleic Acids Research* **32** D23–D26.

Beral V, Million Women Study Collaborators 2003 Breast cancer and hormone replacement therapy in the Million Women Study. *The Lancet* **362** 419–427.

Berg C, Geipel A, Germer U, Pertersen-Hansen A, Koch-Dorfler M and Gembruch U 2001 Prenatal detection of Fraser syndrome without cryptophthalmos: case report and review of the literature. *Ultrasound in Obstetrics and Gynecology* **18** 76–80.

Bergh T, Ericson A, Hillensjo T, Nygren KG and Wennerholm UB 1999 Deliveries and children born after in-vitro fertilisation in Sweden 1982–95: a retrospective cohort study. *The Lancet* **354** 1579–1585.

Bermudez MG, Piyamongkol W, Tomaz S Dudman E, Sherlock JK and Wells D 2003 Single-cell sequencing and mini-sequencing for preimplantation genetic diagnosis. *Prenatal Diagnosis* **23** 669–677.

Bernstein F 1924 Ergebnisse einer biostatistischen zusammenfassen Betrachtung über die erblichen Blutstrukturen des Menschen. *Wiener Klinische Wochenschrift* **3** 1495–1497.

Bernstein F 1925 Zusammenfassende Betrachtungen über die erblichen Blutstrukturen des Menschen. *Zeitschrift für Induktive Abstammungs- und Vererbungslehre* **37** 237–270.

Berrettini S, Neri E, Forli F, Panconi M, Massimetti M, Ravecca F, Sellari-Franceschini S and Bartozolozzi C 2001 Large vestibular aqueduct in distal renal tubular acidosis. High-resolution MR in three cases. *Acta Radiologica* **42** 320–322.

Berry WTC, Cowin PJ and Davies DR 1954 A relationship between body fat and plasma pseudo-cholinesterase. *British Journal of Nutrition* **8** 79–82.

Bertolí A, Girouod J, Vincenso B, Álvarez E, Heredero L, van Duijn C and Heutink P 2003 Suggestive linkage to chromosome 19 in a large Cuban family with late onset Parkinson disease. Movement Disorders, published on line in Wiley InterScience DOI 10.1002 mds.10534. *www.interscience.wiley.com*

Bertoli A, Marcheco B, Llibre J, Gómez N, Borrajero I, Severijen E, Joose M, van Duijn C, Heredero L and Heutink P 2002 A novel presenilin 1 mutation (L174 M) in a large Cuban family with early onset Alzheimer disease. *Neurogenetics* **4** 97–104.

Bertwistle D and Ashworth A 1998 Functions of the BRCA 1 and BRCA 2 genes. *Current Opinion in Genetics and Development* **8** 14–20.

Biben C, Hatzistavrou T and Harvey RP 1998 Expression of NK-2 class homeobox gene Nkx2–6 in foregut endoderm and heart. *Mechanisms of Development* **73** 125–127.

Bielschowsky M and Schofield GC 1962 Studies on megacolon in piebald mice. *Australian Journal of Experimental Biology and Medical Science* **40** 395–403.

Billerbeck AEC, Cavaviere H, Goldberg AC, Kalil J and Medeiros-Neto G 1994 Clinical and genetic studies in Pendred syndrome. *Thyroid* **4** 279–284.

Birch-Jensen A 1949 Congenital deformities of the upper extremities. *Opera ex Domo Biologiae Hereditariae Humanae Universitatis Hafniensis* **33** 19. Munksgaard, Copenhagen.

Birkebaek NH, Cruger D, Hansen J, Nielsen J and Bruun-Petersen G 2002. Fertility and pregnancy outcome in Danish women with Turner syndrome. *Clinical Genetics* **61** 35–39.

Birney E, Andrews D, Bevan P, Caccamo M, Cameron G, Chen Y, Clarke L, Coates G, Cox T, Cuff J, Curwen V, Cutts T, Down T, Durbin R, Eyras E, Fernandez-Suarez XM, Gane P, Gibbins B, Gilbert J, Hammond M, Hotz H, Iyer V, Kahari A, Jekosch K, Kasprzyk A, Keefe D, Keenan S, Lehvaslaiho H, McVicker G, Melsopp C, Meidl P, Mongin E, Pettett R, Potter S, Proctor G, Rae M, Searle S, Slater G, Smedley D, Smith J, Spooner W, Stabenau A, StalkerJ, Storey R, Ureta-Vidal A, Woodwark C, Clamp M and Hubbard T 2004 Ensembl 2004. *Nucleic Acids Research* **32** D468-D470.

Bittles A 2001 Consanguinity and its relevance to clinical genetics. *Clinical Genetics* **60** 89–98.

Bjarnason O, Bjarnason V, Edwards JH, Fridriksson S, Magnusson M, Mourant AE and Tills D 1973 The blood groups of Icelanders. *Annals of Human Genetics* **36** 425–458.

Bladt F, Tafuri A Gelkop S, Langille L and Pawson T 2002 Epidermolysis bullosa and embryonic lethality in mice lacking the multi-PDZ domain protein GRIP1 *Proceedings of the National Academy of Sciences USA* **99** 6816–6821.

Blake DL, Dean NL, Knight C, Tan SL and Ao A 2001 Direct comparison of detection systems used for the development of single-cell genetic tests in preimplantation genetic diagnosis. *Journal of Assisted Reproductive Genetics* **18** 557–565.

Blakemore C and Jennett S (editors) 2001 *The Oxford Companion to the Body* Oxford University Press, Oxford.

Blohmé J and Tornqvist K 1997 On visual impairment in Swedish children. III Diagnoses. *Acta Ophthalmologica Scandinavica* **75** 681–87

Bloom D 1954 Congenital telangiectatic erythema resembling lupus erythematosus in dwarfs. Probably a syndrome entity. *American Journal of Diseases of Children* **88** 754–758.

Blumberg BS 1964 Polymorphisms of serum proteins and the development of isoprecipitins in transfused patients. *Bulletin of the New York Academy of Medicine* **40** 377–386.

Blumberg BS 1972 Australia antigen: The history of its discovery with comments on genetic and family aspects. In: *Viral Hepatitis and Blood Transfusion.* (editors Vyas GN, Perkins HA and Schmid R) Grune & Stratton, New York, pp. 63–83.

Blumberg BS 1976 Bioethical questions related to hepatitis-B antigen. *American Journal of Clinical Pathology* **65** 848–853.

Blumberg BS and Gartler SM 1959 High prevalence of high level beta-amino-isobutyric acid excretors in Micronesians. *Nature* **184** 1990–1992.

Blumberg BS and Millman I 1972 *Vaccine Against Viral Hepatitis and Process, U.S. Patent Office No. 3,636,191.*

Blumberg BS and Robbins J 1961 In: *Advances in Thyroid Research*, (editor Rivers RPH) Pergamon, New York, vol. 2, p. 461.

Blumberg BS and Tombs MT 1958 Possible polymorphism of bovine alpha-lactalbumin. *Nature* **181** 683–684.

Blumberg BS, Allison AC and Gerry B 1959 The haptoglobins and haemoglobins of Alaskan Eskimos and Indians. *Annals of Human Genetics* **23** 349–356.

Blumberg BS, Alter HJ and Visnich S 1965 A 'new' antigen in leukemia sera. *Journal of the American Medical Association* **191** 541–546.

Blumberg BS, Dray S and Robinson JC 1962 Antigen polymorphism of a low-density beta-lipoprotein. Allotypy in human serum. *Nature* **194** 656–658.

Blumberg BS, Gerstley BJS, Hungerford DA, London WT and Sutnick AI 1967 A serum antigen (Australia antigen) in Down's syndrome leukemia and hepatitis. *Annals of Internal Medicine* **66** 924–931.

Blumberg BS, Larouzé B, London WT, Werner B, Hesser JE, Millman I, Saimot G and Payet M 1975 The relation of infection with the hepatitis B agent to primary hepatic carcinoma. *American Journal of Pathology* **81** 669–682.

Blumberg BS, London WT, Lustbader ED, Drew JS and Werner BG 1975 Protection vis-à-vis de l'hépatite B par l'anti-HBs chez des malades hémodialysés. In: *Hépatite à Virus B et Hémodialyse* Flammarion, Paris, pp. 175–183.

Blumberg BS, Millman I, Sutnick AI and, London WT 1971 The nature of Australia antigen and its relation to antigen-antibody complex formation *Journal of Experimental Medicine* **134** 320–329.

Blumberg BS, W. Wills, Millman I and London WT 1973 Australia antigen in mosquitoes. Feeding experiments and field studies. *Research Communications in Chemical Pathology and Pharmocology* **6** 719–732.

Böcker W, Moll R, Poremba C, Holland R, van Diest PJ, Dervan P Burger H, Wai D, Ina Diallo R, Brandt B, Herbst H, Schmidt A, Lerch MM and Buchwallow IB 2002 Common adult stem cells in the human breast give rise to glandular and myoepithelial cell lineages: A new cell biological concept. *Laboratory Investigation* **82** 737–745.

Bodian M and Carter CO 1963 Family study of Hirschsprung's disease. *Annals of Human Genetics* **29** 261–277.

Bodmer W 1999 Familial adenomatous polyposis (FAP) and its gene, APC. *Cytogenetics and Cell Genetics* **86** 99–104.

Boeck AT, Fry DL, Sastre A and Lockridge O 2002 Naturally occurring mutation, Asp70His, in human butyrylcholinesterase. *Annals of Clinical Biochemistry* **39** 154–156.

Bolande R 1973 The neurocristopathies: A unifying concept of disease arising in neural crest maldevelopment. *Human Pathology* **5** 409–429.

Bolk S, Pelet A, Hofstra RMW, Angrist M, Salomon R, Croaker D, Buys CHCM, Lyonnet S and Chakravarti A 2000 A human model for multigenic inheritance: phenotypic expression in Hirschsprung disease requires both the RET gene and a new 9q31 locus. *Proceedings of the National Academy of Science* **97** 268–273.

Bonaldo P, Braghetta P, Zanetti M, Piccolo S, Volpin D and Bressan GM 1998 Collagen VI deficiency induces early onset myopathy in the mouse: an animal model for Bethlem myopathy. *Human Molecular Genetics* **7** 2135–2140.

Bonduelle M, Aytoz A, Van Assche E, Devroey P, Liebaers I and Van Steirteghem A 1998 Incidence of chromosomal aberrations in children born after assisted reproduction through intracytoplasmic sperm injection. *Human Reproduction* **13** 781–782.

Bondurand N, Kuhlbrodt K, Pingault V, Enderich J, Sajus M, Tommerup N, Warburg M, Hennekam RC, Read AP, Wegner M and Gossens M 1998 A molecular analysis of the Yemenite deaf-blind hypopigmentation syndrome: SOX10 dysfunction causes different neurocristopathies. *Human Molecular Genetics* **8** 1785–1789.

Bonney GE and Konotey-Ahulu FID 1977 Polygamy and genetic equilibrium. *Nature* **265** 46–47.

Böök JA 1957 Genetic investigations on a north Swedish population: the offspring of first-cousin marriages. *Annals of Human Genetics* **21** 191–221.

Booth ML, Chey T, Wake M, Norton K, Hesketh K, Dollman J and Robertson I 2003 Changes in the prevalence of overweight and obesity among young Australians, 1969–1997 *American Journal of Clinical Nutrition* 77 29–36.

Borrego S, Saez ME, Ruiz A Gimm O, Lopez-Alonso M, Antinolo G and Eng C 1999 Specific polymorphism in the RET proto-oncogene are over-represented in patients with Hirschsprung disease and may represent loci modifying phenotypic expression. *American Journal of Human Genetics* **65** 1200–1205.

Bosch M, Rajmil O, Martinez-Pasarell O, Egozcue J and Templado C 2001 Linear increase of diploidy in human sperm with age: a four-colour FISH study. *European Journal of Human Genetics* **9** 533–538.

Boswell J 1791 *Life of Johnson* (editors Chapman RW and Fleeman JD) Oxford University Press, 1980. Passages cited, in order: Saturday 15 May 1784 p. 1279, Thursday 21 March 1776 p. 696, Thursday 15 May 1784 p. 1235, 1775 p. 662.

Boucekkine C, Toublanc JE, Abbas N, Semrouni M, Vilain, E, McElreavey K, Mugneret F and Fellous M 1992 The sole presence of the testis-determining region of the Y chromosome (*SRY*) in 46, XX patients is associated with phenotypic variability. *Hormone Research* **37** 236–240.

Boué A and Gallano P 1984 A collaborative study of the segregation of inherited chromosome arrangements in 1356 prenatal diagnoses. *Prenatal Diagnosis* **4** 45–67.

Bourne JC, Collier HOJ and Somers GF 1952 Succinylcholine (succinoyl-choline) muscle-relaxant of short action. *The Lancet* **1** 1225–1229.

Bovet D 1957 The relationships between isosterism and competitive phenomena in the field of drug therapy of the autonomic nervous system and that of the neuromuscular transmission. Nobel Lectures, Physiology or Medicine 1942–1962, Elsevier Publishing Company, Amsterdam, 1964. The Official Web Site of The Nobel Foundation: *http://nobelprize.org/medicine/laureates/1957/bovet-lecture.html.*

Box JF 1978 *R. A. Fisher: The Life of a Scientist* New York, John Wiley.

Boyd PA, Keeling JW and Lindenbaum H 1988 Fraser syndrome (cryptophthalmos-syndactyly syndrome): a review of eleven cases with postmortem findings. *American Journal of Medical Genetics* **31** 159–168.

Brabletz T, Jung A, Spaderna S, Hlubek F and Kirchner T 2005 Opinion: migrating cancer stem cells – an integrated concept of malignant tumour progression. *Nature Reviews Cancer* **5** 744–749.

Brain WR 1927 Hereditary and simple goitre. *Quarterly Journal of Medicine* **20** 303–319.

Bray I, Wright DE, Davies CJ and Hook EB 1998 Joint estimation of Down syndrome risk and ascertainment rates: a meta-analysis of nine published data sets. *Prenatal Diagnosis* **18** 9–20.

Bray IC and Wright DE 1998 Estimating the spontaneous loss of Down syndrome fetuses between the time of chorionic villus sampling and livebirth. *Prenatal Diagnosis* **18** 1045–1054.

Brennan J, Karl J and Capel B 2002 Divergent vascular mechanisms downstream of Sry establish the arterial system of the XY gonad. *Developmental Biology* **244** 418–428.

Breslow JL, Zannis VI, SanGiacomo TR, Third JL, Tracy T and Glueck CJ 1982 Studies of familial type III hyperlipoproteinemia using as a genetic marker the apoE phenotype E2/2. *Journal of Lipid Research* **23** 1224–1235.

Breslow NE, Norris R, Norkool PA, Kang T, Beckwith JB, Perlman EJ, Michael L. Ritchey ML, Green DM and Nichols KE 2003 Characteristics and outcomes of children with Wilms tumor-aniridia syndrome: A report from the National Wilms Tumor Study Group. *Journal of Clinical Oncology* **21** 4579–4585.

Bridges CB 1916 Non-disjunction as proof of the chromosome theory of heredity. *Genetics* **1** 1–52; 107–163.

Bridges CB 1925 Sex in relation to chromosomes and genes. *American Naturalist.* **59** 127–137.

Bridges CB 1939 Cytological and genetic basis of sex. In: *Sex and Internal Secretions* (editor Allen E) Bailliere, London 2nd edn, pp.15–63

Brockbank EM 1944 *John Dalton, Some unpublished letters of personal and scientific interest.* Manchester University Press, Manchester.

Brodsky I and Waddy G 1940 Cryptothalmos or ablepharià: A survey of the condition with a review of the literature and the presentation of a case. *Medical Journal of Australia* 1 894–898.

Broman KW, Murray JC, Sheffield VC, White RL and Weber J 1998 Comprehensive human genetic maps: individual and sex-specific variation in recombination. *American Journal of Human Genetics* 63 861–869.

Bronstein IN, Semendjajew KA and Hirsch KA 2002 *Handbook of Mathematics.* Springer Verlag, Heidelberg.

Brook JD, Gosden RG and Chandley AC 1984 Maternal ageing and aneuploid embryos – Evidence from the mouse that biological and not chronological age is the important influence. *Human Genetics* 66 41–45.

Brose MS, Rebbeck TR, Calzone KA, Stopfer JE, Nathanson KL and Weber BL 2002 Cancer risk estimates for BRCA1 mutation carriers identified in a risk evaluation program. *Journal of the National Cancer Institute* 94 1365–1372.

Brown CB, Wenning JM, Lu MM, Epstein DJ, Meyers EN and Epstein JA 2004 Cre-mediated excision of Fgf8 in the Tbx1 expression domain reveals a critical role for Fgf8 in cardiovascular development in the mouse. *Developmental Biology* 267 190–202.

Brownlee J 1905 Statistical studies in immunity: smallpox and vaccination *Biometrika* 4 313–331.

Brunet J-S, Arnes J, Nielsen T, Huntsman D, Wong N, Begin LR, Akslen LA and Foulkes WD 2004 Extending the basal phenotype of BRCA1-related breast cancer. American Society of Human Genetics. Abstract 527.

Brunner HG and van Driel MA 2004 From syndrome families to functional genomics. *Nature Reviews Genetics* 5 545–551.

Brunner HG, Nelen M, Breakefield XO Ropers HH and van Oost BA 1993 Abnormal behavior associated with a point mutation in the structural gene for monoamine oxidase A. *Science* 262 578–580.

Bruwer A, Bargen JA and Kierland RR 1954 Surface pigmentation and generalized intestinal polyposis (Peutz-Jeghers syndrome). *Proceedings of the Staff Meetings of the Mayo Clinic* 29 168–171.

Buchanan A, Brock DW, Daniels N and Wikler D 2000 *From Chance to Choice.* Cambridge University Press, Cambridge, pp. 30, 42–55, 261.

Buchholz TA, Wu X, Hussain A, Tucker SL, Mills GB, Haffty B, Bergh S, Story M, Geara FB, and Brock WA 2002 Evidence of haplotype insufficiency in human cells containing a germline mutation in BRCA1 or BRCA2. *International Journal of Cancer* 97 557–561.

Buehr M, Gu S and McLaren A 1993 Mesonephric contribution to testis differentiation in the fetal mouse. *Development* 117 273–281.

Bundy S and Alam H 1993 A five-year perspective study of the health of children in different ethnic groups, with particular reference to the effects of inbreeding. *European Journal of Human Genetics* 1199 206–219.

Buratti E, Brindisi A, Pagani F and Baralle FE 2004 Nuclear factor TDP-43 binds to the polymorphic TG repeats in CFTR intron 8 and causes skipping of exon 9: a functional link with disease penetrance. *American Journal of Human Genetics* 74 1322–1325.

Burg MA, Tillet E, Timpl R and Stallcup WB 1996 Binding of the NG2 proteoglycan to type VI collagen and other extracellular matrix molecules. *Journal of Biological Chemistry* 271 26110–26116.

Burgoyne PS, Thornhill AR, Boudrean SK, Darling SM, Bishop CE and Evans EP 1995 The genetic basis of XX-XY difference present before gonadal sex differentiation in the mouse. *Philosophical Transactions of the Royal Society of London Series* B 253–261.

Burzynski GM, Nolte IM, Osinga J, Ceccherini I, Twigt B, Maas S, Brooks A, Verheij J, Menacho IP, Buys, CHCM and Hofstra RMW 2004 Localizing a putative mutation as the major contributor to the development of sporadic Hirschsprung disease to the RET genomic sequence between the promoter region and exon 2. *European Journal of Human Genetics* 12 604–612.

Buynak EB, Roehm RR, Tytell AA, Bertland AU, Lampson GP and Hiiteman MR 1976 Vaccine against human hepatitis B. *Journal of the American Medical Association* 235 2832–2834.

Buyse ML 1990 *Birth defects encyclopedia: the comprehensive, systematic, illustrated reference source for the diagnosis, delineation, etiology, biodynamics, occurrence, prevention, and treatment of human anomalies of clinical relevance.* Blackwell Scientific Publications -Year Book Medical Publications, Massachusetts-Chicago.

Cacheux V, Dastot-Le Moal F, Kääriäinen H, Bondurand N, Rintala R, BoissierB, Wilson M, Mowat D and Gossens M 2001 Loss-of-function mutations in SIP1 Smad interacting protein-1 result in a syndromic Hirschsprung disease. *Human Molecular Genetics* **10** 1503–1510.

Cai CL, Liang X, Shi Y, Chu PH, Pfaff SL, Chen J and Evans S 2003 Isl1 identifies a cardiac progenitor population that proliferates prior to differentiation and contributes a majority of cells to the heart. *Developmental Cell* **5** 877–889.

Callegari-Jacques SM and Salzano FM 1999 Brazilian Indian/non-Indian interactions and their effects. *Ciência e Cultura* **51** 166–174.

Calvari V, Bertini V, DeGrandi A, Peverali G, Zuffardi O, Ferguson-Smith M, Knudtzon J, Camerino G, Borsani G and Guioli S 2000 A new submicroscopic deletion that refines the 9p region for sex reversal. *Genomics* **65** 203–212.

Camera G, Marasini M, Pozzolo S and Camera A 1994 Oral-facial-digital syndrome: report on a transitional type between the Mohr and Varadi syndromes in a fetus. *American Journal of Medical Genetics* **53** 196–198.

Cameron C and Williamson R 2003 Is there an ethical difference between preimplantation genetic diagnosis and abortion? *Journal of Medical Ethics* **29** 90–92.

Cannon-Albright LA, Skolnick MH, Bishop DT, Lee RG and Burt RW 1988 Common inheritance of susceptibility to colonic adenomatous polyps and associated colorectal cancers. *New England Journal of Medicine* **319** 533–537.

Carango P, Noble JE, Marks HG and Funanage VL 1993 Absence of myotonic dystrophy protein kinase (DMPK) mRNA as a result of a triplet repeat expansion in myotonic dystrophy. *Genomics* **18** 340–348.

Carcassi V, Ceppellini R and Pitzus F 1957 Frequenza della talassemia in quattro popolazioni sarde e suoi rapporti con la distribuzione dei gruppi sanguini e della malaria. *Bollettino dell Istituto Sieroterapico Milanese* **36** 206.

Cardon LR and Abecasis GR 2003 Using haplotype blocks to map human complex trait loci. *Trends in Genetics* **19** 135–140.

Cardon LR and Bell JI 2001 Association study designs for complex diseases. *Nature Reviews Genetics* **2** 91–99.

Cardon LR and Palmer LJ 2003 Population stratification and spurious allelic association. *The Lancet* **361** 598–604.

Carothers AD, Hecht CA and Hook EB 1999 International variation in reported livebirth prevalence rates of Down syndrome, adjusted for maternal age. *Journal of Medical Genetics* **36** 386–393.

Carrasquillo MM, McCallion AS, Puffenberger EG, Kashuk CS, Nouri N and Chakravarti A 2002 Genome-wide association study and mouse model identify interaction between RET and EDNRB pathways in Hirschsprung disease. *Nature Genetics* **32** 237–244.

Carriline M, Marshall J and Walker J 1984 *Department of Health and Social Security Report of the committee of inquiry into human fertilisation and embryology: Expression of dissent: B. Use of human embryos in research.* Her Majesty's Stationary Office, London, pp. 90–94.

Carvalho-Silva DR, Santos FR, Rocha J and Pena SDJ 2001 The phylogeography of Brazilian Y chromosome lineages. *American Journal of Human Genetics* **681** 281–286.

Carvalho-Silva DR, Tarazona-Santos E, Rocha J, Pena SDJ and Santos FR 2006 Y chromosome diversity in Brazilians: switching perspectives from slow to fast evolving markers. *Genetica* **126** 251–260.

Casana M, Menéndez F, Quintana J, Quiñones O and Lavista M 1986 Cytogenetic prenatal diagnosis. Experiences and results from May 1984 to January 1985. *Revista Cubana de Pediatría* **58** 745–752.

Caspi A, McClay J, Moffitt TE, Mill J, Martin J, Craig IW, Taylor A and Poulton R 2002 Role of genotype in the cycle of violence in maltreated children. *Science* **297** 851–854

Castilla EE and Sod R 1990 The surveillance of birth defects in South America. II. The search for geographic clusters: Endemics. In: *Advances in Mutagenesis Research*, Springer-Verlag, Berlin, pp. 211–230.

Cavalli-Sforza LL and Bodmer W 1971 *Human Population Genetics.* Freeman, San Francisco.

Cavalli-Sforza LL, Menozzi P and Piazza A 1994 *The History and Geography of Human Genes.* Princeton University Press, Princeton, pp. 273–276.

Center EM and Emery KE 1997 Acidic glycosaminoglycans and laminin-1 in renal corpuscles of mutant blebs (my) and control mice. *Histology and Histopathology* **12** 901–907.

Ceppellini R, Curtoni ES, Mattiuz PL, Miggiano V, Schudeller G and Serra A 1967 Genetics of leucocyte antigens. A family study of segregation and linkage. In: *Histocompatibility Testing* (eds Curtoni ES, Mattiuz PL and Tosi RM) Munksgaard, Copenhagen.

Ceppellini R, Siniscalco M and Smith CAB 1955 The estimation of gene frequencies in a random mating population. *Annals of Human Genetics* **20** 97–115.

Chakravarti A 1996 Endothelin receptor-mediated signaling in Hirschsprung disease. *Human Molecular Genetics* **5** 303–307.

Chakravarti A and Lyonnet S 2001 Hirschsprung disease, p. 6231–55 In: *The Metabolic & Molecular Bases of Inherited Disease.* 8th edition, C.R. Scriver *et al.*, editors, McGraw-Hill Med. Publishing Division, New York.

Chan A, McCaul KA, Keane RJ and Haan EA 1998 Effect of parity, gravidity, previous miscarriage, and age on risk of Down's syndrome: population based study. *British Medical Journal* **317** 923–924.

Chan KC, Knox WF, Gee JM, Morris J, Nicholson RI, Potten CS and Bundred NJ 2002 Effect of epidermal growth factor receptor tyrosine kinase inhibition on epithelial proliferation in normal and premalignant breast. *Cancer Research* **62** 122–128.

Chance PF, Cavalier L, Satran D, Pellegrino JE, Koenig M and Dobyns WB 1999 Clinical nosologic and genetic aspects of Joubert and related syndromes. *Journal of Child Neurology* **10** 660–666.

Chang EH, Mebezes M, Meyer NC, Cucci RA, Vervoort VS, Schwartz CE and Smith RJH 2004 Branchio-Oto-Renal Syndrome: the mutation spectrum in *EYA1* and its phenotypic consequences. *Human Mutation* **23** 582–589.

Chang JC, Wooten EC, Tsimelzon A, Hilsenbeck SG, Gutierrez MC, Elledge R, Mohsin S, Osborne CK, Chamness GC, Allred DC and O'Connell P 2003 Gene expression profiling for the prediction of therapeutic response to docetaxel in patients with breast cancer. *The Lancet* **362** 362–369.

Chautard-Freire-Maia EA, Carvalho RDS, Silva MCBO, Souza MGF and Azevêdo ES 1984a Frequencies of atypical serum cholinesterase in a mixed population of Northeastern Brazil. *Human Heredity* **34** 364–370.

Chautard-Freire-Maia EA, Primo-Parmo SL, Lourenço MA and Culpi L 1984b Frequencies of atypical serum cholinesterase among Caucasians and Negroes from Southern Brazil. Human *Heredity* **34** 388–392.

Chautard-Freire-Maia EA, Primo-Parmo SL, Picheth G, Lourenço MAC and Vieira MM 1991 The C_5 isozyme of serum cholinesterase and adult weight. *Human Heredity* **41** 330–339.

Checler F, Grassi J, Masson P and Vincent JP 1990 Monoclonal antibodies allow precipitation of esterasic but not peptidasic activities associated with butyrylcholinesterase. *Journal of Neurochemistry* **55** 750–755.

Chen C-L, Gilbert TJ and Daling JR 1999 Maternal smoking and Down syndrome: the confounding effect of maternal age. *American Journal of Epidemiology* **149** 442–446.

Chen CY, Croissant J, Majesky M, Topouzis S, McQuinn T, Frankovsky MJ and Schwartz RJ 1996 Activation of the cardiac alpha-actin promoter depends upon serum response factor, Tinman homologue, Nkx-2.5, and intact serum response elements. *Developmental Genetics* **19** 119–130.

Chiari H 1883 Congenitales ankylo- et synblepharon und congenitale atresia laryngis bei einem mit mehrfachen anderweitigen Bildungsanomalien Kind. *Zeitschrift für Heilkunde Prag* **4** 143–154.

Chitayat D, Stalker HJ and Azouz EM 1992 Autosomal recessive oral-facial-digital syndrome with resemblance to OFD types II, III, IV and VI: a new OFD syndrome? *American Journal of Medical Genetics* **44** 567–572.

Christiano AM, Greenspan DS, Hoffman GG, Zhang X, Tamai Y, Lin AN, Dietz HC, Hovnanian A and Uitto J 1993 A missense mutation in type VII collagen in two affected siblings with recessive dystrophic epidermolysis bullosa. *Nature Genetics* **4** 62–66.

Clark AG 1999 The size distribution of homozygous segments in the human genome. *American Journal of Human Genetics* **65** 1489–1492.

Clark AG Hubisz MJ, Bustamente CD, Williamson SH and Nielsen R 2005 Ascertainment bias in studies of human genome-wide polymorphism. *Genome Research* 15 1496–1502.

Clarke AC 1962 *Profiles of the Future: An Inquiry into the Limits of the Possible* Victor Gollancz Ltd, London.

Clayton CL, Kleanthous H, Coates PJ, Morgan DD and Tabaqchali S 1992 Sensitive detection of *Helicobacter pylori* by using polymerase chain reaction. *Journal of Clinical Microbiology* 30 192–200.

Cleper R, Kauschansky A, Varsano I and Frydman M 1993 Váradi syndrome (OFD VI) or Opitz trigono-cephaly syndrome: overlapping manifestations in two cousins. *American Journal of Medical Genetics* 47 451–455.

Coelho LGV, Barros CAS, Lima DCA, Barbosa AJA, Magalhães AFN, Oliveira CA, Queiroz DMM, Cordeiro F, Rezende JM, Castro LP, Tolentino MM, Haddard MT and Zaterka S 1996 National consensus on 'H. pylori' and associated infections. *GED* 15 53–58.

Cohan FM 1996 The role of genetic exchange. *ASM News* 62 631–636.

Cohen JC, Kiss RS, Pertsemlidis A, Marcel YL, McPherson R and Hobbs HH 2004 Multiple rare alleles contribute to low plasma levels of HDL cholesterol. *Science* 305, 869–72.

Cohen MM 1983 Craniofacial disorders. In: *Principles and Practice of Medical Genetics* (editors Emery AEH and Rimoin DI) Churchill Livingstone, Edinburgh-London, pp. 567–621.

Cohen MM 2002 Cranio-facial disorders. In: *Emery and Rimoin's Principles and Practice of Medical Genetics*, 4th edition (Rimoin DL, Connor JM, Pyeritz RE and Korf BR) Churchill Livingstone, London, vol.3, pp. 3689–3727.

Cohen MM Jr. 1989 Perspectives on holoprosencephaly: Part I. Epidemiology, genetics, and syndromology. *Teratology* 40 211–235.

Collaborative Group on Hormonal Factors in Breast Cancer 2001 Familial breast cancer: collaborative reanalysis of individual data from 52 epidemiological studies including 58,209 women with breast cancer and 101,986 women without the disease. *The Lancet* 358 1389–1399.

Collazo T, Magariño C, Chávez R, Suardíaz B, Gispert S, Gómez M, Rojo M and Heredero L 1995 Frequency of Delta-F508 mutation and XV2C/KM19 haplotypes in Cuban cystic fibrosis families. *Human Heredity* 45 55–57.

Collins FS, Green ED, Guttmacher AE and Guyer MS 2003 A vision for the future of genomic research. *Nature* 422 835–847.

Concannon P, Erlich HA, Julier C, Morahan G, Nerup J, Pociot F, Todd JA and Rich SS 2005 Type 1 Diabetes Genetics Consortium. Type 1 diabetes: evidence for susceptibility loci from four genomewide linkage scans in 1435 multiplex families. *Diabetes* 54 2995–3001.

Connor F, Bertwistle D, Mee PJ, Ross GM, Swift S, Grigorieva E, Tybulewicz VL and Ashworth A 1997 Tumourigenesis and a DNA repair defect in mice with a truncating BRCA 2 mutation. *Nature Genetics* 17 423–430.

Conterio F and Barrai I 1966 Effeti della consanguineitá sulla mortalitá e sulla morbiditá nelle populazione della diocese di Parma. *Atti Associazione Genetica Italiana* 11 378–391.

Cook-Deegan RM and McCormack SJ 2001 Patents, secrecy, and DNA. *Science* 293 217.

Corcoran PA, Allen FH Jr., Allison AC and. Blumberg BS 1959 Blood groups of Alaskan Eskimos and Indians. *American Journal of Physical Anthropology* 17 187–193.

Couch FJ and Weber BL 1996 Mutations and polymorphisms in the familial early-onset breast cancer (BRCA1) gene. *Human Mutation* 8 8–18.

Crawford MH 1998 *The Origins of Native Americans. Evidence from Anthropological Genetics.* Cambridge University Press, Cambridge.

Creazzo TL, Godt RE, Leatherbury L, Conway SJ and Kirby ML 1998 Role of cardiac neural crest cells in cardiovascular development. *Annual review of Physiology* 60 267–286.

Cresswell WL and Froggatt P 1963 *The Causation of Bus Driver Accidents.* Oxford University Press, London.

Crookshank FG 1924 *The Mongol in our Midst: A Study of Man and his Three Faces.* EP Dutton, New York.

Crow JF 1958 Some possibilities for measuring selection intensities in Man. *Human Biology.* 30 1–13.

Crow JF 1970 Genetic loads and the cost of natural selection. In: *Mathematical topics in population genetics.* (editor Kojima K) Springer-Verlag, Berlin, pp. 128–177.

Crow JF 1980 The estimation of inbreeding from isonymy. *Human Biology* **52** 1–4.

Crow JF 1987 Muller, Dobzhansky, and overdominance. *Journal of the History of Biology* **20** 351–380.

Crow JF 1997 The high spontaneous mutation rate: Is it a health risk? *Proceedings of the National Academy of Sciences USA* **94** 8380–8386.

Crow JF and Kimura M 1970 *An Introduction to Population Genetics Theory.* Harper and Row, New York.

Crow JF and Kimura M 1979 Efficiency of truncation selection. *Proceedings of the National Academy of Sciences USA* **76** 396–399.

Crow JF and Mange AP 1965 Measurements of inbreeding from the frequency of marriages between persons of the same surnames. *Eugenics Quarterly* **12** 199–203.

Crow JF and Morton NE 1960 The genetic load due to mother-child incompatibility. *American Naturalist* **94** 413–419.

Crow JF and Temin RG 1964 Evidence for the partial dominance of recessive lethal genes in natural populations of Drosophila. *American Naturalist* **98** 21–33.

Cruse A 2004 *Roman Medicine* Tempus Publishers, Stroud.

Cuckle H 1999 Down syndrome foetal loss rate in early pregnancy. *Prenatal Diagnosis* **19** 1177–1179.

Cuckle H 2002 Potential biases in Down syndrome birth prevalence estimation. *Journal of Medical Screening* **9** 192.

Cuckle H 2005 Primary prevention of Down's syndrome. *International Journal of Medical Science* **2** 93–99.

Cuckle HS, Wald NJ and Thompson SC 1987 Estimating a women's risk of having a pregnancy associated with Down's syndrome using her age and serum alpha-fetoprotein level. *British Journal of Obstetrics and Gynaecology* **94** 387–402.

Cucuianu M, Opincaru A and Tapalagă D 1978 Similar behaviour of lecithin:cholesterol acyltransferase and pseudocholinesterase in liver disease and hyperlipoproteinemia. *Clinica Chimica Acta* **85** 73–79.

Cucuianu M, Popescu T A and Haragus ST 1968 Pseudocholinesterase in obese and hyperlipemic subjects. *Clinica Chimica Acta* **22** 151–155.

Culotta E and Koshland DE Jr. 1993 p53 sweeps through cancer research. *Science* **262** 1958–1961.

Cunningham GC and Tompkinson DG 1999 Cost and effectiveness of the California triple marker prenatal screening program. *Genetics in Medicine* **1** 200–207.

Curtin PD 1969 *The Atlantic Slave Trade: a Census.* University of Wisconsin Press, Madison.

Cutting GR 2005 Modifier genetics: cystic fibrosis. *Annual Review of Genomics Human Genetics.* **6** 237–260.

Cuzick J, Powles T, Veronesi U, Forbes J, Edwards R, Ashley S and Boyle P 2003 Overview of the main outcomes in breast-cancer prevention trials. *The Lancet* **361** 296–300.

da Rocha ST and Ferguson-Smith AC 2004 Genomic imprinting. *Current Biology* **14** R646–649.

Dacou-Voutekakis C, Karavanaki K and Tsoka-Gennatas H 1995 National data on the epidemiology of IDDM in Greece. Cases diagnosed in 1992. Hellenic Epidemiology Study Group. *Diabetes Care* **18** 552–554.

Dahl E 2003 Ethical issues in new uses of PGD: should parents be allowed to use PGD to choose the sexual orientation of their children? *Human Reproduction* **18** 1368–1369.

Dahlen P, Liukkonen L, Kwiatkowski M, Dahlen P, Liukkonen L, Kwiatkowski M, Hurskainen P, Iitia A, Siitari H, Ylikoski J, Mukkala VM and Lovgren T 1994 Europium-labeled oligonucleotide hybridization probes: preparation and properties. *Bioconjugate Chemistry* **5** 268–272.

Dale HH 1936 Some recent extensions of the chemical transmission of the effects of nerve impulses. Nobel Lecture, December 12, 1936. The Official Web Site of The Nobel Foundation: *http://nobelprize.org/medicine/laureates/1936/dale-lecture.html.*

Daly MJ, Rioux JD, Schaffner SF, Hudson TJ and Landers ES 2001 High-resolution haplotype structure in the human genome. *Nature Genetics* **29** 229–232.

Dane DS, Cameron CH and Briggs M 1970 Virus-like particles in serum of patients with Australia antigen-associated hepatitis. *The Lancet* **1** 695–698.

Danforth CH 1923 The frequency of mutation and the incidence of hereditary traits in man. I. *Eugenics, genetics and the family.* Papers. 2nd International Congress of Eugenics, New York **1** 120–128.

Danish JM, Tillson JK and Levitan M 1963 Multiple anomalies in congenitally deaf children. *Eugenics Quarterly* **10** 12–21.

Darling S and Gossler A 1994 A mouse model for Fraser syndrome? *Clinical Dysmorphology* **3** 91–95.

Darvesh S, Hopkins DA and Geula C 2003 Neurobiology of butyrylcholinesterase. *Nature Reviews Neuroscience* **4** 131–138.

Darvesh S, Kumar R, Roberts S, Walsh R and Martin E 2001 Butyrylcholinesterase-mediated enhancement of the enzymatic activity of trypsin. *Cellular and Molecular Neurobiology* **21** 285–289.

Darwin C 1868 *The Variation of Animals and Plants under Domestication* John Murray, London.

David B 1953 Über einen dominanten Erbgang bei einen polytopen enchondralen Dysostose Typ Pfaundler-Hurler. *Zeitschrift für Orthopädie* **84** 657–660.

David JB, Edoo BB and Mustaffah JOAF 1971 Adamarobe – A 'deaf' village. *Sound* **5** 70–72.

Davies JL, Kawaguchi Y, Bennett ST, Copeman JB, Cordell HJ, Pritchard LE, Reed PW, Gough SC, Jenkins SC, Palmer SM, *et al.* 1994 A genome-wide search for human type 1 diabetes susceptibility genes. *Nature* **371** 130–136.

Davies JNP 1973 In: *The Liver* (editors Gall EA and Mostofi FK) Williams & Wilkins, Baltimore, pp. 361–369.

Daw EW, Wijsman EM and Thompson EA 2003 A score for Bayesian genome screening. *Genetic Epidemiology* **24** 181–190.

Dawkins R 1999 Foreword. In: *The Genetic Revolution and Human Rights.* University Press, Oxford.

Dawson E, Abecasis GR, Bumpstead S, Chen Y, Hunt S, Beare DM, Pabial J, Dibling T, Tinsley E, Kirby S, Carter D, Papaspyridonos M, Livingstone S, Ganske R, Lõhmussaar E, Zernant J, Tõnisson N, Remm M, Mägi R, Puurand T, Vilo J, Kurg A, Rice K, Deloukas P, Mott R, Metspalu A, Bentley DR, Cardon LR and Dunham I 2002 A first-generation linkage disequilibrium map of chromosome 22. *Nature* **418** 544–548.

de Lonlay-Debeney P, Cormier-Daire V, Amiel J, Abadie V, Odent S, Paupe A, Couderc S, Tellier AL, Bonnet D, Prieur M, Vekemans M, Munnich A and Lyonnet S 1997 Features of DiGeorge syndrome and CHARGE association in five patients. *Journal of Medical Genetics* **34** 986–989.

de Maupertuis PLM 1756 *Oeuvres de M. de Maupertuis* Nouvelle edition, corrigée et augmentée. 4 vols. J-M Bruyset, Lyon.

De Vos A and Van Steirteghem A 2001 Aspects of biopsy procedures prior to preimplantation genetic diagnosis. *Prenatal Diagnosis* **21** 767–780.

De Vriese C, Gregoire F, Lema-Kisoka R, Waelbroeck M, Robberecht P and Delporte C 2004 Ghrelin degradation by serum and tissue homogenates: identification of the cleavage sites. *Endocrinology* **145** 4997–5005.

Deeming DC and Ferguson MWJ 1988 Environmental regulation of sex determination in reptiles. *Philosophical Transactions of the Royal Society, London B* **322** 19–39.

Degner D, Bleich S, Riegel A and Ruther E 1999 Orofaciodigital syndrome. A new variant? Psychiatric, neurologic and neuroradiological findings. *Fortschritte der Neurologie-Psychiatrie* **67** 525–528.

D'Elia AV, Tell G, Paron I, Pellizzari L, Lonigro R and Damante G 2001 Missense mutations of human homeoboxes: A review. *Human Mutation* **18** 361–374.

Della Valle A 1924 Contributo alla conoscenza della forma famigliare del megacolon congenito. *Pediatria* **32** 569–599.

Dempster AP, Laird NM and Rubin DB 1977. Maximum likelihood from incomplete data via the EM algorithm. *Journal of the Royal Statistical Society* Series B **34** 1–38.

Dempster ER and Lerner IM 1950 Heritability of threshold characters. *Genetics* **35** 212–236.

Dermitzakis ET, Reymond A and Antonarakis SE 2005 Conserved non-genic sequences – an unexpected feature of mammalian genomes. *Nature Reviews Genetics* **6** 151–157.

DeStefano AL, Couples LA, Arnos KS, Asher JH Jr, Baldwin CT, Blaton S, Carey ML, da Silva EO, Friedman TB, Greenburg J, Lalwani AK, Milunsky A, Nance WE, Pandya A, Ramesar RS, Read AP, Tassabejhi M, Wilcox ER and Farrer LA 1998 Correlation between Waardenburg syndrome phenotype and genotype in a population of individuals with identified PAX3 mutations. *Human Genetics* **102** 499–506.

Dey DC, Kanno T, Sudo K and Maekawa M 1997 Genetic analysis of 6 patients with no detectable butyrylcholinesterase activity – three novel silent genes and heterogeneities of mutations in *BCHE* gene in Japan. *American Journal of Human Genetics* **61**(supp) 2307.

Dey DC, Maekawa M, Sudo K and Kanno T 1998 Butyrylcholinesterase genes in individuals with abnormal inhibition numbers and with trace activity: one common mutation and two novel silent genes. *Annals of Clinical Biochemistry* (Part 2) **35** 302–310.

Diels H and Kranz W 1964 *Die Fragmente der Vorsokratiker.* Hildesheim Zürich.

Diogenes Laertius 1925 Lives of Eminent Philosophers. Harvard University Press: Boston.

Dipierri JE, Alfaro EL, Scapoli C, Mamolini E, Rodriguez-Larralde A and Barrai I 2005 Surnames in Argentina. *American Journal of Physical Anthropology* **128** 199–209.

Dobzhansky T 1937 *Genetics and the Origin of Species.* Columbia University Press, New York.

Dobzhansky T 1955 A review of some fundamental concepts and problems of population genetics. *Cold Spring Harbor Symposia on Quantitative Biology* **20** 1–15.

Doll R 2001a Cohort studies: history of the method. I. Prospective cohort studies. *Sozial- und Präventivmedizin* **46** 75–86.

Doll R 2001b Cohort studies: history of the method. II. Retrospective cohort studies. *Sozial- und Präventivmedizin* **46** 152–160.

Doll R, Peto R, Boreham J and Sutherland I 2004 Mortality in relation to smoking: 50 years' observations on male British doctors. *British Medical Journal* **328** 1519–1528.

Donahue RP, Bias WB, Renwick JH and McKusick VA 1968 Probable assignment of the Duffy blood group locus to chromosome 1 in man. *Proceedings of the National Academy of Sciences USA* **61** 949–955.

Donnai D 2004 The Carter Lecture. British Society of Human Genetics. York.

Donnai D and Winter RM 1995 *Congenital malformation syndromes.* Chapman and Hall Medical, London.

Donne J 1624 *Devotions* London.

Doray B, Salomon R, Amiel J Pelet A, Touraine R, Billaud M, Attie T, Bachy B, Munnich A and Lyonnet S 1998 Mutation of the *RET* ligand, neurturin, supports multigeneic inheritance in Hirschsprung disease. *Human Molecular Genetics* **8** 1449–1452.

Doria-Rose VP, Kim HS, Augustine ET and Edwards KL 2003 Parity and the risk of Down's syndrome. *American Journal of Epidemiology* **158** 503–508.

Dorticós A, Martin M, Echevarria P, Robaina M, Rodríguez M, Moras F and Granda H 1997 Reproductive behaviour of couples at risk for sickle cell disease in Cuba: a follow-up study. *Prenatal Diagnosis* **17** 737–742.

Doss BJ, Jolly S, Qureshi F, Jacques SM, Evans MI Johnson MP, Lampinen J and Kupsky WJ 1998 Neuropathologic findings in a case of OFDS type VI (Váradi syndrome). *American Journal of Medical Genetics* **77** 38–42.

Down JL 1866 Observations on an ethnic classification of idiots, *London Hospital Reports*, 1866, reprinted in J Langdon Down, *On Some of the Mental Affections of Childhood and Youth* (JA Churchill, 1887), pp. 210–217.

Doyle PE, Beral V, Botting B and Wale CJ 1990 Congenital malformations in twins in England and Wales. *Journal of Epidemiology and Community Health* **45** 43–48.

Dozortsev DI and McGinnis KT 2001 An improved fixation technique for fluorescence in situ hybridization for preimplantation genetic diagnosis. *Fertility and Sterility* **76** 186–188.

Drew JS, London WT, Lustbader ED and Blumberg BS 1977 Cross reactivity *bet*ween hepatitis B surface antigen and an antigen on male cells. *Proceedings of the 1977 March of Dimes Birth Defects Conference.*

Dubbs CA 1966 Ultrasonic effects on isoenzymes. *Clinical Chemistry* **12** 181–186.

Dufier J-L and Kaplan J (editors) 2005 *Œil et Génétique* Masson, Paris.

Duncan MH and Miller RW 1983. Another family with the Li-Fraumeni cancer syndrome. *Journal of the American Medical Association* **249** 195.

Dunning AM, Chiano M, Smith NR, Dearden J, Gore M, Oakes S, Wilson C, Stratton M, Peto J, Easton D, Clayton D and Ponder BA 1997 Common BRCA1 variants and susceptibility to breast and ovarian cancer in the general population. *Human Molecular Genetics* **6** 285–289.

Durocher D, Charron F, Warren R, Schwartz RJ and Nemer M 1997 The cardiac transcription factors Nkx2–5 and GATA-4 are mutual cofactors. *EMBO Journal* **16** 5687–5696.

Duster T 1990 *Backdoor to Eugenics.* Routledge, New York, pp. ix, 76.

Dyck AJ 1997 Eugenics in historical and ethical perspective. In: *Genetic Ethics. Do the Ends Justify the Genes?* (editors Kilner JF, Pentz RD, Young FE) Paternoster Press WB Eerdsmans Publishing Co., Grand Rapids, MI. pp. 25–39.

Eaves IA, Merriman TR, Barber RA, Nutland S, Tuomilehto-Wolf E, Tuomilehto J, Cucca F and Todd JA 2000 The genetical isolated populations of Finland and Sardinia may not be a panacea for linkage disequilibrium mapping of common disease genes. *Nature Genetics* **25** 320–323.

Eberhardt W 1911 Cryptophthalmia. *Ophthalmic Reconstruction* **20** 4–6.

Eddy SR 2001 Non-coding RNA genes and the modern RNA world. *Nature Reviews Genetics* **2** 919–929.

Edery P, Attie T, Amiel J, Pelet A, Pelet A, Eng C, Hofstra RMW, Martelli H, Bidaud C, Munnich A and Lyonnet S 1996 Mutation of the endothelin-3 gene in the Waardenburg-Hirschsprung disease (Shah-Waardenburg syndrome). *Nature Genetics* **12** 442–444.

Editorial 2003 The International HapMap Project. *Nature* **426** 789–796.

Edwards JH 1960 The simulation of Mendelism. *Acta Genetica et Statistica Medica* **10** 63–70.

Edwards JH 1965 The meaning of the association between blood groups and disease. *Annals of Human Genetics* **29** 77–83.

Edwards JH 1980 Allelic association in man. In: *Population Genetics and Genetic Disorders.* (editor Ericksson AW) Academic Press, New York.

Edwards JH 2003 Sib-pairs in multifactorial disorders: the sib-similarity problem. *Clinical Genetics* **63** 1–9.

Edwards SM, Kote-Jarai Z, Meitz J, Hamoudi R, Hope Q, Osin P, Jackson R, Southgate C, Singh R, Falconer A, Dearnaley DP, Arden-Jones A, Murkin A, Dowe A, Kelly J, Williams S, Oram R, Stevens M, Teare DM, Ponder BA, Gayther SA, Easton DF and Eeles RA 2003 Two percent of men with early-onset prostate cancer harbor germline mutations in the BRCA2 gene. *American Journal of Human Genetics* **72** 1–12.

Egger J, Bellman MH, Ross EM and Baraitser M 1982 Joubert-Boltshauser syndrome with polydactyly in siblings. *Journal of Neurology Neurosurgery and Psychiatry* **45** 737–739.

Egginton J 1997 *From Cradle to Grave – Why did a Mother's Nine Babies have to Die?* Virgin Books, London.

Eiberg H and Mohr J 1996 Assignment of genes coding for brown eye colour (BEY2) and brown hair colour (HCL3) on chromosome 15q. *European Journal of Human Genetics* **4** 237–241.

Eichenlaub-Ritter U 1998 Genetics of oocyte ageing. *Maturitas* **30** 143–169.

Eisenmann KM, McCarthy JB, Simpson MA, Keely PJ, Guan JL, Tachibana K, Lim L, Manser E, Furcht LT and Iida J 1999 Melanoma chondroitin sulphate proteoglycan regulates cell spreading through Cdc42, Ack-1 and p130cas. *Nature Cell Biology* **1** 507–513.

Elçioglu HN and Berry AC 2000 Fraser syndrome: Diagnosed in a 50-year-old museum specimen. *American Journal of Medical Genetics* **94** 262–246.

Ellis NA, Ciocci S, Protcheva M, Lennon D, Groden J and German J 1998 The Ashkenazic Jewish Bloom syndrome mutation blmAsh is present in non-Jewish Americans of Spanish ancestry. *American Journal of Human Genetics* **63**:1685–1693.

Ellis NA, Groden J, Ye T-Z, Straughen J, Lennon DJ, Ciocci S, Proytcheva M and German J 1995 The Bloom's syndrome gene product is homologous to RecQ helicases. *Cell* **83** 655–666.

Emery AEH 1988 Portraits in medical genetics: John Dalton (1766–1844). *Journal of Medical Genetics* **25** 422–426.

Emison ES, McCallion AS, Kashuk CS, Bush RT, Grice E, Lin S, Portnoy ME, Cutler DJ, Green ED and Chakravarti A 2005 A common sex-dependent mutation in a *RET* enhancer underlies Hirschsprung disease. *Nature* **434** 857–863.

Engvall E and Wewer UM 2003 The new frontier in muscular dystrophy research: booster genes. *FASEB Journal* **17** 1579–84.

Epstein CJ 2003 Presidential address: Is modern genetics the new eugenics? *Genetics in Medicine* **5** 469–470.

Epstein CJ 2004 Genetic testing: hope or hype? *Genetics in Medicine* **6** 165–172.

Erickson JD 1978 Down's syndrome, paternal age, maternal age and birth order. *Annals of Human Genetics* **41** 289–298.

Erickson RP 1997 Does sex determination start at conception? *BioEssays* **19** 1027–1032.

Ericson A and Kallen B 2001 Congenital malformations in infants born after IVF: a population-based study. *Human Reproduction* **16** 504–509.

Eriksson M, Brown WT, Gordon LB, Glynn MW, Singer J, Scott L, Erdos MR, Robbins CM, Moses TY, Berglund P, Dutra A, Pak E, Durkin S, Csoka AB, Boehnke M, Glover TW and Collins FS 2003 Recurrent de novo point mutations in lamin A cause Hutchinson-Gilford progeria syndrome. *Nature* **423** 293–298.

EURODIAB ACE Study Group 2000 Variation and trends in incidence of childhood diabetes in Europe. *The Lancet* **355** 873–876.

Evans FT, Gray PWS, Lehmann H and Silk E 1952 Sensitivity to succinylcholine in relation to serum-cholinesterase. *The Lancet* **1** 1229–1230.

Everett LA, Glaser B, Beck JC, Idol JR Buchs A, Heyman M, Adawi F, Hazani E, Nassir E, Baxevanis AD, Sheffield VC and Green ED 1997 Pendred syndrome is caused by mutations in a putative sulphate transporter gene. *Nature Genetics* **17** 411–422.

Everett SM, White KLM, Schorah CJ, Calvert RJ, Skinner C, Miller D and Axon ATR 2000 In vivo DNA damage in gastric epithelial cells. *Mutation Research* **468** 73–85.

Ewens WJ 2004 *Mathematical Population Genetics*. Springer, New York.

Ewens WJ 1970 Remarks on the substitutional load. *Theoretical Population Biology* **1** 129–139.

Ewens WJ 1993 Beanbag genetics and after. In: *Human Population Genetics: A Tribute to J.B.S. Haldane*, (editor Majumder PP), Plenum Press, New York and London.

Eyre-Walker A and Keightley PD 1999 High genomic deleterious mutation rates in hominids. *Nature* **397** 344–347.

Fagundes NJR, Salzano FM, Batzer MA, Deininger PL and Bonatto SL 2005 Worldwide genetic variation at the 3'-UTR region of the *LDLR* gene: possible influence of natural selection. *Annals of Human Genetics* **69** 389–400.

Falcinelli C, Iughetti L, Percesepe A, Calabrese G, Chiarelli F, Cisternino M, De Sanctis L, Pucarelli I, Radetti G, Wasniewska M, Weber G, Stuppia L, Bernasconi S and Forabosco A 2002 SHOX point mutations and deletions in Leri-Weill dyschondrosteosis. *Journal of Medical Genetics* **39** E33.

Falconer DS 1960 *Introduction to Quantitative Genetics*. Oliver and Boyd, Edinburgh.

Falconer DS 1965 The inheritance of liability to certain diseases, estimated from the incidence among relatives. *Annals of Human Genetics* **29** 51–76.

Falk R 1961 Are induced mutations in Drosophila overdominant? II Experimental results. *Genetics* **46** 737–757.

Fearnhead NS, Wilding JL and Bodmer WF 2002 Genetics of colorectal cancer: Hereditary aspects and overview of colorectal tumorigenesis. *British Medical Bulletin* **64** 27–43.

Fearnhead NS, Wilding JL, Winney B, Tonks S, Bartlett S, Bicknell DC, Tomlinson IP, Mortensen NJ and Bodmer WF 2004 Multiple rare variants in different genes account for multifactorial inherited susceptibility to colorectal adenomas. *Proceedings of the National Acadamy of Sciences USA* **101** 15992–15997.

Fearnhead NS, Winney B and Bodmer WF 2005 Rare variant hypothesis for multifactorial inheritance: susceptibility to colorectal adenomas as a model. *Cell Cycle* **4** e26-e30.

Ferdman B and Singh G 2003 Persistent truncus arteriosus. *Current Treatment Options in Cardiovascular Medicine* **5** 429–438.

Ferguson-Smith MA 1965 Karyotype-phenotype correlations in gonadal dysgenesis and their bearing on the pathogenesis of malformations. *Journal of Medical Genetics* **2** 142–155.

Ferguson-Smith MA 1966 X-Y chromosomal interchange in the aetiology of true hermaphroditism and of XX Klinefelter's syndrome. *The Lancet* **ii** 475–476.

Ferguson-Smith MA 2002 *In situ* hybridisation. In: *Encyclopedia of Genetics* (editors Brenner S and Miller JH) Academic Press, San Diego, vol. 2 pp. 1002–1004.

Ferguson-Smith MA and Munro IB 1958 Spermatogenesis in the presence of female nuclear sex. *Scottish Medical Journal* **3** 39–42.

Ferrero M, Lotti F, Cendán I and Pérez A 1998 Tendencias del Síndrome Down en Cuba. Su relación con la edad materna y la tasa de fecundidad. *Revista Cubana de Pediatría* **70** 141–147.

Findlay I, Matthews P and Quirke P 1999 Preimplantation genetic diagnosis using fluorescence polimerase chain reaction: results and future development. *Journal of Assisted Reproduction and Genetics* **16** 199–206.

Fiorentino F, Magli MC, Podini D, Ferraretti AP, Nuccitelli A, Vitale N, Baldi M, and Gianaroli L 2003 The minisequencing method: an alternative strategy for preimplantation genetic diagnosis of single gene disorders. *Molecular Human Reproduction* **9** 399–410.

Fisch H, Hyun G, Golden R, Hensle TW, Olsson CA and Liberson GL 2003 The influence of paternal age on Down syndrome. *Journal of Urology* **169** 2275–2278.

Fisher RA 1924 The elimination of mental defect. *Eugenics Review* **16** 114–116.

Fisher RA 1949 *The Theory of Inbreeding*. Oliver and Boyd, Edinburgh.

Fisher RA and Yates F 1963 *Statistical Tables for Use in Agricultural, Biological and Medical Research* (6th edition) Oliver and Boyd, Edinburgh.

Fitze G, Appelt H, König IR, Gorgens H, Stein U, Walther W, Gossen M, Schreiber M, Ziegler A, Roesner D and Schackert HK 2003 Functional haplotypes of the RET proto-oncogene promoter are associated with Hirschsprung disease (HSCR). *Human Molecular Genetics* **12** 3207–3214.

Fitze G, Schreiber M, Kuhlisch E and Schackert HK 1999 Association of *RET* protooncogene codon 45 polymorphism with Hirschsprung disease. *American Journal of Human Genetics* **65** 1469–1473.

Flannery DB 1993 Tale of a nail. *Proceedings of the Greenwood Genetics Center* **12** 90.

Fleischmann RD, Adams MD, White O, Clayton RA, Kirkness E, Kerlavage AR, Bult CJ, Tomb J-F, Dougherty BA, Merrick JM, McKenney K, Sutton G, FitzHugh W, Fields C, Gocayne JD, Scott J, Shirley R, Liu L-I, Glodek A, Kelley JM, Weidman JF, Phillips CA, Spriggs T, Hedblom E, Cotton MD, Utterback TR, Hanna MC, Nguyen DT, Saudek DM, Brandon RC, Fine LD, Fritchman JL, Fuhrmann JL, Geoghagen NSM, Gnehm CL, McDonald LA, Small KV, Fraser CM, Smith HO and Venter JC 1995 Whole-genome random sequencing and assembly of *Haemophilus influenzae* Rd. *Science* **269** 496–512.

Flint J and Knight S 2003 The use of telomere probes to investigate submicroscopic rearrangements associated with mental retardation. *Current Opinion in Genetics and Development* **13** 310–316.

Flint J, Harding RM, Boyce AJ and Clegg JB 1998 The population genetics of the haemoglobinopathies. In: *Baillière's Clinical Haematology; 'Haemoglobinopathies'* (editors D.R. Higgs and D.J. Weatherall), Vol. 11, pp. 1–51. Baillière Tindall and W.B. Saunders, London.

Flint J, Hill AVS, Bowden DK, Oppenheimer SJ, Sill.R, Serjeantson SW, Bana-Koiri J, Bhatia K, Alpers MP, Boyce AJ, Weatherall DJ and Clegg JB 1986 High frequencies of α-thalassaemia are the result of natural selection by malaria. *Nature* **321** 744–749.

Flück CE, Tajima T, Pandey AV, Arlt W, Okuhara K, Verge CF, Jabs EW, Mendonca BB, Fujieda K and Miller WL 2004 Mutant P450 oxidoreductase causes disordered steroidogenesis with and without Antley-Bixler syndrome. *Nature Genetics* **36** 228–230.

Foley TR, McGarrity TJ and Abt AB 1988 Peutz-Jeghers syndrome: a clinicopathologic survey of the 'Harrisburg family' with a 49-year follow-up. *Gastroenterology* **95** 1535–1540.

Foray N, Randrianarison V, Marot D, Perricaudet M, Lenoir G and Feunteun J 1999 Gamma-rays-induced death of human cells carrying mutations of BRCA1 or BRCA2. *Oncogene* **18** 7334–7342.

Forbat A, Lehmann H and Silk E 1953 Prolonged apnoea following injection of succinylcholine. *The Lancet* **2** 1067–1068.

Ford CE, Miller OJ, Polani PE, De Almeida JC and Briggs JH 1959 A sex-chromosome anomaly in a case of gonadal dysgenesis (Turner's Syndrome). *The Lancet* **i** 711–713.

Ford D, Easton DF, Stratton M, Narod S, Goldgar D, Devilee P, Bishop DT, Weber B, Lenoir G, Chang-Claude J, Sobol H, Teare MD, Struewing J, Arason A, Scherneck S, Peto J, Rebbeck TR, Tonin P, Neuhausen S, Barkardottir R, Eyfjord J, Lynch H, Ponder BA, Gayther SA and Zelada-Hedman M 1998 Genetic heterogeneity and penetrance analysis of the BRCA1 and BRCA2 genes in breast cancer families. The Breast Cancer Linkage Consortium. *American Journal of Human Genetics* **62** 676–689.

Ford EB 1940 Polymorphism and taxonomy. pp. 493–516. In: *New International Conference: The Human Genome Biological, Medical and Ethical Prospects.* (editor Huxley JS).

Ford EB 1956 *Genetics for Medical Students* Methuen, London, p. 202.

Forsdyke DR 2001 *'The Origin of Species' Revisited: A Victorian who Anticipated Modern Developments in Darwin's Theory* McGill-Queen's University Press, Montreal.

Forshew T, Johnson CA, Khaliq S, Pasha S, Willis C, Abbasi R, Tee L, Smith U, Trembath RC, Mehdi SQ, Moore AT and Maher ER 2005 Locus heterogeneity in autosomal recessive congenital cataracts: linkage to 9q and germline HSF4 mutations. *Human Genetics* **117** 452–9.

Foster MW and Sharp RR 2004 Beyond race: towards a whole-genome perspective on human populations and genetic variation. *Nature Reviews Genetics* **5** 790–796.

Foulkes WD 2004 BRCA1 functions as a breast stem cell regulator. *Journal of Medical Genetics* **41** 1–5.

Foulkes WD, Brunet JS, Stefansson IM, Straume O, Chappuis PO, Begin LR, Hamel N, Goffin JR, Wong N, Trudel M, Kapusta L, Potter P and Akslen 2004 The prognostic implication of the basal-like (cyclin E high/p27 low/p53+/glomeruloid-microvascular-proliferation+) phenotype of BRCA1-related breast cancer. *Cancer Research* **64** 830–835.

Foulkes WD, Chappuis PO, Wong N, Brunet JS, Vesprini D, Rozen F Yuan ZO, Kuperstein G, Narod SA and Begin LR 2000 Primary node negative breast cancer in BRCA1 mutation carriers has a poor outcome. *Annals of Oncology* **11** 307–313.

Foulkes WD, Metcalfe K, Hanna W, Lynch HT, Ghadirian P, Tung N Olopade O, Weber B, McLennan J, Olivotto IA, Sun P, Chappuis PO, Begin LR, Brunet JS and Narod SA 2003 Disruption of the expected positive correlation between breast tumor size and lymph node status in BRCA1-related breast carcinoma. *Cancer* **98** 1569–1577.

Foulkes WD, Stefansson IM, Chappuis PO, Begin LR, Goffin JR, Wong N, Trudel M and Akslen LA 2003 Germ-line BRCA1 mutations and a basal epithelial phenotype in breast cancer. *Journal of the National Cancer Institute* **95** 1482–1485.

Foulkes WD, Wong N, Brunet JS, Begin LR, Zhang JC, Martinez JJ Rozen F, Tonin PN, Narod SA, Karp SE and Pollak MN 1997 Germ-line BRCA1 mutation is an adverse prognostic factor in Ashkenazi Jewish women with breast cancer. *Clinical Cancer Research* **3** 2465–2469.

Fox JL 2002 Eugenics concerns rekindle with application of gene therapy and genetic counseling. *Nature Biotechnology* **20** 531–531.

Fraccaro M, Glen-Bott M, Salzano FM, Ross Russell RW and Cranston WI 1962 Triple chromosomal mosaic in a woman with clinical evidence of masculinisation. *The Lancet* **1** 1379–1381.

Francannet C, Lefrançois P, Dechelotte P, Robert E, Malpuech G, Robert JM 1990 Fraser syndrome with renal agenesis in two consanginous Turkish families. *American Journal of Medical Genetics* **36** 477–479.

Franco D and Campione M 2003 The role of Pitx2 during cardiac development. Linking left-right signaling and congenital heart diseases. *Trends in Cardiovascular Medicine* **13** 157–163.

Franco RF and Reitsma PH 2001 Genetic risk factors of venous thrombosis. *Human Genetics* **109** 369–384.

François J 1959 *Les cataractes congénitales.* Masson et Cie, Paris. pp. 326–327.

Fraser FC 1956 Heredity counseling: the darker side. *Eugenics Quarterly* **3** 45–51.

Fraser FC 1968 Genetic counselling and the physician (The Blackader Lecture). *Canadian Medical Association Journal* **99** 927–934.

Fraser FC 1974 Genetic counseling. *American Journal of Human Genetics* **26** 636–659.

Fraumeni JF and Glass AG 1970 Rarity of Ewing's sarcoma among US Negro children. *The Lancet* **i** 366.

Fraumeni JF Jr. and Miller RW 1967a Adrenocortical neoplasms with hemihypertrophy, brain tumors, and other disorders. *Journal of Pediatrics* **70** 129–138.

Fraumeni JF Jr. and Miller RW 1967b Epidemiology of human leukemia: Recent observations. *Journal of the National Cancer Institute* **38** 593–605.

Fraumeni JF Jr. and Miller RW 1969 Cancer deaths in the newborn. *American Journal of Diseases of Children* **117** 186–189.

Fraumeni JF Jr. (Ed.) 1975 *Persons at High Risk of Cancer: An Approach to Cancer Etiology and Control.* New York, Academic Press.

Fraumeni JF Jr., Miller RW and Hill JA 1968 Primary carcinoma of the liver in childhood: An epidemiologic study. *Journal of the National Cancer Institute* **40** 1087–1099.

Frayling IM, Beck NE, Ilyas M, Dove-Edwin I., Goodman P, Pack K, Bell JA, Williams CB, Hodgson SV, Thomas HJ, Talbot IC, Bodmer WF and Tomlinson IP 1998 The APC variants I1307K and E1317Q are associated with colorectal tumors, but not always with a family history. *Proceedings of the National Acadamy of Sciences USA* **95** 10722–10727.

Freeman DJ, Samani NJ, Wilson V, McMahon AD, Braund PS, Cheng S, Caslake MJ, Packard CJ and Gaffney D on behalf of the West of Scotland Study Group 2003 A polymorphism of the cholesteryl ester transfer protein gene predicts cardiovascular events in non-smokers in the West of Scotland Coronary Prevention Study. *European Heart Journal* **24** 1833–1842.

Freeman MW 2006 Statins, cholesterol and the prevention of coronary heart disease. *FASEB Journal* **20** 200–201.

Freeman SB, Yang Q, Allran K, Taft LF and Sherman SL 2000 Women with a reduced ovarian complement may have an increased risk for a child with Down syndrome. *American Journal of Human Genetics* **66** 1680–1683.

Freire-Maia A 1968 Genética da Aquiropodia. PhD thesis, Ribeirão Preto Medical School, University of São Paulo.

Freire-Maia A 1970a The handless and footless families of Brazil. *The Lancet* **1** 519–520.

Freire-Maia A 1970b The handless and footless families of Brazil. *The Lancet* **2** 727–728.

Freire-Maia A, Freire-Maia N and Schull WJ 1975 Genetics of acheiropodia ('the handless and footless families of Brazil'). IX. Genetic counseling. *Human Heredity* **25** 329–336.

Frézal J, Baule MS, de Fougerolle T, Kaplan J and Le Merrer M 1989 *Genatlas: a catalog of mapped genes and other markers.* Inserm/Joh Libbey, Paris.

Frias S, Ramos S, Molina B, del Castillo V and Mayen DG 2002 Detection of mosaicism in lymphocytes of parents of free trisomy 21 offspring. *Mutation Research* **520** 25–37.

Froggatt P 1977 A cardiac cause in cot death: A discarded hypothesis? *Journal of the Irish Medical Association* **60** 408–414.

Froggatt P and Adgey AJJ 1978 A case of the cardio-auditory syndrome (long QT interval and profound deafness) diagnosed in the perinatal period and kept under surveillance for two years. *Ulster Medical Journal* **47** 114–133.

Froggatt P and James TN 1973 Sudden unexpected death in infants: Evidence on a lethal cardiac arrhythmia. *Ulster Medical Journal* **42** 136–152.

Froggatt P and Nevin N 1999. The 'Incredible Mr Kavanagh': Arthur MacMurrough Kavanagh, P.C., M.P. (1831–1889). *Familia (Ulster Genealogical Review)* **15** 40–53.

Froggatt P, Lynas MA and MacKenzie G 1971a Epidemiology of sudden unexpected death in infants ('cot death') in Northern Ireland. *British Journal of Preventive and Social Medicine* **25** 119–134.

Froggatt P, Lynas MA and Marshall TK 1968. Sudden death in babies. Epidemiology. *American Journal of Cardiology* **22** 457–468.

Froggatt P, Lynas MA and Marshall TK 1971b Sudden unexpected death in infants ('cot death'). Report of a collaborative study in Northern Ireland. *Ulster Medical Journal* **40** 116–135.

Frota-Pessoa O, Otto PA and Otto PG 1975 *Genética clínica* (Clinical genetics). Livraria Francisco Alves Editora, Rio de Janeiro.

Fryns JP, van Schoubroeck D, Vandenberghe K, Nagels H and Klerckx P 1997 Diagnostic echographic findings in cryptophthalmos syndrome (Fraser syndrome). *Prenatal Diagnosis* **17** 582–584.

Fuchs E 1889 Krankendemonstration. (Session protocol of the K.K. Soc of Physicians of Vienna 29/3/1889.) *Wiener klinische Wochenschrift* **2** 281–282.

Fukushi JI, Makagiansar IT and Stallcup WB 2004 NG2 Proteoglycan promotes endothelial cell motility and angiogenesis via engagement of galectin-3 and {alpha}3{beta}1 integrin. *Molecular Biology of the Cell* **15** 3580–3590.

Fukuzawa N, Breslow I, Morison P, Dwyer T, Kusafuka Y, Kobayashi D, Becroft J, Beckwith E, Perlman A and Reeve 2004 Epigenetic differences between Wilms' tumours in white and east-Asian children. *The Lancet* **363** 446–451.

Gabriel SB, Salomon R, Pelet A, Angrist M, Amiel J, Fornage M, Attie-Bitach T, Olson JM, Hofstra R, Buys C, Steffann J, Munnich A, Lyonnet S and Chakravarti A 2002a Segregation at three loci explains familial and population risk in Hirschsprung disease. *Nature Genetics* **31** 89–93.

Gabriel SB, Schaffner SF, Nguyen H, Moore JM, Roy J, Blumenstiel B, Higgins J, DeFelice M, Lochner A, Faggart M, Liu-Cordero SN, Rotimi C, Adeyemo A, Cooper R, Ward R, Lander ES, Daly MJ and Altshuler D. 2002b The structure of haplotype blocks in the human genome. *Science* **296** 2225–2229.

Galjaard H 1994 Genetic technology in health care. *International Journal of Technology Assessment in Health Care* **10** 527–545.

Galjaard H 2002 New names for old disciplines. *Journal of Inherited Metabolic Disorders* **25** 139–156.

Gamaleia NF 1926 Troubles multiples du développement et glandes endocrines (par le Professeur Schereschewsky) *Revue Française d'Endocrinologie* **4** 181.

Gambetti P, Parchi P and Chen SG 2003 Hereditary Creutzfeldt-Jakob disease and fatal familial insomnia. *Clinical and Laboratory Medicine* **23** 43–64.

Gangloff S, McDonald JP, Bendixen C, Arthur L, and Rothstein R 1994 The yeast type I topoisomerase Top3 interacts with Sgs1, a DNA helicase homolog: a potential eukaryotic reverse gyrase. *Molecular Cell Biology* **14** 8391–8398.

García T, Nordet I, Machín S, González A, Muñiz A, Martínez G, Wade M and Svarch E 1999 Aportes al estudio de la drepanocitosis. Análisis clínico y hematológico en los primeros 5 años de vida. *Revista Cubana de Hematología e Inmunología y Hemoterapia* **15** 96–104.

Gardiner K, Davisson MT, Pritchard M, Patterson D, Groner Y, Crnic LS, Antonarakis S and Mobley W 2005 Report on the 'Expert Workshop on the biology of chromosome 21: towards gene-phenotype correlations in Down syndrome', held 11–14 June 2004, Washington D.C. *Cytogenetic and Genome Research* **108** 269–277.

Garg V, Kathiriya IS, Barnes R, Schluterman MK, King IN, Butler CA, Rothrock CR, Eapen RS, Hirayama-Yamada K, Joo K, Matsuoka R, Cohen JC and Srivastava D 2003 GATA4 mutations cause human congenital heart defects and reveal an interaction with TBX5. *Nature* **424** 443–447.

Garriga S and Crosby WH 1959 The incidence of leukemia in families with hypoplasia of the marrow. *Blood* **14** 1008–1014.

Garrod AE 1908 The Croonian Lectures on inborn errors of metabolism. *The Lancet* ii 1–7; 73–79; 142–148; 214–220. Reprinted with a supplement by Harris H. In: *Garrod's Inborn Errors of Metabolism*, Oxford University Press

Garza-Chapa R and Rojas-Alvarado MA 1996 Risk estimation of ABO and Rho(D) incompatibility in persons with mono- and polyphyletic surnames in Monterrey, Mexico. Comparison with other Mexican populations. *Archives of Medical Research* **27** 243–251.

Gätke MR, Østergaard D, Bundgaard JR, Varin F and Viby-Mogensen J 2001 Response to mivacurium in a patient compound heterozygous for a novel and a known silent mutation in the butyrylcholinesterase gene: genotyping by sequencing. *Anesthesiology* **95** 600–606.

Gaulden ME 1992 Maternal age effect: The enigma of Down syndrome and other trisomic conditions. *Mutation Research* **296** 69–88.

Gausden E, Coyle B, Armour JAL, Coffey R, Grossman A, Fraser GR, Winter RM, Pembrey ME, Kendall-Taylor P, Stephens D, Luxon LM, Phelps PD, Reardon W and Trembath R 1997a Pendred syndrome: evidence for genetic homogeneity and further refinement of linkage. *Journal of Medical Genetics* **34** 126–129.

Gavin AC and Superti-Furga G 2003 Protein complexes and proteome organization from yeast to man. *Current Opinion in Chemical Biology* 7 21–27.

Gayon J 2003 Do the biologists need the expression 'human race'? UNESCO 1950–1951, In: *Bioethical and Ethical Issues Surrounding the Trials and Code of Nuremberg*, (editor Jacques Rozenberg), Edwin Melon Press, New York. pp. 23–48.

Gencik A and Gencikova A 1983 Mohr syndrome in two siblings. *Journal de Génétique Humaine* 31 307–315.

George ST and Balasubramanian AS 1981 The aryl acylamidases and their relationship to cholinesterases in human serum, erythrocyte and liver. *European Journal of Biochemistry* 121 177–186.

German J 1968 Mongolism, delayed fertilization and human sexual behaviour. *Nature* 217 516–518.

German J 1969 Bloom's syndrome. I. Genetical and clinical observations in the first twenty-seven patients. *American Journal of Human Genetics* 21 196–227.

German J 1993 Bloom syndrome: A mendelian prototype of somatic mutational disease. *Medicine* 72 393–406.

German J and Ellis NA 2002 Bloom syndrome. In: *The Genetic Basis of Human Cancer*, 2nd Edition (editors Vogelstein B and Kinzler KW) McGraw-Hill, NewYork. pp. 267–288.

German J and Passarge E 1989 Bloom's syndrome. XII. Report from the Registry for 1987. *Clinical Genetics* 35 57–69.

Geula G and Mesulam M 1989 Special properties of cholinesterases in the cerebral cortex of Alzheimer's disease. *Brain Research* 498 185–189.

Gianaroli L, Magli, MC Ferraretti AP, Tabanelli C Trombetta C and Boudjema C 2001 The role of preimplantation diagnosis for aneuploidies. *Reproductive Biomedicine* Online 4 31–36.

Giardiello FM, Welsh SB, Hamilton SR, Offerhaus GJA, Gittelsohn AM, Booker SV, Krush AJ, Yardley JH and Luk GD 1987 Increased risk of cancer in the Peutz-Jeghers syndrome. *New England Journal of Medicine* 316 1511–1514.

Gibbons A 2004 Tracking the evolutionary history of a 'warrior' gene. *Science* 304 818.

Gilford H 1904 Ateleiosis and progeria: continuous youth and premature old age. *British Medical Journal* 2 914–918.

Gill H, Michaels L, Phelps PD and Reardon W 1999 Histopathological findings suggest the diagnosis in an atypical case of Pendred syndrome. *Clinical Otolaryngology* 24 523–526.

Gillam L 1999 Prenatal diagnosis and discrimination against the disabled. *Journal of Medical Ethics* 25 163–171.

Gispert S, Twells R, Orozco G, Brice A, Weber J, Heredero L, Sceufler K, Riley B, Allotey R, Nothers C, Hillermann R, Lunkes A, Khati C, Stevanin G, Hernández A, Magariño C, Klockgether T, Durr A, Chneiweiss H, Enczmann J, Farral M, Beckmann J, Mullan M, Wernet P, Agid Y, Freund J, Williamson R, Auburger G and Chamberlain S 1993 Chromosomal assignment of the second locus for autosomoal dominant cerebellar ataxia (SCA2) to chromosome 12q23–24.1. *Nature Genetics* 4 295–299.

Glas R, Graves JAM, Toder R, Ferguson-Smith MA and O'Brien PCM 1999 Cross-species chromosome painting between human and marsupial demonstrates the ancient region of the mammalian X. *Mammalian Genome* 10 1115–1116.

Glass AG and Miller RW 1968 U.S. mortality from Letterer-Siwe disease, 1960–1964. *Pediatrics* 42 364–367.

Glass AG, Hill JA and Miller RW 1968 Significance of leukemia clusters. *Journal of Pediatrics* 73 101–107.

Glass HB 1972 Human heredity and ethical problems. *Perspectives in Biology and Medicine* 15 237–253.

Gleeson JG, Keeler LC, Parisi MA, Marsh SE, Chance PF, Glass IA, Graham Jr JM, Maria BL, Barkovich AJ and Dobyns WB 2004 Molar tooth sign of the midbrain-hindbrain junction: occurrence in multiple distinct syndromes. *American Journal of Medical Genetics* 125 125–134

Glick D 1941 Some additional observations on the specificity of cholinesterase. *Journal of Biological Chemistry* 137 357–362.

Gnatt A, Prody CA, Zamir R, Lieman-Hurwitz J, Zakut H and Soreq H 1990 Expression of alternatively terminated unusual human butyrylcholinesterase messenger-RNA transcripts, mapping to chromosome 3q26-ter, in nervous-system tumors. *Cancer Research* 50 1983–1987.

Gocke DJ and Kavey NB 1969 Correlation with disease and infectivity of blood donors. *The Lancet* **1** 1055–1059.

Goffin JR, Straume O, Chappuis PO, Brunet JS, Begin LR, Hamel N, Wong N, Akslen LA and Foulkes WD 2003 Glomeruloid microvascular proliferation is associated with p53 expression, germline BRCA1 mutations and an adverse outcome following breast cancer. *British Journal of Cancer* **89** 1031–1034.

Golowin S 1902 Beiträge zur Anatomie und Pathogenese des Kryptophthalmus congenitus. *Zeitschrift für Augenheilkunde* **8** 175–212.

Goodman M 1999 The genomic record of humankind's evolutionary roots. *American Journal of Human Genetics* **64** 31–39.

Goretzki L, Burg MA, Grako KA and Stallcup WB 1999 High-affinity binding of basic fibroblast growth factor and platelet-derived growth factor-AA to the core protein of the NG2 proteoglycan. *Journal of Biological Chemistry* **274** 16831–16837.

Gorlin RJ, Cohen MM and Levin LS 1990 *Syndromes of the Head and Neck*. Oxford University Press, Oxford.

Gorlin RJ, Pindar L and Cohen MM 1988 *Syndromes of the Head and Neck*. McGraw-Hill Book Co., New York.

Gowers WR 1879 *Pseudo-hypertrophic Muscular Paralysis – a Clinical Lecture*. J & A Churchill, London.

Graham JM Jr 2001 A recognisable syndrome within CHARGE association: Hall-Hittner syndrome. *American Journal of Medical Genetics* **99** 120–123.

Granda H, Gispert S, Dorticós A, Martin M, Cuadras Y, Calvo M, Martinez G, Zayas MA, Oliva JA, and Heredero L 1991 Cuban programme for prevention of sickle cell disease. *The Lancet* **337** 152–153.

Granda H, Gispert S, Martínez G, Gómez M, Ferreira R, Collazo T, Magarino C, and Heredero L 1994 Results from a reference laboratory for prenatal diagnosis of sickle cell disorders in Cuba. *Prenatal Diagnosis* **14** 659–662.

Granda H. Personal communication to Luis Heredero-Baute

Graves JAM 1994 Mammalian sex-determining genes. In: *The Differences between the Sexes* (editors Short RV and Balaban E) Cambridge University Press, Cambridge. pp. 397–418..

Graves JAM 2003 Sex and death in birds: a model of dosage compensation that predicts lethality of sex chromosome aneuploids. *Cytogenetic and Genome Research* **101** 278–282.

Graves JAM and Shetty S 2001 Sex from W to Z: evolution of vertebrate sex chromosomes and sex determining genes. *Journal of Experimental Zoology* **290** 449–462.

Graves JAM and Westerman M 2002 Marsupial genetics and genomics. *Trends in Genetics* **18** 517–521.

Green J and Statham H 1996 Psychosocial aspects of prenatal screening and diagnosis. In: *The Troubled Helix: Social and Psychological Implications of the New Human Genetics* (editors Marteau T and Richards M) Cambridge University Press, Cambridge. pp. 143–165.

Greenberg CP, Primo-Parmo SL, Pantuck EJ and La Du BN 1995 Prolonged response to succinylcholine: a new variant of plasma cholinesterase that is identified as normal by traditional phenotyping methods. *Anesthesia and Analgesia* **81** 419–421.

Greenberg F 1987 Choanal atresia and athelia: Methimazole teratogenicity or a new syndrome? *American Journal of Medical Genetics* **28** 931–934.

Greenberg F, Keenan B, DeYanis V, Feingold M 1986 Gonadal dysgenesis and gonadoblastoma *in situ* in a female with Fraser (cryptophthalmos) syndrome. *Journal of Pediatrics* **108** 952–954

Greenwood TA, Rana BK and Schork NJ 2004 Human haplotype block sizes are negatively correlated with recombination rates. *Genome Research* **14** 1358–1361.

Gregory-Evans CY, Vieira H, Dalton R, Salt A, Adams GGW and Gregory-Evans K 2004 Ocular coloboma and high myopia with Hirschsprung disease associated with a novel ZFHX1B missense mutation and trisomy 21. *American Journal of Medical Genetics* **131A** 86–90.

Griseri P, Pesce B, Patrone G, Osinga J, Puppo F, Sancandi M, Hofstra R, Romeo G, Ravazzolo R, Devoto M and Ceccherini I 2002 A rare haplotype of the RET proto-oncogene is a risk modifying allele in Hirschsprung disease. *American Journal of Human Genetics* **71** 969–674.

Grüneberg H 1952a Genetical studies on the skeleton of the mouse. IV quasi-continuous variation. *Journal of Genetics* **51** 95–114.

Grüneberg H 1952b *The Genetics of the Mouse*, 2nd edn Martinus Nijhoff, The Hague.

Grüneberg H 1963 *The Pathology of Development: a Study of Inherited Skeletal Disorders in Animals*. Blackwell Scientific, Oxford.

Grützner F, Deakin J, Rens W, El-Mogharbel N and Graves JAM 2003 The monotreme genome: a patchwork of reptile, mammal and unique features? *Comparative Biochemistry and Physiology Part A* **136** 867–881.

Grützner F, Rens W, Tsend-Ayush E, EL-Mogharbel N, O'Brien PCM, Jones RC, Ferguson-Smith MA and Graves JAM 2004 In the platypus a meiotic chain of ten sex chromosomes shares genes with the bird Z and mammal X chromosomes. *Nature* **432** 913–917.

Guillaumin C 1972 *L'idéologie raciste. Genèse et langage actuel*. Mouton, Paris et La Haye.

Guldberg P, Straten P, Ahrenkiel V, Seremet T, Kirkin AF and Zeuthen J 1999 Somatic mutation of the Peutz-Jeghers gene, LKB1/STK11, in malignant melanoma. *Oncogene* **18** 1777–1780.

Gustavson KH, Kreuger A and Petersson PO 1971 Syndrome characterized by lingual malformations, polydactyly, tachypnea, and psychomotor retardation (Mohr syndrome) *Clinical Genetics* **2** 261–266.

Gusterson BA, Ross DT, Heath VJ and Stein T 2005 Basal cytokeratins and their relationship to the cellular origin and functional classification of breast cancer. *Breast Cancer Research* **7** 143–148.

Gutiérrez E, Barrios B and Taboada L 2002 Valoración clínica, psicológica y de laboratorio a niños con hiperfenilalaninemia benigna al nacimiento. *Revista Cubana de Pediatría* **74** 4.

Guven MA, Ceylaner S, Prefumo F and Uzel M 2004 Prenatal sonographic findings in a case of Varadi-Papp syndrome. *Prenatal Diagnosis* **24** 989–991.

Haargaard B, Wohlfahrt J, Fledelius HC, Rosenberg T and Melbye M 2004a Incidence and cumulative risk of childhood cataract in cohort of 2.6 million Danish children. *Investigative Ophthalmology and Visual Science.* **45** 1316–1320.

Haargaard B, Wohlfahrt J, Fledelius HC, Rosenberg T and Melbye M 2004b A National Danish study of 1027 cases of congenital/ infantile cataracts. Etiological and clinical classifications. *Ophthalmology* **111** 2292–2298.

Habermas J 2000 *Il futuro della natura umana. I rischi di una genetica liberale.* (translation of *The Future of Human Nature. The Risks of Liberal Genetics*) Giulio Einaudi editore, Turin, pp. 3, 72.

Habets PE, Moorman AF, Clout DE, van Roon MA, Lingbeek M, van Lohuizen M, Campione M and Christoffels VM 2002 Cooperative action of Tbx2 and Nkx2.5 inhibits ANF expression in the atrioventricular canal: implications for cardiac chamber formation. *Genes and Development* **16** 1234–1246.

Hacker A, Capel B, Goodfellow P and Lovell-Badge R 1995 Expression of *Sry*, the mouse sex-determining gene. *Development* **121** 1603–1614.

Hackstein JH, Hochstenbach R and Pearson PL 2000 Towards an understanding of the genetics of human male infertility: lessons from flies. *Trends in Genetics* **16** 565–572.

Hada T, Muratani K, Ohue T, Imanishi H, Moriwaki Y, Ito M, Amuro Y and Higashino K 1992 A variant serum cholinesterase and a confirmed point mutation at Gly-365 to Arg found in a patient with liver cirrhosis. *Internal Medicine* **31** 357–362.

Haim M 2002 Epidemiology of retinitis pigmentosa in Denmark. *Acta Ophthalmologica Scandinavica* Suppl 233 p. 10.

Haldane JBS 1937 The effect of variation on fitness. *American Naturalist* **71** 337–349.

Haldane JBS 1949 The rate of mutation of human genes. *Proceedings of the VIII International Congress of Genetics. Hereditas* **35** 267–273.

Haldane JBS 1957 The cost of natural selection. *Journal of Genetics* **55** 511–524.

Haldane JBS 1961 More precise expressions for the cost of natural selection. *Journal of Genetics* **57** 351–350.

Haldane JBS 1970 Karl Pearson In: *Studies in the History of Probability and Statistics* (editors Pearson ES and Kendall MG) Griffin.

Haldane JBS and Jayakar SD 1965 The nature of human genetic loads. *Journal of Genetics* **59** 143–149.

Hall BD 1979 Choanal atresia and associated multiple anomalies. *Journal of Pediatrics* **95** 395–398.

Hambire SD, Bhavsar PP, Meenakshi B and Tayakar AV 2003 Fraser – cryptophthalmos syndrome with cardiovascular malformation: A rare association. *Indian Pediatrics* **40** 888–890.

Hamosh A, Scott AF, Amberger JS, Bocchini CA and McKusick VA 2005 Online Mendelian Inheritance in Man (OMIM), a knowledgebase of human genes and genetic disorders. *Nucleic Acids Research* **33** D514–517.

Handyside AH, Kantogianni EH, Hardy K and Winston RM 1990 Pregnancies from biopsied human preimplantation embryos sexed by Y-specific DNA amplification. *Nature* **344** 768–770.

Hansen E, Flage T, Rosenberg T, Rudanko S-L, Viggoson G and Riise R 1992 Visual impairment in Nordic children III. Diagnoses. *Acta Ophthalmologica Scandinavica* **70** 597–604

Harding RM and McVean G 2004 A structured ancestral population for the evolution of modern humans. *Current Opinion in Genetics and Development* **14** 667–674.

Hardy GH 1908 Mendelian proportions in a mixed population. *Science* **28** 49–50.

Hardy ICW (ed.) 2002 *Sex Ratios: Concepts and Research Methods.* Cambridge University Press, Cambridge.

Harper JC, Wells D, Piyamongkol W, Abou-Sleiman P, Apessos A, Ioulianos A, Davis M, Doshi A, Serhal P, Ranieri M, Rodeck C and Delhanty JD 2002 Preimplantation genetic diagnosis for single gene disorders: experience with five single gene disorders. *Prenatal Diagnosis* **22** 525–533.

Harper P and Clarke A 1997 *Genetics and Society and Clinical Practice.* BIOS Scientific Publishers, Oxford. pp. 3, 181.

Harper PS, Harley HG, Reardon W and Shaw DJ 1992 Anticipation in myotonic dystrophy: new light on an old problem. *American Journal of Human Genetics* **51** 10–16.

Harrelson Z, Kelly RG, Goldin SN, Gibson-Brown JJ, Bollag RJ, Silver LM and Papaioannou VE 2004 Tbx2 is essential for patterning the atrioventricular canal and for morphogenesis of the outflow tract during heart development. *Development* **131** 5041–5052.

Harris H 1966 Enzyme polymorphisms in man. *Proceedings of the Royal Society Series B* **164** 298–310.

Harris H 1973 Lionel Sharples Penrose. *Biographical Memoirs of Fellows of the Royal Society* **19** 521–561.

Harris H and Whittaker M 1961 Differential inhibition of human serum cholinesterase with fluoride: recognition of two new phenotypes. *Nature* **191** 496–498.

Harris H, Hopkinson DA, Robson EB and Whittaker M 1963 Genetical studies on a new variant of serum cholinesterase detected by electrophoresis. *Annals of Human Genetics* **26** 359–382.

Hartmann A and Speit G. 1997 The contribution of cytotoxicity to DNA-effects in the single cell gel test (comet assay). *Toxicology Letters* **90** 183–188.

Hartmann A, Agurell E, Beevers C, Brender-Schwaab S, Burlinson B, Clay P, Collins, AR, Smith A, Speit G, Thyband V and Tice RR 2003 Recommendations for conducting the in vivo alkaline comet assay. *Mutagenesis* **18** 45–51.

Hartmann LC, Schaid DJ, Woods JE, Crotty TP, Myers JL, Arnold PG, Petty PM, Sellers TA, Johnson JL, McDonnell SK, Frost MH and Jenkins RB 1999 Efficacy of bilateral prophylactic mastectomy in women with a family history of breast cancer. *New England Journal of Medicine* **340** 77–84.

Hartmann LC, Sellers TA, Schaid DJ, Frank TS, Soderberg CL, Sitta DL, Frost MH, Grant CS, Donohue JH, Woods JE, McDonnell SK, Vockley CW, Deffenbaugh A, Couch FJ and Jenkins RB 2001 Efficacy of bilateral prophylactic mastectomy in BRCA1 and BRCA2 gene mutation carriers. *Journal of the National Cancer Institute* **93** 1633–1637.

Harvey RP 1996 NK-2 homeobox genes and heart development. *Developmental Biology* **178** 203–216.

Hassold T and Sherman S 2000 Down syndrome: genetic recombination and the origin of the extra chromosome 21. *Clinical Genetics* **57** 95–100.

Hassold T, Warburton D, Kline J and Stein Z 1984 The relationship of maternal age and trisomy among trisomic spontaneous abortions. *American Journal of Human Genetics* **36** 1349–1356.

Hatch M, Kline J, Levin B, Hutzler M and Warburton D 1990 Paternal age and trisomy among spontaneous abortions. *Human Genetics* **85** 355–361.

Haug K, Khan S, Fuchs S and Konig R 2000 OFD II, OFD VI, and Joubert syndrome manifestations in 2 sibs. *American Journal of Medical Genetics* **91** 135–137.

Haumont D and Pelc S 1983 The Mohr syndrome: are there two variants? *Clinical Genetics* **24** 41–46

He L and Hannon GJ 2004 MicroRNAs: small RNAs with a big role in gene regulation. *Nature Reviews Genetics* **5** 522–531.

Headings H 1998 The present state of genetic knowledge and implications for genetic screening. In: *Genetic Knowledge: Human Values and Responsibility.* (editor Kegley JAK) ICUS, Lexington, KY, pp. 113–122.

Heath AC, Cates R, Martin NG, Meyer J, Hewitt JK, Neale MC, and Eaves LJ 1993 Genetic contribution to risk of smoking initiation: comparisons across birth cohorts and across cultures. *Journal of Substance Abuse* **5** 221–246.

Heath SC 1997 Markov chain segregation and linkage analysis for oligogenic models. *American Journal of Human Genetics* **61** 748–760.

Heath S, Robledo R, Beggs W, Feola G, Parodo C, Rinaldi A, Contu L, Dana D, Stambolian D and Siniscalco M 2001 A novel approach to search for identity by descent in small samples of patients and controls from the same Mendelian Breeding Unit: A pilot study on Myopia. *Cytogenetics and Cell Genetics* **52** 183–190.

Hecht CA and Hook EB 1994 The imprecision in rates of Down syndrome by 1-year maternal age intervals: a critical analysis of rates used in biochemical screening. *Prenatal Diagnosis* **14** 729–738.

Hecht CA and Hook EB 1996 Rates of Down syndrome at livebirth by one-year maternal age intervals in studies with apparent close to complete ascertainment in populations of European origin: a proposed rate schedule for use in biochemical screening. *American Journal of Medical Genetics* **62** 376–385.

Heilstedt HA, Wu YQ, May KM, Bedell JA, Starkey DE, McPherson JD, Shapira SK and Shaffer LG 1998 Bilateral high frequency hearing loss is commonly found in patients with the 1p36 deletion syndrome. *American Journal of Human Genetics* Suppl **63** A106.

Hemminki A, Markie D, Tomlinson I, Avizienyte E, Roth S, Loukola A, Bignell G, Warren W, Aminoff M, Hoglund P, Jarvinen H, Kristo P, Pelin K, Ridanpaa M, Salovaara R, Toro T, Bodmer W, Olschwang S, Olsen AS, Stratton MR, de la Chapelle A and Aaltonen LA 1998 A serine/threonine kinase gene defective in Peutz-Jeghers syndrome. *Nature* **391** 184–187.

Hemminki A, Tomlinson I, Markie D, Jarvinen H, Sistonen P, Bjorkqvist A-M, Knuutila S, Salovaara R, Bodmer W, Shibata D, de la Chapelle A and Aaltonen LA 1997 Localization of a susceptibility locus for Peutz-Jeghers syndrome to 19p using comparative genomic hybridization and targeted linkage analysis. *Nature Genetics* **15** 87–90.

Henderson SA and Edwards RG 1968 Chiasma frequency and maternal age in mammals. *Nature* **218** 22–28.

Henning EMBP 1928 *Ohrenärztliche Untersuchungen von Schülern der Taubstummenschulen Schwedens; nebst Bemerkungen zur Frage des Unterrichts der Schwerhörigen.* Almqvist and Wiksele, Uppsala.

Heredero L 1992 Comprehensive national genetic program in a developing country-Cuba. *Birth Defects: Original Article Series* **28** 52–57.

Heredero L 1997 Los servicios de genética médica en Cuba. *Brazilian Journal of Genetics* **20** 47–53.

Heredero L, Atencio J, Vega E, Gutiérrez E and Damián A 1986 Diagnóstico Precoz de Fenilcetonuria en Cuba (estudio preliminar). *Revista Cubana de Pediatría* **58** 1.

Heredero L, Granda H, Suarez J and Altland K 1974 An economic high speed electrophoretic screening system for hemoglobin S and other proteins. *Humangenetik* **21** 167–177.

Herman R, Bartsocas CS, Soltesz G, Vazeou A, Paschou P, Bozas E, Malamitsi-Puchner A, Simell O, Knip M and Ilonen J 2004 Genetic screening for individuals at high risk for type 1 diabetes in the general population using HLA Class II alleles as disease markers. A comparison between three European populations with variable rates of disease incidence. *Diabetes Metabolism Research Reviews* **20** 322–329.

Hervé G 1912 Maupertuis génétiste. *Revue Anthropologie* **22** 217–230.

Hesser JE, Economidou J and Blumberg BS 1975 Hepatitis B surface antigen (Australia antigen) in parents and sex ratio of offspring in a Greek population. *Human Biology* **47** 415–425.

Hidaka K, Iuchi I, Tomita M, Watanabe Y, Minatogawa Y, Iwasaki K, Gotoh K and Shimizu C 1997a Genetic analysis of a Japanese patient with butyrylcholinesterase deficiency. *Annals of Human Genetics* **61** 491–496.

Hidaka K, Iuchi I, Yamasaki T, Ohhara M, Shoda T, Primo-Parmo SL and La Du BN 1992 Identification of two different genetic mutations associated with silent phenotypes for human serum cholinesterase in Japanese. *Japanese Journal of Clinical Pathology* **40** 535–540.

Hidaka K, Iuchi I, Yamasaki T, Ueda N and Hukano K 1997b Nonsense mutation in exon 2 of the butyryl-cholinesterase gene: a case of familial cholinesterasemia. *Clinica Chimica Acta* **261** 27–34.

Hidaka K, Watanabe Y, Tomita M, Ueda N, Higashi M, Minatogawa Y and Iuchi I 2001 Gene analysis of genomic DNA from stored serum by polymerase chain reaction: identification of three missense mutations in patients with cholinesterasemia and ABO genotyping. *Clinica Chimica Acta* **303** 61–67.

Hill WG 1974 Estimation of linkage disequilibrium in randomly mating populations. *Heredity* **33** 229–239.

Hill WG and Robertson A 1968 Linkage disequilibrium in finite populations. *Theoretical and Applied Genetics* **38** 226–231.

Hilleman MR, Buynak EB, Roehm RR, Tytell AA, Bertland AV and Lampson SP 1975 Purified and inactivated human hepatitis B vaccine: Progress report. *American Journal of Medical Sciences* **270** 401–404.

Hinds DA, Stuve LL, Nilsen GB, Halperin E, Eskin E, Ballinger DG, Frazer KA and Cox DR 2005 Whole-genome patterns of common DNA variation in three human populations. *Science* **307** 1072–1079.

Hingorani SR, Pagon RA, Shepard TH and Kapur RP 1991 Twin fetuses with abnormalities that overlap with three midline malformation complexes. *American Journal of Medical Genetics* **41** 230–235.

Hiraoka M, Berinstein DM, Trese MT and Shastry BS 2001 Insertion and deletion mutations in the dinucleotide repeat region of the Norrie disease gene in patients with advanced retinopathy of prematurity. *Journal of Human Genetics* **46** 178–181.

Hiroi Y, Kudoh S, Monzen K, Ikeda Y, Yazaki Y, Nagai R and Komuro I 2001 Tbx5 associates with Nkx2–5 and synergistically promotes cardiomyocyte differentiation. *Nature Genetics* **28** 276–280.

Hirschsprung H 1889 Stuhlträgheit Neugeborener infolge von Dilatation und Hypertrophie des Colons. *Jahrbuch der Kinderheilkunde* **27** 1–7.

Hittner HM, Hirsch NJ, Kreh GM and Rudolph AJ 1979 Colobomatous microphthalmia, heart disease, hearing loss and mental retardation: a syndrome. *Journal of Pediatric Ophthalmology & Strabismus* **16** 122–128.

Hobbs CA, Sherman SL, Yi P, Hopkins SE, Torfs CP, Hine RJ, Pogribna M, Rozen R and James SJ 2000 Polymorphisms in genes involved in folate metabolism as maternal risk factors for Down syndrome. *American Journal of Human Genetics* **67** 623–630.

Hodor PG, Illies MR, Broadley S and Ettensohn C A 2000 Cell-substrate interactions during sea urchin gastrulation: migrating primary mesenchyme cells interact with and align extracellular matrix fibers that contain ECM3, a molecule with NG2-like and multiple calcium-binding domains. *Developmental Biology* **222** 181–194.

Hoffman JI and Kaplan S 2002 The incidence of congenital heart disease. *Journal of the American College of Cardiology* **19** 1890–1900.

Hoffman JI, Kaplan S and Liberthson RR 2004 Prevalence of congenital heart disease. *American Heart Journal* **147** 425–439.

Hofstra RM, Valdenaire O, Arch E, Osinga J, Kroes H, Loffler, B-M, Hamosh A, Meijers C and Buys CHCM 1999 A loss-of-function mutation in the endothelin-converting enzyme 1 (ECE-1) associated with Hirschsprung disease, cardiac defects, and autonomic dysfunction. *American Journal of Human Genetics* **64** 304–308

Hol FA, Hamel BCJ, Geurds MPA, Mullaart RA, Barr FG, Macina RA and Mariman EC 1995 A frameshift mutation in the gene for PAX3 in a girl with spina bifida and mild signs of Waardenburg syndrome. *Journal of Medical Genetics* **32** 52–56.

Holden C 2003 More men ready for cloning. *Science* **299** 41.

Holz O, Jörres R, Kästner A and Magnussen H 1995 Differences in basal and induced DNA single-strand breaks between human peripheral monocytes and lymphocytes. *Mutation Research* **332** 55–62.

Hook EB 1983 Down syndrome rates and relaxed selection at older maternal ages. *American Journal of Human Genetics* **35** 1307–1313.

Hook EB 1992 In: *Prenatal Diagnosis and Screening* (editors Brock DJH, Rodeck CH and Ferguson-Smith MA) Churchill Livingstone, Edinburgh. pp. 351–392.

Hook EB, Topol BB and Cross PK 1989 The natural history of cytogenetically abnormal fetuses detected at midtrimester amniocentesis which are not terminated electively: New data and estimates of the excess and relative risk of late foetal death associated with 47,+21 and some other abnormal karyotypes. *American Journal of Human Genetics* 45 855–861.

Hoover R, Mason TJ, McKay FW and Fraumeni JF Jr 1975 Cancer by county: new resource for etiologic clues. *Science* 189 1005–1007.

Hosoda H, Kojima M, Matsuo H and Kangawa K 2000 Ghrelin and des-acyl ghrelin: two major forms of rat ghrelin peptide in gastrointestinal tissue. *Biochemical and Biophysical Research Communications* 279 909–913.

Hosoda K, Hammer RE, Richardson JA, Baynash AG, Cheung JC, Giaid A, Yanagisawa M 1994 Targeted and natural (piebald-lethal) mutations of endothelin-B receptor gene produce megacolon associated with spotted coat color in mice. *Cell* 79 1267–1276.

Hoth CF, Milunsky A, Lipsky N, Sheffer R, Clarren SK and Baldwin CT 1993 Mutations in the paired domain of the human PAX3 gene cause Klein-Waardenburg syndrome (WS-III) as well as Waardenburg syndrome type I (WS-I). *American Journal of Human Genetics* 52 455–462.

Houlston RS, Murday V, Harocopos C, Williams CB and Slack J 1990 Screening and genetic counselling for relatives of patients with colorectal cancer in a family cancer clinic. *British Medical Journal* 301 366–368.

Howard FM and Young ID 1988 Unknown syndrome: microcephaly, facial clefting, and preaxial polydactyly. *Journal of Medical Genetics* 25 272–273. *http://www.sigemec.sld.cu\rcgh\esp\revista_esp\rcgh_esp.htm*

Hu T, Yamagishi H, Maeda J, McAnally J, Yamagishi C and Srivastava D 2004 Tbx1 regulates fibroblast growth factors in the anterior heart field through a reinforcing autoregulatory loop involving forkhead transcription factors. *Development* 131 5491–5402.

Huang B, Wang S, Ning Y, Lamb AN and Bartley J 1999 Autosomal XX sex reversal caused by duplication of *SOX9*. *American Journal of Medical Genetics* 87 349–353.

Hubbard R and Wald E 1999 *Exploding the Gene Myth. How Genetic Information is Produced and Manipulated by Scientists, Physicians, Employers, Insurance Companies, Educators, and Law Enforcers.* Beacon Press, Boston, p. 27.

Hume J 1996 Disability, feminism and eugenics: who has the right to decide who should or should not inhabit the world? *http://www.wwda.org.au/eugen.htm.*

Hunt R and Taveau RM 1906 On the physiological action of certain cholin derivatives and new methods for detecting cholin. *British Medical Journal* 2 1788–1791.

Hunter DJ 2005 Gene-environment interactions in human diseases. *Nature Reviews Genetics* 6 287–298.

Hunter J 1787 Observations tending to shew that the wolf, jackal and dog are all of the same species. *Philosophical Transactions of the Royal Society of London* 77 253–266.

Hurskainen P, Dahlen P, Ylikoski J, Kwiatkowski M, Siitari H and Lovgren T 1991 Preparation of europium-labelled DNA probes and their properties. *Nucleic Acid Research* 19 1057–1061.

Hurst JA, Meinecke P and Baraitser M 1991 Balanced t(6;8)(6p8p;6q8q) and the CHARGE association. *Journal of Medical Genetics* 28 54–55.

Hurst LD 1994 Embryonic growth and the evolution of the Y chromosome. I. The Y as an attractor of selfish growth factors. *American Journal of Human Genetics* 73 223–232.

Hussey ND, Donggui H, Froiland DAH, Hussey DJ, Haan EA, Matthews CD and Craig JE 2002 Preimplantation genetic diagnosis for beta-thalassemia using sequencing of single cell PCR products to detect mutation and polymorphic loci. *Molecular Human Reproduction* 8 1136–1143.

Hutchinson J 1886 Case of congenital absence of hair, with atrophic condition of the skin and its appendages, in a boy whose mother had been almost wholly bald from alopecia areata from the age of six. *The Lancet* 1 923

Hutchinson J 1896 Records of demonstrations at the Clinical Museum. Pigmentation of lips and mouth. *Archives of Surgery* 7 286–294.

Hutchinson J 1913 Foreword. *Syphilis.* Cassell and Co., London. (2nd edn).

Hüther W 1954 Die Hirschsprung'sche Krankheit als Folge einer Entwicklungsstörung der intramuralen Ganglien. *Beitrage zur Pathologischen Anatomie und zur Allgemeinen Pathologie* **114** 161–191.

Ibanez P, Bonnet AM, Debarges B, Lohmann E, Tison F, Pollak P, Agid Y, Durr A and Brice A 2004 Causal relation between alpha-synuclein gene duplication and familial Parkinson's disease. *The Lancet* **364** 1169–1171.

Ilonen J, Reijonen H, Herva E, Sjoroos M, Iitia A, Longren T, Veijola R, Knip M and Akerblom HK 1996 Rapid HLA-DQB1 genotyping for four alleles in the assessment of risk for IDDM in the Finnish population. *Diabetes Care* **19** 795–800.

Ilonen J, Sjöroos M, Knip M, Veijola R, Simell O, Akerblom HK, Paschou P, Bozas E, Havarani B, Malamitsi-Puchner A, Thymelli J, Vazeou A and Bartsocas CS 2002 Estimation of genetic risk for type 1 diabetes. *American Journal of Medical Genetics* (Seminars in Medical Genetics) **115** 30–36.

INCA 2002 Estimativas sobre a incidência e mortalidade por câncer no Brasil. *Revista Brasileira de Cancerologia* **48** 175–179.

Ingram VM 1956 A specific chemical difference between the globins of normal human and sickle-cell anaemia haemoglobin. *Nature* **178** 792–794.

Ingram VM 1957 Gene mutations in human haemoglobin: the chemical difference between normal and sickle cell haemoglobin. *Nature* **180** 325–328.

Inoue K, Khajavi M, Ohyama T, Hirabayashi S-I, Wilson J, Reggin JD, Mancias P, Butler IJ, Wilkinson MF, Wegner M and Lupski JR 2004 Molecular mechanism for distinct neurological phenotypes conveyed by allelic truncating mutations. *Nature Genetics* **36** 361–369.

Inoue K, Tanabe Y and Lupski J 1999 Myelin deficiencies in both the central and the peripheral nervous systems associated with a *SOX10* mutation. *Annals of Neurology* **46** 313–318.

Ishihara N, Yamada K, Yamada Y, Miura K, Kato J, Kuwabara N, Hara Y, Kobayashi Y, Hoshino K, Nomura Y, Mimaki M, Ohya K, Matsushima M, Nitta H, Tanaka K, Segawa M, Ohki T, Ezoe T, Kumagai T, Onuma A, Kuroda T, Yoneda M, Yamanaka T, Saeki M, Segawa M, Saji T, Nagaya M and Wakamatsu N 2004 Clinical and molecular analysis of Mowat-Wilson syndrome associated with ZFHX1B mutations and deletions at 2q22-q24.1 *Journal of Medical Genetics* **41** 387–933.

Iwashita T, Kruger GM, Pardal R, Kiel MJ and Morrison SJ 2002 Hirschsprung disease is linked to defects in neural crest stem cell function. *Science* **301** 972–976.

Iwashita T, Murakami H, Asai N and Takahashi M 1996 Mechanism of Ret dysfunction by Hirschsprung mutations affecting its extracellular domain. *Human Molecular Genetics* **5** 1577–80.

Jacobi A 1869 On some important causes of constipation in infants. *American Journal of Obstetrics* **2** 96–113.

Jacobs PA and Ross A 1966 Structural abnormalities of the Y chromosome in man. *Nature* **210** 352–354.

Jacobs PA and Strong JA 1959 A case of human intersexuality having a possible XXY sex-determining mechanism. *Nature* **183** 302–303.

Jadeja S, Smyth I, Pitera JE, Taylor MS, van Haelst M, Bentley E, McGregor L, Hopkins J, Chalepakis G, Philip N, Perez Aytes A, Watt FM, Darling SM, Jackson I, Woolf AS and Scambler PJ 2005 Identification of a new gene mutated in Fraser syndrome and mouse myelencephalic blebs. *Nature Genetics* **37** 520–525.

Jamar M, Lemarchal C, Lemaire V, Koulischer L and Bours V 2003 A low rate of trisomy 21 in twin-pregnancies: a cytogenetic retrospective study of 278 cases. *Genetic Counselling* **14** 395–400.

James RS, Ellis K, Pettay D and Jacobs PA 1998 Cytogenetic and molecular study of four couples with multiple trisomy 21 pregnancies. *European Journal of Human Genetics* **6** 207–212.

James SJ, Pogribna M, Pogribny IP, Melnyk S, Hine RJ, Gibson JB, Yi P, Tafoya DL, Swenson DH, Wilson VL and Gaylor DW 1999 Abnormal folate metabolism and mutation in the methylene-tetrahydrofolate reductase gene may be maternal risk factors for Down syndrome. *American Journal of Clinical Nutrition* **70** 495–501.

James TN 1968 Sudden death in babies. New observations in the heart. *American Journal of Cardiology* **22** 479–506.

Jeghers H, McKusick VA and Katz KH 1949 Generalized intestinal polyposis and melanin spots of the oral mucosa, lips and digits. *New England Journal of Medicine* **241** 993–1005 and 1031–1036.

Jennings HS 1917 The numerical results of diverse systems of breeding, with respect to two pairs of characters, linked or independent, with special relation to the effects of linkage. *Genetics* **2** 97–154.

Jensen FS, Bartels CF and La Du BN 1992 Structural basis of the butyrylcholinesterase H-variant segregating in two Danish families. *Pharmacogenetics* **2** 234–240.

Jensen H, Warburg M, Sjö O and Schwartz M 1995 Duchenne muscular dystrophy: negative electroretinograms and normal dark adaptation. *Journal of Medical Genetics* **32** 348–351.

Jervell A and Lange-Nielsen F 1957 Congenital deaf-mutism, functional heart disease with prolongation of QT interval and sudden death. *American Heart Journal* **54** 59–68.

John Paul II 2003 Address to the plenary session of the Pontifical Academy of Sciences, 28 October 1994, in The Pontifical Academy of Sciences, Vatican City. *Scripta Varia* **100** 358–363.

Johnson S 1759 *The History of Rasselas, Prince of Abissinia* Project Gutenberg *http://www.gutenberg.org/browse/authors/*

Jones C, Ford E, Gillett C, Ryder K, Merrett S, Reis-Filho JS Fulford LG, Hanby A and Lakhani SR 2004 Molecular cytogenetic identification of subgroups of grade III invasive ductal breast carcinomas with different clinical outcomes. *Clinical Cancer Research* **10** 5988–5997.

Jones C, Nonni AV, Fulford L, Merrett S, Chaggar R, Eusebi V and Laklani SR 2001 CGH analysis of ductal carcinoma of the breast with basaloid/myoepithelial cell differentiation. *British Journal of Cancer* **85** 422–427.

Jones KLF 1998 *Smith's recognizable patterns of human malformation.* WB Saunders Co., Philadelphia. p. 262.

Jost A 1953 Problems of fetal endocrinology: the gonadal and hypophyseal hormones. *Recent Progress in Hormone Research* **8** 379–418.

Jost A, Magre S and Agelopoulou R 1981 Early stages of testicular differentiation in the rat. *Human Genetics* **58** 59–63.

Jourdi H, Iwakura Y, Narisawa-Saito M, Ibaraki K, Xiong H, Watanabe M, Hayashi Y, Takei N and Nawa H 2003 Brain-derived neurotrophic factor signal enhances and maintains the expression of AMPA receptor-associated PDZ proteins in developing cortical neurons. *Developmental Biology* **263** 216–230.

Judge DP, Aprhys CMJ, Guerrerio P, Geubtner J, Zhang J, Cheng A and Dietz H A novel cardiocutaneous progeria syndrome caused by mutation in lamin A/C. pers. comm. to VA McKusick.

Just W, Rau W, Vogel W, Akhuerdian M, Fredga K, Graves JAM and Lyapunova E 1995 Absence of *Sry* in species of the vole *Ellobius. Nature Genetics* **11** 117–118.

Juul P 1968 Human plasma cholinesterase isoenzymes. *Clinica Chimica Acta* **19** 205–213.

Kahn J 2005 Misreading race and genomics after BiDil. *Nature Genetics* **37** 655–656.

Kajiwara K, Berson EL and Dryja TP 1994 Digenic retinitis pigmentosa due to mutations at the unlinked peripherin/RDS and ROM1 loci. *Science* **264** 1604–1608.

Kallen K 1997 Parity and Down syndrome. *American Journal of Medical Genetics* **70** 196–201.

Kalow W and Genest K 1957 A method for the detection of atypical forms of human serum cholinesterase. Determination of dibucaine numbers. *Canadian Journal of Biochemistry and Physiology* **35** 339–346.

Kalow W and Gunn DR 1959 Some statistical data on atypical cholinesterase of human serum. *Annals of Human Genetics* **23** 239–250.

Kalow W and Staron N 1957 On distribution and inheritance of atypical forms of human serum cholinesterase, as indicated by dibucaine numbers. *Canadian Journal of Biochemistry and Physiology* **35** 1305–1320.

Kant I 1785 *Grundlegung zur Metaphysik der Sitten (Foundations of the Metaphysics of Morals).* 1986 edition, Reclam, Leipzig.

Karp SE, Tonin PN, Begin LR, Martinez JJ, Zhang JC, Pollak MN and Foulkes WD 1997 Influence of BRCA1 mutations on nuclear grade and estrogen receptor status of breast carcinoma in Ashkenazi Jewish women. *Cancer* **80** 435–441.

Kasahara H, Usheva A, Ueyama T, Aoki H, Horikoshi N and Izumo S 2001 Characterization of homo- and heterodimerization of cardiac Csx/Nkx2.5 homeoprotein. *Journal of Biological Chemistry* **276** 4570–4580.

Katsanis N, Ansley SJ, Badano JL, Eichers ER, Lewis RA, Hoskins BE Scambler PJ, Davidson WS, Beales PL and Lupski JR 2001 Triallelic inheritance in Bardet-Biedl syndrome, a Mendelian recessive disorder. *Science* 293 2256–2259.

Katz MG, Fitzgerald L, Bankier A, Savulescu J and Cram DS 2002 Issues and concerns of couples presenting for preimplantation genetic diagnosis (PGD). *Prenatal Diagnosis* 22 1117–1122.

Kauff ND, Perez-Segura P, Robson ME, Scheuer L, Siegel B, Schluger A, Rapaport B, Frank TS, Nafa K, Ellis NA, Parmigiani G and Offit K 2002 Risk-reducing salpingo-oophorectomy in women with a BRCA1 or BRCA2 mutation. *New England Journal of Medicine* 346 1609–1615.

Kazaura MR and Lie RT 2002. Down's syndrome and paternal age in Norway. *Paediatric and Perinatal Epidemiology* 16 314–319.

Keefe DL, Niven-Fairchild T, Powell S and Buradagunta S 1995 Mitochondrial deoxyribonucleic acid deletions in oocytes and reproductive aging in women. *Fertility and Sterility* 64 577–583.

Keitges E, Rivest M, Siniscalco M and Gartler SM 1985 X-linkage of steroid sulphatase in the mouse is evidence for a functional Y-linked allele. *Nature* 315 226–227.

Kelly EC 1940 Selections from the writing of Sir Jonathan Hutchinson. *Medical Classics* 5 109–245.

Kelly RG and Buckingham ME 2002 The anterior heart-forming field: voyage to the arterial pole of the heart. *Trends in Genetics* 18 210–216.

Kennedy JL, Farrer LA, Andreasen NC, Mayeux R and St George-Hyslop P 2003 The genetics of adult-onset neuropsychiatric disease: Complexities and conundra? *Science* 302 822–826.

Kennedy RD, Quinn JE, Johnston PG and Harkin DP 2002 BRCA 1: Mechanisms of inactivation and implications for management of patients. *The Lancet* 360 1007–1014.

Kerr WE 1975 Population genetic studies in bees (Apidae, Hymenoptera). 1 Genetic load. *Anais da Academia Brasileira de Ciências* 47 319–332.

Kettlewell JR, Raymond CS and Zarkower D 2000 Temperature-dependent expression of turtle DMRT1 prior to sexual differentiation. *Genesis* 26 174–178.

Key SN 1920 Report of a case of cryptophthalmia. *American Journal of Ophthalmology* 3 684–685.

Killian JK, Byrd JC, Jirtle JV, Munday BL, Stoskopf MK, MacDonald RG and Jirtle RL 2000 M6P/IGF2R imprinting evolution in mammals. *Molecular Cell* 5 707–716.

Kimura M 1960 Optimum mutation rate and degree of dominance as determined by the principle of minimum genetic load. *Journal of Genetics* 57 21–34.

Kimura M 1968 Evolutionary rate at the molecular level. *Nature* 217 624–626.

Kimura M 1983 *The Neutral Theory of Molecular Evolution*. Cambridge University Press.

Kimura M and Crow JF 1978 Effect of overall phenotypic selection on genetic change at individual loci. *Proceedings of the National Academy of Sciences USA* 75 6168–6171.

Kimura M and Ohta T 1971 *Theoretical Aspects of Population Genetics*. Princeton University Press, Princeton, NJ.

King JL 1967 Continuously distributed factors affecting fitness. *Genetics* 55 483–492.

King JL and Jukes TH 1969 Non-Darwinian evolution. *Science* 164 788–798.

King MC, Marks JH and Mandell JB 2003 New York Breast Cancer Study Group. Breast and ovarian cancer risks due to inherited mutations in BRCA1 and BRCA2. *Science* 302 643–646.

Kingman JFC 1982 The coalescent. *Stochastic Processes and Their Applications* 13 235–248.

Kingman JFC 2000 Origins of the coalescent: 1974-1982. *Genetics* 156 1461–1463.

Kirk GS, Raven JE and Schofield M 1983 *The Presocratic Philosophers*. (2nd edition) Cambridge University Press

Kirk GS, Raven JE and Schofield M 1988 *The Presocratic Philosophers*. (Greek edition) MIET, Athens.

Kirsch S, Weiss B, Kleiman S, Roberts K, Pryor J, Milunsky A, Ferlin A, Foresta C, Matthijs G and Rappold GA 2002 Localisation of the Y chromosome stature gene to a 700 kb interval in close proximity to the centromere. *Journal of Medical Genetics* 39 507–513.

Kirsch S, Weiß B, Miner TL, Waterston RH, Clark RA, Eichler EE, Münch C, Schempp W and Rappold G 2005 Interchromosomal segmental duplications of the pericentromeric region on the human Y chromosome. *Genome Research* 15 195–204.

Kissinger CR, Liu BS, Martin-Blanco E, Kornberg TB and Pabo CO 1990 Crystal structure of an engrailed homeodomain-DNA complex at 2.8 A resolution: a framework for understanding homeodomain-DNA interactions. *Cell* **63** 579–590.

Kiyozumi D, Osada A, Sugimoto N, Weber CN, Ono Y, Imai T, Okada A and Sekiguchi K 2005 Identification of a novel cell-adhesive protein spatiotemporally expressed in the basement membrane of mouse developing hair follicle. *Experimental Cell Research* **306** 9–23.

Klein D 1983 Historical background and evidence for dominant inheritance of the Klein-Waardenburg syndrome (type III). *American Journal of Medical Genetics* **14** 231–239.

Kline J and Levin B 1992 Trisomy and age at menopause: predicted associations given a link with rate of oocyte atresia. *Paediatric and Perinatal Epidemiology* **6** 225–239.

Kline J, Kinney A, Levin B and Warburton D 2000 Trisomic pregnancy and earlier age at menopause. *American Journal of Human Genetics* **67** 395–404.

Kline J, Kinney A, Reuss ML, Kelly A, Levin B, Ferin M and Warburton D 2004 Trisomic pregnancy and the oocyte pool. *Human Reproduction* **19** 1633–1643.

Knight SJL and Flint J 2000 Perfect endings: a review of subtelomeric probes and their se in clinical diagnosis. *Journal of Medical Genetics* **37** 401–409.

Knouff C, Malloy S, Wilder J, Altenburg MK and Maeda N 2001 Doubling expression of the low density lipoprotein receptor by truncation of the 3'-untranslated region sequence ameliorates type III hyperlipoproteinemia in mice expressing the human ApoE2 isoform. *Journal of Biological Chemistry* **276** 3856–3862.

Kobayashi S, Boggon TJ, Dayaram T, Janne PA, Kocher O, Meyerson M Johnson BE, Eck MJ, Tenen DG and Halmos B 2005 EGFR mutation and resistance of non-small-cell lung cancer to gefitinib. *New England Journal of Medicine* **352** 786–792.

Kohl T, Hering R, Bauriedel G, van de Vondel P, Heep A, Keiner S, Müller A, Franz A, Bartmann P and Gembruch U 2006 Percutaneous fetoscopic and ultrasound-guided decompression of the fetal trachea permits normalization of fetal hemodynamics in a human fetus with Fraser syndrome and congenital high airway obstruction syndrome (CHAOS) from laryngeal atresia. *Ultrasound Obstetrics Gynecology* **27** 84–88.

Kohn M, Kehrer-Sawatzki H, Vogel W, Graves JAM and Hameister H 2004 Wide genome comparisons reveal the origins of the human X chromosome. *Trends in Genetics* **20** 598–603.

Koivurova S, Hartikainen AL, Gissler M, Hemminki E, Sovio U and Jarvelin MR 2002 Neonatal outcome and congenital malformations in children born after in-vitro fertilization. *Human Reproduction* **17** 1391–1398.

Kojima M, Hosoda H, Matsuo H and Kangawa K 2001 Ghrelin: discovery of the natural endogenous ligand for the growth hormone secretagogue receptor. *Trends in Endocrinology & Metabolism* **12** 118–126.

Kolata G 2003 Genetic revolution: how much, how fast? *New York Times* (25 Feb. 2003) F6.

Kondrashov AS and Crow JF 1993 A molecular approach to estimating the human deleterious mutation rate. *Human Mutation* **2** 229–234.

Konotey-Ahulu FID 1970 Maintenance of high sickling rate in Africa: Role of polygamy. *Journal of Tropical Medicine and Hygiene* **73** 19–21.

Konotey-Ahulu FID 1977 Male procreative superiority in African populations: The fact established and quantified. In: *Medical Genetics, Excerpta Medica* (editors Szabo G and Papp Z) pp. 600–607.

Konotey-Ahulu FID 1980 Male procreative superiority index (MPSI): The missing co-efficient in African anthropogentics. *British Medical Journal* **281** 1700–1702.

Konotey-Ahulu FID 1990 The genetics of Ghanaian high blood pressure. (Editorial) *Ghana Medical Journal* **24** 160–163.

Konotey-Ahulu FID 1996 *The sickle cell disease patient: natural history from a clinico-epidemiological study of the first 1550 patients of Korle Bu Hospital Sickle Cell Clinic.* Tetteh-A'Domeno, Watford.

Konotey-Ahulu FID 2004 Some aspects of the macro-genetics of hypertension in African people. Ethnic Health Symposium. *Royal Society of Medicine* (*In preparation*)

Koopman P 2005 Sex determination: a tale of two *Sox* genes. *Trends in Genetics* **21** 367–370.

Kopp P, Arseven OK, Sabacan L, Kotlar T Dupuis J, Cavaliere H, Santos CLS, Jameson JL and Medeiros-Neto G 1999 Phenocopies for deafness and goiter development in a large inbred Brazilian kindred with Pendred's syndrome associated with a novel mutation in the PDS gene. *Journal of Clinical Endocrinology and Metabolism* 84 336–341.

Korner H, Rodríguez L, Fernández JL, Schulze M, Horn A, Heredero L, Witkowski R, Tinschert S, Oliva JA, Sommer D, Zwahr Ch, Prenzlau P, Cobet G and Gunter H 1986 Maternal serum alpha-feto-protein screening for neural tube defects and other disorders using an ultramicro-ELISA. *Human Genetics* 73 60–63.

Korsching E, Packeisen J, Agelopoulos K, Eisenacher M, Voss R, Isola J, van Diest PJ, Brandt B, Boecker W and Buerger H 2002 Cytogenetic alterations and cytokeratin expression patterns in breast cancer: Integrating a new model of breast differentiation into cytogenetic pathways of breast carcinogenesis. *Laboratory Investigation* 82 1525–1533.

Krejci E, Thomine S, Boschetti N, Legay C, Sketelj J and Massoulié J 1997 The mammalian gene of acetyl-cholinesterase-associated collagen. *Journal of Biological Chemistry* 272 22840–22847.

Kringlebach J and Wennevold A 1971 Prolonged QT and cardiac syncope (Ward's syndrome). *Acta Pediatrica Scandinavica* 60 248–249.

Krivit W and Good RA 1956 The simultaneous occurrence of leukemia and mongolism; report of four cases. *American Journal of Diseases of Children* 91 218–222.

Krob G, Braun A. and Kuhnle U 1994 True hermaphroditism: geographical distribution, clinical findings, chromosomes and gonadal histology. *European Journal of Pediatrics* 153 2–10.

Kruger LM and Kumar A 1994 Acheiropody. A report of two cases. *Journal of Bone and Joint Surgery (American Volume)* 76 1557–1560.

Krugman S, Giles JP and Hammond J 1971 Viral hepatitis, type B (MS2 strain). Studies on active immunization. *Journal of the American Medical Association* 217 41–45.

Kucherlapati RD, McKusick VA and Ruddle FH 1996 Genomics: an established discipline, a commonly used name, a mature journal. (Editorial) *Genomics* 31 1–2.

Kuehn MR, Bradley A, Robertson EJ and Evans MJ 1987 A potential animal model for Lesch-Nyhan syndrome through introduction of HPRT mutations into mice. *Nature* 326 295–298.

Kuhl CK, Schmutzler RK, Leutner CC, Kempe A, Wardelmann E, Hocke A, Maringa M, Pfeifer U, Krebs D and Schild HH 2000 Breast MR imaging screening in 192 women proved or suspected to be carriers of a breast cancer susceptibility gene: preliminary results. *Radiology* 215 267–279.

Kuliev A and Verlinsky Y 2002 Current features of preimplantation genetic diagnosis. *Reproductive BioMedicine* Online 5 294–299.

Kumar S and Hedges SB 1998 A molecular timescale for vertebrate evolution. *Nature* 392 917–920.

Kumar S, Deffenbacher K, Marres HAM, Cremers CWRJ and Kimberling WJ 2000 Genomewide search and genetic localization of a second gene associated with autosomal dominant branchio-oto-renal syndrome: clinical and genetic implications. *American Journal of Human Genetics* 66 1715–1720.

Kutty KM, Rowden G and Cox AR 1973 Interrelationship between serum α-lipoprotein and cholinesterase. *Canadian Journal of Biochemistry* 51 883–887.

Ladeira MSP, Rodrigues MA, Salvadori DMF, Neto PP, Achilles P, Lerco MM, Rodrigues PA, Goncalves Jr I, Queiroz DMM and Freire-Maia DV 2004a Relationships between cagA, vacA, and iceA genotypes of *Helicobacter pylori* and DNA damage in the gastric mucosa. *Environmental and Molecular Mutagenesis* 44 91–98.

Ladeira MSP, Rodrigues MAM, Salvadori DMF, Queiroz DMM and Freire-Maia DV 2004b DNA damage in patients infected by Helicobacter pylori. *Cancer Epidemiology and Biomarkers Prevention* 13 631–637.

Ladeira MSP, Salvadori DMF, Bueno RCA, dos Santos BF, Achilles P, Kobayasi S, Rodrigues PA, Lerco MM and Rodrigues M 2004c Oxidative DNA damage in patients infected by *H. pylori*. *Proceedings of II Intercontinental Congress of Pathology* 5–9.

Ladeira MSP, Rodrigues MAM, Freire-Maia DV and Salvadori DMF 2005 Use of comet assay to assess DNA damage in patients infected by *Helicobacter pylori*. *Mutation Research* 586 76–86.

Lahn BT and Page DC 1999 Four evolutionary strata on the human X chromosome. *Science* 286 964–967.

Laken SJ, Petersen GM, Gruber SB, Oddoux C, Ostrer H, Giardiello FM, Hamilton SR, Hampel H, Markowitz A, Klimstra D, Jhanwar S, Winawer S, Offit K, Luce MC, Kinzler KW and Vogelstein B 1997 Familial colorectal cancer in Ashkenazim due to a hypermutable tract in APC. *Nature Genetics* 17 79–83.

Lakhani SR, Jacquemier J, Sloane JP, Gusterson BA, Anderson TJ, van de Vijver MJ, Farid LM, Venter D, Antoniou A, Storfer-Isser A, Smyth E, Steel CM, Haites N, Scott RJ, Goldgar D, Neuhausen S, Daly PA, Ormiston W, McManus R, Scherneck S, Ponder BA, Ford D, Peto J, Stoppa-Lynnet D, Bignon YJ, Stuewing JP, Spurr NK, Bishop DT, Klijn JG, Devilee P, Cornelisse CJ, Lasset C, Lenoir G, Barkardottir RB, Egilsson V, HamannU, Chang-Claude J, Sobol H, Weber B, Stratton MR, and Easton DF 1998 Multifactorial analysis of differences between sporadic breast cancers and cancers involving BRCA1 and BRCA2 mutations. *Journal of the National Cancer Institute* 90 1138–1145.

Lakhani SR, Reis-Filho JS, Fulford L, Penault-Llorca F, van der Vijver M, Parry S, Bishop T, Benitez J, Rivas C, Bignon YJ, Chang-Claude J, Hamann U, Cornelisse CJ, Devilee P, Beckmann MW, Nestle-Kramling C, Daly PA, Haites N, Varley J, Lalloo F, Evans G, Maugard C, Meijers-Heijboer H, Klijn JG, Olah E, Gusterson BA, Pilotti S, Radice P, Scherneck S, Sobol H, Jacquemier J, Wagner T, Peto J, Stratton MR, Mcguffog L, Easton DF andBreast Cancer Consortium 2005 Prediction of BRCA1 status in patients with breast cancer using estrogen receptor and basal phenotype. *Clinical Cancer Research* 11 5175–5180.

Lakhani SR, van de Vijver MJ, Jacquemier J, Anderson TJ, Osin PP, McGuffog L and Easton DF 2002 The pathology of familial breast cancer: predictive value of immunohistochemical markers estrogen receptor, progesterone receptor, HER-2, and p53 in patients with mutations in BRCA1 and BRCA2. *Journal of Clinical Oncology* 20 2310–2318.

Lam WWK, Keng WT, Metcalfe K, Stewart F *et al.*, 2004 Fetal carbimazole – characteristic facial features. 11th Manchester Birth Defects Conference.

Lamb NE, Feingold E, Savage A, Avramopoulos D, Freeman SB, Gu Y, Hallberg A, Hersey J, Karadima G, Pettay D, Saker D, Shen J, Taft L, Mikkelsen M, Petersen MB, Hassold T and Sherman S 1997 Characterization of susceptible chiasma configurations that increase the risk for maternal nondisjunction of chromosome 21. *Human Molecular Genetics* 6 1391–1399.

Lamb NE, Freeman SB, Savage-Austin A, Pettay D, Taft L, Hersey J, Gu Y, Shen J, Saker D, May KM, Avramopoulos D, Petersen MB, Hallberg A, Mikkelsen M, Hassold TJ and Sherman S 1996 Susceptible chiasmata configurations of chromosome 21 predispose to non-disjunction in both maternal meiosis I and meiosis II. *Nature Genetics* 14 400–405.

Lamb NE, Yu K, Shaffer J, Feingold E and Sherman SL 2005 Association between maternal age and meiotic recombination for trisomy 21. *American Journal of Human Genetics* 76 91–99.

Lancaster HO 1995 Mathematicians in medicine and biology; Genetics before Mendel, Maupertuis and Réaumur. *Journal of Medical Biography* 3 84–89.

Lander ES 1996 The new genomics: global views of biology. *Science* 274 586–589.

Lander E and Kruglyak L 1995 Genetic dissection of complex traits: guidelines for interpreting and reporting linkage results. *Nature Genetics* 11 241–247.

Lane PW 1966 Association of megacolon with two recessive spotting genes in the mouse. *Journal of Heredity* 57 29–31.

Lansac J, Thepot F, Mayaux MJ, Czyglick F, Wack T, Selva J and Jalbert P 1997 Pregnancy outcome after artificial insemination or IVF with frozen semen donor: a collaborative study of the French CECOS Federation on 21,597 pregnancies. *European Journal of Obstetrics Gynecology and Reproductive Biology* 74 223–228.

Lantigua A, Moras F, Arechaederra M, Rojas I, Morales E, Rodríguez H, Viñas C, Noa C and Barrios B 1999 Etiological characterization of 512 severely mentally retarded institutionalized patients in Havana. *Community Genetics* 2 184–189.

Larouzé B, Blumberg BS, London WT, Lustbader ED, Sankale M and Payet M 1977 Forecasting the development of primary hepatic carcinoma by the use of risk factors. Studies in West Africa. *Journal of the National Cancer Institute* 58 1557–1561.

Larouzé B, London WT, Saimot G, Werner BG, Lustbader ED, Payet M and Blumberg BS 1976 Host responses to hepatitis B infection in patients with primary hepatic carcinoma and their families. A case/control study in Senegal, West Africa. *The Lancet* ii 534–538.

Larsen PH, Wells JE, Stallcup WB, Opdenakker G and Yong W 2003 Matrix metalloproteinase-9 facilitates remyelination in part by processing the inhibitory NG2 proteoglycan. *Journal of Neuroscience* 23 11127–11135.

Lasker GW 1983 *Surnames and Genetic Structure*. Cambridge University Press, Cambridge.

Lavery SA, Aurell R, Turner C, Castello C, Veiga A, Barri PN and Winston RM 2002 Preimplantation genetic diagnosis: patients' experiences and attitudes *Human Reproduction* 17 2464–2467.

Lawrence SH and Melnick PJ 1961 Enzymatic activity related to human serum beta-lipoprotein: histochemical, immuno-electrophoretic and quantitative studies. *Proceedings of the Society for Experimental Biology and Medicine* 107 998–1001.

Layer PG 1983 Comparative localization of acetylcholinesterase and pseudocholinesterase during morphogenesis of the chick embryo. *Proceedings of the National Academy of Sciences USA* 80 6413–6417.

Layer PG and Sporns O 1987 Spatiotemporal relationship of embryonic cholinesterases with cell proliferation in chicken brain and eye. *Proceedings of the National Academy of Sciences USA* 84 284–288.

Layer PG, Weikert T and Albert R 1993 Cholinesterases regulate neurite growth of chick nerve cells in vitro by means of a non-enzymatic mechanism. *Cell and Tissue Research* 273 219–226.

LDDB *London dysmorphology database*. Oxford University Press, Oxford http//www.oup.com/uk/omd

Le Douarin N and Kalcheim C 1999 *The Neural Crest*, 2nd edn Cambridge University Press, Cambridge.

Leach MO, Boggis CR, Dixon AK, Easton DF, Eeles RA, Evans DG, Gilbert FJ, Griebsch I, Hoff RJ, Kessar P, Lakhani SR, Moss SM, Nerurkar A, Padhani AR, Pointon LJ, Thompson D, Warren RM and MARIBS study group 2005 Magnetic resonance imaging and mammography of a UK population at high familial risk of breast cancer: a prospective multicentre cohort study (MARIBS). *The Lancet* 365 1769–1778.

Ledbetter DH and Engel E 1995 Uniparental disomy in humans – development of an imprinting map and its implications for prenatal diagnosis. *Human Molecular Genetics* 4 1757–1764.

Legg J, Jensen UB, Broad S, Leigh I and Watt FM 2003 Role of melanoma chondroitin sulphate proteoglycan in patterning stem cells in human interfollicular epidermis. *Development* 130 6049–6063.

Lehmann DJ 2002 The genetics of butyrylcholinesterase and Alzheimer's disease. *Alzheimer Insights* 7 1–5.

Lehmann DJ, Nagy Z, Litchfield S, Borja MC and Smith AD 2000 Association of butyrylcholinesterase K variant with cholinesterase-positive neuritic plaques in the temporal cortex in late-onset Alzheimer's disease. *Human Genetics* 106 447–452.

Lehmann H and Ryan E 1956 The familial incidence of low pseudocholinesterase level. *The Lancet* ii 124.

Lehur P-A, Madarnas P, Devroede G, Perey BJ, Menard DB and Hamade N 1984 Peutz-Jeghers syndrome: association of duodenal and bilateral breast cancers in the same patient. *Digestive Diseases and Sciences* 29 178–182.

Lejeune J 1981 Vingt Ans Après. In: *Trisomy 21: An International Symposium*. (editors Burgio GR, Fraccaro M, Tiepolo L and Wolf U) pp. 91–102, Springer Verlag, Berlin.

Lejeune J, Gautier M and Turpin MR 1959 Étude des chromosomes somatiques de neuf enfants mongoliens. *Les Comptes rendus de l'Académie des sciences* (Paris) 248 1721–1722.

Lentner C (ed.) 1981 *International Medical and Pharmaceutical Information*, Vol.1. Ciba-Geigy, Basle.

Lerner IM 1954 *Genetic Homeostasis*. Oliver and Boyd, London.

Levano S, Ginz H, Siegemund M, Miodrag F, Voronkov E, Urwyler A and Girard T 2005 Genotyping the butyrylcholinesterase in patients with prolonged neuromuscular block after succinylcholine. *Anesthesiology* 102 531–535.

Levin M 2005 Left-right asymmetry in embryonic development. *Mechanisms of Development* 122 1–25.

Levine JM and Card JP 1987 Light and electron microscopic localization of a cell surface antigen (NG2) in the rat cerebellum: association with smooth protoplasmic astrocytes. *Journal of Neuroscience* 7 2711–2720.

Levine SA and Woodworth CR 1958. Congenital deaf-mutism, prolonged QT interval, syncopal attacks and sudden death. *New England Journal of Medicine* 259 412–417.

Levy-Marchal C, Patterson C and Green A 1995 Variation by age group and seasonality at diagnosis of childhood IDDM in Europe. The EURODIAB ACE Study Group. *Diabetologia* **38** 823–830.

Lewes GH 1875 'The Eleatics. I. Xenophanes' pp. 36–42 in *The Biographical History of Philosophy.* 2nd edition. Appleton and Company, New York.

Lewis CM, Pinel T, Whittaker JC and Handiside AH 2001 Controlling misdiagnosis errors in preimplantation genetic diagnosis: a comprehensive model encompassing extrinsic and intrinsic sources of error. *Human Reproduction* **16** 43–50.

Lewontin R 1964 The interaction of selection and linkage. I. General considerations of heterotic models. *Genetics* **49** 49–67.

Lewontin RC and Hubby JL 1966 A molecular approach to the study of genic heterozygosity in natural populations. II. Amount of variation and degree of heterozygosity in natural populations of *Drosophila pseudoobscura. Genetics* **54** 595–609.

Li FP and Fraumeni JF Jr. 1969 Soft-tissue sarcomas, breast cancer, and other neoplasms. A familial syndrome? *Annals of Internal Medicine* **71** 747–752.

Li L, Eng C, Desnick R, German J and Ellis NA 1998 Frequency of *blm^{Ash}*, the mutation responsible for the high frequency of Bloom's syndrome in the Ashkenazim. *Molecular Genetics and Metabolism* **64** 286–290.

Lin AE, Garver KL, Diggans G, Clemens M Wenger SL, Steele MW, Jones MC and Israel J 1988 Interstitial and terminal deletions of the long arm of Chromosome 4: further delineation of phenotypes. *American Journal of Medical Genetics* **31** 533–548.

Lindsay EA, Morris MA,Gos A, Nestadt G, Wolyniec PS, Lasseter VK, Shprintzen R, Antonarakis SE, Baldini A and Pulver AE 1995 Schizophrenia and chromosomal deletions within 22q11.2. *American Journal of Human Genetics* **56** 1502–1503.

Lippman-Hand A and Fraser FC 1979 Genetic counseling – provision and reception of information. *American Journal of Medical Genetics* **3** 113–127.

Little People of America 2003 *http://www.LPAOnline.org/resources_faq.html.*

Liu J, Aoki M, Illa I, Wu C, Fardeau M, Angelini C, Serrano C, Urtizberea JA, Hamida MB, Bohlega S, Culper EJ, Amato AA, Bossie K, Oeltjen J, McKenna-Yasek D, Hosler BA, Schurr E, Arahata K, de Jond PJ and Brown RH Jr 1998 Dysferlin, a novel skeletal muscle gene, is mutated in Miyoshi myopathy and limb girdle muscular dystrophy. *Nature Genetics* **20** 31–36.

Liu W, Cheng J, Iwasaki A, Imanishi H and Hada T 2002 Novel mutation and multiple mutations found in the human butyrylcholinesterase gene. *Clinica Chimica Acta* **326** 193–199.

Llanusa C, Sánchez R and Rodríguez L 1998 Alpha-fetoprotein Screening program: organizational aspects and psycho-social impact (Havana City). *Revista Cubana de Medicina General Integral* **14** 43–47. On line in English available in MEDICC Review. *www.medicc.org*

Lockridge O, Bartels CF, Vaughan TA, Wong CK, Norton SE and Johnson LL 1987 Complete amino acid sequence of human serum cholinesterase. *Journal of Biological Chemistry* **262** 549–557.

Loenen JHMM 1954 Was Anaximander an evolutionist? *Mnemosyme* **4** 215–232.

Loewi O 1936 The chemical transmission of nerve action. Nobel Lecture, 12 December 1936. The Official Web Site of The Nobel Foundation: *http://nobelprize.org/medicine/laureates/1936/loewi-lecture.html.*

Loft A, Petersen K, Erb K, Mikkelsen AL, Grinsted J, Hald F, Hindkjaer J, Nielsen KM, Lundstrom P, Gabrielsen A, Lenz S, Hornnes P, Ziebe S, Ejdrup HB, Lindhard A, Zhou Y and Nyboe Andersen A 1999 A Danish national cohort of 730 infants born after intracytoplasmic sperm injection (ICSI) 1994–1997. *Human Reproduction* **14** 2143–2148.

London WT, Drew JS, Blumberg BS, Grossman RA and Lyons PS 1977b Association of graft survival with host response to hepatitis B infection in patients with kidney transplants. *New England Journal of Medicine* **296** 241–244.

London WT, Drew JS, Lustbader ED, Werner BG and Blumberg BS 1977a Host response to hepatitis B infection among patients in a chronic hemodialysis unit. *Kidney International* **12** 51–58.

London WT, Sutnick AI and Blumberg BS 1969 Australia antigen and acute viral hepatitis. *Annals of Internal Medicine* **70** 55–59.

London WT, Sutnick AI, Millman IV Coyne, Blumberg BS and Vierucci A 1972 Australia antigen and hepatitis: Recent observations on the serum protein polymorphism, infectious agent hypotheses. *Canadian Medical Association Journal* **106** 480–485.

Lovell-Badge R, Canning C and Sekido R 2002 Sex-determining genes in mice: building pathways. In: *The Genetics and Biology of Sex Determination* (Novartis Foundation Symposium 244) John Wiley & Sons, Chichester, pp. 4–22.

Lu WD, Hada T, Fukui K, Imanishi H, Matsuoka N, Iwasaki A and Higashino K 1997 Familial hypocholinesterasemia found in a family and a new confirmed mutation. *Internal Medicine* **36** 9–13.

Lunkes A, Hartung U, Magariño C, Rodríguez M, Palmero A, Rodríguez L, Heredero L, Weisenbachu J, Weber J and Auburger G 1993 Refinement of the OPA I locus on chromosome 3q28-q29 to a region of 2–8 cM, in one Cuban pedigree with autosomal dominant optic atrophy type Kjer. *American Journal of Human Genetics* **57** 968–970.

Lurie IW and Cherstvoy ED 1984 Renal agenesis as a diagnostic feature of the cryptophthalmos – syndactyly syndrome. *Clinical Genetics* **25** 528–532.

Lurie IW, Supovitz KR, Rosenblum-Vos LS and Wulfsberg EA 1994 Phenotypic variability of del (2)(q22-q23): report of a case and review of the literature. *Genetic Counseling* **5** 11–14.

Lustbader ED, London WT and Blumberg BS 1976 Study design for a hepatitis B vaccine trial. *Proceedings of the National Academy of Sciences USA* **73** 955–959.

Lynch TJ, Bell DW, Sordella R, Gurubhagavatula S, Okimoto RA, Brannigan BW, HarrisPL, Haserlat SM, Supko JG, Haluska FG, Louis DN, Christiani DC, Settleman J and Haber DA 2004 Activating mutations in the epidermal growth factor receptor underlying responsiveness of non-small-cell lung cancer to Gefitinib. *New England Journal of Medicine* **350** 2129–2139.

Lyon M 1961 Gene action in the mammalian X-chromosome of the mouse (*Mus musculus* L.). *Nature* **190** 372–373.

Lyonnet S, Bolino A, Pelt A, Abel L, Nihoul-Fekete C, Briard ML, Mok-Siu V, Kääriäinen H, Martucciello G, Lerone M, Puliti A, Luo Y, Weissenbach J, Devoto M, Munnich A and Romeo G 1993 A gene for Hirschsprung disease maps to the proximal long arm of chromosome 10. *Nature Genetics* **4** 346–350.

Lyons I, Parsons LM, Hartley L, Li R, Andrews JE, Robb L and Harvey RP 1995 Myogenic and morphogenetic defects in the heart tubes of murine embryos lacking the homeo box gene Nkx2-5. *Genes and Development* **9** 1654–1666.

Macatee TL, Hammond BP, Arenkiel BR, Francis L, Frank DU and Moon AM 2003 Ablation of specific expression domains reveals discrete functions of ectoderm- and endoderm-derived FGF8 during cardiovascular and pharyngeal development. *Development* **130** 6361–6374.

Madigan MP, Ziegler RG, Benichou J, Byrne C and Hoover RN 1995 Proportion of breast cancer cases in the United States explained by well-established risk factors. *Journal of the National Cancer Institute* **87** 1681–1685.

Madsen CM 1964 *Hirschsprung's disease*. Munksgaard, Copenhagen.

Maekawa M, Sudo K, Kanno T, Kotani K, Dey DC, Ishikawa J and Izume M 1997 Genetic mutations of butyrylcholine esterase identified from phenotypic abnormalities in Japan. *Clinical Chemistry* **43** 924–929.

Maekawa M, Sudo K, Kanno T, Kotani K, Dey DC, Ishikawa J, Izume M and Etoh K 1995 Genetic basis of the silent phenotype of serum butyrylcholinesterase in three compound heterozygotes. *Clinica Chimica Acta* **235** 41–57.

Maekawa M, Taniguchi T, Ishikawa J, Toyoda S and Takahata N 2004 Problem with detection of an insertion-type mutation in the *BCHE* gene in a patient with butyrylcholinesterase deficiency. *Clinical Chemistry* **50** 2410–2411.

Magnius LO 1975 Characterization of a new antigen-antibody system associated with hepatitis B. *Clinical and Experimental Immunology* **20** 209–216.

Majumdar M, Vuori K and Stallcup WB 2003 Engagement of the NG2 proteoglycan triggers cell spreading via rac and p130cas. *Cellular Signalling* **15** 79–84.

Malécot G 1966 *Probabilités et hérédité*. Presses universitaires de France, Paris.

Malécot G 1969 *The Mathematics of Heredity*. WH Freeman, New York.

Malkin D, Li FP, Strong LC, Fraumeni Jr JF, Nelson CE, Kim DH, Kassel J, Gryka MA, Bischoff FZ, Tainsky MA and Friend SH 1990 Germ line p53 mutations in a familial syndrome of breast cancer, sarcomas, and other neoplasms. *Science* **250** 1233–1238.

Mangion J, Chang-Claude J, Eccles D, Eeles R, Evans D G, Houlston R, Murday V, Narod S, Peretz T, Peto J, Phelan C, Zhang HX, Szabo C, Devilee P, Goldgar D, Futreal PA, Nathanson KL, Weber B, Rahman N and Stratton MR 2002 Low-penetrance susceptibility to breast cancer due to CHEK2(*)1100delC in noncarriers of BRCA1 or BRCA2 mutations. *Nature Genetics* **31** 55–59.

Maniatis N, Collins A, Xu C-F, McCarthy LC, Hewett DR, Tapper W, Ennis S, Ke X and Morton NE 2002 The first linkage disequilibrium (LD) maps: Delineation of hot and cold blocks by diplotype analysis. *Proceedings of the National Academy of Science* USA **99** 2228–2233.

Marles SL and Chudley AE 1990 Another case of microcephaly, facial clefting, and preaxial polydactyly. *Journal of Medical Genetics* **27** 593–594.

Marquez C, Sandalinas M, Bahce M, Alikani M and Munne S 2000 Chromosome abnormalities in 1255 cleavage-stage embryos. *Reproductive Biomedicine Online* **1** 17–26.

Marshall BJ and Warren JR 1984 Unidentified curved bacilli in the stomach of patients with gastritis and peptic ulceration. *The Lancet* **1** 1311–1315

Martin-DeLeon PA and Williams MB 1987 Sexual behaviour and Down syndrome: the biological mechanism. *American Journal of Medical Genetics* **27** 693–700.

Martinez-Frias ML, Bermejo SE, Felix V, Calvo CR, Ayala GA and Hernandez RF 1998 [Fraser syndrome: frequency in our environment and clinical-epidemiological aspects of a consecutive series of cases]. *Anales Españoles de Pediatria* **48** 634–638.

Mascie-Taylor CGN and Lasker GW 1984 Geographic distribution of surnames in Britain: the Smiths and Joneses have clines like blood group genes. *Journal of Biosocial Science* **16** 301–308.

Mascie-Taylor CGN and Lasker GW 1985 Geographical distribution of common surnames in England and Wales. *Annals of Human Biology* **12** 397–401.

Masson P 1989 A naturally occurring molecular form of human plasma cholinesterase is an albumin conjugate. *Biochimica et Biophysica Acta* **988** 258–266.

Masson P 1991 Molecular heterogeneity of human cholinesterase. In: *Cholinesterase: structure, function, mechanisms, genetics and cell biology.* (editors Massoulié J, Bacou F, Barnard E, Chatonnet A, Doctor BP and Quinn DM) American Chemical Society, Washington, D.C.

Massoulié J and Bon S 1982 The molecular forms of cholinesterase and acetylcholinesterase in vertebrates. *Annual Review of Neurosciences* **5** 57–106.

Matera I, Bachetti T and Puppo F 2004 PHOX2B mutations and polyalanine expansions correlate with the severity of the respiratory phenotype and associated symptoms in both congenital and late onset central hypoventilation syndrome. *Journal of Medical Genetics* **41** 373–380.

Matoba S, Kanai Y, Kidokoro T, Kanai-Azuma M, Kawakami H, HayashiY and Kurohmaru M 2005 A novel *Sry* downstream cellular event which preserves the readily available energy source of glycogen in mouse sex differentiation. *Journal of Cell Science* **18** 1449–1459.

Matsuda Y, Nishida-Umehara C, Tarui H, Kuroiwa A, Yamada K, Iosobe T, Ando J, Fujiwara A, Hirao Y, Nishimura O, Ishijima J, Hayashi A, Murakami T, Murakami Y, Kuratani S and Agata K 2005 Highly conserved linkage homology between birds and turtles: bird and turtle chromosomes are precise counterparts of each other. *Chromosome Research* **13** 601–615.

Mattei JF and Aymé S 1983 Syndrome of polydactyly, cleft lip, lingual hamartomas, renal hypoplasia, hearing loss, and psychomotor retardation: variant of the Mohr syndrome or a new syndrome? *Journal of Medical Genetics* **20** 433–435.

Maubaret C and Hamel C 2005 Génétique des retinitis pigmentaires: classification métabolique et correlations phénotype/génotype *Journal Français d'Ophtalmologie* **28** 71–92.

Mauceri L, Greco F, Baieli S, Sorge G 2000 Varadi-Papp syndrome: report of a case. *Clinical Dysmorphology* **9** 289–290.

Maugh TH 1975 Hepatitis B: A new vaccine ready for human testing. *Science* **188** 137–138.

Maupas P, Coursaget P, Goudeau A, Drucker J and Bagros P 1976 Immunisation against hepatitis B in man. *The Lancet* **1** 1367–1370.

Maynard-Smith J 1968 'Haldane's Dilemma' and the rate of evolution. *Nature* **219** 1114–1116.

Mayo O 1978 Polymorphism, selection and evolution. In: *The Biochemical Genetics of Man* (2nd edition) (editors DJH Brock and O Mayo) Academic Press, London.

Mayo O 1994 Genetic analysis of a complex disease. *Alcohol & Alcoholism* (Supp. 2) 9–18.

Mayo O 2004 Interaction and quantitative trait loci. *Australian Journal of Experimental Agriculture* **44** 1135–1140.

Mayo O and Brock DJH 1978 The uses of polymorphism. In: *The Biochemical Genetics of Man.* (2nd edition) (editors Brock DJH and Mayo O) Academic Press, London.

Mayr E 1957 Species concepts and definitions In: *The Species Problem.* (editor E Mayr) *Bulletin of the American Society for the Advancement of Science* **50** 1–22.

McCabe L, Griffin LD, Kinzer A, Chandler M, Beckwith JB, McCabe ERB 1990 Overo lethal white foal syndrome: equine model of aganglionic megacolon (Hirschsprung disease). *American Journal of Medical Genetics* **36** 336–40.

McCallion AS and Chakravarti A 2004 *RET* and Hirschsprung disease, p. 421–32. In: *Inborn Errors of Development.* (editors Epstein CJ, Erickson RP, Wynshaw-Boris A) Oxford University Press, Oxford.

McCallion AS, Emison ES, Kashuk CS, Bush RT, Kenton M, Carrasquillo MM, Jones KW, Kennedy GC, Portnoy ME, Green ED and Chakravarti A 2003 Genomic variation in multigenic traits: Hirschsprung disease. *Cold Spring Harbor Symposia on Quantitative Biology* **68** 373–81.

McCarthy MI and Froguel P 2002 Genetic approaches to the molecular understanding of type 2 diabetes. *American Journal of Physiology, Endocrinology and Metabolism* **283** E217–225.

McElhinney DB, Geiger E, Blinder J, Benson DW and Goldmuntz E 2003 NKX2.5 mutations in patients with congenital heart disease. *Journal of the American College of Cardiology,* **42** 1650–1655.

McGahan JP, Nyberg DA and Mack LA 1990 Sonography of facial features of alobar and semilobar holoprosencephaly. *American Journal of Roentgenology* **154** 143–148.

McGregor L, Makela V, Darling SM, Vrontou S, Chalepakis G, Roberts C, Smart N, Rutland P, Prescott N, Hopkins J, Bentley E, Shaw A, Roberts E, Mueller R, Jadeja S, Philip N, Nelson J, Francannet C, Perez-Ayles A, Mégarbané A, Kerr B, Wainwright B, Woolf AS, Winter RM and Scambler PJ 2003 Fraser syndrome and mouse blebbed phenotype caused by mutations in *FRAS1*/Fras1 encoding a putative extracellular matrix protein. *Nature Genetics* **34** 203–208.

McGuire MC, Nogueira CP, Bartels CF, Lightstone AH, Hajra A, van der Spek AFL, Lockridge O and La Du BN 1989 Identification of the structural mutation responsible for the dibucaine-resistant (atypical) variant form of human serum cholinesterase. *Proceedings of the National Academy of Sciences USA* **86** 953–957.

McKusick VA 1952 The clinical observations of Jonathan Hutchinson. *American Journal of Syphilis, Gonorrhoea and Venereal Disease* **36** 101–126.

McKusick VA 1955 The cardiovascular aspects of Marfan's syndrome: a heritable disorder of connective tissue. *Circulation* **11** 321–342.

McKusick VA 1956 *Heritable Disorders of Connective Tissue.* CV Mosby, St. Louis 1956 (1st edn), 1960 (2nd edn), 1966 (3rd edn), 1972 (4th edn), with P Beighton, 1993 (5th edn).

McKusick VA l963 Frederick Parkes Weber: 1863–1962. *Journal of the American Medical Association* **l83** 45–59.

McKusick VA 1969 On lumpers and splitters, or the nosology of genetic disease. *Perspectives in Biology and Medicine* **12** 298–318.

McKusick VA 1978 *Medical Genetic Studies of the Amish. Selected Papers with Commentary.* Johns Hopkins University Press, Baltimore.

McKusick VA *Mendelian Inheritance in Man.* Johns Hopkins University Press, Baltimore, 1966 (1st edn), 1998 (12th edn). Online version, OMIM (*www.ncbi.nlm.nih.gov/omim*).

McKusick VA 1998 *Mendelian Inheritance in Man. A Catalog of Human Genes and Genetic Disorders.* 12th edn Johns Hopkins University Press, Baltimore.

McKusick VA 2005 *OMIM (On Line Mendelian Inheritance in Man).* *http://www.ncbi.nlm.nih.gov:80/entrez/query.fcgi?CMD=Details&DB=OMIM*

McKusick VA and Ruddle FH 1987 A new discipline, a new name, a new journal. (Editorial) *Genomics* **1** 1–2.

McKusick VA, Eldridge R, Hostetler JA, Egeland JA and Ruangwit U 1965 Dwarfism in the Amish. II. Cartilage-hair hypoplasia. *Bulletin of the Johns Hopkins Hospital* **116** 285–326.

McLaren A 1986 Prelude to embryogenesis. In: *Human embryo research, yes or no?* The Ciba Foundation Tavistock Publications, London, p. 12.

McQueen HA, McBride D, Miele G, Bird AP and Clinton M 2001 Dosage compensation in birds. *Current Biology* **11** 253–257.

McTiernan C, Adkins S, Chatonnet A, Vaughan TA, Bartels CF, Kott M, Rosenberry TL, La Du BN and Lockridge O 1987 Brain cDNA clone for human cholinesterase. *Proceedings of the National Academy of Sciences USA* **84** 6682–6686.

Meadow R 1977 Munchausen Syndrome by proxy. The hinterland of child abuse. *The Lancet* **2** 343–345.

Meadow R 1982 Munchausen Syndrome by proxy. *Archives of Disease in Childhood* **57** 92–98.

Meijers-Heijboer H, van den Ouweland A, Klijn J, Wasielewski M, de Snoo A, Oldenburg R, Hollestelle A, Houben M, Crepin E, van Veghel-Plandsoen M, Elstrodt F, van Duijn C, Bartels C, Meijers C, Schutte M, McGuffog I, Thompson D, Easton D, Sodha N, Seal S, Barfoot R, Mangion J, Chang-Claude J, Eccles D, Eeles R, Evans DG, Houlston R, Murday V, Narod S, Peretz T, Peto J, Phelan C, Zhang HX, Szabo C, Devilee P, Goldgar D, Futreal PA, Nathanson KL, Weber B, Rahman N and Stratton MR: CHEK2-Breast Cancer Consortium 2002 Low-penetrance susceptibility to breast cancer due to CHEK2 1100delC in noncarriers of BRCA1 or BRCA2 mutations. *Nature Genetics* **31** 55–59.

Meijers-Heijboer H, Van Geel B, Van Putten WL, Henzen-Logmans SC, Seynaeve C, Menke-Pluymers MB, Bartels CC, Verhoog LC, van den Ouweland AM, Niermeijer MF, Brekelmans CT and Klijn JG 2001 Breast cancer after prophylactic bilateral mastectomy in women with a BRCA1 or BRCA2 mutation. *New England Journal of Medicine* **345** 159–164.

Mein CA, Esposito L, Dunn MG, Mein CA, Esposito L, Dunn MG, Johnson GC, Timms AE, Goy JV, Smith AN, Sebag-Montefiore L, Merriman ME, Wilson AJ Pritchard LE, Cucca F, Barnett AH, Bain SC and Todd JA *et al.* 1998 A search for type 1 diabetes susceptibility in families from the United Kingdom. *Nature Genetics* **19** 297–300.

Meissner FL 1856 *Taubstummenheit und Taubstummenbildung.* [Deaf – Mutism and the Education of Deaf Mutes] Winter, Leipzig and Heidelberg, p. 119.

Mendel B and Rudney H 1943 Studies on cholinesterase. I. Cholinesterase and pseudo-cholinesterase. *Biochemical Journal* **37** 59–63.

Meredith S and Doolin D 1997 Timing of activation of primordial follicles in mature rats is only slightly affected by foetal stage at meiotic arrest. *Biology of Reproduction* **57** 63–67.

Meyers C, Adam R, Dungan J and Prenger V 1997 Aneuploidy in twin gestations: when is maternal age advanced? *Obstetrics and Gynecology* **89** 248–251.

Mikami LR, Souza RLR, Chautard-Freire-Maia EA and Lockridge O 2004 New genetic variants of human butyrylcholinesterase in the Brazilian population. VIII International Meeting on Cholinesterases, Abstract 60 23.

Miki Y, Swensen J, Shattuck-Eidens D, Futreal PA, Harshman K, Tavtigian S, Liu Q, Cochran C, Bennett LM, Ding W, Bell R, Rosenthal J, Hussey CH, Tran T, McClure M, Frye Ch, Hattier T, Phelps R, Haugen-Strano A, Katcher H, Yakumo K, Gholami Z, Shaffer D, Stone S, Bayer S, Wray CH, Bogden R, Dayananth P, Ward J, Tanin P, Narod S, Bristow PK, Norris FH, Helvering L, Morrison P, Rosteck P, Lai M, Barrett JC, Lewis C, Neuhausen S, Cannon-Albright L, Goldgar D, Wiseman R, Kamb A and Skolnick MH 1994 A strong candidate for the breast and ovarian cancer susceptibility gene BRCA1. *Science* **266** 66–71.

Milkman R 1967 Heterosis as a major cause of heterozygosity in nature. *Genetics* **55** 493–495.

Milkman R 1978 Selection differentials and selection coefficients. *Genetics* **88** 391–403.

Miller LH, Mason SJ, Clyde DF and McGinniss MH 1976 The resistance factor to *Plasmodium vivax* in Blacks. *New England Journal of Medicine* **295** 302–304.

Miller NE 1987 Association of high density lipoprotein subclasses apolipoproteins with ischaemic heart disease and coronary atherosclerosis. *American Heart Journal* **113** 589–597.

Miller RW 1956 Delayed effects occurring within the first decade after exposure of young individuals to the Hiroshima atomic bomb. *Pediatrics* **18** 1–18.

Miller RW 1966 Medical Progress. Relation between cancer and congenital defects in man. *New England Journal of Medicine* 275 87–93.

Miller RW 1968 Deaths from childhood cancer in sibs. *New England Journal of Medicine* 279 122–126.

Miller RW 1979 Cancer epidemics in the People's Republic of China (and addendum). *National Cancer Institute Monographs.* **52** 31–39.

Miller RW 1998 Autobiography: clinical genetics: key to cancer etiology. *Am. J. Med. Genet.* 76 9–20.

Miller RW and Dalager NA 1974 U.S. childhood cancer deaths by cell type, 1960–68. *Journal of Pediatrics* **85** 664–668.

Miller RW, Fraumeni JF Jr. and Manning MD 1964 Association of Wilms' tumor with aniridia, hemihypertrophy and other congenital malformations. *New England Journal of Medicine* 270 922–927.

Miller RW, Watanabe S and Fraumeni JF Jr. eds. 1988 Unusual occurrences as clues to cancer etiology. In *Proceedings of the 18th Princess Takamatsu Cancer Research Fund Symposium.* Taylor and Francis, Ltd, Philadelphia.

Millman I, Hutanen H, Merino F, Bayer ME and Blumberg BS 1971 Australia antigen: Physical and chemical properties *Research Communications in Chemical Pathology and Pharmacology* 2 667–686.

Mittwoch U 1996 Unilateral manifestations of bilateral structures: which phenotype matches the genotype? *Frontiers of Endocrinology* 16 121–129.

Mittwoch U 2000 Three thousand years of questioning sex differentiation. *Cytogenetics and Cell Genetics* 91 186–101.

Mittwoch U 2004 The elusive action of sex-determining genes: mitochondria to the rescue? *Journal of Theoretical Biology* 228 359–365.

Mittwoch U, Delhanty JDA and Beck 1969 Growth of differentiating testes and ovaries. *Nature* 224 1323–1332.

Modell B and Kuliev A 1998 The history of community genetics: The contribution of the haemoglobin disorders. *Community Genetics* 1 3–11.

Moll R, Franke WW, Schiller DL, Geiger B and Krepler R 1982 The catalog of human cytokeratins: patterns of expression in normal epithelia, tumors and cultured cells. *Cell* 31 11–24.

Moller P, Borg A, Evans DG, Haites N, Reis MM, Vasen H, Anderson E, Steel CM, Apold J, Goudie D, Howell A, Lalloo F, Maehle L, Gregory H and Heimdal K 2002 Survival in prospectively ascertained familial breast cancer: analysis of a series stratified by tumour characteristics, BRCA mutations and oophorectomy. *International Journal of Cancer* 101 555–559.

Moore W 2005 *The Knife Man.* Bantam Press, London.

Moran CN, Scott RA, Adams SM, Warrington SJ, Jobling MA, Wilson RH, Goodwin WH, Georgiades E, Wolde B and Pitsiladis YP 2004 Y chromosomal haplogroups of elite Ethiopian endurance runners. *Human Genetics* 115 492–497.

Morell R, Spritz RA, Ho L, Pierpont J, Guo W, Friedman TB and Asher JH Jr 1997 Apparent digenic inheritance of Waardenburg syndrome type 2 (WS2) and autosomal recessive ocular albinism (AROA). *Human Molecular Genetics* 6 659–664.

Mörner M 1967 *Race Mixture in the History of Latin America.* Little, Brown and Company, Boston.

Morquio L 1901 Sur une maladie infantile et familiale caracterisée par des modifications permanentes du poul, des attaques syncopales et epileptiformes et la mort subite. *Archives de Medicine des Infants* 4 467–475.

Morris JK, De Vigan C, Mutton D and Alberman E 2005 Risk of Down syndrome live birth in women 45 years of age and older. *Prenatal Diagnosis* 25 275–278.

Morris JK, Wald NJ and Watt HC 1999 Foetal loss in Down syndrome pregnancies. *Prenatal Diagnosis* 19 142–145.

Morris JK, Wald NJ, Mutton D and Alberman E 2003 Comparison of models of maternal age-specific risk for Down syndrome live births. *Prenatal Diagnosis* 23 252–258.

Morrow AC and Motulsky AG 1968 Rapid screening method for the common atypical pseudo-cholinesterase variant. *Journal of Laboratory and Clinical Medicine* 71 350–356.

Morson B 1974 President's address. The polyp-cancer sequence in the large bowel. *Proceedings of the Royal Society of Medicine* **67** 451–457.

Morton NE 1955 Sequential tests for the detection of linkage. *American Journal of Human Genetics* **7** 277–318.

Morton NE 1960 The mutational load due to detrimental genes in man. *American Journal of Human Genetics* **12** 348–364.

Morton NE 1963 The components of genetic variability. In: *The genetics of migrant and isolate populations.* (editor Goldschmidt E) pp. 226–236. Williams and Wilkins, Philadelphia.

Morton NE 1975 Theory of inbreeding effect on diploid bees. Appendix to Kerr WE *Anais da Academia Brasileira de Ciências* **47** 332–334.

Morton NE 1981 Mutation rates for human autosomal recessives. In: *Population and biological aspects of human mutation* (editors Hook EB and Porter IJ) pp. 65–90. Academic Press, New York.

Morton NE 1991 Genetic epidemiology of hearing impairment. *Annals of the New York Academy of Sciences* **630** 16–30.

Morton NE, Chung CS and Mi M-P 1967 *Genetics of Interracial Crosses in Hawaii.* S. Karger, Basel.

Morton NE, Chung CS, Friedman LD, Morton NT, Miki C and Yee S 1968 Relation between homozygous viability and average dominance in *Drosophila melanogaster. Genetics* **60** 601–614.

Morton NE, Crow JF and Muller HJ 1956 An estimate of the mutational damage in man from data on consanguineous marriages. *Proceedings of the National Academy of Sciences USA* **42** 855–863.

Mowat DR, Croaker GDH, Cass DT, Kerr BA, Chaitow J, Ades LC, Chia NL and Wilson MJ 1998 Hirschsprung Disease, microcephaly, mental retardation, and characteristic facial features: delineation of a new syndrome and identification of a locus at chromosome 2q22-q23. *Journal of Medical Genetics* **35** 617–623.

Mowat DR, Wilson MJ and Goossens M. 2003 Mowat-Wilson syndrome. *Journal of Medical Genetics* **40** 305–310

Müller E 1922 Über einen Fall von Kryptophthalmus congenitus des einen und Oberlid-kolobom des anderen Auges. *Klinische Monatsblätter für Augenheilkunde* **68** 247 (brief abstract only).

Muller F, Rebiffe M, Taillandier A, Oury JF and Mornet E 2000 Parental origin of the extra chromosome in prenatally diagnosed foetal trisomy 21. *Human Genetics* **106** 340–344.

Muller HJ 1948 Mutational prophylaxis. *Bulletin of the New York Academy of Medicine* **24** 447–469.

Muller HJ 1950 Our load of mutations. *American journal of Human Genetics* **2** 111–176.

Muller HJ 1967 What genetic course will man steer? In: *Proceedings of the III International Congress on Human Genetics* (editors Crow JF and Neel J) Johns Hopkins, Baltimore, pp. 521–535.

Muller U, Wang D, Denda S, Meneses JJ, Pedersen RA and Reichardt LF 1997 Integrin alpha8beta1 is critically important for epithelial-mesenchymal interactions during kidney morphogenesis. *Cell* **88** 603–613.

Müller-Hill B 1997 The spectre of kakogenics: a personal view. *The Genetical Society of Great Britain Newsletter* **33** 13–14.

Müller-Hill B 1984 *Tödliche Wissenschaft. Die Aussonderung von Juden, Zigeunern und Geisteskranken 1933 – 1945.* Rowohlt Taschenbuch Verlag, Reinbek

Müller-Hill B 1988 *Murderous Science. Elimination by Scientific Selection of Jews, Gypsies, and others in Germany 1933 – 1945.* Translated by George Fraser, Oxford University Press, Oxford. With an afterword by JD Watson reprinted by Cold Spring Harbor Laboratory Press 1998

Mulvihill JJ, Miller RW and Fraumeni JF Jr. eds. 1977 *Genetics of Human Cancer.* Raven Press, New York.

Muniz FJ and Micks DW 1973 The persistence of hepatitis B antigen in *Aedes aegypti. Mosquito News* **33** 509–511.

Münke M 1989 Clinical, cytogenetic, and molecular approaches to the genetic heterogeneity of holoprosencephaly. *American Journal of Medical Genetics* **34** 237–245.

Münke M, McDonald DM, Cronister A, Stewart JM, Gorlin RJ and Zackai EH 1990 Oral-facial-digital syndrome type VI (Váradi syndrome): further clinical delineation. *American Journal of Medical Genetics* **35** 360–369.

Münke M, Ruchelli ED, Rorke LB, McDonald-McGinn DM, Orlow MK, Isaacs A, Craparo FJ, Dunn LK and Zackai EH 1991 On lumping and splitting: a fetus with clinical findings of the oral-facial-digital syndrome type VI, the hydrolethalus syndrome, and the Pallister-Hall syndrome. *American Journal of Medical Genetics* **41**:548–556.

Munne S, Bahce M, Sandalinas M, Escudero T, Marquez C, Velilla E, Colls P, Oter M, Alikani M and Cohen J 2004 Differences in chromosome susceptibility to aneuploidy and survival to first trimester. *Reproductive Biomedicine Online* **8** 81–90.

Muratani K, Hada T, Yamamoto Y, Kaneko T, Shigeto Y, Ohue T, Furuyama J and Higashino K 1991 Inactivation of the cholinesterase gene transposition. *Proceedings of the National Academy of Sciences USA* **88** 11315–11319.

Murphy EA and Chase GA 1975 *Principles of Genetic Counseling.* Year Book Medical Publishers, Inc., Chicago.

Mutton D, Alberman E and Hook EB 1996 Cytogenetic and epidemiological findings in Down syndrome, England and Wales 1989 to 1993. *Journal of Medical Genetics* **33** 387–384.

Myers AK and Reardon W 2005 Choanal atresia – a recurrent feature of fetal carbimazole syndrome. *Clinical Otolaryngology* **30** 375–377.

Mykytun K, Nishimura DY, Searby CC, Beck G, Bugge K, Haines HL, Cornier AS, Cox GF, Fulton AB, Carmi R, Iannaccone A, Jacobson SG, Weleber RG, Wright AF, Riise R, Hennekam RC, Lulici G, Berker-Karauzum S, Biesecker LG, Stone EM and Sheffield VC 2003 Evaluation of complex inheritance involving the most common Bardet-Biedl syndrome locus (BBS1). *American Journal of Human Genetics* **72** 429–437.

Nachman MW and Crowell SL 2000 Estimate of the mutation rate per nucleotide in humans. *Genetics* **156** 297–304.

Nance WE and Grove J 1972 Genetic determination of phenotypic variation in sickle cell trait. *Science* **177** 716–718.

Nanda I, Zend-Ajusch E, Shan Z, Grützner F, Schartl M, Burt DW, Koehler M, Fowler VM, Goodwin G, Schneider WJ, Mizuno S, Dechant G, Haaf T and Schmid M 2000 Conserved synteny between the chicken Z sex chromosome and human chromosome 9 includes the male regulatory gene DMRT1: a comparative (re)view on avian sex determination. *Cytogenetics and Cell Genetics* **89** 67–78.

Narod SA and Foulkes WD 2004 BRCA1 and BRCA2: 1994 and beyond. *Nature Reviews Cancer* **4** 665–676.

Narod SA, Brunet JS, Ghadirian P, Robson M, Heimdal K, Neuhausen SL, Stoppa-Lyonnet D, Lerman C, Pasini B, de los Rios P, Weber B and Lynch H; Hereditary Breast Cancer Clinical Study Group 2000 Tamoxifen and risk of contralateral breast cancer in BRCA1 and BRCA2 mutation carriers: a case-control study. Hereditary Breast Cancer Clinical Study Group. *The Lancet* **356** 1876–1881.

Narod SA, Dube MP, Klijn J, Lubinski J, Lynch HT, Ghadirian P, Provencher D, Heimdal K, Moller P, Robson M, Offit K, Isaacs C, Weber B, Friedman E, Gershoni-Baruch R, Rennert G, Pasini B, Wagner T, Daly M, Garber JE, Neuhausen SL, Ainsworth P, Olsson H, Evans G, Osborne M, Couch F, Foulkes WD, Warner E, Kim-Sing C, Olopade O, Tung N, Saal HM, Weitzel J, Merajver S, Gauthier-Villars M, Jernstrom H, Sun P and Brunet JS 2002 Oral contraceptives and the risk of breast cancer in BRCA1 and BRCA2 mutation carriers. *Journal of the National Cancer Institute* **94** 1773–1779.

Narod SA, Risch H, Moslehi R, Dorum A, Neuhausen S, Olsson H, Provencher D, Radice P, Evans G, Bishop S, Brunet JS and Ponder BA 1998 Oral contraceptives and the risk of hereditary ovarian cancer. Hereditary Ovarian Cancer Clinical Study Group. *New England Journal of Medicine* **339** 424–428.

Nasse CF 1820 Von einer erblichen Neigung zu tödlichen Blutungen. *Archiv der medizinischen Polizei und der gemeinnutzigen Arzneikunde* (*Horn's Archives*). G Reimer, Berlin. pp. 385–434.

Nasseri A, Mukherjee T, Grifo JA, Noyes N, Krey L and Copperman AB 1999 Elevated day 3 serum follicle stimulating hormone and/or estradiol may predict foetal aneuploidy. *Fertility and Sterility* **71** 715–718.

National Institute of Health Consensus Statement. 1994 12 April 5–7.

National Research Council. 1988 *Report.* Committee on Mapping and Sequencing the Human Genome. National Academy of Science Press, Washington, D.C.

National Statistics 2000 *Cancer registrations in England, 2000 HMSO,* London

Nedambale TL, Dinnyés A, Yang X and Tian XC 2004 Bovine blastocyst development in vitro: timing, sex, and viability following vitrification. *Biology of Reproduction* 71 1671–1676.

Neel JV 1962 Diabetes mellitus: a 'thrifty' genotype rendered detrimental by 'progress'? *American Journal of Human Genetics* 14 353–362.

Nef S, Verma-Kurvari S, Merenmies J, Vassalli J-D, Efstratiadis A and Accili D 2003 Testis determination requires insulin receptor family function in mice. *Nature* 426 421–425.

Nei M 2005 Selectionism and neutralism in molecular evolution. *Molecular Biology and Evolution* 22 2318–2342.

Nelkin D and Lindee MS 1995 *The DNA Mystique. The Gene as a Cultural Icon.* Freeman, New York. p. 171.

Nelson HD, Huffman LH, Fu R and Harris EL 2005 Genetic risk assessment and BRCA mutation testing for breast and ovarian cancer susceptibility: systematic evidence review for the U.S. Preventive Services Task Force. *Annals of Internal Medicine* 143 362–379.

New England Regional Genetics Group Social and Ethical Concerns Committee. 1999 Statement on cost-effectiveness and cost-benefit analysis. *http://www.nergg.org/pop cost benefit.html*

Newkirk MM, Downe AER and Simon JB 1975 Fate of ingested hepatitis B antigen in blood-sucking insects. *Gastroenterlogy* 69 982–987.

Newman CS and Krieg PA 1998 Abstract, tinman-related genes expressed during heart development in Xenopus. *Developmental Genetics* 22 230–238.

Newson AJ and Humphries SE 2005 Cascade testing in familial hypercholesterolaemia: how should family members be contacted? *European Journal of Human Genetics* 13 401–408.

Nieuwenhuis B, Van Assen-Bolt AJ, Van Waarde-Verhagen MA, Sijmons RH, Van der Hout AH, Bauch T, Streffer C and Kampinga HH 2002 BRCA1 and BRCA2 heterozygosity and repair of X-ray-induced DNA damage. *International Journal Radiation Biology* 78 285–295.

Nikolova M, Chen X and Lufkin T 1997 Nkx2.6 expression is transiently and specifically restricted to the branchial region of pharyngeal-stage mouse embryos. *Mechanisms of Development* 69 215–218.

Ninomiya K and Kaneko M 1970 Australia antigen, transfusion and hepatitis. *Vox Sanguinis* 18 289–300.

Nirenberg MW and Matthaei JH 1961 The dependence of cell-free protein synthesis in *E. coli* upon naturally occurring or synthetic polyribonucleotides. *Proceedings of the National Academy of Science USA* 47 1588–1602.

Nogueira CP, Bartels CF, McGuire MC, Adkins S, Lubrano T, Rubinstein HM, Lightstone H, van der Spek AFL, Lockridge O and La Du BN 1992 Identification of two different point mutations associated with the fluoride-resistant phenotype for human butyrylcholinesterase. *American Journal of Human Genetics* 51 821–828.

Nogueira CP, McGuire MC, Graeser C, Bartels CF, Arpagaus M, van der Spek AFL, Lightstone H, Lockridge O and La Du BN 1990 Identification of a frameshift mutation responsible for the silent phenotype of human serum cholinesterase, GLY 117 (GG*T* → GG*A*G). *American Journal of Human Genetics* 40 934–940.

Nybo Anderson M-A, Wohlfahrt J, Christens P, Olsen J and Melbye M 2000 Maternal age and foetal loss: population based register linkage study. *British Medical Journal* 320 1708–1712.

O'Malley MA, Bostanci A and Calvert J 2005 Whole-genome patenting. *Nature Reviews Genetics* 6 502–507.

Ohbayashi A, Okochi K and Mayumi M 1972 Familial clustering of asymptomatic carriers of Australia antigen and patients with chronic liver disease or primary liver cancer. *Gastroenterology* 62 618–625.

Ohno S 1967 *Sex Chromosomes and Sex-Linked genes.* Springer-Verlag, Berlin.

Ohta T 1973 Slightly deleterious mutation substitutions in evolution. *Nature* 241 96–98.

Okabe M, Ikawa M and Ashkenas J 1998 Gametogenesis '98 – Male infertility and the genetics of spermatogenesis. *American Journal of Human Genetics* 62 1274–1281.

Okamoto E and Ueda T 1967 Embryogenesis of intramural ganglion of the gut and its relation to Hirschsprung's disease. *Journal of Pediatric Surgery* 2 437–443.

Okochi K and Murakam S 1968 Observations on Australia antigen in Japanese. *Vox Sanguinis* 15 374–385.

O'Leary VB, Parle-McDermott A, Molloy AM, Kirke PN, Johnson Z, Conley M, Scott JM and Mills JL 2002 MTRR and MTHFR polymorphism: link to Down syndrome? *American Journal of Human Genetics* **107** 151–155.

Omenn GS and McKusick VA 1979 The association of Waardenburg syndrome and Hirschsprung megacolon. American Journal of Medical Genetics *3 217–23.*

Onishi 1911 Über Kryptophthalmos. *Klinische Monatsblätter für Augenheilkunde* **50** 271–272.

Onkano P, Vaanamen S, Karvonen M and Tuomilehto J 1999 Worldwide increase in incidence of type 1 diabetes: the analysis of the data on published incidence trends. *Diabetologia* **42** 1395–1403

On-Kei Chan A, Lam CW, Tong SF, Man Tung C, Yung K, Chan YW, Au KM, Yuen YP, Hung CT, Ng KP and Shek C 2005 Novel mutations in the *BCHE* gene in patients with no butyrylcholinesterase activity. *Clinica Chimica Acta* **351** 155–159.

Online Mendelian Inheritance in Man, OMIM (TM). McKusick-Nathans Institute for Genetic Medicine. Johns Hopkins University (Baltimore, Maryland) and National Center for Biotechnology Information, National Library of Medicine (Bethesda, Maryland), 2000, at World Wide Web URL: (*http://www.ncbi.nlm.nih.gov/Omim/*).

Orozco G, Nodarse A, Cordovés M and Auburger G 1990 Autosomal dominant cerebellar ataxia. Clinical analysis of 263 patients from a homogeneous population in Holguín, Cuba. *Neurology* **40** 1369–1375.

Osborn F 1968 *The Future of Human Heredity: An Introduction into Eugenics in Modern Society.* Weybright and Talley, New York.

Osborn FH and Robinson A 1987 Eugenics. In: *The New Encyclopedia Britannica,* 15th edn Encyclopedia Britannica, Chicago, vol 19 726–727.

Osler W 1900 The importance of post-graduate study. *The Lancet* **ii** 73.

Osler W 1915 The iconography of Jonathan Hutchinson. *Bulletin Johns Hopkins Hospital* **26** 82.

Osorio MG, Kopp P, Marui S, Latronico AC, Mendonca BB and Arnhold IJ 2000 Combined pituitary hormone deficiency caused by a novel mutation of a highly conserved residue (F88S) in the homeodomain of PROP-1. *Journal of Clinical Endocrinology and Metabolism* **85** 2779–2785.

Otto JC 1803 An account of an haemorrhagic disposition existing in certain families. *Medical Repository* **6** 1–4.

Otto PA 1989 Estimativas de riscos de doença genética na prole de primos em primeiro grau (Estimates of genetic disease in the offspring of first cousins). *Ciencia e Cultura* **41** 471–474.

Otto PA and Frota-Pessoa O 1979 Estimativas de riesgos genéticos (Estimates of genetic risks). *Anales de la Academia Ciencias Exactas Físicas y Naturales Buenos Aires* **31** 271–289.

Oxford English Dictionary 1970, Clarendon Press, Oxford. Volume III, p. 319.

Paabo S 2003 The mosaic that is our genome. *Nature* **421** 409–412.

Paez JG, Janne PA, Lee JC, Tracy S, Greulich H, Gabriel S, Herman P, Kaye FJ, Lindeman N, Boggon TJ, Naoki K, Saski H, Fujii Y, Eck MJ, Sellars WR, Johnson BE and Meyerson M 2004 EGFR mutations in lung cancer: correlation with clinical response to gefitinib therapy. *Science* **304** 1497–1500.

Pagon RA, Graham JM, Zonana J and Young SL 1981 Congenital heart disease and choanal atresia with multiple anomalies. *Journal of Pediatrics* **99** 223–237.

Palmer S, Perry J and Ashworth A 1995 A contravention of Ohno's law in mice. *Nature Genetics* **10** 472–476.

Paneque M, Santos N, Tamayo H, Reinaldo R, Velásquez L, Almagur L, Hechavarría R, Mourino T and Cruz T 2001 Type 2 Spinocerebellar ataxia: acceptance of oprenatal diagnosis in descents at risk. *Revista de Neurología* **33** 904–908.

Pangalos CG, Talbot CC Jr, Lewis JG, Adelsberger PA, Petersen MB, Serre JL, Rethore MO, de Blois MC, Parent P, Schinzel AA Binkert F, Boue J, Corbin E, Croquette MF, Gilgenkrantz S, de Grouchy J, Bertheas MF, Prieur M, Raoul O, Serville F, Siffroi JP, Thepot F, Lejeune J and Antonarakis SE. 1992. DNA polymorphism analysis in families with recurrence of free trisomy 21. *American Journal of Human Genetics* **51** 1015–1027.

Papp Z and Váradi V 1985 A further case of Váradi-Papp syndrome. Personal communication to V. McKusick. Oxford, Debrecen.

Parens E and Asch A 1999 The disability rights critique of prenatal genetic testing. Reflections and recommendations. *Hasting Center Report* 5 S1-S22.

Parra EJ, Marcini A, Akey J, Martinson J, Batzer MA, Cooper R, Forrester T, Allison DB, Deka R, Ferrell RE and Shriver MD 1998 Estimating African-American admixture proportions by use of population-specific alleles. *American Journal of Human Genetics* 63 1839–1851.

Parra FC, Amado RC, Lambertucci JR, Rocha J, Antunes CM and Pena SDJ 2003 Color and genomic ancestry in Brazilians. *Proceedings of the National Academy of Sciences USA* 100 177–182.

Paschou P, Bozas E, Dokopoulou M, Havarani B, Malamitsi-Puchner A, Ylli A, Ylli Z, Thymelli I, Gerasimidi-Vazeou A and Bartsocas CS 2004 HLA alleles and type 1 diabetes mellitus in low disease incidence populations of Southern Europe: a comparison of Greeks and Albanians. *Journal of Pediatric Endocrinology and Metabolism* 17 173–182.

Passarge E 1967a The genetics of Hirschsprung's disease: evidence for heterogeneous etiology and a study of sixty three families. *New England Journal of Medicine* 276 138–143.

Passarge E 1967b Quelques considérations étiologiques et génétiques sur la maladie de Hirschsprung. *Médecine et Hygiéne* 25 240–241.

Passarge E 1972 Genetic heterogeneity and recurrence risk of congenital intestinal aganglionosis. *Birth Defects: Original Article Series* 8 63–7.

Passarge E 1991 Bloom's syndrome: the German experience. *Annales de Génétique* 34 179–197.

Passarge E 2002 Dissecting Hirschsprung disease. *Nature Genetics* 31 11–12.

Passarge E 2003 Gastrointestinal tract: Molecular genetics of Hirschsprung disease. *Nature Encyclopedia of the Human Genome* 2 578–83.

Passarge E 2006 Gastrointestinal tract and hepatobiliary duct system. In: *Emery & Rimoin's Principles and Practice of Medical Genetics*, 5th edn (editors Rimoin DL, Connor JM, Pyeritz RE and Korf B) Churchill Livingstone, London.

Patrinos GP and Brookes AJ 2005 DNA, diseases and databases: disastrously deficient. *Trends in Genetics* 21 333–338.

Patterson CP Boyer KM Maynard JE and Kelly PC 1974 Epidemic hepatitis in a clinical laboratory. *Journal of the American Medical Association* 230 854–857.

Paul DB 1998 *The Politics of Heredity. Essays on Eugenics, Biomedicine, and The Nature-Nurture Debate.* State Univ of NY Press, Albany, p. 97.

Pauling L, Itano HA, Singer SJ and Wells IC 1949 Sickle cell anemia: a molecular disease. *Science* 110 543–548.

Payet M, Camain R and Pene P 1956 Le cancer primitif du foie, étude critique à propos de 240 cas. *Revue Internationale d'hepatologie* 4 1–20.

Peacock W 1929 Hereditary absence of hands and feet. *Eugenical News* 14 46–47.

Pearson CE, Edamura KN and Cleary JD 2005 Repeat instability: mechanisms of dynamic mutations. *Nature Reviews Genetics* 6 729–742.

Pearson K 1901 Talk. *The London Edinburgh and Dublin Philosophical Magazine and Journal* 6 566.

Pearson K and Elderton EM 1923 On the variate difference method. *Biometrika* 14 281–310.

Pearson PL 2002 Chromosome Mapping. In: *Encyclopedia of Genetics* (editors Brenner S and Miller JH) Academic Press, San Diego, vol. 1, pp. 361–363.

Peek RM Jr., Miller GG, Tham KT, Perez-Perez GI, Zhao X, Atherton JC and Blaser MJ 1995 Heightened inflammatory response and cytokine expression in vitro to *cagA+ Helicobacter pylori* strains. *Laboratory Investigation* 7 1760–1770.

Pellestor F, Andreo B, Arnal F, Humeau C and Demaille J 2003 Maternal aging and chromosomal abnormalities: new data drawn from in vitro unfertilized human oocytes. *Human Genetics* 112 195–203.

Pena SDJ and Bortolini MC 2004 Pode a genética definir quem deve se beneficiar das cotas universitárias e demais ações afirmativas? *Estudos Avançados* 181 31–50.

Penchaszadeh V 2002 Preventing congenital anomalies in developing countries. *Community Genetics* 5 61–69.

Penchaszadeh VB, Heredero L, Punales-Morejón D, Rojas I and Pérez ET 1997 Genetic counselling training in Cuba. *American Journal of Human Genetics* 6 Supplement A 1098.

Pendred V 1896 Deaf-mutism and goitre. *The Lancet* **ii** 532–533.

Pennisi E 2003 A low number wins the genesweep pool. *Science* **300** 1484.

Penrose LS 1932 The blood grouping of Mongolian imbeciles. *The Lancet* **7** 394–395.

Penrose LS 1933 The relative effect of paternal and maternal age in mongolism. *Journal of Genetics* **27** 219–224.

Penrose LS 1934a The relative aetiological importance of birth order and maternal age in Mongolism. *Proceedings of the Royal Society* **115** 431–450.

Penrose LS 1934b *Mental Defect* Sidgwick and Jackson, London.

Penrose LS 1938 *The Colchester Survey: A Clinical and Genetical Study of 1280 Cases of Mental Defect.* H. M. Stationery Office, London, Privy Council of Medical Research Council, No. 229.

Penrose LS 1946 Phenylketonuria: A problem in eugenics. *The Lancet* **1** 949–951.

Penrose LS 1949 *The Biology of Mental Defect* (Sidgwick and Jackson, London; 2nd edition 1953, 3rd ed. 1963, 4th ed. 1972).

Penrose LS 1953 The genetical background to common diseases. *Acta Genetica et Statistica Medica* **4** 257–265.

Penrose LS 1956 A note on the prevalence of genes for deleterious recessive traits in man. *Annals of Human Genetics* **21** 222–223.

Penrose LS 1966 Human chromosomes, normal and aberrant. *Proceedings of the Royal Society of London Series B* **164** 311–319.

Penrose LS and Smith GF 1966 *Down's Anomaly.* Churchill, London.

Penrose LS 1967 Presidential Address – the influence of the English tradition in human genetics. In: *Proceedings of the Third International Congress of Human Genetics* (editors Crow JF and Neel JV) pp. 13–25. Johns Hopkins University Press, Baltimore.

Penrose LS, Ellis JR and Delhanty JDA 1960 Chromosomal translocations on mongolism and in normal relatives. *The Lancet* **2** 409.

Penrose O 1998 *Lionel S Penrose, human geneticist and human being,* talk given to the Wisbech Society on 11 May 1998 (unpublished booklet).

Perou CM, Sorlie T, Eisen MB, van de Rijn M, Jeffrey SS, Rees CA, Pollack JR, Ross DT, Johnsen H, Asklen LA, Fluge O, Pergamenschikov A, Williams c, Zhu SX, Lonning PE, Borresen-Dale AL, Brown PO and Botstein D 2000 Molecular portraits of human breast tumours. *Nature* **406** 747–752.

Perrier AL, Massoulié J and Krejci E 2002 PRIMA: The membrane anchor of acetylcholinesterase in the brain. *Neuron* **33** 275–285.

Perry EK, Perry RH, Blessed G and Tomlinson BE 1978 Changes in brain cholinesterases in senile dementia of Alzheimer type. *Neuropathology and Applied Neurobiology* **4** 273–277.

Peto J and Mack T 2000 High constant incidence in twins and other relatives of women with breast cancer. *Nature Genetics* **26** 411–414.

Peto J, Collins N, Barfoot R, Seal S, Warren W, Rahman N, Easton DF, Evans C, Deacon J and Stratton MR 1999 Prevalence of BRCA1 and BRCA2 gene mutations in patients with early-onset breast cancer. *Journal of the National Cancer Institute* **91** 943–949.

Petrou P, Pavlakis E, Dalezios Y, Galanopoulos VK and Chalepakis G. 2005 Basement membrane distortions impair lung lobation and capillary organization in the mouse model for Fraser syndrome. *Journal of Biological Chemistry* **280** 10350–10356. (Epub 2004 Dec 28)

Peutz JLA 1921 Very remarkable case of familial polyposis of mucous membrane of intestinal tract and nasopharynx accompanied by peculiar pigmentations of skin and mucous membrane. (Dutch). *Nederland Maandschrift voor Geneeskunde* **10** 134–146.

Pfeiffer RA, Henkel KE and Stoss H 1993 Genetisch-morphologische Letalsyndrome. *Pathologe* **14** 365–370.

Pharoah PD, Dunning AM, Ponder BA and Easton DF 2004 Association studies for finding cancer-susceptibility genetic variants. *Nature Reviews Cancer* **4** 850–860.

Phelps PD 1996 Large vestibular aqueduct: large endolymphatic sac? *Journal of Laryngology and Otology* **110** 1103–1104.

Phelps PD, Coffey RA, Trembath RC, Luxon LM, Grossman AB, Britton KE, Kendall-Taylor P, Graham JM, Cadge BC, Stephens SG, Pembrey ME and Reardon W 1998 Radiological malformations of the ear in Pendred Syndrome. *Clinical Radiology* **53** 268–273.

Phillips ED 1973 *Aspects of Greek Medicine* Thames & Hudson, London.

Phillips KA, Andrulis IL and Goodwin PJ 1999 Breast carcinomas arising in carriers of mutations in BRCA 1 or BRCA 2: are they prognostically different? *Journal of Clinical Oncology* **17** 3653–3663.

Phillips OP, Cromwell S, Rivas M, Simpson JL and Elias S 1995 Trisomy 21 and maternal age of menopause: Does reproductive age rather than chronological age influence risk of nondisjunction? *Human Genetics* **95** 117–118.

Phinney AL, Horne P, Yang J, Janus C, Bergeron C and Westaway D 2003 Mouse models of Alzheimer's disease: the long and filamentous road. *Neurology Research* **25** 590–600.

Picheth G, Fadel-Picheth C, Primo-Parmo SL, Chautard-Freire-Maia EA and Vieira MM 1994 An improved method for butyrylcholinesterase phenotyping. *Biochemical Genetics* **32** 83–89.

Pike MG, Holmstrom G, de Vries LS, Pennock JM, Drew KJ, Sonksen PM and Dubowitz LMS 1994 Patterns of visual impairment associated with lesions of the preterm infant brain. *Developmental Medicine and Child Neurological* **36** 849–862.

Pingault V, Guiochon-Mantel A, Bondurand N, Faure C, Lacroix C, Lyonnet S, Gossens M and Landrieu P 2000 Peripheral neuropathy with hypomyelination, chronic intestinal pseudo-obstruction and deafness: a developmental 'neural-crest syndrome' related to a SOX10 mutation. *Annals of Neurology* **48** 671–676.

Pluschke G, Vanek M, Evans A, Dittmar T, Schmid P, Itin P, Filardo EJ and Reisfeld RA 1996 Molecular cloning of a human melanoma-associated chondroitin sulfate proteoglycan. *Proceedings of the National Academy of Sciences USA* **93** 9710–9715.

Pool-Zobel BL, Lotzmann N, Knoll M, Kuchenmeister F, Lambertz R, Leucht U, Scroeder GU and Schmezer P 1994 Detection of genotoxic effects in human gastric and nasal mucosa cells isolated from biopsy samples. *Environmental and Molecular Mutagenesis* **24** 23–45.

POSSUM Murdoch Childrens Research Institute, *http//www.possum.net.au*

Possum System F. 1988–2005 *Pictures of standard syndromes and undiagnosed malformations.* CP Export Pty Ltd, Melbourne.

Primo-Parmo SL, Bartels CF, Wiersema B, van der Spek AFL, Innis JW, La Du BN 1996 Characterization of 12 silent alleles of the human butyrylcholinesterase (*BCHE*) gene. *American Journal of Human Genetics* **58** 52–64.

Primo-Parmo SL, Lightstone H and La Du BN 1997 Characterization of an unstable variant (BChE115D) of human butyrylcholinesterase. *Pharmacogenetics* **7** 27–34.

Prince AM 1968 An antigen detected in the blood during the incubation period of serum hepatitis. *Proceedings of the National Academy of Sciences USA* **60** 814–827.

Prince AM, Metselaar D, Kafuko GW, Mukwaya LG, Ling CM and Overby LR 1972 Hepatitis B antigen in wild-caught mosquitoes in Africa. *The Lancet* **ii** 247–250.

Pritchard JK 2001 Are rare variants responsible for susceptibility to complex diseases? *American Journal of Human Genetics* **69** 124–137.

Pritchard JK, Stephens M and Donnelly P 2000 Inference of population structure using multilocus genotype data. *Genetics* **155** 945–959.

Prody CA, Zevin-Sonkin D, Gnatt A, Goldberg O and Soreq H 1987 Isolation and characterization of full-length cDNA clones coding for cholinesterase from fetal human tissues. *Proceedings of the National Academy of Sciences USA* **84** 3555–3559.

Pryor SP, Madeo AC, Reynolds JC, Sarlis NJ, Arnos KS, Nance WE, Yang Y, Zalewski CK, Brewer CC, Butman JA and Griffith AJ 2000 *LC26A4/PDS* genotype-phenotype correlation in hearing loss with enlargement of the vestibular aqueduct (EVA): evidence that Pendred syndrome and non-syndromic EVA are distinct clinical and genetic entities *Journal of Medical Genetics* **42** 159–165.

Puffenberger EG, Hosoda K, Washington SS, Nakao K, deWit D, Yanagisawa M and Chakravarti A 1994a A missense mutation of the Endothelin-B receptor gene in multigenic Hirschsprung's disease. *Cell* **79** 1257–1266.

Puffenberger EG, Hu-Lince D, Parod JM, Craig DW, Dobrin SE, Conway AR, Donarum EA, Strauss KA, Dunckley T, Cardenas JF, Melmed KR, Wright CA, Liang W, Stafford P, Flynn CR, Morton DH and Stephen DA 2004 Mapping of sudden infant death with the dysgenesis of the testes syndrome (SIDDT) by a SNP genome scan and identification of TSPYL loss of function. *Proceedings of the National Academy of Sciences USA* **101** 11689–11694.

Puffenberger EG, Kauffman ER, Bolk S *et al.* 1994b Identity-by-descent and association mapping of a recessive gene for Hirschsprung disease on human chromosome 13q22. *Human Molecular Genetics* **3** 1217–1225.

Pulst S, Nechiporuk A, Nechiporuc T, Sispert S, Xiao Ning C, Lopez I, Pearlman S, Starkman S, Orozco G, Lunkes A, DeJong P, Rouleau G, Auburger G, Korenberg J, Figueroa C and Sahba S 1996 Moderate expansion of a normally biallelic trinucleotide repear in spinocerebellar ataxia type 2. *Nature Genetics* **14** 269–276.

Purcell RH and Gerin JL 1975 Hepatitis B subunit vaccine: A preliminary report of safety and efficacy tests in chimpanzees. *American Journal of Medical Sciences* **270** 395–399.

Quintana J 1999 Clinical cytogenetics in Cuba. *European Cytogeneticists Association Newsletter* **4** 23–24.

Quintana J, Quiñones O, Méndez L, Lavista M, Gómez M and Dieppa N 1999 Resultados del diagnóstico prenatal citogenético en las provincias occidentales de Cuba, 1984–1998. *Revista Cubana de Genética Humana* **1** 3. Electronic journal.

Rando TA 2004 Artificial sweeteners – enhancing glycosylation to treat muscular dystrophies. *New England Journal of Medicine* **35** 1254–1256.

Ranson M and Wardell S 2004 Gefitinib, a novel, orally administered agent for the treatment of cancer. *Journal of Clinical Pharmacy and Therapeutics* **29** 95–103.

Raunio VK, London WT, Sutnick AI, Millman I and Blumberg BS 1970 Specificities of human antibodies to Australia antigen. *Proceedings of the Society for Experimental Biology and Medicine* **134** 548–557.

Raymond CS, Kettlewell JR, Hirsch B, Bardwell VJ and Zarkower D 1999 Expression of DMRT1 in the genital ridge of mouse and chicken embryos suggests a role in vertebrate sexual development. *Developmental Biology* **215** 208–220.

Raymond CS, Shamu CE, Shen MM, Seifert HJ, Hirsch B, Hodgkin J and Zarkower D 1998 Evidence for evolutionary conservation of sex-determining genes. *Nature* **391** 691–695.

Read AP 2005 *Genes Hearing and Deafness – from molecular biology to clinical practice.* Caserta, Italy.

Read AP and Newton V 1996 Waardenburg Syndrome. In: *Genetics and hearing Impairment.* Editors, Martine, Read, Stephens. Whurr publications.

Reardon W 1997 Historical note: Dr. George Fraser. *Journal of Audiological Medicine* **6** 185–189.

Reardon W and Hall CM 2003 Broad thumbs and halluces with deafness: a patient with Keipert syndrome. *American Journal of Medical Genetics* **118A** 86–89.

Reardon W, Mahoney CF, Trembath R, Jan H and Phelps PD 2000 Enlarged vestibular aqueduct: a radiological marker of Pendred syndrome, and mutation of the PDS gene. *Quarterly Journal of Medicine* **93** 99–104.

Reardon W, Smith A, Honour JW, Hindmarsh P, Das, Rumsby G, Nelson I, Malcolm S, Adès L, Sillence D, Kumar D, DeLozier-Blanchet C, McKee S, Kelly T, McKeehan WL, Baraitser M and Winter RM 2000 Evidence for digenic inheritance in some cases of Antley-Bixler syndrome? *Journal of Medical Genetics* **37** 26–32.

Rebbeck TR, Levin AM, Eisen A, Snyder C, Watson P, Cannon-Albright L, Isaacs C, Olopade O, Garber JE, Godwin AK, Daly MB, Narod SA, Neuhausen SL, Lynch HT and Weber BL 1999 Breast cancer risk after bilateral prophylactic oophorectomy in BRCA1 mutation carriers. *Journal of the National Cancer Institute* **91** 1475–1479.

Rebbeck TR, Lynch HT, Neuhausen SL, Narod SA, Van't Veer L, Garber JE, Evans G, Isaacs C, Daly MB, Matloff E, Olopade OI and Weber BL 2002 Prevention and Observation of Surgical End Points Study Group. Prophylactic oophorectomy in carriers of BRCA1 or BRCA2 mutations. *New England Journal of Medicine* **346** 1616–1622.

Rechitsky S, Storm C, Verlinsky O, Amet T, Ivakhnenko V, Kukharenko V Kuliev A and Verlonsky Y 1999 Accuracy of preimplantation diagnosis of single gene disorders by polar body analysis of oocytes. *Journal of Assisted Reproductive Genetics* 16 192–198.

Reddy MA, Francis PJ Berry V, Battacharya SS and Moore AR 2004 Molecular genetic basis of inherited cataract and associated phenotypes. *Survey of Ophthalmology* 49 300–315.

Reich DE and Lander ES 2001 On the allelic spectrum of human disease. *Trends in Genetics* 17 502–510.

Reik W and Lewis A 2005 Co-evolution of X-chromosome inactivation and imprinting in mammals. *Nature Reviews, Genetics* online.

Rens W, Grützner F, O'Brien PCM, Fairclough H, Graves JAM and Ferguson-Smith MA 2004 Resolution and evolution of the duck-billed platypus karyotype with an $X_1Y_1X_2Y_2X_3Y_3X_4Y_4X_5Y_5$ male sex chromosome constitution. *Proceedings of the National Academy of Sciences USA* 101 16257–16261.

Renwick JH 1969 Progress in mapping human autosomes. *British Medical Bulletin* 25 655–673.

Retina International's Scientific Newsletter – protein database (*http://www.retina-international.org/sci-news/protdat.hum*).

Retwein P, Yokoyama S, Didier DK and Chirgwin JM 1986 Genetic analysis of the hypervariable region flanking the human insulin gene. *American Journal of Human Genetics* 39 291–299.

Reya T, Morrison SJ, Clarke MF and Weissman IL 2001 Stem cells, cancer, and cancer stem cells. *Nature* 414 105–111.

Riccardi VM and Mulvihill JJ eds. 1981 *Neurofibromatosis (von Recklinghausen Disease)*. Advances in Neurology. Raven Press, New York.

Rickard S, Parker M, van't Hoff W, Barnicot A, Russell-Eggit I, Winter RM and Bitner-Glindzicz M 2001 Oto-facial-cervical (OFC) syndrome is a contiguous gene deletion syndrome involving EYA1: molecular analysis confirms allelism with BOR syndrome and further narrows the Duane syndrome critical region to 1 cM. *Human Genetics* 108 398–403.

Riise R, Flage T, Hansen E, Rosenberg T, Rudanko S-L, Viggoson G and Warburg M 1992 Visual impairment in Nordic children. I. Nordic registers and prevalence data. *Acta Ophthalmologica Scandinavica* 70 145–154

Riordan JR, Rommens JM, Kerem B, Alon N, Rozmahel R, Grzelczak Z, Zielenski J, Lok S, Plavsic N, Chou JL, Drumm ML, Iannuzzi MC, Collin FS and Tsui L-C 1989 Identification of the cystic fibrosis gene: Cloning and characterization of complementary DNA. *Science* 245 1066–1073.

Risch N and Zhang H 1995 Extreme discordant sib pairs for mapping quantitative trait loci in humans. *Science* 268 1584–1589.

Robbins RB 1918 Some applications of mathematics to breeding problems III. *Genetics* 3 375–389.

Roberts SA, Spreadborough AR, Bulman B, Barber JB, Evans DG and Scott D 1999 Heritability of cellular radiosensitivity: a marker of low-penetrance predisposition genes in breast cancer? *American Journal of Human Genetics* 65 784–794.

Robertson JA 2003 Extending preimplantation genetic diagnosis: the ethical debate: Ethical issues in new uses of PGD. *Human Reproduction* 18 465–471.

Robinson WS and Lutwick LI 1976 The virus of hepatitis type B. *New England Journal of Medicine* 295 1168–1175.

Robson M, Levin D, Federici M, Satagopan J, Bogolminy F, Heerdt A, Borgen P, McCormick B, Hudis C, Norton L, Boyd J and Offit K 1999 Breast conservation therapy for invasive breast cancer in Ashkenazi women with BRCA gene founder mutations. *Journal of the National Cancer Institute* 91 2112–2117.

Rodríguez L Personal communication to L Heredero Baute.

Rodríguez L, Heredero L and Padrón L 1979 Incidencia de los defectos de cierre del tubo neural. *Revista Cubana de Pediatría* 51 125–131.

Rodríguez L, Heredero L, López J, Fernández J and Oliva JA 1982 Neural tube defects in Havana City incidence and prenatal diagnosis. *Zentralblatt fur Gynakologie* 104 1325–1327.

Rodríguez L, Heredero L, Oliva JA and Zaldivar JO 1987 Prenatal diagnosis by measurement of alpha-feto-protein in Havana City. *Prenatal Diagnosis* 7 657–661.

Rodríguez L, Sánchez R, Hernández J, Carrillo L, Oliva JA and Heredero L 1997 Results of 12 year's combined maternal serum alpha-fetoprotein and ultrasound monitoring for prenatal detection of foetal malformations in Havana City, Cuba. *Prenatal Diagnosis* **17** 301–304.

Rodriguez-Larralde A, Gonzales-Martin A, Scapoli C and Barrai I 2003 The names of Spain: a study of the isonymy structure of Spain. *American Journal of Physical Anthropology* **121** 280–292.

Rodriguez-Larralde A, Scapoli C, Beretta M, Nesti C, Mamolini E and Barrai I 1998a Isonymy and the genetic structure of Switzerland. II. Isolation by distance. *Annals of Human Biology* **25** 533–540.

Rodriguez-Larralde A, Scapoli C, Beretta M, Nesti C, Mamolini E and Barrai I 1998b Isonymy and isolation by distance in Germany. *Human Biology* **70** 1041–1076.

Rogers AR 1991 Doubts about isonymy. *Human Biology* **63** 663–668.

Rojas-Alvarado MA and Garza-Chapa R 1994 Relationships by isonymy between persons with monophyletic and polyphyletic surnames from the Monterrey metropolitan area, Mexico. *Human Biology* **66** 1021–1036.

Romano C 1965 Congenital cardiac arrhythmia. *The Lancet* **1** 658–659.

Romano C, Gemme G and Pongiglione R 1963 Aritmie cardiache rare dell'ete pediatrica: il accessi sincopali per fibrillazione ventriculari parossistica. *Clinica Pediatrica (Bologna)* **45** 656–661.

Romeo G, Ronchetto P, Luo Y, Barone Y, Seri M, Ceccherini I, Pasini B, Bocciardi R, Lerone M, Kääriäinen H and Martuciello G 1994 Point mutations affecting the tyrosine kinase domain of the *RET* proto-oncogene in Hirschsprung's disease. *Nature* **367** 377–378.

Ronningen KS and Keiding N 2001 Correlations between the incidence of childhood-onset type 1 diabetes in Europe and HLA genotypes. *Diabetologia* **44** (Suppl 3) B51-B59.

Ronningen KS and Thorsby E 1993 Particular HLA-DQ molecules play a dominant role in determining susceptibility or resistance to type 1 (insulin-dependent) diabetes mellitus. *Diabetologia* **36** 371–377.

Rosen EM, Fan S, Pestell RG and Goldberg ID 2003 BRCA1 gene in breast cancer. *Journal of Cell Physiology* **196** 19–41

Rudnicka AR, Wald NJ, Huttly W and Hackshaw AK 2002 Influence of maternal smoking on the birth prevalence of Down syndrome and on second trimester screening performance. *Prenatal Diagnosis* **22** 893–897.

Ruf RG, Berkman J, Wolf MTF, Nurnberg P, Gattas M, Ruf EM, Hyland V, Kromberg J, Glass I, Macmillan J, Otto E, Nurnberg G, Lucke B, Hennies HC and Hildebrandt F 2003 A gene locus for branchio-otic syndrome maps to 14q21.3–24.3. *Journal of Medical Genetics* **45** 515–519.

Ryhänen RJJ, Jauhianen MS, Laitinen MV and Puhakainen EV 1982 The relationships between human serum pseudocholinesterase, lipoproteins, and apolipoproteins (APOHDL). *Biochemical Medicine* **28** 241–245.

Rzewuska-Lech E and Lubinski J 2004 Breast cancer and oestrogen. *Hereditary Cancer in Clinical Practice* **2** 31–36.

Sabry MA, Al Saleh Q, Farah S, Obenbergerova D, Simeonov S, Al Awadi SA and Farag TI 1997 Another Arab patient with overlap of Váradi-Papp/Opitz trigonocephaly syndromes? *American Journal of Medical Genetics* **68** 54–57.

Sakamoto N, Hidaka K, Fujisawa T, Maeda M and Iuchi I 1998 Identification of a point mutation associated with a silent phenotype of human serum butyrylcholinesterase – a case of a familial cholinesterasemia. *Clinica Chimica Acta* **274** 159–166.

Salzano FM 2002 Molecular variability in Amerindians: widespread but uneven information. *Anais da Academia Brasileira de Ciências* **74** 223–263.

Salzano FM 2007 The prehistoric colonization of the Americas. In: *Anthropological Genetics: Theory, Methods and Applications.* (editor Crawford MH) Cambridge University Press, Cambridge (*in press*).

Salzano FM and Bortolini MC 2002 *The Evolution and Genetics of Latin American Populations.* Cambridge University Press, Cambridge.

Salzano FM and Callegari-Jacques SM 1988. *South American Indians. A Case Study in Evolution.* Clarendon Press, Oxford.

Salzano FM and Callegari-Jacques SM 2006 Amerindian and non-Amerindian autosome molecular variability – a test analysis. *Genetica* **126** 237–42

Salzano FM and Freire-Maia N 1970 *Problems in Human Biology: a Study of Brazilian Populations*. Wayne State University Press, Detroit.

Salzano FM and Hurtado AM 2004 *Lost Paradises and the Ethics of Research and Publication*. Oxford University Press, New York.

Sandler I 1983 Pierre Louis Moreau de Maupertuis – a precursor of Mendel? *Journal of Historical Biology* **16** 101–136.

Savulescu J 2001 Procreative beneficence: why we should select the best children. *Bioethics* **15** 413–426

Scambler PJ 2003 In: *Nature Encyclopedia of the Human Genome* (editor Cooper D) Macmillan, London.

Schabath MB, Spitz MR, Grossman HB, Zhang K, Dinney CP, Zheng PJ and Wu X 2003 Genetic instability in bladder cancer assessed by the comet assay. *Journal of National Cancer Institute* **95** 540–547.

Schauer GM, Dunn LK, Godmilow L, Eagle RC Jr. and Knisely AS 1990 Prenatal diagnosis of Fraser syndrome at 18.5 weeks gestation, with autopsy findings at 19 weeks. *American Journal of Medical Genetics* **37** 583–591.

Scherer G 2002 The molecular genetic jigsaw puzzle of vertebrate sex determination and its missing pieces. In *The Genetics and Biology of Sex Determination* (Novartis Foundation Symposium 244) pp. 225–239. John Wiley & Sons, Chichester.

Schmahl J, Eicher EM, Washburn LL and Capel B 2000 *Sry* induces cell proliferation in the mouse gonad. *Development* **127** 65–73.

Schmahl J, Kim Y, Colvin JS, Ornitz DM and Capel B 2004 *Fgf9* induces proliferation and nuclear localization of FGFR2 in Sertoli precursors during male sex determination. *Development* **131** 3627–3636.

Schottenfeld D and Fraumeni JF eds. 2006 *Cancer Epidemiology and Prevention*. 2nd edition Oxford University Press, New York.

Schull WJ and Neel JV 1965 *The Effects of Inbreeding on Japanese Children* Harper and Row, New York.

Schupf N, Kapell D, Lee JH, Ottman R and Mayeux R 1994 Increased risk of Alzheimer's disease in mothers of adults with Down's syndrome. *The Lancet* **344** 353–356.

Schupf N, Kapell D, Nightingale B, Lee JH, Mohlenhoff J, Bewley S, Ottman R and Mayeux R 2001 Specificity of the fivefold increase in AD in mothers of adults with Down syndrome. *Neurology* **57** 979–984.

Schwartz PJ 1983 Autonomic nervous system, ventricular fibrillation, and SIDS. In: *Sudden Infant Death Syndrome* (editors Tildon JT, Roeder LM and Steinschneider A) Academic Press, New York, pp.319–339.

Schwarz EM and Benzer S 1997 Calx, a Na-Ca exchanger gene of *Drosophila melanogaster*. *Proceedings of the National Academy of Sciences USA* **94** 10249–10254.

Scott D, Spreadborough AR, Jones LA, Roberts SA and Moore CJ 1996 Chromosomal radiosensitivity in G2-phase lymphocytes as an indicator of cancer predisposition. *Radiation Research* **145** 3–16.

Scriven PN, Flinter FA., Braude PR, Mackie Ogilvie C 2001 Robertsonian translocations – reproductive risks and indications for preimplantation genetic diagnosis. *Human Reproduction* **16** 2267–2273.

Sebat J, Lakshmi B, Troge J, Alexander J, Young J, Lundin P, Maner S, Massa H, Walker M, Chi M, Navin N, Lucito R, Healy J, Hicks J, Ye K, Reiner A, Gilliam TC, Trask B, Patterson N, Zetterberg A and Wigler M 2004 Large-scale copy number polymorphism in the human genome. *Science* **305** 525–528.

Senior JR, Sutnick AI, Goeser E, London WT, Dahlke MD and Blumberg BS 1974 Reduction of post-transfusion hepatitis by exclusion of Australia antigen from donor blood in an urban public hospital. *American Journal of Medical Science* **267** 171–177.

Sepulveda L, Belaguli N, Nigam V, Chen Y, Nemer M and Schwartz J 1998 GATA-4 and Nkx-2.5 coactivate Nkx-2 DNA binding targets: role for regulating early cardiac gene expression. *Molecular and Cellular Biology* **18** 3405–3415.

Serra A 1999 Riflessioni sulle «Tecnologie di Riproduzione Assistita» a 21 anni dalla nascita della prima bambina concepita in vitro. *Medicina e Morale* **5** 861–883.

Serville F, Carles D and Broussin B 1989 Fraser syndrome: Prenatal ultrasound detection. *American Journal of Medical Genetics* **32** 561–563.

Sever JL and Miller RW 1971 Safety of vaccines: Short and long-term evaluations. In: *Proceedings— International Conference on the Application of Vaccines Against Viral, Rickettsial, and Bacterial Diseases of Man*. Pan American Health Organisation Scientific Publications No. 226, pp. 432–435.

Shah KN, Dalal SJ, Desai MP, Sheth PN, Joshi NC and Ambani LM 1981 White forelock, pigmentary disorder of irides, and long segment Hirschsprung disease: possible variant of Waardenburg syndrome. *Journal of Pediatrics* **99** 432–435.

Shapira S, McCaskill C, Northrup H, Spikes AS, Elder FF, Sutton VR, Korenberg JR, Greenberg F and Shaffer LG 1997 Chromosome 1p36 deletions: the clinical phenotype and molecular characterization of a common newly delineated syndrome. *American Journal of Human Genetics* **61** 642–650.

Sharp AJ, Locke DP, McGrath SD, Cheng Z, Bailey JA, Vallente RU, Pertz LM, Clark RA, Schwartz S, Segraves R, Oseroff VV, Albertson DG, Pinkel D and Eichler EE 2005 Segmental duplications and copy-number variation in the human genome. *American Journal of Human Genetics* **77** 78–88.

Shashi V, Clark P, Rogol AD and Wilson WG 1995 Absent pituitary gland in two brothers with an oral-facial-digital syndrome resembling OFD II and VI: a new type of OFDS? *American Journal of Medical Genetics* **57** 22–26.

Shastry BS 2006 Pharmacogenetics and the concept of individualized medicine. *Pharmacogenomics Journal* **6** 16–21.

Sherrod PH 1997 NLRG: nonlinear regression analysis program. *http://www.sandh.com/sherrod*.

Shetty S, Griffin DK and Graves JAM 1999 Comparative painting reveals strong homology over 80 million years of bird evolution. *Chromosome Research* **7** 289–295.

Shorrocks J, Tobi SE, Latham H, Peacock JH, Eeles R, Eccles D and McMillan TJ 2004 Primary fibroblasts from BRCA1 heterozygotes display an abnormal G1/S cell cycle checkpoint following UVA irradiation but show normal levels of micronuclei following oxidative stress or mitomycin C treatment. *International Journal of Radiation Oncology Biology Physics* **58** 470–478.

Shriver MD, Parra EJ, Dios S, Bonilla C, Norton H, Jovel C, Pfaff C, Jones C, Massac A, Cameron N, Baron A, Jackson T, Argyropoulos G, Jin L, Hoggart CJ, McKeigue PM and Kittles RA 2003 Skin pigmentation, biogeographical ancestry and admixture mapping. *Human Genetics* **112** 387–399.

Shriver MD, Smith MW, Jin L, Marcini A, Akey JM, Deka R and Ferrell RE 1997 Ethnic-affiliation estimation by use of population-specific DNA markers. *American Journal of Human Genetics* **60** 957–964.

Shull GH 1911 Experiments with maize. *American Naturalist* **45** 234–252.

Sickle Cell Anemia in Cuba and its Prevention. 1974 *Revista Cubana de Pediatría* **46** 2 and 3.

Silengo MC, Bell GL, Biagioli M and Franceschini P 1987 Oro-facial-digital syndrome II. Transitional type between the Mohr and the Majewski syndromes: report of two new cases. *Clinical Genetics* **31** 331–336

Simpson NE 1966 Factors influencing cholinesterase activity in a Brazilian population. *American Journal of Human Genetics* **18** 243–252.

Sinclair AH, Berta P, Palmer MS, Hawkins JR, Griffiths BL, Smith MJ, Foster JW, Frischauf AM, Lovell-Badge R and Goodfellow PN 1990 A gene from the human sex determining region encodes a protein with homology to a conserved DNA-binding motif. *Nature* **346** 240–244.

Singh NP, McCoy MT, Tice RR and Schneider EL 1988 A simple technique for quantification of low levels of DNA damage in individual cells. *Experimental Cell Research* **175** 184–191.

Siniscalco M, Bernini L, Latte B and Motulsky AG 1961 Favism and thalassaemia in Sardinia and their relationship to malaria. *Nature* **190** 1179–1180.

Siniscalco M, Robledo R, Bender PK, Carcassi C, L. Contu L and Beck JC 1999 Population genomics in Sardinia: a novel approach to hunt for genomic combinations underlying complex traits and diseases. *Cytogenetics and Cell Genetics* **86** 148–152 and **87** 296.

Siniscalco MLB, Filippi G, Latte B, Khan M, Piomeli S and Rattazi M 1966 Population genetics of haemoglobin variants, thalassaemia and glucose-6-phosphate dehydrogenase deficiency, with particular reference to the malaria hypothesis. *Bulletin of the World Health Organisation* **34** 379–393.

Sjöroos M, Iitiä A, Ilonen J, Reijonen H and Lovgren T 1995 Triple-label hybridization assay for type-1 diabetes related HLA alleles. *Biotechniques* **18** 870–877.

Skaletsky H, Kuroda-Kawaguchi, Minx TPJ, Cordum JHS, Hillier LaD, Brown LG, Repping S, Pyntikova T, Ali J, Bieri T, Chinwalla A, Delehaunty A, Delehaunty K, Du H, Fewell G, Fulton L, Fulton R, Graves T, Hou S-H, Latrielle P, Leonard S, Mardis E, Maupin R, McPherson J, Miner T, Nash W, Nguyen C, Ozersky P, Pepin K, Rock S, Rohlfing T, Scott K, Schultz B, Strong C, Tin-Wollam A, Yang S-P, Waterston RH, Wilson RK, Rozen S and Page DC 2003 The male specific region of the human Y chromosome is a mosaic of discrete sequence classes. *Nature* **243** 825–837.

Slatis HM, Reis RH and Hoene RE 1958 Consanguineous marriages in the Chicago region. *American Journal of Human Genetics* **10** 446–464.

Slavotinek A, Shaffer LG and Shapira SK 1999 Monosomy 1p36. *Journal of Medical Genetics* **36** 657–663.

Slavotinek AM and Tifft CJ 2002 Fraser syndrome and cryptophthalmos: review of the diagnostic criteria and evidence for phenotypic modules in complex malformation syndromes. *Journal of Medical Genetics* **39** 623–633.

Slim R, Graia F, de Lurdes Pereira M, Nivelon A, Croquette M-F, Lacombe D, Vigneron J, Helias J, Broyer M, Callen DF, Haan EA, Weissenbach J, Lacroix B, Bellané-Chantelot C, Le Paslier D, Cohen D and Petit C 1994 A proposed new contiguous gene syndrome on 8q consists of branchio-oto-renal (BOR) syndrome, Duane syndrome, a dominant form of hydrocephalus and trapeze aplasia; implications for the mapping of the BOR gene. *Human Molecular Genetics* **3** 1859–1866.

Smith C 1975 Quantitative inheritance. Chapter 9 in *Textbook of Human Genetics.* (editors GR Fraser & O Mayo) Blackwell Scientific, Oxford.

Smith CA and Sinclair AH 2004 Sex determination: insights from the chicken. *Bio Essays* **26** 120–132.

Smith CAB 1957 Counting methods in genetical statistics. *Annals of Human Genetics* **21** 254–276.

Smith GF and Berg JM 1976 *Down's Anomaly.* Churchill Livingstone, London.

Smith JA, Ogunba EO and Francis TI 1970 Transmission of Australia Au(l) antigen by Culex mosquitoes. *Nature* **237** 231–232.

Smith RA and Gardner-Medwin D 1993 Orofaciodigital syndrome type III in two sibs. *American Journal of Medical Genetics* **30** 870–872.

Smith TR, Miller MS, Lohman KK, Case LD and Hu JJ 2003 DNA damage and breast cancer risk. *Carcinogenesis* **5** 883–889.

Smyth AL 1966 *John Dalton 1766–1844, A bibliography of works by and about him.* Manchester University Press, Manchester.

Smyth I, Du X, Taylor MS, Justic MJ, Beutler B and Jackson IJ 2004 The extracellular matrix gene Frem1 is essential for the normal adhesion of the embryonic epidermis. *Proceedings of the National Academy of Sciences USA* **101** 13560–13565.

Snijders RJ, Sundberg K, Holzgreve W, Henry G and Nicolaides KH 1999 Maternal age- and gestation-specific risk for trisomy 21. *Ultrasound in Obstetrics and Gynecology* **13** 167–170.

Sokal RR and Rohlf FJ 1995 *Biometry.* Third edition. WH Freeman & Co., New York.

Solis R, Heredero L, Carlos N, Quintana J, Casanave J, Días W, Martínez A, Orraca M, Becker D, Coto R, Estrada R, Rosquete R, Bencomo F, Barrios M, Rodríguez L, Fernández J 2001 Detección sistemática de riesgo prenatal de cromosomopatías (Syndrome Down) mediante la evaluación combinada de HCG y AFP en suero materno y edad de la gestante. Unpublished. Final report of a research project.

Sommer A and Bartholomew DW 2003 Craniofacial-deafness-hand Syndrome Revisited. *American Journal of Medical Genetics* **123A** 91–94.

Sorlie T, Perou CM, Tibshirani R, Aas T, Geisler S, Johnsen H, Hastie T, Eisen MB, van de Rijn M, Jeffery SS, Thorsen T, Quist H, Matese JC, Brown PO, Botstein D, Eystein Lonning P, and Borresen-Dale AL 2001 Gene expression patterns of breast carcinomas distinguish tumor subclasses with clinical implications. *Proceedings of the National Academy of Sciences USA* **98** 10869–10874.

Sorlie T, Tibshirani R, Parker J, Hastie T, Marron JS, Nobel A, Deng S, JohnsenH, Pesich R, Geisler S, Dementer J, Perou CM, Lonning PE, Borresen-Dale AL and Botstein D 2003 Repeated observation of breast tumor subtypes in independent gene expression data sets. *Proceedings of the National Academy of Sciences USA* **100** 8418–8423.

Sotiriou C, Neo SY, McShane LM, Korn EL, Long PM, Jazaeri A, Martiat P, Fox SB, Harris AL and Liu ET 2003 Breast cancer classification and prognosis based on gene expression profiles from a population-based study. *Proceedings of the National Academy of Sciences USA* **100** 10393–10398.

Souza RLR, Fadel-Picheth C, Allebrandt KV, Furtado L and Chautard-Freire-Maia EA 2005a Possible influence of *BCHE* locus of butyrylcholinesterase on stature and body mass index. *American Journal of Physical Anthropology* **126** 329–334.

Souza RLR, Mikami LR, Maegawa ROB and Chautard-Freire-Maia EA 2005b Four new mutations in the *BCHE* gene of human butyrylcholinesterase in a Brazilian blood donor sample. *Molecular Genetics and Metabolism* **84** 349–353.

Spandorfer SD, Davis OK, Barmat LI, Chung PH and Rosenwaks Z 2005 Relationship between maternal age and aneuploidy in in vitro fertilization pregnancy loss. *Fertility and Sterility* **81** 1265–1269.

Spencer B and Gillen FJ 1899 *The Native Tribes of Central Australia*. Macmillan & Co, London.

Spencer JA, Sinclair AH, Watson JM and Graves JA 1991 Genes on the short arm of the human X chromosome are not shared with the marsupial X. *Genomics* **11** 339–345.

Stallcup WB 2002 The NG2 proteoglycan: past insights and future prospects. *Journal of Neurocytology* **31** 423–435.

Staub E, Hinzmann B and Rosenthal A 2002 A novel repeat in the melanoma-associated chondroitin sulfate proteoglycan defines a new protein family *FEBS Letters* **527** 114–118.

Stedman E, Stedman E and Easson LH 1932 CCXLV. Choline-esterase. An enzyme present in the blood-serum of the horse. *Biochemical Journal* **26** 2056–2066.

Stegmuller J, Werner H, Nave KA and Trotter J 2003 The proteoglycan NG2 um, R. H. 1988 Fraser syndrome is complexed with alpha-amino-3-hydroxy-5-methyl-4-isoxazolepropionic acid (AMPA) receptors by the PDZ glutamate receptor interaction protein (GRIP) in glial progenitor cells. Implications for glial-neuronal signalling. *Journal of Biological Chemistry* **278** 3590–3598.

Steiner PD and Davies JNP 1957 Cirrhosis and primary liver carcinoma in Uganda Africans. *British Journal of Cancer* **11** 523–534.

Stene E, Stene J and Stengel-Rutkowski S 1987 A reanalysis of the New York State prenatal diagnosis data on Down's syndrome and paternal age effects. *Human Genetics* **77** 299–302.

Stenson PD, Ball EV, Mort M, Phillips AD, Shiel JA, Thomas NS, Abeysinghe S, Krawczak M, Cooper DN 2003 Human Gene Mutation Database (HGMD): 2003 update. *Human Mutation* **21** 577–581.

Stephan MJ, Brooks KL, Moore DC, Coll EJ and Goho C 1994 Hypothalamic hamartoma in oral-facial-digital syndrome type VI (Váradi syndrome). *American Journal of Medical Genetics* **51** 131–136.

Stern C 1943 The Hardy-Weinberg law. *Science* **97** 137–138.

Stern C 1973 *Principles of Human Genetics*. W. H. Freeman and Company, San Francisco (3rd edition).

Stern C, Carson G, Kinst M, Novitsky E and Uphoff D 1952 The viability of heterozygotes for lethals. *Genetics* **37** 413–449.

Stevens CA, McClanahan C, Steck A, Shiel F O'M, Carey JC 1994 Pulmonary hyperplasia in the Fraser cryptophthalmos syndrome. *American Journal of Medical Genetics* **52** 427–431.

Stevens NM 1905 Studies in spermatogenesis, with special reference to the accessory chromosome. *Carnegie Institute of Washington Publications* **36** 1–32.

Stevenson AC and Davison BCC 1976 *Genetic Counseling*. J.B. Lippincott Company, Philadelphia.

Stickler GB and Pugh DG 1967 Hereditary progressive arthro-ophthalmopathy II. Additional observations on vertebral abnormalities, a hearing defect, and a report of a similar case. *Mayo Clinic Proceedings* **42** 495–500.

Stickler GB, Belau PG, Farrell FJ, Jones ID, Pugh DG, Steinberg AG and Ward LE 1965 Hereditary progressive arthro-ophthalmopathy. *Mayo Clinic Proceedings* **40** 433–455.

Stoppa-Lyonnet D, Ansquer Y, Dreyfus H, Gautier C, Gauthier-Villars M, Bourstyn E, Clough KB, Magdelenat H, Pouillart P, Vincent-Salomon A, Fourquet A and Asselain B 2000 Familial invasive breast cancers: worse outcome related to BRCA1 mutations. *Journal of Clinical Oncology* **18** 4053–4059

Stoutjesdijk MJ, Boetes C, Jager GJ, Beex L, Bult P, Hendriks JH, Laheij RJ, Massuger L, van Die LE, Wobbes T and Barentsz JO 2001 Magnetic resonance imaging and mammography in women with a hereditary risk of breast cancer. *Journal of the National Cancer Institute* **93** 1095–1102.

Strachan T, Abitbol M, Davidson D and Beckmann JS 1997 A new dimension for the human genome project: towards comprehensive expression maps. *Nature Genetics* **16** 126–132.

Strom CM, Levin R, Strom S, Masciangelo C. Kuliev A and Verlinsky Y 2000 Neonatal outcome of preimplantation genetic diagnosis by polar body removal: the first 109 infants. *Pediatrics* **106** 650–653.

Strong LC 2003 The two-hit model for Wilms' tumor: where are we 30 years later? *Genes Chromosomes and Cancer* **38** 294–299.

Stueber-Odebrecht N, Chautard-Freire-Maia EA, Primo-Parmo SL and Carrenho JMX. 1985 Studies on the *CHE1* locus of serum cholinesterase and surnames in a sample from Santa Catarina (Southern Brazil). *Revista Brasileira de Genética* **8** 535–543.

Sturtevant AH 1913 The linear arrangement of six sex-linked factors in Drosophila, as shown by their mode of association. *Journal of Experimental Zoology* **14** 43–59.

Sturtevant AH 1965 *A History of Genetics.* Harper and Row, New York.

Sudo K, Maekawa M, Akizuki S, Magara T, Ogasawara H and Tanaka T 1997 Human butyrylcholinesterase L330I mutation belongs to a fluoride-resistant gene, by expression in human fetal kidney cells. *Biochemical and Biophysical Research Communications* **240** 372–375.

Sudo K, Maekawa M, Kanno T, Akizuki S and Magara T 1996 Three different point mutations in the butyrylcholinesterase gene of three Japanese subjects with a silent phenotype: possible Japanese type alleles. *Clinical Biochemistry* **29** 165–169.

Summers J, O'Connell A and Millman I 1975 Genome of hepatitis B virus: Restriction enzyme cleavage and structure of DNA extracted from Dane particles. *Proceedings of the National Academy of Sciences USA* **72** 4597–4601.

Suresh S, Rajesh K, Suresh I, Raja V, Gopish D and Gnanasoundari S 1995 Prenatal diagnosis of orofaciodigital syndrome: Mohr type. *Journal of Ultrasound Medicine* **14** 863–866.

Sutter J and Tabah L 1952 Effets de la consanguinité et de l'endogamie. Une enquête dans le Morbihan et le Loir-et-Cher (Effects of consanguinity and inbreeding. A survey in the regions of Morbihan and Loir-et-Cher). *Population* **7** 249–266.

Sved JA 1968 Possible rates of gene substitution in evolution. *American Naturalist* **102** 283–293.

Sved JA 1976 The relationship between genotype and fitness for heterotic models. In: *Population Genetics and Ecology.* (editors Karlin S and. Nevo E) Academic Press, NY.

Sved JA, Reed TE and Bodmer WF 1967 The number of balanced polymorphisms that can be maintained in a natural population. *Genetics* **55** 469–481.

Swarbrick MM and Vaisse C 2003 Emerging trends in the search for genetic variants predisposing to human obesity. *Current Opinion in Clinical Nutrition & Metabolic Care.* **6** 369–375.

Swiergiel JJ, Funderburgh JL, Justice MJ and Conrad GW 2000 Developmental eye and neural tube defects in the eye blebs mouse. *Developmental Dynamics* **219** 21–27.

Syrris P, Carter ND and Patton MA 1999 Novel missense mutation of the endothelin-B receptor gene in a family with Waardenburg-Hirschsprung disease. *American Journal of Medical Genetics* **87** 69–71.

Takagi H, Narahara A, Takayama H, Shimoda R, Nagamine T and Mori M 1997 A new point mutation in cholinesterase: relationship between multiple mutation sites and enzyme activity. *International Hepatology Communications* **6** 288–293.

Takamiya K, Kostourou V, Adams S, Jadeja S, Chalepakis G, Scambler PJ, Huganir RL and Adams RH 2004 A direct functional link between the multi-PDZ domain protein GRIP1 and the Fraser syndrome protein Fras1. *Nature Genetics* **36** 172–175.

Talks SJ, Ebenezer N, Hykin P, Adams G, Yanf F, Schulenberg E, Gregory-Evans K and Gregory-Evans CY 2001 De novo mutations in the 5' regulatory region of the Norrie disease gene in retinopathy of prematurity. *Journal of Medical Genetics* **38** E46

Tanaka M, Schinke M, Liao HS, Yamasaki N and Izumo S 2001 Nkx2.5 and Nkx2.6, homologs of Drosophila tinman, are required for development of the pharynx. *Molecular and Cellular Biology* **21** 4391–4398.

Tanaka M, Yamasaki N and Izumo S 2000 Phenotypic characterization of the murine Nkx2.6 homeobox gene by gene targeting. *Molecular and Cellular Biology* **20** 2874–2879.

Tarani L, Lampariello S, Raguso G, Colloridi F, Pucarelli I, Pasquino AM and Bruni LA 1998 Pregnancy in patients with Turner's syndrome: six new cases and review of literature. *Gynecological Endocrinology* **12** 83–87.

Tassabehji M, Read AP, Newton V, Harris R, Balling R, Gruss P and Strachan T 1992 Waardenburg syndrome patients have mutations in the human homologue of the *Pax-3* paired box gene. *Nature* **355** 635–636.

Tay JS and Yip WC 1984 The estimation of inbreeding from isonymy: relationship to the average inbreeding coefficient. *Annals of Human Genetics* **48** 185–194.

Tease C and Fisher G 1991 The influence of maternal age on radiation-induced chromosome aberrations in mouse oocytes. *Mutation Research* **262** 57–62.

Tellier AL, Cormier-Daire V, Abadie V, Amiel J, Siguady S, Bonnet D, de Lonlay-Debeney P, Morrisseau MP, Hubert P, Michel JL, Jan D, Dollfus H, Baumann C, Labrune P, Lacombe D, Philip N, LeMerrer M, Briard ML, Munnich A and Lyonnet S 1998 CHARGE syndrome: report of 47 cases and review. *American Journal of Medical Genetics* **76** 402–409

Teranishi M, Shimada Y, Hori T, Nakabayashi O, Kikuchi T, Macleod T, Pym R, Sheldon B, Slovei I, Macgregor H and Mizuno S 2001 Transcripts of the MHM regions on the chicken Z chromosome accumulate as non-coding RNA in the nucleus of female cells and adjacent to the DMRT1 locus. *Chromosome Research* **9** 147–165.

Terwilliger JD and Weiss KM 1998 Linkage disequilibrium mapping of complex disease: fantasy or reality? *Current Opinion in Biotechnology* **9** 578–594.

Testart J and Séle B 1995 Towards an efficient medical eugenics: is the desirable always the feasible? *Human Reproduction* **10** 3086–3090.

The Breast Cancer Linkage Consortium 1999 Cancer risks in BRCA2 mutation carriers. *Journal of the National Cancer Institute* **91** 1310–1316.

The Huntington's Disease Collaborative Research Group. 1993 A novel gene containing a trinucleotide repeat that is expanded and unstable on Huntington's disease chromosomes. *Cell* **72** 971–983.

The International HapMap Consortium 2005 A haplotype map of the human genome. *Nature* **437** 1299–1320.

The Report of the Chief Medical Officer's Expert Group on Therapeutic Cloning – 'Stem Cell Research: Medical Progress with Responsibility' Department of Health 16 August 2000 *http://www.dh.gov.uk*

Thomas G 2002 Furin at the cutting edge: from protein traffic to embryogenesis and disease. *Nature Reviews Molecular Cellular Biology* **3** 753–766.

Thomas IT, Frías JL, Felix V, Sanchez de Leon L, Hernandez RA and Jones MC 1986 Isolated and syndromic cryptophthalmos. *American Journal of Medical Genetics* **25** 85–98.

Thompson D and Easton D 2002b Breast Cancer Linkage Consortium. Variation in BRCA1 cancer risks by mutation position. *Cancer Epidemiology Biomarkers & Prevention* 2002 **11** 329–336.

Thompson D and Easton DF 2002a Breast Cancer Linkage Consortium. Cancer incidence in BRCA1 mutation carriers. *Journal of the National Cancer Institute* **94** 1358–1365.

Tice RR, Agurell E, Anderson D, Burlinson B, Hartmann A, Kobayashi H, Miyamae Y, Rojas E, Ryu JC and Sasaki YF 2000 Single Cell Gel/Comet Assay: Guidelines for in vitro and in vivo genetic toxicology testing. *Environmental and Molecular Mutagenesis* **35** 206–221.

Tierney P 2000 *Darkness in El Dorado: How Scientists and Journalists Devastated the Amazon*. Norton, New York.

Tilanus-Linthorst MM, Kriege M, Boetes C, Hop WC, Obdeijn IM, Oosterwijk JC, Peterse HL, Zonderland HM, Meijer S, Eggermont AM, de Koning HJ, Klijn JG and Brekelmans CT 2005 Hereditary breast cancer growth rates and its impact on screening policy. *European Journal of Cancer* **41** 1610–1617.

Tillet E, Gential B, Garrone R and Stallcup WB 2002 NG2 proteoglycan mediates beta1 integrin-independent cell adhesion and spreading on collagen VI.. *Journal of Cellular Biochemistry* **86** 726–736.

Tiret L, Nicaud V, Ehnholm C, Havekes L, Menzel H, Ducimetiere P and Cambien F 1993 Inference of the strength of genotype-disease association from studies comparing offspring with and without parental history of disease. *Annals of Human Genetics* **57** 141–149.

Tishkoff SA and Verrelli BC 2003 Role of evolutionary history on haplotype block structure in the human genome: implications for disease mapping. *Current Opinion in Genetics and Development* **13** 569–575.

Tittel K 1901 Über eine angeborene Mißbildung des Dickdarms. *Klinische Wochenschrift* **14** 903–907.

Tjio JH and Levan A 1956 The chromosome number of man. *Hereditas* **42** 1–6.

Todd JA, Bell JL and McDevitt HO 1987 HLA-DQ beta gene contributes to susceptibility and resistance to insulin-dependent diabetes mellitus. *Nature* **329** 599–604.

Tomasetti M, Alleva R, Borghi B and Collins AR 2001 *In vivo* supplementation with coenzyme Q10 enhances the recovery of human lymphocytes from oxidative DNA damage. *FASEB Journal* **15** 1425–1427.

Tomatis L, Aitio A, Day N, Heseltine E, Kaldor J, Miller A, Parkin D and Riboli E eds. 1990 *Cancer: Causes, Occurrence and Control*. International Agency for Research on Cancer, Lyon. IARC Scientific Publications No. 100.

Tomlinson IP and Bodmer WF 1995 The HLA system and the analysis of multifactorial genetic disease. *Trends in Genetics* **11** 493–498.

Toomes C, Downey M, Bottomley HM, Mintz-Hittner H and Inglehearn CF 2005 Further evidence ofgenetic heterogeneity in familial exudative vitreoretinopathy; exclusion of EVR1, EVR3, and EVR4 in a large autosomal dominant pedigree. *British Journal of Ophthalmology* **89** 194–197.

Topaloglu H 2005 An Autosomal recessive limb girdle muscular dystrophy (LGMD2) with mild mental retardation is allelic to Walker-Warburg syndrome(WWS) caused by a mutation in the POMT1 gene. *Neuromuscular disorders*. **25** 271–275.

Toriello HV 1988 Heterogeneity and variability in the oral-facial-digital syndromes. *American Journal of Medical Genetics* Suppl 4 149–159.

Toriello HV 1993 Oral-facial-digital syndromes, 1992. *Clinical Dysmorphology* **2** 95–105.

Tot T 2000 The cytokeratin profile of medullary carcinoma of the breast. *Histopathology* **37** 175–181.

Tournamille C, Colin Y, Cartron JP and Le Van Kim C 1995 Disruption of a GATA motif in the *Duffy* gene promoter abolishes erythroid gene expression in Duffy-negative individuals. *Nature Genetics* **10** 224–228.

Towbin JA and Friedman RA 1998 Prolongation of the QT interval and the sudden infant death syndrome. *New England Journal of Medicine* **338** 1760–1761.

Tranebjaerg L, Sjö O and Warburg M 1986 Retinal cone dysfunction and mental retardation associated with a de novo balanced translocation 1;6(q44;q27). *Ophthalmic Paediatrics and Genetics* **7** 167–173.

Treacher Collins E 1900 Case with symmetrical congenital notches in the outer part of each lid and defective development of the malar bones. *Transactions of the Ophthalmology Society UK* **20** 190.

Tsao MS, Sakurada A, Cutz JC, Zhu CQ, Kamel-Reid S, Squire J, Lorimer I, Zhang T, Liu N, Daneshmand M, Marrano P, da Cunha Santos G, Lagrade A, Richardson F, Seymour L, Whitehead M, Ding K, Pater J and Shepherd FA 2005 Erlotinib in lung cancer – molecular and clinical predictors of outcome. *New England Journal of Medicine* **353** 133–144.

Tschöp M, Smiley DL and Heiman ML 2000 Ghrelin induces adiposity in rodents. *Nature* **407** 908–913.

Tschöp M, Weyer C, Tataranni PA, Devanarayan V, Ravussin E and Heiman ML 2001 Circulating ghrelin levels are decreased in human obesity. *Diabetes* **50** 707–709.

Tsuda H, Takarabe T, Hasegawa F, Fukutomi T and Hirohashi S 2000 Large, central acellular zones indicating myoepithelial tumor differentiation in high-grade invasive ductal carcinomas as markers of predisposition to lung and brain metastases. *American Journal of Surgical Pathology* **24** 197–202.

Tupler R and Gabellini D 2004 Molecular basis of facioscapulohumeral muscular dystrophy. *Cellular and Molecular Life Sciences* **61** 557–566.

Turney J 1998 *Frankenstein's Footsteps. Science, Genetics, and Popular Culture*. New Haven, Yale University Press, 192–193.

Tutt A and Ashworth A 2002 The relationship between the roles of BRCA genes in DNA repair and cancer predisposition. *Trends in Molecular Medicine* **8** 571–576.

Twining P, McHugo JM and Pilling DW 2000 *Textbook of Fetal Abnormalities*. Churchill Livingstone, London.

Vahteristo P, Bartkova J, Eerola H, Syrjakoski K, Ojala S, Kilpivaara O, Tamminen A, Kononen J, Aittomak, K, Heikkila P, Holli K, Blomqvist C, Bartek J, Kallioniemi OP and Nevanlinna H 2002 A CHEK2 genetic variant contributing to a substantial fraction of familial breast cancer. *American Journal of Human Genetics* **71** 432–438.

Vaidya B, Coffey R, Coyle B, Trembath R, San Lazaro C, Reardon W and Kendall-Taylor P 1999 Concurrence of Pendred syndrome, autoimmune thyroiditis and simple goiter in one family. *Journal of Clinical Endocrinology and Metabolism* **84** 2736–2738.

van de Velde H, De Vos A, Sermon K, Staessen C, De Rycke M, Van Assche E, Lissens W, Vandervorst M, Van Ranst H, Liebaers I and Van Steirteghem A 2000 Embryo implantation after biopsy of one or two cells from cleavage-stage embryos with a view to preimplantation genetic diagnosis. *Prenatal Diagnosis* **20** 1030 – 1037.

van der Groep P, Bouter A, van der Zanden R, Menko FH, Buerger H, Verheijen RH, van der Wall E and van Diest PJ. 2004 Re: Germline BRCA1 mutations and a basal epithelial phenotype in breast cancer. *Journal of the National Cancer Institute* **96** 712–713.

Van Doorn L, Figueiredo C Sanna R, Plaisier A, Schneeberger P, De Boer W and Quint W 1998 Clinical relevance of the *cagA, vacA, iceA* status of *H. pylori*. *Gastroenterology* B 58–66.

Van Driessche N, Demsar J, Booth EO, Hill P, Juvan P, Zupan B, Kuspa A and Shaulsky G 2005 Epistasis analysis with global transcriptional phenotypes. *Nature Genetics* **37** 471–477.

van Montfrans JM, Dorland M, Oosterhuis GJ, van Vugt JM, Rekers-Mombarg LT and Lambalk CB 1999 Increased concentrations of follicle-stimulating hormone in mothers of children with Down's syndrome. *The Lancet* **353** 1853–1854.

van Niekerk WA and Retief AE 1981 The gonads of human true hermaphrodites. *Human Genetics* **58** 117–122.

Van Reeuwijk J, Janssen M, van den Elzen C, Beltran-Valero de Bernabe D, Sabatelli P, Merlini L, Boon M, Scheffer, Brockington M, Muntoni F, Huynen M, Verrips A, Walsh C, Barth P, Brunner H and van Bokhoven H 2005 POMT2 mutations cause alpha-dystrogloycan hypoglycosylation and Walker-Warburg syndrome. *Journal of Medical Genetics* **42** 907–912.

Váradi V, Szabó L and Papp Z 1980 Syndrome of polydactyly, cleft lip/palate or lingual lump, and psychomotor retardation in endogamic gypsies. *Journal of Medical Genetics* **17** 119–122.

Varnum DS and Fox SC 1981 Head blebs: a new mutation on chromosome 4 of the mouse. *Journal of Heredity* **72** 293.

Veldhuyzen van Zanten SJ, Pollak PT, Best LM, Bezanson GS and Marrie T 1994 Increasing prevalence of *Helicobacter pylori* with age: continuous risk of infection in adults rather than cohort effect. *Journal of Infection Disease* **169** 434–437.

Velluz L 1969 *Maupertuis* Librairie Hachette, Paris.

Venkitaraman AR 2002 Cancer susceptibility and the functions of BRCA1 and BRCA2. *Cell* **108** 171–182.

Venkitaraman AR 2003 A growing network of cancer-susceptibility genes. *New England Journal of Medicine* **348** 1917–1919.

Vergel G, Sánchez R, Heredero L, Rodríguez L and Martínez A 1990 Primary prevention of neural tube defects with folic acid supplementation: Cuban experience. *Prenatal Diagnosis* **10** 140–152.

Verheij JBG, Kunze J, Osinga J van Essen AJ and Hofstra RMW 2002 ABCD syndrome is caused by a homozygous mutation in the EDNRB gene. *American Journal of Medical Genetics* **108** 223–225.

Verloes A 1995 Numerical syndromology: a mathematical approach to the nosology of complex phenotypes. *American Journal of Medical Genetics* **55** 433–443.

Vidal VPI, Chaboissier M-C, de Rooij D and Schedl A 2001 *Sox9* induces testis development in XX transgenic mice. *Nature Genetics* **28** 216–217.

Vieira FJ 1992 *Riscos de doença genética na prole de consangüíneos* (Risks of genetic disease in the offspring of consanguineous parents). PhD Thesis, DB IB USP, São Paulo.

Vierucci A, Bianchini AM, Morgese G, Bagnoli F and Messina G 1968 L'antigen Australia. 1. Rapporti con l'epatite infettiva e da siero. Una ricerca in pazienti pediatrici. *Pediatria Internazione* **18** No. 4.

Vilain E 2002 Anomalies of human sexual development: clinical aspects and genetic analysis. In: *The Genetics and Biology of Sex Determination* (Novartis Foundation Symposium 244) pp. 43–56. John Wiley & Sons, Chichester.

Vincent C, Kalatzis V, Compain S, Levilliers J, Slim R, Graia F, de Lurdes Pereira M, Nivelon A, Croquette M-F, Lacombe D, Vigneron J, Helias J, Broyer M, Callen DF, Haan EA, Weissenbach J, Lacroix B, Bellané-Chantelot C, Le Paslier D, Cohen D and Petit C 1994 A proposed new contiguous gene syndrome on 8q consists of branchio-oto-renal (BOR) syndrome, Duane syndrome, a dominant form of hydrocephalus and trapeze aplasia; implications for the mapping of the BOR gene. *Human Molecular Genetics* **3** 1859–1866.

Viscardi H Jr 1959 *Give Us the Tools; with an introduction by Eleanor Roosevelt.* Eriksson-Taplinger Co., New York.

Viscardi H Jr 1962 ' … *a letter to Jimmy*'. Paul S. Eriksson, New York.

Vissers LELM, van Ravenswaaij CMA, Admiraal R, Hurst JA, de Vries BBA, Janssen IM, van der Vliet WA, Huys EHLPG, de Jong PJ, Hamel BC, Schoenmakers EFPM, Brunner HG, Veltman JA, van Kessel AG 2004. Mutations in a new member of the chromodomain gene family cause CHARGE syndrome. *Nature Genetics* **36** 955–7.

Vlad MO, Cavalli-Sforza LL, and Ross J 2004 Enhanced (hydrodynamic) transport induced by population growth in reaction-diffusion systems with application to population genetics. *Proceedings of the National Academy of Sciences USA* **101** 10249–10253.

Vogelstein B and Kinzler KW eds. 2002 *The Genetic Basis of Human Cancer.* McGraw-Hill, New York.

Volff JN, Kondo M and Schartl M 2003 Medaka dmY/dmrtY is not the universal primary sex-determining gene in fish. *Trends in Genetics* **19** 196–199.

Volpe P, Paladini D, Marasini M, Buonadonna AL, Russo MG, Caruso G, Marzullo A, Vassallo M, Martinelli P and Gentile M 2003 Common arterial trunk in the fetus: characteristics, associations, and outcome in a multicentre series of 23 cases. *Heart* **89** 1437–1441.

von Hippel E 1906 (Abstract presented in 1904). Kryptophthalmus congenitus. *Albrecht Von Graefes Archiv für klinische und experimentelle Ophthalmologie* **63** 25–38.

von Mering C, Krause R, Snel B, Cornell M, Oliver SG, Fields S and Bork P 2002 Comparative assessment of large-scale data sets of protein-protein interactions. *Nature* **417** 399–403.

Vrontou S, Petrou P, Meyer BI, Galanopoulos VK, Imai K, Yanagi M, Chowdhury K, Scambler PJ, Chalepakis G 2003 *Fras1* deficiency results in cryptophthalmos, renal agenesis and blebbed phenotype in mice. *Nature Genetics* **34** 209–214.

Waardenburg PJ 1932 *Das menschliche Auge and seine Erblangen*, pp. 47–48 Martinus Nijhoff, the Hague. (also published as *Bibliographia Genetica* VII, 1932)

Waardenburg PJ 1951 A new syndrome combining developmental anomalies of the eyelids, eyebrows and nose root with pigmentary defects of the iris and head hair and with congenital deafness. Dystopia canthi medialis et punctorum lacrimalium lateroversa, hyperlasia supercilii nedialis et radicis nasi, heterochromia iridium totalis sive partialis, albinismus circumscriptus (surdimutitas). *American Journal of Human Genetics* **3** 195–253.

Wagner T, Wirth J, Meyer J, Zabel B, Held M, Zimmer J, Pasantes J, Dagna Bricarelli F, Keutel J, Hustert E, Wolf U, Tommerup N, Schempp W and Scherer G 1994 Autosomal sex reversal and campomelic dysplasia are caused by mutations in and around the *SRY*-related gene *SOX9*. *Cell* **79** 1111–1120.

Wakamatsu N, Yamada Y and Yamada K 2001 Mutations in SIP1, encoding Smad interacting protein-1, cause a form of Hirschsprung disease. *Nature Genetics* **27** 369–370.

Wald NJ and Cuckle HS 1987 Recent advances in screening for neural tube defects and Down's syndrome. In: *Baillière's Clinical Obstetrics and Gynaecology* (editor Rodeck CH) Foetal Diagnosis of Genetic Defects, Vol.1(3). Baillière Tindall, London pp. 649–676.

Wall JD and Pritchard JK 2003 Haplotype blocks and linkage disequilibrium in the human genome. *Nature Reviews Genetics* **4** 587–597.

Wallace B 1970 *Genetic Load, its Biological and Conceptual Aspects.* Prentice-Hall, New York.

Wang D, Johnson AD, Papp AC, Kroetz DL and Sadee W 2005 Multidrug resistance polypeptide 1 (MDR1, ABCB1) variant 3435C>T affects mRNA stability. *Pharmacogenetics and Genomics* **15** 693–704.

Wang WYS, Barratt BJ, Clayton DG and Todd JA 2005 Genome wide association studies: Theoretical and practical concerns. *Nature Reviews Genetics* **6** 109–118

Warburg M 2001 Visual impairment in adult people with moderate, severe, and profound intellectual disability. *Acta Ophthalmologica Scandinavica* **79** 450–454.

Warburg M 1978 Hydrocephaly, congenital retinal nonattachment, and congenital falciform folds. *American Journal of Ophthalmology* **85** 88–94.

Warburg M, Frederiksen P and Rattleff J 1979 Blindness among 7700 mentally retarded children in Denmark. *Clinics in Developmental Medicine* **73** 68–75.

Warburg M, Sjö O and Tranebjerg L 1991a Tapetoretinal dystrophy and mental retardation. In: *Degenerative retinopathies: Advances in Clinical and Genetic Research.* (editors Humphries P, Bhattacharya S and Bird A) CRC Press, Chicago. pp. 163–169.

Warburg M, Sjo O, Fledelius HC and Pedersen SA 1993 Autosomal recessive microcephaly, microcornea, congenital cataract, mental retardation, optic atrophy, and hypogenitalism. Microsyndrome. *American Journal of Diseases of Childhood* **147** 1309–1312.

Warburg M, Sjö O, Tranebjerg L and Fledelius H 1991b Deletion mapping of a retinal cone-rod dystrophy: assignment to 18q211 *American Journal of Medical Genetics* **39** 288–293.

Warburton D, Dallaire L, Thangavelu M, Ross L, Levin B and Kline J 2004 Trisomy recurrence: a reconsideration based on North American data. *American Journal of Human Genetics* **75** 376–385.

Ward OC 1964 A new familial cardiac syndrome in children. *Journal of the Irish Medical Association* **54** 103–106.

Ward OC 2005 Long QT syndromes: The Irish dimension. *Irish Medical Journal* **98** 120–122.

Warkany J and Schraffenberger E 1944 Congenital malformation of the eyes induced in rats by maternal vitamin A deficiency. *Proceedings of the Society for Experimental Biology and Medicine* **57** 49–52.

Warner E, Plewes DB, Shumak RS, Catzavelos GC, Di Prospero LS, Yaffe MJ, Goel V, Ramsay E, Chart PL, Cole DE, Taylor GA, Cutrara M, Samuels TH, Murphy JP, Murphy JM and Narod SA 2001 Comparison of breast magnetic resonance imaging, mammography, and ultrasound for surveillance of women at high risk for hereditary breast cancer. *Clinical Oncology* **19** 3524–3531.

Warren AC, Chakravarti A, Wong C, Slaugenhaupt SA, Halloran SL, Watkins PC, Metaxotou C and Antonarakis SE 1987 Evidence for reduced recombination on the nondisjoined chromosomes 21 in Down syndrome. *Science* **237** 652–654.

Warren M, Lord CJ, Masabanda J, Griffin D and Ashworth A 2003 Phenotypic effects of heterozygosity for a BRCA2 mutation. *Human Molecular Genetics* 2003 **12** 2645–2656.

Waters PD, Delbridge ML, Deakin JE, El-Mogharbel N, Kirby PJ, Calvalho-Silva DR and Graves JAM 2005 Autosomal location of genes from the conserved mammalian X in the platypus (*Ornithorhynchus anatinus*): implications for mammalian sex chromosome evolution. *Chromosome Research* **13** 401–410.

Watson JD and Crick FHC 1953a A structure for deoxyribose nucleic acids. *Nature* **171** 737–738.

Watson JD and Crick FHC 1953b Genetical implications of the structure of deoxyribonucleic acid. *Nature* **171** 964–967.

Wayne S, Robertson NG, DeClau F, Chen N, Verhoeven K, Prasad S, Tranebjarg L, Morton CC, Ryan AF, Van Camp G and Smith RJ 2001 Mutations in the transcriptional activator EYA4 cause late-onset deafness at the DFNA10 locus. *Human Molecular Genetics* **10** 195–200.

Weatherall DJ 2001 Phenotype-genotype relationships in monogenic disease: lessons from the thalassaemias. *Nature Reviews Genetics* **2** 245–55.

Weatherall DJ and Clegg JB 2001 *The Thalassaemia Syndromes.* Blackwell Science, Oxford.

Weatherall DJ and Clegg JB 2002 Genetic variability in response to infection. Malaria and after. *Genes and Immunity* **3** 331–337.

Webb S, Qayyum SR, Anderson RH, Lamers WH and Richardson MK 2003 Septation and separation within the outflow tract of the developing heart. *Journal of Anatomy* **202** 327–342.

Weber F, Fukino K, Sawada T, Williams N, Sweet K, Brena RM, Plass C, Caldes T, Mutter GL, Villanona-Caler MA and Eng C 2005 Variability in organ-specific EGFR mutational spectra in tumour epithelium and stroma may be the biological basis for differential responses to tyrosine kinase inhibitors. *British Journal of Cancer* **92** 1922–1926.

Weber FP 1919 Patches of deep pigmentation of oral mucous membrane not connected with Addison's disease. *Quarterly Journal of Medicine* **12** 404.

Webster's New Collegiate Dictionary 1981 G & C Merriam Co., Springfield, MA.

Weinberg W 1908 Über den Nachweiss der Vererbung beim Menschen. *Jahreshefte des Vereins fur Vaterlandische Naturkunde in Wurttemburg* **64** 368–382.

Weir BS, Cardon LR, Anderson AD, Nielsen DM and Hill WG 2005 Measures of human population structure show heterogeneity among genomic regions. *Genome Research* **15** 1468–1476.

Weismann A 1889 *Essays upon Heredity and Kindred Biological Problems*. Clarendon Press, Oxford.

Weismann A 1893 *The Germ-Plasm: A Theory of Heredity*. Charles Scribner's Sons, New York.

Werner B and London WT 1975 Host responses to hepatitis B infection: Hepatitis B surface antigen and host proteins. *Annals of Internal Medicine* **83** 113–114.

Wertz DC 1998 Eugenics is alive and well: a survey of genetic professionals around the world. *Science in Context* **11** 493–510.

Westergaard HB, Johansen AM, Erb K and Andersen AN 1999 Danish National In-Vitro Fertilization Registry 1994 and 1995: a controlled study of births, malformations and cytogenetic findings. *Human Reproduction* **14** 1896–1902.

Wey PD, Neidich JA, Hoffman LA and LaTrenta GS 1994 Midline defects of the orofaciodigital syndrome type VI (Váradi syndrome). *Cleft Palate-Craniofacial Journal* **31** 397–400.

Whitehouse FR and Kernohan JW 1948 Myenteric plexus in congenital megacolon. Study of eleven cases. *Archives of Internal Medicine* **82** 75–111.

Whitfield JB, Zhu G, Nestler JE, Heath AC,and Martin NG 2002 Genetic covariation between serum gamma glutamyl transferase activity and cardiovascular risk factors. *Clinical Chemistry* **48** 1426–1431.

Whittaker M, Britten JJ and Wicks RJ 1981 Inhibition of the plasma cholinesterase variants by propranolol. *British Journal of Anaesthesia* **53** 511–516.

Whittaker M 1986 *Cholinesterase*. Monographs in Human Genetics, vol. 11, (editor Beckman L) Karger, Basel.

Whittemore AS, Gong G and Itnyre J 1997 Prevalence and contribution of BRCA1 mutations in breast cancer and ovarian cancer: results from three U.S. population-based case-control studies of ovarian cancer. *American Journal of Human Genetics* **60** 496–504.

Whitteridge G 1976 *An Anatomical Disputation concerning the Movement of the Heart and Great Vessels in Animals* (Translation of *Ezercitatio Anatomica de Motu Cordis et Sanguinis in Animalibus*. William Harvey 1628). Blackwell Scientific Publications, Oxford.

Wietnauer E, Ebert C, Hucho F, Robitzki A, Weise C and Layer PG 1999 Butyrylcholinesterase is complexed with transferrin in chicken serum. *Journal of Protein Chemistry* **18** 205–214.

Wikler W and Palmer E 1992 Neo-eugenics and disability rights in philosophical perspective. In: *Human Genome Research and Society. Proceedings of the Second International Bioethics Seminar in Fukui* (editors Fujuki N and Macer DRJ) http://zobell.biol.tsukuba.ac.jp/-macer/HGR/HGRDW.html

Wildervanck LS 1952 Een geval van aandoening van Klippel-Feil, gecombineerd med abducensparalyse, retractio bulbi en doofstomheid. (Kippel-Feil syndrome associated with abducens paralysis, bulbar retraction and deaf-mutism) *Nederlands tijdschrift voor geneeskunde* **96** 2752.

Wildervanck LS 1963 Perceptive deafness associated with split-hand and -foot -a new syndrome. *Acta Genetica et Statistica Medica* **13** 161–169.

Williams TN, Maitland K, Bennett S, Ganczakowski M, Peto TEA, Newbold CI, Bowden DK, Weatherall DJ and Clegg JB 1996 High incidence of malaria in α-thalassaemic children. Nature **383** 522–525.

Williams TN, Mwangi TW, Roberts DJ, Alexander ND, Weatherall DJ, Wambua S, Kortok M, Snow RW and Marsh K 2005 An immune basis for malaria protection by the sickle cell trait. *PLoS Medicine* **2** e128.

Willis R 1847 *The Works of William Harvey MD*. Translated from the Latin with a Life of the Author. Sydenham Society, London.

Wills C 1978 Rank-order selection is capable of maintaining all genetic polymorphisms. *Genetics* **89** 403–417.

Wills W, Larouzé B, London WT, Blumberg BS, Millman I, Pourtaghra M and Coz J 1976a Hepatitis B surface antigen in West African mosquitoes and bedbugs. Abstract *25th Annual Joint Meeting of the American Society of Tropical Medicine and Hygiene and the Royal Society of Tropical Medicine and Hygiene*, Philadelphia, Pennsylvania, 3 to 5 November 1976.

Wills W, Saimot G, Brochard C, Blumberg BS, London WT, Dechene R and Millman I 1976b Hepatitis B surface antigen (Australia antigen) in mosquitoes collected in Senegal, West Africa. *American Journal of Tropical Medicine* 25 186–190.

Wilson EB 1909 Recent researches on the determination and heredity of sex. *Science* 29 52–70.

Wilson EB 1911 The sex chromosomes. *Archiv für Anatomie und Entwicklungsmechanismus* 77 249–271.

Wilson GM, Vasa MZ and Deeley RG 1998 Stabilization and cytoskeletal-association of LDL receptor mRNA mediated by distinct domains in its 3' untranslated region. *Journal of Lipid Research* 39 1025–1032.

Wilson M, Mowat D, Dastot-Le Moal F, Cacheux V, Kaariainen H, Cass D, Donnai D, Clayton-Smith J, Townshend S, Curry C, Gattas M, Braddock S, Kerr B, Aftimos S, Zehnwirth H, Barrey C and Goossens M 2003 Further delineation of the phenotype associated with heterozygous mutations in ZFHX1B. *American Journal of Medical Genetics* 119 257–265.

Winawer SJ, Zauber AG, Ho MN, O'Brien MJ, Gottlieb LS, Sternberg SS, Waye JD, Schapiro M, Bond JH, Panish JF, Ackroyd F, Shike M, Kurtz RC, Hornsby-Lewis L, Gerdes H, Stewart ET and The National Polyp Study Workgroup. 1993 Prevention of colorectal cancer by colonoscopic polypectomy. The national polyp study workgroup. *New England Journal of Medicine* 329 1977–1981.

Winston RML and Hardy K 2002 Are we ignoring potential dangers of in vitro fertilization and related treatments? *Nature Cell Biology* 4 S14-S18.

Winter RM 1988 Malformation syndromes: a review of mouse/human homology. *Journal of Medical Genetics* 25 480–487.

Winter RM 1990 Fraser syndrome and mouse 'bleb' mutants. *Clinical Genetics* 37 494–495.

Winter RM and Baraitser M 1991 *Multiple Congenital Anomalies. A Diagnostic Compendium.* Chapman and Hall Medical Publications, London.

Wittbrodt J, Shima A and Schartl M 2002 Medaka – a model organism from the far east. *Nature Reviews in Genetics* 3 53–64.

Wollnik B, Tukel T, Uyguner O, Ghanbari A, Kayserili H, Emiroglu M, Yuksel-Apak M 2003 Homozygous and heterozygous inheritance of PAX3 mutations causes different types of Waardenburg syndrome. *American Journal of Medical Genetics* 122A 42–45.

Wolstenholme J and Angell RR 2000 Maternal age and trisomy – a unifying mechanism of formation. *Chromosoma* 109 435–438. Erratum in: 2001. *Chromosoma* 110 130.

Woolf B 1955 On estimating the relation between blood group and disease. *Annals of Eugenics* 19 251–253.

Woolfe A, Goodson M, Goode DK, Snell P, McEwen GK, Vavouri T, Smith SF, North P, Callaway H, Kelly K, Walter K, Abnizova I, Gilks W, Edwards YJK, Cooke JE and Elgar G 2005 Highly conserved non-coding sequences are associated with vertebrate development. *PLoS Biology* Jan;3(1):e7

Wooster R, Bignell G, Lancaster J, Swift S, Seal S, Mangion J, Collins N, Gregory S, Gumbs C and Micklem G 1995 Identification of the breast cancer susceptibility gene BRCA2. *Nature* 378 789–792.

World Atlas of Birth Defects 2003 2nd edition. International Centre for Birth Defects of the International Clearinghouse for Birth Defects Monitoring Systems. Geneva, Switzerland.

Wright CI, Geula C and Mesulam MM 1993 Protease inhibitors and indoleamines selectively inhibit cholinesterases in the histopathologic structures of Alzheimer disease. *Proceedings of the National Academy of Sciences USA* 90 683–686.

Wright S 1930 RA Fisher *The Genetical Theory of Natural Selection* – a Review. *Journal of Heredity* 21 349–356.

Wright S 1934a An analysis of variability in number of digits in an inbred strain of guinea pigs. *Genetics* 19 503 536.

Wright S 1934b The results of crosses between inbred strains differing in numbers of digits. *Genetics* 19 537–551.

Wright S 1960 On the appraisal of genetic effects of radiation in man. In: *The Biological Effects of Atomic Radiation* Summary reports. *National Academy of Science – National Research Council* pp. 18–24.

Wright S 1968 *Evolution and the Genetics of Populations.* Vol. 1. *Genetic and biometric foundations.* University of Chicago Press, Chicago.

Wright S 1977 *Evolution and the Genetics of Populations. Vol 4. Variability Within and Among Natural Populations.* University of Chicago Press, Chicago.

www.tecnosuma.com Sistema Ultramicroanalítico. Programas Nacionales. Tecnología SUMA.

Xenophon *Anabasis* Translations by Dakyns HG Project Gutenberg
  http://www.gutenberg.org/browse/authors/

Xu H, Morishima M, Wylie JN, Schwartz RJ, Bruneau BG, Lindsay EA and Baldini A 2004 Tbx1 has a dual role in the morphogenesis of the cardiac outflow tract. *Development* **131** 3217–3227.

Yasuda N, Cavalli-Sforza LL, Skolnick M and Moroni A 1974 The evolution of surnames: an analysis of their distribution and extinction. *Theoretical Population Biology* **5** 123–142.

Yen T, Nightingale BN, Burns JC, Sullivan DR and Stewart PM 2003 Butyrylcholinesterase (BCHE) genotyping for post-succinylcholine apnea in an Australian population. *Clinical Chemistry* **49** 1297–1308.

Yildirim S, Akan M, Deviren A and Akoz T 2002 Penile agenesis and clavicular anomaly in a child with an oral facial digital syndrome. *Clinical Dysmorphology* **11** 29–32.

Young S, Gooneratne S, Straus FH II, Zeller WP, Bulun SE and Rosenthal IM 1995 Feminizing Sertoli cell tumours in boys with Peutz-Jeghers syndrome. *American Journal of Surgical Pathology* **19** 50–58.

Yu N, Chen FC, Ota S, Jorde LB, Pamilo P, Patthy L, Ramsay M, Jenkins T, Shyue SK and Li WH 2002 Larger genetic differences within africans than between Africans and Eurasians. *Genetics* **161** 269–274.

Yuasa Y 2003 Control of gut differentiation and intestinal-type gastric carcinogenesis. *Nature Review Cancer* **3** 592–600.

Zak NB, Shifman S, Shalom A and Darvasi A 2001 Population-based gene discovery in the post-genomic era. *Drug Discovery Today* **6** 1111–1115.

Zehender W 1872 Eine Missbildung mit hautüberwachsenen Augen oder Kryptophthalmus. *Klinische Monatsblätter für Augenheilkunde* **10** 225–249.

Zei G, Matessi RG, Siri E, Moroni A and Cavalli-Sforza LL 1983 Surnames in Sardinia. I. Fit of frequency distributions for neutral alleles and genetic population structure. *Annals of Human Genetics* **47** 329–352.

Zei G, Piazza A and Cavalli-Sforza LL 1984 Geographic analysis of surname distributions in Sardinia: a test for neutrality. *Atti Associazione Genetica Italiana* **30** 247.

Zhang K, Akey JM, Wang N, Xiong M, Chakraborty R and Jin L 2003 Randomly distributed crossovers may generate block-like patterns of linkage disequilibrium: an act of genetic drift. *Human Genetics* **113** 51–59.

Zhang W, Collins A, Maniatis N, Tapper W and Morton NE 2002 Properties of linkage disequilibrium (LD) maps. *Proceedings of the National Academy of Science USA* **99** 17004–17007.

Zheng C-J, Guo S-W and Byers B 2000 Modelling the maternal-age dependency of reproductive failure and genetic fitness. *Evolution & Development* **22** 203–207.

Zhu G, Evans DM, Duffy DL, Montgomery GW, Medland SE, Nathan A, Gillespie NA, Ewen KR, Jewell M, Liew YW, Hayward NK, Sturm RA, Trent JM and Martin NG 2004 A genome scan for eye color in 502 twin families: most variation is due to a QTL on chromosome 15q. *Twin Research* **7** 197–210.

Zuelzer WW and Wilson JL 1948 Functional intestinal obstruction of congenital neurogenic basis in infancy. *American Journal of Diseases of Childhood* **75** 40–64.

Index of Authors

Subject Index